SINGLE- AND MULTI-PHASE FLOWS IN AN ELECTROMAGNETIC FIELD
ENERGY, METALLURGICAL, AND SOLAR APPLICATIONS

Edited by
Herman Branover
Ben-Gurion University of the Negev
Beer-Sheva, Israel

Paul S. Lykoudis
Purdue University
West Lafayette, Indiana

Michael Mond
Ben-Gurion University of the Negev
Beer-Sheva, Israel

Volume 100
PROGRESS IN
ASTRONAUTICS AND AERONAUTICS

Martin Summerfield, Series Editor-in-Chief
Princeton Combustion Research Laboratories, Inc.
Monmouth Junction, New Jersey

Technical papers from the Proceedings of the Fourth Beer-Sheva International Seminar on Magnetohydrodynamic Flows and Turbulence, Ben-Gurion University of the Negev, Beer-Sheva, Israel, February 27-March 2, 1984, and subsequently revised for this volume.

Published by the American Institute of Aeronautics and Astronautics, Inc.
1633 Broadway, New York, N.Y. 10019

American Institute of Aeronautics and Astronautics, Inc.
New York, New York

Library of Congress Cataloging in Publication Data

Bat-Sheva Seminar on MHD Flows and Turbulence
 (4th : 1984 : Ben-Gurion University of the Negev)
 Single- and multi-phase flows in an electromagnetic field.

 (Progress in astronautics and aeronautics; v. 100)
 Technical papers from the Proceedings of the Fourth Beer-Sheva
International Seminar on Magnetohydrodynamic Flows and
Turbulence, Ben-Gurion University of the Negev, Beer-Sheva, Israel,
February 27-March 2, 1984, and subsequently revised for this volume.
 Includes index.
 1. Magnetohydrodynamics—Congresses. 2. Turbulence—Congresses.
3. Two-phase flow—Congresses. 4. Liquid metals—Congresses.
I. Branover, G.G. (German Gertsevich). II. Lykoudis, P.S. (Paul S.),
1926-   . III. Mond, Michael. IV. American Institute of Aeronautics
and Astronautics. V. Title. VI. Series.
TL507.P75   vol. 100   629.1 s   [538'.6]   85-19979
[QC155.7]
ISBN 0-930403-04-5

Copyright © 1985 by the American Institute of Aeronautics and Astronautics, Inc. All rights reserved. Reproduction or translation of any part of this work beyond that permitted by Sections 107 and 108 of the U.S. Copyright Law without the permission of the copyright owner is unlawful. The code following this statement indicates the copyright owner's consent that copies of articles in this volume may be made for personal or internal use, on condition that the copier pay the per-copy fee ($2.00) plus the per-page fee ($0.50) through the Copyright Clearance Center, Inc., 21 Congress Street, Salem, Mass. 01970. This consent does not extend to other kinds of copying, for which permission requests should be addressed to the publisher. Users should employ the following code when reporting copying from this volume to the Copyright Clearance Center:

0-930403-04-5/85 $2.00 + .50

Progress in Astronautics and Aeronautics

Series Editor-in-Chief
Martin Summerfield
Princeton Combustion Research Laboratories, Inc.

Series Associate Editors

Burton I. Edelson
*National Aeronautics
and Space Administration*

Allen E. Fuhs
Naval Postgraduate School

Jack L. Kerrebrock
Massachusetts Institute of Technology

Assistant Series Editor
Ruth F. Bryans
Ocala, Florida

Norma J. Brennan
Director, Editorial Department
AIAA

Jeanne Godette
Series Managing Editor
AIAA

Table of Contents

Preface xi

Chapter I.   Laminar MHD Flows 1

Laminar Duct Flows in Strong Magnetic Fields 3
    J.S. Walker, *University of Illinois, Urbana, Illinois*

**Approximate Side Layer Solutions for a Liquid Metal Flow
in a Rectangular Duct with a Strong Nonuniform Magnetic Field** 17
    J.S. Walker and J.C. Petrykowski, *University of Illinois, Urbana, Illinois*

Applications of MHD Flows Between Rotating Disks 32
    P. Herve, *Université de la Réunion, France,* and C. Vives,
    Centre Universitaire d'Avignon, France

**Unsteady Magnetoaerodynamic Supersonic Flows Past Oscillating
Thin Bodies and Lifting Surfaces** 55
    L. Librescu, *Tel Aviv University, Israel*

Chapter II.   MHD and HD Turbulence 75

**Two-Dimensional Behavior of Electrically Driven
Flows at High Hartmann Numbers** 77
    J. Sommeria, *Groupement d'Interet Scientifique, Madylam,
Saint Martin d'Heres, France*

**Transition from Three-Dimensional to Quasi-Two-Dimensional
MHD Grid Turbulence** 89
    P. Caperan and A. Alemany, *Institut de Mecanique de Grenoble,
Saint Martin d'Heres, France*

Direct Numerical Simulation of Two-Dimensional Turbulence 100
    M.E. Brachet, *Observatoire de Nice, France,* and P.L. Sulem,
    Tel Aviv University, Israel, and Observatoire de Nice, France

**Experiments in Duct Flows with Reversed Turbulent
Energy Cascades** 111
    S. Sukoriansky, I. Zilberman, and H. Branover, *Ben-Gurion
University of the Negev, Beer-Sheva, Israel*

Direct Numerical Simulation of Three-Dimensional Convection
in Liquid Metals..125
    P.L. Sulem, *Tel Aviv University, Israel, and Observatoire de Nice,
France,* C. Sulem, *Ben-Gurion University of the Negev, Israel,
and Universite' de Nice, France,* and O. Thual, *Meteorologie Nationale,
Toulouse, France*

Magneto-Fluid-Mechanic Turbulent Vortex Streets.......................152
    D.D. Papailiou, *University of Patras, Greece*

Numerical Simulation of Homogeneous Turbulence Submitted
to Two Successive Plane Strains and Solid Body Rotation..............174
    P. Roy, *Office National d'Etudes et de Recherches Aerospatiales,
Chatillon, France*

Sensitivity of Turbulent Channel Flow to the Interactions
at the Perimeter..202
    D. Naot, *Center for Technological Education, Holon, Israel*

Electrogasdynamic and Kinetic Phenomena in Diffuse Electrical
Discharges in Turbulent Gas Flows..213
    Y.L. Khait, *Ben-Gurion University of the Negev, Beer-Sheva, Israel*

Homotopic Structural Invariants in HD and MHD Turbulence.........241
    D. Weil, *Racah Institute of Physics, The Hebrew University,
Jerusalem, Israel*

Chapter III.   Two-Phase Flows..253

Liquid Metal Magneto-Fluid-Mechanic Turbubblence...................255
    P.S. Lykoudis, *Purdue University, West Lafayette, Indiana*

Bubble Growth in a Superheated Liquid Metal
in a Uniform Magnetic Field..280
    P.S. Lykoudis, *Purdue University, West Lafayette, Indiana*

Analysis of Two-Phase MHD Flow
in Converging-Diverging Ducts...304
    S.I. Kamiyama, *Institute of High-Speed Mechanics,
Tohoku University, Sendai, Japan*

Stability of Two-Phase Liquid Metal MHD Channel Flow..............317
    S. Morioka and T. Toma, *University of Tsukuba, Sakura, Ibaraki, Japan*

An Analytical Model for Bubbly Flow.......................................329
    M. Mond and S. Sukoriansky, *Ben-Gurion University of the Negev,
Beer-Sheva, Israel*

Computer Modeling for Single-Phase Reacting Flow Patterns...........340
    E-D. Cristea, D. Mihai, and N. Lemnean, *Icsitee, Bucharest, Romania*

Two-Phase Flow Measurement Using a Modified Laser Doppler
Anemometry System..355
    Y. Levy and Y.M. Timnat, *Technion, Israel Institute of Technology, Haifa, Israel*

Chapter IV. MHD Power Generation and Application to Fission and Fusion Reactors...................369

Liquid Metal MHD Power Generation—Its Evolution and Status......371
    M. Petrick, *Argonne National Laboratory, Argonne, Illinois* and H. Branover, *Ben-Gurion University of the Negev, Beer-Sheva, Israel*

Tin-Water Faraday Generator..401
    J.P. Thibault, F. Joussellin, R. Laborde, and A. Alemany, *Institut de Mecanique de Grenoble, Saint Martin d'Heres, France,* and F. Werkoff, *Centre d'Etudes Nucleaires de Grenoble, France*

The ETGAR Liquid Metal MHD Project..................................413
    A. El-Boher and H. Branover, *Ben-Gurion University of the Negev, Beer-Sheva, Israel,* and M. Petrick, *Argonne National Laboratory, Argonne, Illinois*

Investigation of a Lithium-Caesium Faraday Converter..................435
    F. Joussellin, R. Laborde, A. Alemany, and J.P. Thibault, *Institut de Mecanique de Grenoble, Saint Martin d'Heres, France,* and F. Werkoff, *Centre d'Etudes Nucleaires de Grenoble, France*

The Feasibility of Remote Power Generation Based on LMMHD
and Biomass Energy..449
    D.G. Malcolm, *D.G. Malcolm & Associates, Inc., Saskatoon, Canada*

Interaction of Hall Currents and Turbulent Boundary Layers
in Closed-Cycle MHD Experiments...455
    W.F.H. Merck and J.G.A. Arts, *Eindhoven University of Technology, The Netherlands*

Streamer Dynamics in MHD Generators....................................475
    H.J. Flinsenberg and J. Uhlenbusch, *Eindhoven University of Technology, The Netherlands*

Magneto-Fluid-Dynamic Issues for Fusion First-Wall
and Blanket Systems...496
    B. Picologlou, C.B. Reed, R. Nygren, and J. Roberts, *Argonne National Laboratory, Argonne, Illinois*

Experiments on a Large Thin-Wall Duct..516
    C.C. Alexion and A.R. Keeton, *Westinghouse Electric Corporation, Pittsburgh, Pennsylvania*

Demonstration of Flow Couplers for the LMFBR..........................533
    R.D. Nathenson, C.C. Alexion, and A.R. Keeton, *Westinghouse Electric Corporation, Pittsburgh, Pennsylvania,* and O.E. Gray III, *Electric Power Research Institute, Naperville, Illinois*

Disk Generator Performance Prospects..548
    H.K. Messerle, *School of Electrical Engineering, University of Sydney, Australia*

High-Temperature Liquid Metal MHD Solar Thermal Systems..........562
    E.S. Pierson and W.D. Jackson, *HMJ Corporation, Washington, D.C.,* and G. Berry, M. Petrick, and C. Dennis, *Argonne National Laboratory, Argonne, Illinois*

Chapter V.   Metallurgical Applications..................587

Metallurgical Applications of MHD..589
    M. Garnier, *Groupement d'Interet Scientifique, Madylam, Saint Martin d'Heres, France*

Current Paths and MHD in Vacuum Arc Remelting......................617
    L.A. Bertram, F.J. Zanner, and B.M. Marder, *Sandia National Laboratories, Albuquerque, New Mexico*

Electromagnetic Modelization of Cold Crucibles..........................634
    A. Gagnoud, D. Delage, and M. Garnier, *Groupement d'Interet Scientifique, Madylam, Saint Martin d'Heres, France*

Shaping of Liquid Metal Cylinders...652
    J-P. Brancher, R. de Framond, and O. Sero-Guillaume, *Groupement d'Interet Scientifique, Madylam, Saint Martin d'Heres, France*

Shield Effects in Continuous Electromagnetic Casting....................667
    R. Ricou and C. Vives, *Centre Universitaire d'Avignon, France*

Investigation of the Turbulent Flow in an Induction Furnace Supplied with Various Frequencies..680
    E. Taberlet and Y.R. Fautrelle, *Groupement d'Interet Scientifique, Madylam, Saint Martin d'Heres, France*

The Electromagnetic Force of Narrow Stirring Inductors.................694
    F.R. Block and E. Julius, *Technical University, Aachen, Federal Republic of Germany*

Study of the Electromagnetic Features in Channel Induction Furnaces...........706
    A. Moros and J.C.R. Hunt, *University of Cambridge, Great Britain,* and D.C. Lillicrap, *Electricity Council Research Centre, Capenhurst, Chester, Great Britain*

Hartmann Layers in Slowly Solidifying Liquids...........716
    F.S. Hall and G.S.S. Ludford, *Cornell University, Ithaca, New York,* and J.S. Walker, *University of Illinois, Urbana, Illinois*

Stirring an Aluminum Ingot Mold with a Linear Motor: Electromagnetic, Hydrodynamic, and Thermal Effects...........736
    J-L. Meyer, R. Ernst, and F. Durand, *Groupement d'Interet Scientifique, Madylam, Saint Martin d'Heres, France*

Author Index for Volume 100...........756
List of Series Volumes...........757

Preface

The variety of physical phenomena and technological areas referred to in the title of this volume belong to a discipline known for thirty years or so as magnetohydrodynamics (MHD). To the minds of many researchers and engineers, however, MHD still means something strange and unfamiliar, even when they deal with problems which come directly within the purview of MHD. So for this volume we have chosen a title that underscores the classical physics of the work presented and its wide implications for engineering. *Single- and Multi-Phase Flows in an Electromagnetic Field* presents fundamentals for laminar and turbulent flow (Chapters I and II), but then emphasizes two-phase flows in the presence of magnetic fields (Chapter III)—an area of great complexity, both experimentally and theoretically, with implications for many applications. Then, Chapter IV deals with MHD power generation, with special attention to solar liquid-metal MHD power generation; Chapter V, with MHD problems in fission and fusion reactors; and Chapter VI, with metallurgical applications.

Since 1975, four international seminars on MHD flows and turbulence have been held at Beer-Sheva, Israel, drawing scholars from many countries interested in diverse aspects of MHD and especially in its dynamics with liquid metals. John Wiley and Sons published the proceedings of the 1975 and 1978 seminars. AIAA published proceedings of the 1981 seminar as Volume 84, *Liquid Metal Flows and Magnetohydrodynamics,* in the Progress in Astronautics and Aeronautics series. This volume comprises the proceedings of the 1984 Beer-Sheva Conference.

MHD theory and methods play a crucial role in the investigation of a great variety of innovative technological systems. New ideas and projects based on MHD theory and technology will provide important tools and momentum to future technological advance. Three areas, in particular, exemplify the large extent to which MHD theory can be applied to technological problems: MHD power generation (this book reflects mainly liquid-metal MHD power generation); lithium blankets in proposed fusion reactors; and metallurgical applications, where a whole new world is opened by the possibility of producing metals of high purity in unusual geometric shapes.

All three areas have exhibited substantial development over the last few years, but especially liquid-metal MHD power generation. Indeed, liquid-metal power conversion systems—discussed and analyzed for years—have now become a reality, at least on a pilot-plant scale. A fully engineered pilot plant has been designed and constructed in Israel with the active participation of American specialists. Another pilot plant has been developed in France. Two-phase-flow, liquid-metal power conversion systems offer great advantages in utilization of heat sources—conventional, nuclear, or solar. A few years from now, the first commercial power plants of this type will probably be built. This development in turn will demand that much greater effort be put into research dealing with single- but especially two-phase flows of liquid metals with and without magnetic fields. In addition, entire new research fields will become important in materials engineering, surface phenomena, power conditioning and controls, etc.

The history and status of liquid-metal MHD power generation are covered in this volume in the review paper by *M. Petrick* and *H. Branover*. Details of pilot plants can be found in papers by *A. El-Boher et al., J.P. Thibault et al., E.S. Pierson et al.,* and others. Two-phase-flow phenomena characteristic of liquid-metal MHD systems are treated in papers by *P.S. Lykoudis, S. Morioka* and *T. Toma, S.I. Kamiyama,* and *M. Mond* and *S. Sukoriansky*.

More than twenty years ago, the very early studies of liquid-metal MHD power generation were related to space applications (D. Elliott and others). Then, for a number of years, the field attracted little interest. Now interest in MHD applications in space seems to be reviving, and will probably be energized by space-platform programs and the U.S. Strategic Defense Initiative (SDI). The paper by *F. Joussellin et al.* reflects this aspect of MHD power generation.

Development of various concepts of thermonuclear-fusion reactors has revived interest in lithium blankets, in which liquid lithium flows in ducts of extremely complicated geometry, in the presence of very strong magnetic fields. Development of such blankets impels the study of laminar and turbulent flows of a liquid metal under these conditions. *B. Picologlou et al.* give a comprehensive review of problems related to the MHD aspects of fusion reactors, their present status, and future research programs. *J.S. Walker* and *J.C. Petrykowski* discuss specific flow phenomena in lithium blankets.

Regarding metallurgical applications of MHD, the progress achieved during recent years seems at first glance less dramatic, since various MHD methods and devices have been used for a long time for stirring and transporting molten metals. However, new ideas have been generated in this area also, and additional studies of laminar and turbulent flows in both constant and (particularly) pulsating magnetic fields have become necessary. Nine papers on metallurgical applications of MHD included in the present volume reflect theoretical and experimental investigations of a very wide range of problems and phenomena, such as the shaping of liquid-metal cylinders, continuous electromagnetic casting, stirring, and solidification in magnetic fields. *M.Garnier* reviews this area.

Studies of the more basic and universal topics of magnetohydrodynamic and hydrodynamic turbulence also receive extensive treatment in this volume. Great attention is given to the influence of magnetic fields on turbulent transfer processes and to the possibility of understanding better the nature of turbulence in general through analyzing the effects of magnetic fields on transfer processes.

The extensive references in the papers in this volume should greatly aid the reader, helping to introduce him quickly to the field. And, of course, the previous three volumes represent an excellent source of information on MHD problems, particularly for liquid metals (an area for which there is no appropriate single source).

The authors and editors are pleased to have AIAA publish the proceedings of the 1984 (4th) Beer-Sheva Seminar on MHD Flows and Turbulence. It was through the various AIAA journals, we should note, that the early MHD papers found a forum and a home. In fact, they still do, much to the credit of aerospace engineers.

Finally, the editors would like to acknowledge the valuable guidance and enthusiasm of Martin Summerfield, Editor-in-Chief of the AIAA Progress in Astronautics and Aeronautics series, Jeanne Godette, Managing Editor of the Progress Series, and Myra Rabinovitch and Judy Copeland of Ben-Gurion University, without whose invaluable assistance this volume would never have been published. Finally, we thank all of the contributors for their cooperation and care in the preparation of their papers.

We now look forward to the fifth volume of this series, and we invite the reader to participate in the 5th MHD Conference, which we expect to be held in Beer-Sheva in 1987.

Herman Branover
Paul S. Lykoudis
Michael Mond
September 1985

Chapter I. Laminar MHD Flows

Laminar Duct Flows in Strong Magnetic Fields

John S. Walker*
University of Illinois, Urbana, Illinois

Abstract

This paper reviews a number of studies of three-dimensional, inertialess, liquid metal duct flows with uniform and nonuniform magnetic fields. The solutions are summarized in a matrix of different combinations of cross-sectional geometries, wall conductivities, and types of magnetic fields. Current efforts to treat cases that have not been solved to date are discussed. Recent numerical results support a tentative hypothesis that the inertialess approximation provides reliable predictions of much lower values of the interaction parameter than indicated by an order-of-magnitude momentum balance. A possible interactive, computer aided design approach to applications such as tokamak blankets is presented.

Introduction

This paper reviews some analytical studies of three-dimensional, inertialess, laminar, liquid metal duct flows with strong, transverse magnetic fields. For liquid metal MHD flows, the four principal parameters are the magnetic Reynolds number, Hartmann number, interaction parameter, and wall conductance ratio, which are defined by

$$R_m = \mu \sigma V L \qquad M = B L (\sigma/\eta)^{1/2}$$

$$N = \sigma B^2 L/\rho V \qquad c = \sigma_w T/\sigma L$$

respectively. Here, μ, σ, η, and ρ are the liquid metal's magnetic permeability, electrical conductivity, viscosity, and density, respectively; V, L, and B are the character-

---

Paper presented at the Fourth Beer-Sheva Seminar on MHD Flows and Turbulence, Ben-Gurion University of the Negev, Beer-Sheva, Israel, Feb. 27-March 2, 1984. Copyright © 1985 by the American Institute of Aeronautics and Astronautics, Inc. All rights reserved.

*Professor, Department of Theoretical and Applied Mechanics.

istic velocity, length, and magnetic field strength respectively; σ_w and T are the electrical conductivity and thickness of the duct's wall respectively. For the tokamak liquid lithium blanket treated by Holroyd and Mitchell,[1] typical values are $R_m = 0.015$, $M = 27,000$, $N = 145,000$, and $c = 0.0085$.

The studies reviewed here all involve the same three assumptions. First, $R_m \ll 1$, so that the induced magnetic field produced by the electric currents in the liquid metal and in the duct's walls can be neglected. Actually, if the dimensionless electric current is small, then neglect of the induced magnetic field involves a much weaker restriction. For example, if the current is $O(c^{1/2})$ in a duct with thin metal walls and with $c \ll 1$, or if the current is $O(M^{-1/2})$ in a duct with insulating walls and with $M \ll 1$, then the restriction is simply $R_m \ll c^{-1/2}$ or $R_m \ll M^{1/2}$, respectively.

The second assumption is that $M \ll 1$, so that viscous effects are confined to thin boundary or free shear layers. The third assumption is that the magnetic field is sufficiently strong that inertial effects are negligible everywhere. The inertialess assumption involves a restriction on the interaction parameter that is different for each combination of wall conductivities and geometries, and these restrictions are presented in Table 1.

In addition, each reviewed study includes one of three assumptions on the wall conductance ratio for each wall:
1) perfectly conducting wall: $c \gg M^{1/2}$,
2) thin conducting wall: $M^{-1} \ll c \ll 1$ and $T \ll L$,
3) insulating wall: $c \ll M^{-1}$.

Some authors have wrongly stated that a wall can be treated as a perfect conductor if $c \gg 1$, and this error led to some confusion. For example, Holroyd[2] finds experimentally that the flow pattern for a rectangular duct with $c = 13$ and $M = 505$ involves M-shaped velocity profiles, which are characteristic of the analytical solutions for thin conducting walls rather than those for perfectly conducting walls. Since he had expected to find good agreement between experimental results for $c = 13$ and analytical results for perfectly conducting walls, he concludes that the analysis is somehow deficient. In fact, the correct analysis is in complete agreement with Holroyd's experimental results. Walker[3] describes the correct analytical solutions for various values of c: For $M^{-1} \ll c \ll 1$, the M-shaped velocity profiles involve sufficiently strong sheet jets adjacent to the walls parallel to the magnetic field that part of the mass flux is carried by these jets; for $c = O(1)$, the same description applies; for $1 \ll c \ll M^{1/2}$, the M-shaped velocity profiles persist and the velocity in the sheet jets is still large, but not large enough to carry any of the mass flux; for $c = O(M^{1/2})$, the M-shaped velocity profiles and sheet jets are gone and the velocity everywhere is $O(1)$, but the flow is still significantly different from that for perfectly conducting wall; finally, for $c \gg M^{1/2}$, the solution for perfectly conducting walls is achieved. Since Holroyd's experiments fall into the range $1 < c < M^{1/2}$, the observed M-shaped velocity profiles are precisely those predicted by the correct analysis. The analysis also shows that, to first order, the pressure gradient is the same as that for perfectly conducting walls for all $c \gg 1$, even though the flow is quite different, and this fact is verified by Holroyd's experiments.

Matrix of Cases

The studies reviewed here consider specific cases in a matrix of different combinations of geometries, wall conductivities, and types of magnetic fields. Part of this

Table 1 Cases of fully three-dimensional MHD duct flows

	Restriction for inertialess approximation	Variable-area ducts with uniform magnetic fields	Any ducts with non-uniform magnetic fields
Circular ducts:			
Perfectly conducting wall	$N \gg 1$	Ref. 17	Unpublished approximate Solution (Ref. 23)
Thin conducting wall	$N \gg c^{-1/2}$	Ref. 18	Ref. 15
Insulating wall	$N \gg M^{1/2}$	Ref. 19	Ref. 15
Rectangular ducts:			
All perfectly conducting walls	$N \gg M$	Ref. 20	No solution
Perfectly conducting sides, insulating top and bottom	$N \gg M$	Ref. 4	Ref. 6
Perfectly conducting top and bottom, insulating sides	$N \gg M^{3/2}$	Ref. 21	Current research
All insulating walls	$N \gg M^{3/2}$	Ref. 22	Current research
All thin conducting walls	$N \gg M^{3/2}$	Ref. 3	Current research

matrix is presented in Table 1 along with the restriction on N for the inertialess approximation in each case and with the reference for each case that has been successfully analyzed. Circular and rectangular cross sections are prototypes for sections without and with flat surfaces joined at sharp corners, respectively. For rectangular ducts, the transverse magnetic fields intersect one pair of walls (top and bottom) and are parallel to the other pair (sides); the sides are parallel, and L is half the distance between the sides. For the one case of a rectangular duct with insulating top and bottom and perfectly conducting sides, Walker et al.[4] find that diverging or converging the sides, as well as the top and bottom, radically change the flow and change the restriction for the inertialess approximation from $N \gg M$ to $N \gg M^{3/2}$. For circular ducts, L is the radius at some reference cross section, while V and B are the average axial velocity and the magnetic field strength at some reference cross section for all ducts.

Table 1 only includes fully three-dimensional flows. For the variable-area ducts with uniform, transverse magnetic fields, the circular ducts involve expansions or contractions with O(1) slopes, while the rectangular ducts have tops and bottoms that diverge or converge with O(1) slopes. However, these solutions also cover the three-dimensional evolution of the flow from some entrance or exit conditions to fully developed flow in a constant-area duct with a uniform magnetic field.[5] For ducts with nonuniform magnetic fields, the ducts may be constant-area or variable-area ducts, while the magnetic fields change over axial distances that are comparable to L. The nonuniform fields considered are planar and are symmetric about a plane coinciding with one of the duct's planes of symmetry.

For cases with complete solutions in print, the references for these solutions are given in Table 1. For the perfectly conducting circular duct with a nonuniform magnetic field, Walker and Holroyd developed an approximate solution in 1978, but the results have not been published. The extension of the analysis of Petrykowski and Walker[6] to a rectangular duct with all perfectly conducting walls and with a nonuniform magnetic field would be straightforward. Current efforts to treat rectangular ducts with nonuniform magnetic fields and with either insulating sides or all thin conducting walls are discussed in the next two sections.

Current Approximate Studies

The solutions for circular ducts have generally involved less difficulty than those for rectangular ducts because the former do not involve flat surfaces that are parallel to the magnetic field. The most difficult part of the solutions for rectangular ducts is always the treatment of the boundary layers adjacent to the sides. This treatment cannot be bypassed, because the side layers frequently involve high-velocity sheet jets that carry all or part of the mass flux and frequently determine the pressure gradients throughout the duct. For fully developed flows, the side layers can be treated with separation of variable methods or with Laplace transforms, but these methods could not be extended to side layers in fully three-dimensional duct flows. All of the side layers in the rectangular duct solutions with references in Table 1 were treated with a method first developed by Walker et al.[4]: Fourier sine and cosine transforms in the coordinate normal to the side reduce the boundary value problem to a pair of coupled ordinary differential equations with two unknown functions of the coordinates along the side; the coupled equations are solved with Green's functions or variation of parameter methods; Fourier inversion leads to two integral equations in the coordinate parallel to the magnetic field or to integro-differential equations that also include derivatives in the axial direction; the integral or integro-differential equations are treated analytically to properly account for singularities in the kernels; separation of variable methods reduces the integro-differential equations to eigenvalue problems with homogeneous integral equations; the integral equations are reduced to matrix equations using several Guass quadrature approximations, and the matrix equations are solved numerically. This essentially analytical method reached its limit in the treatment of ducts with perfectly conducting sides and nonuniform magnetic fields.[6] While the extension of this method to ducts with thin conducting or insulating walls and nonuniform fields is theoretically possible, it is impossible in practice because of the extraordinary amount of algebra that would be involved. Current efforts are focused on two approaches: a method that involves less analytical processing and more numerical analysis, and a simplification of the problems by considering slowly varying magnetic fields.

For rectangular ducts with thin conducting walls and with fully three-dimensional nonuniform magnetic fields, there are two boundary layers on each side. An outer layer carries a large axial electric current, and an inner layer carries half the total mass flux. Walker[7] presents solutions for the outer layer, but the flow-carrying inner layer remains untreated. For insulating walls, there is only one boundary layer on each side that carries both an axial electric current and half the mass flux.

While treatment of these side layers should involve more numerical analysis than previous side layer solutions, a fully numerical finite-difference or finite-element solution of each boundary value problem with no analytical processing would be undesirable for several reasons. The governing equations involve double diffusion operators: One operator is intrinsically unstable when marching upward, while the other operator is intrinsically unstable when marching downward. Singularities at the corners play a key role in the physics of the side layers but would be difficult to incorporate properly into a fully numerical scheme. The analytical approach provides physical insights into the complex interaction of electric currents, voltages, velocities, and pressures in the side layers; and purely numerical approaches might not provide the same insights. Therefore, it seems more profitable to look for a method that combines the benefits of the analytical and numerical approaches. A Galerkin-type approach using trial functions generated from the boundary value problem with guidance from the characteristics of known analytical solutions is proposed in another presentation here.[8]

Slowly Varying Magnetic Fields

There are three important reasons for the study of MHD duct flows with transverse magnetic fields that change over an axial distance much greater than L. First, the boundary value problems for the side layers in rectangular ducts with thin conducting or insulating walls are much simpler for slowly varying magnetic fields than for fully three-dimensional ones and can be treated with established analytical methods. Therefore, current research involves two approaches for these difficult problems: a combined analytical and numerical approach for fully three-dimensional fields and an analytical approach for slowly varying fields. Comparison of results from the two approaches should provide physical insights into the flows and should indicate the limitations of each approach.

The second reason for studying slowly varying magnetic fields is that many important applications involve gradual rather than abrupt field changes. For example, in a recent Argonne National Laboratory design for a tokamak blanket, the magnetic field strength changes from 3.0 T to 7.5 T in an axial distance of 7 m, while L = 0.2 m. The third reason for studying slowly varying fields is that such studies provide the transitions between sometimes quite different flows for uniform and fully three-dimensional magnetic fields. For variable-area ducts with O(1) slopes, the slowly varying field solutions would provide the transitions between the two columns in Table 1. On the other hand, for ducts with constant or slowly varying areas, the slowly varying field solutions provide the transitions between fully three-dimensional flows and slowly varying or fully developed flows. These transitions are more dramatic, since they involve the transitions from sheet jets with stagnant cores to uniform velocity profiles in many cases; and from the strong restriction on N for the inertialess approximation to no restriction, since fully developed flow is intrinsically inertialess for any N. This reduction of the inertialess restriction for slowly varying fields means that many important applications involving slowly varying fields and geometries can be treated with the inertialess approximation at much lower values of N than indicated in Table 1.

Only one slowly varying analysis has been completed to date: the flow in an insulating rectangular duct with parallel sides, with a uniform transverse magnetic field, and with a small slope b for the diverging or converging top and bottom. Walker and Ludford[9] present solutions for various values of b and find the following restrictions on N for the inertialess approximation:
1) For $b = O(1)$, $N \gg M^{3/2}$;
2) for $M^{-1/2} \ll b \ll 1$, $N \gg b^2 M^{3/2}$;
3) for $b = O(M^{-1/2})$, $N \gg M^{1/2}$;
4) for $b \ll M^{-1/2}$, $N \gg bM$.

For case 4, the flow is locally fully developed, and the restriction on N is satisfied by any N as $b \to 0$. El-Consul and Walker[10] present inertial perturbation solutions for cases 2 and 3.

The analysis for slowly varying magnetic fields is different from that for slowly varying areas with uniform magnetic fields, but the results and the relaxation of the inertialess restriction should be similar. In the matrix of cases in Table 1, there should be an additional column for slowly varying magnetic fields, but there are currently no references to put into the missing column. Current research with attention focused on tokamak blanket flows should yield a number of solutions to begin to fill this gap in the theory.

Range of Validity for the Inertialess Approximation

The conditions on N given in Table 1 are certainly sufficient to insure that inertial effects are negligible in each fully three-dimensional flow, but they may not be necessary. Indeed, inertialess solutions may provide reliable flow predictions for much smaller values of N. Hunt[11] found exact agreement between the predictions of the inertialess solution for the drag of an insulated sphere in a transverse magnetic field and the results of experiments, even though the inertialess restriction was $N \gg M^{3/2}$, while N and $M^{3/2}$ were comparable in the experiments.

The restriction on N insures that inertial effects are negligible everywhere. Generally a much weaker restriction is required for inertial effects to be negligible in the core regions, but not in the thin boundary and free shear layers. For example, only $N \gg M^{1/2}$ is required to neglect inertial effects in core regions for the drag of the insulated sphere. Hunt and Leibovich[12] present solutions for a free shear layer between two core regions in which inertial effects are negligible in the cores, but dominate in the free shear layer, that require that $1 \ll N \ll M^{3/2}$. On the other hand, the inertialess solution for the same problem requires that $N \gg M^{3/2}$ and involves a totally different free shear layer structure, but this different shear layer matches the same two inertialess cores. Outside the free shear layer, the flows are the same for all $N \gg 1$.

Until recently, it was generally believed that 1) the excellent agreement found by Hunt[11] between predictions of the inertialess approximation and results of experiments at values of N below the range required by the inertialess restriction and 2) the similarity of inertial and inertialess flows found by Hunt and Leibovich[12] represented aberrant cases. It was assumed that most flows would involve a progressive emergence of inertial effects as N was reduced from values satisfying the inertialess restriction and that the inertialess approximation would cease to be valid as soon as the inertialess restriction was not satisfied.

However, recent numerical solutions indicate that, as N is reduced, inertial effects remain essentially negligible until some critical value of N and that, below this critical value, inertial effects emerge abruptly and immediately overwhelm the previously dominant electromagnetic effects. The numerical solutions treat the Czochralski growth of silicon crystals from a crucible of molten silicon with a uniform axial magnetic field applied in order to suppress thermal convection and thermocapillary motions in the molten silicon.[13] The numerical analysis includes all inertial terms in the momentum equation, as well as all convective terms in the energy equation.

For the magnetic Czochralski problem, the inertialess approximation would require that $N \gg M^{3/4}$. The fully inertial results for the streamlines and electric current lines for $B = 0.2$ T ($N = 13.7$, $M^{3/4} = 188$) are identical to those predicted by an inertialess analysis, even though $N = 0.07\ M^{3/4}$. The numerical results for $B = 0.1$ T ($N = 3.4$, $M^{3/4} = 112$) are still very similar to the inertialess results. There is some blending of the cores and the free shear layer, but this blending appears to result from the fact that the free shear layer thickness is now 5% of the crucible radius, which is not small, rather than from any inertial effects. We are currently computing streamlines and current lines from a composite inertialess solution that allows for the blending of adjacent regions at finite M. Comparison of these results and the fully inertial numerical results is expected to demonstrate that the inertial approximation is still valid, even though $N = 0.031\ M^{3/4}$. Finally, the numerical results for $B = 0.05$ T ($N = 0.86$, $M^{3/4} = 66$) are totally different from the predictions of the inertialess approximations. Indeed, the numerical results look very much like the results without a magnetic field, except that the magnitude of the convective circulation is reduced by the electromagnetic body force. At $N = 0.031\ M^{3/4}$, the flow is dominated by electromagnetic effects, while inertial effects are apparently negligible or minor, but, at $N = 0.013\ M^{3/4}$, the flow is dominated by inertial effects with modest modifications by the electromagnetic effects. The critical N at which effects emerge abruptly lies between these two cases. There is an error in the results presented by Langlois and Walker,[13] while correct and more extensive numerical results are presented by Langlois and Lee.[14]

These numerical results need to be supported by more studies before we can state that the inertialess approximation provides reliable predictions at much lower values of N than those required by the conditions in Table 1, and before we can accurately estimate the critical values of N for various cases. Nevertheless, this new hypothesis is more compatible with the exact agreement found by Hunt.[11]

Designs with Thin Conducting or Insulating Walls

The ultimate objective of the reviewed research is the technology transfer from the basic research focused on simple prototypes to the applied analysis of the complex and contorted geometries, wall thicknesses, and magnetic fields of real designs such as tokamak blankets. For the past few years, a concept of certain characteristic surfaces has been recognized as the tool that must be used in such a technology transfer, but this concept still requires development before reliable design calculations can be done on a routine basis. For a general three-dimensional magnetic field, let s be the distance measured along a magnetic field line and B(s) be the local magnetic field strength at each point along this line. If each field line has two intersections with insulating or thin conducting walls confining the liquid metal, say, at s_1 and s_2, then each magnetic field line has a value for a scalar quantity given by

$$K = \int_{s_1}^{s_2} B^{-1} \, ds$$

If K has the same value for every magnetic field line in a region, then there are no distinct characteristic surfaces, and the flow is "free" in this region. In this case, the core flow is a two-dimensional potential flow in planes perpendicular to the field, the core solutions easily accomodate perturbations from the boundary layers, and the boundary layers merely match the core and satisfy the non-slip condition in most cases.

If K varies, then a set of magnetic field lines with the same value of K defines a characteristic surface. Three different flows with distinct characteristic surfaces are possible:

1) "Guided" flows with axial characteristic surfaces. If the characteristic surfaces extend far upstream and downstream, then the fluid can flow along the duct by following the characteristic surfaces and with no O(1) electric currents. The long surfaces also allow very slow migration of the flow across surfaces over long distances. This is the case for variable-area circular ducts in uniform or nonuniform magnetic fields and for circular pipes in nonuniform fields.[15]

2) "Guided" flows with transverse characteristic surfaces. If the characteristic surfaces are essentially perpendicular to the flow direction and if an O(1) electric current can flow along the surfaces because they intersect highly conducting sides, then there can be an O(1) core velocity perpendicular to the characteristic surfaces, i.e., in the flow direction. While this core flow can carry all or part of the mass flux, it is a much more restricted core flow that that for "free" flows. The restricted core cannot accomodate perturbations from the side layers, and this produces a more severe side layer structure.

3) "Blocked" flows with transverse characteristic surfaces. If characteristic surfaces at right angles to the flow direction intersect thin conducting or insulating sides, there can be no O(1) core velocity, so that the core is essentially stagnant and the mass flux is carried by

sheet jets with very large velocities adjacent to the sides, i.e., an M-shaped velocity profile.

The terms "free," "guided," and "blocked" are borrowed from geostrophic rotating flows that are analogous to MHD flows with strong magnetic fields. This classification is described in greater detail by Petrykowski and Walker.[6] The transition between guided or blocked flows and free flows occurs when the variation of K is $O(M^{-1})$. If the variations of K are much less than M^{-1}, then the flows are essentially free; if the variations of K are much greater than M^{-1}, then the flows are essentially guided or blocked.

With this classification, the picture of a design process is beginning to emerge. The first step in any design calculation will be the complete and accurate definition of all characteristic surfaces. This step cannot be done "by hand" for a tokamak blanket for three reasons: The combination of spatially variable toroidal and poloidal magnetic fields is complex; the flow path geometry with manifolds, changing cross-sectional shapes and thicknesses, and possibly modules is complex; and the difference between truly parallel and almost parallel can be very important in the presence of a strong magnetic field. However, the purely geometric task of accurately defining all the characteristic surfaces is an almost trivial one for a computer. With a complete description of the characteristic surfaces, the flow in each part of the blanket can be classified, and these classifications provide information about many important aspects of the flows with no detailed flow analyses, e.g., are there stagnant regions with jets along some surfaces or is the flow guided or free; is thermal convection suppressed by the magnetic field; what are the orders of magnitude of the pressure gradient, heat and mass transfer, wall stresses, etc.? With the computer prepared to determine the characteristic surfaces and classify flows for given geometries and magnetic fields, the designer can alter blanket designs interactively until the flow classifications change from, say, blocked flows with stagnant regions and jets to guided flows. This change in the classification of the flow implies changes in the effect on thermal convection, order of magnitude of the pressure gradient, etc., all without any detailed flow analysis beyond mapping characteristic surfaces.

For example, interactive design modification can lead to improved thermal convection. Most of the energy from fusion is deposited on the first wall or in a relatively thin layer of lithium adjacent to the first wall in a tokamak blanket. Without thermal convection, an acceptable liquid temperature at the first wall can only be maintained either with very high flow rates or with complex modules or manifolds that force heat and mass transfer away from the first wall. Both alternatives imply large pressure drops and large pumping powers. With thermal convection, mixing maintains a more uniform liquid temperature, so that modest flow rates and no modules or manifolds are required. The mapping of the characteristic surfaces and the classification of the flows for an initial design might show that convection is suppressed because the flow is blocked or because the guided flow's direction does not coincide with the convection's direction or because the potential flow planes for free flow are perpendicular to the convection planes. The design can be modified and the computer can provide revised characteristic surface maps and flow classifications until thermal convection is not suppressed because the convection follows the characteristic surfaces for guided flow or because planes for thermal convection and free flows coincide. Thus, interactive computer aided design can lead to free thermal convection and more modest flow rates, pressure drops, and pumping powers. Once a design is optimized in its general aspects, detailed thermal hydraulic analyses would be carried out numerically using an intrinsic, curvilinear coordinate system associated with the characteristic surfaces,[16] and these detailed flow analyses would lead to more refined optimizations.

Acknowledgment

This research was supported by the U. S. National Science Foundation under Grant CPE-8108952.

References

[1] Holroyd, R. J. and Mitchell, J. T. D., "Liquid Lithium as a Coolant for Tokomak Fusion Reactors," Culham Laboratory, Abingdon, Oxon, U.K., Report CLM-R231, 1982.

[2] Holroyd, R. J., "MHD Flow in a Rectangular Duct with Pairs of Conducting and Nonconducting Walls in the Presence of a Nonuniform Magnetic Field," Journal of Fluid Mechanics, Vol. 96, No. 2, 1980, pp. 335-353.

[3] Walker, J. S., "Magnetohydrodynamic Flows in Rectangular Ducts with Thin Conducting Walls. Part 1. Constant-Area and Variable-Area Ducts with Strong Uniform Magnetic Fields," Journal de Mecanique, Vol. 20, No. 1, 1981, pp. 79-112.

[4] Walker, J. S., Ludford, G. S. S., and Hunt, J. C. R., "Three-Dimensional MHD Duct Flows with Strong Transverse Magnetic Fields. Part 2. Variable-Area Rectangular Ducts with Conducting Sides," Journal of Fluid Mechanics, Vol. 46, No. 4, 1971, pp. 657-684.

[5] Ludford, G. S. S. and Walker, J. S., "On Establishing Fully Developed Duct Flow in Strong Magnetic Fields," MHD Flows and Turbulence: Proceedings of the First Bat Sheva Seminar, Vol. 1, H. Branover, editor, John Wiley and Sons, New York, 1976, pp. 7-15.

[6] Petrykowski, J. C. and Walker, J. S., "Liquid-Metal Flow in a Rectangular Duct with a Strong, Nonuniform Magnetic Field," Journal of Fluid Mechanics, Vol. 139, 1984, pp. 309-324.

[7] Walker, J. S., "Three-Dimensional Laminar MHD Flows in Rectangular Ducts with Thin Conducting Walls and Strong, Transverse, Nonuniform Magnetic Fields," Liquid-Metal Flows and Magnetohydrodynamics: Progress in Astronautics and Aeronautics, Vol. 84, H. Branover, P. S. Lykoudis and A. Yakhot, editors, AIAA, New York, 1983, pp. 3-19.

[8] Walker, J. S. and Petrykowski, J. C., "Approximate Side Layer Solutions for a Liquid-Metal Flow in a Rectangular Duct with a Strong, Nonuniform Magnetic Field," published elsewhere in this volume.

[9] Walker, J. S. and Ludford, G. S. S., "Three-Dimensional MHD Duct Flows with Strong Transverse Magnetic Fields. Part 4. Fully Insulated, Variable-Area Rectangular Ducts with Small Divergences," Journal of Fluid Mechanics, Vol. 56, No. 3, 1972, pp. 481-496.

[10] El-Consul, A. M. and Walker, J. S., "Inertial Perturbation to Inertialess MHD Flows in Insulating, Rectangular Ducts with Small Divergences," Developments in Mechanics: Proceedings of the Sixteenth Midwestern Mechanics Conference, Vol. 10, Kansas State University, Manhattan, Kansas, 1979, pp. 167-171.

[11] Hunt, J. C. R., "The Resistance of a Blunt Body in a Strong Transverse Magnetic Field," Magnitnaya Gidrodinamika, Vol. 6, No. 1, 1970, pp. 35-38.

[12] Hunt, J. C. R. and Leibovich, S., "Magnetohydrodynamic Flow in Channels of Variable Cross-Section with Strong Transverse Magnetic Field," Journal of Fluid Mechanics, Vol. 28, No. 2, 1967, pp. 241-260.

[13] Langlois, W. E. and Walker, J. S., "Czochralski Crystal Growth in an Axial Magnetic Field," Computational and Asymptotic Methods for Boundary and Interior Layers: Proceedings of the BAIL II Conference, Dublin, J. J. H. Miller, editor, Boole Press Limited, Dublin, Ireland, 1982, pp. 299-304.

[14] Langlois, W. E. and Lee, K. J., "Digital Simulation of Magnetic Czochralski Flow Under Various Laboratory Conditions for Silicon Growth," IBM Journal of Research and Development, Vol. 27, No. 3, 1983, pp. 281-284.

[15] Holroyd, R. J. and Walker, J. S., "A Theoretical Study of the Effects of Wall Conductivity, Nonuniform Magnetic Fields and Variable Areas on Liquid-Metal Flows at High Hartmann Number," Journal of Fluid Mechanics, Vol. 84, No. 3, 1978, pp. 471-495.

[16] Ludford, G. S. S. and Walker, J. S., "Current Status of MHD Duct Flows," MHD Flows and Turbulence: Proceedings of the Second Bat Sheva Seminar, Vol. 2, H. Branover and A. Yakhot, editors, Israel Universities Press, Jerusalem, 1980, pp. 83-95.

[17] Walker, J. S. and Ludford, G. S. S., "MHD Flow in Conducting Circular Expansions with Strong Transverse Magnetic Fields," International Journal of Engineering Science, Vol. 12, No. 3, 1974, pp. 193-204.

[18] Walker, J. S. and Ludford, G. S. S., "MHD Flow in Circular Expansions with Thin Conducting Walls," International Journal of Engineering Science, Vol. 13, No. 3, 1975, pp. 361-369.

[19] Walker, J. S. and Ludford, G. S. S., "MHD Flow in Insulating Circular Expansions with Strong Transverse Magnetic Fields," International Journal of Engineering Science, Vol. 12, No. 12, 1974, pp. 1045-1061.

[20] Walker, J. S. and Ludford, G. S. S., "MHD Flow in Variable-Area Conducting Rectangular Ducts," Developments in Mechanics: Proceedings of the 13th Midwestern Mechanics Conference, Vol. 7, University of Pittsburgh, Pittsburgh, Pennsylvania, 1973, pp. 265-276.

[21] Walker, J. S. and Ludford, G. S. S., "MHD Flows in Variable-Area Rectangular Ducts with Mixed Wall Conductivities," Developments in Theoretical and Applied Mechanics: Proceedings of the Seventh Southeastern Conference on Theoretical and Applied Mechanics, Vol. 7, Catholic University of America, Washington, 1974, pp. 45-60.

[22] Walker, J. S., Ludford, G. S. S., and Hunt, J. C. R., "Three-Dimensional MHD Duct Flows with Strong Transverse Magnetic Fields. Part 3. Variable-Area Rectangular Ducts with Insulating Walls," Journal of Fluid Mechanics, Vol. 56, No. 1, 1972, pp. 121-141.

[23] Holroyd, R. J. and Walker, J. S., unpublished results, 1978.

Approximate Side Layer Solutions for a Liquid Metal Flow in a Rectangular Duct with a Strong Nonuniform Magnetic Field

John S. Walker* and John C. Petrykowski†
University of Illinois, Urbana, Illinois

Abstract

This paper treats the boundary layers on the walls of a rectangular duct that are parallel to an applied transverse magnetic field. An essentially analytical method has been successfully used to treat these side layers in many different ducts with uniform magnetic fields and in a duct with perfectly conducting sides and with a particular nonuniform magnetic field. However, further extensions of this method to other ducts with this particular nonuniform magnetic field or to other nonuniform fields appears to be impractical. The present paper proposes a Galerkin method that optimizes an approximate series solution with respect to certain trial functions and proposes a separation of variable method to generate trial functions that reflect the important physical phenomena in the side layers and that closely resemble the correct solutions. With such trial functions, excellent results can be obtained with a small number of trial functions, while each trial function reveals some particular aspects of the flow. Trial functions are compared with the results of the one successful analytical solution for the side layers with a nonuniform magnetic field in order to demonstrate the close resemblance between the analytical results and the trial functions generated by the proposed method.

Paper presented at the Fourth Beer-Sheva Seminar on MHD Flows and Turbulence, Ben-Gurion University of the Negev, Beer-Sheva, Israel, February 27 - March 2, 1984. Copyright © American Institute of Aeronautics and Astronautics, Inc., 1985. All rights reserved.
*Department of Theoretical and Applied Mechanics.
†Department of Theoretical and Applied Mechanics. Present address: Oak Ridge National Laboratory, Oak Ridge, Tennessee.

Introduction

The magnetic field in a magnetic-confinement fusion reactor consists of a toroidal field that varies inversely with distance from the reactor's centerline and a poloidal field that varies inversely with distance from the plasma current producing it. For any liquid lithium circuit through a fusion reactor, the magnetic field is everywhere nonuniform. For large dc electromagnetic (EM) pumps, there is a central, uniform field region, but there are also nonuniform, fringing magnetic field regions at the ends of the magnet's pole faces. These fringing field regions coincide with the ends of the electrodes, so that the three-dimensional end effects involve a coupling between fringing electric currents and fringing magnetic fields. Therefore, reliable design calculations for fusion reactor blankets and for EM pumps require a complete theory of three-dimensional liquid metal flows in strong nonuniform magnetic fields.

For duct cross sections with no corners, no discontinuities in electrical conductivity, and no straight walls parallel to the magnetic field, the solutions involve core regions, Hartmann layers, and side regions (at points of wall-field tangency). The circular duct is the prototype here. During the past decade, efforts to develop a theory for these ducts have been reasonably successful. In particular, the concept of certain characteristic surfaces composed of magnetic field lines has led to an analogy between a duct of this type in a particular nonuniform magnetic field and an "equivalent" duct in a uniform magnetic field with the same distribution of characteristic surfaces. Since methods of treating three-dimensional MHD flows in uniform magnetic fields are well established, this analogy provides many solutions with modest effort.[1]

For duct cross sections with corners or with discontinuities in wall conductivity or with at least one straight wall parallel to the magnetic field, the solution also involves side layers on the walls parallel to the magnetic field and/or free shear layers emanating from the corners or wall conductivity discontinuities and lying along magnetic field lines. The rectangular duct is the prototype here: If the sides are exactly parallel to the magnetic field, the solution involves side layers, while, if the sides are not parallel to the field, the solution involves free shear layers emanating from two opposite corners.[2] If one increases the angle between the sides and the magnetic field from zero: a) each side layer is initially symmetric about the midpoint of the side at zero angle; b) the side layer becomes asymmetric with a greater thickness near the top than at the bottom if the field is slanting away at the top; and c) the side layer detaches from the side at some small angle and becomes a free shear layer emanating from the bottom corner for this case. The simiinfinite side layers are intrinsically more difficult to treat than the doubly infinite free shear layers, so the former are the focus here.

The analogy based on characteristic surfaces that solved the circular duct problem only provides a qualitative similarity for rectangular ducts: The flow in a rectangular duct with a nonuniform magnetic field is similar to that in a variable-area rectangular duct with a uniform field. For given wall conductivities, both involve essentially stagnant core regions and high velocities in the side layers that carry all of the volume flux, or both involve overshoots in the side layer velocity profiles so that the side layer displacement thickness is everywhere zero. However, the side layer boundary value problem for nonuniform magnetic fields involve a coupling between the voltage and the pressure that is absent from the side layer problems for uniform fields. This coupling precludes any exact or approximate predictions for nonuniform field problems based on uniform field solutions. Because of this coupling, nonuniform field problems are much more difficult to solve than the corresponding uniform field problems. While a number of uniform field side layer solutions for arbitrary geometries are known, only one nonuniform field side layer solution for a particular magnetic field has been found to date.[3] For both cases, the boundary value problem is reduced to an integral equation using Fourier transforms and Green's function methods; the integral equation is integrated and modified in other ways to properly represent various aspects of the singular kernel; the modified integral equation is solved numerically; and the side layer profiles of velocity, electric current, voltage, and pressure are generated from the integral equation solution. This approach is 90% analytical and 10% numerical. The extension from general cases for uniform fields to the one special case for a particular nonuniform field involved an enormous increase in the amount of algebra in the analytical part of the solution. The further extensions of this approach to other nonuniform magnetic fields or to other ducts in the same nonuniform field are impossible from a practical point of view because the algebraic efforts involved would be unreasonable. An approach that is more numerical and less analytical has a greater promise of success with reasonable effort. However, a total role reversal with, for example, finite-difference or finite-element solutions of the three-dimensional side layer boundary value problems with no analytical reductions is also undesirable, because the analytical component frequently reveals aspects of the physical phenomena that might be missed in purely numerical approaches.[4]

The present paper represents a preliminary step toward a more balanced approach to solving nonuniform magnetic field side layer problems. The ultimate objective is a Galerkin method to find an optimal solution for a given set of trial functions, where the trial functions are carefully chosen to reflect the analytical character of the boundary value problems and the important physical phenomena in the problems. In the nonuniform field coupling, the voltage is associated with the primary axial velocity, and the pressure is associated with a secondary flow with circulation in surfaces perpendicular to the primary flow direction. From the mathematical character of the coupling, one might expect the primary and secondary flows to be comparable. However, for the one known, exact, special case solution, the secondary velocity turns out to be less than 2% of the primary velocity. While this secondary flow turns out to be small, it still plays a key role in the physics of the side layers. If we simply set the secondary flow equal to zero, we find that the primary velocity does not have the overshoot, which is the most important characteristic of these side layers. For other wall conductivities, setting the secondary flow equal to zero would eliminate the large axial velocity that must carry all of the volume flux. Thus, even an approximate solution is impossible without the secondary flow.

Since the secondary flow is small, perhaps the important primary flow is relatively insensitive to changes in the details of the secondary flow. This lead suggests a method of generating Galerkin trial functions that reflect the basic physics. First, we choose a set of preliminary trial functions for the voltage that have the correct general characteristics and that satisfy most of the boundary conditions, but ignore the coupling. Second, we solve the boundary value problem for the pressure and secondary flow with this approximate voltage introduced into one of the two coupling terms. Third, we solve the boundary value problem for the voltage and primary flow with the results of the second step introduced into the second coupling term. The result is a set of improved trial functions for the voltage that satisfy the governing equations and boundary conditions and that incorporate a good approximation of the coupling through the successive rather than simultaneous treatment of the two coupling terms. In theory, one could iterate through the second and third steps to further improve the trial functions. With a careful choice of the preliminary trial functions, the first improvement is relatively

SIDE LAYER SOLUTIONS IN NONUNIFORM MAGNETIC FIELD   21

straightforward, but the second or successive improvement would involve almost as much algebra as the exact solution. Since the objective is to find an approach with a balance between its numerical and analytical components, increasing the number of trial functions with one improvement is better than successive improvements of a given set of trial functions. The solution is given by the sum of the trial functions times coefficients that are chosen to minimize some measure of the error. The objectives of the present paper are to develop a logical method of deriving trial functions that reflect the basic mathematics and physics, and to compare a few simple trial functions with the results of the one known, exact solution.

Ludford and Walker[5] formulate the governing equations for side layers in rectangular ducts with any combination of wall conductivities and with any planar magnetic field parallel to the sides. The magnitudes of the side layer variables and the boundary conditions are different for different wall conductivities. For ducts with thin conducting walls, there are two boundary layers on each side: an outer layer that carries an axial electric current and an inner layer that carries half the mass flux. Walker[6] presents solutions for the outer layer. The side layer formulation of Ludford and Walker[5] applies to the inner layer, but this inner layer has not been successfully treated to date. For ducts with insulating walls, there is only one side layer on each side, and this layer carries both an axial current and half the mass flux. This side layer is also governed by the formulation of Ludford and Walker[5] and has not been successfully treated to date. For ducts with perfectly conducting sides and with a polar magnetic field, Petrykowski and Walker[3] present side layer solutions that represent the practical limit of the essentially analytical approach originally developed by Walker et al.[7] For perfectly conducting sides, the side layer problems can be solved independently in each cross section, while, for thin conducting or insulating sides, the boundary conditions introduce axial derivatives that change the integral equations for the former side conductivity to integro-differential equations for the latter ones. This addition of the axial derivative makes the further extension of the essentially analytical approach impossible in practice.

The present objective is to forumlate an approximate method of treating side layer problems that can be practically applied to ducts with thin conducting or insulating walls and to ducts with nonpolar magnetic fields. We will only consider the problem already solved by Petrykowski and Walker[3], so that we can compare our trial functions with the known solutions. For ducts with thin conducting or insulating sides, the constant coefficients multiplying the trial functions in the present Galerkin solution are replaced by coefficient functions of the axial coordinate. The boundary conditions with the axial derivatives, the corresponding error minimization, and the introduction of separation of variable solutions[8] yield an eigenvalue problem on a matrix in place of the present set of linear, algebraic equations.

For nonpolar magnetic fields, the governing equations have coefficients which vary along magnetic field lines.[5] The present separation of variable method used to generate trial functions then yields the same ordinary differential equation in the boundary-layer coordinate, but the ordinary differential equation in the coordinate along magnetic field lines involves variable coefficients. The sines and cosines of θ in the present trial functions are replaced by the special functions for the ordinary differential equation with variable coefficients.

Basic Side Layer Equations

Petrykowski and Walker[3] treat liquid metal flows in rectangular ducts with parallel perfect conductors at $z = \pm 1$ (sides), with straight diverging insulators at $\theta = \pm\theta_0$ (top and bottom), and with a polar magnetic field, $\underset{\sim}{B} = r^{-1}\hat{\theta}$, where r, θ, z are cylindrical coordinates, $\hat{\theta}$ is a unit vector, the characteristic length L is half the distance between the conductors, the characteristic velocity V_0 is the average radial velocity at $r = 1$, and the characteristic magnetic field strength B_0 is the field strength at $r = 1$. To leading order, the only nonzero dimensionless core variables are the radial velocity, transverse electric current density, electric potential function, and pressure, which are given by

$$v_r = r^{-1} \quad j_z = r^{-2} - \phi_0$$

$$\phi = \phi_0 z \quad p = \phi_0 \ln r + 1/2r^2$$

respectively, where ϕ_p is half the dimensionless voltage difference between the conductors.

The boundary value problem for the side layer at $z = -1$ consists of the governing equations

$$\frac{\partial^2 \phi}{\partial \theta^2} - \frac{\partial^4 \phi}{\partial z^4} = 2\frac{\partial p}{\partial z} \tag{1a}$$

$$\frac{\partial V_\Theta}{\partial \Theta} - \frac{\partial^2 P}{\partial Z^2} = -2\frac{\partial \Phi}{\partial Z} \qquad (1b)$$

$$\frac{\partial P}{\partial \Theta} - \frac{\partial^2 V_\Theta}{\partial Z^2} = 0 \qquad (1c)$$

and the boundary conditions

$$\frac{\partial \Phi}{\partial Z} = 1 \quad \text{at} \quad Z = 0 \qquad (2a)$$

$$\frac{\partial P}{\partial Z} = \Phi \quad \text{at} \quad Z = 0 \qquad (2b)$$

$$\frac{\partial P}{\partial Z} = V_\Theta \quad \text{at} \quad Z = 0 \qquad (2c)$$

$$\frac{\partial P}{\partial Z} = 0 \quad \text{at} \quad Z = 0 \qquad (2d)$$

$$V_\Theta = 0 \quad \text{at} \quad \Theta = \pm\Theta_o \qquad (3a)$$

$$\frac{\partial \Phi}{\partial \Theta} = \frac{\partial^2 \Phi}{\partial Z^2} \quad \text{at} \quad \Theta = \pm\Theta_o \qquad (3b)$$

$$\Phi \to 0 \quad \text{as} \quad Z \to \infty \qquad (4a)$$

$$V_\Theta \to 0 \quad \text{as} \quad Z \to \infty \qquad (4b)$$

$$P \to 0 \quad \text{as} \quad Z \to \infty \qquad (4c)$$

The other side layer variables are given by

$$V_r = 1 + \frac{\partial \Phi}{\partial Z} \qquad (5a)$$

$$V_z = S(\Phi) - \frac{\partial P}{\partial Z} \qquad (5b)$$

$$J_r = \frac{\partial P}{\partial Z} \qquad (5c)$$

$$J_\Theta = -\frac{\partial \Phi}{\partial \Theta} \qquad (5d)$$

$$J_z = P + S(P) + \frac{\partial^3 \Phi}{\partial Z^3} \tag{5e}$$

where

$$S(P) = P + Z\frac{\partial P}{\partial Z} \quad Z = \frac{M^{1/2}(z+1)}{r}$$

Here, $M = B_0 L(\sigma/\eta)^{1/2}$ is the large Hartmann number, while σ and η are the liquid metal's electrical conductivity and viscosity. The capital letters denote the leading-order side layer variables with core values subtracted and with factors involving multiplicative powers of r and M removed, so that all variables denoted by capital letters are independent of r, and all except V_r vanish as $Z \to \infty$, while $V_r \to 1$.

The boundary value problem (1a, 2a, 2c, 3b, 4a) governing Φ is coupled to the problem (1b, 1c, 2b, 2d, 3a, 4b, 4c) governing P and V_Θ through the right-hand sides of Eqs. (1a, 1b). If these right-hand sides were replaced by zeros, then $P = V_\Theta = 0$, and, as we will see, Φ could not satisfy the matching condition (4a). The coupling makes the solution of this side layer problem much more difficult than the corresponding problem for a uniform magnetic field.[7] Nevertheless, Petrykowski and Walker[3] succeed in applying the same essentially analytical method to this problem, and some typical velocity profiles from their solution are presented as the solid lines in Fig. 1. Since further extension of

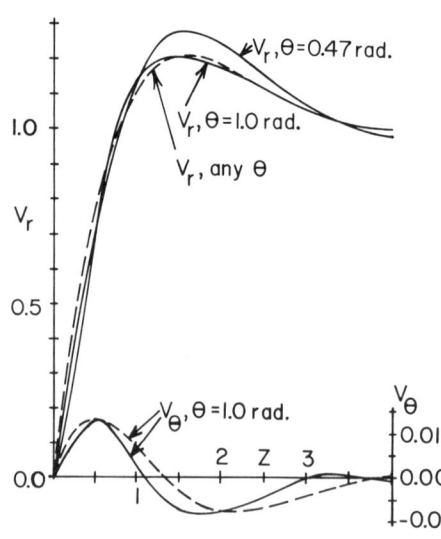

Fig. 1 Side layer velocity profiles for $\Theta_0 = 1.1071$ rad (tan $\Theta_0 = 2.0$). Solid lines present the results of Petrykowski and Walker.[3] Dashed lines present the first trial function in each case.

this method is impractical, we begin development of an approximate method and test it against the results of the more exact method.

The first step in the approximate method is the analytical solution of the boundary value problem (1a, 2a, 2c, 3b, 4a), treating $P(\Theta,Z)$ as a known function. Fourier sine and cosine transforms are introduced for Φ and P, respectively. These transforms reduce Eq. (1a) to an ordinary differential equation in Θ, introduce the unknown value of the dimensionless radial wall shear stress

$$F(\Theta) = \frac{\partial^2 \Phi}{\partial Z^2} \quad \text{at} \quad Z = 0$$

ignore the boundary condition (2a), and reduce the boundary conditions (3b) to simple point boundary conditions at $\Theta = \pm\Theta_0$. The inhomogeneous ordinary differential equation is solved using Green's functions to obtain the transform of Φ in terms of integrals of Green's functions times F or the transforms of P. Fourier inversion and a convolution integral for the term involving the transform of P give

$$\Phi = \int_{-\Theta_0}^{\Theta_0} \left(\int_0^\infty P(\Theta^*,t) \left[E(t + Z) - E(t - Z) \right] dt - F(\Theta^*) \, E(Z) \right) d\Theta^* \tag{6}$$

where

$$E(Z) = 0.5 \, \text{erf} \, (Z/2|\Theta - \Theta^*|^{1/2})$$

The unknown function F is governed by an integral equation obtained by introducing the solution (6) into the previously ignored boundary condition (2a),

$$\int_{-\Theta_0}^{\Theta_0} F(\Theta^*) |\Theta - \Theta^*|^{-1/2} d\Theta^* = 2\pi^{1/2}$$

$$+ 2 \int_{-\Theta_0}^{\Theta_0} \int_0^\infty P(\Theta^*,t) |\Theta - \Theta^*|^{-1/2} \tag{7}$$

$$\times \exp(-t^2/4|\Theta - \Theta^*|) dt \, d\Theta^*$$

If $P = 0$, then Eq. (7) is precisely the integral equation solved independently by Hunt and Stewartson[9] and by Chiang and Lundgren,[10] and Eq. (6) with their solution for F gives their solution for Φ. Since

$$\Phi \to -0.5 \int_{-\Theta_0}^{\Theta_0} F(\Theta^*) \, d\Theta^* \quad \text{as} \quad Z \to \infty$$

if $P = 0$, and since the solution of Eq. (7) with $P = 0$ gives $F > 0$ for all Θ, a solution with $P = 0$ cannot satisfy the matching condition (4a). Correspondingly, the side layer velocity profile given by Hunt and Stewartson[9] is monotonic and has no overshoot where $V_r > 1$, as in Fig. 1. The overshoot means that the side layer's displacement thickness is zero at every point on the side, and this feature is an important one for nonuniform field side layers. Therefore, at least an approximation of the correct P is needed to capture the essential physics of these side layers.

Carleman[11] presents explicit solutions for the integral equation (7) with an arbitrary function of Θ on the right-hand side, while Lundgren and Chiang[12] extend Carleman's solution to more general fractional powers of $|\Theta - \Theta^*|$. Therefore, once an approximate $P(\Theta,Z)$ is known, the solution of the integral equation (7) to obtain F and the generation of voltage, velocity and electric current profiles from the solutions (5,6) with this F are both straightforward.

Generation of Trial Functions

The steps to generate trial functions follow: i) Select a series of trial functions for $\Phi(\Theta,Z)$ that satisfy as much of the boundary value problem (1a, 2a, 2c, 3b, 4a) as possible, especially important features such as the overshoots in the radial velocity profiles, but that also lead to simple explicit solutions of the boundary value problem (1b, 1c, 2b, 2d, 3a, 4b, 4c) for the corresponding trial functions for V_Θ and P. ii) Introduce the trial functions for P into the integral equations (7) and the solution (6) in order to obtain the improved trial functions for Φ that satisfy all features of the problem (1a, 2a, 2c, 3b, 4a). In order to insure simple explicit trial functions for V_Θ and P, we obtain the fundamental elements of our trial functions from separation of variable solutions for the Eqs. (1b, 1c) with $\Phi = 0$ and for the boundary conditions (3a, 4b, 4c); namely,

$$V_\Theta = \sin(n\pi\Theta/\Theta_0) \, (A_n C_n + B_n S_n) \quad (8a)$$

$$P = \cos(n\pi\Theta/\Theta_0)(B_n C_n - A_n S_n) \qquad (8b)$$

where

$$C_n(Z) = \exp(-\alpha_n Z) \cos(\alpha_n Z)$$

$$S_n(Z) = \exp(-\alpha_n Z) \sin(\alpha_n Z)$$

$$\alpha_n = (n\pi/2\Theta_0)^{1/2}$$

and n is any positive integer, while A_n and B_n are arbitrary constants. For the inhomogeneous version of Eqs. (1b, 1c) with Φ included, the homogeneous solution (8a) suggests a particular solution of the form

$$V_\Theta = \sin(n\pi\Theta/\Theta_0) Z (a_n C_n + b_n S_n) \qquad (9)$$

which leads to trial functions for Φ, namely,

$$\Phi = 2 \cos(n\pi\Theta/\Theta_0)(b_n C_n - a_n S_n) \qquad (10)$$

When we combine the homogeneous solutions (8) and the particular solutions (9, 10) (and the corresponding solution for P), we obtain the solutions

$$V_\Theta = \sum_{n=1}^{\infty} a_n \sin(n\pi\Theta/\Theta_0) Z C_n \qquad (11a)$$

$$P = g + \sum_{n=1}^{\infty} a_n \cos(n\pi\Theta/\Theta_0)(C_n + S_n - \alpha_n Z S_n)/\alpha_n \qquad (11b)$$

$$\Phi = 0.5 \, g' - 2 \sum_{n=1}^{\infty} a_n \cos(n\pi\Theta/\Theta_0) S_n \qquad (11c)$$

which satisfy the Eqs. (1b, 1c) and the boundary conditions (2b-d, 3a, 4), as long as the arbitrary function g(Z) satisfies

$$g'(0) = 0 \quad g \to 0 \text{ as } Z \to \infty \qquad (12)$$

In addition, the solution (11c) with g = 0 satisfies the homogenous version of Eq. (1a) with P = 0. If we introduce the solution (11c) and the average of the solution (11b) over $-\Theta_o < \Theta < \Theta_o$ into Eq. (1a), we obtain an ordinary differential equation governing g. The solution of this equation that satisfies the conditions (12) and the condition

$$g''(0) = -2 \qquad (13)$$

is

$$g = \exp(-Z)\left[\cos(Z) + \sin(Z)\right] \qquad (14)$$

The condition (13) is obtained by introducing the average of the solution (11c) over $-\Theta_o < \Theta < \Theta_o$ into the boundary condition (2a). Therefore, the solutions (11) with g given by the solution (14): i) satisfy the equations (1b, 1c) and the boundary conditions (2b-d, 3a, 4) exactly; ii) satisfy Eq. (1a) with the average P; iii) satisfy boundary condition (2a) with the average ϕ; and iv) only ignore boundary condition (3b).

The next step is to solve the integral equation (7) with the solution (11b) for P to obtain

$$F = f_o + \sum_{n=1}^{\infty} a_n f_n \qquad (15)$$

where f_o combines the solution for P = 0 (Ref. 9) and the solution for the second term on the right-hand side of Eq. (7) with P_o = g, while f_n is the solution for the second term with

$$P_n = \cos(n\pi\Theta/\Theta_o)(C_n + S_n - \alpha_n Z S_n)/\alpha_n \qquad (16)$$

When the solution (15) is introduced into the solution (6), we obtain

$$\phi = \phi_o + \sum_{n=1}^{\infty} a_n \phi_n \qquad (17)$$

where ϕ_o is given by Eq. (6) with P_o = g and F = f_o, while ϕ_n is given by Eq. (6) with expression (16) and F = f_n.

The solutions ϕ_j, for j = 0, 1, 2, ..., represent improved trial functions that satisfy the boundary value problem (1a, 2c, 3b, 4a) with the trial functions P_j on the right-hand side of Eq. (1a) and that satisfy

$$\frac{\partial \phi_j}{\partial Z}(\Theta,0) = -1 \quad \text{for} \quad j = 0$$

$$= 0 \quad \text{for} \quad j > 0$$

SIDE LAYER SOLUTIONS IN NONUNIFORM MAGNETIC FIELD   29

If solution (11c) with some set of values for the constants a_n is the correct solution, then the solutions (11a, 11b, 17) are also correct, and the solutions (11c, 17) are the same. The best possible final solution (17) is obtained from the best possible initial solution (11c), and both are achieved by minimizing the difference between the solutions (11c, 17). If we define the error as the integral of the square of the difference over the side layer, i.e., over $-\Theta_0 \leq \Theta \leq \Theta_0$, $0 \leq Z < \infty$, and if we minimize this error with respect to a_n, then we obtain an infinite set of simultaneous, linear, algebraic equations. In these equations, each a_n is multiplied by a coefficient given by integrals of $\Phi_m \Phi_n$ or of $\Phi_m \cos(n\pi\Theta/\Theta_0) S_n$, since these functions are not orthogonal, while the inhomogeneous terms are given by integrals of the products of g' or Φ_0 with either Φ_n or $\cos(n\pi\Theta/\Theta_0) S_n$. We must truncate this infinite set of simultaneous equations with N equations, and we can then solve for a_n, for n = 1, 2, 3, ..., N. These a_n give the best possible solution (17) with these N trial functions.

If the first few trial functions in the expressions (11) closely resemble the correct solutions, then the error minimization gives excellent results with small values for N. The dashed lines in Fig. 1 present the trial functions

$$V_r = 1 + 0.5 g'' \tag{18a}$$

$$V_\Theta = a_1 \sin(n\pi\Theta/\Theta_0) Z C_1 \tag{18b}$$

with a_1 chosen to give the same maximum value of V_Θ in order to demonstrate the similarity of the shapes of the graphs. The excellent agreement of V_r at $\Theta = 1.0$ rad is a coincidence, and comparison with the solid line for $\Theta = 0.47$ rad is more typical.

Figure 1 demonstrates that the present separation of variable method generates trial functions that closely resemble the correct solutions, so that truncation after a few trial functions should give excellent results with minimal effort. Figure 1 only shows the first trial function in each case, but the roles of the second trial functions are easily identified. The expression (18a) gives the same profile and the same maximum V_r of 1.208 for all Θ, while the correct solution has a maximum value of V_r that varies from 1.29 at $\Theta = 0$ to 1.20 at $\Theta = 1$, for $\Theta_0 = 1.1071$ rad. On the other hand, if we use the first two trial functions in expression (11c) with a_1 still determined by matching the maximum V_Θ (not at all optimal), then the two trial functions give a maximum V_r that varies from 1.28 at $\Theta = 0$ to 1.14 at

$\Theta = 1$. The expression (18b) gives $V_\Theta = 0$ at $Z = 1.319$ for all Θ, while the actual position where $V_\Theta = 0$ varies from $Z = 1.05$ to $Z = 1.45$. The second trial function in expression (11a) gives a similar variation of this position with Θ, which is absent in the first trial function. Therefore, the second trial functions begin to represent weak variations with Θ that are missed by the first trial functions.

Acknowledgment

This research was supported by the U.S. National Science Foundation under Grant CPE-8108952.

References

[1] Holroyd, R. J. and Walker, J. S., "A Theoretical Study of the Effects of Wall Conductivity, Nonuniform Magnetic Fields and Variable-Area Ducts on Liquid-Metal Flows at High Hartmann Numbers," Journal of Fluid Mechanics, Vol. 84, No. 3, February 1978, pp. 471-495.

[2] Alty, C. J. N., "MHD Duct Flow in Uniform Transverse Magnetic Fields at Arbitrary Orientation," Journal of Fluid Mechanics, Vol. 48, No. 3, 1971, p. 429.

[3] Petrykowski, J. C. and Walker, J. S., "Liquid-Metal Flow in a Rectangular Duct with a Strong Nonuniform Magnetic Field," Journal of Fluid Mechanics, Vol. 139, February 1984, pp. 309-324.

[4] Walker, J. S., "Laminar Duct Flows with Strong Magnetic Fields," Liquid-Metal Flows and Magnetohydrodynamics, Proceedings of the Fourth Beer-Sheva International Seminar, Vol. 2, AIAA, New York, 1984.

[5] Ludford, G. S. S. and Walker, J. S., "Current Status of MHD Duct Flow," MHD-Flows and Turbulence II, Proceedings of the Second Bat-Sheva International Seminar, Beer-Sheva, Israel Universities Press, Jerusalem, 1980, pp. 83-95.

[6] Walker, J. S., "Three-Dimensional Laminar MHD Flows in Rectangular Ducts with Thin Conducting Walls and Strong Transverse Nonuniform Magnetic Fields," Liquid-Metal Flows and Magnetohydrodynamics, Proceedings of the Third Beer-Sheva International Seminar, New York, 1982, pp. 3-19.

[7] Walker, J. S., Ludford, G. S. S., and Hunt, J. C. R., "Three-Dimensional MHD Duct Flows with Strong Transverse Magnetic Fields, Part 2: Variable-Area Rectangular Ducts with Conducting Sides," Journal of Fluid Mechanics, Vol. 46, No. 4, April 1971, pp. 657-684.

[8] Walker, J. S., Ludford, G. S. S., and Hunt, J. C. R., "Three-Dimensional MHD Duct Flows with Strong Transverse Magnetic Fields, Part 3: Variable-Area Rectangular Ducts with Insulating Walls," Journal of Fluid Mechanics, Vol. 56, No. 1, November 1972, pp. 121-141.

[9] Hunt, J. C. R. and Stewartson, K., "Magnetohydrodynamic Flow in Rectangular Ducts. II," *Journal of Fluid Mechanics*, Vol. 23, No. 3, November 1965, pp. 563-581.

[10] Chiang, D. and Lundgren, T., "Magnetohydrodynamic Flow in a Rectangular Duct with Perfectly Conducting Electrodes," *Zeitschrift fuer Angewandte Mathematik und Physik*, Vol. 18, No. 1, 1967, pp. 92-104.

[11] Carleman, T., "Uber die Abelsche Integralgleichung mit Konstanten Integrationsgrengen," *Mathematische Zeitschrift*, Vol. 15, 1922, pp. 111-120.

[12] Lundgren, T. and Chiang, D., "Solution of a Class of Singular Integral Equations," *Quarterly of Applied Mathematics*, Vol. 24, No. 4, 1967, pp. 303-313.

Applications of MHD Flows Between Rotating Disks

P. Herve*
Université de la Réunion, Ile de la Réunion, France

and

C. Vives†
Centre Universitaire d'Avignon, Avignon, France

Abstract

In this study, a liquid metal is contained between two parallel disks of very different electrical conductivities in the presence of a stationary, uniform, axial magnetic field. A theoretic study is made of the motion produced both by an outward radial flow from a central source and the rotation of a conducting, or insulating, upper disk. This paper presents an analysis of the velocity and pressure distribution, and also of the viscous friction torque upon the disks, when a single disk is rotating and when the two disks are rotating with different velocities. The results are compared with the results from experiments in the absence of a source. An example of practical use is presented: the magnetomechanic pump.

Introduction

As far as classical fluid mechanics is concerned, numerous papers have been published on the study of a viscous liquid flow between two rotating disks with or without a central source. Beyond its fundamental aspect, this study can lead to interesting applications, for example, the effects of convective diffusion in the liquids, particularly the action on the speeds of electrochemical reactions upon moving electrodes (rotating disks) immersed in electrolytes.[1]

Paper presented at the Fourth Beer-Sheva Seminar on MHD Flows and Turbulence, Ben-Gurion University of the Negev, Beer-Sheva, Israel, Feb. 27-March 2, 1984. Copyright © 1985 by the American Institute of Aeronautics and Astronautics, Inc. All rights reserved.

*Teaching Assistant, Faculté des Sciences, Laboratoire de Physique.

†Professor, Faculté des Sciences, Laboratoire de Magnetohydrodynamique.

The methods of the rotating electrode and crystal[2] used to control the growth of crystals are closely linked to this problem, too. Moreover, the features and performances of the viscously driven pumps and compressors have also been analyzed.[3,4]

The radial flow between two disks rotating with a constant angular velocity has been studied by Peube and Kreith[5]; the results were obtained by successive approaches in the form of series expansion.

The great number of possible combinations between mechanic and electromagnetic boundary conditions (wall electrical conductivity and wall magnetic permeability) increases the problems when an axial, stationary, uniform magnetic field is applied. Analytic results have been obtained when one (or two) insulating or conducting disks rotate(s) without source, in a space limited by insulating coaxial cylinders,[6] or electroconducting and short-circuiting cylinders.[7] A theoretical study[8] shows that, to adopt the simplifying assumption of an infinite conductivity of the rotating disk, its thickness must be greater than one-tenth of the channel width and its electrical conductivity must exceed the fluid conductivity. Moreover, the Hartmann number M must be greater than 50, which holds for most laboratory experiments. Finally, theoretical results compared with experimental ones for M > 50 (Ref. 7) show that if the disks rotate about the axis of symmetry, the fluid trajectories are circles centered on this axis.

Only a few papers that take into account the presence of a central source have been published, and none include experimental work. The influence of an application of a transverse magnetic field on a strictly radial laminar flow between two insulating, fixed, parallel planes has been studied by Reddy[9]; the induced magnetic fields being neglected, the solutions were proposed in power series expansion.

The more complex case, with the superpositions of a radial flow due to a central source and the motion of the liquid metal caused by the rotation of two insulating disks in the presence of an axial magnetic field, has been treated by Khan.[10] Restrictive assumptions were made: The induced magnetic field is neglected, the Reynolds number and especially the Hartmann number are very small (the numeric results were only given for M = 1), and electroconducting walls are not considered. The solutions show that the magnetic field decreases the radial component of the velocity and increases the azimuthal component.

Metallurgists show an increasing interest relating to the industrial applications of magnetodynamic, especially the improvement of the structure and fineness of the grain of solidification by means of control of the forced convection inside the molten metal. Thus, our purpose was to extend the former studies by removing limitations in the assumptions and on disk conductivity, to compare the theoretical results to experiments, and to apply the results to the design of clutches, brakes and pumps of a new type: the magnetomechanic devices.

Flows Between Parallel Disks in the Absence of a Source

Theoretical Study

We consider a liquid metal within a toroidal duct made by two insulating coaxial cylinders with axis OZ, height 2a, radii R_1, R_2, and limited up and down by two disks with radius R_2. The slow motion, around OZ, of the upper disk, laminarly drives the fluid in the presence of a steady magnetic field that is uniform and parallel to the axis of rotation of the apparatus. Two cases are considered:
   1) The rotating disk is electroconducting; the fixed disk is insulating (Fig. 1).
   2) The rotating disk is insulating; the fixed disk is electroconducting.
   The theoretical results for the different boundary conditions, for laminar flow, and for any value of M have been obtained analytically through long, but classical calculations.[6,8] Thus, only the main results are presented here.

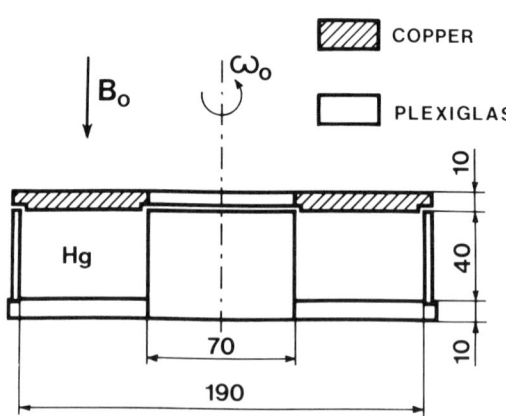

Fig. 1 Setup of experimental device. Annular vessel with upper rotating disk.

Using the Navier-Stokes and Maxwell's equations added to the equation of continuity, we obtain

$$\frac{\partial^2 V}{\partial X^2} + \frac{1}{X}\frac{\partial V}{\partial X} - \frac{V}{X^2} + \gamma^2 \frac{\partial^2 V}{\partial Z^2} + M\gamma^2 \frac{\partial h}{\partial Z} = 0 \quad (1)$$

$$\frac{\partial^2 h}{\partial X^2} + \frac{1}{X}\frac{\partial h}{\partial X} - \frac{h}{X^2} + \gamma^2 \frac{\partial^2 h}{\partial Z^2} + M\gamma^2 \frac{\partial V}{\partial Z} = 0 \quad (2)$$

with the following dimensionless parameters:

$$X = \frac{r}{R_1} \quad Z = \frac{z}{a} \quad V(X,Z) = \frac{V^+}{\omega_0 R_1} \quad h(X,Z) = \frac{h_\theta^+}{\sqrt{\sigma\eta}\omega_0 R_1}$$

$$\alpha = \frac{R_2}{R_1} \quad \gamma = \frac{R_1}{a} \quad M = \sqrt{\frac{\sigma}{\eta}} B_0 a \quad (3)$$

(the exponent + denotes a physical parameter).

Using electromagnetic and mechanic boundary conditions and following classical calculations, the expressions for $V(X,Z)$ are

$$V_1(X,Z) = 2 \sum_n H(k_n) t_1(k_n, X) \frac{\{sh(r_1-r_2)/2\, Z\}}{(r_1-r_2)sh2(r_1-r_2)}$$

$$\times \left[r_1 ch\left(\frac{r_1+r_2}{2} Z - 2r_1\right) - r_2 ch\left(\frac{r_1+r_2}{2} Z - 2r_2\right)\right] \quad (4)$$

when the moving disk has an infinite conductivity, and

$$V_2(X,Z) = \sum_n H(k_n) T_1(k_n, X)$$

$$\frac{ch2r_2 shr_1 Z - ch2r_1 shr_2 Z}{sh2(r_1 - r_2)} \quad (5)$$

when the moving disk has zero conductivity, and where

$$H(k_n) = \frac{2k_n J_1^2(k_n)}{J_1^2(k_n\alpha) - J_1^2(k_n)} \{J_1(k_n\alpha)[\alpha^2 Y_2(k_n\alpha) - Y_2(k_n)]$$

$$- Y_1(k_n\alpha)[\alpha^2 J_2(k_n\alpha) - J_2(k_n)]\}$$

$$T_1(K_n, X) = J_1(k_n X) Y_1(k_n\alpha) - J_1(k_n\alpha) Y_1(k_n X)$$

$$r_1 = -(M/2) + \sqrt{(M/2)^2 + (k_n/\gamma)^2} > 0$$

$$\text{and} \quad r_2 = -M - r_1 < 0 \tag{6}$$

k_n are the positive solutions of $T_1(k_n, 1) = 0$, and J_1, Y_1 and Y_2, Y_2 are the Bessel functions and modified Bessel functions, respectively.

Thus, we can calculate the viscous friction torques upon the disks. The tangential viscous stress in the fluid is $\tau = \eta(V^+/z)$, and the viscous force moment $||\vec{M}_A^+||$ about the axis is

$$C = \int_1^\alpha X^2 \frac{V}{Z} \frac{V}{Z} dX$$

where

$$C = \frac{||M_A^+|| a}{2\pi \eta R_1^4 \omega_0} \tag{7}$$

is the dimensionless viscous force moment. Relations (4) and (5) justify a simplifying development of $V(X,Z)$:

$$V(X,Z) = \sum_n H(k_n) T_1(k_n, X) \psi(Z)$$

and we obtain

$$C = \sum_n K(k_n) \frac{d\psi(Z)}{dZ} \quad \text{with} \quad K(k_n) = H^2(k_n) \frac{J_1^2(k_n\alpha) - J_1^2(k_n)}{2k_n^2 J_1^2(k_n)} \tag{8}$$

When the moving disk is electroconducting, the moment of the torque is

$$C_1(Z) = \sum_n K(k_n)\{r_1^2 \text{ch } r_1(Z-2) + r_2^2 \text{ch } r_2(Z-2) - r_1 r_2$$

$$\times [\text{ch}(2r_1-r_2 Z) + \text{ch}(2r_2-r_1 Z)]/(r_1-r_2)\text{sh}2(r_1-r_2)\} \quad (9)$$

The solution for the electroconducting disk (moving) is

$$C_1(2) = \sum_n K(k_n) \frac{r_1^2 + r_2^2 - 2r_1 r_2 \text{ch}2(r_1-r_2)}{(r_1-r_2)\text{sh}2(r_1-r_2)} \quad (10)$$

and for the insulating disk (motionless),

$$C_1(0) = \sum_n K(k_n) \frac{r_1 \text{ch}2r_1 - r_2 \text{ch}2r_2}{\text{sh}2(r_1-r_2)} \quad (11)$$

For large values of M, $r_1 \to 0^+$ and $r_2 \to -M$, while the moments become $C_1(2) \to 0$ and $C_1(0) \to M \sum_n K(k_n)$. Figure 2 presents $C_1(0) = f(M)$ and $C_1(2) = f'(M)$ when increased from 0 to 350. We note that $C_1(2) \to 0$ rapidly and becomes very small, compared with $C_1(0)$, when $M > 150$. The function $C_1(0)$ becomes linear, when $M > 10$, which is explained by $M\sum_n K(k_n) = 13.3$ M for the values in Fig. 1. If M is equal to

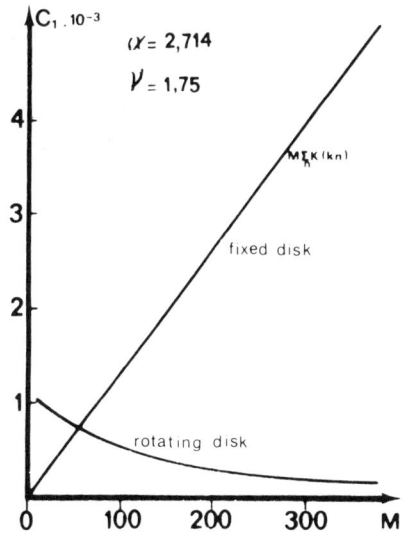

Fig. 2 Influence of the Hartmann number upon the moment of the viscous friction torques (first case).

zero, $C_1(0) = 2.39$, and the ratio $C_1(0)_{M \neq 0}/C_1(0)_{M=0}$ is equivalent to 5.54 M. Consequently, when M increases from 0 to 200, the moment of the viscous friction torque upon the fixed disk is multiplied by 1100, whereas, at the same time, this moment is strongly decreased upon the rotating disk.

It should be noted that this does not mean that there are different total moments on the two disks. The circuit for the radial electric current in the Hartmann layer on the insulating disk is completed by an axial current in the boundary layer on the cylinder wall, a potential current flow inside the conducting disk.[11] This current and the axial magnetic field produce an azimuthal body force inside the disk. The total moment on the conducting disk (viscous and electromagnetic body force) exactly equals the total moment on the insulating disk (only viscous).

When the moving disk is insulating, the moment of the torque is

$$C_2(Z) = \sum_n K(k_n) \frac{r_1 \text{ch} 2r_2 \text{ch} \, r_1 Z - r_2 \text{ch} 2r_1 \text{ch} \, r_2 Z}{\text{sh} 2(r_1 - r_2)} \quad (12)$$

the solution for the insulating disk (moving) is

$$C_2(2) = \sum_n K(k_n)(r_1 - r_2) \frac{\text{ch} 2r_1 \text{ch} 2r_2}{\text{sh} 2(r_1 - r_2)} \quad (13)$$

and for the electroconducting disk (motionless):

$$C_2(0) = \sum_n K(k_n)(r_1 - r_2) \frac{r_1 \text{ch} 2r_2 - r_2 \text{ch} 2r_1}{\text{sh} 2(r_1 - r_2)} \quad (14)$$

For large values of M, $C_2(0) \to 0$ and $C_2(2) \to M \Sigma K(k_n)$.

Figure 3 shows the variation of $C_2(2)$ as a function of M. The function $C_2(0) = f(M)$ is not represented because its numerical values are too small compared to $C_2(2)$ and decrease from 2.39 to 0 as the Hartmann number increases. In this case, the very important viscous torque, observed upon the moving disk, enables us to envisage a MHD speed reducer (or brake) presenting a progressive adjustment.

Finally, when the insulating disk can rotate under the action of the fluid movement, we introduce a new parameter $m = \omega_0/\omega$ (≥ 1), where ω_0 is the angular velocity of the driving electroconducting ring and ω is the angular velocity of the driven insulating ring. The viscosity force moment

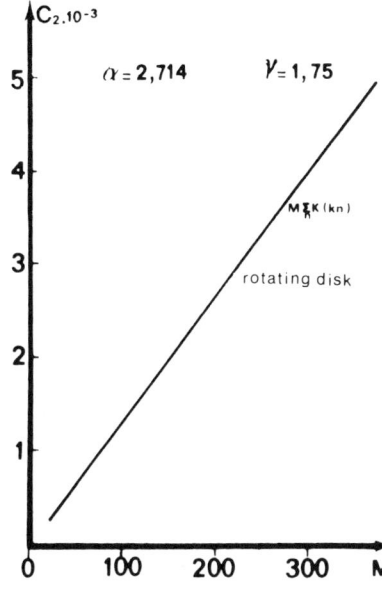

Fig. 3 Influence of the Hartmann number upon the moment of the viscous friction torques (second case).

upon the disks can be expressed by

$$C_1^*(Z) = \frac{1}{2m} \sum_n \frac{K(k_n)}{1+B_n} \left\{ [r_1 e^{r_1 z} + B_n r_2 e^{r_2 z}] \right.$$
$$\left. - (a_n + B_n b_n) r_2 e^{-r_2 z} + (c_n + d_n B_n) r_2 e^{-r_1 z} \right\} \quad (15)$$

and the solution for the insulating disk is

$$C_1^*(0) = \frac{1}{2m} \sum_n \frac{K(k_n)}{1+B_n} \left[r_1(1-c_n-B_n d_n) \right.$$
$$\left. + r_2(-B_n b_n + B_n - a_n) \right] \quad (16)$$

where

$$a_n = \frac{2[ch(2r_1) - m]}{e^{-2r_1} - e^{-2r_2}} \qquad b_n = \frac{e^{2r_2} + e^{-2r_1} - 2m}{e^{-2r_1} - e^{-2r_2}}$$

$$c_n = \frac{-e^{2r_1} - e^{-2r_2} + 2m}{e^{-2r_1} - e^{-2r_2}} \qquad d_n = \frac{2[m - ch(2r_2)]}{e^{-2r_1} - e^{-2r_2}}$$

(continued)

$$B_n = \frac{r_1(e^{2r_1} + c_n e^{-2r_1}) + a_n r_2 e^{-2r_2}}{r_2(e^{2r_2} + b_n e^{-2r_2}) + d_n r_1 e^{-2r_1}} \quad (17)$$

We note that if the parameter M is large, then

$$a_n \approx b_n \approx 0 \quad c_n \approx d_n \approx 1 \quad B_n \approx 2(m-1)/(1-2m)$$

Hence,

$$C_1^*(0) \to \frac{m-1}{m} M \sum_n K(k_n)$$

Therefore, the moment of the torque upon the free disk tends to a linear function of M, for a specific m, and an hyperbolic function of m for a specific M. Figure 4 clearly illustrates this point and this finding. It is important to note that the curves $C_1^*(0)$ are, whatever m, very close to their asymptotic value for $M \to \infty$. So, we remark that if $m = 1.5$ ($\omega = 2/3 \omega_0$), the ratio $C_1^*(0)_{M=200}/C_1(0)_{M=0}$ is close to 300 (1100, for $M \to \infty$, when the insulating disk is fixed).

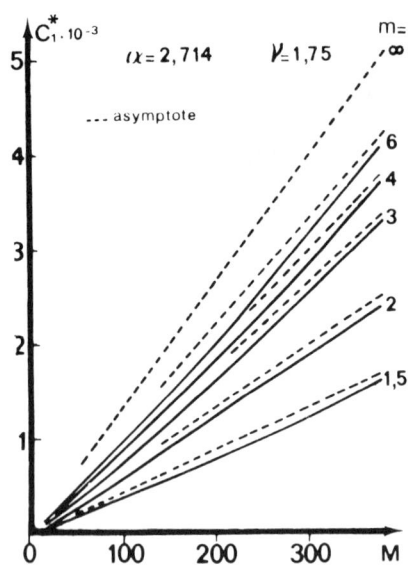

Fig. 4 Influence of the $m = \omega_0/\omega$ parameter upon the function C_1^*(M) with Z = 0.

Consequently, the effect of the magnetic field upon the viscous couples remains very important. It enables us to envisage, from such an apparatus, the working of liquid metal electromagnetic clutches offering the advantage of being progressively adjusted by a separate or simultaneous action on ω_0 and B_0 and with a very slight wear since there is no solid-solid friction.

Experimental Study

Two annular vessels of different geometric factors were made; their sizes were respectively

$2a = 40$ mm $R_1 = 35$ mm $R_2 = 95$ mm

$2a = 40$ mm $R_1 = 45.6$ mm $R_2 = 85.6$ mm

All the walls of the container were made of Plexiglas, except one of the disks, which was made of copper with a thickness e = 10 mm. As the liquid used was mercury, the conditions $e/2a = 0.25$ and $\sigma_{Hg}/\sigma_{Cu} = 0.018$ enabled us to assume, in the theoretical study, the hypothesis of an infinite conductivity.[8] The magnetic field B_0 varied between 0 and 0.35 T, and the period for one rotation varied between 3.2 and 31.7 s.

The measurements of local velocities were led in a cross-sectional area by means of different technique[12]: a miniaturized double Pitot tube connected with an electromagnetic micromanometer (M = 0 and M ≠ 0), an incorporated magnet probe (M = 0), and an electromagnetic conduction probe (M ≠ 0).

Electroconducting Rotating Disk and Insulating Fixed Disk. Figure 5 shows a velocity profile, plotted at midheight of the tank and along a radius (d = r - R_1), for different values of the magnetic field. It occurs that the liquid is notably accelerated. The velocity distributions, for M > 50 (B_0 > 0.1 T), and in accordance with the theory, tend to an asymptote V = $r\omega_0$; that means, towards a flow for which all the points of the fluid and those of the rotating disk have the same angular velocity. In Fig. 6, for the d given, it is clear that the velocity is not a function of Z, except in the very thin boundary layer, in the vicinity of the fixed disk.

Insulating Rotating Disk and Electroconducting Fixed Disk. Examples of velocity distributions, always plotted at midheight of the vessel, are depicted in Fig. 7. It may be

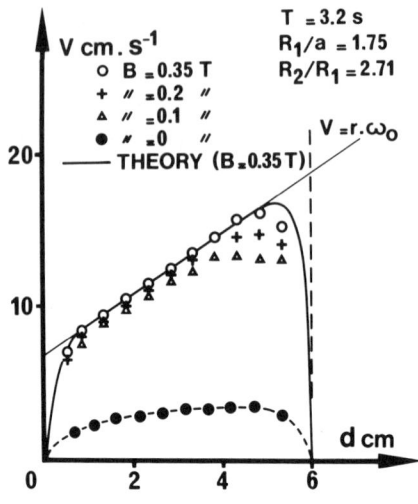

Fig. 5 Velocity profiles plotted along a radius for different values of the magnetic field.

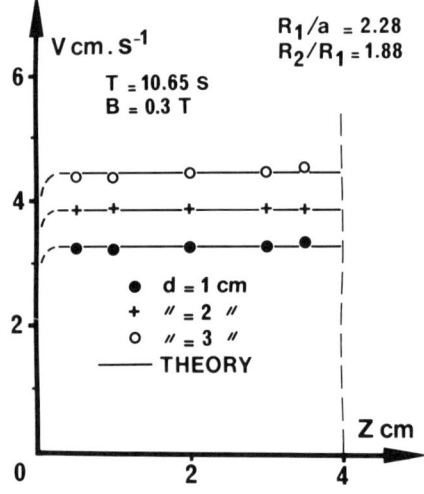

Fig. 6 Velocity profiles plotted along the OZ axis.

seen that, for large values of B_0, the velocity profile is characterized by two maximums very close to the cylinders, of R_1 and R_2 radii, and by a nearly complete stopping of the flow within the duct. Figure 8, which represents the velocity profile along Z, depends on conditions different from those of the theoretical assumptions (short period and weak magnetic field). A clear slowdown of the fluid flow is still observed, and it does not depend on Z, except near the disks, where there is a boundary layer with a large gradient. (V_p indicates the fluid velocity with a distance d from the moving disk.)

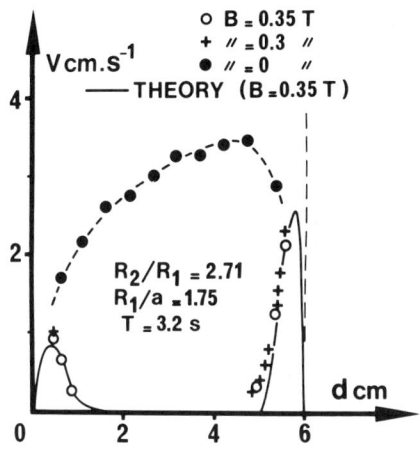

Fig. 7 Velocity profiles plotted along a radius for different values of the magnetic field.

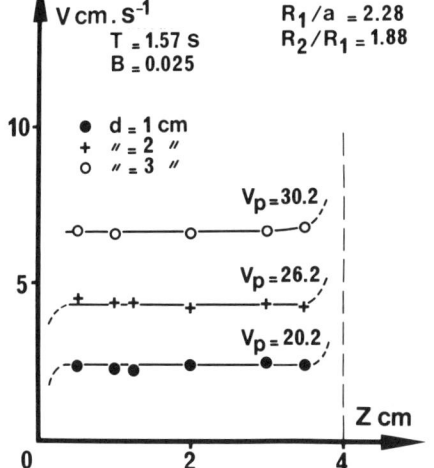

Fig. 8 Velocity profiles plotted along the OZ axis.

Remarks. Although the theory is partly based upon restrictive assumptions already enumerated, the confrontation of numerical calculations with experiments is satisfying (Figs. 2-4). The Stuart number $N^* = B_0 \sigma/\rho \omega_0$ (the ratio of electromagnetic body forces to initial forces) is here the best dimensionless parameter to fix the validity limit of the theory. A great number of experiments show that the theory is in agreement with measurements, when $N^* > 5$ (although the Taylor number $R_e^2 \omega/\nu$ reaches 177,000 in the cases presented in Figs. 3 and 4), and the discrepancy does not exceed 10% for $N^* = 0.4$.

Flows with a Central Source

Basic Equations

One insulating or electroconducting disk rotates with a uniform angular velocity ω_0. The second disk, parallel to the former one, is fixed with an infinite or zero conductivity. In each case, the two walls have very different electric properties, and they are separated by a distance 2a. A uniform and stationary field B_0 is applied perpendicular to the two disks.

The steady laminar flow of a liquid metal is produced by an outward radial flow from a central source and also from the rotation of the moving disk (Fig. 9). The following dimensionless parameters are used:

$$x = r/a \quad u = u^+/V \quad v = v^+/V \quad w = w^+/V \quad M = \sqrt{\sigma/\eta}\mu H_0 a$$

$$Z = z/a \quad h_i = h_i^+/\sqrt{\sigma\eta}V \quad i = \{r,\theta,z\} \quad p = p^+/\rho V^2$$

$$Q = q/2\pi a^2 V \quad B_t = \sigma\mu\eta/\rho \quad \beta = \omega_0 a^2/\nu = \omega_0 a^2 \rho/\eta \quad (18)$$

$Va = \eta/\rho = \nu$, where V is a characteristic velocity. The flow rate q is defined by

$$q = 2\pi r \int_0^{2a} u^+ dz$$

After simplification (B_t being neglected), a set of seven differential equations appears:

$$\frac{\partial^2 u}{\partial x^2} + \frac{1}{x}\frac{\partial u}{\partial x} - \left(\frac{1}{x^2}+M^2\right)u + \frac{\partial^2 u}{\partial z^2} - \left(u\frac{\partial u}{\partial x} + w\frac{\partial u}{\partial x} - \frac{v^2}{x} + \frac{\partial P}{\partial x}\right) = 0 \quad (19)$$

$$\frac{\partial^2 v}{\partial x^2} + \frac{1}{x}\frac{\partial v}{\partial x} + \frac{\partial^2 v}{\partial z^2} - \left[u\left(\frac{\partial v}{\partial x} + \frac{v}{x}\right) + w\frac{\partial v}{\partial z}\right] + M\frac{\partial h_\theta}{\partial z} = 0 \quad (20)$$

$$\frac{\partial^2 w}{\partial x^2} + \frac{1}{x}\frac{\partial w}{\partial x} + \frac{\partial^2 w}{\partial z^2} - u\frac{\partial w}{\partial x} + w\frac{\partial w}{\partial z} + \frac{\partial P}{\partial z} = 0 \quad (21)$$

$$\frac{\partial^2 h_r}{\partial x^2} + \frac{1}{x}\frac{\partial h_r}{\partial x} - \frac{h_r}{x^2} + \frac{\partial^2 h_r}{\partial z^2} + M\frac{\partial u}{\partial z} = 0 \quad (22)$$

MHD FLOWS BETWEEN ROTATING DISKS

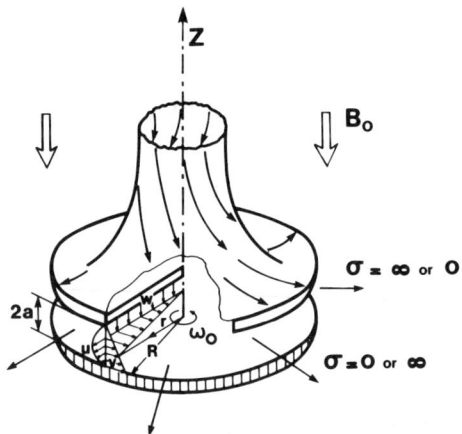

Fig. 9 Sketch of the flowfield developed in a fluid both by a source and a rotating disk.

$$\frac{\partial^2 h_\theta}{\partial x^2} + \frac{1}{x}\frac{\partial h_\theta}{\partial x} - \frac{h_\theta}{x^2} + \frac{\partial^2 h_\theta}{\partial z^2} + M\frac{\partial v}{\partial z} = 0 \tag{23}$$

$$\frac{\partial^2 h_z}{\partial x^2} + \frac{1}{x}\frac{\partial h_z}{\partial x} + \frac{\partial^2 h_z}{\partial z^2} + M\frac{\partial w}{\partial z} = 0 \tag{24}$$

$$\frac{u}{x} + \frac{\partial u}{\partial x} + \frac{\partial w}{\partial z} = 0 \tag{25}$$

Solutions in the Two Cases

In order to solve Eq. (25) we can write

$$u(x,z) = \frac{1}{x}\frac{\partial F(x,z)}{\partial z} \quad w(x,z) = -\frac{1}{x}\frac{\partial F(x,z)}{\partial x} \tag{26}$$

referring to the methods already used[8,10]; the solutions of these differential equations can be expressed as (parity problems are solved by Peube[5]):

$$F(x,z) = x^2 f_{-1}(z) + \frac{f_3(z)}{x^2} + \ldots \tag{27}$$

$$u(x,z) = x f'_{-1}(z) + \frac{f'_1(z)}{x} + \frac{f'_3(z)}{x^3} + \ldots \tag{28}$$

$$v(x,z) = x g_{-1}(z) + \frac{g_1(z)}{z} + \frac{g_3(z)}{x^3} + \ldots \tag{29}$$

$$w(x,z) = -2f_{-1}(z) + \frac{2}{x^4} f_3(z) + \ldots \tag{30}$$

$$P(x,z) = x^2 h_{-2}(z) + h_0(z) + h_1(z) \log x + \frac{h_2(z)}{x^2} + \ldots \tag{31}$$

$$h_r(x,z) = xJ_{-1}(z) + \frac{J_1(z)}{x} + \frac{J_3(z)}{x^3} + \ldots \tag{32}$$

$$h_\theta(x,z) = xk_{-1}(z) + \frac{k_1(z)}{x} + \frac{k_3(z)}{x^3} + \ldots \tag{33}$$

$$h_z(x,z) = \iota_{-1}(z) + \frac{\iota_3(z)}{x^4} + \ldots \tag{34}$$

Considering the boundary conditions, and the assumption that $X \gg 1$ ($r \gg a$), we obtain after lengthy calculations the solutions for the two following cases.

Electroconducting Rotating Disk. Using classical boundary conditions, we can demonstrate that $(\partial h_\theta/\partial x)(x,z) = -M(v - \beta x)$ [see Eq. (20)]. Hence, the system to be solved is

$$\frac{\partial^2 u}{\partial x^2} + \frac{1}{x}\frac{\partial u}{\partial x} - \left(M^2 + \frac{1}{x^2}\right)u + \frac{\partial^2 u}{\partial x^2} - \left(u\frac{\partial u}{\partial x} + w\frac{\partial u}{\partial z} - \frac{v^2}{x} + \frac{\partial P}{\partial x}\right) = 0 \tag{35}$$

$$\frac{\partial^2 v}{\partial x^2} + \frac{1}{x}\frac{\partial v}{\partial x} - \left(M^2 + \frac{1}{x^2}\right)v + \frac{\partial^2 v}{\partial z^2} - \left[u\left(\frac{\partial v}{\partial x} + \frac{v}{x}\right) + w\frac{\partial v}{\partial z}\right] + \beta M^2 x = 0 \tag{36}$$

$$\frac{\partial^2 w}{\partial x^2} + \frac{1}{x}\frac{\partial w}{\partial x} + \frac{\partial^2 w}{\partial z^2} - \left(u\frac{\partial w}{\partial x} + w\frac{\partial w}{\partial z} + \frac{\partial P}{\partial z}\right) = 0 \tag{37}$$

$$\frac{u}{x} + \frac{\partial u}{\partial x} + \frac{\partial w}{\partial z} = 0 \quad \text{with} \quad u = \frac{1}{x}\frac{\partial F}{\partial z} \quad \text{and} \quad w = -\frac{1}{x}\frac{\partial F}{\partial x} \tag{38}$$

which leads to an infinite system of differential equations, with one independent variable z. The equations for the first three terms in each of the series (21-34) are

$$M^2(\beta - g_{-1}) + g'''_{-1} - 2g_{-1}f'_{-1} + 2f_{-1}g'_{-1} = 0$$

$$-M^2 f'_{-1} + F'''_{-1} + 2f_{-1}f''_{-1} - f'^2_{-1} + g^2_{-1} - 2h_{-2} = 0$$

$$2f'''_{-1} + 4f_{-1}f'_{-1} + h'_0 = 0 \qquad h_{-2} = \text{cste} \tag{39}$$

$$-M^2 g_1 + g_1'' - 2g_{-1} f_1' + 2f_{-1} g_1' = 0$$

$$-M^2 f_1' + f_1''' + 2f_{-1} f_1'' + 2g_1 g_{-1} - h_1 = 0$$

$$h_1 = \text{cste} \qquad (40)$$

$$-M^2 g_3 + g_3'' + 2g_3 f_{-1}'' - 2g_{-1} f_3'$$

$$- 2f_3 f_{-1}' + 2f_{-1} g_3' = 0$$

$$-M^2 f_3' + f_3''' + 2f_{-1}' f_3' + 2f_1'^2 + 2f_{-1} f_{-1}'' f_3$$

$$+ 2(g_3 g_{-1} + g_1^2) - 2h_2 = 0$$

$$h_2 = \text{cste} \qquad (41)$$

The difference between these equations and those found by Khan[10] for two insulating disks in a uniform rotation is the new term βM^2 in the first equation. The function can be expanded in Taylor's series in terms of β or β^2.

Only small values of β will be considered in this study. Using classical integration methods, we get the expressions:

$$g_{-1}(z) = \beta(A_1 e^{Mz} + B_1 e^{-Mz} + 1) + O(\beta^3)$$

$$g_1(z) = \beta \Big[A_6 e^{2Mz} + B_6 e^{-2Mz} + (A_5 + C_5 z) e^{Mz}$$

$$+ (B_5 + C_5' z) e^{-Mz} + C_6 \Big] + O(\beta^3)$$

$$f_{-1}(z) = \beta^2 \Big[(A_2 + C_2 z) e^{Mz} + (B_2 + C_2' z) e^{-Mz}$$

$$+ A_3 e^{2Mz} + B_3 e^{-2Mz} + C_3 z + C_1 \Big] + O(\beta^4)$$

$$f_1(z) = A_4 e^{Mz} + B_4 e^{-Mz} - \frac{QM}{2} \frac{\text{ch } M}{\text{sh}M - M\text{ch}M} z + C_4 + \beta^2$$

(continued)

$$x\left[A_7 e^{3Mz} + B_7 e^{-3Mz} + (C_7 + D_7 z)e^{2Mz} + (E_7 + F_7 z)e^{-2Mz}\right.$$

$$+ (G_7 + H_7 z + K_7 z^2)e^{Mz}$$

$$+ (L_7 + M_7 z + N_7 z^2)e^{-Mz}$$

$$\left. + P_7 z^2 + Q_7 z + R_7 \right] + O(\beta^3) \quad (42)$$

$$h_{-2}(z) = \beta^2 (\tfrac{1}{2} + A_1 B_1) - \frac{C_3 M^2}{2} + O(\beta^4)$$

$$h_0(z) = -2\beta^2 F_1'(z) + \text{cste} + O(\beta^4)$$

$$h_1(z) = \frac{QM^3}{2} \frac{\operatorname{ch} M}{\operatorname{sh} M - M \operatorname{ch} M} + \beta^2 \left[-M^2 Q_7 - 2(M^2(A_2 B_4 + A_4 B_2)\right.$$

$$\left. - C_6 - A_5 B_5 - B_5 A_1)\right] + O(\beta^3)$$

where the coefficients A_1, B_1, C_1, ... are easily determined by applying the boundary conditions. Finally, definitions (28-31) give the u, v, w velocity components and the pressure inside the fluid. The analysis of these functions (we do not present these profiles for various M in order to keep this paper clear) especially shows that for M > 20,

$$g_{-1}(z) \simeq \beta \quad g_1(z) \simeq f_{-1}(z) \simeq f'_{-1}(z) \simeq 0 \quad \text{and} \quad f'_1(z) \simeq Q/2$$

(with the assumption that the Taylor number β associated with the rotating disk, is small).

So, for M > 20, the velocity components are

$$u = Q/2x \quad v = \beta x \quad w = 0$$

Hence, the vertical component of the velocity disappears, and the whole fluid is driven with the angular velocity ω_0 of the rotating wall; the flow is practically orthoradial for large x values. So, the moving disk tends, when M increases, to become the main cause of the liquid metal transport, thus generating a pump effect quite similar to the one observed in the absence of a source (Figs. 5 and 6).

Insulating Rotating Disks. Using classical boundary conditions, we can demonstrate that $\partial h_\theta/\partial X(X,Z) = -MV$ and only the differential equation of V Eq. (36) is modified. The first equation of the infinite system (39) becomes

$$-M^2 g_{-1} + g'''_{-1} + 2g_{-1} f'_1 + 2f_{-1} g'_1 = 0$$

Classical integration methods yield

$$g_{-1}(z) = \beta \frac{\text{sh } Mz}{\text{sh } 2M} + O(\beta^3)$$

$$g_1(z) = \beta \left[A_6 e^{2Mz} + B_6 e^{-2Mz} + (A_5 + C_5 z) e^{Mz} \right.$$
$$\left. + (B_5 + C_5 z) e^{-Mz} + C_6 \right] + O(\beta^3)$$

$$f_{-1}(z) = \beta^2 (A_2 \text{sh} 2Mz + B_2 e^{Mz} + C_2 e^{-Mz} + D_2 z + E_2) + O(\beta^4)$$

$$f_1(z) \text{ remains unchanged} \qquad (43)$$

$$h_{-2}(z) = \beta^2 \left[M(C_2 - B_2) - 2A_1^2 (1 - \frac{1}{3M^2}) \right] + O(\beta^4)$$

$h_0(z)$ remains unchanged

$$h_1(z) = \frac{QM^3}{2} \frac{\text{ch } M}{\text{sh} M - M\text{ch} M} + \beta^2 \left[-M^2 M_7 \right.$$
$$\left. - 2(A_1(A_5 - B_5) + M(B_2 B_4 - A_4 C_2)) \right] + O(\beta^3)$$

Once again, the new coefficients, A_1, B_1, C_1, ... are easily found.

When the rotating disk is insulating and the fixed disk conducting, the flow becomes strictly radial for M > 20. The magnetic field contributes to practically cancel the fluid motion due to the rotation of the disk. As in the case without a source, the flow rate through a radial cross section tends to zero everywhere (Fig. 7) and becomes similar to the one of radial MHD flow between two fixed disks.[9]

The Magnetomechanical Pump

Description

A magnetomechanical pump with an axial inlet (central source) has been fashioned in order to take advantage of the notable increase of the liquid metal flow rate caused by the magnetic field and clearly seen in the section entitled Flows with a Central Source.

The pump (Fig. 10) was mainly made up of two fixed coaxial cylinders made of altuglas, with diameters D_1 = 70 mm and D_2 = 190.5 mm. The lower fixed disk, 190.5 mm diameter, was also insulating. A ring, made of copper, with thickness e = 12 mm and with diameters D_1 = 70.5 mm, D_2 = 190 mm, was moved with a constant angular velocity ω_0.

A stationary and nearly uniform axial magnetic field B_0, ranging between 0 and 0.40 T, was produced by an electromagnet for which the diameter of the pole was 27 cm. The pump discharged mercury in a loop having a circular cross

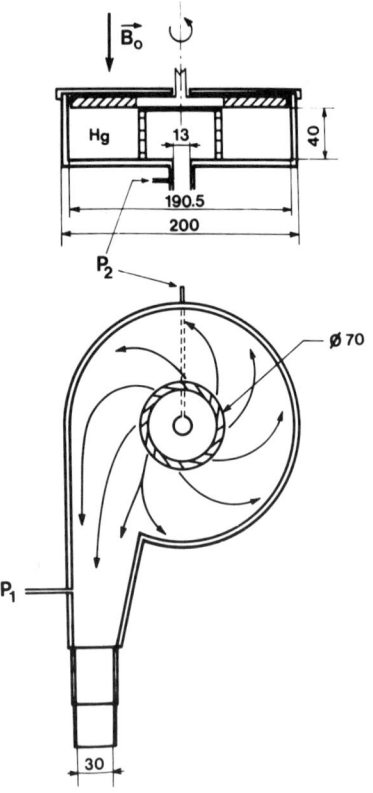

Fig. 10 Sketch of the magnetomechanic pump.

section, 40 mm in diameter and 4 m in length. The flow rate Q, measured with an electromagnetic flowmeter, was modulated at will by means of a gate valve R. Two pressure taps (P_1, P_2), respectively situated at the inlet and the outlet of the pump, were connected to manometric tubes (Fig. 10). The hydraulic power provided by the pump to the external circuit was $W_h = Q\Delta P$, with $P = P_1 - P_2$.

Operating Features and Efficiency

The performance curves of the pump are first plotted with a constant speed N, for different values of B_0 (Fig. 11), and second with a constant magnetic field, for several values of N (Fig. 12). Variations of $\Delta P(Q)$ are approximately linear with a rather weak negative slope. For example, $\Delta P(Q)$ varies from $\Delta P_0 = 57$ cm Hg, for $Q = 0$, to $\Delta P = 32$ cm Hg, for a flow rate of 21.6 tuns/h mercury.

In this last case, the hydraulic power W_h is 19 w, for a power $W_e = 172$ w dissipated by the driving motor. Besides, the static pressure rise ($Q = 0$) is increased by a factor of 17, for $N = 400$ rpm, when B_0 increases from 0 to 0.32 T (Fig. 12).

Performances and efficiencies depend upon the Hartmann and Taylor numbers of the pump, or more exactly, upon one of their combinations, the Stuart number $N^* = B_0^2 \sigma/\rho\omega_0$, which denotes the elctromagnetic and inertial forces ratio.

The $r^* = W_{h,B}/_{h,0}$ parameter is the ratio of the hydraulic powers (for given loop and rotating speed) with and without magnetic field. It expresses an amplification power factor of the pump, arising from the B_0 application. Figure 13 shows that r is proportional to N^*; so, W_h varies as

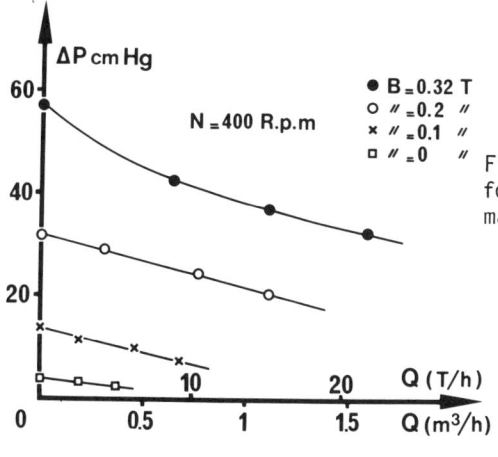

Fig. 11 Pump characteristics for different values of the magnetic field.

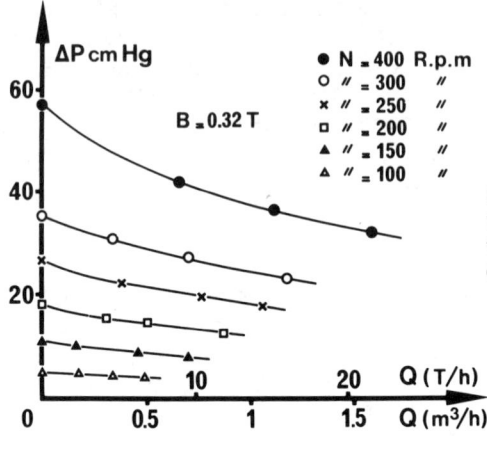

Fig. 12 Pump characteristics in the presence of a magnetic field, for different values of the rotational speed.

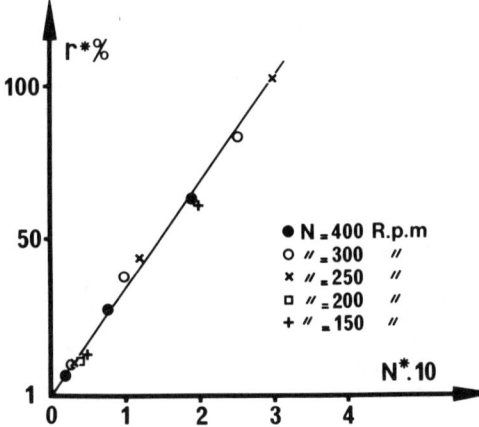

Fig. 13 Variation of the amplification coefficient of the pumps as a function of the interaction parameter.

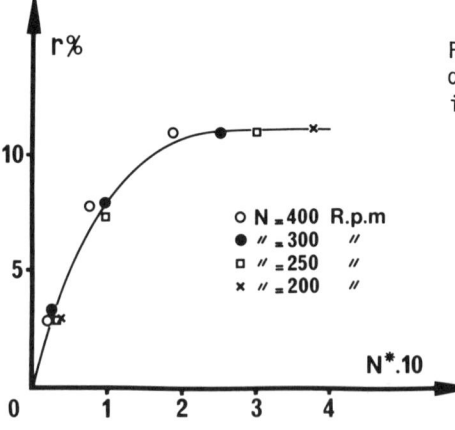

Fig. 14 Mechanical efficiency of the pump as a function of the interaction parameter.

B_0^2/N and is increased by a factor of over 100, when N evolves from 0 to 0.3.

The mechanic efficiency $r = W_h/W_e$ (W_e is the electric power dissipated by the motor) tends to a limit of about 12%, for N > 0.25 (Fig. 14).

Remarks. Because of its electric conductivity ($\sigma = 1.05 \times 10^6 \ \Omega^{-1}$), mercury is not the ideal liquid to get the best efficiency, which can reach 30% with molten aluminum ($\sigma = 5 \times 10^6 \ \Omega^{-1}$), which is a satisfying achievement for liquid metal pumps.

The magnetomechanic pump permits independent modulation of the pressure ΔP and the flow rate Q by changing, separately or simultaneously, upon N and B_0. This pump avoids the use of electrodes feeding an electromagnetic pump, as well as expensive rotors with blades or vanes in a conventional pump, and can be multistage.

The existence of powerful permanent magnets with a magnetic field in the direction of the smaller dimension and keeping half magnetization at 700°C (Curie's point at 860°C) makes possible immersion working in liquid metals or alloys such as sodium, tin, lead, zinc, aluminum, etc....

Conclusion

This paper shows the presence of strong magnetohydrodynamic effects, particularly upon the velocity and pressure profiles. If the rotating disk is electroconducting, the whole liquid metal flow tends, for M > 50, to rotate with a constant angular velocity ω_0. For large values of the Hartmann number M, the viscous torque upon the disks is very strong upon the motionless disk and tends to zero upon the rotating disk. So, it becomes possible to yield a uniform rotation of the lower disk.

If the rotating disk is insulating, the flow becomes strictly radial, for M > 20, and the moving disk is submitted to a very large viscous torque.

These properties allow the development of new type of couplers, clutches, brakes, and magnetomechanical pumps, offering the advantage of being progressively adjusted (action on ω_0 and B_0) and with a very slight wear of mechanical parts.

References

[1] Levich, V. G., Physicochemical Hydrodynamics, Prentice-Hall, Englewood Cliffs, N.J., 1962, pp. 61-138.

[2] Plamplin, B. R., Crystal Growth, Pergamon Press, New York, 1980, pp. 65-103.

[3] Hansinger, S. H. and Kehrt, L. G., "Investigation of a Shear-Force Pump," Transactions of the ASME, Vol. 85, No. 3, 1963, pp. 201-212.

[4] Rice, W., "An Analytical and Experimental Investigation of Multiple Disk Pumps and Compressors," Transactions of the ASME, Vol. 85, No. 3, 1963, pp. 191-200.

[5] Peube, J. L. and Kreith, F., "Steady Flow of a Viscous and Incompressible Fluid Between Two Parallel Rotating Disks," Journal de Mecanique, Vol. 5, No. 2, 1966, pp. 261-286.

[6] Bendaoud, M. and Vives, C., "Theoretical Study of the Vortical Motion of a Conducting Fluid Driven by the Uniform Rotation of the Wall of an Annular Vessel in the Presence of an Axial Magnetic Field," Comptes Rendus de l'Academie des Sciences, Serie A, Vol. 273, 1971, pp. 642-645.

[7] Bas, J., Bendaoud, M. and Vives, C., "Theoretical and Experimental Study of the Laminar Flow of an Incompressible and Conducting Fluid Driven by the Uniform Rotation of a Disk in a Confined Medium and in the Presence of an Axial Magnetic Field," Comptes Rendus de l'Academie des Sciences, Series A, Vol. 273, 1973, pp. 73-76.

[8] Herve, P., "Theoretical Study of the Laminar Flow of a Conducting Fluid Driven by the Uniform Rotation of a Ring of Finite Conductivity in the Presence of an Axial Magnetic Field," Comptes Rendus de l'Academie des Sciences, Series B, Vol. 286, 1978, pp. 69-72.

[9] Reddy, P., "Effect of Magnetic Field on the Laminar Radial Flow Between Parallel Plates," Journal of the Physics of Japan, Vol. 21, No. 12, 1966, pp. 2710-2715.

[10] Khan, M. A., "Laminar Source Flow Between Rotating Disks in Presence of a Transverse Magnetic Field," Journal de Mecanique, Vol. 9, No. 1, 1970, pp. 90-110.

[11] Herve, P., "Contribution to the Study of Magnetohydrodynamic Flows Provoked by Rotating Disks," Thesis Universite de Provence, Aix-Marseille, Sept. 1963, pp. 43-45.

[12] Ricou, R. and Vives, C., "Local Velocity and Mass Transfer Measurements in Molten Metals Using an Incorporated Magnet Probe," International Journal of Heat and Mass Transfer, Vol. 25, No. 10, 1982, pp. 1579-1588.

Unsteady Magnetoaerodynamic Supersonic Flows Past Oscillating Thin Bodies and Lifting Surfaces

Liviu Librescu*
Tel Aviv University, Israel

Abstract

This paper is concerned with the determination of the aerodynamic pressure field on elastic thin bodies and lifting surfaces that undergo oscillatory motions in a supersonic ionized flow environment. The presence of an ambient magnetic field has been included within the analysis, and, in this respect, the cases of aligned and crossed fields are treated separately. Attention is paid to some special cases that have wide applicability in aeroelastic stability investigations. Finally, some flutter problems of two-dimensional lifting surfaces oscillating in an ionized flowfield are analyzed, and the strong influence played by the electromagnetic effects is pointed out.

Introduction

A great deal of research activity in aeroelasticity has been directed during the past few decades toward the development and refinement of unsteady lifting surface theories (LST).

In spite of all refinements and generalizations in the field, it must be emphasized that the basic assumptions idealizing the properties of the flowing gas generally used throughout these investigations are, in fact, classical in their character. However, during the high-speed flight of space vehicles, there are many important instances when some of these assumptions, traditionally used both in the sub-

---
Paper presented at the Fourth Beer-Sheva Seminar on MHD Flows and Turbulence, Ben-Gurion University of the Negev, Beer-Sheva, Israel, Feb. 27-March 2, 1984. Copyright © 1985 by the American Institute of Aeronautics and Astronautics, Inc. All rights reserved.
*Professor, Faculty of Engineering, Department of Solid Mechanics, Materials and Structures.

stantiation of LST and in their associated aeroelastic analyses, are violated.

Such instances arise, e.g., as a result of the aerodynamic heating during launch and re-entry, as a result of nuclear and solar radiation, and under the influence of the plume exhaust of rocket engines.

Under such conditions, the gas surrounding the space vehicle will be ionized. It should be noted that it is the emergence of this phenomenon that has suggested, together with the use of onboard magnetic fields, new prospects of guidance and flight control, control of lift and drag of flying bodies, and control of heat transfer from hot gas streams to adjacent bodies, etc.[1,2] In this context, a complete analysis of the aeroelastic stability of lifting surfaces and thin elastic bodies should include in the governing equations the appropriate unsteady air loads. They are to be determined by taking into account the ionization of the high-speed flow environment and the presence of magnetic fields.

This paper is concerned with the development of this topic† and, in this context, some flutter problems of two-dimensional lifting surfaces are analyzed. The results obtained reveal some new and interesting features, underlining the great influence played by the electromagnetic field interacting with the aerodynamic and the elastic ones.‡

General Equations

Let us consider the case of planar lifting surfaces (or of elastic thin panels) extending to infinity in the x_2 direction and flown in the x_1 direction by a supersonic ionized gas. Let us consider in addition the existence of an ambient magnetic field. In this context, two different instances are envisaged. They are referred to as case I and case II and correspond to the aligned (characterized by $\vec{U} || \vec{H}$) and the crossed fields (characterized by $\vec{U} \perp \vec{H}$), respectively. Here, $\vec{U}$ and $\vec{H}$ denote the vectors of the undisturbed stream velocity and the applied magnetic field, respectively. In order to evaluate the unsteady aerodynamic loads, we shall stipulate that 1) the thermoelectric and viscosity effects are disregarded; 2) the gas processes are isentropic; 3) the electrical conductivity of the gas is infinite; 4) the Hall effect is absent; and 5) the electromagnetic ef-

---

†For some of the earlier developments in the field, see Refs. 3-7.

‡A similar field interaction may also occur in some nonaeronautical devices, e.g., of ionic engines, plasmatrons, etc.

fects in the solid medium are disregarded. On this basis, and by using in addition the small disturbance concept from Ref. 5, the linearized field equations appropriate to the two envisaged field configurations read as follows. For case I:

$$\frac{D\vec{h}}{Dt} = H_1 \frac{\partial \vec{v}}{\partial x_1} + \frac{H_1}{\rho_0} \frac{D\hat{\rho}}{Dt} \vec{I}_1 \tag{1a}$$

$$\frac{D\vec{v}}{Dt} = -\frac{1}{\rho_0} \text{grad} \left(a_0^2 \hat{\rho} + \frac{1}{4\pi} H_1 h_1 \right) + \frac{1}{4\pi\rho_0} H_1 \frac{\partial \vec{h}_1}{\partial x_1} \tag{1b}$$

$$\text{div } \vec{h} = 0 \tag{1c}$$

$$\frac{D\hat{\rho}}{Dt} + \rho_0 \text{ div } \vec{v} = 0 \tag{1d}$$

$$\hat{p} = a_0^2 \hat{\rho} \quad p = p_0 + \hat{p} \tag{1e}$$

For case II:

$$\frac{D\vec{h}}{Dt} = H_2 \frac{\partial \vec{v}}{\partial x_2} + \frac{H_2}{\rho_0} \frac{D\hat{\rho}}{Dt} \vec{I}_2 \tag{2a}$$

$$\frac{D\vec{v}}{Dt} = -\frac{1}{\rho_0} \text{grad} \left(a_0^2 \hat{\rho} + \frac{1}{4\pi} H_2 h_2 \right) + \frac{1}{4\pi\rho_0} H_2 \frac{\partial \vec{h}_1}{\partial x_2} \tag{2b}$$

Equations (1c-1e) maintain their validity in case II as well. Consequently, in the forthcoming developments they are to be used also in conjunction with Eqs. (2). In Eqs. (1-2), $\vec{h}$ ($\equiv h_i \vec{I}_i$); $\vec{v}$ ($\equiv v_i \vec{I}_i$), $\hat{\rho}$ and $\hat{p}$ stand for the small perturbations of the undisturbed primary fields, i.e., of the applied magnetic field H, of the velocity U, of the density ρ_0, and of the pressure p_0, respectively; a_0 denotes the speed of sound in the undisturbed gas; $\vec{I}_i$ denotes the in-plane orthogonal unit vectors parallel to the coordinates x_i ($i = 1,2$) in which x_1 ($0 \le x_1 \le \ell$, where ℓ denotes the wing chord) coincides with flow direction, while x_2 (extending from $-\infty$ to $+\infty$) coincides with the leading edge of the lifting surface (LS) (see Fig. 1). It is pointed out that the primary fields are considered time- and space-independent quantities, whereas the disturbance quantities (due to the two-dimensional approach) are considered func-

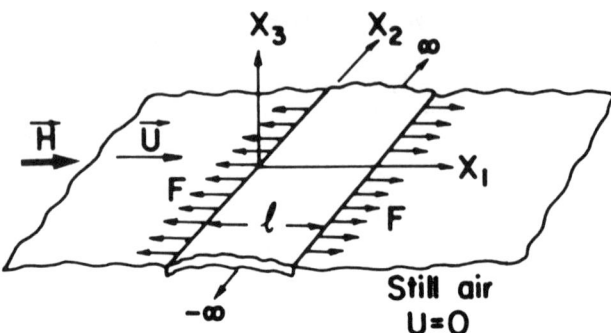

Fig. 1 Two-dimensional thin panel placed in an aligned-fields flow.

tions of x_1, x_3, and t. In both sets of equations, $D/Dt \equiv \partial/\partial t + U \, \partial/\partial x_1$, where $U \equiv \vec{U}_1$. It is also mentioned that in the two envisaged cases, $\vec{H}$ has the representations $\vec{H} \equiv H_1 \vec{i}_1$ (in case I) and $\vec{H} \equiv H_2 \vec{i}_2$ (in case II).

The solution to the field equations is to fulfill the impenetrability condition

$$v_3|_{x_3 = 0} = - \frac{DZ_a}{Dt}, \quad (x_1, x_2) \in R_a \qquad (3)$$

and the finiteness condition at infinity. $Z_a = Z_a(x_1, x_2, t)$ denotes the transverse deflection of the panel (or of LS) immersed in the flowing gas. Its mean position is considered placed as close as possible to the $x_3 = 0$ plane; R_a denotes the region belonging to the two-dimensional lifting surface (bounded by $x_1 = 0, \ell; \ x_2 = \pm\infty$).

The pressure difference across the lifting surface required in flutter analyses may be expressed as follows:

$$P|_{x_3 = 0} = (\delta p - \delta t_{33}) \qquad (4)$$

where $\delta(\ ) \equiv (\ )|_{x_3 = 0_+} - (\ )|_{x_3 = 0_-}$ denotes the jump across the surface $x_3 = 0$; $x_3 = 0_\pm$ refers to upper and lower surfaces, respectively, while t_{33} denotes the transverse component of the Maxwell stress tensor. Its linearized expression for the two considered cases reads

$$t_{33} = - h_1 H_1/(4\pi) \quad \text{in case I}$$

$$= - h_2 H_2/(4\pi) \quad \text{in case II} \qquad (5)$$

With all the preliminaries in view, we may proceed to the explicit determination of the unsteady aerodynamic pressure.

Pressure Evaluation

Its Representation in Terms of $Z_a(x_1,t)$

This problem is treated in the framework of the purely hyperbolic flow regime. Referring to Sears-Resler diagrams of the associated steady flow regimes,[1,8] in case I, this domain is restricted to that corresponding to $M > 1$, $M^2 > \lambda_1^2$, while in case II, it is defined by $M^2 > 1 + \lambda_2^2$. The considered purely hyperbolic flow regime, characterized by the nonoccurence of disturbance ahead of the airfoil (i.e., for $x_1 \leq 0$), permits us to use a Laplace transform (LT), with respect to the streamwise coordinate, i.e.,

$$f^L(s,\bar{x}_3) = \int_0^\infty f(\bar{x}_1,\bar{x}_3) e^{-s\bar{x}_1} d\bar{x}_1 \tag{6}$$

By considering simple harmonic motion, which implies the following representation of field variables,

$$f(\bar{x}_1,\bar{x}_3,\bar{t}) = \overset{o}{f}(\bar{x}_1,\bar{x}_3)\exp(j\bar{\omega}\bar{t}) \quad j = (-1)^{1/2} \tag{7}$$

and applying LT to Eqs. (1) and (2), all these yield the following representation of $\overset{o}{h}{}_i^L(s,\bar{x}_3)$ and $\overset{o}{v}{}_i^L(s,\bar{x}_3)$ in terms of $\overset{o}{\rho}{}^L(s,\bar{x}_3)$:

$$\overset{o}{h}{}_1^L = \frac{H_1}{\rho_0} \frac{\frac{v^2}{\mu} - s^2}{\frac{v^2}{\mu}} \overset{o}{\rho}{}^L \quad \overset{o}{h}{}_2^L = \overset{o}{v}{}_2^L = 0 \quad \overset{o}{v}{}_1^L = -\frac{a_0}{\rho_0} \frac{s}{\frac{v}{\mu}} \overset{o}{\rho}{}^L$$

$$\overset{o}{h}{}_3^L = -\frac{H_1 s}{\rho_0 v^2} \frac{\frac{v^2}{\mu} + \lambda_1^2 (\frac{v^2}{\mu} - s^2)}{\frac{v^2}{\mu} - \lambda_1^2 s^2} \overset{o}{\rho}{}^L_{,3} \tag{8}$$

$$\overset{o}{v}{}_3^L = -\frac{a_0}{\rho_0 \frac{v}{\mu}} \frac{\frac{v^2}{\mu} + \lambda_1^2 (\frac{v^2}{\mu} - s^2)}{\frac{v^2}{\mu} - \lambda_1^2 s^2} \overset{o}{\rho}{}^L_{,3}$$

while in case II,

$$\overset{o}{h}{}_1^L = \overset{o}{h}{}_3^L = 0 \quad \overset{o}{h}{}_2^L = \frac{H_2}{\rho_0} \overset{o}{\rho}{}^L$$

$$\overset{o}{v}{}^L_1 = -\frac{a_0}{\rho_0}\frac{s(1+\lambda_2^2)}{v_\mu}\overset{o}{\rho}{}^L$$

$$\overset{o}{v}{}^L_2 = 0 \qquad \overset{o}{v}{}^L_3 = -\frac{a_0(1+\lambda_2^2)}{\rho_0 v_\mu}\overset{o}{\rho}{}^L_{,3} \qquad (9)$$

In Eqs. (8) and (9), $M(\equiv U/a)$ and $\lambda_i^2\;[\equiv H_i^2/(4\pi\rho_0 a_0^2)]$ ($i = 1,2$), stand for the Mach and Alfvén numbers,§ respectively (the last one being also referred to as magnetic Mach number); $v_\omega(\equiv \omega\ell/U)$ denotes the reduced frequency; $v_\mu \equiv M(jv_\omega+s)$; $(\ )_{,3} \equiv \partial(\ )/\partial\bar{x}_3$; s denotes the LT variable; $\bar{x}_1(\equiv \tilde{x}_1/\ell)$ ($i = 1,3$) denotes the nondimensional coordinates; $\overset{o}{\rho}{}^L(s,\bar{x}_3)$ $[\equiv L(\overset{o}{\rho}(\bar{x}_1,\bar{x}_3))]$ denotes the LT of the spatial part of the disturbance quantities denoted generically by $f\;[\equiv f(\bar{x}_1,\bar{x}_3,\bar{t})]$. The quantities $\overset{o}{\rho}{}^L_{h_i}$ and $\overset{o}{v}{}^L_{v_i}$ as expressed through Eqs. (8) and (9) satisfy identically the Eqs. (1a,b) and (2a,b), respectively (these ones being transformed through LT).

Substitution of Eqs. (8) and (9) into the remaining equations not yet used yields the governing equation, which has a similar form for both cases, i.e.,

$$\overset{o}{\rho}{}^L_{,33} - \phi^2\overset{o}{\rho}{}^L = 0 \qquad (10)$$

where

$$\phi^2 = \frac{(v_\mu^2 - s^2)(v_\mu^2 - \lambda_1^2 s^2)}{(1+\lambda_1^2)v_\mu^2 - \lambda_1^2 s^2} \qquad \text{for case I}$$

$$= \frac{v_\mu^2 - s^2(1+\lambda_2^2)}{1+\lambda_2^2} \qquad \text{for case II} \qquad (11)$$

$\overset{o}{\rho}$ plays the role of a potential function. [The representation in terms of $\hat{\rho}$ appears preferable in the aeroelastic problems to the one implying the small-disturbance velocity potential (see Refs. 6, 7, 9).] The general solution to Eq. (10) that fulfills the impenetrability condition and

---

§ Some authors refer to $A_i^2[\equiv U^2/(H_i^2/4\pi\rho_0)]$ as the Alfvén number. The two considered expressions are interrelated by $A_i^2=M^2/\lambda_i^2,(i=1,2)$.

the finiteness condition at infinity is

$$\left. p^{oL} \right|_{\bar{x}_3=0_+} = \frac{\rho_0 \mu^{v2}(\mu^{v2} - \lambda_1^2 s^2)^{1/2}}{\ell(\mu^{v2} - s^2)^{1/2}\left[(1+\lambda_1^2)\mu^{v2} - \lambda_1^2 s^2\right]^{1/2}}$$

$$\times \overset{o}{Z}_a(s) \text{ sign } \bar{x}_3 \quad \text{for case I}$$

$$= - \frac{\rho_0 \mu^{v2}}{\ell(1 + \lambda_2^2)^{1/2}\left[\mu^{v2} - s^2(1 + \lambda_2^2)\right]^{1/2}}$$

$$\times \overset{o}{Z}_a(s) \text{ sign } \bar{x}_3 \quad \text{for case II} \qquad (12)$$

where $Z_a(\bar{x}_1,\bar{t}) = \bar{Z}_a(\bar{x}_1)\exp(j\bar{\omega}\bar{t})$, while $\overset{o}{Z}_a(s) = L[\bar{Z}_a(\bar{x}_1)]$; sign $\bar{x}_3$ (=1 for $\bar{x}_3 > 0$ and = -1 for $\bar{x}_3 < 0$) denotes the signum distribution. Equation (12) shows that p^{oL} is antisymmetrical with respect to $\bar{x}_3=0$. This property proves very useful in the evaluation of the pressure jump on lifting surfaces. The appropriate use of Eqs. (12) and (1e) results in the following representation (in the image space) of the pressure field:

$$\left. p^L \right|_{\bar{x}_3=0} = \bar{C} C \frac{P}{Q} \overset{o}{Z}_a(s) e^{j\bar{\omega}\bar{t}} \quad \bar{t} = tU/\ell \qquad (13a)$$

where for case I,

$$P = \left(s^2 + 2j \frac{M^2(1 + \lambda_1^2)\bar{\omega}s}{M^2(1 + \lambda_1^2) - \lambda_1^2} - \frac{M^2(1 + \lambda_1^2)\bar{\omega}^2}{M^2(1 + \lambda_1^2) - \lambda_1^2} \right)^{1/2}$$

$$\times \left(s^2 + 2j \frac{M^2 \bar{\omega}s}{M^2 - \lambda_1^2} - \frac{M^2 \bar{\omega}^2}{M^2 - \lambda_1^2} \right)^{1/2}$$

$$Q = \left[\left(s + j \frac{M^2 \bar{\omega}}{M^2 - 1} \right)^2 + \frac{M^2 \bar{\omega}^2}{(M^2 - 1)^2} \right]^{1/2} \qquad (13b)$$

$$C = \frac{[(M^2(1 + \lambda_1^2) - \lambda_1^2)(M^2 - \lambda_1^2)]^{1/2}}{(M^2 - 1)^{1/2}}$$

while for case II,

$$P = (s + j\overset{v}{\omega})^2$$

$$Q = \left[\left(s + j\frac{M^2\overset{v}{\omega}}{M^2 - 1 - \lambda_2^2}\right)^2 + \frac{M^2(1 + \lambda_2^2)\overset{v2}{\omega}}{(M^2 - 1 - \lambda_2^2)^2}\right]^{1/2}$$

$$C = \frac{M^2(1 + \lambda_2^2)^{1/2}}{(M^2 - 1 - \lambda_2^2)^{1/2}} \qquad (13c)$$

For both cases, the coefficient $\bar{C}$ is given by $\bar{C} = -\gamma a_0^2 \rho_0 / \ell$. Here, the tracer γ takes the value 1 or 2 whether the flow takes place either on the upper surface of the panel or on both the upper and lower surfaces (this instance is characteristic of LS theory), respectively. The inverse LT of Eq. (13a) yields the two-dimensional unsteady pressure distribution. In order to obtain closed-form solutions, two spe-

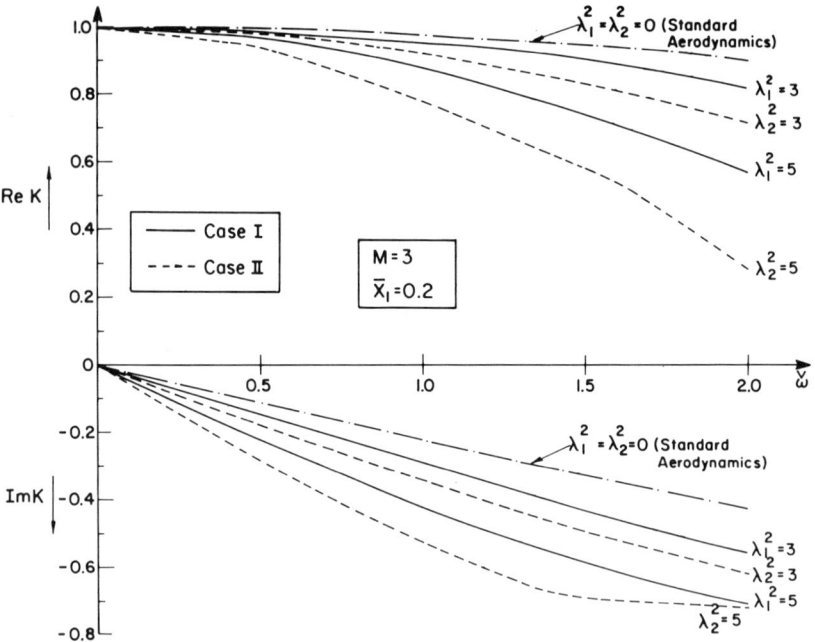

Fig. 2 Depiction of the kernel function K ($K \equiv \text{Re}K + j\text{Im}K$) for case I, case II, and classical nonconducting aerodynamics.

cial instances belonging to the case I are considered. They consist of $\lambda_1^2 \gg 1$ and $\lambda_1^2 \ll 1$.

Concerning case II, it is considered in full generality. Making use in Eqs. (13) of the well-known properties of LT, the unsteady two-dimensional pressure distribution results under the following unitary form:

$$P\big|_{\bar{x}_3=0} = C\left[c_1 K(\bar{x}_1)\left(\frac{\partial Z_a}{\partial x_1} + \Omega \frac{\partial Z_a}{\partial t}\right)\right|_{\bar{x}_1 = 0_+}$$

$$+ \int_0^{\bar{x}_1} K(\bar{x}_1 - \bar{\xi}_1)\left(c_1 \frac{\partial^2}{\partial \bar{\xi}_1^2} + 2M^2 \frac{\partial^2}{\partial \bar{\xi}_1 \partial t} + M^2 \frac{\partial^2}{\partial t^2}\right) Z_a(\bar{\xi}_1, \bar{t}) d\bar{\xi}_1\right] \quad (14)$$

In Eq. (14), $\bar{\xi}_1$ is a dummy variable (associated to $\bar{x}_1$), while K denotes the aerodynamic kernel expressed by

$$K \equiv K(\bar{x}_1, \overset{v}{\omega}, M, \lambda_1^2) = J_0(\Gamma \bar{x}_1)\exp(-j\overline{\omega}\bar{x}_1) \quad (15)$$

The coefficients entering into Eqs. (14) and (15) are defined as follows. In case I,

$$\bar{\omega} = \frac{M^2 \overset{v}{\omega}}{M^2 - \lambda_1^2} \quad \Gamma = \frac{\overline{\omega} \lambda_1}{M} \quad C_1 = \frac{\lambda_1}{(M^2 - \lambda_1^2)^{1/2}}$$

$$c_1 = M^2 - \lambda_1^2 \quad \Omega = \frac{M^2}{M^2 - \lambda_1^2} \quad \text{for } \lambda_1^2 \gg 1,\, M^2 < \lambda_1^2 \quad (16)$$

and

$$\bar{\omega} = \frac{M^2 \overset{v}{\omega}}{M^2 - 1} \quad \Gamma = \frac{\overline{\omega}}{M} \quad C_1 = \frac{1}{(M^2 - 1)^{1/2}} \quad c_1 = M^2 - \lambda_1^2$$

$$\Omega = \frac{M^2(M^2 - 2 + \lambda_1^2)}{(M^2 - 1)(M^2 - \lambda_1^2)} \quad \text{for } \lambda_1^2 \ll 1,\, M^2 > 1$$

while in case II,

$$\bar{\omega} = \frac{M^2 v}{M^2-1-\lambda_2^2} \qquad \Gamma = \frac{\bar{\omega}(1+\lambda_2^2)^{1/2}}{M} \qquad C_1 = \frac{(1+\lambda_2^2)^{1/2}}{(M^2-1-\lambda_2^2)^{1/2}}$$

$$c_1 = M^2 \qquad \Omega = \frac{M^2-2-2\lambda_2^2}{M^2-1-\lambda_2^2} \qquad \text{for } M^2 > 1 + \lambda_2^2 \qquad (17)$$

In both cases we have $C = CC_1$, while J_0 denotes the Bessel function of the first kind and zero order. It is readily seen that the specialization for $\lambda_1^2 \to 0$ of the equations appropriate to case I (for $\lambda_1^2 \ll 1$, $M > 1$) and case II, provides the standard supersonic kernel function (see, e.g., Ref. 10). Several depictions of the kernel function, in the cases envisaged in the paper, are given in Fig. 2.

An Alternative Representation of the Pressure Field

In some instances, it is preferable to represent P in terms of the downwash velocity. In the image space such a representation may be obtained by making use of Eq. (3) (transformed through LT) into Eq. (13a). This way, the required relationship results formally by multiplying the right-hand side of Eq. (13a) by $-\ell/[a_0 M(s+j\bar{\omega})]$ and by replacing there Z_a^L by v_3^L. In order to convert the obtained relationships into the real space, the convolution rule will be applied twice. This yields the result

$$P\Big|_{\bar{x}_3=0} = \frac{\gamma a_0 \rho_0}{M} C_1 c_1 \Bigg[K(\bar{x}_1) \, v_3(0,\bar{t})$$

$$+ \int_0^{\bar{x}_1} \left(\frac{\partial v_3(\bar{\xi}_1,\bar{t})}{\partial \bar{\xi}_1} + \sigma \frac{\partial v_3(\bar{\xi}_1,\bar{t})}{\partial \bar{t}} \right) K(\bar{x}_1-\bar{\xi}_1) d\bar{\xi}_1$$

$$- \Lambda \int_0^{\bar{x}_1} \tilde{K}(\bar{x}_1-\bar{\xi}_1) \, v_3(\bar{\xi}_1,\bar{t}) d\bar{\xi}_1 \Bigg] \qquad (18)$$

In Eq. (18), along with the denotations (16) and (17), we have also

$$\sigma \equiv \frac{M^2+\lambda_1^2}{M^2-\lambda_1^2} \quad \text{valid for } \lambda_1^2 \gg 1 \text{ and } \lambda_1^2 \ll 1 \text{ of case I}$$

$$\equiv 1 \quad \text{in case II} \tag{19a}$$

$$\Lambda \equiv \frac{\lambda_1^2(j\overset{v}{\omega})^2}{M^2-\lambda_1^2} \quad \text{valid for } \lambda_1^2 \gg 1 \text{ and } \lambda_1^2 \ll 1 \text{ of case I}$$

$$\equiv 0 \quad \text{in case II} \tag{19b}$$

while

$$\tilde{K}(\bar{x}_1, \overset{v}{\omega}, M, \lambda_1^2) \equiv \int_0^{\bar{x}_1} e^{-j\overset{v}{\omega}\bar{\zeta}_1} K(\bar{x}_1-\bar{\zeta}_1) \, d\bar{\zeta}_1$$

denotes a modified aerodynamic kernel, where $\bar{\zeta}_1$ is a dummy variable associated to $\bar{x}_1$ (and $\bar{\xi}_1$). The appropriate specialization of Eq. (18), for $\lambda^2 \to 0$, yields the unsteady pressure expression valid for standard supersonic flows (see, e.g., Ref. 10).

Note that in case I, when $\lambda_1^2 \gg 1$, and $M \to \lambda_1$ (or equivalently when $A_1 \to 1$) or when $\lambda_1^2 \ll 1$ and $M \to 1$ and, in case II, when $M^2 \to 1 + \lambda_2^2$, the corresponding arguments of the Bessel function J_0 (intervening in the aerodynamic kernel K) become infinite.

In these instances, the critical borderline between the purely hyperbolic (H) and the elliptic (E) flow regimes (for case I) and between the purely hyperbolic and hyper-liptic (H-L) flow (for case II) are respectively reached. (The H-L regime is a superposition of H and E flow regimes; for more details, see Ref. 8.)

These instances may easily be identified in the body of Sears-Resler diagrams, where the various regimes of flow are represented in the coordinates (M, A_i), for aligned and crossed fields (see Figs. 1 and 3 in Ref. 8).

Using the asymptotic representation of Bessel function J_0, $J_0(x) \sim [2/(\pi x)]^{1/2} \cos(x-\pi/4)$, and paralleling the classical procedure,[10] a single limiting expression for $C_1 K(\bar{x}_1)$ is obtained. It is $C_1 K(\bar{x}_1) = (2\pi j \overset{v}{\omega} \bar{x}_1)^{-1/2} \exp(-j\overset{v}{\omega}\bar{x}_1/2)$. The result entails the conclusion that the unsteady pressure loads along transcritical boundaries are still finite quantities. In addition, it is worth remarking that this expression of $C_1K(\bar{x}_1)$ coincides with the one in the classical unsteady transonic aerodynamics (see Refs. 10 and 11).

Special Cases of the Pressure Loads

Some special cases playing a great role in flutter analysis will be considered next.

Very-Low Reduced-Frequency Approximation

This case involves the approximation $\bar{\omega} \ll 1$. Consequently, J_0 entering the kernel function may be considered very close to unity (whenever its argument is less than 0.2), while J_1 (intervening at a later stage) may be assimilated to zero. In addition, taking advantage of the series expansion for the exponential (entering the kernel function), integrating by parts in Eq. (14) whenever possible, and retaining only the terms that are first order in frequency, all yield the result:

$$P(\bar{x}_1,\bar{x}_3 = 0,\bar{t}) = \frac{\gamma a_0 \rho_0}{M} C_1 c_1 \left[v_3(\bar{x}_1) \right.$$

$$\left. + \int_0^{\bar{x}_1} (-j\bar{\omega} + j\overset{v}{\omega}\sigma) v_3(\bar{x}_1 - \bar{\zeta}_1) d\bar{\zeta}_1 \right] e^{j\overset{v}{\omega}\bar{t}} + O(\overset{v}{\omega}^2) \quad (21)$$

Equation (21) adequately specialized for $\lambda^2 \to 0$ yields the classical pressure expression valid for nonconducting supersonic flows and very-low reduced frequencies (see, e.g., Ref. 12).

Quasistatic Supersonic Approximation

For this case, the assumption of the very small frequency $\overset{v}{\omega}$ is applicable. Due to its smallness it is considered explicitly zero. As a result, from Eq. (14) [or Eq. (21)], the pertinent pressure approximation reads

$$P(\bar{x}_1,\bar{t})\big|_{\bar{x}_3=0} = -\frac{\gamma a_0^2 \rho_0}{\ell} \frac{\{[M^2(1+\lambda_1^2)-\lambda_1^2](M^2-\lambda_1^2)\}^{1/2}}{(M^2-1)^{1/2}} \times \frac{\partial Z_a(\bar{x}_1,\bar{t})}{\partial \bar{x}_1}$$

(22a)

UNSTEADY MAGNETOAERODYNAMIC SUPERSONIC FLOWS 67

for case I, when $M > 1$, $M^2 > \lambda_1^2$;

$$P(\bar{x}_1,\bar{t})\big|_{\bar{x}_3=0} = - \frac{\gamma a_0^2 \rho_0}{\ell} \frac{M^2(1+\lambda_2^2)^{1/2}}{(M^2-1-\lambda_2^2)^{1/2}} \frac{\partial Z_a(\bar{x}_1,\bar{t})}{\partial \bar{x}_1} \qquad (22b)$$

for case II, when $M^2 > 1 + \lambda_2^2$. Equation (22a), obtained in an alternative manner by Sears,[1] has been used in panel flutter analyses in Ref. 13.

High-Frequency Approximation

Another case deserving attention is that corresponding to high frequencies. By making use in Eqs. (13) of the following pertinent replacement:

$$-\omega^{V2} + 2j\omega^V s + Ss^2 \rightarrow (j\omega^V + s)^2 \qquad (23)$$

where S may represent any of the following quantities:

$$S: \quad \frac{M^2-\lambda_i^2}{M^2} \quad \frac{M^2-1}{M^2} \quad \frac{M^2(1+\lambda_i^2)-\lambda_i^2}{M^2(1+\lambda_i^2)} \quad \frac{M^2-1-\lambda_i^2}{M^2} \qquad (24)$$

the modified pressure expression reads

$$P(x_1,t)\big|_{x_3=0} = - \gamma a_0 \rho_0 (1+\lambda_i^2)^{1/2} \left(\frac{\partial Z_a}{\partial t} + U \frac{\partial Z_a}{\partial x_1} \right) \qquad (25)$$

($i = 1$ or 2 for case I or case II, respectively).

A simple inspection of the expressions of S given by Eq. (24) allows one to infer that high reduced frequency and high Mach number approximations yield identical results.

Equation (25) constitutes the generalized counterpart of the standard linearized piston theory aerodynamics[14] widely used in flutter analyses.

Airfoil Flutter

In the remaining sections of the paper, two different cases that concern the supersonic flutter of airfoils will be briefly treated. This will permit one to get an idea of the influence played by the electromagnetic field in the flutter behavior of two-dimensional lifting surfaces.

Bending-Torsion Flutter at High Supersonic Mach Numbers

The problem may adequately be treated by using the pressure approximation (25). For the airfoil undergoing a plunging $h(\bar{t})$ and a torsional motion consisting of a turning [with the instantaneous angle $\alpha \equiv \alpha(\bar{t})$], about the elastic axis (EA), whose dimensionless position with respect to the leading edge is $\bar{X}_1 (\equiv X_1/\ell)$, $Z_a(\bar{x}_1,\bar{t})$ may be represented as

$$Z_a(\bar{x}_1,\bar{t}) = h(\bar{t}) + (\bar{x}_1 - \bar{X}_1)\ell\alpha(\bar{t}) \qquad (26)$$

where $h = h_0 \exp(j\bar{\omega}\bar{t})$ and $\alpha = \alpha_0 \exp(j\bar{\omega}\bar{t})$, α_0 and h_0 being constants (generally complex). Replacement in the bending-torsion equations of motion of the lift force L and of the torsional moment M (about the EA), defined by

$$L = \ell \int_0^1 P d\bar{x}_1 \qquad M = \ell \int_0^1 (\bar{x}_1 - \bar{X}_1) P d\bar{x}_1 \qquad (27)$$

[expressions which are rendered explicit with the aid of Eqs. (25) and (26)] and employment in the resulting governing equations of standard methods allowing determination of the stability boundary (see, e.g., Ref. 10) yield the following closed-form expression of the dimensionless flutter speed:

$$\frac{U_F}{\ell\omega_\alpha} = \frac{\mu M (1+\lambda_1^2)^{-1/2}}{\chi^{1/2}} \frac{N_1}{N_2} \qquad (28)$$

where

$$N_1 = \{x_\alpha^2 - [(\omega_h/\omega_\alpha)^2 \chi - 1] r_\alpha^2 (\chi-1)\}^{1/2} \qquad (29)$$

$$N_2 = \{\mu M (1+\lambda_i^2)^{-1/2} [x_\alpha + (1-x_0)(\omega_h/\omega_\alpha)^2 \chi - 1)] - 1/3\}^{1/2}$$

while

$$\chi \equiv \left(\frac{\omega_\alpha}{\omega}\right)^2 = \frac{r_\alpha^2 - 2x_\alpha(1-x_0) + (4/3 - 2x_0 + x_0^2)}{r_\alpha^2 + (\omega_h/\omega_\alpha)^2 (4/3 - 2x_0 + x_0^2)} \qquad (30)$$

denotes the flutter frequency; the index $i = 1$ identifies case I, available for $M^2 \gg 1$, $M^2 > \lambda_1^2$; while $i = 2$ identifies case II, available for $M^2 \gg 1$, $M^2 > 1 + \lambda_2^2$.

UNSTEADY MAGNETOAERODYNAMIC SUPERSONIC FLOWS 69

Equations (28-30) represent the generalized counterpart of the classical solution obtained in Ref. 14.

Here, ω_h, ω_α stand for the bending and torsion uncoupled natural frequencies; r_α denotes the dimensionless radius of gyration about the EA; b ($\equiv \ell/2$) denotes the semichord; $\mu \ [\equiv m/(\rho_0 \ \ell^2)]$ is the wing density parameter; x_0 is the modified dimensionless distance from leading edge to EA ($x_0 = 2\bar{x}_1$); x_α is the dimensionless static unbalance about the EA; and m is the mass of wing (per unit span).

Figure 3 depicted on the basis of Eqs. (28-30) shows the influence played by the magnetic Mach number on the flutter critical boundary. At large Mach numbers, for moderate λ_i^2, the influence of the magnetic Mach number appears to be destabilizing. However, at large λ_i^2, this tendency is declining.

The present analysis, based on the linear piston-theory aerodynamics, allows determination of the flutter instability boundary. For an investigation of the stability in the vicinity of the flutter boundary, a full representation of piston-theory aerodynamics must be used, requiring the inclusion in its expression of appropriate nonlinear terms. For such an analysis, developed in the classical framework (and incorporating also the pertinent structural nonlinearities), see Ref. 15.

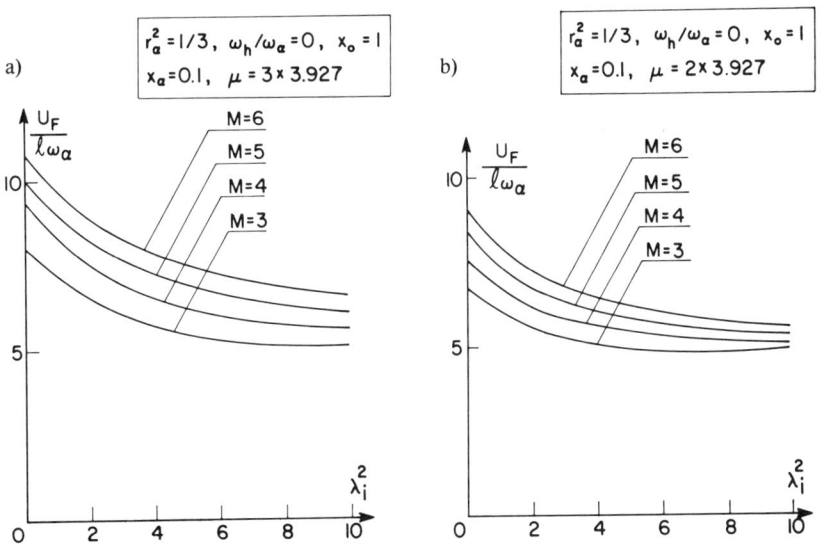

Fig. 3 Flutter velocity coefficient ($U_F/\ell\omega_\alpha$) as a function of magnetic Mach number λ_i^2 (i = 1,2) for two different values of the wing density parameter μ. The points afferent to $\lambda_i^2 = 0$ correspond to classical aeroelasticity theory (Ref. 10).

Torsional Flutter

As is shown in the classical aerelasticity (see, e.g., Refs. 10 and 11), a two-dimensional airfoil oscillating at very low frequencies exhibits the possibility of a single-degree-of-freedom torsional instability in the Mach-number range $1 < M < 1.58$ (which depends on the location of the EA). Let us re-examine the problem in the context of the field interaction considered here.

The concept of pure torsional oscillations about $\overline{x}_1 = \overline{X}_1$ implies the following representations of the displacement and downwash velocity:

$$Z_a = \ell(\overline{x}_1-\overline{X}_1)\,\alpha(\overline{t})$$

$$v_3(\overline{x}_1,\overline{t}) = -U\left[1+j\omega^\vee (\overline{x}_1-\overline{X}_1)\right]\alpha(\overline{t}) \qquad (31)$$

The appropriate use of Eqs. (31), (21), and (27) provides the aerodynamic torsional moment $M \equiv M_\alpha$ (associated with pitching motion only). As it may easily be shown, the condition of torsional instability may be expressed, as in the classical framework,[12] by stating that $\mathrm{Im}M_\alpha < 0$. Positive values of $\mathrm{Im}M_\alpha$ imply stable conditions, while $\mathrm{Im}M_\alpha = 0$ defines the borderline between damped and undamped torsional oscillations. Here, Im denotes "imaginary part."

Under an explicit form, the last condition reads

$$2F(1-3\overline{X}_1+3\overline{X}_1^2) - G(2-3\overline{X}_1) = 0 \qquad (32)$$

where, for case I,

$$F \equiv M^2-1 \quad \text{for } M>1,\ \lambda_1^2 \ll 1$$

$$\equiv M^2-\lambda_1^2 \quad \text{for } M>\lambda_1,\ \lambda_1^2 \gg 1$$

and, for case II,

$$F \equiv M^2-1-\lambda_2^2$$

and where, for case I,

$$G \equiv \frac{M^2+\lambda_1^2-2M^2\lambda_1^2}{M^2-\lambda_1^2} \quad M>1,\ \lambda_1^2 \ll 1$$

$$\equiv -\lambda_1^2 \quad M>\lambda_1,\ \lambda_1^2 \gg 1$$

UNSTEADY MAGNETOAERODYNAMIC SUPERSONIC FLOWS 71

and, for Case II,

$$G \equiv 1 + \lambda_2^2$$

Equation (32) shows that, in addition to the Mach number M and location of the axis of rotation $\bar{X}_1$, in this instance the Alfvén number enters into play.

Examination of Eq. (32) defining the torsional flutter boundary allows one to infer the following:

1) In case I, for $\lambda_1^2 \ll 1$ ($M > 1$), the flutter boundary does not differ appreciably from the one obtained in the classical framework (i.e., for $\lambda_1^2 = 0$). However, a slight increase of the stability domain appears in the range $\bar{X}_1 > 0.2$ (see Fig. 4). In the instance corresponding to $\lambda_1^2 \gg 1$, ($M > \lambda_1$), the torsional flutter is completely eliminated. In this case, the magnetic field appears to have a strong beneficial influence.

2) In case II, the instability domain has a similar extent to the one in the classical case (formally corresponding to $\lambda_2^2 = 0$). The instability domain for $\lambda_2^2 \neq 0$ may be obtained formally from its classical counterpart (which corresponds to $\lambda_1^2 = 0$ in Fig. 3 by translating it as a whole

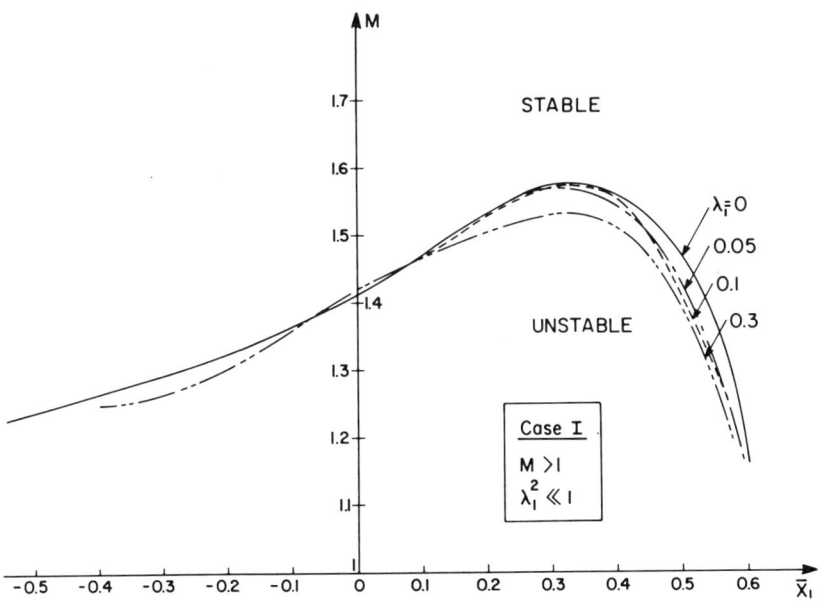

Fig. 4 Domain of unstable pitching oscillation for two-dimensional airfoils and $\lambda_1^2 \ll 1$, $M > 1$.

along the M axis. An arbitrary point of the classical flutter boundary generically denoted as $(M_f, \tilde{X}_1)$, for the case $\lambda_2 \neq 0$, becomes $[(1+\lambda_2^2)^{1/2} M_f, \tilde{X}_1]$. The same is valid for the bottom line of the classical flutter domain, i.e., the line M=1, which goes into $M = (1+\lambda_2^2)^{1/2}$ (i.e., into the transcritical line pertaining to this case). This shows that in the case of the torsional flutter, the electromagnetic field may play an exceptional stabilizing role.

Conclusions

In the present paper, some basic results which concern the behavior of elastic thin bodies and lifting surfaces harmonically oscillating in a supersonic ionized flow have been presented. The peculiarities of the obtained results and some of their implications in the associated flutter problems are pointed out. It is hoped that the obtained results could constitute a basis for further developments in the field and for an appropriate approach to the aeroelastic phenomena, arising during the flight of advanced space vehicles in an ionized field environment.

References

[1] Sears, W. R., "Magnetohydrodynamic Effects in Aerodynamic Flows," ARS Journal, Vol. 29, June 1958, pp. 397-407.

[2] Resler, E. L. Jr. and Sears, W. R., "The Prospects for Magneto-Aerodynamics," Journal of Aeronautical Science, Vol. 25, No. 4, 1958, pp. 235-246.

[3] Selezhov, I. T. and Selezhova, L. V., Waves in Magnetohydroelectric Media, Naukova Dumka, Kiev, 1975, pp. 83-166.

[4] Ambartsumian, S. A., Bagdasarian, E. A., and Belubekian, M. V., Magnetoelasticity of Shells and Plates, Nauka, Moscow, 1977, pp. 204-271.

[5] Librescu, L., "Recent Contributions Concerning the Flutter Problems of Elastic Thin Bodies in an Electrically Conducting Gas Flow, A Magnetic Field Being Present," Solid Mechanics Archives, Vol. 2, No. 1, pp. 1-108.

[6] Librescu, L. and Badoiu, T., "Sur la Stabilité Magneto-Aeroelastique des Paneaux Minces de Longueur Infinie et de Largeur Finie," Comptes Rendus de l'Academie des Sciences, Ser. A, Jan 1977, pp. 203-206.

[7] Librescu, L., "Unsteady Magnetoaerodynamic Forces on an Oscillating Circular Cylindrical Shell of Finite Length," Atti della Accademia Nazionale dei Lincei. Rendiconti della Classe di Scienze fisiche, matematiche e naturali, Part I., Vol. LXII, No. 5, May 1977, pp. 641-646; Part II, Vol. LXIII, No. 6, Dec. 1977, pp. 538-543; Part III, Vol. LXIV, No. 1, June, 1978, pp. 82-87.

[8]McCune, J. E. and Resler, E. L. Jr., "Compressibility Effects in Magnetoaerodynamic Flows Past Thin Bodies," Journal of Aerospace Science, Vol. 27, July 1960, pp. 493-503. (See also corrections by Sears, W. R., "Sub-Alfvénic Flow in Magnetoaerodynamics," Journal of Aerospace Sciences, Vol. 28, March 1961, pp. 249-250.)

[9]Librescu, L., "Unsteady Aerodynamics of Chemically Reacting Flows Past Oscillating Thin Bodies," AIAA Progress in Astronautics and Aeronautics, AIAA, New York, Vol. 95, "Dynamics of Flames and Reactive Systems," pp. 593-609.

[10]Bisplinghoff, R. L. and Ashley, H., Principles of Aeroelasticity, John Wiley and Sons, New York, 1962.

[11]Nelson, H. C. and Berman, J. H., "Calculations of the Forces and Moments for the Oscillating Wing-Aileron Combination in Two-Dimensional Potential Flow at Sonic Speed," NACA Report 1128, 1953.

[12]Miles, J. W., The Potential Theory of Unsteady Supersonic Flow, Cambridge University Press, Cambridge, England, 1959.

[13]Librescu, L. and Malaiu, E., "Aeroelastic Stability of Plane Sandwich-Type Structures Placed in a Current of Supersonic Gas," NASA TTF-13778, Aug. 1971.

[14]Ashley, H. and Zartarian, G., "Piston Theory--A New Aerodynamic Tool for the Aeroelastician," Journal of Aeronautical Science, Vol. 23, Dec. 1956, pp. 1109-1118.

[15]Librescu, L., "Aeroelastic Stability of Orthotropic Heterogeneous Thin Panels in the Vicinity of the Flutter Critical Boundary," Journal de Mécanique, Part I, Vol. 4, No. 1, 1965, pp. 51-76; Part II, Vol. 6, No. 1, 1967, pp. 133-152.

Chapter II. MHD and HD Turbulence

Two-Dimensional Behavior of Electrically Driven Flows at High Hartmann Numbers

Joël Sommeria
Groupement d'Interet Scientifique, Madylam, Saint Martin d'Heres, France

Abstract

Novel experiments are presented on two-dimensional electrically driven flows of mercury in a closed domain, in which three-dimensional perturbations are suppressed by the effect of a uniform magnetic field. The main cause of energy dissipation is a linear friction associated with the Hartmann boundary layer. The validity of the two-dimensional approximation is first tested for isolated vortices, then with a four vortex steady forcing. Then, a 36-electrode network steadily generates a corresponding vortex network, which is very unstable; and a pairing process leads to an inverse energy cascade, which is limited at large scales either by the Hartmann friction or by the size of the domain. In the latter case, the flow is organized in a global rotation, with weak turbulent fluctuation, in a way similar to that for the four-vortex case.

I. Introduction

The two-dimensional behavior of MHD turbulence has been the subject of very impressive work, essentially experimental, during the last 20 years. References 1-3 are recent reviews. Most of it has been performed in channel flows placed in a uniform magnetic field. These investigations have brought some understanding of a complicated situation that occurs in several industrial devices, such as MHD power generators, and have contributed to providing some experimental foundation to the concept of two-dimensional

---

Paper presented at the Fourth Beer-Sheva Seminar on MHD Flows and Turbulence, Ben-Gurion University of the Negev, Beer-Sheva, Israel, Feb. 27-March 2, 1984. Copyright © 1985 by the American Institute of Aeronautics and Astronautics, Inc. All rights reserved.

turbulence. However, no really new ideas concerning two-dimensional turbulence itself have been obtained from these experiments. One of the reasons for this is that three-dimensional perturbations are never completely negligible, and their influence is difficult to estimate. Incidentally, such a problem is encountered in all the practical fields of application of two-dimensional turbulence, such as rotating fluids of geophysical flows. Another drawback of these previous works is the difficulty of knowing the effect of the magnetic field on the turbulence generation, even when a grid is used. Otherwise, a lot of experimental problems are encountered, for controlling the mean flow and suppressing perturbations from upstream, and for velocity measurements.

In this paper, these problems are avoided by generating two-dimensional flows in a closed domain, with a free upper surface to allow visualization. Motion is driven by injecting electric currents in a network of bottom electrodes, a kind of "electric grid" (Figs. 1 and 2). The vorticity production is proportional to the current density, Eq. (1), so that it is very well controlled, and versatile forcings are possible by changing only the electrical device. The three-dimensional perturbations are reduced to a minimum, following the criteria of Ref. 4, by using a small depth a in the direction of the magnetic field and relatively large horizontal turbulent scales. Details of the experimental procedure are given in Sec. II.

The two-dimensional equation of motion for the vertical component of vorticity is obtained by including the effect of injected electric currents density $\tilde{j}(x,y)$ in Eq. (18) of Ref. 4.

$$\frac{d\tilde{\omega}}{dt} = \frac{\beta \tilde{j}}{\rho a} - \frac{\tilde{\omega}}{t_h} + \nu \nabla^2 \tilde{\omega} \qquad (1)$$

where ρ stands for the mercury density, σ its conductivity, and ν its viscosity, d/dt is the substantial derivative, and $t_h = (a/B)(\rho/\sigma\nu)^{1/2}$ is the typical time of Hartmann bottom friction. If the upper surface is rigid, there is an upper Hartmann layer and t_h is doubled. Furthermore, all the electric currents are confined in the Hartmann layer, and Ohm's law is written in the two-dimensional core:

$$\nabla \tilde{\phi} = \vec{V} \times \vec{B} \qquad (2)$$

so that the potential gradients are always perpendicular to the velocity and the electric potential ϕ is proportional to the stream function. The impermeability condition at

BEHAVIOR OF ELECTRICALLY DRIVEN FLOWS 79

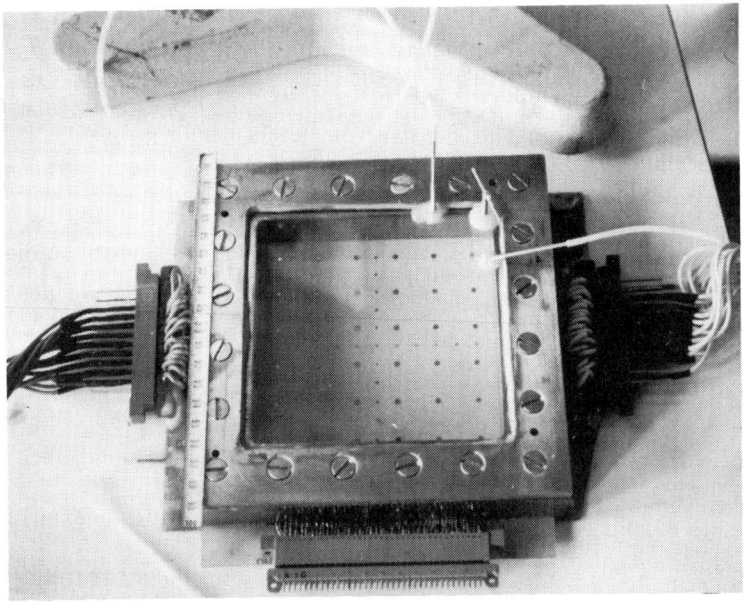

Fig. 1 Photograph of the 36 electrode mercury container.

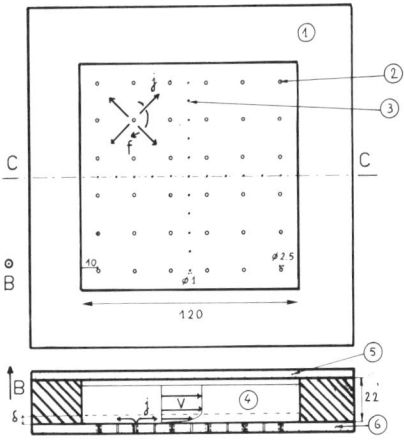

Fig. 2 Apparatus, current distribution near one electrode, and velocity profile are schematized. The Hartmann layer depth is denoted by 1) copper frame, 2) electrodes for current injection and electric potential measurements, 3) electrodes for electric potential measurements only, 4) mercury, 5) glass cover, 6) electrically insulating bottom plate in which electrodes are embedded.

the lateral walls and the electric condition of infinite
conductivity (copper is much more conducting than mercury)
are consistent with Eq. (2) and impose ϕ = const on the
frame. One could associate with it the no-slip condition,
but it seems, from the comparison with the numerical
simulations of Sec. III, that the zero vorticity condition
(perfect slip) is at least as suitable, when one considers
the flow outside the parallel boundary layers. However,
this simplified description misrepresents possible boundary
layer detachment at large velocity.

It is suitable to use a nondimensional form of Eq. (1),
in which the forcing is of the order of unity, by means of
the new variables defined by

$$\tilde{t} = (\rho a L^2/BI)^{\frac{1}{2}} t, \quad \tilde{\vec{r}} = L\vec{r}, \quad \tilde{\vec{j}} = (I/L^2)\vec{j} \qquad (3)$$

where L is an horizontal length scale, and I is the total
current provided by the power supply so that
$\int_0^1 \int_0^1 |j(x, y)|\ dx\ dy = 2$. (Electric current entering the
box through the bottom must exit the same way, contributing
twice to the current density.) Then the nondimensional
equation is written

$$\frac{d\omega}{dt} = j(x,y) - \omega/Rh + 1/Re\ \nabla^2\omega \qquad (4)$$

The experimental conditions are characterized by the
two nondimensional numbers:

$$Re = L/\nu\ (BI/\rho a)^{\frac{1}{2}}, Rh = (Ia/B\sigma\nu)^{\frac{1}{2}}\ L^{-1} \qquad (5)$$

Re is the Reynolds number, and Rh is the ratio of the bottom
friction time t_h to a typical turnover time. The main
cause of energy dissipation is the bottom friction, and the
effect of lateral viscosity, characterized by Re, seems not
to be very pertinent for the dynamics, according to the
experimental results, Secs. III and IV.

The two-dimensional approximation is tested in Sec. III.
The results, reported in Ref. 5, for the relatively simple
case of isolated vortices are first summarized. Then, in
a more complex situation, it is investigated in Ref. 6,
where vorticity is steadily generated in a four-vortex mode,
and experimental results are compared with a numerical
integration of Eq. (1). In Sec. IV, new results, concerning
a small-scale forcing and the development of an inverse
energy cascade toward large scales, are presented.

II. The Experimental Procedure

The facility is a closed box containing a horizontal layer of mercury (whose depth a is 2 cm) located in the gap of an electromagnet of vertical magnetic field < 1 Ta. The box is made of an insulating bottom plate in which copper electrodes are embedded, a copper frame, and a glass cover. The upper mercury surface is generally free, and velocity is kept small enough (< 10 cm/s) so that the effect of wave propagation is negligible. In some experiments, a rigid upper boundary condition is provided by a thin oxide skin, formed at the air contact. One set of vortex experiments is achieved in a circular box with one central electrode, while the others are made in a square box containing 36 electrodes (Figs. 1 and 2). Great care is necessary to avoid mercury pollution and to keep a really free upper surface. For this purpose, the apparatus is filled with pure nitrogen (U quality), the insulating parts are covered polypropylene, and the copper parts (electrodes and frame) by a nickel layer (50 μm thick) followed by a very thin gold coating to improve electric contact with mercury. The walls are highly polished to avoid perturbations of the boundary layers. A constant current is provided to each electrode by a set of electrical resistances (3 Ω each) and a dc power supply (0 → 20 A).

The flow investigation is made by two methods: The first one is direct photography of the trajectory of small particules (∅ 50 μm) following the free upper surface. The exposure time is adjusted to get an instantaneous velocity field from the small, approximately straight photographic traces. Because of the presence of the electromagnet above the mercury, a 45-deg mirror must be used, and a 2.5-cm-wide band can be visualized at a given time. The second method is the measurement of the electric potential at the bottom, by means of the electrodes which are used for injecting electric currents, or best by a separate array of electrodes. By means of relation (2), this method provides a precise measurement of the stream function.

III. Testing the Two-Dimensional Dynamics

A. One Steady Vortex

A simple solution of Eq. (1) is obtained when the current is injected through a small electrode at the center of a circular box (and returns through the copper frame). It is a vortex in which vorticity is localized above the electrode and in which total circulation is $BIt_h/\rho a$. The experimental nondimensional velocity distribution at the free surface (from photographs) is represented in Fig. 3. The discrepancy with theory is limited to the central core region and comes from two sources: the effect of viscosity in the horizontal, which produces a parallel boundary layer near the electrode, and the nonlinear three-dimensional convective effects. The correction for the former effect is calculated in Ref. 5 and represented in Fig. 3. The agreement with experiments is then very good for small injected currents (for which nonlinear effects are negligible). A very precise agreement ($\sim 1\%$) is also obtained for the electric potential. Since the electric currents are confined in the Hartmann layer, this result is a good test of its thickness by means of the measure of its electrical resistance. The nonlinear effects consist mainly in secondary flows which are confined to the vortex central core.

B. Free Vortices

More generally, one or several localized two-dimensional vortices can be generated by electric pulses. The interaction between two vortices, simultaneously produced in this manner, is studied in Ref.5, as well as the interaction of one vortex with the wall. These results confirm the validity of the two-dimensional equation (1), when vorticity is convected.

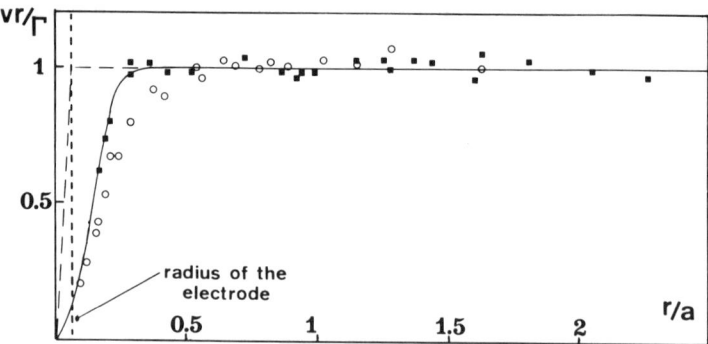

Fig. 3 Nondimensional product rv/Γ vs the nondimensional distance to the center r/a ($\Gamma = I/2\pi \sqrt{\sigma\rho\nu}$). ■:I = 50 mA; o:I = 200 mA; ---: theoretical curve from Eq. (1); ———: theoretical curve taking into account viscous effects in the horizontal directions near the electrode.

C. A Steady Four-Vortex Forcing

The square box with the 36 electrode bottom is used and the injected current at the electrode which position is x, y is proportional to sin 2π x/L sin 2π y/L in order to get a four-vortex steady forcing. For small values of Rh (small electric currents), a linear four-vortex regime is obtained. There is an instability threshold Rh = 1.52 above which a global steady rotation occurs. The rotation rate can be characterized by the stream function at the box center, which is represented in Fig. 4 versus Rh. Some turbulent fluctuations appear for Rh>5, but for still larger (>20), they again become very weak. So this indicates that convection has a stabilizing effect at large Rh, probably because of the rotation, somewhat in contradiction with classical ideas about transition to turbulence.

The good similarity of the nondimensional rotation rate for different experimental conditions is a good verification of the validity of Eq. (4) and of the fact that Rh is the main pertinent parameter. The most convincing verification is provided by the comparison between the results obtained with a free upper surface and that obtained with a rigid one, for which the friction is found to be precisely

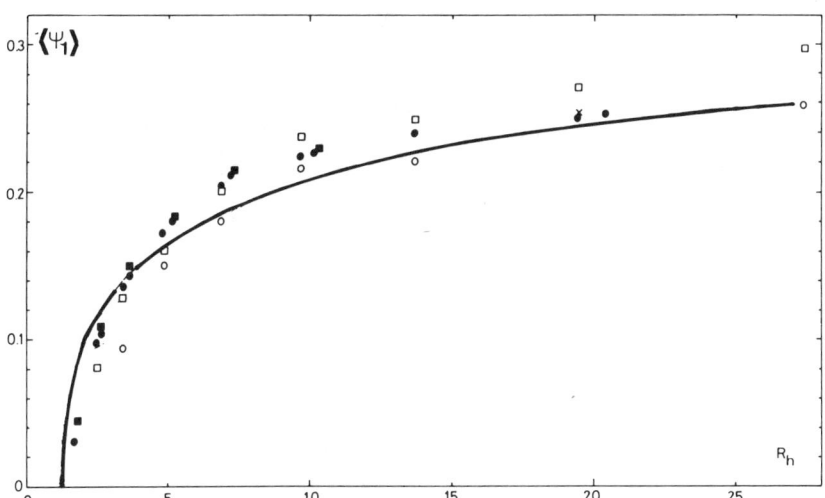

Fig. 4 Mean stream function at the box center vs the friction parameter Rh, compared with numerical simulations (from Ref. 6).
●: Experiments with a free upper surface, B = 0.48 T; □ : idem, B = 0.24 T; O:idem, B = 0.12 T; ■ :rigid upper surface, B = 0.48 T; ———:numerical simulation of Eq. (5) by a finite-difference scheme with 100x100 grid points.

doubled, as is expected if the velocity is constant along a vertical line. A more direct proof of the validity of the two-dimensional approximation is provided by a numerical finite-difference integration of Eq. (1) (see Fig. 4). Note that this calculation is made with a spatially continuous forcing, so the vorticity created at each electrode is well spread by convection, and the global effect is close to that of a continuous creation.

IV. The 36 Vortex Steady Forcing

The injected electric current is such that its modulus is the same for each electrode, but it changes sign from one electrode to its nearest neighbor. Thus, a regular network of alternating sign vortices is created (Fig. 5a); this network is unstable and generates turbulence by pairing processes. The study of the flow consists of visualization and potential measurements. Three main kinds of

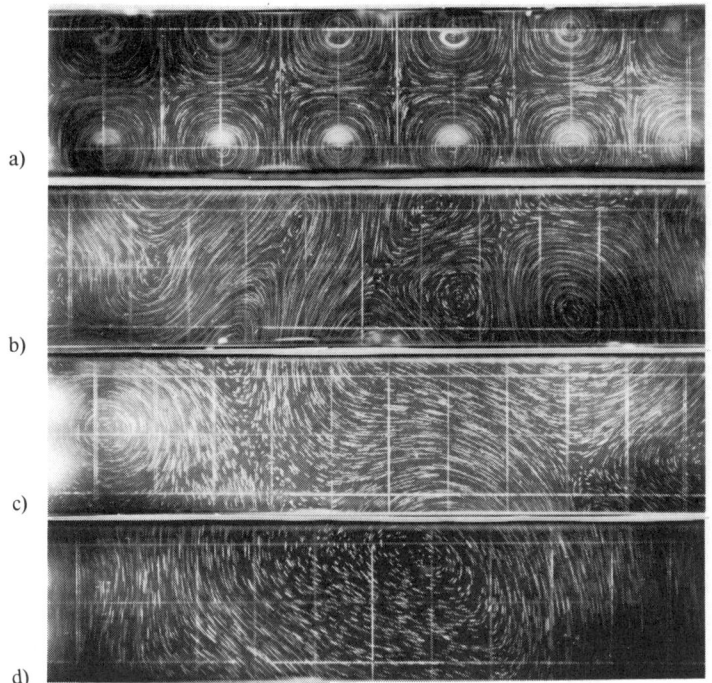

Fig. 5 Photographs of a central band of the upper surface using a 45-deg mirror (1 graduation = 1 cm). t is the time of exposure; a) the initial motion when the current is switched on; b) Rh = 6.86, B = 0.48 T (t = 195 ms); c) Rh = 19.4, B = 0.48 T (t = 30 ms); d) Rh = 38.8, B = 0.12 T (t = 195 ms).

BEHAVIOR OF ELECTRICALLY DRIVEN FLOWS   85

behavior can be distinguished as the number Rh is increased.
The linear 36 vortex regime ends at a very small Rh and has
not been studied as well as the transition to turbulence.
A regime of nearly isotropic turbulence then occurs, characterized by large chaotic fluctuations and a zero mean
value of the stream function at the box center (i.e., no
mean rotation). For Rh>40, an organized global rotation

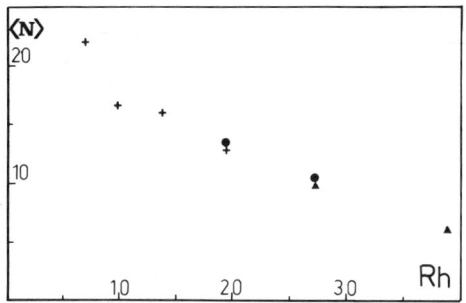

Fig. 6 Decrease of the mean number $\langle N \rangle$ of vortex cores in the whole
flow as Rh is increased, due to inverse energy transfers. +:B =
0.48 T; ●:B = 0.24 T; ▲:B = 0.12 T.

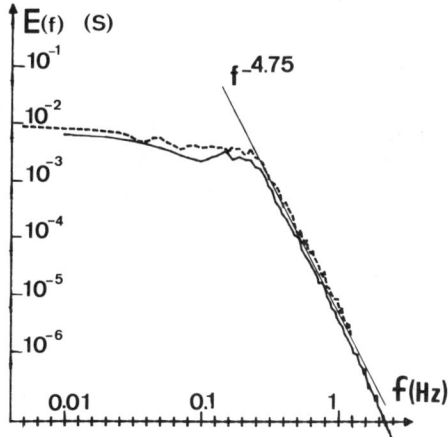

Fig. 7 Time spectra of the central stream function ψ_1 in log-log
coordinates [normalized in such a way that $\int_0^\infty E(f)\, df = \langle \psi_1^2 \rangle$] for
Rh = 3.56. ———:B = 1 T, I = 1 A, free upper surface; ----:B =
0.5 T, I = 2 A, rigid upper surface. The good similarity between
these two spectra is a test of the two-dimensional approximation.
Spectra are obtained from samples of 50,000 successive data. The
average is calculated from 210 Fast Fourier Transforms realized
on 512 points. Each sample overlaps the preceding by 50% and is
multiplied by a Blackmann Harris window.

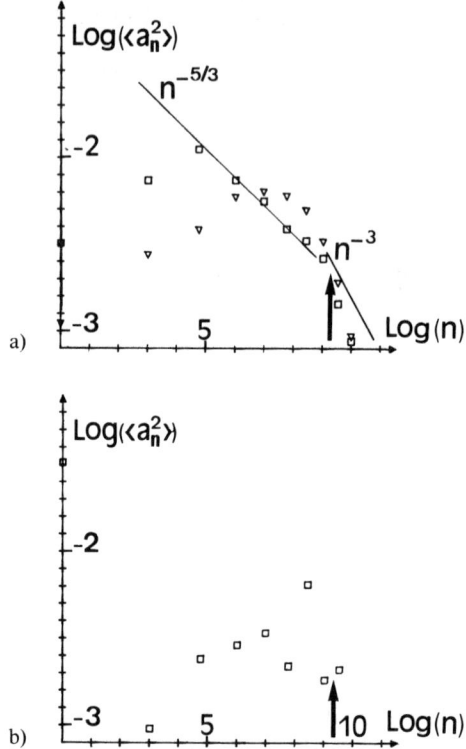

Fig. 8 One-dimensional spectra of the transverse velocity component in log-log coordinates. The points correspond to the 12 first terms of the Fourier series computed from a periodic function which is an extension of the actual velocity profile. The normalization is such that $\frac{1}{2} \sum_{1}^{\infty} \langle a_n^2 \rangle = \langle v_2^2 \rangle$. The wave number is expressed in the unit π/L. The slopes $K^{-5/3}$ and K^{-3} are indicated as well as the injection wave number. a) ▽:Rh = 5.04; □:Rh = 14.24. b) □:Rh = 40.3; notice that most of the energy is concentrated in the smallest available wave number π/L.

appears, analogous to that obtained in the four-vortex case, and associated with an important decrease of the turbulent fluctuation (Fig. 5d).

The nearly isotropic regime is definitely chaotic, as is revealed by smooth time spectra of the central stream function (Fig. 7). Otherwise, a $k^{-4.7}$ decay seems to occur at high frequencies, although it is difficult to measure such a steep slope with precision. The eddies are larger and larger as Rh increases, as is clearly seen in the visualizations of Fig. 5. This effect can be quantified by counting the mean number <N> of vortex cores in the whole

box (Fig. 6). This mean is calculated from 25 photographs of
of each 2-cm-wide band that can be observed at a given time.
From this number, a turbulent scale L <N>$^{-\frac{1}{2}}$ can be obtained
that is smaller than the integral scale derived from correlation functions, since small eddies are counted with the
same weight as larger ones. A more precise description is
provided by one-dimensional velocity spectra, obtained from
an array of 11 electrodes (Fig. 8). The electric potential
measured at these points is interpolated by cubic spline
functions and is differentiated to give the transverse
velocity profile by means of relation (2). Spectra are then
obtained as an averaged squared Fourier transform of 800
such profiles.

The results are in good agreement with the standard
description of two-dimensional steady isotropic homogeneous
turbulence (see Refs. 7 or 8 for a recent review). Two
modifications of these theories must be noted here. First,
a drain in the inverse energy cascade and in the enstrophy
cascade is due to the bottom friction. This drain is weaker
as Rh gets larger, so that the integral scale, corresponding
roughly to the maximum of the spectrum, increases. Second,
for large enough values of Rh ($\sim$40), the inverse energy
cascade reaches the size of the box and energy accumulates
in the largest available scale, as in truncated equilibrium
systems. However, the description in terms of isotropic
turbulence is no longer valid for this state of global
organized rotation.

V. Conclusions

Different kinds of two-dimensional flows were produced
in mercury by means of a uniform magnetic field. The two-
dimensional dynamics was checked directly by comparisons
with analytical results and numerical computations, and
indirectly by the observation that the nondimensional
results depend only on the friction parameter Rh. Using a
fairly thin mercury layer is essential to efficiently
suppress three-dimensional perturbations. The effect of the
walls perpendicular to the magnetic field is then limited
to a linear Hartmann friction whose time scale t_h depends
only on the fluid properties and on the magnetic field.

The electrically driven network of vortices is
strongly unstable and turns into a nearly isotropic homogeneous two-dimensional turbulence. The presence of a $k^{-5/3}$
inverse energy cascade is in good agreement with standard
theories. When this cascade is limited by the finite size
of the box, a mean rotating flow spontaneously appears, for
which most of the kinetic energy is in the largest available

scale. This behavior, which seems to be quite general in a square box, is fairly surprising, since no global vorticity is brought by the electric forcing.

Acknowledgments

The author has benefitted from many useful discussions with J. Verron, who performed the numerical calculations, and P. Tabeling, who suggested the surface treatment of the copper electrodes.

References

[1] Branover, H., Magnetohydrodynamics Flows in Ducts, Halsted, 1978.

[2] Moreau, R., "Why, How and When MHD Turbulence Becomes Two-Dimenional," MHD Flows and Turbulence: AIAA Progress in Astronautics and Aeronautics, Vol. 84, AIAA, New York, 1983; pp. 20-29.

[3] Sommeria, J., "Two-Dimensional Behavior of MHD Fully Developed Turbulence (Rm << 1)," J.M.T.A (special issue), 1983, pp. 169-190.

[4] Sommeria, J. and Moreau, R., "Why, How and When MHD Turbulence Becomes Two-Dimensional," Journal of Fluid Mechanics, Vol. 118, 1982, pp. 507-518.

[5] Sommeria, J., "Electrically Driven Quasi-Two-Dimensional Vortices in a Uniform Magnetic Field," Journal of Fluid Mechanics (to be published).

[6] Sommeria, J. and Verron, J., "An Investigation of Nonlinear Interactions in a Two-Dimensional Recirculating Flow," Physics of Fluids (to be published).

[7] Lesieur, M., "Introduction to Two-Dimensional Turbulence," J.M.T.A. (special issue), 1983, pp. 5-20.

[8] Kraichnan, R. H. and Montgomery, D., "Two-Dimensional Turbulence," Report on Progress in Physics, Vol. 43, 1980, p. 547.

Transition from Three-Dimensional to Quasi-Two-Dimensional MHD Grid Turbulence

P. Caperan* and A. Alemany†

Institut de Mecanique de Grenoble, Saint Martin d'Heres, France

Abstract

This paper gives the results of an experiment conducted on homogeneous MHD turbulence in mercury, at low magnetic Reynolds number. The first part describes how an initial power spectrum decreasing as $k^{-5/3}$ is slowly invaded from the highest wave number to the lowest one by a k^{-3} law. The transition between these two laws occurs at a rather constant level of energy in the spectrum. The second part of the paper deals with the asymptotic quasi-two-dimensional phase of this kind of turbulence. In order to characterize the anisotropy at this stage, the energy density function of the velocity parallel to B has been mapped in the plane parallel to B. This map of energy in Fourier space clearly demonstrates the existence of a conical zone, as expected from the phenomenology.

Introduction

Experimental works on MHD turbulence have been undertaken to get a better insight into the MHD duct flows that are involved in some industrial cases, like two-phase flow MHD generators. An idealization of this turbulence, as homogeneous grid turbulence, was first attempted by Kolesnikov and Tsinober,[1] measured a k^{-3} decrease of the energy

---

Paper presented at the Fourth Beer-Sheva Seminar on MHD Flows and Turbulence, Ben-Gurion University of the Negev, Beer-Sheva, Israel, Feb. 27-March 2, 1984. Copyright © 1985 by the authors. Published by the American Institute of Aeronautics and Astronautics, Inc. with permission.

*Assistant to INPG.
†CMRS Research Head.

spectrum in the direction perpendicular to the magnetic field $\vec{B}$. That property has been related to the two-dimensional behavior of such flows. Further experiments conducted by Alemany[2] complete these results with the exhibition of the spectra in the direction parallel to B, again showing a k^{-3} decrease law. The decay along time at a fixed wave number, occurs in t^{-2}. An explanation of this $k^{-3} \, t^{-2}$ law has been proposed using the dissipation cone concept, which was first introduced by Moreau[3]. In fact, the local equilibrium between angular transfer of energy toward the conical dissipation zone axial to $\vec{B}$ and the Joule dissipation fully explains the behavior of free decaying MHD turbulence submitted to a steady magnetic field $\vec{B}$.

The anisotropy of these flows has been characterized today, quite indirectly, by means of passive scalar diffusion (indium[4] and thermal plume[5]). These experiments exhibit higher diffusion in the direction perpendicular to the B plane than in the parallel one. Some theoretical investigations have been performed by Moffatt[6] and Schumann[7] on the setting of the k^{-3} spectrum from an initial isotropic three-dimensional turbulence. But at present, no experiment has been attempted to settle this problem.

The experimental works described in this paper attempt to answer two questions:

1) How does the settlement of an energy density spectrum decreasing as k^{-3} occur from an initial isotropic stage in $k^{-5/3}$?

2) Can the conical dissipation zone phenomenology be supported by some experimental evidence? This question leads to the measurement of the two-dimensional energy density in Fourier space, in order to exhibit any conical structure in it.

Experimental Apparatus

The experimental facility is sketched in Fig. 1. A tank of mercury 2 m high and 20 cm in diameter is disposed in the core of a solenoid delivering an axial homogeneous magnetic field in the range of 0 to 0.4 T. A grid descending at controlled velocity generates a turbulence that is studied by means of cylinder hot film probes at different distances behind the grid. The solidity ratio of the different grids used in this study is 0.44, the mesh size of the bar and the characteristic velocity fluctuation just behind the grid, corresponding to 30% of the mean velocity.

The acquisition chain is shown in Fig. 2. After applying a bucking voltage to reduce the dc component, band-pass filters are used in order to respect the Shannon acquisition condition and to eliminate any dc components from the anemometer signal. The data are sampled and processed on a Norsk 100 computer. The energy density spectra are obtained by the periodogram method.

Some problems arise from mechanical vibrations caused by the suspension chain. In order to resolve them, a Wiener filter has been used, which has necessitated the use of a supplementary probe to measure the pure vibrations in front of the grid.

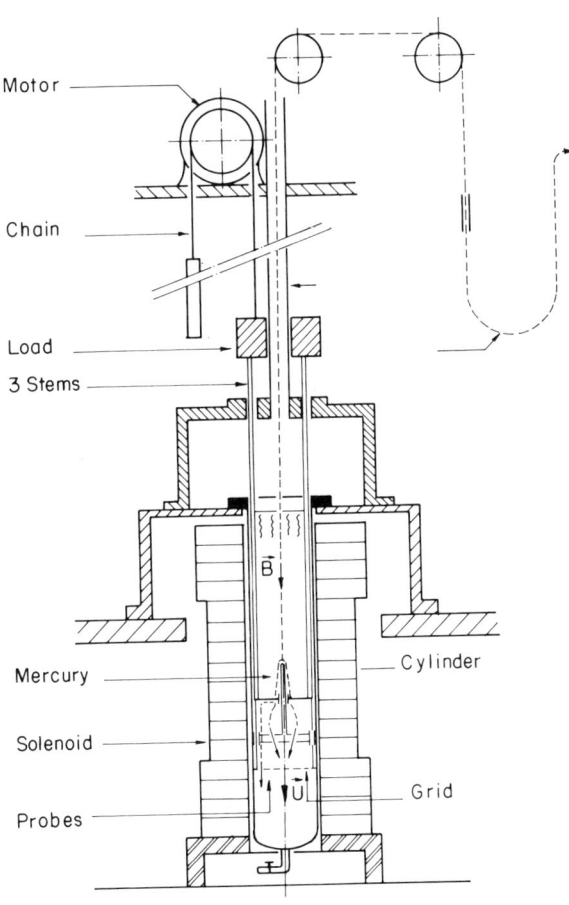

Fig. 1 Experimental facility.

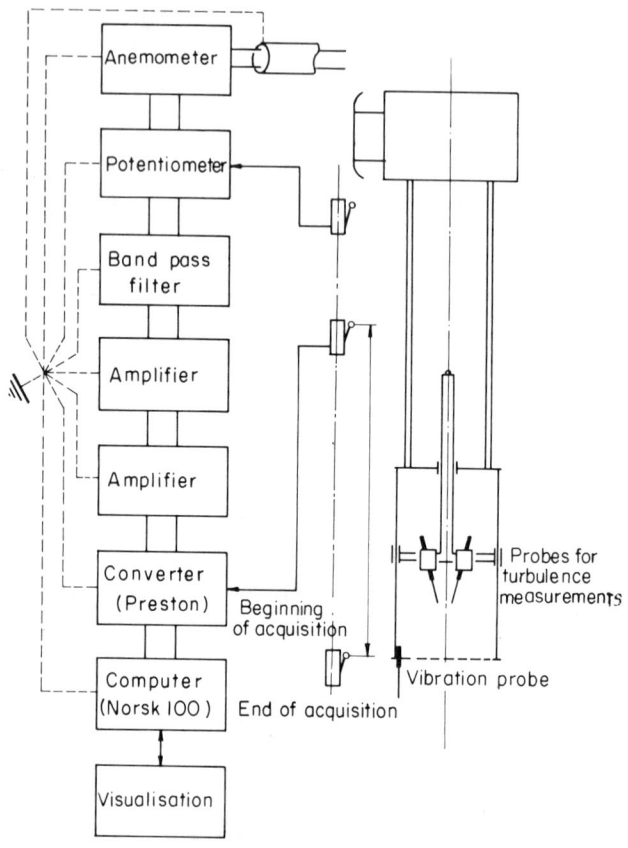

Fig. 2   Signal processing chain.

Experimental Results

Transition Phase

In order to characterize the behavior of the turbulence during the transitional phase between a $k^{-5/3}$ law and a k^{-3} one, the one-dimensional spectra behind the grid have been measured. The presentation of the spectra in a compensated way allows a better segregation between the two characteristic zones (Fig. 3).

A first experiment has been carried out with an initial interaction parameter of moderate value 0.5 (Re=1800). The results, plotted on Fig. 3, show a decrease of $k^{-5/3}$, so the dynamic of the turbulence in this case is mainly three-dimensional with a usual energy cascade.

TRANSITION TO QUASI-TWO-DIMENSIONAL TURBULENCE 93

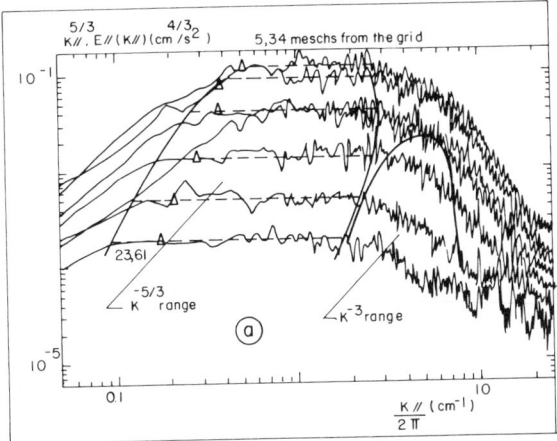

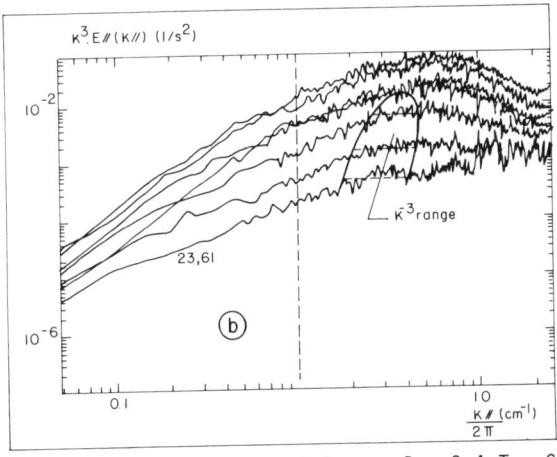

Fig. 3 $N = 0.6$; Re = 1800; $m = 1.2$ cm; $B = 0.4$ T. Compensated spectra at different distances from the grid, $E// (k//)k//^n$. (a) $n = 5/3$; (b) $n = 3$.

The second experiment corresponds to an initial N_0, equal to 1, and a Reynolds number of 900. The results, plotted on Fig. 4, exhibit two different zones in the energy spectra. On Fig. 4a, the $k^{-5/3}$ law appears on one decade (0.1<k<2 cm^{-1}) at 5.34 meshes from the grid. This inertial zone is slowly invaded by the k^{-3} law (Fig. 4b) which occupies a decade at the last experimental point at 34.2 meshes from the grid. A transition wave number k_t between these two zones of different decrease can be defined by taking the midpoint between them. The energy on k_t is shown on Fig. 5, which plots along the distance from the grid of the energy

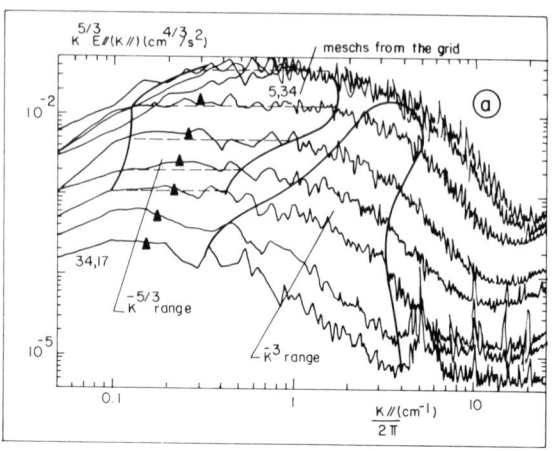

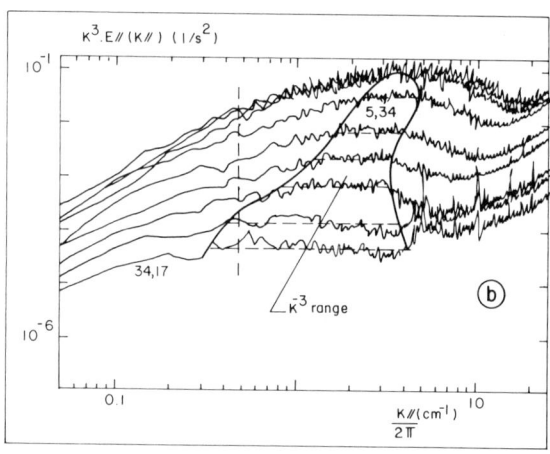

Fig. 4 N = 1.05; Re = 900; m = 1.2 cm; B = 0.4 T. Compensated spectra at different distances from the grid, $E_{//}(k_{//}) k_{//}^n$.
(a) n = 5/3; (b) n = 3.

at different fixed wave numbers. The straight line corresponding to $E_{//}(k_t)$ distinctly separates two zones of decay: one, at a low wave number, corresponds to a $t^{-1.2}$ decay, whereas the other one, at a wave number higher than k_t, corresponds to a t^{-2} decay. These two regions are respectively identified with the $k^{-5/3}$ and k^{-3} decrease zones of Fig. 4. Thus, one can infer a $k^{-5/3} t^{-1.2}$ decay zone at wave numbers lower than k_t and a $k^{-3} t^{-2}$ decay otherwise. This behavior corresponds with the ones found respectively in usual turbulence and in MHD turbulence at high interac-

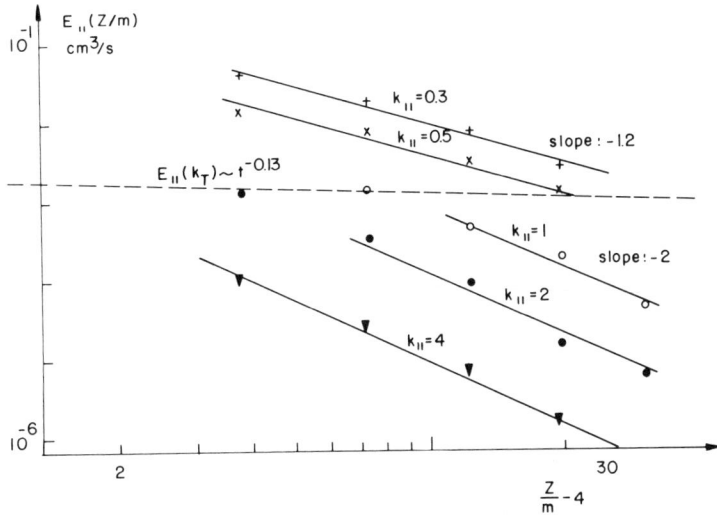

Fig. 5   Decay of energy along the distance from the grid at different wave numbers.

tion parameters. From the continuity of the spectrum, the transition wave number should take place where

$$E_{//}(k_t) \sim t^{-1.2} \, k^{-5/3} \sim t^{-2} \, k^{-3} \qquad (1)$$

which gives us

$$E_{//}(k_t) \sim t^{-0.2} \qquad (2a)$$

$$k_t \sim t^{-0.6} \qquad (2b)$$

The evolution of k_t (plotted on Fig. 6) is in fairly good agreement with the behavior predicted by Eq.(2b.).

These results will be further developed in a future work,[8] but some observations can be made about them now. It is quite surprising that a k^{-3} law occurs first at high wave numbers, since the interaction parameter is higher on the lower ones. However, this damping of the big structure energy will exhaust the energy cascade toward little scales. As these cease to be fed, their interaction parameter will increase on the one hand, and, on the other hand, their faster initial turnover time will favor a quick restructuring toward a k^{-3} decrease law.

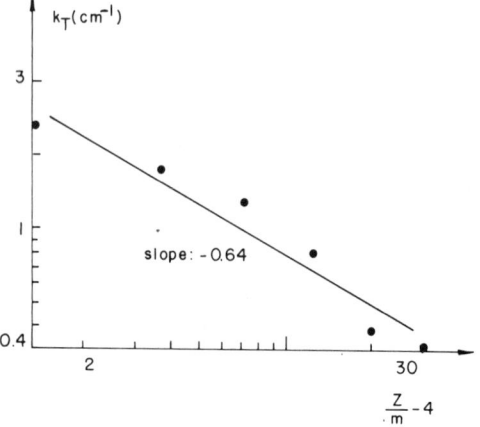

Fig. 6 Decay of energy along the distance from the grid at the transition wave number k_i.

All these measurements have been realized using a one-dimensional investigation method. This has been useful in a first stage to detect modifications of behavior in the turbulence. Nevertheless, in order to get a better insight into such an anisotropic phenomenon, a three-dimensional investigation method is needed. We have used a new procedure that permits mapping the correlation field and the energy density in physical and Fourier three-dimensional spaces. These first tests have been made on a MHD turbulence in a quasi-bidimensional phase, at an interaction parameter of value 3.5.

Two-Dimensional Results on Asymptotic Phase

In order to map the correlation field on chosen points behind the grid, we have measured the set of frequency interspectral functions between two probes at the same distance from the grid on 32 different transverse spacings between them (on a regular step). The use of the Taylor hypothesis allows conversion of the frequency wave number in the direction parallel to the magnetic field. The transverse Fourier transform of the set of interspectral data at fixed parallel wave numbers gives the energy density in Fourier space.

Whereas the energy, in the absence of a magnetic field (Fig. 7a) is confined in a roughly squared zone of the Fourier space, it is located in the MHD case (Fig. 7b) in a rectangular region perpendicular to the magnetic field B. This phenomenon occurs very near to the grid (four meshes), which means that the structure of the turbulence is strongly influenced by B, even during its generation.

TRANSITION TO QUASI-TWO-DIMENSIONAL TURBULENCE

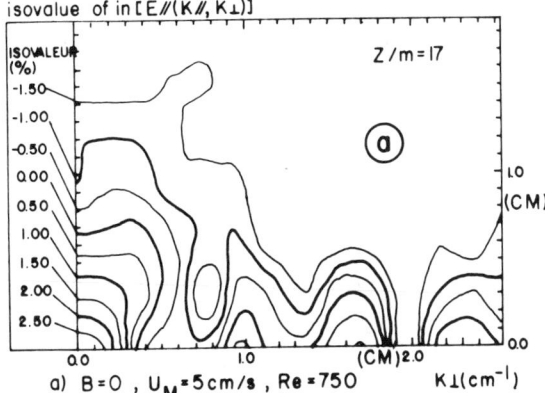

Fig. 7 Isologarithm of the energy in Fourier space. $U_M = 5$ cm/s; Re = 750; m = 2 cm.

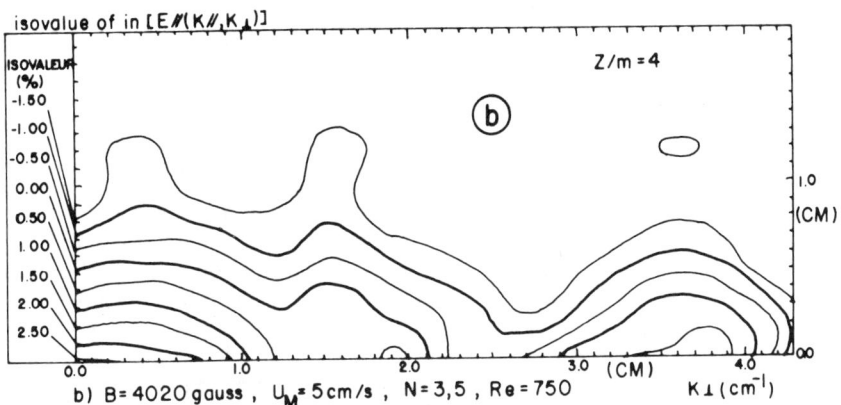

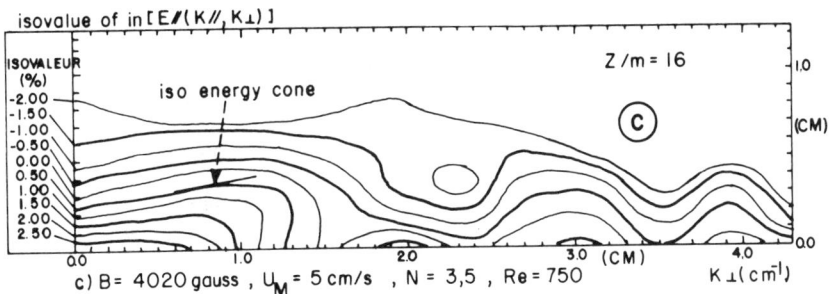

In fact, an aspect ratio can be defined for the region containing the energy of value $k_{//}/k_{\perp} \sim 0.32$ (Fig. 7b).

The low value of this coefficient means that the eddies are elongated in the direction of the magnetic field. The evolution of the energy density between 4 meshes and 16 meshes corresponds to a decrease of the aspect ratio to the value 0.18 (correlative increase of the length of the eddies along B). Also note the modification of shape of the isoenergy lines. Indeed, these lines become straighter from 4 to 16 meshes on their upper part. We have the apparition of a conical zone in the Fourier space, which must be related to the phenomenology of the dissipation cone.

In another experiment the transverse correlation of u// has been measured at different distances from the grid. Whereas the characteristic transverse scale increases along the distance from the grid for a usual turbulence (B=0, Fig. 8a), the action of B (Fig. 8b) is to block the evolution as soon as eight meshes from the grid. This effect is the same as that observed on Fig. 7b and 7c, where the perpendicular wave number does not seem to evolute any more. This blocking has been predicted by the model described in Ref. 2. It can be compared to the similar effect found in a turbulence submitted to a strong rotation.

Conclusion

The main results obtained in this experimental study follow:

1) Evidence of two drastically different behaviors in the one-dimensional spectra obtained during the transitional phase from "isotropic" to quasi-bidimensional turbulence. It is supposed that this is the trace on the one-dimensional spectra of a complex energy exchange, which must be precisely established by means of three-dimensional spectra.

2) The proof of the existence of a conical shaped isoenergy density surface in the Fourier space, which supports

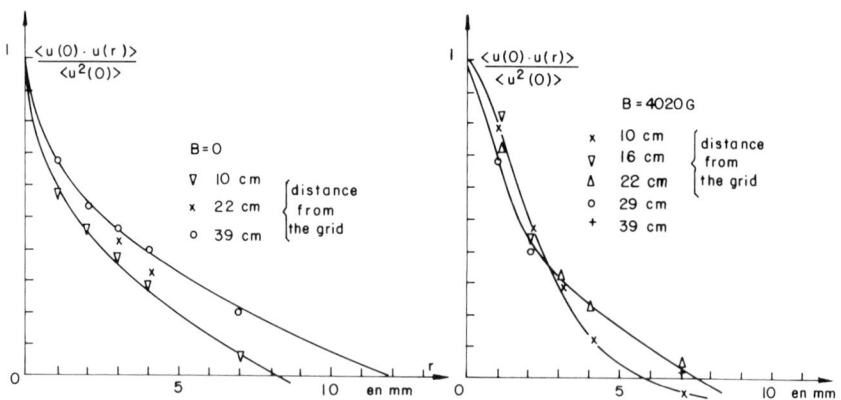

Fig. 8 Evolution of the perpendicular A to B correlation of u// along different distances from the grid. U_M = 15 cm/s; m = 2 cm; solidity of the grid = 0.44.

the phenomenology of Joule dissipation cone in the asymptotic phase. Further studies on this phase would be to evaluate the inertial transfer by bispectral analysis.[9]

References

[1] Kolesnikov, B.Y. and Tsinober, A.B., "MHD Turbulence Behind One-Dimensional Grid," Magn-Gridodynamica, Vol. 3.33, 1972.

[2] Alemany, A., Sulem, P.L., Frisch, U., and Moreau, R., "Influence of an External Magnetic Field on Homogeneous MHD Turbulence," Journal de Mecanique, Vol. 18, No. 2, 1979.

[3] Moreau, R., "On Magnetohydrodynamic Turbulence," Proceedings of the Symposium on Turbulence of Fluid and Plasma, Polytechnic Institute of Brooklyn, New York, 1968, p. 359.

[4] Kolesnikov, B.Y. and Tsinober, A.B., "Experimental Study of Two-Dimensional Grid Turbulence," Isvestia Academii, Nauk Gidkostivy Gaza, No. 4, 1974, p. 146.

[5] Sommeria, J., "Two-Dimensional Behavior of MHD Fully Developed Turbulence (Rm << 1)", J.M.T.A., Special Issue, 1983, pp. 169-190.

[6] Moffatt, H.F., "On the Suppression of the Turbulence by a Uniform Magnetic Field," Journal of Fluid Mechanics, Vol. 28, No. 3, 1967, pp. 571-592.

[7] Schumann, U., "Numerical Simulation of the Transition from the Three- to Two-Dimensional Turbulence under a Uniform Magnetic Field," Journal of Fluid Mechanics, Vol. 74, NO. 1, 1976, p. 31.

[8] Caperan, P., and Alemany, A., "Turbulence Homogene MHD a Petit Nombre de Reynolds Magnetique. Etude de la Transition 3D-2D et de l'Anisotropie en Phase Asymptotique," J.M.T.A., to be published, 1984.

[9] Lü, K.S., Rosenblatt, M., and Van Atta, C., "Bispectral Measurements in Turbulence," Journal of Fluid Mechanics, Vol. 77, No. 1, 1976, pp. 45-62.

Direct Numerical Simulation of Two-Dimensional Turbulence

M.E. Brachet*
Observatoire de Nice, France
and
P.L. Sulem†
Tel Aviv University, Israel, and Observatoire de Nice, France

Abstract

The small-scale dynamics of decaying two-dimensional turbulence is investigated by direct numerical simulations (with up to 1024 × 1024 modes) of Navier-Stokes equations for random flows with Taylor-Green-like symmetries. At high Reynolds number, the following scenario is observed. At early times, large-scale straining generates quasirectilinear vorticity gradient sheets with thickness decaying exponentially in time until dissipation becomes relevant. In Fourier space, the energy spectrum displays a k^{-n} range, with n ≈ 4.0, in agreement with the Saffman theory. Close to the time of maximum entropy dissipation, we observe a sharp transition to a n ≈ 3.2 inertial range, consistent with the Batchelor-Kraichnan theory of entropy cascade. In this regime, vorticity gradients are distributed on highly convoluted secondary dissipative structures resulting from folding and reconnection of early time sheets.

I. Introduction

We have performed direct numerical simulations of freely decaying high Reynolds number turbulence in two dimen-

---

Paper presented at the Fourth Beer-Sheva Seminar on MHD Flows and Turbulence, Ben-Gurion University of the Negev, Beer-Sheva, Israel, Feb. 27-March 2, 1984. Copyright © 1985 by the American Institute of Aeronautics and Astronautics, Inc. All rights reserved.
*Centre National de la Recherche Scientifique.
†School of Mathematical Sciences, Tel Aviv, and Centre National de la Recherche Scientifique.

sional flows governed by the Navier-Stokes equations

$$\frac{\partial \vec{u}}{\partial t} + \vec{u}.\vec{\nabla}u = -\vec{\nabla}p + \nu\nabla^2\vec{u}$$
$$\vec{\nabla}.\vec{u} = 0$$
(1)

with periodic boundary conditions. Although two-dimensional turbulence is hardly realized in nature and experiments, it may idealize some large-scale phenomena in the atmosphere[1] and also in an electrostatic guiding-center plasma.[2] A specific property of two-dimensional turbulence is the absence of vortex stretching (a fundamental feature of three-dimensional turbulence). This is due to the conservation of vorticity ω = curl u in the inviscid limit. Vorticity gradient stretching is the basic mechanism for small-scale generation in two-dimensional flows.

A controversial question in high Reynolds number two-dimensional turbulence is the behavior of the energy spectrum. Saffman[3] suggests that advection of vorticity will bring close together different values of ω, producing thin regions across which vorticity jumps. Such quasidiscontinuities of vorticity lead to an inertial range with a k^{-4} energy spectrum. In contrast, the theory of the entrophy cascade[4-6] predicts a k^{-3} energy spectrum with a possible logarithmic correction due to nonlocal interactions.[7] Furthermore, Kraichnan[8] predicts that because of this nonlocality, intermittency will not affect the energy spectrum. This point has been questioned by Basdevant et al.,[9] who claim that intermittency will restore the predominance of nonlocal interactions and steepen the energy spectrum.

It has long been recognized that the small-scale behavior of two-dimensional turbulence could be investigated by direct numerical simulations.[10-13] Later, Herring et al.,[14] using resolutions up to $(128)^2$ concluded that at least $(512)^2$ modes were required to simulate property and inertial range. Preliminary calculations at this resolution were presented by Orszag.[15] He observed that when the large-scale Reynolds number is increased from 1100 to 25,000, a distinct change is observed from a k^{-4} energy spectrum to a spectrum roughly proportional to k^{-3}.

The present paper is mainly concerned with simulations of spatially periodic solutions at $(1024)^2$ resolution. To achieve such a resolution on a GRAY 1 computer, the so called sparse mode technique has been implemented. It consists in dealing with random initial conditions with Taylor-Green-like symmetries.[16]

II. Initial Data and Numerical Method

For all the runs reported here, the stream function has a Fourier representation

$$\psi(x,y) = \sum_{\ell,m=0}^{N/2} a_{\ell m} \sin\ell x \, \sin my \qquad (2)$$

where $a_{\ell m}$ vanishes unless ℓ and m are both even or odd integers jointly. As in the Taylor-Green vortex, this representation implies flow symmetries, including reflectional invariance on the sides of an impermeable box x = 0 or π, y = 0 or π; vorticity and normal velocity vanish on these sides. The nonlinear terms of the Navier-Stokes equations are evaluated in the form $v\omega$ to insure energy conservation. Entropy is also preserved because aliasing is suppressed by spectral truncation at a maximum wave number k_M = N/3. Time marching is done by leapfrog from the nonlinear terms and Crank-Nicolson for the viscous term.

Runs with deterministic initial conditions are reported in Brachet.[17] We report here on runs with random initial data: Each Fourier mode is a zero-mean, isotropic, Gaussian random variable with dispersion such that the energy spectrum has the form

$$E_0(k) = c \, k \, e^{-(k/k_0)^2} \qquad (3)$$

This choice of initial data minimizes the inverse energy transfer (absolute equilibrium at small k). In order to permit spatial averaging, conditions with a sufficiently large range of excited scales must be used. However, when the integral scale is decreased, the accessible Reynolds numbers are also reduced (see Table 1). For the run at resolution $(1024)^2$ described in Sec. III, we have used k_0 = 3.5, c = 0.02 (corresponding to an energy Σ = 0.132 and an entropy Ω = 1.632), and a viscosity 1/256,000. With the time step of 0.00125 we used, the integration up to t = 20 required five hours CPU on a CRAY 1 machine.

III. High Reynolds Number Simulation

The short time behavior of the flow is observed to be dominated by the formation of vorticity gradient sheets (see Fig. 1). A simple model of this phenomenon may be derived from the (inviscid) equation for the curl of vortici-

Table 1 Influence of integral scale $1/k_0$ [see Eq. (3)] on time of maximum enstrophy dissipation t_{max}. $R_L = \Sigma/(\nu\eta^{1/3})$, $R_\lambda = \Omega^{3/2}/\eta$, $k_d = (\eta/\nu^3)^{1/6}$, where Σ is the energy, Ω the enstrophy, and η the enstrophy dissipation. The viscosity is $\nu = 1/6000$, and the resolution is $(512)^2$

k_0	t_{max}	R_λ	R_L	k_D
5	4	22	2376	102
7	2	21	1490	124
10	1.5	16	950	150

ty (curlω)

$$\frac{\partial}{\partial t}\,\text{curl}\vec{\omega} + \vec{u}\cdot\vec{\nabla}\,\text{curl}\vec{\omega} = (\text{curl}\vec{\omega})\cdot\vec{\nabla}\vec{u} \qquad (4)$$

Note that both the symmetric (straining) and antisymmetric (rotation) parts of the velocity gradient contribute to the dynamics of curlω. This contrasts with the somewhat analogous three-dimensional vortex stretching where only straining contributes. Following Weiss,[18] if we assume that strain and vorticity are slowly varying compared with curlω, Eq. (4) can be viewed as an essentially linear problem in Lagrangian coordinates, leading to either an exponential growth of curlω if strain dominates vorticity, or an oscillatory behavior if rotation dominates. For a two-dimensional incompressible flow, the eigenvalues of ∇u are both imaginary or both real with opposite signs. In the latter case, the curl of vorticity is locally elongated in the direction of the eigenvector associated with the positive eigenvalue and contracted in the other direction. This mechanism appears to dominate the early time small-scale generation.

A diagnostic of small-scale generation is provided by the evolution of the energy spectrum

$$E(k,t) = \frac{1}{2}\sum_{k \leq |\vec{k}'| > k+1}|\hat{u}(\vec{k}',t)|^2 \qquad (5)$$

where $\hat{u}$ is the velocity in Fourier space. In order to extract quantitative information on power law exponents and high k exponential decay, we have resorted to analyzing the data in terms of an assumed functional form for E(k,t). We

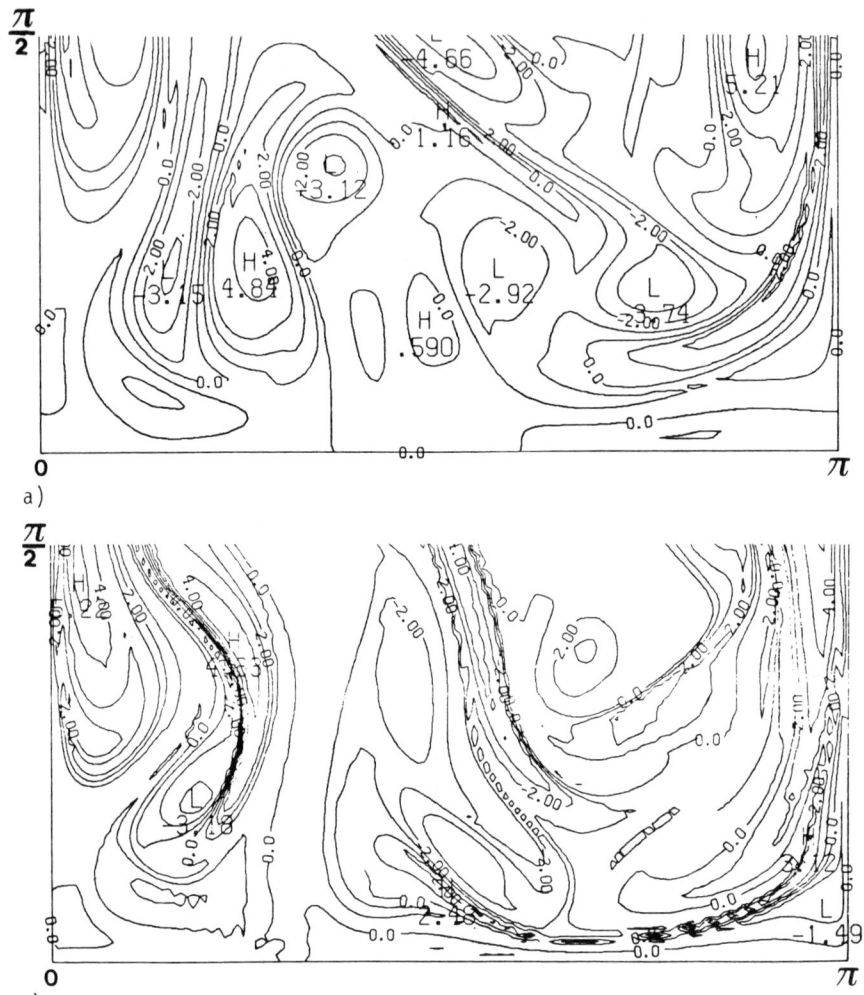

Fig. 1 Vorticity contours in the slab $0<x<\pi$, $0<y<\pi/2$; the complete flow is reconstructed by a rotation of π around the point ($\pi/2$, $\pi/2$) followed by mirror symmetries on the axes $x=0$, $x=\pi$, $y=0$, $y=\pi$. a) $t = 2$; b) $t = 4$.

fit log $E(k,t)$ by a function of the form (see Fig. 2)

$$E(k,t) = \Lambda(t) \, k^{n(t)} \, e^{-\beta(t)k} \qquad (6)$$

In this way, we estimate both the smallest excited scales by the logarithmic decrement β,[16,19,20] and the inertial exponent n.

Figure 3 shows that the early time sheet formation is associated with an exponential decay of the logarithmic

TWO-DIMENSIONAL TURBULENCE

a)
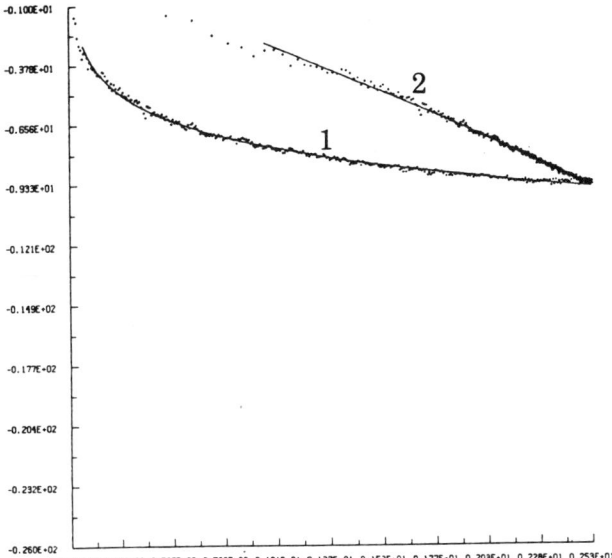

b)
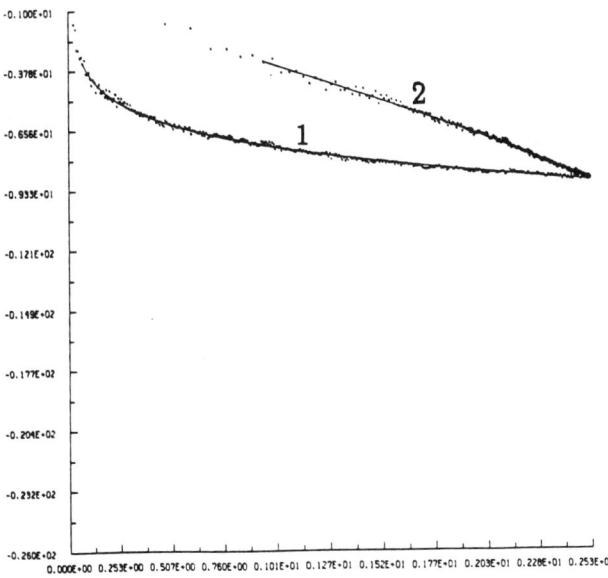

Fig. 2 Energy spectrum for the run described in Sec. III. a) t = 4; b) t = 16. The solid lines show a fit of the equation type (6) for $10 \leq k \leq 341$, giving $n = -4.05$, $\beta = 6.10^{-2}$ for $t = 4$ and $n = -3.2$, $\beta = 4.10^{-2}$ for $t = 16$. Curve 1: lin-log coordinates; curve 2: log-log coordinates.

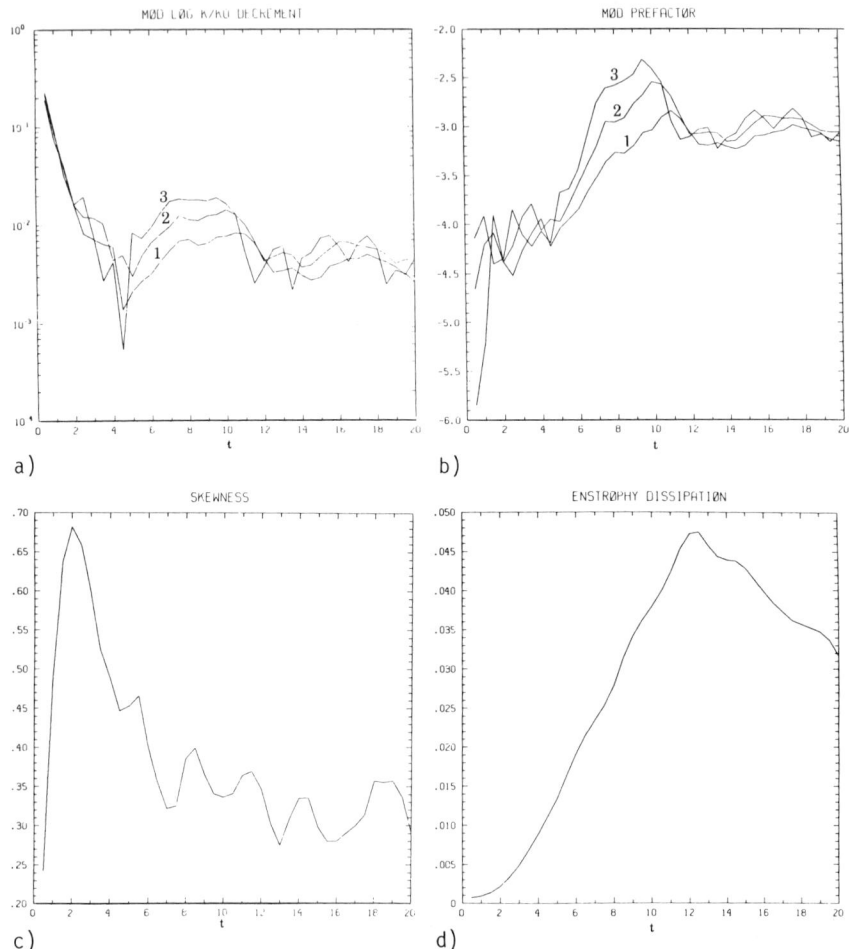

Fig. 3 Spectral behavior of the flow described in Sec. III. a) and b) Plots vs time of β and n obtained by fitting E(k) with the function $A \left[\log(k/k_0)\right]^{-1/3} k^n e^{-\beta k}$ on the interval $10 \leq k \leq 341$ for curve 1, $10 \leq k \leq 256$ for curve 2, and $10 \leq k \leq 170$ for curve 3.
c) The two-dimensional skewness § (same definition as in Ref. 14).
d) Gives the enstrophy dissipation η.

decrement. This process, consistent with the model presented above[††], stops around t ≈ 2 when scales small enough for the viscosity to act are reached. Around this time, the skewness (a nondimensional measure of the rate of production of mean-square vorticity gradients by nonlinearities) reaches its maximum. The spectral exponent of enstrophy transfer reaches its maximum. The spectral expo-

---

[††]A similar effect is discussed in appendix 4 of Ref. 1.

nent n is then close to -4, in agreement with the Saffman[3] theory (see Table 2). Indeed, flow visualizations show conspicuous vorticity gradient sheets (Fig. 1). Later, around the time t ≈ 12 of maximum entrophy dissipation, a transition to a new regime characterized by a n ≈ -3.2 ±0.1 inertial spectral exponent is observed (see Table 2). This developed regime persists until the end of our run (t = 20). The n ≈ -3.2 spectral exponent is consistent with the theory of entrophy cascade.[4-6]

Concerning the k^{-3} range, it may be noticed that, like the k^{-1} range for a passive scalar, it has probably a statistical origin. Unlike the k^{-4} range, which is of deterministic origin, a k^{-3} range may therefore go unnoticed in individual realization unless some (ensemble, temporal, or spatial) averaging is performed, as is the case when, for example, the integral scale is a fraction of the total box size.[21]

Furthermore, if we take into account the Kraichnan's log $(k/k_0)^{-1/3}$ correction,[7] by fitting $[\log(k/k_0)]^{1/3} E(k,t)$ with $\Lambda k^{-n} e^{-\delta k}$, we get n ≈ 3.0 ±0.1 (see Fig. 3b). Flow visualizations in physical space show that vorticity gradient sheets are still present but have developed a rolled-up convoluted structure (Fig. 4). This leads to the formation of isolated vortices somehow analogous to those observed by Basdevant et al.[9] and McWilliams[22] at lowest resolution with super viscosities. It is still unclear whether the formation of these "secondary structures" is produced by a purely inviscid mechanism or if viscous effects contribute in a nontrivial way. This point is currently under investigation by using a combination of runs at higher resolution and/or modifications of the functional form of dissi-

Table 2 Fit of the energy spectrum $E(k) \propto k^{-n} e^{-\beta k}$ on the interval $10 \le k \le 341$ for the run described in Sec. III

t	n	β
2	-4.49	1.59×10^{-2}
4	-4.05	0.578×10^{-2}
6	-3.96	0.297×10^{-2}
8	-3.38	0.686×10^{-2}
10	-3.15	0.739×10^{-2}
12	-3.17	0.431×10^{-2}
14	-3.32	0.283×10^{-2}
16	-3.20	0.380×10^{-2}

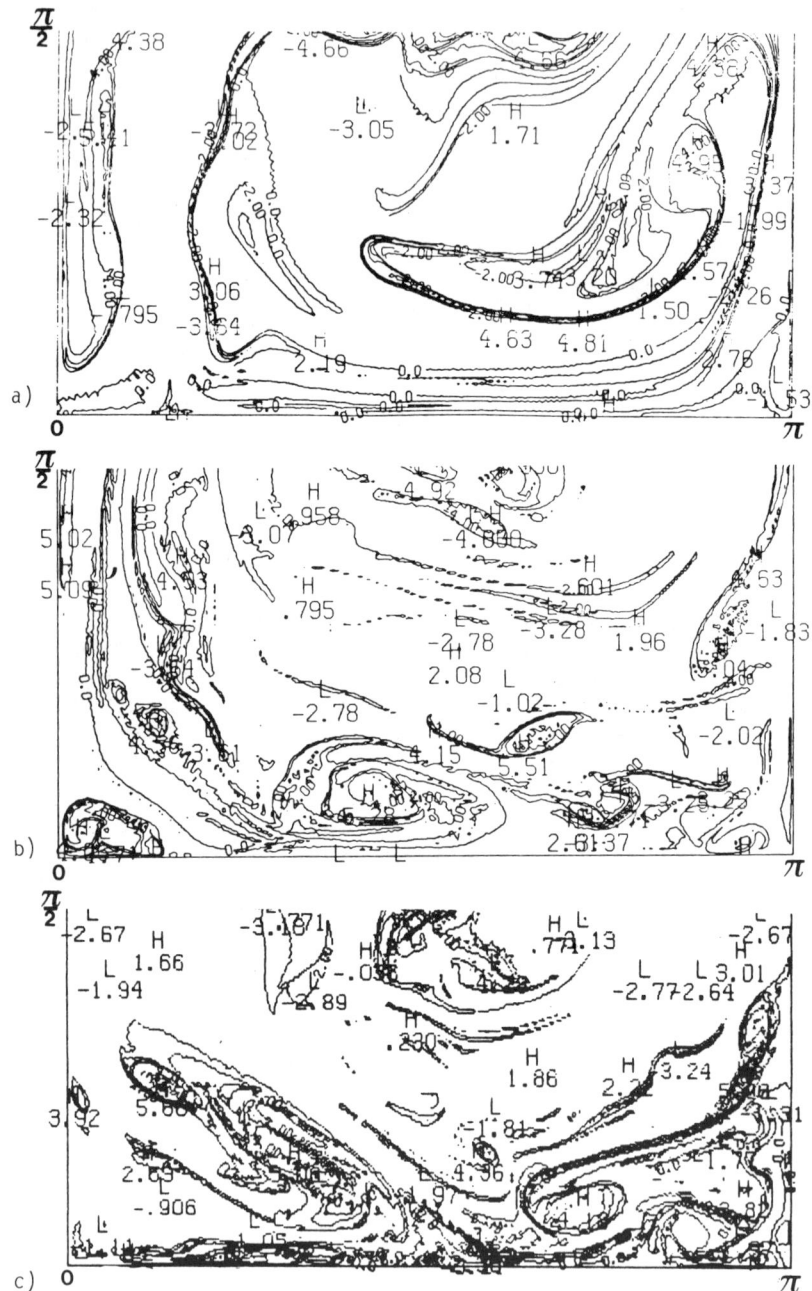

Fig. 4 Same as Fig. 1 for a) t = 8, b) t = 16, and c) t = 20. Convoluted secondary structures have emerged and the energy spectrum now displays a k^{-3} inertial behavior. The small islands are the consequence of insufficient precision in the contouring procedure.

pative term. We are also testing the effects of large-scale symmetries.

Acknowledgments

We would like to thank U. Frisch, J. Herring, R. H. Kraichnan, M. Meneguzzi, and S. A. Orszag for very useful discussions and suggestions. The computations were performed partly on the CCVR CRAY 1 and partly on the CISI CRAY 1 with the support of a DRET contract.

References

[1] McWilliams, J. C., "On the Relevance of Two-Dimensional Turbulence to Geophysical Fluid Motions," Journal de Mecanique Theorique et Appliquee, Special Issue on Two-Dimensional Turbulence, 1983, p. 83.

[2] Kraichnan, H. R. and Montgomery, D., "Two-Dimensional Turbulence," Report on Progress in Physics, Vol. 43, 1980, p. 547.

[3] Saffman, R. G., "On the Spectrum and Decay of Random Two-Dimensional Vorticity Distributions at Large Reynolds Number," Studies in Applied Mathematics, Vol. 50, 1971, p. 377.

[4] Kraichnan, H. R., "Inertial Ranges in Two-Dimensional Turbulence," Physics of Fluids, Vol. 10, 1967, p. 1417.

[5] Leith, C., "Diffusion Approximation for Turbulent Scalar Fields," Physics of Fluids, Vol. 11, 1968, p. 671.

[6] Batchelor, G. K., "Computation of the Energy Spectrum in Homogeneous Two-Dimensional Turbulence," Physics of Fluids, Vol. 12, 1969, p. 233.

[7] Kraichnan, H. R., "Inertial Range Transfer in Two and Three Dimensional Turbulence," Journal of Fluid Mechanics, Vol. 47, 1971, p. 525.

[8] Krachnan, H. R., "Statistical Dynamics of Two-Dimensional Flows," Journal of Fluid Mechanics, Vol. 67, 1975, p. 155.

[9] Basdevant, C., Legras, B., Sadourny, R., and Béland, M., "A Study of Barotropic Model Flows: Intermittency, Wave and Predictability," Journal of the Atmospheric Sciences, Vol. 38, 1981, p. 2305.

[10] Lilly, D. K., "Numerical Simulation of Two-Dimensional Turbulence," Physics of Fluids Supplement, Vol. 12, II, 1969, p. 240.

[11] Lilly, D. K., "Numerical Simulation of Developing and Decaying Two-Dimensional Turbulence," Journal of Fluid Mechanics, Vol. 45, 1971, p. 395.

[12] Lilly, D. K., "Numerical Simulation Studies of Two-Dimensional Turbulence: II Stability and Predictability Studies," Geophysics of Fluid Dynamics, Vol. 3, 1972, p. 289, Vol. 4, p. 1.

[13] Deem, G. S. and Zabusky, N. J., "Vortex Wave: Stationary V States, Interactions, Recurrence and Breaking," Physical Review Letters, Vol. 27, 1971, p. 396.

[14] Herring, J. R., Orszag, S. A., Kraichnan, H. R., and Fox, D. G., "Decay of Two-Dimensional Homogeneous Turbulence," Journal of Fluid Mechanics, Vol. 66, 1974, p. 417.

[15] Orszag, S. A., "Turbulence and Transition: A Progress Report," Proceedings of the Fifth International Conference on Numerical Methods in Fluid Dynamics: Lecture Notes in Physics, Vol. 59, edited by A. I. van de Vooren and P. J. Zandbergen, Springer-Verlag, New York, p. 32.

[16] Brachet, M. E., Meiron, D. I., Orszag, S. A., Nickel, B. G., Morf, R. H., Frisch, U., "Small Scale Structure of the Taylor-Green Vortex," Journal of Fluid Mechanics, Vol. 130, 1983, p. 411.

[17] Brachet, M. E., "Simulation Numerique Directe d'écoulements Turbulents Tridimensionnels," These D'Etat, Université de Nice, France, 1983.

[18] Weiss, J., The Dynamics of Entrophy Transfer in Two-Dimensional Hydrodynamics, La Jolla Institute, La Jolla, California, Report LJI-TN-81-121, 1981.

[19] Sulem, C., Sulem, P. L., and Frisch, H., "Tracing Complex Singularities with Special Methods," Journal of Computational Physics, Vol. 50, 1983, p. 138.

[20] Frisch, U., Pouquet, A., Sulem, P. L., and Meneguzzi, M., "The Dynamics of Two-Dimensional Ideal MHD," Journal de Mecanique Theorique et Appliquee, Special Issue on Two-Dimensional Turbulence, 1983, p. 191.

[21] Frisch, U., private communication, 1983.

[22] McWilliams, J. C., "The Emergence of Isolated, Coherent Vortices in Turbulent Flow," Journal of Fluid Mechanics, Vol. 146, 1984, pp. 21-43.

Experiments in Duct Flows with Reversed Turbulent Energy Cascades

S. Sukoriansky,* I. Zilberman,† and H. Branover‡
Ben-Gurion University of the Negev, Beer-Sheva, Israel

Abstract

Further experimental evidence has been obtained on the existence of reversed turbulent energy cascades in duct flows subjected to magnetic fields. Turbulence intensity and energy spectra were investigated experimentally in mercury flows in a 2 x 4.8-cm² rectangular cross-section duct placed into a transverse magnetic field. Velocity fluctuations were measured by means of a quartz-coated hot film probe at $Re = 78.5 \times 10^3$, $Re = 63 \times 10^3$, and $0 < Ha < 785$. Changes of structure and energy transfer mechanisms are discussed.

Nomenclature

B = magnetic field induction
D = hydraulic diameter of the channel's cross-section
Ha = $BD(\sigma/\rho\nu)^{1/2}$ = Hartmann number
f = pulsation frequency
k = wave number
N = $\sigma B^2 D/\rho U$ = Stuart number (interaction parameter)
Re = UD/ν = Reynolds number
U = mean velocity
u' = velocity fluctuation
ε = energy transfer rate
ν = kinematic viscosity
ρ = density
σ = electrical conductivity

Paper presented at the Fourth Beer-Sheva Seminar on MHD Flows and Turbulence, Ben-Gurion University of the Negev, Beer-Sheva, Israel, Feb. 27-March 2, 1984. Copyright © 1985 by the American Institute of Aeronautics and Astronautics, Inc. All rights reserved.
*Lecturer, Dept. of Mechanical Engineering.
†Head of Electronics and Electrics on Etgar-3 Project.
‡Professor, Dept. of Mechanical Engineering.

Introduction

Some striking peculiarities related to the structure of turbulence and transfer phenomena in duct flows with strong magnetic fields were observed experimentally earlier.[1-4] It was found in particular that while the mean flow in a duct becomes laminar -- judging from the pressure drop -- when Ha/Re is sufficiently high, turbulence intensity level remains essentially nonzero. Later, it was found that turbulence intensity may, in some cases, even increase.[5,6] It was also established that velocity fluctuations become highly correlated in distant points located on a line parallel to the magnetic field.[7,8]

The first theoretical study in which appearance of strong anisotropy of turbulence in magnetic fields was established was published by Moffatt[9] in 1967. This tendency has been used to account for the fact that strong residual velocity fluctuations have been observed in flows "laminarized" by the magnetic field in the sense that pressure drop along the duct corresponds to laminar flow theory. Indeed, this situation resembles that of two-dimensional turbulence[10,12], where there is no energy transfer toward higher wave number fluctuations. In two-dimensional turbulence, energy spectra in the inertial range can have the conventional $-5/3$ slope (however, corresponding to an inverse energy cascade toward low wave numbers) in addition to a -3 slope (corresponding to enstrophy cascade). Because of this, it was important to study experimentally not only the turbulent intensity but also the turbulent energy spectra under the influence of magnetic fields. In view of the above, turbulence spectra have been measured in a number of cases of shear and grid generated turbulence[3,6,13,14]. The appearance of -3 slope regions in the spectra have been clearly demonstrated. More detailed local measurements in flows past grids with aligned magnetic fields[13] showed that in the -3 slope range fluctuations persist in all three directions and are all of the same order of magnitude. The integral length scale in the magnetic field direction increases slowly with the increase in field strength. Analysis of transfer and dissipation mechanisms by Alemany et al.[13], led to a conclusion about the importance of angular energy transfer from the inside to the outside of a cone (in wave number space) with axis parallel to the magnetic field. Supposing an equilibrium between this angular energy transfer and Joule dissipation, it can be shown that the energy spectrum must vary as k^{-3} in the self-similar range. Whether the velocity field becomes two-dimensional or not depends, however, on the presence of solid boundaries transverse to the magnetic field.[15] A number of reviews and summarizing papers have been written on the subject under consideration.[15,16] Nevertheless, a clear and complete picture of the physical phenomena (especially in shear turbulence) have not as yet been presented.

Regarding flows in ducts, the role of different turbulizing factors, including the wall created shear and phenomena at the entrances to the channel and the magnetic field, remain especially unclear.

There are three basic situations of a duct flow in a magnetic field: 1) flow in an aligned field, 2) flow in a transverse field, and 3) flow in an azimuthal field (Fig. 1). All other practical situations can be regarded as combinations of these three cases. Obviously, the only case when the flow can ever become ideally two-dimensional is case 3. In case 1 components of velocity in the field direction should always persist and vary with variation in the distance from the wall. In case 2 the velocity component in field direction can vanish but the two other components are dependent on the distance from the wall. In cases 2 and 3, the entrance of the flow in the magnetic field leads to strong perturbation of the mean flow, namely, formation of M-shaped mean velocity profiles[17,18] (Fig. 2). This perturbation can eventually strongly influence the turbulence structure in the magnetic field region and could even be regarded as a possible reason for the mere existence of residual velocity fluctuations in a magnetically "laminarized" flow. Indeed, one of the experimental studies, where residual fluc-

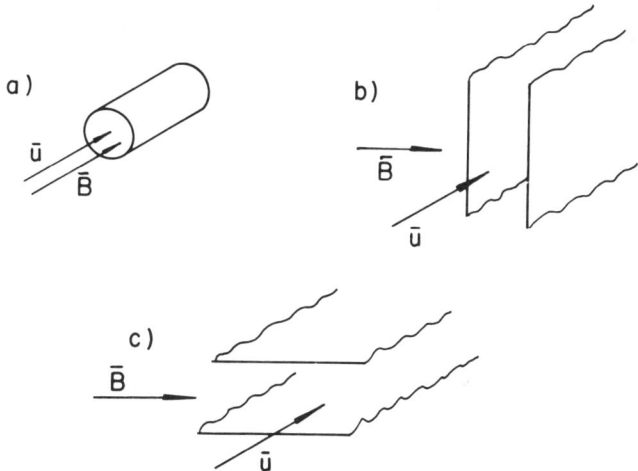

Fig. 1 Three basic situations of a duct flow in a magnetic field: a) aligned field, b) transverse field, c) azimuthal field.

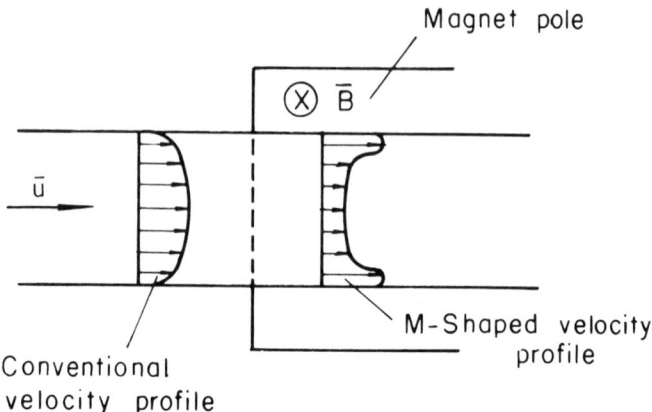

Fig. 2 Formation of M-shaped velocity profile at flow entrance.

tuations were found to approach zero,[6] was performed with the flow in case 3 while special measures for elimination of entrance perturbation were arranged. Unfortunately, this study was performed only at rather low Reynolds numbers (up to Re = 2.5×10^4).

One of the main purposes of the present work was to study how a gradually increasing magnetic field changes different regions of the energy spectra and how it correlates with the changes in turbulence intensity. This enables clarification of the influence of the magnetic field on different processes in the transfer of turbulent energy. A number of experimental situations (all corresponding to the case designated above as 2) were arranged. In some of these situations there was a honeycomb in the inlet section of the channel, while the magnetic field could either cover the honeycomb region or it could begin just downflow of the honeycomb. In other situations the honeycomb was removed while the magnetic field could still begin at different distances from the channel inlet.

The purpose of the honeycomb was not only to eliminate the level of turbulence entering the magnetic field but also to prevent the formation of an M-shape velocity profile.

Facility and Instrumentation

The schematic diagram of the experimental facility is presented in Fig. 3, while the photograph in Fig. 4 gives the general view of this facility. The facility has a pump, an overflow constant level tank with a number of dense meshes for quieting the entering flow disturbances, and a

constant level sump tank. The tanks are provided with water cooling which keeps the mercury temperature constant within a 0.5°C accuracy.
The experimental channel, made from perspex, has a rectangular 2x4.8-cm^2 on cross section. Magnetic field is directed transverse to the longer side of the channel cross section. In one part of the experiments a honeycomb made of an assembly of electrically insulated tubes with 0.24 cm external diameter and 0.025-cm-thick walls. The length of the honeycomb in the flow direction was 26 cm. Solidity of the honeycomb was ~0.25. It was assumed that in cases when the entrance edge of the magnet was in the honeycomb region, no M-shaped velocity profiles have been formed. Electromagnet pole length is 90 cm, and magnetic field can vary from 0 to 1.1 T. The magnet can easily be moved along the channel. The mercury flow rate is measured by a venturi flow meter with liquid manometers. Experiments have been performed at two mean velocity values U_1=0.25 m/s and U_2=0.32 m/s corresponding to Re_1=63x10^3 and Re_2=78.5x10^3, respectively.
Local velocity measurements have been performed by means of TSI quartz-coated hot film probes (1212-20 HG model with a 0.05-mm sensor diameter and a 1-mm sensor length). The signal was processed through a Disa 55M system with a 55M10 constant temperature standard anemometric bridge. The usual probe overheating ratio was 5%. Spectral functions were obtained by means of FFT Spectrum Analyzer (Spectral Dynamics Model SD 340). The probe was moved by means of a specially built traversing mechanism and could be placed in any point of the longer axis of the channel cross section up to 4 mm distance from the wall.

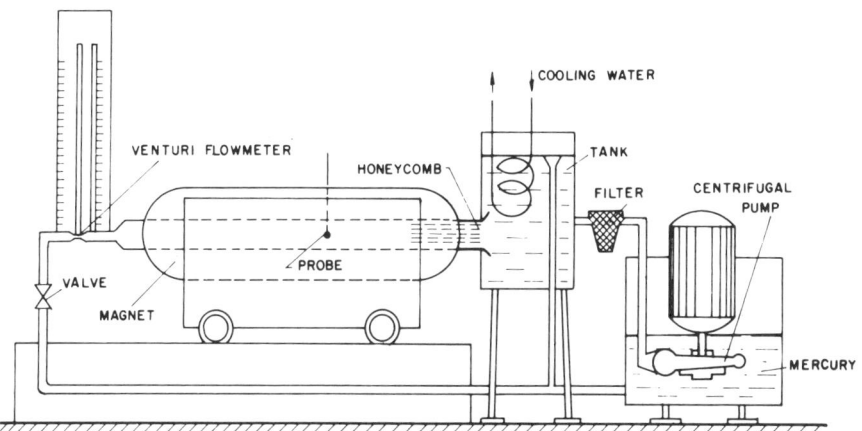

Fig. 3   Schematic diagram of the experimental facility.

Fig. 4   General view of the experimental facility.

Results and Conclusions

Measurements of Turbulence Intensity

There is a striking difference in the influence of the magnetic field on intensity of turbulence in cases when there is no honeycomb (regardless of the position of the magnet) and when there is a honeycomb but the magnetic field begins at a distance of 5.5 D (where D is the hydraulic diameter of the channel's cross section), on the one hand, and when the honeycomb is covered by the magnetic field, on the other hand (Figs. 5-7). Changes of turbulence intensity when Ha/Re is increasing are identical in all cases belonging to the first group mentioned above. Namely, u'/U reaches approximately 30% to 40% of the initial level without magnetic field at Ha/Re ≈ $(2-3) \times 10^{-3}$ and then remains practically constant up to the highest experimental value of Ha/Re (about 9×10^{-3}, which is approximately twice the critical value corresponding to flow laminarization according to friction measurements). The above relates equally to turbulence intensity measured on the axis of the channel and near the wall parallel to the magnetic field. However, in the cases when the field is covering the honeycomb, intensity reaches a minimal value, exactly as in the cases above, but after that increases strongly with further increase of the Ha/Re value, reaching a level two or more times higher than

REVERSED TURBULENT ENERGY CASCADES   117

the initial level without magnetic field. A careful analysis of possible explanations of the phenomena described above leads to the following conclusions.

As established in numerous previous experimental studies, friction in an initially turbulent flow, in a channel similar to ours, corresponds to laminar theory when the value of Ha/Re reaches approximately 4×10^{-3}.[16] The velocity profile in the plane perpendicular to the magnetic field becomes clearly M-shaped when the interaction parameter reaches the value $N \approx 1$ (in all of our cases this is close to the region of "laminarization" according to friction). The residual turbulence intensity seen in Figs. 5 and 6, cases II and III, and in case II of Fig. 7 could originate in the instability of M-shaped mean velocity profiles and/or also in other turbulizing factors. Persistence of this residual turbulence, together with the experimental fact that friction is laminar, can be viewed as evidence for its close to two-dimensional structure and elimination of the conventional energy cascade.

In cases when the magnetic field covers the honeycomb (Fig. 6, case I; Fig. 7, cases I and III) instead of a non-

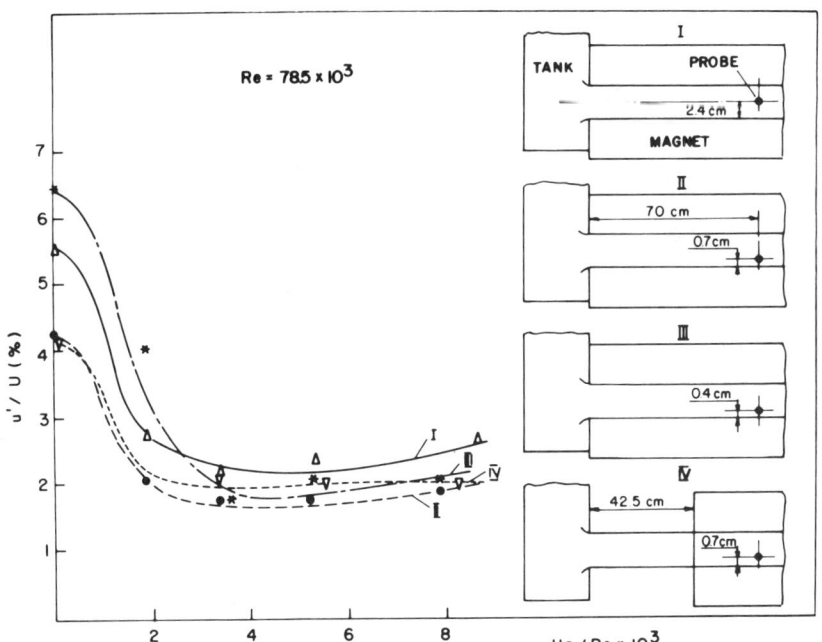

Fig. 5   Turbulence intensities as a function of Ha/Re at different distances from wall and different positions of the magnet (without honeycomb).

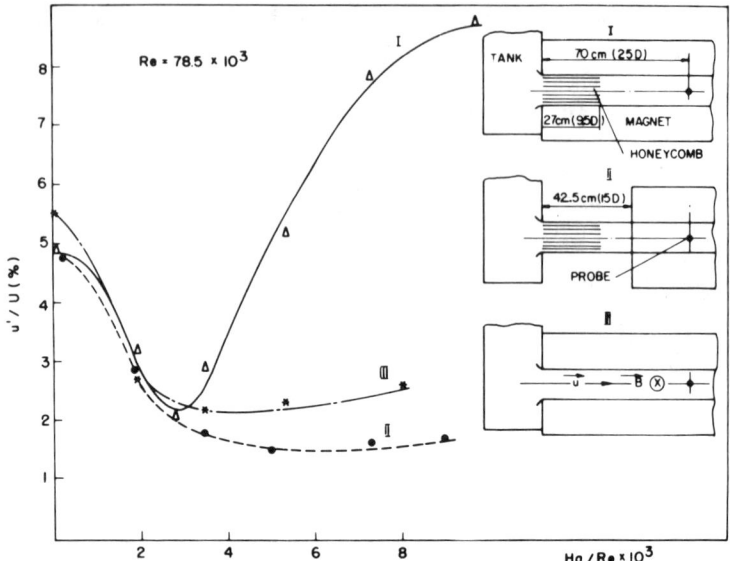

Fig. 6 Turbulence intensities on flow axis as a function of Ha/Re at different positions of the magnet with and without honeycomb.

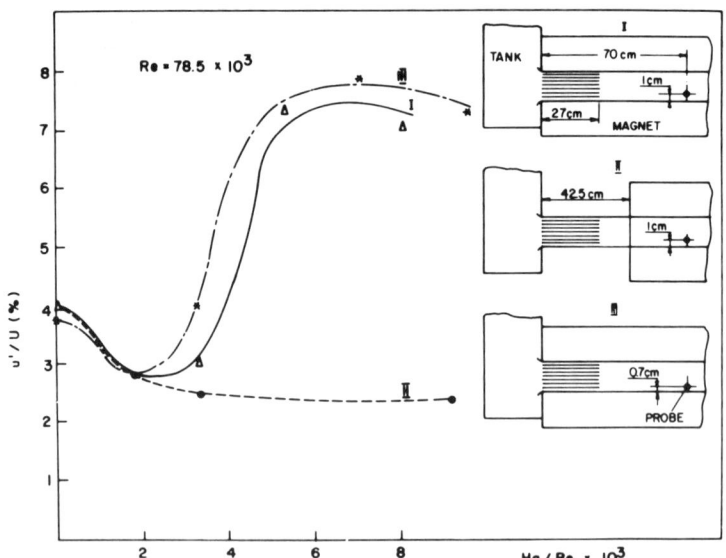

Fig. 7 Turbulence intensities as a function of Ha/Re at different distances from the wall and different positions of the magnet (with honeycomb).

disturbed flow at high Ha/Re values, as one could expect, u'/U becomes, as already mentioned, twice as high or more than without magnetic field. The explanation of this is that the tail parts of the honeycomb's fine pipes are acting as a strong source of forced turbulence. When the magnetic field is weak (i.e., that for k value under consideration the local magnetic interaction parameter $N_k < 1$), the energy of this turbulence is conventionally transferred to higher k and dissipated. If, however, $N_k > 1$ disturbances become two-dimensional, the conventional energy cascade does not exist any more and an inverse cascade takes its place, supplying energy to the low wave number region. All the above can be additionally verified through the analysis of spectra.

Experimental Spectra

Experimental spectra are presented in Figs. 8-10. The spectrum analyzer performed the calculations for frequencies

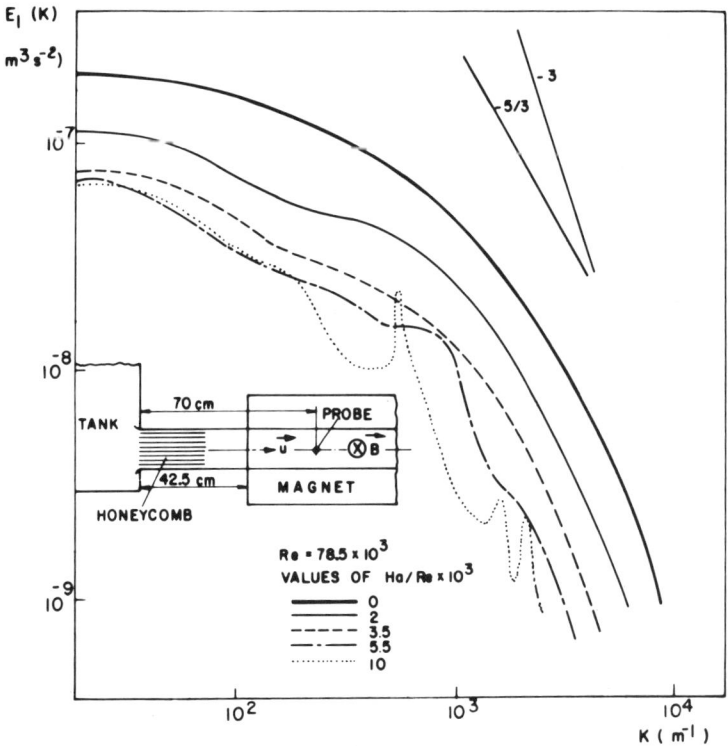

Fig. 8 Turbulence spectra measured on flow axis at different Ha/Re values with the magnet positioned downflow of the honeycomb.

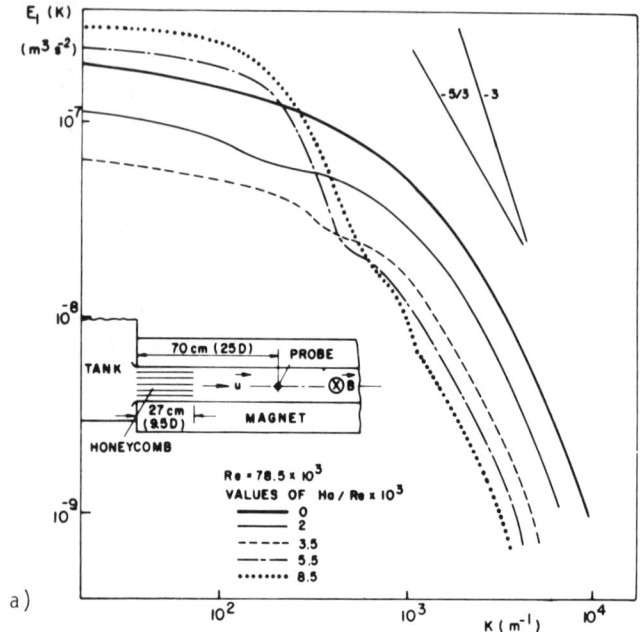
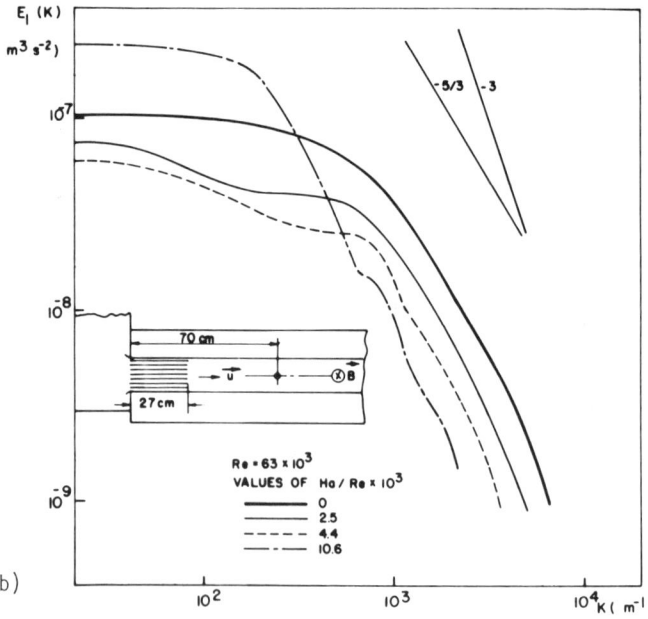

Fig. 9 Turbulence spectra measured on flow axis at different Ha/Re values with the magnet covering the honeycomb: a) Re = 78.5x 10^3, b) Re = 63x10^3.

up to 500 Hz with 1.25 Hz resolution (time window 0.8 s), using 64 ensembles for averaging. Using the Taylor hypothesis, corresponding wave numbers $k=2\pi f/U$ have been established. When the honeycomb is not covered by the magnetic field (Fig. 8), energy density in the low wave number region gradually decreases with the increase in the value of Ha/Re. There is a slight tendency in the low wave number region toward steeper slope, and, simultaneously, the energy density at higher wave numbers decreases while the $-5/3$ slope is preserved (conventional energy cascade with lower ε). At $Ha/Re \approx 5\times 10^{-3}$ and higher low wave numbers, the energy density remains almost constant, but at $k>5\times 10^2$ (former inertial range) the slope becomes clearly k^{-3}.

Finally, at $k \approx 5\times 10^2$ there is a strong energy peak which is probably caused by the turbulizing effect of the M-shaped mean velocity profile (the peak occurs when $N \approx 1$). At higher k the slope is -3, and therefore energy transfer to cascade higher k cannot exist. On the other hand, energy can be transferred to lower k (in Fig. 8 there is a $-5/3$ slope region left of the peak on the curve for $Ha/Re = 10^{-2}$) and that probably explains why energy density there remains constant despite the increase of Ha/Re.

When the field covers the honeycomb (Figs. 9, 10), both on flow axis and near the wall, energy density at low k initially goes down when Ha/Re increases as in the previous case. However, at higher Ha/Re values, energy density increases in the low k region. This coincides with intensity measurements for the same case presented above and indicates the reversal of energy cascade. This statement deserves some more detailed discussion.

Characteristic value of wave number k for honeycomb's tail generated pulsations may be estimated from the expression $k \sim 2\pi/d$, where d is the diameter of the honeycomb pipes. In our case, d = 2.4 mm, hence k is in the order of $3\times 10^3 m^{-1}$. For comparison, it could be mentioned here that shear generated turbulence (and also pulsations induced by the entrance of the flow in the magnetic field and formation of M-shaped velocity profile) has characteristic wave numbers of the order of $2\pi/D \sim 5\times 10^2 m^{-1}$. Thus, the spectra presented in Figs. 9 and 10 show that at $Ha/Re \gtrsim 5\times 10^{-3}$ the influence of the magnetic field on velocity fluctuations (which is proportional to the magnetic interaction parameter based on characteristic length of pulsation; i.e., $\sigma B^2/k\rho u'$ reaches the region of pulsations generated by the tails of the honeycomb's pipes and the energy of this pulsation cascaded to the smallest scales. This coincides with all the available experimental evidence, namely, 1) decrease of wall friction to the laminar level, 2) increase of turbulence intensity,

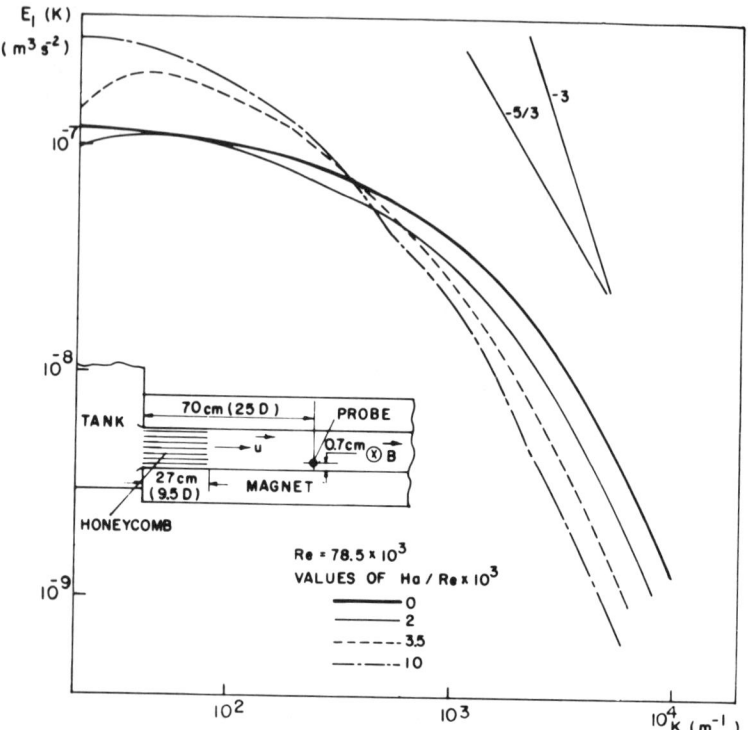

Fig. 10 Turbulence spectra measured near the wall at different Ha/Re values with the magnet covering the honeycomb.

3) increase of energy density in the low k region, and 4) preservation of $-5/3$ slope of the spectral curve in the inertial region (5×10^2 m^{-1} ⩽k⩽ 5×10^3 m^{-1}) even at highest Ha/Re values, while the energy input to turbulence occurs at $k \sim 5 \times 10^3$ m^{-1}. Interestingly enough, in the case when energy input occurs only at lower k, as in Fig. 8 (energy input at $k < 5 \times 10^2$ m^{-1}), the slope in the inertial region is -3 (enstrophy cascade), since there is no turbulence generation in the $k > 10^3$ region. Clean -3 and $-5/3$ slopes at $k ⩽ 5 \times 10^2$ m^{-1} in Figs. 9 and 10 are not achieved probably because of superposition of the shear generated turbulence effects and the effects of honeycomb generated turbulence. In the close to two-dimensional structure situation, a $-5/3$ slope should occur to the left of turbulent energy input frequency and a -3 slope to the right of it. Since in our case the two regions of k where generation occurs differs by an order of magnitude, the slope in the intermediate k region becomes a superposition of $-5/3$ and -3 slopes.

In conclusion, it should be said that while the explanations given above seem to fit with the experimental results most satisfactorily, further experimental verification is most desirable. Different turbulizing factors should be investigated, preferably at higher Reynolds numbers. Changes of spectra along the flow in the magnetic field should also be studied. Other flow situations -- namely, flow in azimuthal and aligned magnetic fields -- should be investigated. The former case is of special interest, since, as mentioned in the introduction, it is the only situation when a duct flow in a magnetic field could eventually become completely two-dimensional. Moreover, in one of the previous works,[6] a completely nondisturbed flow was observed in strong magnetic fields when all entrance disturbances were eliminated. However, these results were obtained only at relatively low Reynolds numbers. Finally, going back to the present case, one should conclude that for this geometry (Fig. 1, case b), there is no way to avoid the persistence of high residual velocity fluctuations (at least 30% to 40% of the initial intensity but usually much higher), since an introduction of a honeycomb either does not influence this effect or even strongly amplifies the residual fluctuations. The latter fact is of great practical importance, especially because the "residual" disturbances obviously can have a great impact on heat and mass transfer.

Moreover, further studies of the phenomena described in this article are desirable not only for the sake of better understanding of the specific case when electroconductive liquid is flowing in the presence of a magnetic field, but also -- and even more so -- for understanding of much more general cases of flows with three- and two-dimensional turbulence, and transfer mechanisms in these respective cases.

References

[1] Branover, H.H., Slyusarev, N.M., and Scherbinin, E.V., "Some Results of Measuring Turbulent Pulsation of Velocity in a Mercury Flow in the Presence of a Transverse Magnetic Field," Magnitnaya Gidrodinamika, Vol. 1, 1965, p. 33.

[2] Branover, H.H., Gelfgat, Yu. M., et al., "Effect of a Transverse Magnetic Field on the Intensity Profiles of Turbulent Velocity Fluctuations in a Channel of Rectangular Cross Section," Magnitnaya Gidrodinamika, Vol. 3, 1970, p. 41.

[3] Gardner, R.A. and Lykoudis, P.S., "Magneto-Fluid-Mechanic Pipe Flow in a Transverse Magnetic Field. Part I, Isothermal Flow," Journal of Fluid Mechanics, Vol. 47, 1971, p. 737.

[4] Hua, H.M. and Lykoudis, P.S., "Turbulence Measurements in a Magneto-Fluid-Mechanic Channel," Nuclear Science and Engineering, Vol. 54, 1974, p. 445.

[5] Kolesnikov, Yu. B. and Tsinober, A.B., "Two-Dimensional Turbulent Flow behind a Circular Cylinder," Magnitnaya Gidrodinamika, Vol. 3, 1972, p. 23.

[6] Branover, H.H. and Gershon, P., "Experimental Investigation of the Origin of Residual Disturbances in Turbulent MHD Flows after Laminarization," Journal of Fluid Mechanics, Vol. 94, 1979, p. 629.

[7] Branover, H.H. and Platnieks, I.A., "Measurement of Spatial Correlations of Velocity Pulsations in an MHD Channel," 1971, unpublished.

[8] Votsish, A.D. and Kolesnikov, Yu. B., "Transition from Three-Dimensional to Two-Dimensional Turbulence in a Magnetic Field," Magnitnaya Gidrodinamika, Vol. 3, 1975, p. 141.

[9] Moffatt, H.K., "On the Suppression of Turbulence by a Uniform Magnetic Field," Journal of Fluid Mechanics, Vol. 28, 1967, p. 571.

[10] Kraichnan, R.H., "Inertial Ranges in Two-Dimensional Turbulence," Physics of Fluids, Vol. 10, 1967, p. 1417.

[11] Leith, C.E., "Diffusion Approximation for Two-Dimensional Turbulence," Physics of Fluids, Vol. 11, p. 671.

[12] Batchelor, G.K., "Computation of the Energy Spectrum in Homogeneous Two-Dimensional Turbulence," Physics of Fluids (Suppl. 2), Vol. 12, 1969, p.233.

[13] Alemany, A., Moreau, R., Sulem, P.L., and Frish, V., "Influence of an External Magnetic Field on Homogeneous MHD Turbulence," Journal de Mecanique, Vol. 18, 1979, p. 277.

[14] Sommeria, J., "Two-Dimensional Behaviour of MHD Fully-Developed Turbulence," Journal de Mecanique Theorique et Appliquee, Numero Special, 1983, p. 169.

[15] Sommeria, J. and Moreau, R., "Why, How, and When, MHD Turbulence Becomes Two-Dimensional," Journal of Fluid Mechanics, Vol. 118, 1982, p. 507.

[16] Branover, H.H., Magnetohydrodynamic Flow in Ducts, Halsted Press, 1978.

[17] Shercliff, J.A., The Theory of Electromagnetic Flow Measurement, Cambridge University Press, Cambridge, England, 1962.

[18] Bocheninskii, V.P., Branover, H.H., Tanaev, A.V. and Chernyaev, Yu. P., "Experimental Study of Resistance to Flow of Conducting Fluid on Plane Insulated Channels in the Presence of a Transverse Magnetic Field taking into account End Effects and Wall Roughness," Mekhanika Zhidkosti i Gaza, Vol. 4, 1971, p. 10.

Direct Numerical Simulation of Three-Dimensional Convection in Liquid Metals

P.L. Sulem*
Tel Aviv University, Israel, and Observatoire de Nice, France
and
C. Sulem†
Ben-Gurion University of the Negev, Israel and Université de Nice, France
and
O. Thual‡
Meteorologie Nationale, Toulouse, France

Abstract

A direct numerical simulation of three-dimensional Boussinesq convection in a horizontal layer of mercury heated from below is presented. Periodicity is assumed in the horizontal directions (k_x = 3.117, k_y = 2.5). Different regimes are observed when the Rayleigh number R is increased. Convection appears when R exceeds 1708, leading to the formation of steady two-dimensional rolls. At higher Rayleigh numbers (1887 < R < 1895), three-dimensional oscillatory instability occurs, leading to time periodic solutions where horizontal disturbances propagate along the rolls. For R > 1900, we observe a modulation of the amplitude of the oscillation and vertical oscillations of the rolls. When a magnetic field is prescribed in the direction of the rolls, the oscillatory instability occurs at a higher Rayleigh number, and the amplitude of the oscillations is reduced. For the aspect ratios we have considered (two rolls in a period), the frequency of the oscillation is not appreciably affected.

Paper presented at the Fourth Beer-Sheva Seminar on MHD Flows and Turbulence, Ben-Gurion University of the Negev, Beer-Sheva, Israel, Feb. 27-March 2, 1984. Copyright © 1985 by the American Institute of Aeronautics and Astronautics, Inc. All rights reserved.

*School of Mathematical Sciences, Tel Aviv, and Centre National de la Recherche Scientifique.
†Centre National de Recherche Météorologique.
‡Department of Mathematics and Centre de la Recherche Scientifique.

Introduction

This paper is concerned with a direct numerical simulation of three-dimensional convection in a layer of low Prandtl number fluid (Pr-0.025) confined between two perfectly conducting horizontal plates. Low Prandtl number fluids are encountered both in industrial applications (liquid metals) and in astrophysical and geophysical convection. Indeed, in the range of Rayleigh numbers between the onset of convection and the first time-dependent instability (oscillatory instability), the convective structures are steady two-dimensional rolls. At Pr > 1, in contrast, the steady convective structures generally become three dimensional before any bifurcation to a time-dependent regime.

Convection at low Prandtl number has recently been addressed both experimentally and theoretically. Experiments in parallelepipedic boxes with small aspect ratios in the presence of a horizontal magnetic field have been performed by Fauve et al,[1] Libchaber et al.[2] Various routes to chaos were observed according to the intensity of the magnetic field. In numerical and analytical studies, periodic boundary conditions are generally assumed in the horizontal directions. In this case, Clever and Busse[3] computed steady solutions corresponding to two-dimensional rolls. They also analyzed the linear stability of these solutions for three-dimensional disturbances and determined the onset of the oscillatory instability. More recently, they investigated the stabilizing effect of a magnetic field parallel to the rolls.[4] It appears that vertical walls have an important stabilizing effect in boxes with small aspect ratios. For example, in mercury (Pr=0.025), when no magnetic field is present, the oscillatory instability appears when the Rayleigh number exceeds by a factor 2 the critical Rayleigh number in the experiments, while only 10% are required with periodic boundary conditions.

Dynamical Equations

We consider a horizontal layer of electrically conducting fluid confined between two horizontal plates and heated from below in the presence of a horizontal magnetic field (assumed in the y direction). The plates, separated by a distance d, are assumed to act as rigid, perfectly conducting boundaries. The temperatures of the lower and upper plates are T_1 and $T_2 < T_1$, respectively. Periodic boundary conditions are assumed in the horizontal directions. In the x direction, we choose a period $L_x = 2\pi/3.117 \simeq 2.0$, which corresponds to the most unstable mode. In the y direction, we choose a period $L_y = 2\pi/2.5 \simeq 2.5$. Numerical simulations of convection in analogous geometries were previously done by Lipps[5] and McLaughlin and Orszag[6] for the (thermal) Prandtl number $Pr = 0.71$. Here, we consider the case of $Pr = 0.025$ (mercury).

In the Boussinesq approximation, the MHD equations read

$$\frac{\partial V}{\partial t} + V \cdot \nabla V = -\nabla \pi + \nu \nabla^2 V + \alpha g \theta \hat{e}_3 + \frac{1}{\rho_0} B \cdot \nabla B$$

$$\frac{\partial B}{\partial t} + V \cdot \nabla B = B \cdot \nabla V + \lambda \nabla^2 B$$

$$\frac{\partial \theta}{\partial t} + V \cdot \nabla \theta + \frac{T_1 - T_2}{d} w = \kappa \nabla^2 \theta$$

$$\nabla \cdot V = 0 \quad \nabla \cdot B = 0$$

(1)

where $V = (u,v,w)$ is the velocity field; $\pi = p + B^2/2$ with p the pressure, and θ the deviation of temperature from the diffusive profile. The magnetic field $B = B_0 + b$ is the sum of the external magnetic field B_0 (in the y direction) and an induced magnetic field b. The parameters ν, λ, and κ respectively denote kinematic viscosity, magnetic diffusivity, and thermal diffusivity. g is the intensity of the gravity field, and α the coefficient of thermal dilatation. $\hat{e}_3$ is a unit vector in the vertical direction, and ρ_0 the mass of the unit volume.

We shall use the following units: d for length, d^2/ν for time, ν/d for velocity, $B_0 \nu/\lambda$ for induced magnetic field, $(T_1-T_2)\nu/\kappa$ for deviation of temperature from the diffusive profile. The equations of motion (1) are then rewritten:

$$\frac{\partial V}{\partial t} + V \cdot \nabla V = -\nabla \pi + \nabla^2 V + R \theta \hat{e}_3 + Q \left(\frac{\partial b}{\partial y} + P_M b \cdot \nabla b \right)$$

$$P_M \left(\frac{\partial b}{\partial t} + V \cdot \nabla b - b \cdot \nabla V \right) = \frac{\partial V}{\partial y} + \nabla^2 b$$

$$\frac{\partial \theta}{\partial t} + V \cdot \nabla \theta = \frac{1}{Pr} (\nabla^2 \theta + w)$$

(2)

$$\nabla \cdot V = \nabla \cdot b = 0$$

where $R = g(T_1-T_2)d^3/\nu\kappa$ is the Rayleigh number; $Q = B_0^2 d^2/\rho_0\lambda\nu$ is the Chandrasekhar parameter; $Pr = \nu/\kappa$, the (thermal) Prandtl number; and $P_M = \nu/\lambda$, the magnetic Prandtl number. When the magnetic Prandtl number is very small ($P_M \sim 10^{-6}$ in mercury), Eq. (2) reduces to

$$\frac{\partial V}{\partial t} + V \cdot \nabla V = -\nabla p + \nabla^2 V + Ra\theta\hat{e}_3 + Q\frac{\partial b}{\partial y}$$

$$\nabla^2 b = -\frac{\partial V}{\partial y}$$

$$\frac{\partial \theta}{\partial t} + V \cdot \nabla\theta = \frac{1}{Pr}(\nabla^2\theta + w) \tag{3}$$

$$\nabla \cdot V = 0$$

On the horizontal plates, velocity and temperature deviation vanish, together with the normal component of the induced magnetic field and the tangential component of the electric current $j = \text{curl } b$ (Ref. 7). One easily checks that the condition $\nabla \cdot b = 0$ is preserved in the above asymptotics.

Computational Technique

Time marching is made using a second-order Adams-Bashforth-Crank-Nicolson scheme of the following form. Stage i:

$$\frac{V^* - V^n}{\delta t} + \tfrac{1}{2}(3V^n \cdot \nabla V^n - V^{n-1} \cdot \nabla V^{n-1})$$

$$= \frac{R}{2}(3\theta^n - \theta^{n-1}) + \frac{Q}{2}\frac{\partial b^n}{\partial y} + \tfrac{1}{2}\nabla^2 V^n$$

stage ii:

$$\frac{\theta^* - \theta^n}{\delta t} + \tfrac{1}{2}(3V^n \cdot \nabla\theta^n - V^{n-1} \cdot \nabla\theta^{n-1}) = \frac{1}{2Pr}(\nabla^2\theta^n + w^n)$$

$$\frac{V^{n+1} - V^*}{\delta t} = -\nabla p + \tfrac{1}{2}\nabla^2 V^{n+1} + \frac{Q}{2}\frac{\partial b^{n+1}}{\partial y} \tag{4}$$

stage iii:

$$\nabla \cdot V^{n+1} = 0 \qquad \nabla^2 b^{n+1} = -\frac{\partial V^{n+1}}{\partial y}$$

and stage iv:

$$\frac{\theta^{n+1}-\theta^*}{\delta t} = \frac{1}{2Pr}(\Delta^2\theta^{n+1}+w^{n+1})$$

The Stages i and ii are explicit. The boundary conditions for the implicit stages iii and iv read

$$V^{n+1} = 0 \qquad b_3^{n+1} = 0 \qquad \theta^{n+1} = 0$$

$$\frac{\partial b_1^{n+1}}{\partial z} = \frac{\partial b_2^{n+1}}{\partial z} = 0$$

(5)

for $z=\pm\frac{1}{2}$. In Eq. (5), b_1, b_2, b_3 denote the components of the induced magnetic field b.

We rewrite stage iii in the form:

$$\nabla^2 p = \text{div} \frac{V^*}{\delta t}$$

$$\frac{V^{n+1}-V^*}{\delta t} = -\nabla p + \frac{1}{2}\nabla^2 V^{n+1} + \frac{Q}{2}\frac{\partial b^{n+1}}{\partial y}$$

(6)

$$\nabla^2 b = -\frac{\partial V^{n+1}}{\partial y}$$

This system is solved as in Kleiser and Schumann[8] by prescribing $\nabla \cdot V^{n+1} = 0$ for $z = \pm\frac{1}{2}$ as an additional boundary condition required to compute the pressure. This condition, which replaces the inviscid boundary condition $\partial p/\partial z=0$ used in Orszag and Kells[9] and McLaughlin and Orszag[6] insures an exact conservation of the flow incompressibility by the time discretization.

Velocity, magnetic field, and temperature are expanded in Fourier series in the horizontal directions and in Chebyshev polynomial series in the vertical direction. We write

$$\theta(x,y,z,t) = \sum_{|\ell|<\frac{L}{2}} \sum_{|m|<\frac{M}{2}} \sum_{n=0}^{N} \theta_{\ell mn}(t) \, e^{2\pi i(\frac{\ell x}{L_x}+\frac{my}{L_y})} T_n(2z)$$

(7)

and similar expressions for the other fields. In the runs reported here, we have used L=16, M=10, N=33.

Time marching is made in spectral space. The nonlinear terms are computed using a collocation method: multiplications are performed in physical space, while derivatives are computed in the spectral space. Transformations from physical to spectral space and the reverse are done by using fast fourier transform algorithms. Actually, the advection term in the velocity equation is not computed in the form $V \cdot \nabla V$. It is replaced by

$$\frac{1}{3}\frac{\partial}{\partial x}(u^2-v^2) + \frac{1}{3}\frac{\partial}{\partial x}(v^2-w^2) + \frac{\partial}{\partial y}(uv) + \frac{\partial}{\partial z}(uw)$$

in the equation for $\partial u/\partial t$ and by analogous expressions obtained by cyclic permutation in the equations for $\partial v/\partial t$ and $\partial w/\partial t$. The pressure p must then be replaced by $p+(V^3/3)$. With this procedure,[10] eight transformations between physical and spectral spaces are needed, instead of the nine transformations required when the more usual representations $V \cdot \nabla V$ or curl $V \times V$ are used. Indeed, instead of computing u^2, v^2, w^2, one computes only u^2-v^2 and v^2-w^2; their difference gives u^2-w^2.

To perform the implicit step iii, we first consider the subsystem

$$\nabla^2 p = \text{div}\,\frac{V^*}{\delta t}$$

$$\frac{w^{n+1}-w^*}{\delta t} = -\frac{\partial p}{\partial z} + \frac{1}{2}\nabla^2 w^{n+1} + Q\,\frac{\partial b_3^{n+1}}{\partial y} \qquad (8)$$

$$\nabla^2 b_3^{n+1} = \frac{\partial w^{n+1}}{\partial y}$$

with the boundary conditions

$$w^{n+1} = 0 \qquad \frac{\partial w^{n+1}}{\partial z} = 0 \qquad b_3^{n+1} = 0 \quad \text{for} \quad z = \pm\tfrac{1}{2} \qquad (9)$$

By Fourier transformation in the horizontal variables, denoted by the superscript caret (^), Eq. (8) is rewritten

$$-\frac{\partial^2 \hat{p}}{\partial z^2} + k^2 \hat{p} = \text{div}\,\frac{\hat{V}^*}{\delta t}$$

$$-\frac{\partial^2 \hat{w}^{n+1}}{\partial z^2} + \beta\hat{w}^{n+1} - iQ\,\frac{2\pi m}{M}\,\hat{b}_3^{n+1} + 2\frac{\partial \hat{p}}{\partial z} = \frac{2}{\delta t}\,\hat{w}^* \qquad (10)$$

$$-\frac{\partial^2 \hat{b}_3^{n+1}}{\partial z^2} + k^2 \hat{b}_3^{n+1} + i\,\frac{2\pi m}{M}\,\hat{w}^{n+1} = 0$$

where

$$k^2 = \left(\frac{2\pi\ell}{L}\right)^2 + \left(\frac{2\pi m}{M}\right)^2$$

and

$$\beta = \frac{2}{\delta t} + k^2$$

Denoting by X^{n+1}, the vector $(\hat{p}, \hat{w}^{n+1}, \hat{b}_3^{n+1})$ and by F^n, the right-hand side of Eq. (10) is rewritten in the compact form

$$LX^{n+1} = F^n \tag{11}$$

Because of the linearity of the operator L, the solution of Eq. (11) can be written in the form

$$X^{n+1} = \tilde{X} + \delta^+ X^{(1)} + \delta^- X^{(2)} \tag{12}$$

where $\tilde{X}$, $X^{(1)}$, and $X^{(2)}$ respectively satisfy

$$L\tilde{X} = F^n$$

$$\tilde{b}_3 = \tilde{w} = \tilde{p} - 0 \quad \text{for} \quad z = \pm\tfrac{1}{2} \tag{13}$$

$$LX^{(1)} = 0$$

$$b_3^{(1)} = w^{(1)} = 0 \quad \text{for} \quad z = \pm\tfrac{1}{2} \tag{14}$$

$$p^{(1)}(z = \tfrac{1}{2}) = 1 \quad p^{(1)}(z = -\tfrac{1}{2}) = 0$$

and

$$LX^{(2)} = 0$$

$$b_3^{(2)} = w^{(2)} = 0 \quad \text{for} \quad z = \pm\tfrac{1}{2} \tag{15}$$

$$p^{(2)}(z = \tfrac{1}{2}) = 0 \quad p^{(2)}(z = -\tfrac{1}{2}) = 0$$

Equations (14) and (15) are solved analytically. Only Eq. (13) is solved numerically at each time step. This is done in spectral space by using a tau method.[11] The coefficients δ^+ and δ^- are then derived from the boundary condition div $V^{n+1}=0$ for $z = \pm\frac{1}{2}$ by solving

$$\frac{\partial w}{\partial z}(z=\tfrac{1}{2}) + \delta^{+}\frac{\partial w^{(1)}}{\partial z}(z=\tfrac{1}{2}) + \delta^{-}\frac{\partial w^{(2)}}{\partial z}(z=\tfrac{1}{2}) = 0$$

$$\frac{\partial w}{\partial z}(z=-\tfrac{1}{2}) + \delta^{+}\frac{\partial w^{(1)}}{\partial z}(z=-\tfrac{1}{2}) + \delta^{-}\frac{\partial w^{(2)}}{\partial z}(z=-\tfrac{1}{2}) = 0 \qquad (16)$$

After completion of this step, the subsystems for $(\hat{u}_1^{n+1}, \hat{b}_1^{n+1})$, $(\hat{v}^{n+1}, \hat{b}_2^{n+1})$, and $\hat{\theta}^{n+1}$ in Eqs. (6) are solved separately by using a tau method.

To conclude this section, we briefly review the algorithm used to solve the scalar equation

$$-\frac{\partial^2 f}{\partial z^2}(z) + \lambda f(z) = g(z) \qquad (17)$$

satisfied by the temperature deviation. We consider the problem in the interval (-1, +1) and assume Dirichlet boundary conditions

$$f(-1) = h_1 \qquad f(1) = h_2 \qquad (18a)$$

or Newmann boundary conditions

$$\frac{df}{dz}(-1) = h_1 \qquad \frac{df}{dz}(1) = h_2 \qquad (18b)$$

The functions f and g are developed in Chebyshev polynomial series:

$$f(z) = \sum_{n=0}^{N} a_n T_n(z)$$

$$\frac{d^2 f}{dz^2} = \sum_{n=0}^{N} a_n^{(2)} T_n(z) \qquad (19)$$

$$g(z) = \sum_{n=0}^{N} b_n T_n(z)$$

In spectral space, Eq. (17) is rewritten

$$-a_n^{(2)} + \lambda a_n = b_n \qquad (20)$$

THREE-DIMENSIONAL CONVECTION IN LIQUID METALS

with the boundary conditions

$$\sum_{i=0}^{N} \xi_i a_i = h_1 \quad \sum_{i=0}^{N} \eta_i a_i = h_2 \qquad (21)$$

where

$$\xi_i = (-1)^i \quad \eta_i = 1 \qquad (22a)$$

in the case of Dirichlet boundary conditions, and

$$\xi_i = (-1)^{i-1} i^2 \quad \eta_i = i^2 \qquad (22b)$$

for Neumann conditions. Using the identity

$$a_n = \frac{c_{n-2}}{4n(n-1)} a_{n-2}^{(2)} - \frac{e_{n+2}}{2n(n^2-1)} a_n^{(2)} + \frac{e_{n+4}}{4n(n+1)} a_{n+2}^{(2)} \qquad (23)$$

where

$$e_n = 1 \quad \text{for } n < N$$
$$e_n = 0 \quad \text{for } n > N \qquad (24a)$$

and

$$c_0 = 2 \quad c_n = 1 \quad \text{for } n > 0 \qquad (24b)$$

we rewrite Eq. (20) in the form

$$\lambda p_n a_{n-2} - (1+\lambda q_n) a_n + \lambda r_r a_{n+2}$$
$$= p_n g_{n-2} - q_n g_n + r_n g_{n+2} \qquad (25)$$

with, for Dirichlet conditions,

$$\sum_{n \text{ even}} a_n = \tfrac{1}{2}(h_1+h_2)$$
$$\sum_{n \text{ odd}} a_n = \tfrac{1}{2}(h_2-h_1) \qquad (26a)$$

and for Neumann conditions,

$$\sum_{n \text{ even}} n^2 a_n = \tfrac{1}{2}(h_1+h_2)$$
$$\sum_{n \text{ odd}} n^2 a_n = \tfrac{1}{2}(h_2-h_1) \qquad (26b)$$

Equations (26a) and (26b) are thus decoupled in two subsystems for the Chebyshev coefficients with odd and even indices. Each subsystem is of the form

$$LX = F \tag{27}$$

The components of the vector X are the even or odd Chebyshev coefficients a_n. The components of F are the corresponding coefficients of the right-hand side of Eq. (25). The matrix L has the form (M=N/2):

$$L = \begin{bmatrix} \lambda_0 & \mu_0 & \nu_0 & & & & \\ & \ddots & \ddots & \ddots & & & \\ & & & \ddots & & \nu_{M-2} & \\ & & & & \lambda_{M-1} & & \\ & & & & \ddots & & \\ & & & & & \lambda_{M-1} & \mu_{M-1} \\ \ell_0 & \ell_1 & \cdots & \cdots & \cdots & \ell_{M-1} & \ell_M \end{bmatrix} \tag{28}$$

The elements x_j of X are computed by induction[12] by writing

$$x_{j+1} = \alpha_j x_j + \beta_j \qquad j = 0,\ldots,M-1$$

with

$$\alpha_j = \frac{-\lambda_j}{\nu_j \alpha_{j+1} + \mu_j}$$

$$\beta_j = \frac{-\nu_j \beta_{j+1} + \beta_j}{\nu_j \alpha_{j+1} + \mu_j} \qquad j = M-2,\ldots,0 \tag{29}$$

To initialize the procedure, we have to compute x_0. This is done by eliminating $x_i (i>1)$ from the system

$$x_{j+1} = \alpha_J x_J + \beta \qquad \ell_0 x_0 + \cdots + \ell_M x_M = 0 \tag{30}$$

To this purpose, we write

$$x_n = \psi_n x_0 + \Psi_n \qquad n=0,\ldots,M \tag{31}$$

THREE-DIMENSIONAL CONVECTION IN LIQUID METALS 135

where
$$\psi_0 = 1 \quad \Psi_0 = 0$$
and
$$\psi_{n+1} = \alpha_n \psi_n \quad \Psi_{n+1} = \alpha_n \Psi_n + \beta_n \quad n=0,\ldots,M-1 \quad (32)$$

We then obtain

$$\sum_{n=0}^{M} \ell_n x_n = \sum_{n=0}^{M} \ell_n (\psi_n x_0 + \Psi_n) = f_M \quad (33)$$

and thus

$$x_0 = \left(h - \sum_{n=0}^{M} \ell_n \Psi_n\right) / \sum_{n=0}^{M} \ell_n \psi_n \quad (34)$$

A similar method is used to solve vector equations of the form

$$-\binom{w''}{b''_3} + \Lambda \binom{w}{b_3} = \binom{g_1}{g_2} \quad (35)$$

by replacing scalar coefficients by 2x2 matrices in the previous algorithm.

Two-Dimensional Solutions

It is well known that when the Rayleigh number exceeds a critical value R_{crit}, convection occurs. For an horizontal periodic layer with fundamental wave number $k_x=3.117$ confined between two rigid conductive plates, $R_{crit} \simeq 1708$ (Ref. 13). There is a range of Rayleigh numbers for which the system evolves to a steady state where the convective patterns are two-dimensional rolls. This range shrinks when the Prandtl number is decreased. It is extended in the presence of a magnetic field parallel to the rolls that dissipates three-dimensional disturbances. The two-dimensional problem was considered by Clever and Busse[3,14]. They computed two-dimensional steady solutions of Eq. (4) and then studied their stability relative to infinitesimal three-dimensional disturbances. In the absence of an external magnetic field, they obtained that, for $Pr=0.025$, $k_x=3.117$ and $2<k_y<3$, the two-dimensional rolls become unstable when the Rayleigh number exceeds $R_{osc} \simeq 1885$. As a test of our numerical code, we have reproduced some of their results and obtained a very good agreement. As an example, the Nusselt number and the kinetic energy of the rolls are given in Table 1. The two-dimensional regime is illustrated

Table 1 Kinetic energy and convective heat transport as functions of the Rayleigh number for the steady two-dimensional solutions at Pr=0.025 and k_x=3.117

Rayleigh number, R	$\frac{1}{2} V^2/(R-R_{cr})$, in thermal units (k^2/d)	Nusselt number Simulation of the evolution problem	Nusselt number Stationary solu- solution of Ref. 3
3000	0.11	1.435	1.437
2000	4.83x10^{-3}	1.0606	1.0610
1900	3.38x10^{-3}	1.0286	...
1800	1.88x10^{-3}	1.0077	...

in Fig. 1, which displays contours of temperature deviation from the diffusive profile, velocity field, and vorticity contours in a plane perpendicular to the rolls at Rayleigh number R=1800.

Time-Dependent Regimes

Steady two-dimensional solutions are unstable relative to three-dimensional disturbances when the Rayleigh number exceeds a threshold R_{osc} that depends on the geometry, on the Prandtl number, and also on the Chandrasekhar parameter when a magnetic field in the direction of the rolls is present. In this section, Q=0. For periodic boundary conditions in the horizontal directions with k_x=3.117 and k_y=2.5 and for Prandtl number Pr=0.025, the two-dimensional rolls become linearly unstable when R>R_{osc} ∿ 1885 (Ref. 4). For R>R_{osc}, small three-dimensional discrepancies are not damped: They oscillate with a constant frequency and an amplitude that grows exponentially. This instability is generally referred to as the oscillatory instability.

We start with the steady state corresponding to two-dimensional rolls at R=2000. We destabilize this solution by means of random disturbances on the fundamental Fourier mode in the y direction (m = ±1), with an amplitude that in the mean is 10% of the corresponding m=0 mode.

At early times, the amplitude of the temperature disturbance at a given point of the flow grows exponentially in time (Fig. 2). The period of the oscillation is T_{osc}=0.065 viscous time, or equivalently 2.6 thermal times. This corresponds to a frequency $\sigma_i=2\pi/T_{osc}$ ∿ 2.4 thermal unit, in agreement with Fig. 12 of Ref. 3. At later times

THREE-DIMENSIONAL CONVECTION IN LIQUID METALS

a)
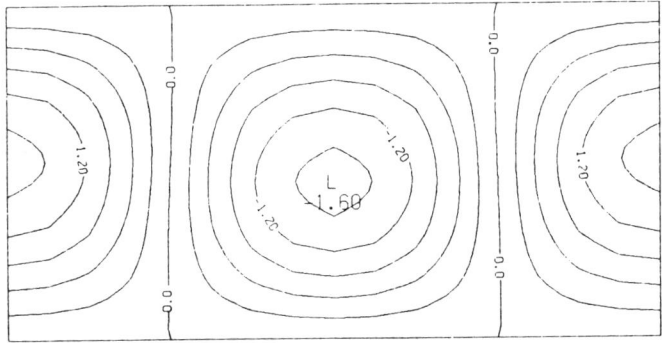

b)
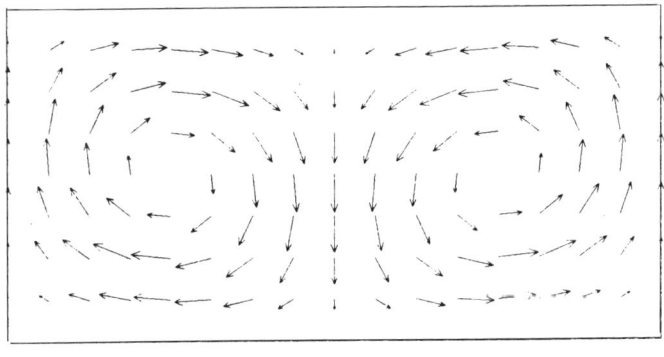

c)
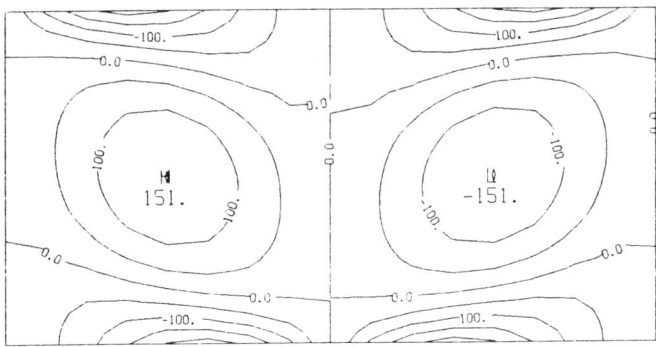

Fig. 1 Steady two-dimensional solution at R=1800: a) Contours of the deviation of temperature from the diffusive profile. b) Velocity field (maximum vector 33.7). c) Contours of vorticity in a plane y=const.

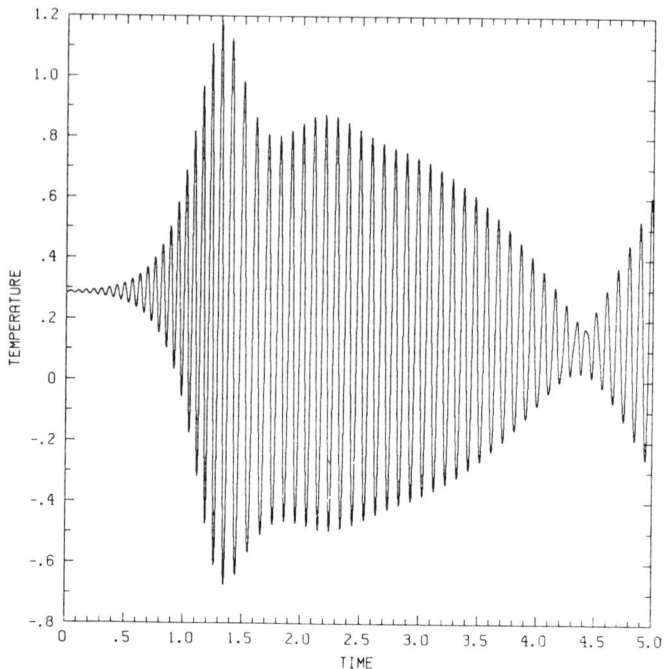

Fig. 2 Temperature deviation at the point ($x=L_x/4$, $y=L_y/5$, $z=\frac{1}{4}$) vs time at R=2000.

owing to nonlinear effects, the amplitude of the oscillation saturates. However, it does not tend to a constant value. This indicates that, at R=2000, the solution has already bifurcated from the periodic state that is expected to be observed for Rayleigh numbers slightly in excess of R_{osc} to another time-dependent regime.

In the following runs, we thus decrease the Rayleigh number. The modulation is still important at R=1925. It becomes very slow at R=1900. At R=1895, a time periodic solution seems to be established. Figure 3 shows the temperature deviation from the diffusive profile and the velocity component in the direction of the rolls at a given point as functions of time. The amplitudes of the oscillations relax to constant values.

To sum up, we obtain that in a layer of low Prandtl number fluid (here Pr=0.025) with periodic conditions in the horizontal directions, bifurcations occur in a rather narrow range of Rayleigh numbers. For aspect ratios corresponding to k_x=3.117 and k_y=2.5, the critical Rayleigh number for two-dimensional steady convection is $R_{crit} \simeq 1708$. At R=1880, we observed stable two-dimensional rolls, while,

THREE-DIMENSIONAL CONVECTION IN LIQUID METALS

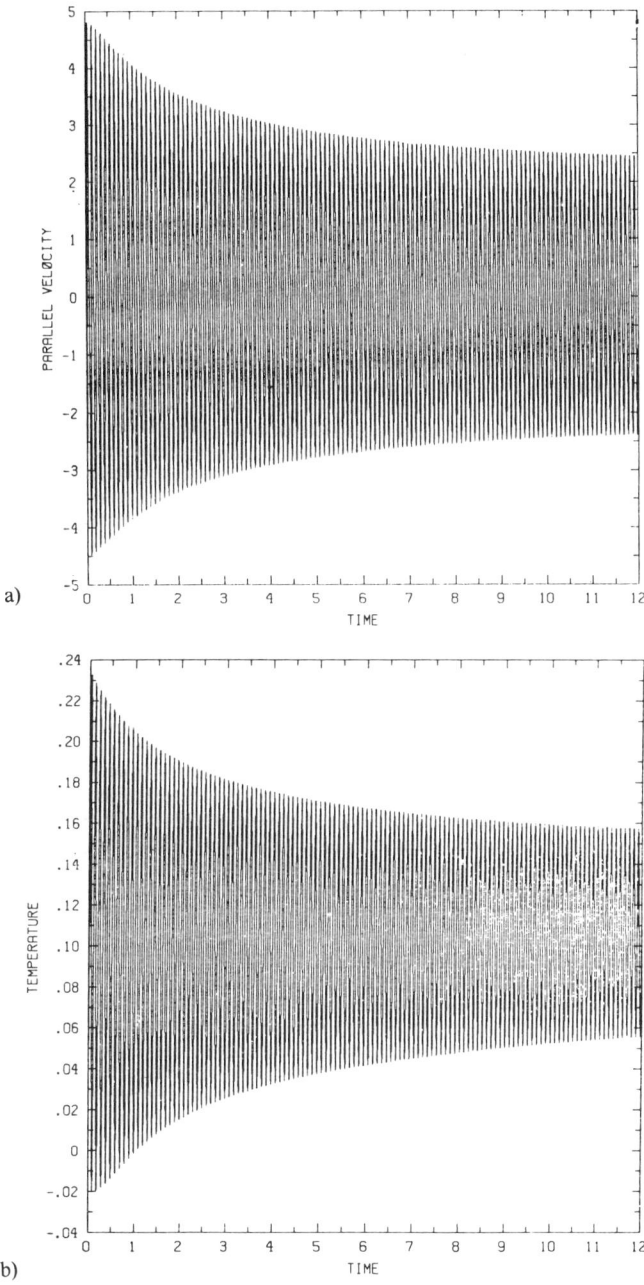

Fig. 3 Temperature deviation (a) and velocity in the direction of the roll axis (b) at the point ($x=L_x/4$, $y=L_y/5$, $z=\frac{1}{4}$) vs time at R=1895. Initial conditions correspond to oscillating solution at R=1900.

at R=1895, the solution evolves to a periodic regime, in agreement with the threshold value $R_{osc} \approx 1885$ for the oscillatory instability.[3,4] When R reaches 1900, we have observed that the solution has already performed another bifurcation to an oscillatory regime where the amplitude of this oscillation is modulated with a very long time scale.

Geometry of the Three-Dimensional Patterns

When two-dimensional rolls become unstable, new patterns are generated by modes that were damped at lower Rayleigh numbers and become coupled with basic two-dimensional rolls as the Rayleigh number is increased. For low Prandtl number, the dominant effect is due to the inertial term $V \cdot \nabla V$, which couples vertical vorticity modes with basic two-dimensional rolls and generates the oscillatory instability. This instability corresponds to waves traveling in the direction of the rolls and consequently to transverse periodic oscillation of the rolls. Figure 4 displays the convective patterns at t=12 and Rayleigh number R=1895 for which a periodic solution has been obtained. Figure 4a shows the (u,v) field and the isotachs of w in the plane $z=\frac{1}{2}$. Distortion of the roll is visible but small because of the proximity of the threshold R_{osc}. It is more easily seen on Fig. 4b, which corresponds to a surface of constant temperature deviation (θ= -1.2) intersected by planes x = const, y = const, z = const. Associated with the transverse oscillation of the rolls, a velocity component develops in the direction of the roll axis, as seen in Fig. 4c [isotachs of v and (u,w) field in the plane $y=L_y/2$]. As noticed by Brachet[15], Fig. 4c is very similar to that observed at the onset of the secondary instability in a plane shear flow.[16] The basic mechanism responsible for the time dependency in low Prandtl number convection is indeed a shear instability [17]. Figure 4d displays the vorticity component in the plane $y=L_y/2$.

Figures 5 and 6 correspond to R=2000 where the amplitude of the oscillation is strongly modulated. Figures 5a and 5b show the velocity field in the horizontal plane $z=\frac{1}{2}$ at t=1.5 and 4.5, respectively. Figures 5c and 5d show a surface of constant temperature deviation (θ= -1.2) at t=1.5 and 4.5, respectively. Figures 6a and 6b display the velocity field in the plane $y=L_y/2$ at t=1.5 and 4.5, respectively. Finally, Figures 6c and 6d are the vorticity component parallel to the roll axis in the same vertical plane at t=1.5 and 4.5, respectively. The main observation is that the rolls oscillate not only transversally but also vertically.

THREE-DIMENSIONAL CONVECTION IN LIQUID METALS 141

a)

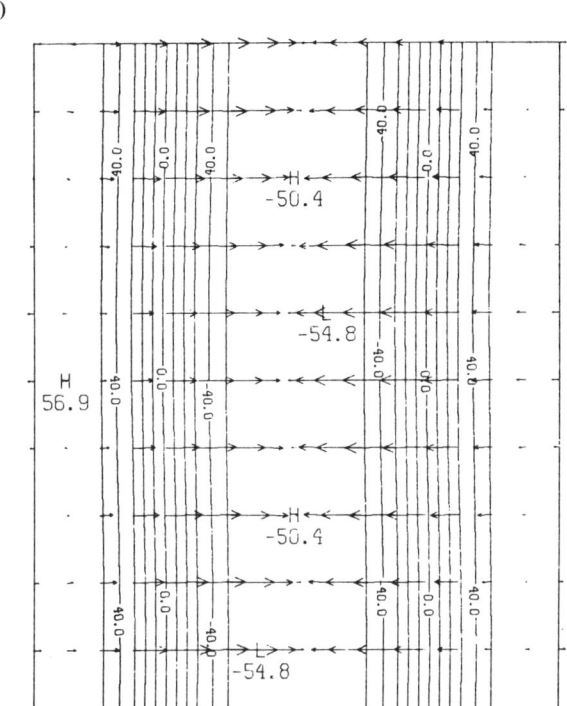

b)

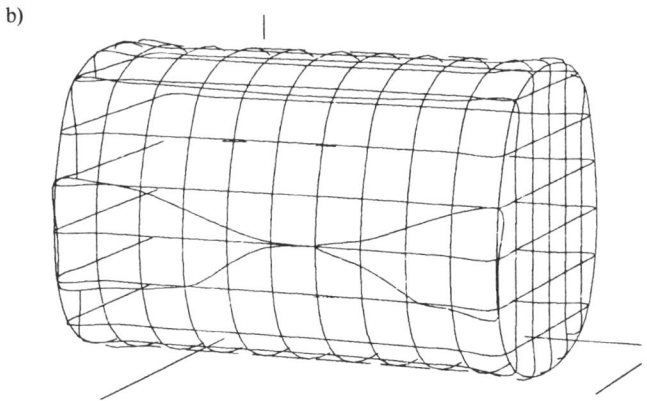

Fig. 4   R=1895, t=12:   a) (u,v) field and isotachs of w in the plane z=½ (maximum vector 20.2).   b) Surface of constant temperature deviation (θ=-1.2) intersected by planes x=const, y=const, z=const. c) (u,w) field and isotachs of v in the plane y=L_y/2, (maximum vector 65.4).   d) y component of vorticity in the plane y=L_y/2.

c)

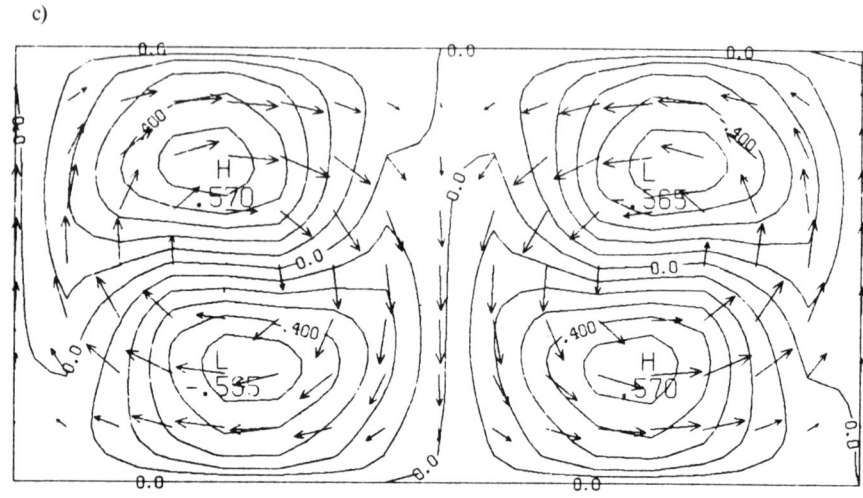

d)

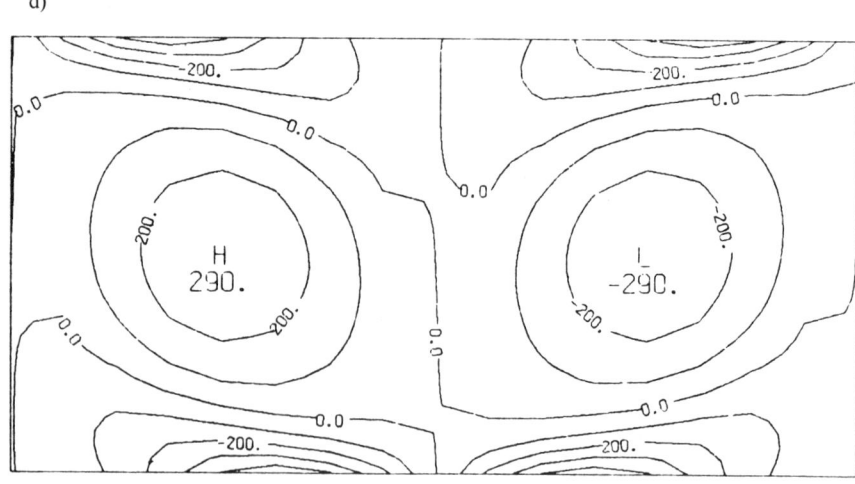

Fig. 4 R=1895, t=12: a) (u,v) field and isotachs of w in the plane $z=\frac{1}{2}$ (maximum vector 20.2). b) Surface of constant temperature deviation ($\theta=-1.2$) intersected by planes x=const, y=const, z=const. c) (u,w) field and isotachs of v in the plane $y=L_y/2$, (maximum vector 65.4). d) y component of vorticity in the plane $y=L_y/2$.

THREE-DIMENSIONAL CONVECTION IN LIQUID METALS

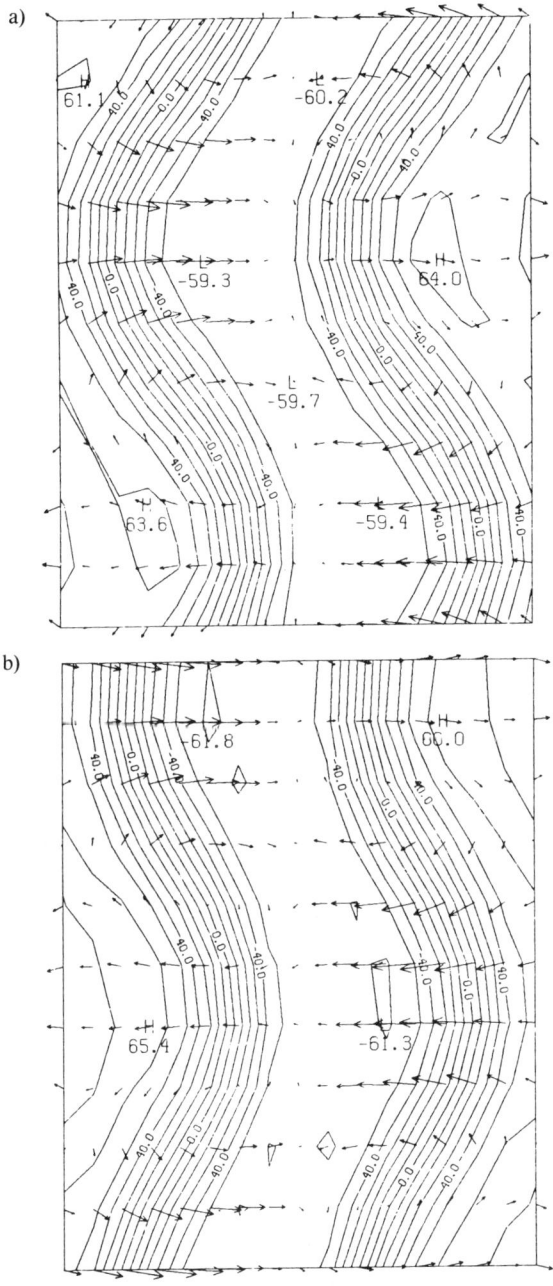

Fig. 5 R=2000: a) (u,v) field and isotachs of w in the plane $z=\frac{1}{2}$ at t=1.5 (maximum vector 60.12). b) Same at t=4.5 (maximum vector 42.4). c) Surface of constant temperature deviation ($\theta=-.1.2$) at t=1.5. d) Same as at t=4.5.

c)

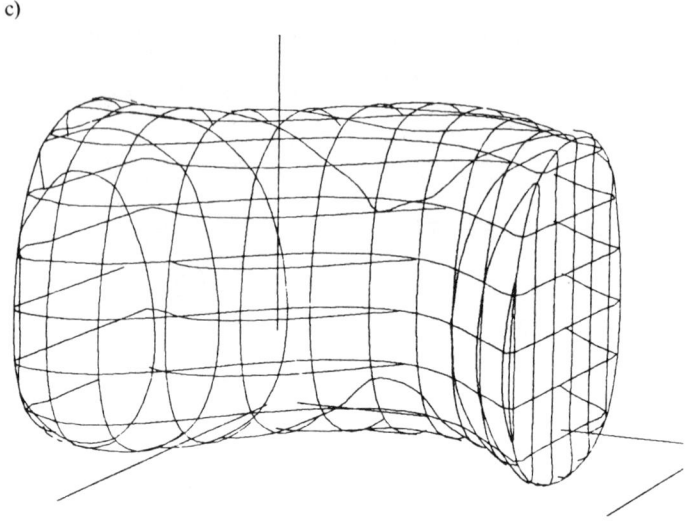

d)

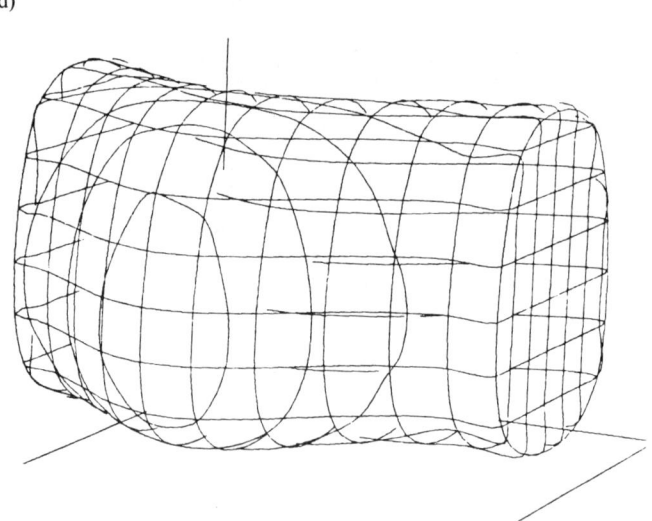

Fig. 5   R=2000:   a) (u,v) field and isotachs of w in the plane $z=\frac{1}{2}$ at t=1.5 (maximum vector 60.12).   b) Same at t=4.5 (maximum vector 42.4).   c) Surface of constant temperature deviation ($\theta=-.1.2$) at t=1.5 . d) Same as at t=4.5.

THREE-DIMENSIONAL CONVECTION IN LIQUID METALS

a)

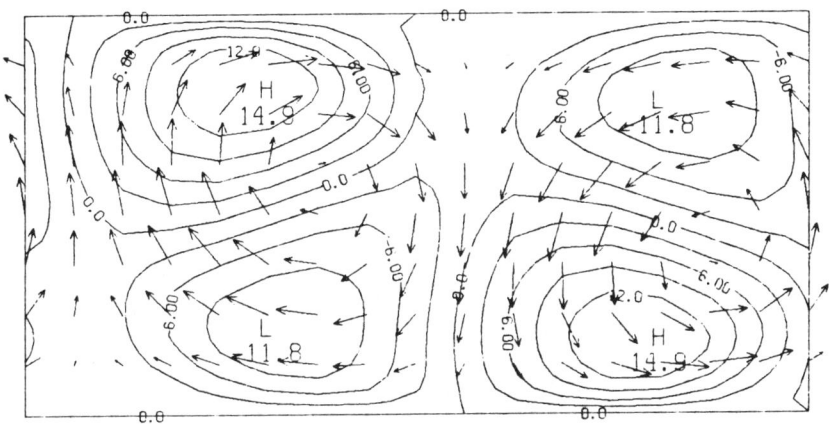

b)

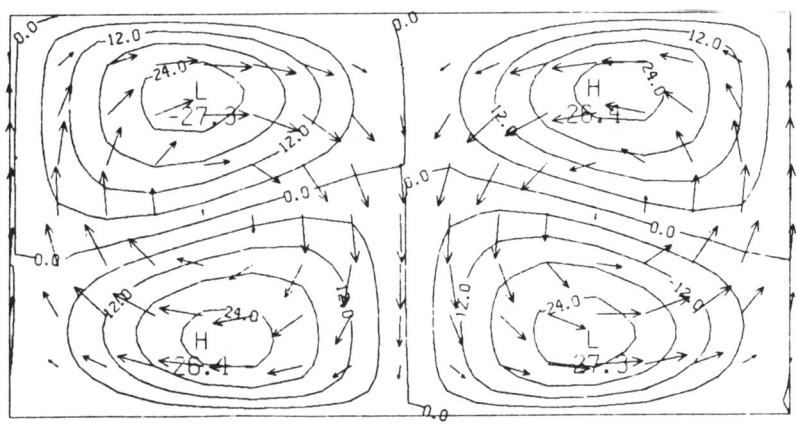

Fig. 6 R=2000: a) (u,w) field and isotachs of v in the plane $y=L_y/2$ at t=1.5 (maximum vector 65.0). b) Same at t=4.5 (maximum vector 64.3). c) y component of vorticity in the plane $y=L_y/2$ at t=1.5. d) Same as at t=4.5.

c)

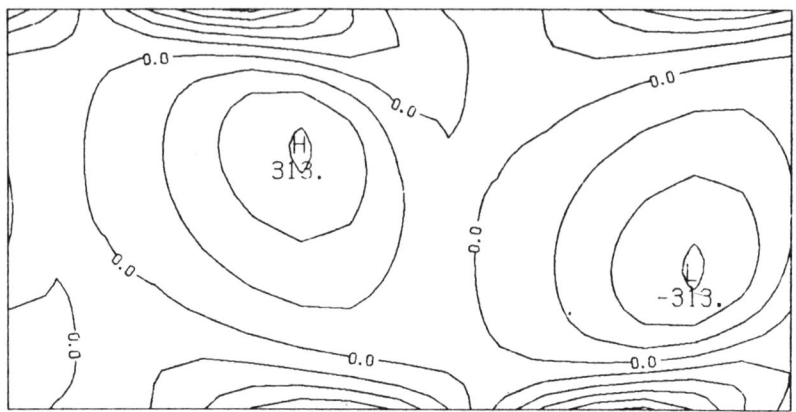

d)

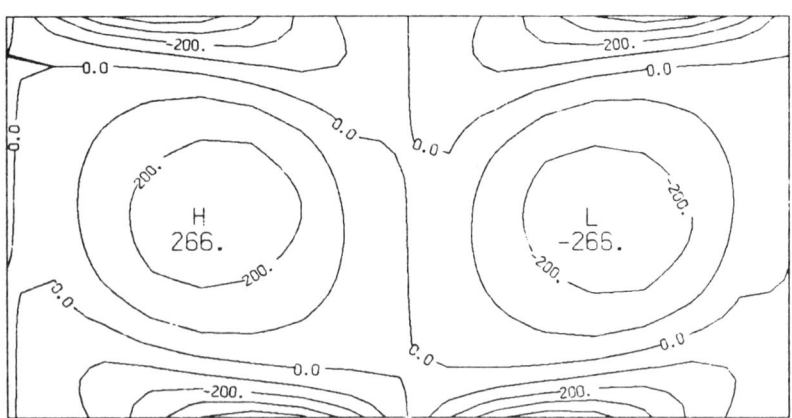

Fig. 6   R=2000:   a) (u,w) field and isotachs of v in the plane y=L_y/2 at t=1.5 (maximum vector 65.0).   b) Same at t=4.5 (maximum vector 64.3).   c) y component of vorticity in the plane y=L_y/2 at t=1.5.   d) Same as at t=4.5.

THREE-DIMENSIONAL CONVECTION IN LIQUID METALS   147

Influence of a Magnetic Field Parallel to the Roll Axis

As expected from the linear analysis of Busse and Clever[4], the stability region of two-dimensional solution is enhanced in the presence of a magnetic field parallel to the roll axis. When an oscillating flow at R=1900 is abruptly subject to a magnetic field corresponding to a Chandrasekhar number Q=50, the oscillations are immediately suppressed, indicating that the relaxation time, is, in this case, smaller than the period of oscillation. When Q=5, we observe an essentially exponential decay of the oscillation whose frequency is not affected (Fig. 7a). When the intensity of the magnetic field is still decreased (Q = 0.3125), the solution tends to become time periodic. (Fig. 7b). The stability region of periodic solutions is thus also increased by a magnetic field. An observation also made in the experiments[18]. Finally, Fig. 8 displays the evolution of temperature deviation and velocity in the roll direction for R=2000 and Q=5. An oscillating regime with a modulated amplitude is established in this case.

In conclusion, for the Rayleigh numbers and Chandrasekhar parameter we have considered a magnetic field in the roll direction does not produce drastic modification of the dynamics. It inhibits the oscillation of the rolls, and its effect is thus essentially equivalent to a reduction of the Rayleigh number.

Summary and Further Developments

A numerical simulation of convection in a horizontal layer of mercury has been performed, and transition to time-dependent regimes has been observed. When the Rayleigh number exceeds by about 10% the critical Rayleigh number $R_{crit} \approx 1708$, the steady solution corresponding to two-dimensional rolls becomes unstable with respect to the oscillatory instability. At R=1895, we obtain a time-dependent three-dimensional solution where the rolls oscillate periodically in a horizontal plane as a consequence of generation of a vertical vorticity. At R=1900, a new oscillating regime is observed where the amplitude of the oscillation is slowly modulated. When the Rayleigh number is still increased, the modulation is amplified. We have observed that in the regime where the amplitude modulation is present, the rolls oscillate both horizontally and vertically.

The same bifurcations were recently observed in a numerical simulation with free-slip boundary conditions at Pr=0.2 (Ref. 19) and also in experiments in mercury in boxes

a)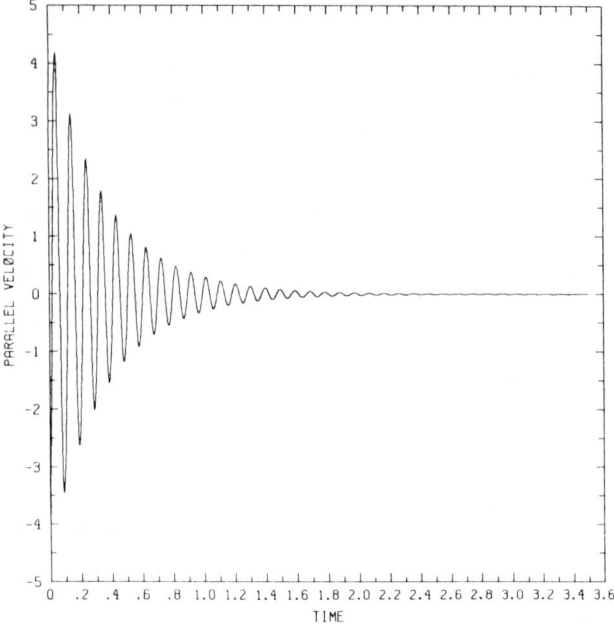

b)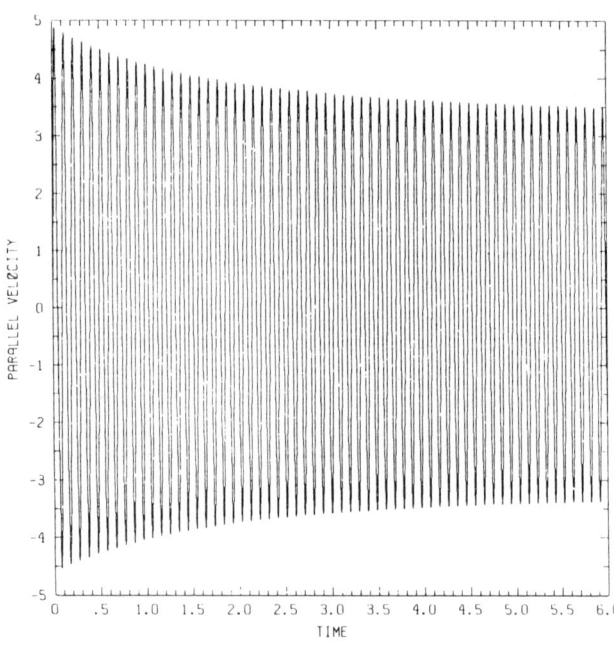

Fig. 7 Velocity in the direction of the roll axis v time at ($x=L_x/4$, $y=L_y/5$, $z=\frac{1}{4}$) vs time for R=1900. a) Q=5, b) Q=0.3125.

THREE-DIMENSIONAL CONVECTION IN LIQUID METALS 149

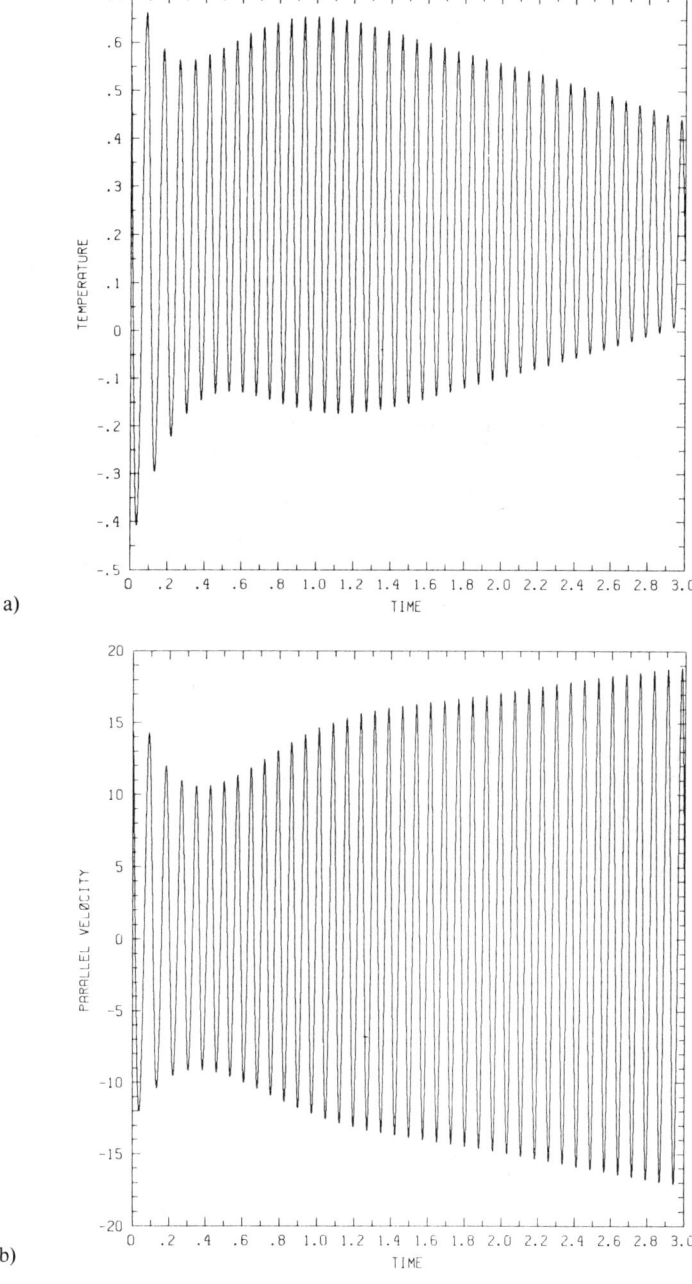

Fig. 8 R=2000, Q=5: a) Temperature deviation v time at ($x=L_x/4$, $y=L_y/5$, $z=\frac{1}{4}$). b) Velocity in the roll direction at the same point as in a).

with aspect ratios 6x6 (Ref. 20). The narrowness of the Rayleigh number range where these bifurcations take place at low Prandtl number suggests that they may be possibly described in terms of multibifurcations.[21,22]

Finally, in the regime we have considered, an external magnetic field does not lead to new time-dependent regimes. This is possibly due to the fact that at the Rayleigh numbers we have considered, the persistence of a time-dependent regime requires a relatively weak magnetic field. Richer dynamics could possibly be expected at higher Rayleigh numbers, as observed in experiments in boxes with small aspect ratios.[1]

Acknowledgments

We acknowledge very useful discussions with A. Arneodo, M.E. Brachet, P. Coullet, S. Fauve, U. Frisch, M. Meneguzzi, S. Orszag, and E. Spiegel. The computations were done on the CRAY 1 of the Centre de Calcul Vectorial pour la Recherche (CCVR). We also benefit from the support of a contract of the Direction des Recherches et Techniques (DRET). Figures were done using subroutines from the National Center for Atmospheric Research (NCAR) graphic library.

References

[1]Fauve,S., Laroche, C., and Libchaber, A., "Rayleigh-Bernard Experiments and Dynamical Systems," Bifurcation Theory, Mechanics and Physics, edited by C.P. Bruter et al, Reidel Publishing Company, 1983, pp. 257-276.

[2]Libchaber, A., Fauve, S., and Laroche, C., "Two-Parameter Study of the Routes to Chaos," Physica, Vol. 70, 1983, pp. 73-84.

[3]Clever, R.M., and Busse, F.H., "Transition to Time-Dependent Convection," Journal of Fluid Mechanics, Vol. 65, 1974, pp. 625-645.

[4]Busse, F.H., and Clever,R.M.,"Stability of Convection Rolls in the Presence of an Horizontal Magnetic Field," Journal de Mechanique Theorique et Appliquee, Vol. 2., 1983, pp. 495-502.

[5]Lipps, F.B., "Numerical Simulation of Three-Dimensional Convection in Air," Journal of Fluid Mechanics, Vol. 75, 1976, pp. 113-148.

[6]McLaughlin, J.B., and Orszag, S.A.O., "Transition from Periodic to Chaotic Thermal Convection," Journal of Fluid Mechanics, Vol. 122, 1982, pp. 123-142.

[7]Roberts, P.H., An Introduction to Magnetohydrodynamics, Longmans, Green, London, 1967.

THREE-DIMENSIONAL CONVECTION IN LIQUID METALS 151

[8]Kleiser, L., and Schumann, U., "Treatment of Incompressibility and Boundary Conditions in Three-Dimensional Numerical Spectral Simulations of Plane Channel Flows," Notes on Numerical Fluid Mechanics: Proceedings in Fluid Mechanics, Vol. 2, edited by E.H. Hirschel, DFVLR, Cologne, 1979, pp. 165-173.

[9]Orszag, S.A.O., and Kells, L.C., "Transition to Turbulence in Plane Poiseuille and Plane Couette Flow," Journal of Fluid Mechanics, Vol. 96, 1980, pp. 159-205.

[10]Basdevant, C., "Technical Improvements for Direct Numerical Simulation of Homogeneous Three-Dimensional Turbulence," Journal of Computational Physics, Vol. 50, 1983, pp. 209-214.

[11]Gottlieb, D., and Orszag, S.A.O., Numerical Analysis of Spectral Methods, SIAM, Philadelphia, 1977.

[12]Richmyer, R., and Morton, K.W., "Difference Methods for Initial Value Problems," Interscience Tracts in Pure and Applied Mathematics, 2nd ed., 1967, John Wiley and Sons, New York, Chichester, Brisbane, Toronto.

[13]Chandrasekhar, S., Hydrodynamics and Magnetohydrodynamic Stability, Oxford University Press, Oxford, 1961.

[14]Clever, R.M., and Busse, F.H., "Low-Prandtl Number Convection in a Layer Heated from Below," Journal of Fluid Mechanics, Vol. 102, 1981, pp. 61-74.

[15]Brachet, M.E., Private Communication, 1983.

[16]Brachet, M.E., and Orszag, S.A.O., "Secondary Instability of Free Shear Flows," Journal of Fluid Mechanics, to be published.

[17]McLaughlin, J.B., and Martin, P.C., "Transition to Turbulence in a Statically Stressed System," Physical Review A, Vol. 12, 1975, pp. 186-203.

[18]Fauve, S., Laroche, C., and Libchaber, A., "Horizontal Magnetic Field and the Oscillatory Instability Onset," GPS/ENS, Paris 1983, preprint.

[19]Meneguzzi, M., Private communication, 1983.

[20]Fauve, S., Private communciation, 1984.

[21]Coullet, P., and Spiegel, E., "Amplitude Equations for Systems with Competing Instabilities," SIAM Journal of Applied Mathematics, Vol. 43, 1983, pp. 776-821.

[22]Guckenheimer, J., and Knoblock, E., "Nonlinear Convection in a Rotating Layer: Amplitude Expansions and Normal Forms," Geophysical and Astrophysical Fluid Dynamics, Vol. 23, 1983, pp. 247-272.

Magneto-Fluid-Mechanic Turbulent Vortex Streets

D.D. Papailiou*

University of Patras, Greece

Abstract

An experimental and theoretical investigation has been conducted regarding the influence of a magnetic field on the geometry and shedding mechanism of a turbulent vortex street. Two configurations were used in the experiments, namely, a mercury open channel with a cylinder placed at the bottom wall and a mercury tray with a cylinder moving in it. In both cases, the applied magnetic field was aligned with the axis of the cylinder. The obtained results showed some difference between the two experiments with respect to the influence of the magnetic field on the shedding mechanism and the vorticity. The moving cylinder experiment indicated that the change of the vortex street goemetry in the presence of the magnetic field is mainly caused by the turbulent suppression. In this case, the developed theory appears to explain adequately the obtained experimental data.

Introduction

The ordinary-fluid-mechanic (OFM) vortex street and its associated wake flow have been the subject of both experimental and theoretical investigation for almost a century. Although the phenomenon is associated with some of the most important basic and applied problems in fluid mechanics,[1,2] it still remains not well understood. This is mainly due to the complexity of some of the participating mechanisms, among which, vortex formation and vortex periodic shedding in the near wake, as well as the influence of turbulence on the vortex street geometry, are more pertinent to the pre-

---

Paper presented at the Fourth Beer-Sheva Seminar on MHD Flows and Turbulence, Ben-Gurion University of the Negev, Beer-Sheva, Israel, Feb. 27-March 2, 1984. Copyright © 1985 by the American Institute of Aeronautics and Astronautics, Inc. All rights reserved.

*Professor, School of Engineering.

sent investigation. As a consequence of the above limitations, existing theoretical attempts related to the problem introduce in the solution a number of parameters to be defined experimentally.[2,3] These inadequacies are transferred to the magneto-fluid-mechanic (MFM) problem, where, in addition, the influence of the magnetic field on the flow must be taken into account.

The present investigation was initiated as a result of certain problems related to measuring turbulent characteristics with a hot-film anemometer in a magneto-fluid-mechanic flow.[4] These measurements, performed in a MFM pipe flow in the presence of a constant magnetic field, showed that when the probe axis was aligned with the magnetic field, no substantial change in the measured Strouhal number related to vortex shedding behind the cylindrical probe was observed. In the case of the probe placed at a 90-deg angle to the magnetic field, a 24% increase in the Strouhal number was found.

Based on these observations, the purpose of the present work was, initially, to investigate the influence of the applied magnetic field on the vortex shedding and the geometry of the turbulent vortex street. However, the obtained results offer some information regarding the changes occurring in the structure of the turbulent free shear flow and the damping of turbulence due to the presence of the magnetic field.

The present work consists of an experimental part and a theoretical study attempting to explain the obtained experimental results. The main body of the experimental data was obtained from the study of a turbulent vortex street formed behind a cylinder moving in a mercury tray. Some

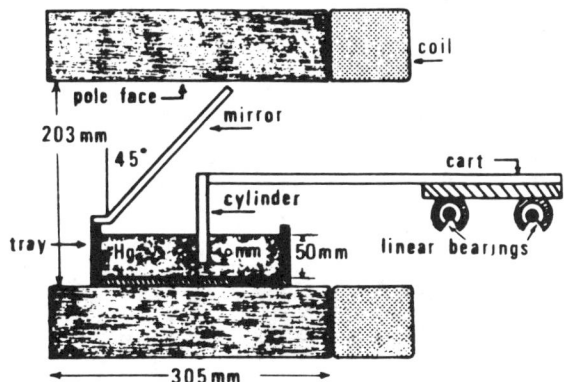

Fig. 1 General experimental setup (the motion is perpendicular to the paper).

inconclusive experiments were also conducted in an open channel. In both cases, the magnetic field was parallel to the axis of the cylinder.

The obtained results revealed a number of basic differences between the two experiments. The observed differences, concerning the influence of the magnetic field on the shedding mechanism and the vorticity, as well as the role of turbulence on the geometry of the vortex street, are discussed in the following sections of this paper.

Finally, the theoretical model developed in Ref. 2 for an OFM turbulent vortex street was extended to cover the MFM case studied in the present experiments. This was achieved by introducing a factor of the form e^{-kN} to act on the eddy diffusivity, in order to take into account the suppression of turbulence by the magnetic field.

The Experiments

Two experimental configurations were used in the study of the turbulent MFM vortex street. Initially, the vortex street was generated by a cylinder attached to the bottom wall of an open channel. As discussed in the following, this experimental setup presented certain difficulties; therefore, a second configuration was employed in which the vortex street was produced by a cylinder towed in a mercury tray. In both cases, the experimental setup was placed between the pole faces of an electromagnet so that the applied magnetic field was parallel to the axis of the cylinder (Fig. 1). A detailed description of the experiments is given in Ref. 2.

In the experiment conducted in the mercury tray, two plexiglass cylinders of diameter $d = 3.17$ mm and $d = 6.35$ mm were used to generate the vortex street by their motion. They were towed in the mercury tray with speeds ranging from 4 to about 10 cm/s, corresponding to a Reynolds number from 1080 to 5200. The applied magnetic field varied from zero in the nonmagnetic case to a maximum of about 6000 G. The interaction parameter $N = \sigma B^2 d/\rho U_o$ (ponderomotive forces over the inertial forces), calculated on the basis of the diameter of the cylinder and its velocity, was in the range of 2×10^{-2} to about 4. The Reynolds number was checked to be constant for each set of experiments.

From photographs of the vortex street, taken at different Reynolds numbers and magnetic fields, it was possible to measure the variation of the distance between the rows h and the longitudinal spacing α in the downstream direction for different magnetic fields. The distances h and α were measured from the apparent center of vorticity as defined

MAGNETO-FLUID-MECHANIC VORTEX STREETS 155

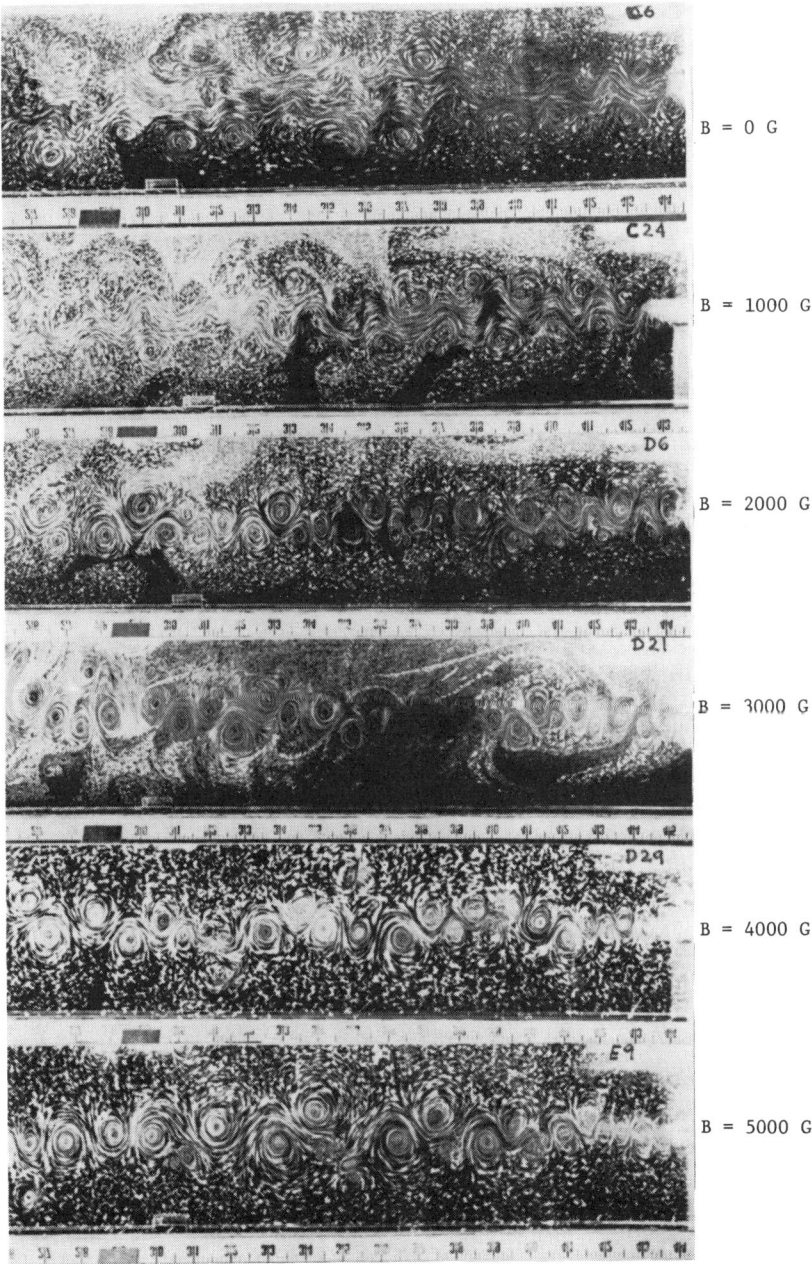

Plate 1 Photographs of the vortex street for varying magnetic fields at Re = 3900 (d = 6.35 mm).

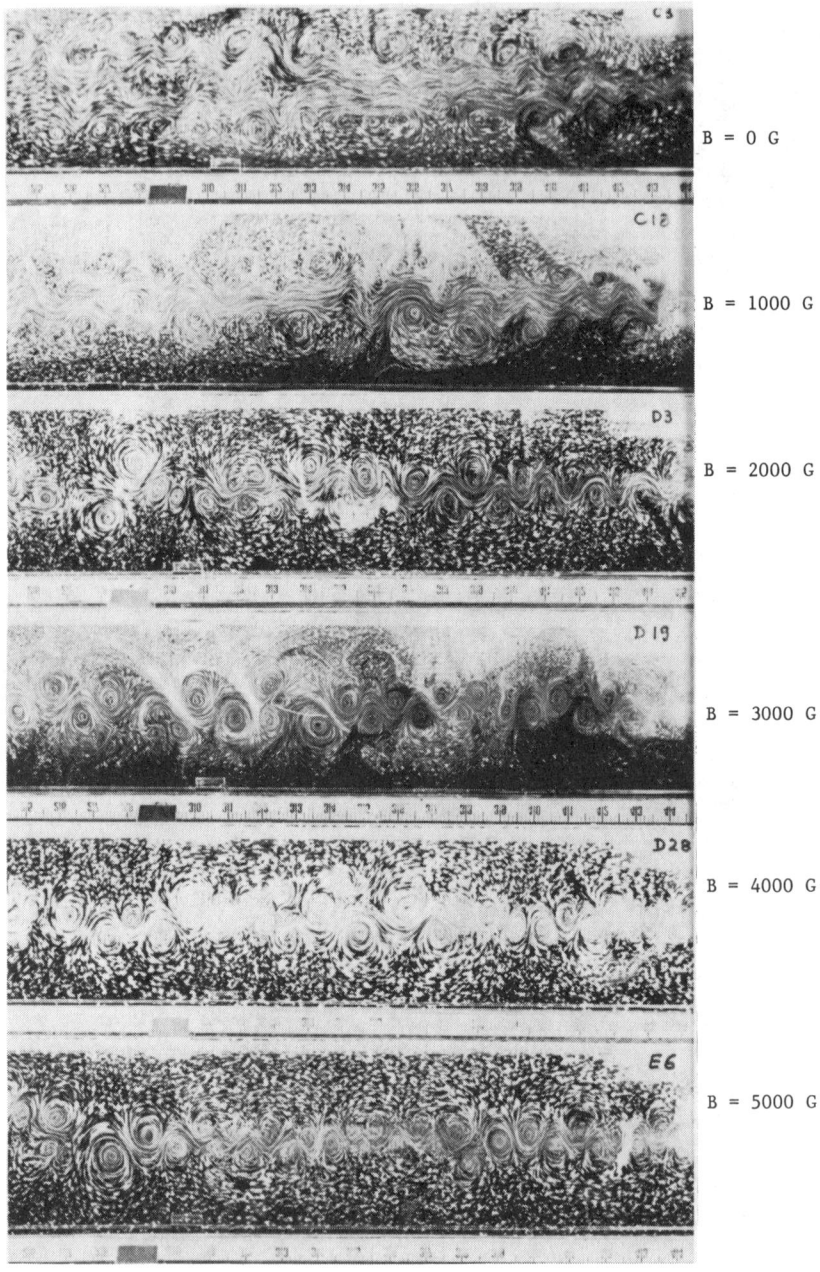

Plate 2 Photographs of the vortex street for varying magnetic fields at Re = 5215 (d = 6.35 mm).

in Refs. 2 and 5. As can be seen from the photographs presented here (Plates 1 and 2), the turbulent motion distorts the vortices, therefore prohibiting the measurement of a vortex velocity profile from which to locate the real center of vorticity.[5]

Typical measurements of h and α as functions of the downstream distance x for different magnetic fields and Reynolds numbers are shown in Figs. 2 through 5. In Fig. 6, the change of h with increasing magnetic field at different sections of the street (x = 10, 15, 20, 30, 40, 50 cm) is shown in the form of the ratio h_m/h_0 versus the magnetic field, for different Reynolds numbers, where h_m and h_0 are values with and without the presence of a magnetic field, respectively. The values presented in this figure were taken from the best-fit curves through the data points of h versus x plots. The change of the longitudinal distance α with the magnetic field is presented in Fig. 7.

At high magnetic fields, the measurements presented certain difficulties due to the fact that the two rows approached very close to each other. In these cases, interaction of certain pairs of vortices of opposite vorticity occurred, resulting in the formation of dipoles traveling away from the street, as shown in Plate 3. This phenomenon will be discussed in the following.

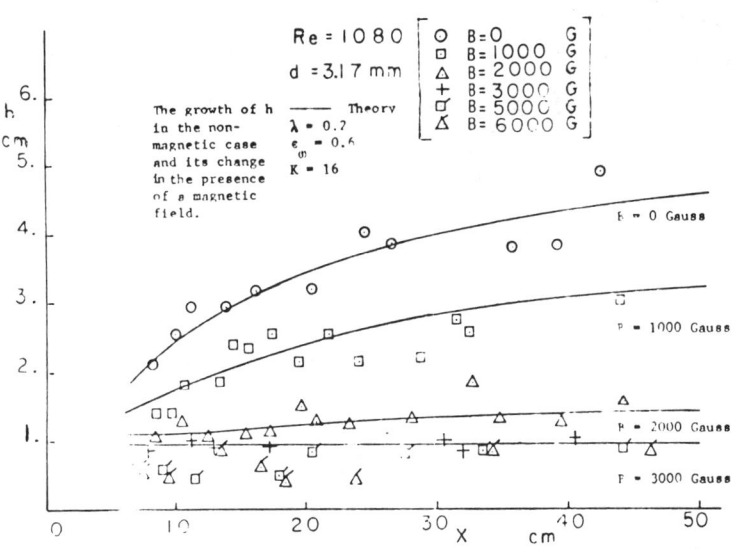

Fig. 2 Growth of h in the nonmagnetic case and its change in the presence of a magnetic field.

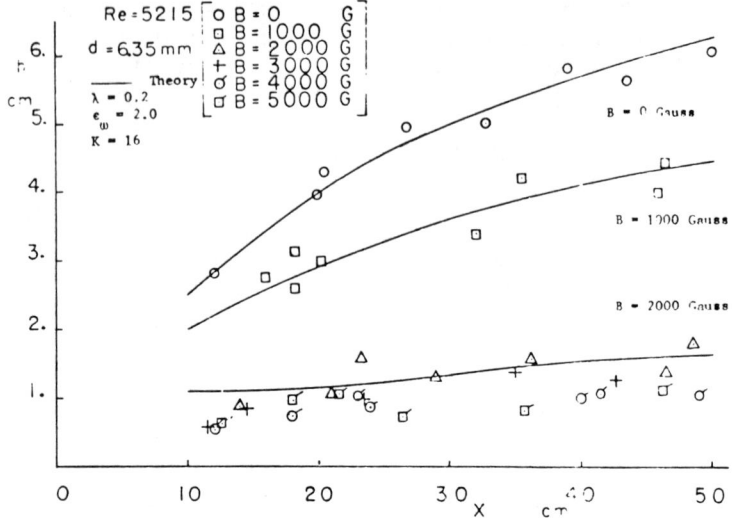

Fig. 3 Growth of h in the nonmagnetic case and its change in the presence of a magnetic field.

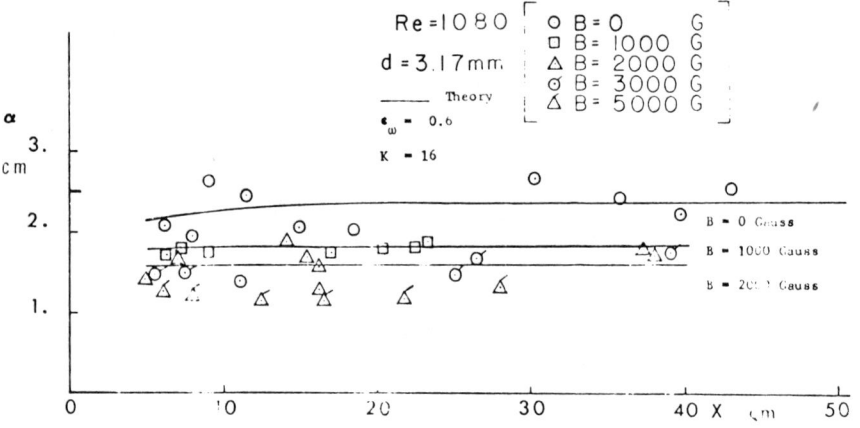

Fig. 4 Downstream change of α due to the turbulent action in the absence and in the presence of a magnetic field.

Measurements of the translational velocity u_T, obtained from consecutive photographs of the vortex street, gave values of u_T which differed by as much as 40%. Therefore, only qualitative information can be drawn from these measurements. It was found that the translational velocity measured near the cylinder was about 5% to 8% of the velocity U_0, and it dropped downstream. It was also found that the applied magnetic field slightly increased the translational velocity, which is also reflected in the decrease of the longitudinal spacing α. The drop of the translational velocity u_T in the downstream direction has little effect on the spacing α due to its small magnitude compared with the U_0.

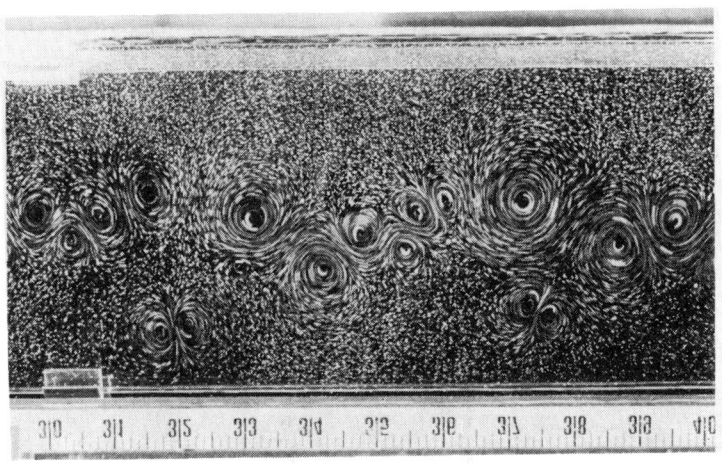

Plate 3 Formation of dipole-like pair of vortices at high magnetic fields.

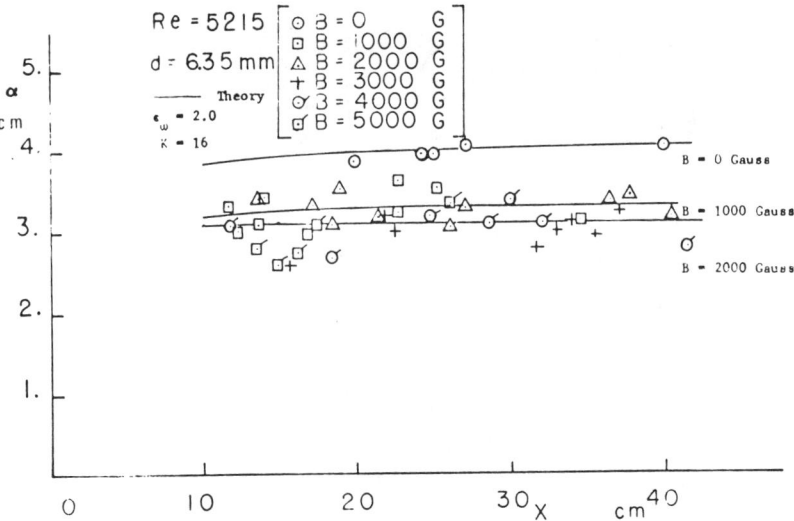

Fig. 5 Downstream change of α due to the turbulent action in the absence and in the presence of a magnetic field.

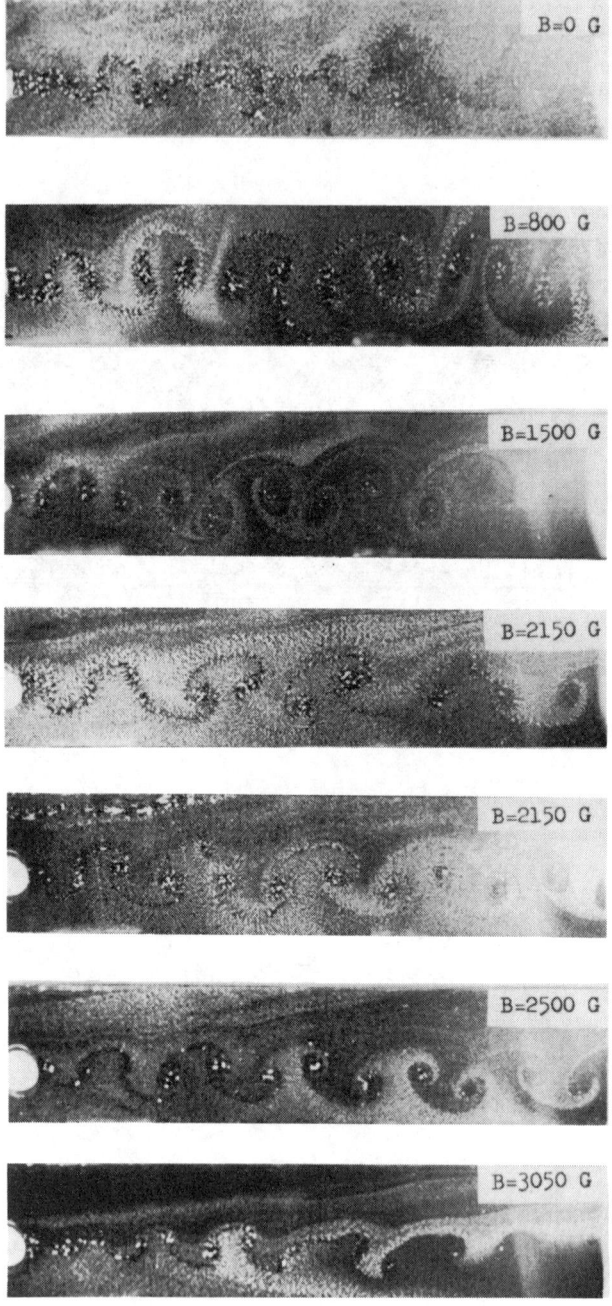

Plate 4 Effect of the magnetic field on the vortex street at Re = 1800 (open channel).

MAGNETO-FLUID-MECHANIC VORTEX STREETS 161

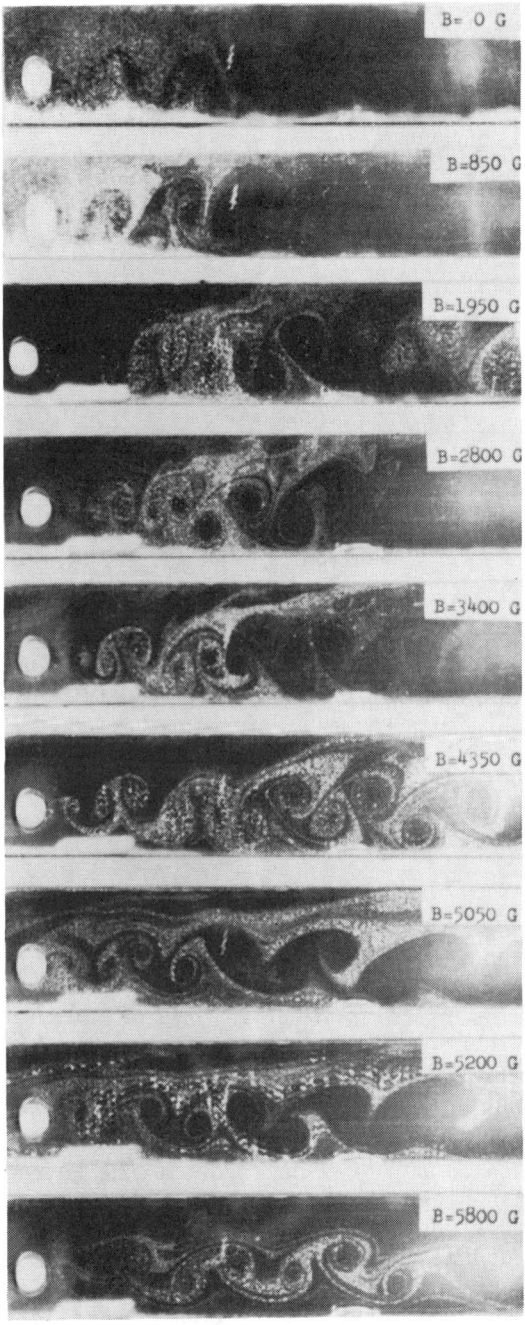

Plate 5 Effect of the magnetic field on the vortex street at Re = 12,500 (open channel).

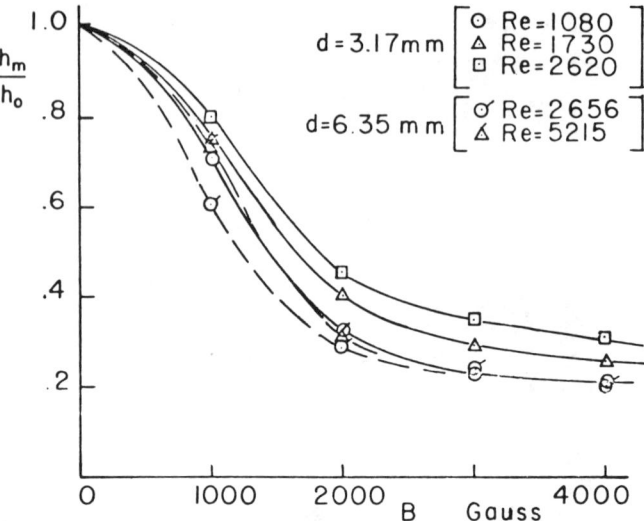

Fig. 6 Experimentally measured ratios h_m/h_o for different Reynolds numbers and distances downstream (x = 10, 15, 20, 30, 50 cm).

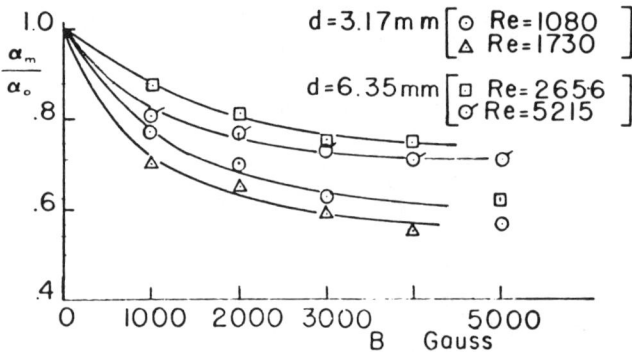

Fig. 7 Change of the longitudinal distance α with increasing magnetic field.

The experiments conducted in the open channel were inconclusive due to an observed decrease in the Reynolds number with increasing magnetic field. This change in the measured Reynolds number was as high as 25% of its initial value in the absence of the magnetic field.

Photographs of the street taken at Reynolds number 1800 and 12,500 are shown in Plates 4 and 5. The formation of the vortices and their subsequent shedding into the street is shown in Plate 6.

MAGNETO-FLUID-MECHANIC VORTEX STREETS

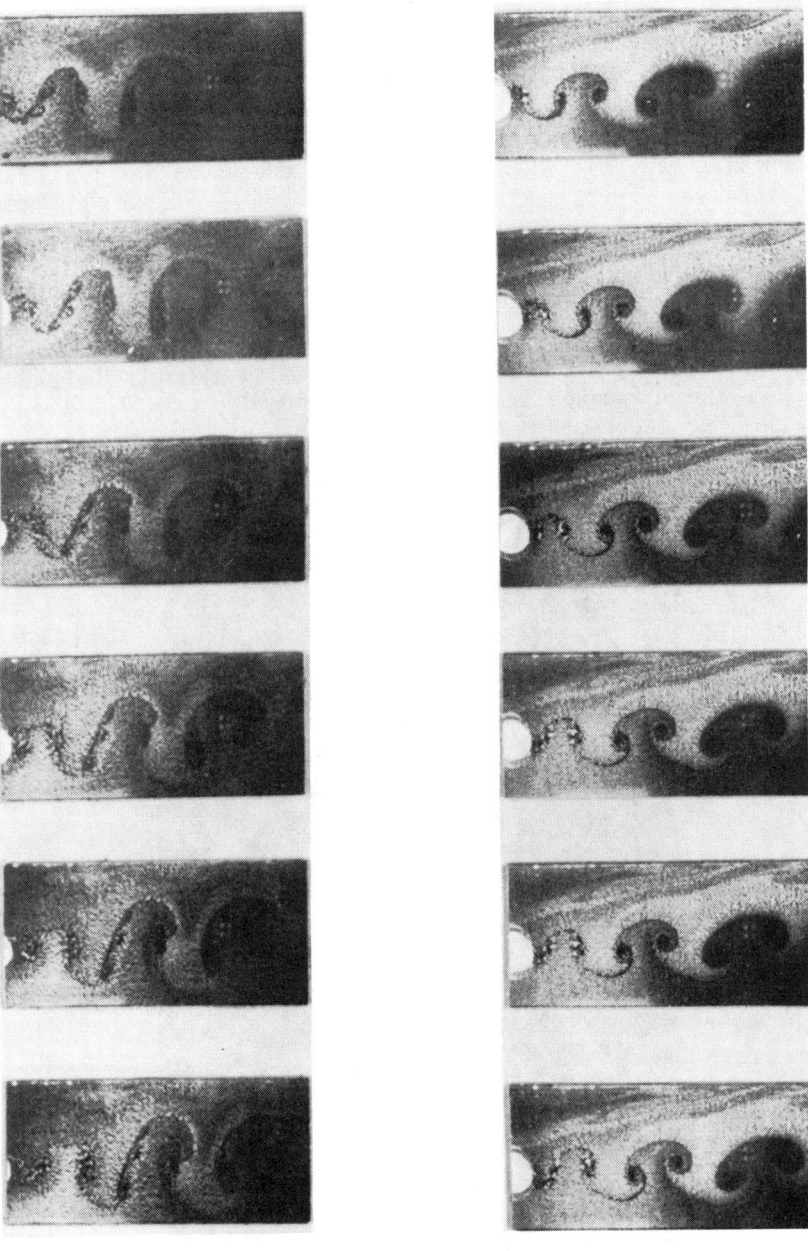

Plate 6 Formation of vortices in the near wake region (open channel).

In the photographs, the return of the wake to an organized vortex street in the presence of a magnetic field can be observed. It was also seen that the vortices stopped rotating in the far wake when a high enough magnetic field was applied, and that they moved with the velocity of the main stream.

The shedding frequency was found to decrease with increasing magnetic field due to the decrease of the stream velocity. It is important to note that the Strouhal number also decreased with increasing magnetic field as shown in Fig. 8. This occurred in a Reynolds number region, for which the Strouhal number remains constant in the absence of a magnetic field. This result differs from what was found in the wake of the moving cylinder, in which the shedding frequency and the Strouhal number remained unchanged under the influence of the magnetic field (Plate 7), and also from the results described in Ref. 4.

The differences found in the two sets of experiments can qualitatively be explained by considering the character

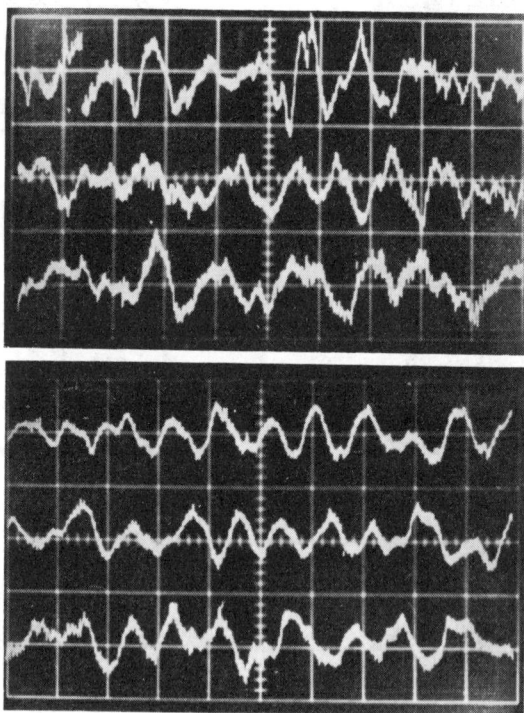

Plate 7   Hot-film signals showing that the shedding frequency remains unchanged under the influence of the magnetic field.

of the freestream flow in the open channel experiment as it is discussed in the following.

Theoretical Study of a Turbulent MFM Street

In this study, an attempt was made to explain the changes in the geometry of the street due to the presence of the applied magnetic field, which have been detected in the above-described experiments.

The developed theoretical study is based on the work of Lin[3] which explains the effect of viscosity on the geometry of a laminar von Karman vortex street. In this work, Lin succeeded in describing the change of the spacing ratio h/a of the street with time by using a linearized vorticity equation superimposed on a uniform stream. In the periodic solution of this equation, the points of maximum vorticity are interpreted as the centers of the vortices. The theory introduces a costant of integration λ, as well as an initial distance h_o, of the two rows of vortices which must be derived from the experiment. This theory has been verified by the experimental work of Wille and Timme.[6]

The above theory has been extended in Ref. 2 to include the effects of turbulence, while the present work examines the effects of an applied magnetic field on the geometry and the shedding mechanism of the street.

The averaged linearized MFM vorticity equation for turbulent flows is

$$\frac{\partial \bar{\omega}}{\partial t} + U \frac{\partial \bar{\omega}}{\partial x} + \frac{\partial}{\partial x} \overline{u'\omega'} + \frac{\partial}{\partial y} \overline{v'\omega'} = \nu \nabla^2 \bar{\omega} - \frac{\sigma B^2}{\rho} \bar{\omega} \quad (1)$$

where

$$\bar{\omega} = \frac{\partial \bar{v}}{\partial x} - \frac{\partial \bar{u}}{\partial y} \qquad \omega' = \frac{\partial v'}{\partial x} - \frac{\partial u'}{\partial y} \qquad U = U_o - u_T$$

The terms $\overline{u'\omega'}$ and $\overline{v'\omega'}$ represent turbulent transport of vorticity on planes parallel to the (x,y) and through planes perpendicular to x and y, respectively.

In the nonmagnetic cases presented in Ref. 2, Taylor's vorticity transport theory[7,8] and a constant transport vorticity coefficient ε_ω were used to express the turbulent vorticity transport terms in Eq. (1) as follows:

$$\overline{u'\omega'} = -\varepsilon_\omega \frac{\partial \bar{\omega}}{\partial x} \qquad \overline{v'\omega'} = -\varepsilon_\omega \frac{\partial \bar{\omega}}{\partial y} \quad (2)$$

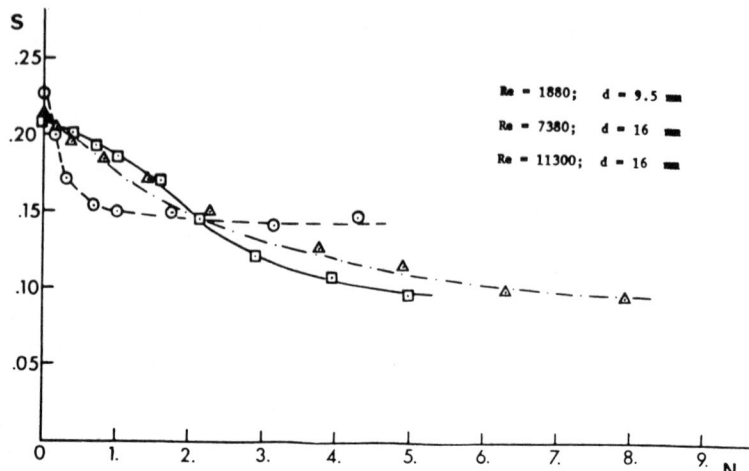

Fig. 8 Change of the Strouhal number with increasing magnetic field in the open channel.

Taking into account the above expressions, Eq. (1) becomes

$$\frac{\partial \bar{\omega}}{\partial t} + U_o \frac{\partial \bar{\omega}}{\partial x} = (\nu + \varepsilon_\omega) \nabla^2 \bar{\omega} - \frac{\sigma B^2}{\rho} \bar{\omega} \qquad (3)$$

In the above equation, the velocity $U = U_o - u_T$ has been replaced by U_o, since measurements of u_T showed that it represents a small percentage of U_o and also that it changes very little with increasing magnetic field.

The last term in Eq. (3) representing the influence of the magnetic field on the vorticity $\bar{\omega}$, could be disregarded on the grounds that, in the case under consideration, the magnetic field is aligned with the axis of vorticity of the traveling vortices.[9] This argument could raise certain objections, since the flow in the wake is not strictly two dimensional, and therefore the vorticity is not expected to be parallel to the applied field. However, the obtained photographs show no apparent vorticity suppression; on the contrary, the vortex street appears far better organized under the influence of the magnetic field. Furthermore, as already mentioned, measured values of the translational velocity u_T indicated no significant change in the presence of the magnetic field. These observations support the above argument, which has been adopted in this study. In this case, Eq. (3) reduces to the form

$$\frac{\partial \bar{\omega}}{\partial t} + U_o \frac{\partial \bar{\omega}}{\partial x} = (\nu + \varepsilon_\omega) \nabla^2 \bar{\omega} \qquad (4)$$

which is similar to that of a turbulent[2] or laminar[3] OFM vortex street.

It should be noted that the same form can be obtained by applying the transformation

$$\bar{\omega} = \bar{\omega}_o e^{-KN} \tag{5}$$

which describes vorticity damping due to the magnetic field.

The form of the vorticity transport coefficient for the OFM case has been discussed in Ref. 2. It is based on the model proposed by Squire[10] in which the transport coefficient ε_ω is assumed to be proportional to the circulation around a single vortex or to the amount of vorticity it possesses. In a two-dimensional nonmagnetic motion, this vorticity remains constant; therefore ε_ω takes the form

$$\varepsilon_\omega = c\bar{\omega}_o \tag{6}$$

where c is a constant, and $\bar{\omega}_o$ is the vorticity of a single vortex.

As a next step, an expression should be found to account for the decrease of ε_ω due to the presence of the magnetic field. Since a vorticity transport model has been adopted to represent turbulent diffusion, the use of an expression describing the damping of vorticity by the magnetic field, similar to that of Eq. (5), appears appropriate. According to this model, the imposed magnetic field suppresses the turbulent motion by damping the vorticity contained in the eddies of various size existing in the flow. Therefore, the transport coefficient in the presence of a magnetic field $\varepsilon_{\omega m}$ is expressed as follows:

$$\varepsilon_{\omega m} = c\bar{\omega}_o e^{-kN} \tag{7}$$

Following the theoretical approach developed in Ref. 2 and Ref. 3 for OFM laminar and turbulent vortex streets, an expression for h can be obtained:

$$h = 2\eta \sqrt{\frac{2(\nu + \varepsilon_{\omega o} e^{-kN})x}{U_o}} \tag{8}$$

in which the effect of the magnetic field on turbulence has been taken into account.

The values of η, in the downstream direction, for different values of magnetic field, are defined by the transcendental equation

$$(\eta_o - \lambda\eta_o)\cosh\eta_o\eta = (\eta-\lambda\eta_o)\sinh\eta_o\eta \qquad (9)$$

where

$$\eta_o = \frac{h'_o}{2}\sqrt{\frac{U_o}{2(\nu+\varepsilon_{\omega o}e^{-kN})x}} \qquad (10)$$

To obtain an expression for the change of α due to the magnetic field, the equation of motion of a turbulent two-dimensional wake is considered:

$$U_o \frac{d\bar{u}}{dx} = \varepsilon_\omega \frac{\partial^2\bar{u}}{\partial y^2} - \frac{\sigma B^2}{\rho}\bar{u} \qquad (11)$$

In this equation, it was assumed that the eddy diffusivity ε is equal to that of vorticity transport ε_ω. Dropping the ponderomotive term for reasons explained above, the velocity profile obtained is of the following Gaussian form:

$$\frac{\bar{u}}{U_o} = A(\frac{x}{d})^{-\frac{1}{2}} \exp\left[\frac{U_o y^2}{4\varepsilon_{\omega o}e^{-kN}x}\right] \qquad (12)$$

The spacing distance α is given by the expression

$$\alpha = \frac{U_o - \bar{u}_T}{f} \qquad (13)$$

It can be assumed that the velocity u_T with which a vortex located at a downstream distance x moves can be obtained from the Gaussian profile [Eq. (12)] by substituting y with the distance h/2 of its center from the axis x.
In this case, u_T is given by the equation

$$\bar{u}_T = AU_o(x/d)^{-\frac{1}{2}} \exp\left[\frac{U_o h^2}{16\varepsilon_{\omega o}e^{-kN}x}\right] \qquad (14)$$

Substituting Eq. (14) in Eq. (13) gives

$$\alpha = \frac{U_o \{1-A(x/d)^{-\frac{1}{2}} \exp\left[(U_o h^2)/(16\varepsilon_{\omega o} e^{-kN} x)\right]\}}{f} \quad (15)$$

The obtained theoretical results regarding the change of the geometrical characteristics h and α of the vortex street under the influence of the magnetic field are compared with the experimental data in Figs. 2 through 5.

Discussion of the Experimental and Theoretical Results

The experimental data show that the flow in the absence of the magnetic field is turbulent, as expected.
The applied magnetic field does not destroy similarity, as can be seen in Fig. 6, where the measured ratio h_m/h_o remains unchanged at different distances downstream for each Reynolds number. This statement is correct insofar as the turbulent eddy diffusivity is strong enough to produce a widening of the vortex street in the downstream direction. It is obvious that when the magnetic field has suppressed the turbulence to the point that the vortex street forms two parallel rows, the ratio h_m/h_o is not constant downstream of the cylinder.

In the presence of the magnetic field, there is a decrease in both h and α with increasing field. This decrease is larger for the distance h than for α. At about 3000 G, the vortices form almost parallel rows for all cases examined.

The same observation can be made from the study of the change of the longitudinal distance α, where the decrease of the ratio α_m/α_o takes place mainly within values of the magnetic field ranging from 0 to 3000 G. At high magnetic field, the initial distance between the rows h_o' decreases slightly but there is no widening of the street. It can thus be concluded that the decrease in both h and α is due to the damping of turbulence with increasing magnetic field. As already mentioned, this conclusion is also supported by measurements of the translational velocity u_T, which shows a slight increase with increasing magnetic field, indicating the strengthening of the vortices due to the reduction of the turbulent diffusion. We must note here that turbulence is present even at high magnetic fields, as is shown in the photographs of the hot-film signals (Plate 7), but its influence on the characteristics h and α seems to be very small. This could indicate that the surviving turbulence consists of eddies with vorticity aligned with the applied magnetic field, and therefore is not affected by it.

The results of the mercury tray experiment showed no influence of the imposed magnetic field on the shedding frequency of the vortices or the Strouhal number. This is in agreement with the experiment reported in Ref. 4 in which the constant magnetic field was aligned to the hot-film probe axis. On the other hand, the experiments with the magnetic field perpendicular to the hot-film probe axis, also described in Ref. 4, and those in the open channel showed that the Strouhal number can be changed in the presence of a magnetic field and that it can either be increased[4] or decreased (open channel experiment). It should be noted that the decrease of the Strouhal number in the presence of a magnetic field, observed in the open channel experiment, cannot be attributed to the deceleration of the main flow, since the measured Reynolds numbers, for all values of the imposed magnetic field, were within the region corresponding to constant values of the Strouhal number for the OFM case.

Although the data reported in the above experiments are not sufficient to allow a complete explanation, the observed changes of the Strouhal number in the pipe and the open channel cases might be attributed to the existing velocity gradients in the main flow, which are absent in the case of the moving cylinder. As Gardner[4] pointed out, the vorticity in the near wake depends on the existing velocity gradients in the main flow as it is fed from these gradients. The presence of the magnetic field will alter the velocity profiles in the main flow and thus will possibly affect the shedding mechanism of the vortices. It might be then concluded that the influence of the shedding mechanism in the presence of a magnetic field depends on the character of the flow in the OFM case and the orientation of the applied magnetic field.

In the nonmagnetic case,[2] the value 0.2 for the integration parameter λ gives the best fit with the experimental data for all Reynolds numbers attained in the experiments. The eddy diffusivity ranged from $\varepsilon_\omega = 0.6$ for the highest one (Re = 5215), and it constantly increased with each increasing Reynolds number.

In the presence of a magnetic field, the same value of the integration parameter ($\lambda=0.2$) was found to give the best fit to the experimental data. This is in agreement with the fact that the ratio h_m/h_o remains unchanged in the downstream direction for a given magnetic field, since according to the theory,[3] λ is a factor pertinent to the shape of the curves $h = h(x)$.

MAGNETO-FLUID-MECHANIC VORTEX STREETS 171

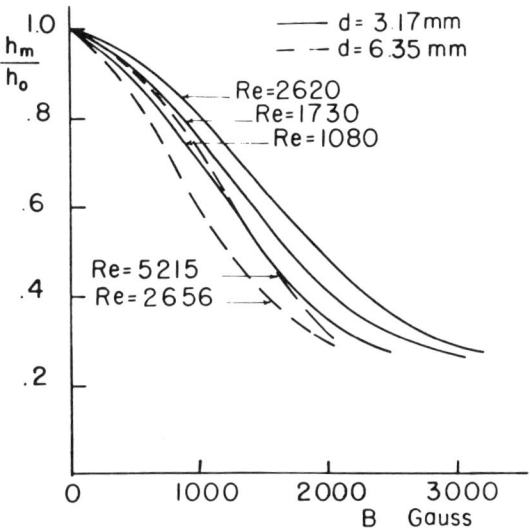

Fig. 9 Theoretically predicted ratios h_m/h_o for different Reynolds numbers and different distances downstream.

In the theoretical part of this work, the form $\varepsilon_{\omega m}e^{-kN}$ was assumed to describe the damping of the turbulent eddy diffusivity in the presence of the magnetic field. This expression was found to be adequate to produce the decrease of the distance h in the presence of the magnetic field. The experimentally determined value for the constant k was equal to 16 for all the attained Reynolds numbers. This value was determined from the calculations of the change of h with the distance x from the cylinder in the presence of the magnetic field.

Due to the inability of the theory to determine the origin of the street, the curves had to be shifted to fit the data. This shift, corresponding to a delay of the broadening of the street, was persistently increased with increasing magnetic field. This can be physically interpreted as a delay of the turbulent action with increasing magnetic field.

It must be noted also that the preservation of similarity in the presence of the magnetic field, which was indicated from the experiment and also found in the theory, was used as a condition to be fulfilled when choosing the set of parameters h_o', λ, and ε_ω. In Fig. 9, similarity is shown for values of h_m/h_o taken from the theoretical curves.

For the theoretical prediction of the variation of the longitudinal distance α far from the cylinder, the values of ε_ω and h used in the calculation of u_T [Eq. (14)] were taken from the theoretical curves $h = h(x)$ for the corresponding Reynolds number.

The constant A appearing in Eq. (12) was determined from the experimental observation that the translational velocity u_T at 20 diameters downstream was found to be approximately 8% of the velocity U_0.

Finally, the value of α estimated from Eq. (15) at a distance from the cylinder equal to 20 diameters was adjusted to match the experimentally obtained value of α at this point. This was necessary, since the arguments presented here in order to explain the variation of α downstream do not account for its change in the near wake region. The lack of information on the flow conditions in the near wake region is reflected also in the lack of explanation of the experimentally found more drastic decrease of α with increasing magnetic field, which occurs mainly in this region (Fig. 7).

It should be stated, that at very high magnetic fields, where turbulence is suppressed, a return to laminar configuration should be expected. However, close observation of the related photographs and the $h = h(x)$ curves shows that the two rows of vortices remain parallel. It also shows that a close approach of the two rows occurs, resulting in the interaction of certain pairs of vortices, as has already been described and is documented in Plate 3.

These vortex dipoles observed at high magnetic fields were first reported in Ref. 11. The existence of vortex dipoles with a translational motion has been predicted in several theoretical works on two-dimensional turbulent free flows.[12-14] Lately, the two-dimensional turbulent wake behind a disk towed in a liquid film has been studied experimentally.[15] These experiments also showed the existence of fast-moving "solitary vortex couples" which escape the central zone of the wake.

Based on the above-mentioned evidence, it is possible to explain the formation of the observed vortex dipoles. The presence of a strong magnetic field suppresses the existing three-dimensional[†] flow, thus establishing a two-dimensional flow which, according to the preceding discussion, is a necessary condition for the appearance of the moving vortex dipoles observed in the present experiments.

It appears, therefore, that the vortex street never returns, under the influence of the magnetic field, to its laminar state corresponding to the related value of the Reynolds number.

In conclusion, a two-dimensional linearized vorticity equation similar to the one used for the OFM turbulent vor-

---
†(effects in the)

MAGNETO-FLUID-MECHANIC VORTEX STREETS

tex street adequately describes the MFM case when one more constant k is introduced (which should be taken from the experiment) in the exponential expression [Eq. (7)] representing turbulence damping due to the magnetic field.

References

[1] Morkovin, M.V., "Flow Around Circular Cylinder: A Kaleidoscope of Challenging Fluid Phenomena," ASME Fluids Engineering Division Conference Symposium on Fully Separated Flows, 1964.

[2] Papailiou, D.D., and Lykoudis, P.S., "Turbulent Vortex Streets and the Entrainment Mechanism of the Turbulent Wake," Journal of Fluid Mechanics, Vol. 62, No. 1., 1974, pp. 11-31.

[3] Lin, C.C., Studies Presented to R.V. Mases, Academic Press, New York, 1954, pp. 170-176.

[4] Gardner, R.A., "Hot Film Anemometry in a Liquid Mercury MFM Pipe Flow," Symposium on Flow -- Its Measurement and Control in Science and Industry, Pittsburgh, PA, May 1971.

[5] Schaeffer, J.W. and Eskinazi, J., Journal of Fluid Mechanics, Vol. 6, No. 2, 1959, p. 241.

[6] Willie, R., Timme, A., Jahrbuch der Schiffbantechn, Gesellschaft, Vol. 51, 1957, pp. 215-221.

[7] Taylor, G.I., Philosophical Transactions of the Royal Society of London, Series A, Vol. 215, 1915, p. 1.

[8] Hinze, J.O, Turbulence, McGraw Hill, New York, 1968.

[9] Shercliff, J.A., A Textbook of Magnetohydrodynamics, Pergamon Press, New York, 1965.

[10] Squire, H.B., Aeronautical Research Council,16,666, FM 2053, 1954.

[11] Papailiou, D.D., Ph.D. Thesis, School of Aeronautics, Astronautics and Engineering Sciences, Purdue University, Lafayette, Ind., 1971.

[12] Larichev, V. and Reznik, G., USSR Academy of Sciences, Vol. 231, No. 5, 1976, pp. 1077-1079.

[13] Flierl, G., Larichev, V., McWilliams, J., and Reznik, G., Dyn. Atm. and Oceans, Vol. 5, 1980, pp. 1-41.

[14] Aref, H., and Siggia, E.D., Journal of Fluid Mechanics, Vol. 109, 1981, pp. 435-463.

[15] Couder, Y., Basdevant, C., and Thome, H., Groupe de Physique des Solides de l'Ecole Normale Superieure and Laboratoire de Meteorologie Dynamique, Ecole Normale Superieure, France (private communication).

Numerical Simulation of Homogeneous Turbulence Submitted to Two Successive Plane Strains and to Solid Body Rotation

P. Roy*

Office National d'Etudes et de Recherches Aerospatiales, Chatillon, France

Abstract

In order to provide new information on fundamental mechanisms of turbulence, we have chosen to simulate numerically two particular constant mean velocity gradient flows: pure strain and solid body rotation. Comparisons are made with available experiments. The numerical method is pseudospectral for space scheme and second-order finite difference for time scheme. We present here full turbulence simulations results, i.e., simulations without any subgrid scale model, obtained with 64^3 discretization points. In the case of two successive plane strains, the main results concern the reorientation process of the tensors (Reynolds stress, pressure velocity correlation, etc.) during the second strain. Rotation effect concerns the transfer rate of energy. Attention has been paid to the spectral domain where linear interactions dominate nonlinear interactions. The visualizations show clearly the shape of this domain.

Introduction

Formulation and Numerical Scheme

We study the flow of an incompressible fluid submitted to uniform strain. Let $\overline{U}$ be the main flow

$$U = \overline{U} + u \qquad \overline{U} = A(t) X$$

$$A(t) = (a_{ij}) = (3, 3) \text{ matrix}$$

Paper presented at the Fourth Beer-Sheva Seminar on MHD Flows and Turbulence, Ben-Gurion University of the Negev, Beer-Sheva, Israel, Feb. 27-March 2, 1984. Copyright © 1985 by the American Institute of Aeronautics and Astronautics, Inc. All rights reserved.
*Aerodynamics Direction Head, Research Group.

NUMERICAL SIMULATION OF HOMOGENEOUS TURBULENCE 175

with $a_{ii}=0$ (incompressible flow).
The velocity U verifies the Navier-Stokes equations. It is easily shown that the turbulent fluctuation u is homogeneous if and only if $(da/dt + A^2)$ is a symmetrical matrix[12]. This restriction means that either the mean flow is irrotational, or the rotation rate and axis are fixed and the irrotational strain is normal to the rotation axis. When we replace U by $\bar{U} + u$, X appears explicitly in the equations. This is not convenient for the use of periodic pseudospectral methods. To get round this difficulty, X is eliminated performing a change of variable and function

$$X = B X'$$
$$u = Bu'$$
with $\frac{dB}{dt} = AB$ $B(o)$ given

The resulting equations for u' and p are

$$\frac{\partial u'}{\partial t} + (u'\cdot\nabla')u' + 2B^{-1}ABu' + B^{-1}\left[B^{-1}\right]^T \nabla'p = \nu\left(\left[B^{-1}\right] \nabla'\right)^2 u'$$

$$\nabla'\cdot u' = 0$$

The change of variable means that we have to compute in a reference frame relative to the variable X', corresponding to a frame in the physical space X moving with the mean flow. To solve the equations, we use the classical pseudospectral method (see Orszag and Patterson[9]). The time scheme is a second-order finite-difference scheme, using quadrature for the diffusion term and leapfrog for the convection term.

$$\hat{u}'^{n+1}_j = -2\Delta t \; (jk'_i \; u'^n_i \; u'^n_j + q_{i1}\hat{u}'^n_j)e^{-\nu\Delta t \; r_{lm}k'_l k'_m}$$

$$- jr_{i1}k'_j \; \hat{p} + \hat{u}^{n-1} \; e^{-2\nu\Delta t r_{lm}k'_l k'_m}$$

$$k'_j \cdot \hat{u}^{n+1}_j = 0$$

with

$$(q_{i1}) = 2 \; (B^{-1}AB)^n \quad \text{and} \quad (r_{ij}) = (B^{-1}[B^{-1}]^T)^n$$

All the simulations have been done with 64 points in each space direction (262, 144 points) using a parallel computer system (a Gould SEL 32/77 minicomputer managing two AP 120 B). (See Fig. 1). The code and the system are described in Ref. 7. They allow numerical simulations at very low cost (the computing time is 75 s per time step).

Numerical Problems

These problems are due to the quasi-Lagrangian method. As the physical frame follows the mean flow, the computing grid is stretched in some directions, shrunk in others, introducing not only a geometric anisotropy in the mesh, but also a numerical anisotropy in all the results. The solution of remeshing has already been used for the shear flow case (see Refs. 10 and 12). In the case of the plane strain, the solution developed at ONERA in collaboration with Dang[2] is the judicious use of predistorted grids ("diamond"-shaped boxes instead of the rectangular boxes usually used). This is done by the choice of B(0) in the change of variables, X-BX'. Not only is the anisotropy induced by the mesh reduced, but the "diamond" shape allows the treatment of higher Reynolds number flows. This has been completely reported by Dang[2]. See the Appendix for details.

In the rotation case, these remarks are not relevant: The frame rotates but is not distorted.

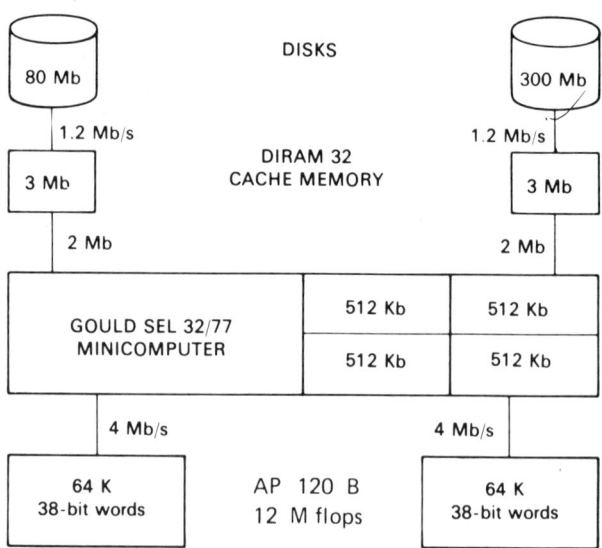

Fig. 1   Parallel computer system.

Initial Conditions

The first step is to create a random velocity field with given square-pulse spectrum[11]. This can be done for any shape of the computing grid [i.e., any matrix $B(0)$]. The obtained field has isotropy property and satisfies the continuity equation. Either two-dimensional or three-dimensional isotropic fields can be created. Another possibility is for the field to be two-dimensional for a range of scales and three-dimensional for another range, to study the interaction between the two kinds of turbulence. If one is interested by the evolution of energy located in a precise spectral domain, the initial condition will be this random field. However, if one is interested by a real turbulent flow, the initial random field must be run until the spectrum becomes smooth and the maximum dissipation rate is reached.

Experiments of Two Successive Strains

Description

This experiment was coducted by Gence[3] in 1979. The apparatus was a wind tunnel made of two distorting ducts (Fig. 2). Their distortion corresponds to a plane strain, normal to the mean flow direction. The connecting section is circular, so it is possible to rotate the second duct and then to modify the principal directions of the second strain. Let α be the angle between the principal directions of the two strains.

This experiment had two main axes: 1) $\alpha=\pi/2$, the study of the possibility of a return to isotropy in presence of a deformation, and 2) $0<\alpha<\pi/2$, the study of the reorientation process of the Reynolds stress tensor (concept of "memory" of turbulence).

Validation of the Simulation

The initial condition is computed in a "diamond"-shaped box. The initial Taylor microscale Reynolds number is equal to 2.5. The strain rate matrix during the first and second strain are respectively

$$A = \begin{pmatrix} 0 & 0 & 0 \\ 0 & D & 0 \\ 0 & 0 & -D \end{pmatrix} \qquad A = \begin{pmatrix} 0 & 0 & 0 \\ 0 & D\cos 2\alpha & D\sin 2\alpha \\ 0 & D\sin 2\alpha & -D\cos 2\alpha \end{pmatrix}$$

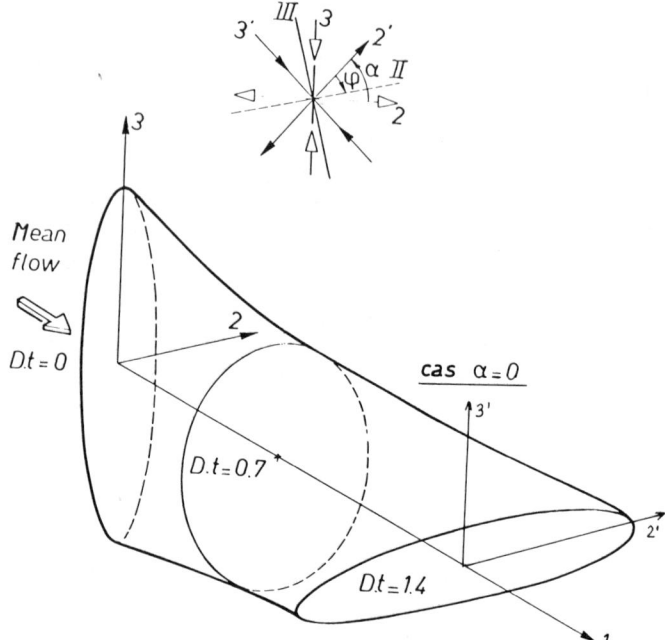

Fig. 2 Gence's experiment wind tunnel (from Ref. 3).

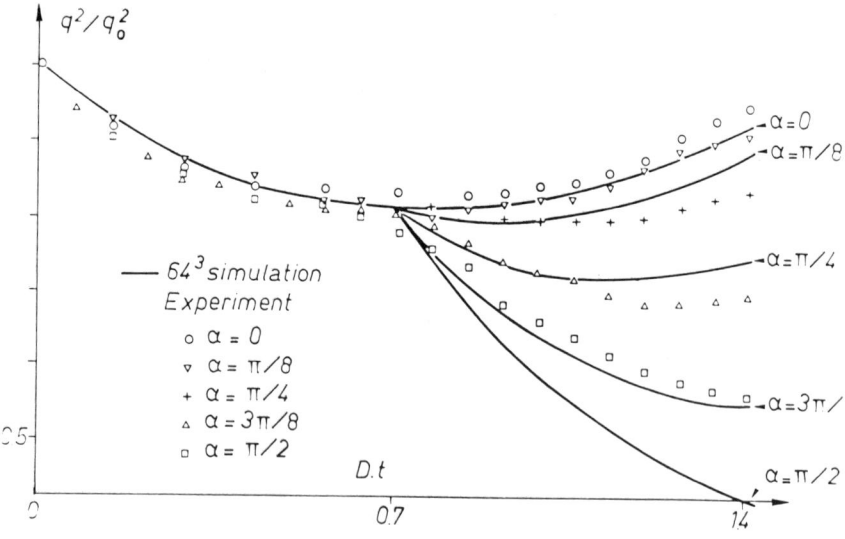

Fig. 3 Time evolution of the energy.

The strain rate D has been first evaluated equalling the ratio of the linear and nonlinear characteristic times ratio with the experimental one:

$$\tau_L/\tau_{NL} = 0.7 \quad \tau_L = 1./D \quad \tau_{NL} = q^2/2\varepsilon$$

$$q^2 = <u_i \cdot u_i> \quad \varepsilon = <u_{i,j} \cdot u_{i,j}>$$

The final choice of D is made adjusting the decay rate of energy with the experimental rate at the end of the first plane strain (Fig. 3). The second strain has been carried for the five values of α: 0, $\pi/8$, $\pi/4$, $3\pi/8$, $\pi/2$.
The comparisons with the experiment show that the decay rate of the energy is greater in the simulation than in the experiment. Seemingly, the main reason is that the Reynolds number of the simulation is 25 instead of 250. Other explanations could be the anisotropy of the experimental initial turbulent flow or its inhomogeneity. We can see, for instance, on Fig. 3 that the effects of the first plane strain on q^2/q^2_0 are different for the different values of α.
These reasons can also explain the discrepancies in the evolution of the structure parameter K (Fig. 4).

$$K = \frac{<u^2_{III}> - <u^2_{II}>}{<u^2_{III}> + <u^2_{II}>}$$

where u_{II} and u_{III} are the components of the turbulent velocity in the directions of the principal axis of the Reynolds stress tensor in the strain plane. Gence[3] has shown by a linear analysis that the slope $[dK/d(Dt)]$ must be equal to zero at the beginning of the second strain for $\alpha = \pi/4$. This is better verified by simulation than by experiment.

Reorientation Process

For $0 < \alpha < \pi/2$, the principal directions of the plane strain and of the Reynolds stress differ when the second strain is imposed. Then, the angle ψ between these two tensors tends to decrease. This evolution of ψ is well followed by the direct simulation (Fig. 5). Furthermore, the simulation also provides the reorientation of two tensors: $<p^{(1)} S_{ij}>$, slow pressure stress tensor, and the deviator of the dissipation tensor.

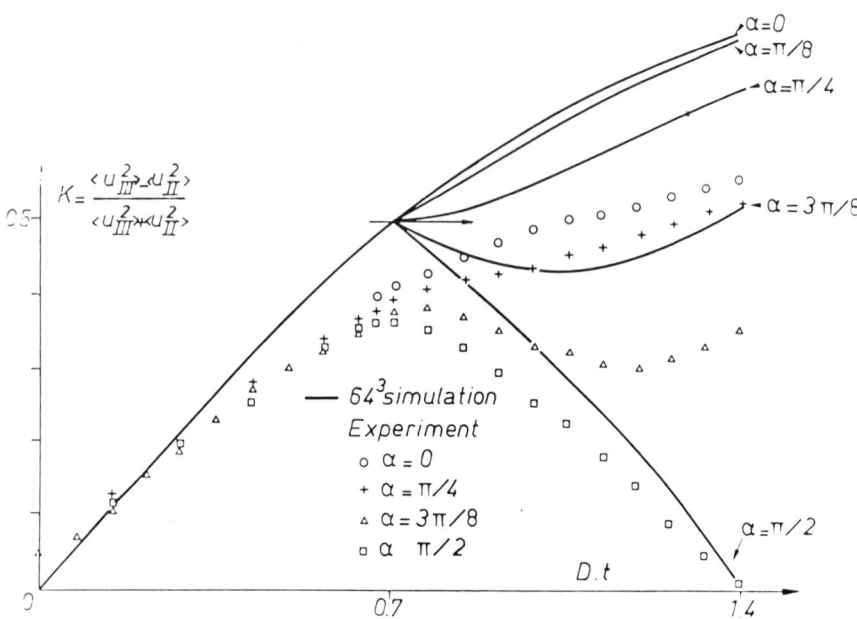

Fig. 4 Time evolution of the structure parameter K.

In Gence's paper[3], it was suggested that $<u_i u_j>$ was aligned with $<p^{(1)} S_{ij}>$. The simulation provides the result that, at low Reynolds number, the alignment is better verified by the tensor ϕ_{ij}:

$$\phi_{ij} = -2 < p^{(1)} S_{ij} >/\varepsilon + (\varepsilon_{ij} - \frac{2\varepsilon \delta_{ij}/3}{\varepsilon}$$

If we study the return to isotropy after the second strain, ψ and ψ'' [ψ'' is the angle between (2',3') and ϕ_{ij}] are stationary while ψ' is still evolving [ψ' is the angle between (2',3') and $<p^{(1)} S_{ij}>$].
This result shows that, as suggested by Lumley and Newman[8], ϕ_{ij} can be related linearly to the anisotropy tensor $b_{ij} = <u_i u_j>/<u_k u_k> - \delta_{ij}/3$).

Visualizations

Some visualizations corresponding to Gence's experiment[3] have been made. Figure 6 is a black-and-white reproduction of color slides. Isovorticity surfaces are represented. The value of the vorticity used as a threshold is the number in the upper right-hand corner ("SEUIL"). It is equal to $2.5 \sqrt{<curlu^2>}$. The different intensities of the grays are related to the distance to the observer position, allowing some deepness effect. In black and white, the closest surfaces are the most black.

Figures 6a through 6e show the isovorticity surfaces at dimensionless times Dt = 0, 0.7, and 1.4 corresponding to the initial condition, between the two strains, and after the second strain for values of the angle 0, $\pi/4, \pi/2$, respectively.

These surfaces are orientated by the strains. For the second plane strain, case $\alpha = \pi/2$, the visualization shows that the flow returned to isotropy, overstepped this state, and then the surfaces became orientated by the second strain, despite all statistic quantities (Reynolds stress tensor spectra, structure parameters, etc.) indicating that the flow is isotropic. This phenomenon could also be observed on the isocorrelation curves provided by Gence[3]. However, we have checked that this effect was not due to a numerical anisotropy problem: A simulation has been carried out starting from the same initial condition, no strain being imposed. After the same period of time, the result (Fig. 6f) shows that in this case, no anisotropy has appeared. So the anisotropy on Fig. 6e is more physical than numerical.

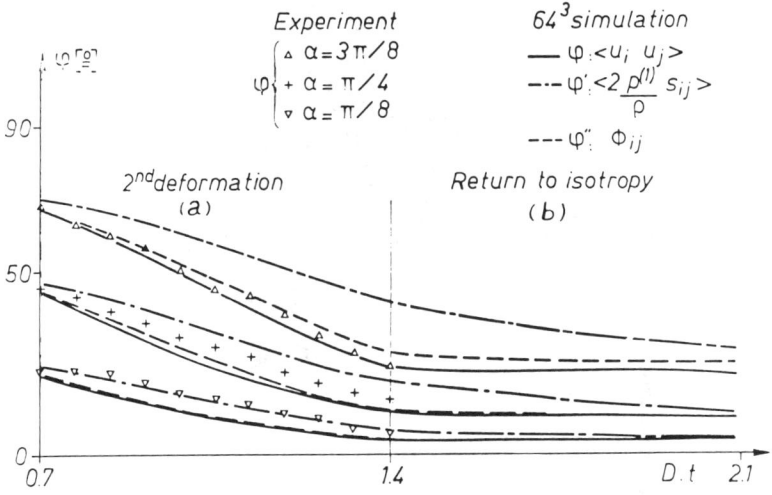

Fig. 5   Time evolution of the reorientation of tensors.

a) **T= 0.00 E+00   SEUIL= 2.79 E-01**

D*T= 0.00 E+00   R*T=

b) **T= 1.73 E-01   SEUIL= 2.62 E+01**

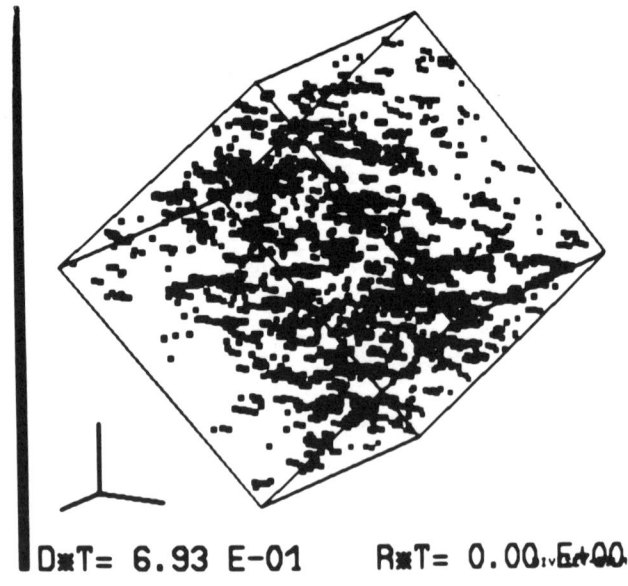

D*T= 6.93 E-01   R*T= 0.00 E+00

(continued)

NUMERICAL SIMULATION OF HOMOGENEOUS TURBULENCE 183

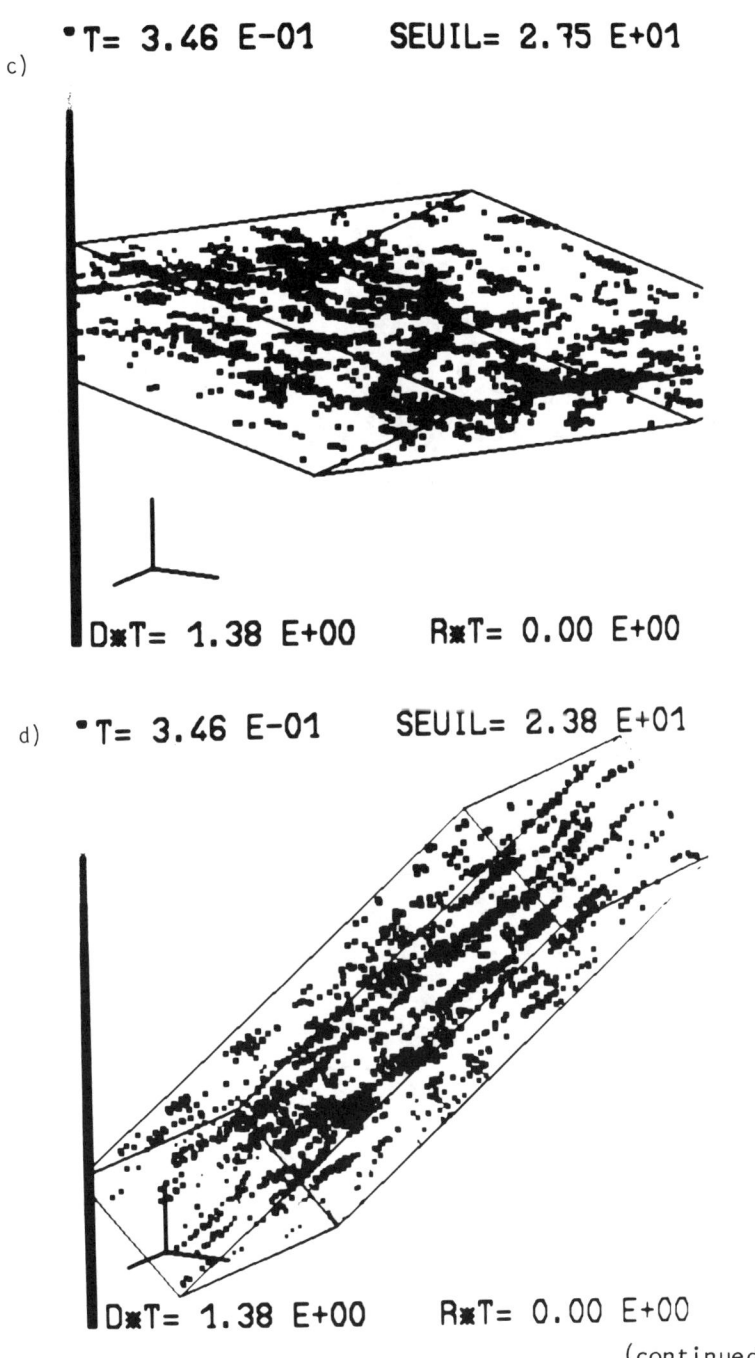

(continued)

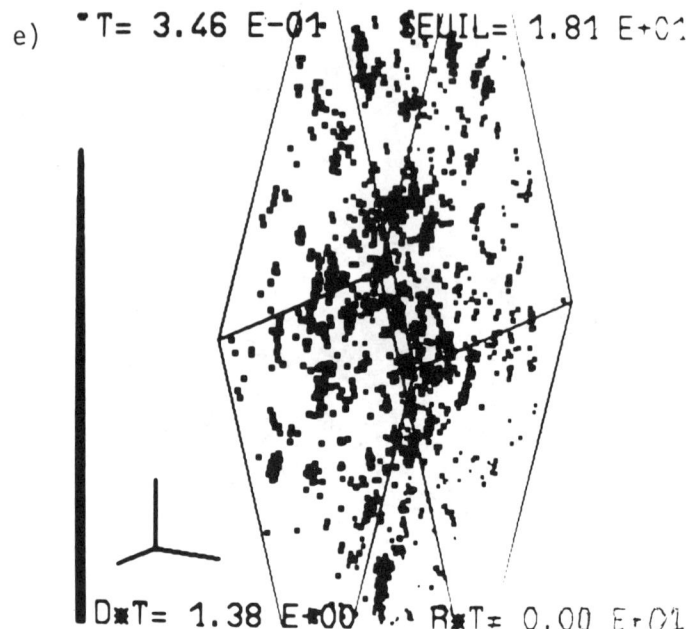

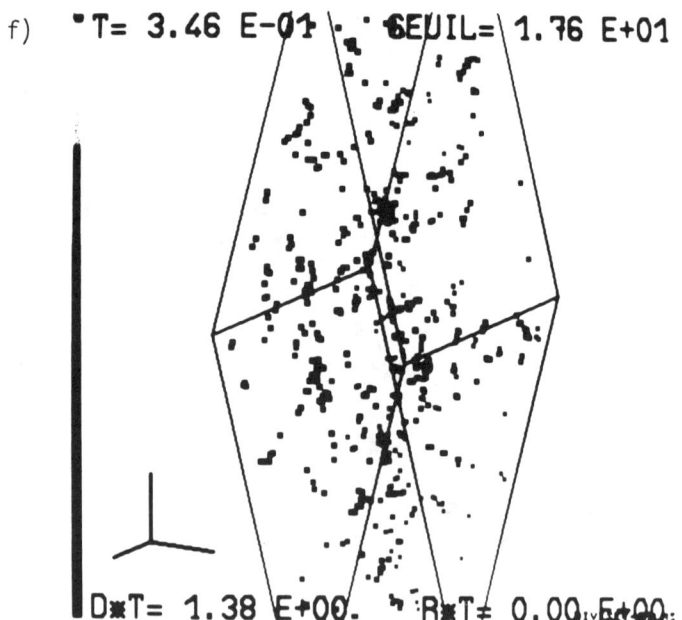

Fig. 6a) Isovorticity surfaces: initial condition. b) Isovorticity surfaces: after the first plane strain. c) Isovorticity surfaces: after the second plane strain ($\alpha = 0$). d) Isovorticity surfaces: after the second plane strain ($\alpha = \pi/4$). e) Isovorticity surfaces: after the second plane strain ($\alpha = \pi/2$). f) Isovorticity surfaces: simulation without strain.

Rotation Experiments

In this case, four experiments have been carried out. In the experiment of Ibbetson and Tritton[6], the walls seem to have too much influence on the results for comparisons with homogeneous turbulence simulations. Traugott[13] and Wigeland and Nagib[14] use similar apparati which seem more compatible with the homogeneity hypothesis. The results of Wigeland and Nagib are more complete, including several mesh sizes, rotation rates, and mean velocity values. However, these results seem to be influenced by an imperfect creation of the turbulent flow. For instance, the decay rate of energy decreases or increases according to the rotation rate. A phenomenology of this variation has been given by Bardina[1]. The last experiment, by Hopfinger et al,[5] is devoted to the influence of three-dimensional perturbations upon a rotating two-dimensional turbulent flow. No statistical results allowing comparisons are given. For these reasons, in this section, no quantitative comparisons with experiments will be given.

Phenomenology of Rotation Effect

Let $\vec{R}$ be the rotation vector. At small Rossby numbers, an analysis can be done to evaluate the spectral domain where rotation has some effect. Plane waves[4,10] $e^{i\vec{k}.\vec{x}}$ propagate with phase speed $2\vec{R}.\vec{k}/\vec{k}$. This linear effect is in competition with the nonlinear interactions.
When the linear characteristic time $(2\vec{R}.\vec{k}/\vec{k})^{-1}$ is much smaller than the nonlinear characteristic time $(kq)^{-1}$, the nonlinear interaction effectiveness is reduced. Let θ be the angle between $\vec{R}$ and $\vec{k}$; This happens if

$$k/\cos \theta \ll 2R/q$$

So linear effects are preponderant for wave number vectors $\vec{k}$ located within two spheres centered on the $\vec{R}$ axis at ordinates $+k^*$ and $-k^*$ with $2k^* \ll 2R/q$.

Effect of Rotation on an Isotropic Turbulent Flow

As for Gence's simulation, the initial condition corresponds to an established isotropic turbulent flow. Then a rotation is imposed to this flow with rotation rate R. This kind of simulation violates the Helmholtz's theorem. But this is the only way to compare different rotation rates. The statistics are compared only after a transitional time, when the flow has incorporated the rotation in its structure.

The rotation rate range is 0-100, allowing the ratio between the linear and the nonlinear times to vary from infinity to 0.03. The Taylor microscale Reynolds number of the initial condition is $R_\lambda = q\lambda/\nu = 32$.
The maximum time is 1.6. The main effects of rotation are described by time evolution of statistical quantities and by means of visualizations.

1) For all rotation rates, the decay rate of energy decreases. The time evolution of q^2/q_0^2 is given for four values of R (0,5,10,50). See Fig. 7.

2) The shape of the energy spectra is modified. In presence of rotation, the slopes of the spectra are much steeper and the peak of energy is situated at higher wave numbers. (k=7 for R=100 instead of k=5 for R=0) (Fig. 8).

3) The Taylor microscales λ and Reynolds number R_λ increase (at t=1.6, $R_\lambda = 26$ for R = 100, $R_\lambda = 15$ for R=0).

4) In the visualizations, the threshold is 2. $\sqrt{<(curl\ u)^2>}$. The rotation modifies the shape of the isovorticity surfaces (Fig. 9). The vorticity is more and more concentrated as R increases. (R=0, 100 for Fig. 9b and 9c, respectively.)

Effect of Rotation on Random Quasi-Two-Dimensional Flows

The simulation of the Wigeland and Nagib experiment[4] confirms the modification of the transfer rate. But no confirmation has been given on the shape of the concerned domain in spectral space. A way of observing it is to take as an initial condition a flow that is the superposition of a random two-dimensional flow and a random isotropic

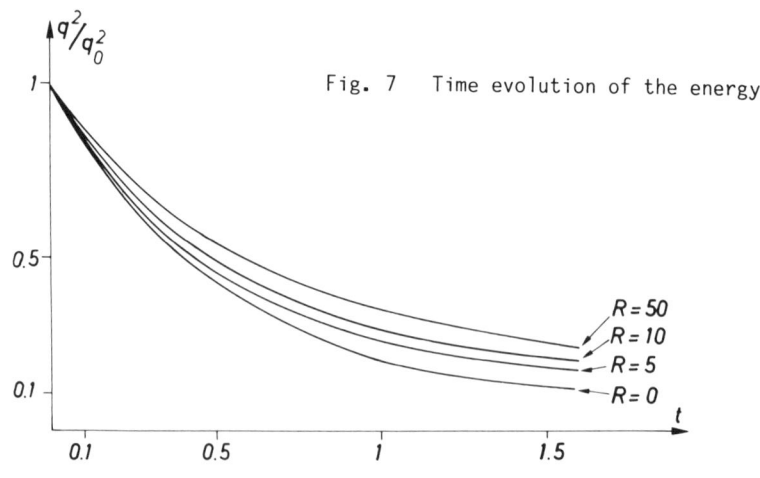

Fig. 7 Time evolution of the energy.

NUMERICAL SIMULATION OF HOMOGENEOUS TURBULENCE 187

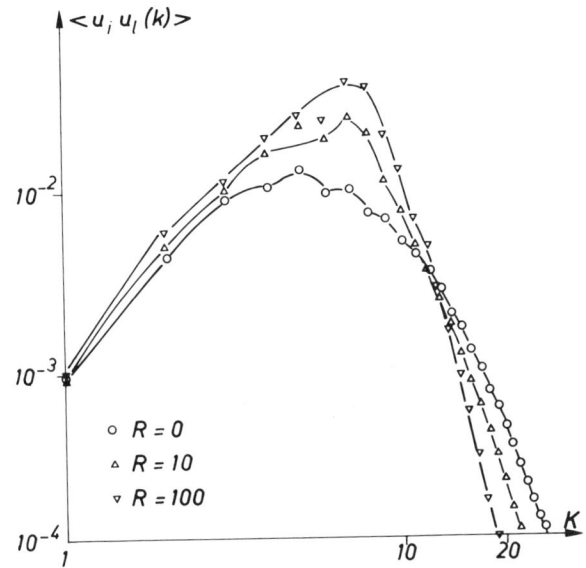

Fig. 8   Energy spectra at T=1.6 for R=0, 10; and 100.

three-dimensional flow. The respective excited special scales ranges are 5 <k<8 and 12<k<15 (Fig. 9a).

We have chosen to restrict the initial modes to k>5 in order to observe the inverse energy cascade related to two-dimensional turbulence. At t=0, there is 15% of three-dimensional energy, i.e. $3 <u_3^2>/q^2 = 0.15$. It is expected that according to the rotation rate (rotation axis normal to the two-dimensional flow), the three-dimensional perturbation will propagate or will be dissipated. The ranges and the energy levels have been chosen to satisfy two criteria:

1) Visualizations must clearly indicate if the flow is evolving toward a two-dimensional or three-dimensional flow.

2) The three-dimensional energy can be dissipated fast enough.

Evolution Toward a Three-Dimensional Flow. When R=0, the three-dimensional energy is first dissipated, its percentage decreases down to 14.9%, but at the same time it is transferred to larger eddies. Then the nonlinear transfer becomes preponderant and the percentage increases. At the time 1.8, its value is 24.3% and the three velocity components spectra are equal when $k \geqslant 13$ (Fig. 10b). This

a)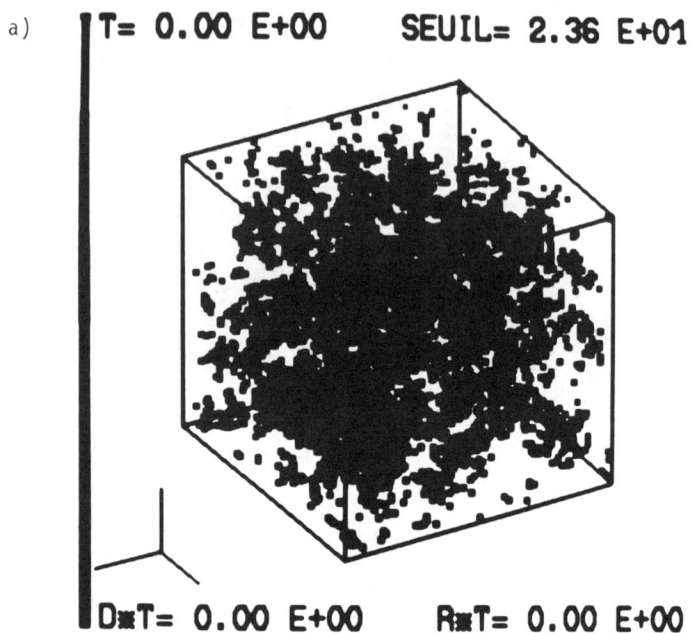

b)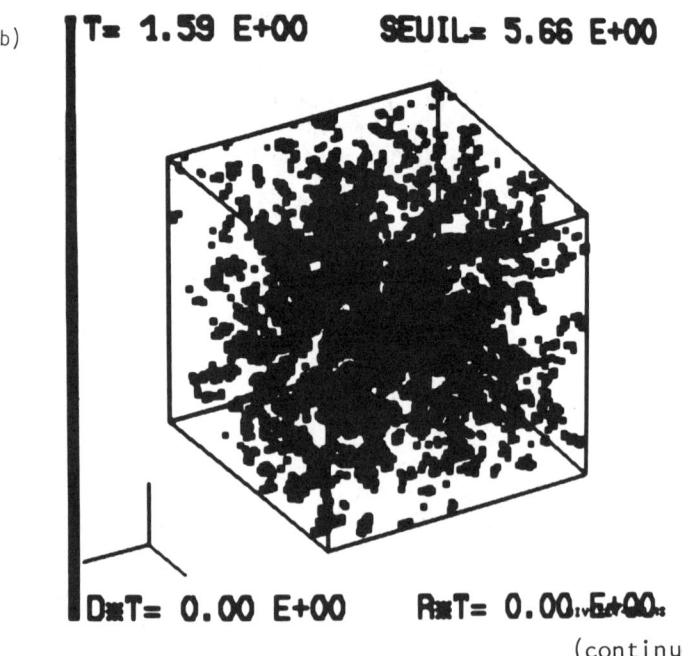

(continued)

NUMERICAL SIMULATION OF HOMOGENEOUS TURBULENCE   189

c)

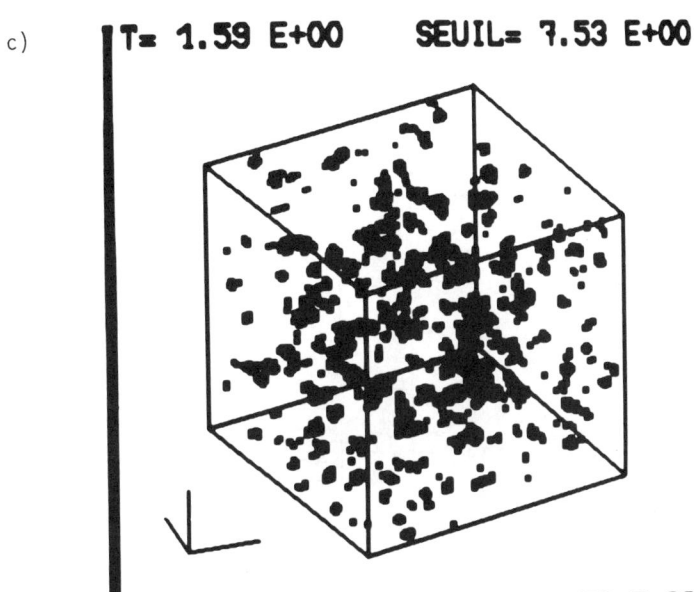

Fig. 9a) Isovorticity surface — the rotation axis is vertical: initial condition T=0. b) Isovorticity surface — the rotation axis is vertical: T=1.6-R=0. c) Isovorticity surface — the rotation axis is vertical: T=1.6-R=100.

simulation has been continued to time 5.4, then the percentage reaches 36% and the velocity is isotropic for $k \geqslant 7$. The visualizations show that the columnar structure of the initial condition has been destroyed (Fig. 11a: t=0; Fig. 11b: t=1.8). This phenomenon is confirmed if we look at the energies in the three spectral planes $k_1=0$, $k_2=0$, $k_3=0$. For the three velocity components, the energy occupies the whole spectral domain (Fig. 12a).

Return to Two-Dimensional Flow. The rotation rates R are 10, 25, 50, 100. In each case, the percentage of three-dimensional energy decreases. At the time 1.8, its value is 1.5% (respectively, 1%, 0.6%, 0.6%) for R=10 (respectively, 25, 50, 100). The visualization at t=1.8 and R=100 shows that the columnar structure has been stabilized (Fig. 11c). More interesting are the energies in the planes $k_i=0$. The spectral domain where the transfer is inhibited appears clearly (Fig. 12b) and its shape is the expected one. We also notice that the blocking is identical for the three velocity components.

a)

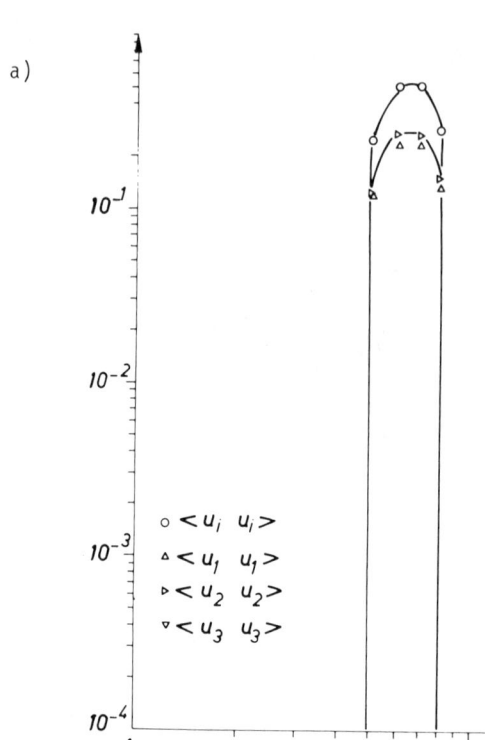

b)

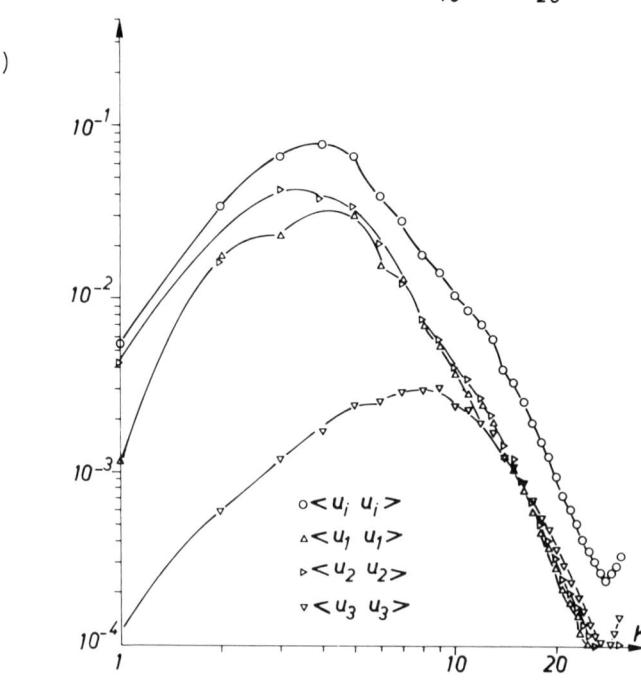

(continued

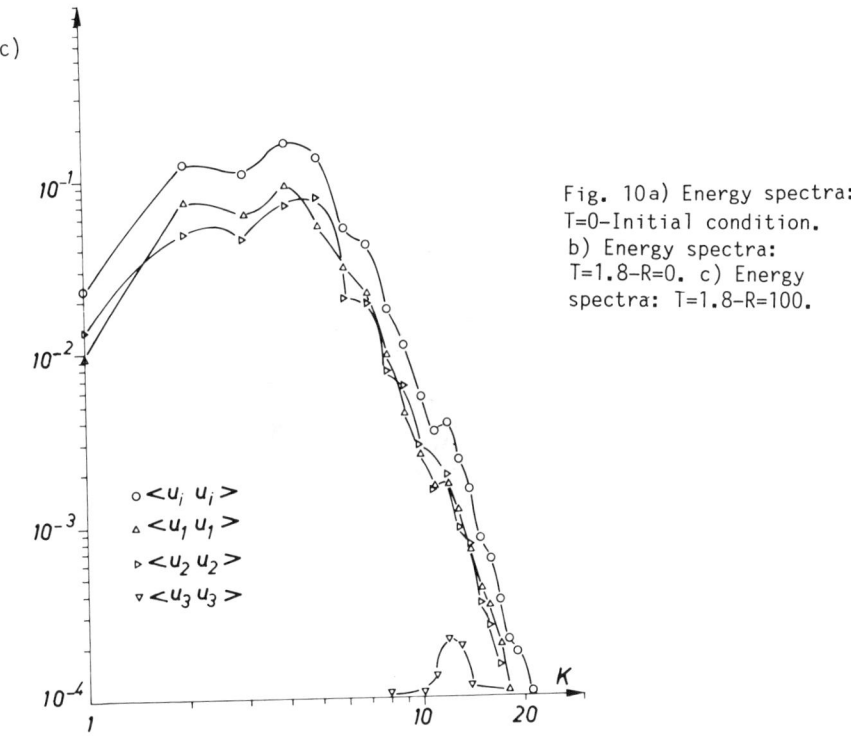

Fig. 10a) Energy spectra: T=0-Initial condition. b) Energy spectra: T=1.8-R=0. c) Energy spectra: T=1.8-R=100.

Conclusion

Three experiments have been studied using full turbulence simulation. The major contributions concern the reorientation process of the Reynolds stress tensor in Gence's experiment[3], and the evidence given for inhibition of the energy transfer by rotation.

We consider that other studies must be done in these cases, to obtain better comprehension of these phenomena:
1) Study of the one-dimensional and three-dimensional transfer energy spectra. 2) Obtain a better separation between the viscous effect and the transfer interactions. This point will soon be solved by repeating some of the previous simulations using either 64x128x128 points FTS or 64^3 points LES with a subgrid scale model settled by Dang[2]. 3) Use new types of visualization to try to study the olitons observed by Hopfinger et al,[5] and the evolution of the velocity phases during roation simulations.

When the rotation effect is better understood, simulations will be done for the superposition of strain and rotation to study their combined effect (shear, for instance).

a) •T= 0.00 E+00   SEUIL= 2.43 E+01

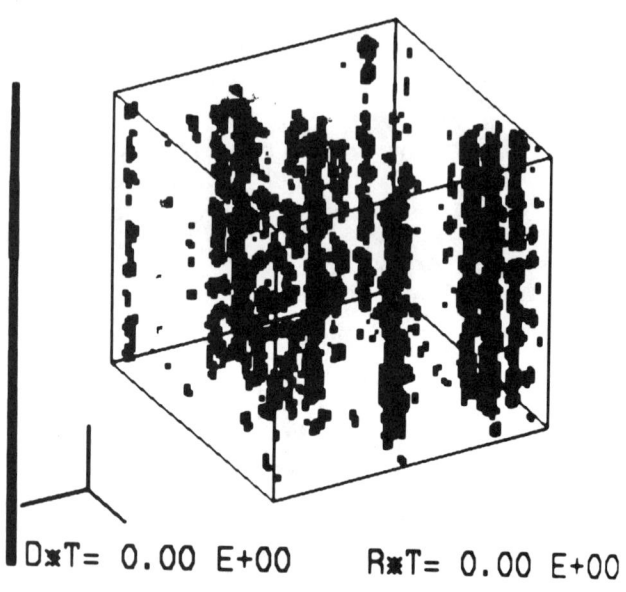

D∗T= 0.00 E+00   R∗T= 0.00 E+00

b) •T= 1.80 E+00   SEUIL= 9.70 E+00

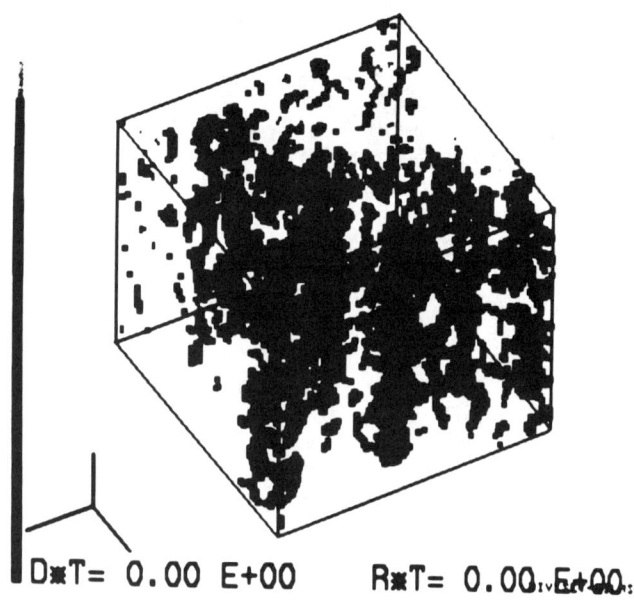

D∗T= 0.00 E+00   R∗T= 0.00 E+00

(continued)

c) **T= 1.80 E+00   SEUIL= 9.20 E+00**

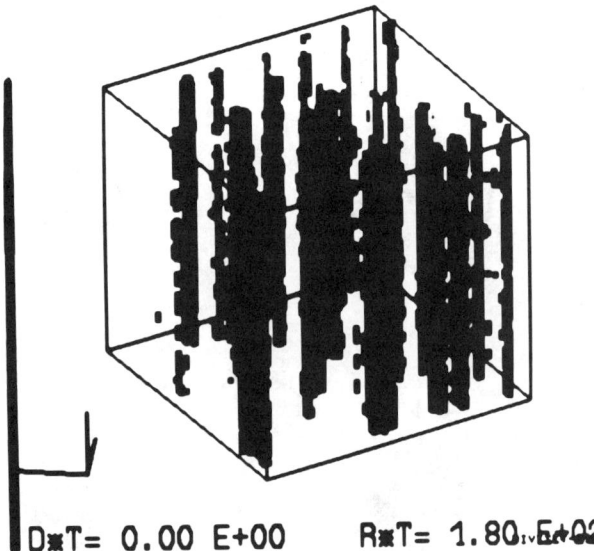

D∗T= 0.00 E+00   R∗T= 1.80 E+02

Fig. 11a) Isovorticity surfaces — the rotation axis is vertical: T=0-initial condition. b) Isovorticity surfaces — the rotation axis is vertical: T=1.8-R=0. c) Isovorticity surfaces — the rotation axis is vertical: T=1.8-R=100.

Appendix

We describe here a numerical experiment that shows the influence of the mesh distortion upon the results, and the advantage of the "diamond-shaped" mesh. For this we simulate the turbulent decay of similar random velocity fields generated on different grids. We use three grids defined by their deformation matrices, B_1, B_2, and B_3:

$$B_1 = \begin{pmatrix} 1 & 0 & 0 \\ 0 & 1 & 0 \\ 0 & 0 & 1 \end{pmatrix} \quad B_2 = \begin{pmatrix} 1 & 0 & 0 \\ 0 & \sqrt{2}/4 & -\sqrt{2}/4 \\ 0 & \sqrt{2} & \sqrt{2} \end{pmatrix}$$

$$B_3 = \begin{pmatrix} 1 & 0 & 0 \\ 0 & 0.5 & 0 \\ 0 & 0 & 2 \end{pmatrix}$$

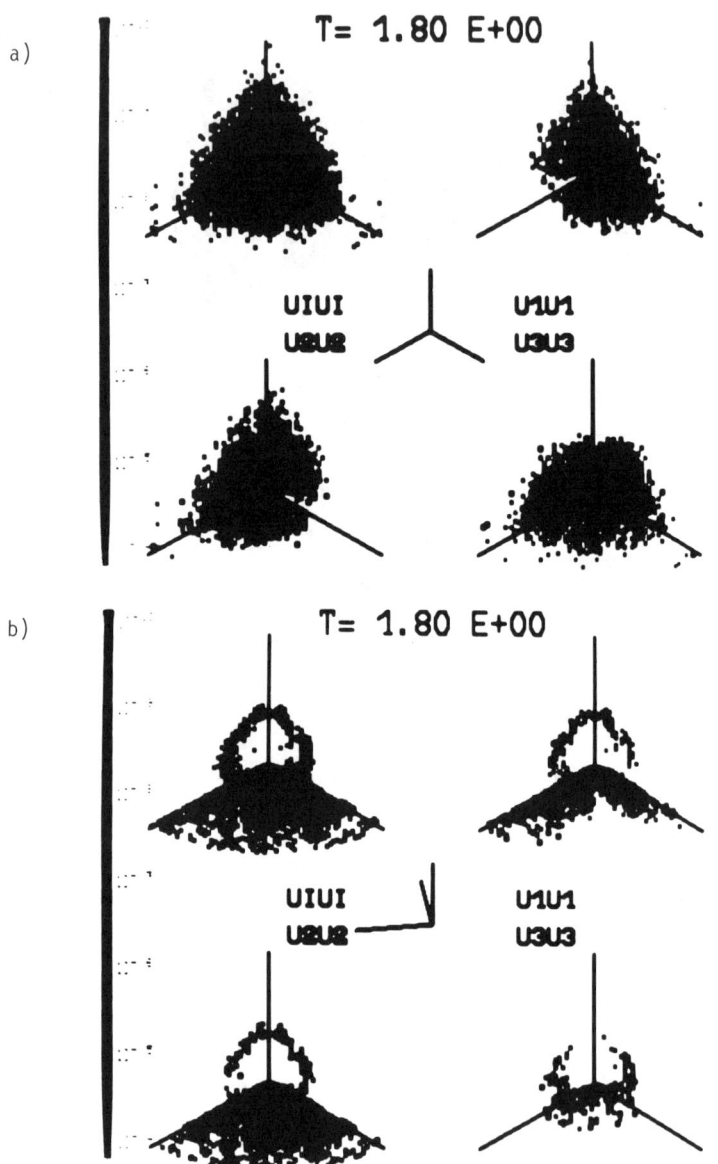

Fig. 12a) Energies in the planes $k_i=0$ (k_3 vertical) — the rotation axis is vertical: R=1.8-R=0. b) Energies in the planes $k_i=0$ (k_3 vertical) — the rotation axis is vertical: T=1.8-R=100.

NUMERICAL SIMULATION OF HOMOGENEOUS TURBULENCE

The matrix B_1 corresponds to the cubic box, and the results computed with B_1 will be considered as the reference ones. B_2 and B_3 have been constructed as follows: They are solutions of the same differential equation at the same time T with respective initial values B_{20} and B_{30}.

$$\frac{dB}{dt} = A\,B \qquad T = (\log 2)/D \qquad D \neq 0$$

$$A = \begin{pmatrix} 0 & 0 & 0 \\ 0 & -D & 0 \\ 0 & 0 & D \end{pmatrix} \qquad B_{20} = \begin{pmatrix} 1 & 0 & 0 \\ 0 & \sqrt{2}/2 & -\sqrt{2}/2 \\ 0 & \sqrt{2}/2 & \sqrt{2}/2 \end{pmatrix}$$

$$B_{30} = \begin{pmatrix} 1 & 0 & 0 \\ 0 & 1 & 0 \\ 0 & 0 & 0 \end{pmatrix}$$

The matrix B_{20} corresponds to a cubic box (matrix B_1) rotated around the first direction with an angle $\pi/4$. So the boxes defined by B_2 and B_3 correspond to the same total irrotational strain applied to B_1, DT= log 2.

In the initial conditions, the modes are excited only for $8 \leq k < 11$ with constant given energy spectra E(k)=0.35. As the computed wave numbers depend on the mesh, the three random generated fields are different but have comparable statistics. For instance, the variation of total energy is less than 2.5×10^{-3}.

For each matrix, 200 time steps have been done (dt = 2.15×10^{-3}). The kinematic viscosity is $\nu = 8 \times 10^{-3}$. In Table 1, we give the results for the value of q^2/q_0^2, ε, z, II, and III.

$q^2 = \langle u_i\, u_i \rangle \qquad\qquad \varepsilon = \nu \langle u_{i,l e}\, u_{i,l e} \rangle$

$Z = \langle \text{rot } u^2 \rangle \qquad\qquad b_{ij} = \langle u_i u_j \rangle / q^2 - \delta_{ij}/3$

$II = b_{ij} b_{ji} \qquad\qquad III = b_{ij} b_{jk} b_{ki}$

We see that the use of B_1 and B_2 gives a relative error on q^2/q_o^2, ε, Z better than 3% to be compared with an estimated error for a 64^3 simulation of the order of 5%. The values of the Reynolds stress tensor invariant II and III show a very good isotropy in the two cases. The use of B_3 gives a much higher distortion on the results of q^2, ε, and Z and a worse isotropy. The energy spectra plotted on Figs. 13a-c furnish the same kind of conclusions. For the

Table 1 Mean value results

Matrix	B_1	B_2	B_3
q^2/q_o^2	0.40	0.41	0.59
ε	2.14	2.07	4.40
Z	11.6	11.4	16.6
II	2.3×10^{-4}	4.2×10^{-5}	5.8×10^{-3}
III	-1.8×10^{-7}	4.2×10^{-8}	-1.7×10^{-4}

a)

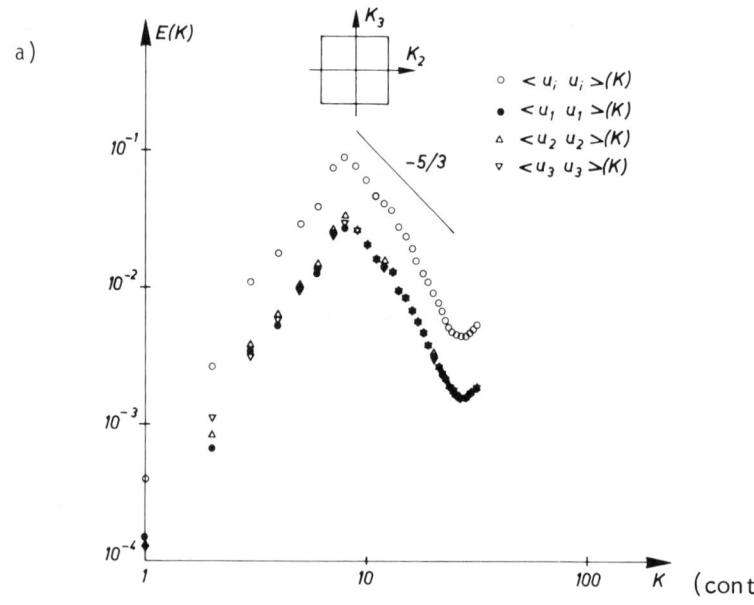

(continue

b)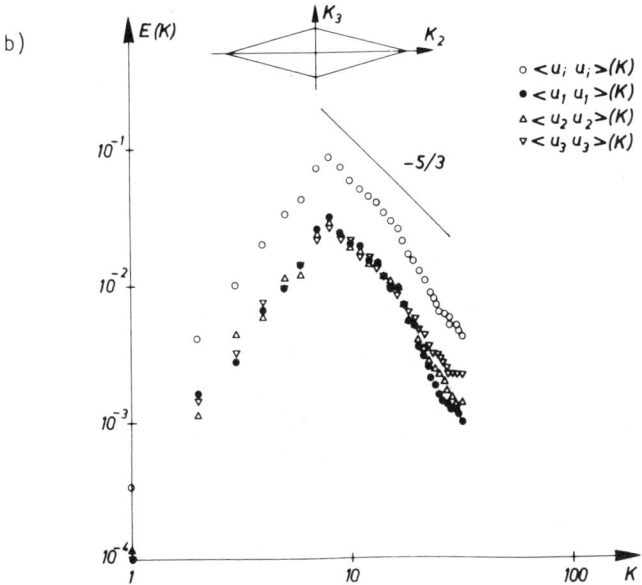

c)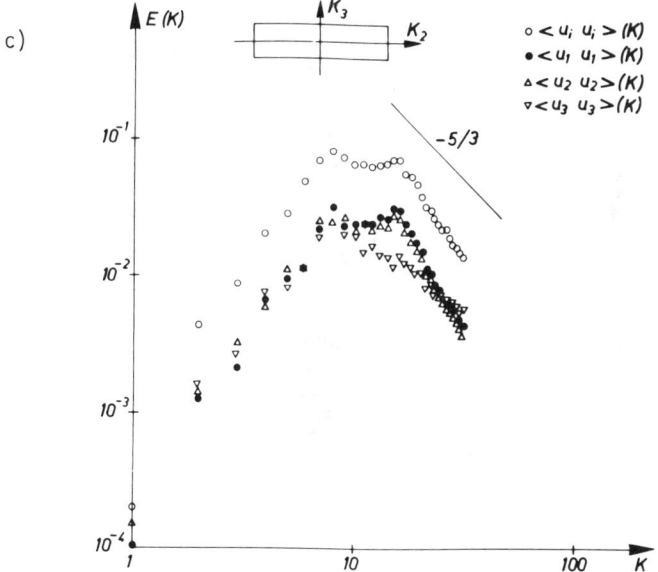

Fig. 13a) Energy spectra (matrix B_1). b) Energy spectra (matrix B_2). c) Energy spectra (matrix B_3).

a)

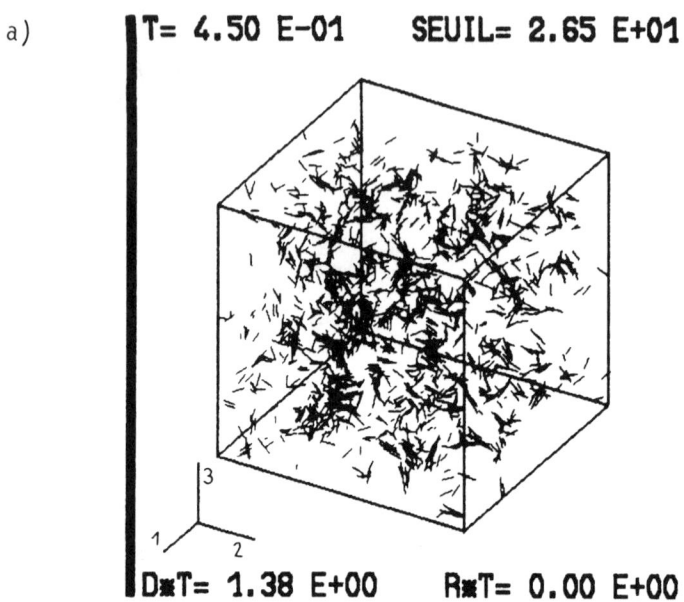

b)

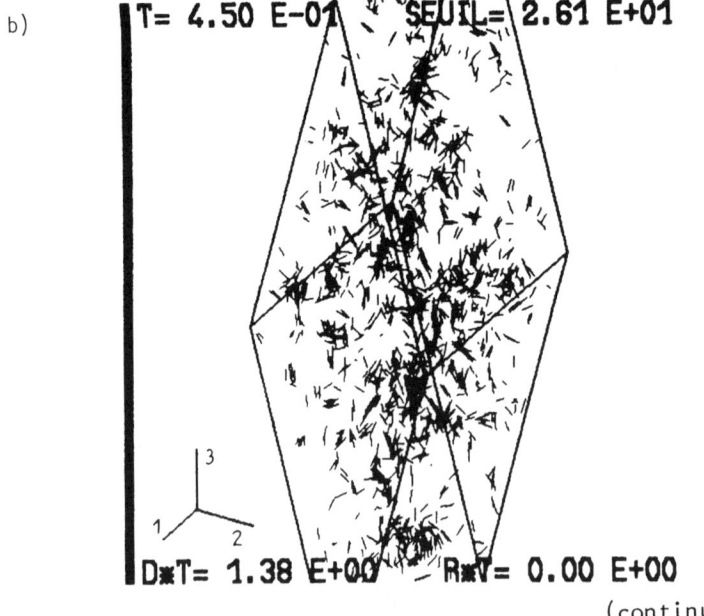

(continued)

c)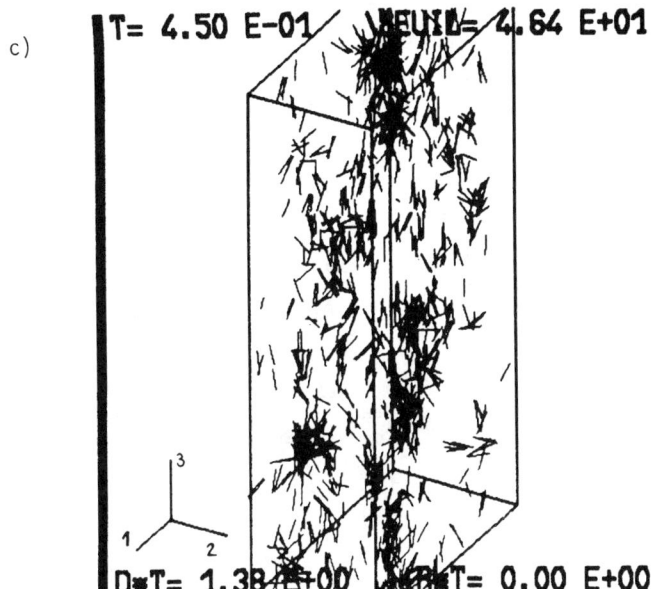

Fig. 14a) Vorticity vectors (matrix B_1) threshold = 2.3<rot u^2>.
b) Vorticity vectors (matrix B_2) threshold = 2.3 <rot u^2>.
c) Vorticity vectors (matrix B_3) threshold = 2.8 <rot u^2>.

matrix B_1, the isotropy is very good at every scale, and the slight accumulation of energy near the cutoff is due to the fact that the whole dissipation range is not included in the computing spectral domain. For the matrix B_2, the isotropy is good up to k=20 (radius of the largest sphere included in the discretization spectral domain) and grows slowly after. For the case B_3, the isotropy is good only if k<10 and there is an accumulation of energy near k=16 (cutoff in the direction 3).
 At last, if we plot vorticity vectors (Fig. 14), they seem to have a privileged direction only for the matrix B_3: A high proportion of vectors seem parallel to the direction 3. In conclusion, for the same irrotational strain, "diamond-shaped" boxes are a more judicious choice than rectangular-shaped boxes.

Acknowledgments

I wish to acknowledge the substantial contribution to this work made by K. Dang and the very efficient collaboration of P. Leca in developing the code on the parallel computer system. I would like to thank C. Basdevant, Y. Morchoisne, and R. Sadourny for their suggestions and contributions to the success of this work, and also thank R. Peyret for helpful comments. A special thanks is also due to Diviset-Films (L.M.D./Lactamme).

References

[1] Bardina, J., Ferziger, J.H., and Reynolds, W.C., "Improved Turbulence Models based on Large Eddy Simulation of Homogeneous Incompressible, Turbulent Flows," Stanford University, Stanford, Calif., Report TF-19, May 1983.

[2] Dang, K., "Evaluation of Simple Subgrid Scale Models for the Simulation of Homogeneous Isotropic and Anisotropic Turbulence," AIAA 16th Fluid and Plasma Dynamics Conference, Danvers, Mass, 12-14 July, 1983, pp. 83-1692.

[3] Gence, J.N., "Action de Deux Deformations Planes Successives sur une Turbulence Isotrope," These de Doctorat d'Etat, Universite Claude Bernard, Lyon, France, 1979.

[4] Greenspan, H.P., The Theory of Rotating Flows, Cambridge University Press, Cambridge, Mass., 1968.

[5] Hopfinger, E.J., Browand, F.K., and Gagne, Y., "Turbulence and Waves in a Rotating Tank," Journal of Fluid Mechanics, Vol. 125, Dec. 1982, pp. 505-534.

[6] Ibbetson, A., and Tritton, D.J., "Experiments on Turbulence in a Rotating Fluid," Journal of Fluid Mechanics, Vol. 68, April 1975, pp. 639-672.

[7] Leca, P. and Roy, P., "Numerical Simulation of Turbulence on Mini-Systems with Attached Processors (in Mono- or Multi-Processor Configuration)," La Recherche Aerospatiale (English and French translations), No. 1983-4, 1983, pp. 107-115.

[8] Lumley, J.L., and Newman, G.R., "The Return to Isotropy of Homogeneous Turbulence," Journal of Fluid Mechanics, Vol. 82, Aug. 1977, pp. 161-178.

[9] Orszag, S.A., and Patterson, G.S., "Numerical Simulation of the Three-Dimensional Homogeneous Isotropic Turbulence," Physical Review Letters, Vol. 28, No. 2, 1972, p. 76.

[10] Rogallo, R.S., "Numerical Experiments in Homogeneous Turbulence," NASA-Ames, Calif., NASA TM 81 315, 1981.

[11] Roy, P., "Solution of Navier-Stokes Equations by a Method of High Accuracy in Space and Time," La Recherche Aerospatiale (French and English translations), No. 1980-6, 1980, pp. 3-15.

[12] Roy, P., "Numerical Simulation of Homogeneous Anisotropic Turbulence," 8th International Conference on Numerical Methods in Fluid

Dynamics, Proceedings, Lecture Notes in Physics, No,170, Springer Verlag, New York, ed. E. Krause, 1982, pp. 440-447.

[13]Traugott, S.C., "Influence of Solid-Body Rotation on Screen Produced Turbulence," NACA TM 4135, 1958.

[14]Wigeland, R.A., and Nagib, H.M., "Grid Generated Turbulence with and without Rotation about the Streamwise Direction," Illinois Institute of Technology, Chicago, ILL, ILT Fluids and Heat Transfer Report, R78-1, 1978.

Sensitivity of Turbulent Channel Flow to the Interactions at the Perimeter

Dan Naot*
Center for Technological Education, Holon, Israel

Abstract

Numerical simulations were performed in order to study the potentiality of the algebraic stress model in integrating the influence of the local interactions at various conditions that may prevail at the channel perimeter emphasizing the variety of the situations in which the flow becomes three-dimensional with apparent effects on the streamwise velocity contours and the wall shear stress distribution.

Introduction

The first use of an algebraic stress model, made by Launder and Ying[1] in a square duct flow, was followed by Naot et al[2] and Reece,[3] who performed conservative calculations of the full Reynolds stress closure equations with no term disregarded and provided a comparative base for the development of simplified models. The algebraic stress model is based on the assumption that the flow is undirectional and that turbulence is in the state of local equilibrium.[4] The transport equations for the turbulent stresses are simplified, and it is possible to express the stresses in terms of the turbulence energy, the dissipation, and the mean velocity gradients. The calculations of channel flow with secondary currents made with this model by Tatchel,[5] Gessner and Emery,[6] and others justified the use of the simplified model, with some reservations made by Naot and Rodi,[7] who modified it to cope with weak lateral velocity gradients.
 The turbulence model used here consists of the following three groups of equations that together form a well-

---

Paper presented at the Fourth Beer-Sheva Seminar on MHD Flows and Turbulence, Ben-Gurion University of the Negev, Beer-Sheva, Israel, Feb. 27-March 2, 1984. Copyright © 1985 by the American Institute of Aeronautics and Astronautics, Inc. All rights reserved.
*Research Associate.

TURBULENT CHANNEL FLOW 203

defined closed parabolic system amenable to the Patankar-Spalding[8] algorithm: 1) momentum and continuity equations governing the three-dimensional mean motion; 2) transport model equations for the turbulent energy and the turbulence dissipation rate; and 3) algebraic stress model specifying the turbulent stresses in terms of the mean velocity gradients.

A full description of the model is given by Naot and Rodi[9] for open channel flow and by Naot and Emrani[10] for the cylindrical closed duct.

Unfortunately, focusing efforts on modeling accurately the interactions inside the flow domain is not enough. Noting that the turbulence energy is mainly produced adjacent to the channel walls, we realize that modeling the turbulent interactions at the channel perimeter are of no lesser importance. The present paper describes efforts made to extend the predictability of the model to describe the hydrodynamic response of channel flow to the following interactions of the turbulent eddies: 1) interactions at a free surface; 2) interactions with roughness discontinuity; and 3) interactions at a gap in the channel wall. Attention will be focused on the flow three-dimensionality, the wall shear stress distribution, and the turbulence energy.

Interaction of the Eddies Near Open Surface

To adjust the turbulence model to open channel flow, it was necessary to consider two additional interactions that are induced by the proximity of a free open surface. Being rejected from the open surface by gravity and surface tension, the turbulent eddies break down, enhancing a process of velocity redistribution combined with high dissipation rate. Following the example of Reece[3] applied to the full Reynolds stress calculations, Naot and Rodi[11] developed a new algebraic stress model that can cope with the velocity redistribution effect. Adopting the suggestion of Hossain and Rodi[12] that expresses the breakdown process via the boundary conditions for the dissipation equation, Naot and Rodi[9] demonstrated the use of the algebraic model for the calculation of secondary currents in open channel flow and obtained reasonable agreement with the experimental data measured in channels with perfect geometry.

Calculated results for open channel and closed square duct are shown in Fig. 1. The use of an identical set-up apart from the inclusion of the free surface interactions shows the sensitivity of the hydrodynamic features to the existence of an open surface. This mainly applies to the location of the streamwise velocity maximum shifted by the strong secondary currents enhanced by the presence of the open surface. The increased dissipation rate shows up in the description of the energy near the free surface. Both effects influence the distribution of the wall shear stress.

Response of Open Channel to Roughness Discontinuity

Since the sand experiments of Nikuradze[13], it is recognized that rough walls are associated with high shear stress and a logarithmic profile that is different from the

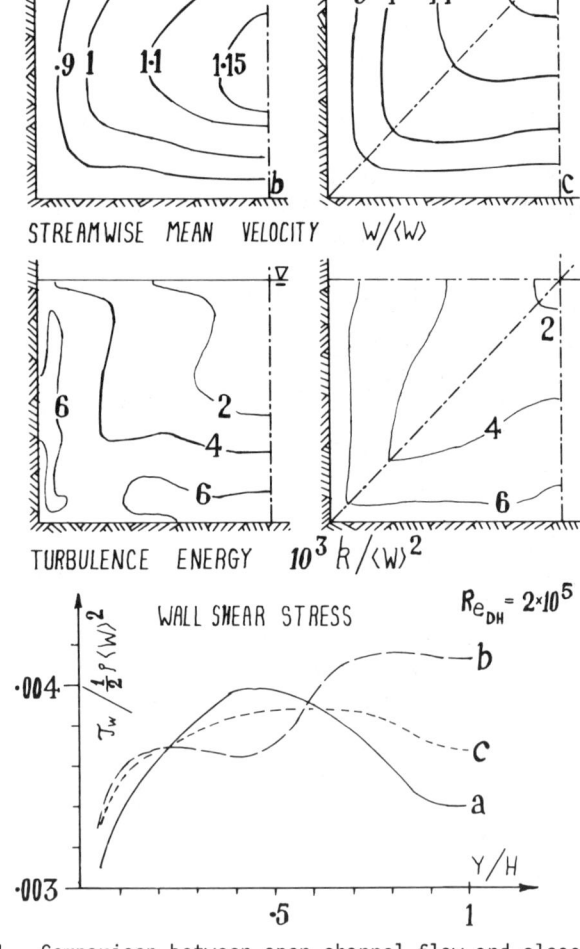

Fig. 1 Comparison between open channel flow and closed square duct flow at Re_{Dh} = 200,000.

profile that characterizes smooth walls. Obviously, the
turbulence energy and the turbulent normal stresses are
also augmented by the roughness. Thus, roughness heterogeneity is expected to induce secondary currents, as was
observed by Hinze[14,15] Grotzbach[16], and others.

To extend the applicability of the algebraic model to
cases with wall roughness heterogeneity, a suggestion was
made to adjust the assumption that the streamwise velocity
at the first grid node follows the logarithmic law and to
introduce a suitable logarithmic law that is the function of
the local roughness height as well as the local shear velocity. The turbulent stresses, the energy, and the dissipation are assumed to show local equilibrium characteristics
and are affected by the roughness height via the local shear
velocity used to calculate them. These assumptions, inspi-
red by the observation of the data of Hua Wang and Nickerson,[17] could be tested only recently when the data of Muller
and Studerus[18,19] became available.

Indeed, with the exclusion of the close vicinity of the
discontinuity essential to avoid local singularity, the
application of these assumptions yields the calculated
results shown in Fig. 2, which compare favorably[20] with the
experimental data.

Typical results for open channel flow with bed roughness discontinuity studied by Naot[20] are shown in Fig. 3.
Although the effect of the bed roughness distribution on the
mean streamwise velocity is less pronounced, the effect on
the secondary currents is dramatic, suggesting a possible
explanation for the differences between the secondary
currents measured by Chiu et al.[21,22] in real operating
channels and those calculated[9] for homogeneously smooth
channels.

Response of Semiopen Duct Flow to Roughness Heterogeneity

The influence of roughness discontinuity at the corners
of cylindrical semiopen subchannels shown by Naot and Em-
Rani[10] is further demonstrated in Figs. 4 and 5 for two
ducts, one with o.d. = 2 x i.d. and 12 fins and the other
with o.d. = 4/3 x i.d. and 4 fins. Two situations are compared for each duct, one with rough external tube and smooth
fin and the other with rough fin and smooth tube. The
pronounced difference in the secondary currents structure
is due to the formation of a relatively stagnant zone adjacent to the rough walls accompanied by the ejection of the
fluid toward the smooth walls. The reflection of the secondary currents structure on the distribution of wall shear
stress at the smooth walls is shown in Fig. 6; note that

lateral flow toward the wall convects high streamwise momentum and enhances the wall shear stress.

Response of a Wide Closed Duct Flow to a Gap

Finally, the possibility of inducing lateral motion by a discontinuity in the boundary conditions due to a gap between the upper end of the vertical wall and the channel ceiling discussed by Emrani and Naot[23] is further illustrated in Fig. 7 for a wide duct with dimensions ratio of 5.2 x 1. The gap is simulated by a symmetry condition that applies zero shear stress, permits the existence of turbulence energy convected to the gap, and allows a lateral flow parallel to the gap. Generally, a perfect corner induces two symmetric vortices. However, now that a part of the

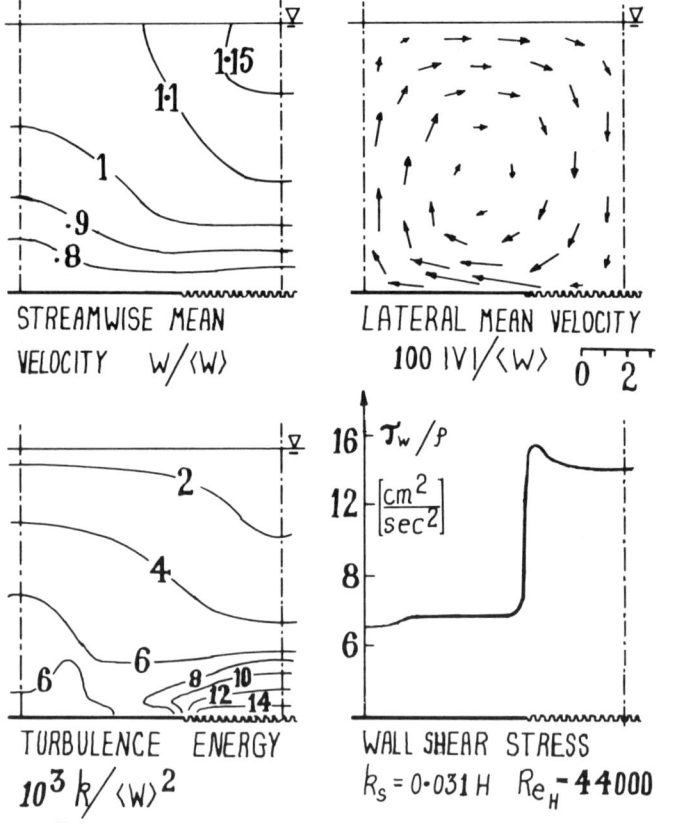

Fig. 2 Response of wide-open channel flow to periodic roughness steps at the channel bed at Re_{Dh} = 176,000.

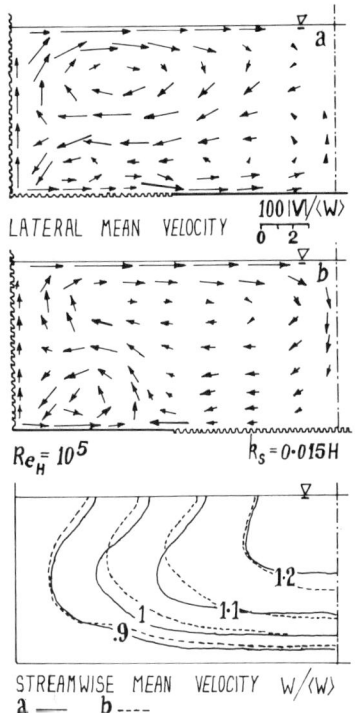

Fig. 3 Response of a 4 × 1 open channel flow to roughness steps at the channel bed at Re_{Dh} = 267,000.

wall is missing, lateral motion is induced flowing from the wall to the gap. In consequence, the vortex adjacent to the wall shrinks, the vortex near the channel ceiling is substantially enlarged, and the shear stress shown in Fig. 7 is enhanced close to the end of the shortened vertical wall.

Discussion

The representative situations here demonstrated where the flow becomes three dimensional, affecting the streamwise velocity contours, and the wall shear stress distribution shows the sensitivity of channel flow to the interactions that take place along its perimeter. The main question raised in this context is about the motivation to deal with such theoretical exercises.

It is believed that properly posed simulations may define the uncertainty margin in the description of the hydrodynamic parameters, which stems from an inability to specify the boundary conditions in particular cases, and justify the use of a simpler model for such a situation.

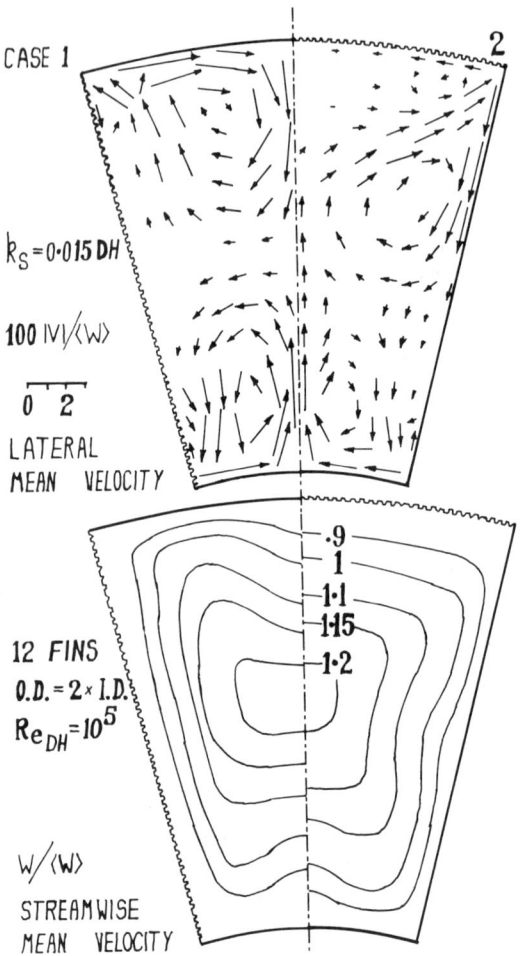

Fig. 4 Response of a semiopen cylindrical channel with 12 fins and o.d. = 2 × i.d. to roughness discontinuity at corners at Re_{Dh} = 100,000.

With a similar approach coupled with the application of local Reynolds analogy, it is possible to estimate the uncertainty margins in the cooling capability for safety studies associated with local damage that is due to partial loss of cooling.

Research aimed at studying sediments transport, organic growth, and other processes that change the surface structure may use properly designed simulations in order to estimate the feedback mechanism of these processes on the flow, which, in turn, control their propagation.

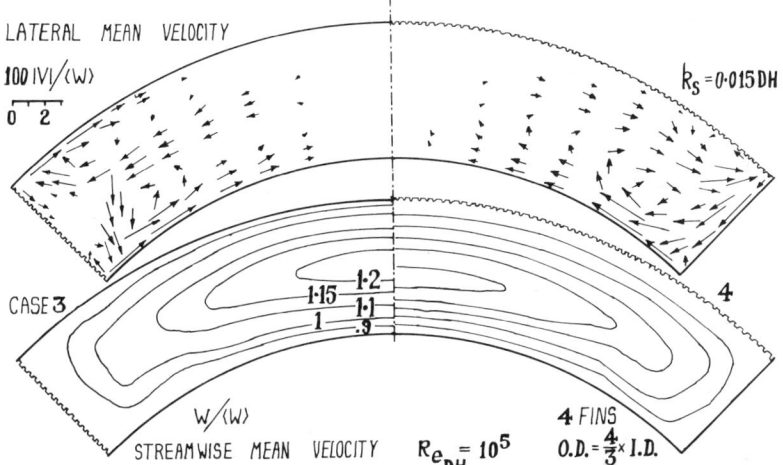

Fig. 5 Response of a semiopen cylindrical channel with 4 fins and o.d. = 4/3 × i.d. to roughness discontinuity at corners at Re_{Dh} = 100,000.

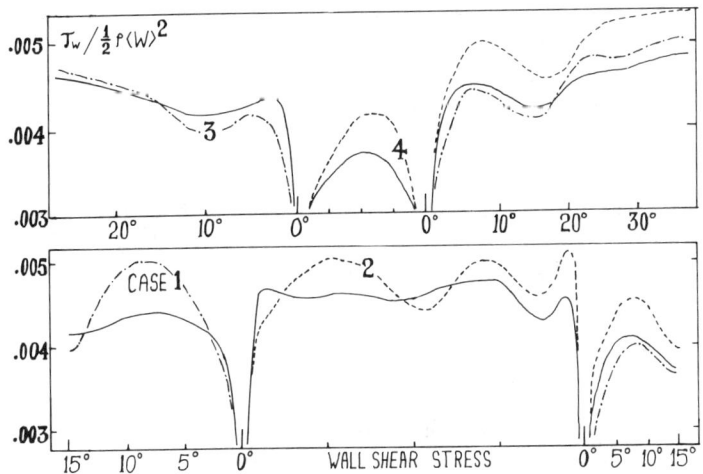

Fig. 6 Shear stress at the smooth walls compared with stress calculated for homogeneously smooth channel (———).

It also seems that the efforts made to improve the modeling of turbulence continuously intrigued by modern experiments should be balanced, as there is no point in improving the prediction capability of internal interactions without devoting similar efforts to the interactions at the channel perimeter.

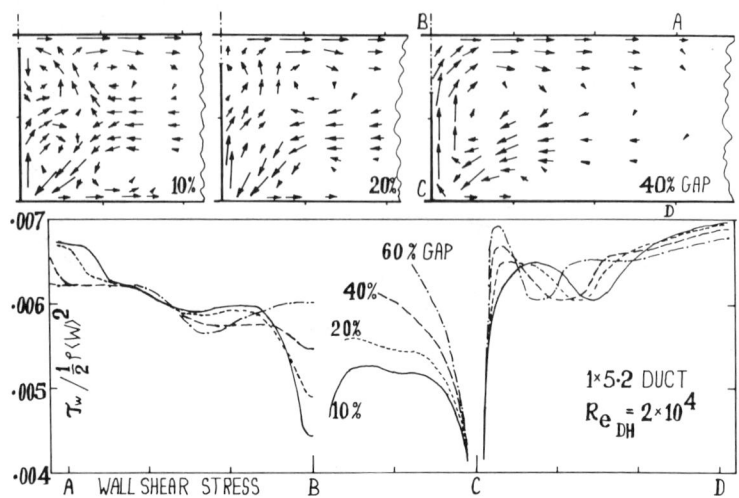

Fig. 7   Response of a wide 5.2 × 1 closed duct flow to the existence of a gap at the corner at Re_{Dh} = 20,000.

Finally, it is important the community of the MHD researchers be aware of the possibility for estimating the contributions of the surface interactions that are not induced by magnetic forces to their findings.

References

1.   Launder, B.E. and Ying, W.M., "Prediction of Flow and Heat Transfer in Ducts of Square Cross Section", Proceedings of the Institution of Mechanical Engineering, Vol. 187, 37/73, 1973, pp. 455-461.

2.   Naot, D., Shavit, A., and Wolfshtein, M., "Numerical Calculations of Reynolds Stresses in Square Duct with Secondary Flow," Warme und Stoffubertrugung, Vol. 7, No. 3, 1974, pp. 151-161.

3.   Reece, G.J., "A Generalized Reynolds Stress Model of Turbulence", Ph.D. Thesis, University of London, Imperial College, England, 1976.

4.   Naot, D., "Two-Dimensional Unidirectional Turbulent Flow in a Local Equilibrium," Physics of Fluids, Vol. 18, No. 12, December 1975, pp. 1813-1814.

5.   Tatchell, D.G., "Convective Processes in Confined Three-Dimensional Boundary Layers," Ph.D. Thesis, University of London, Imperial College, England, 1975.

6.   Gessner, F.B. and Emery, A.F., "The Numerical Prediction of Developing Turbulent Flow in Rectangular Ducts," Proceedings: Turbulent Shear Flow 2, Imperial College, London, England, July 1979.

7. Naot, D. and Rodi, W., "Applicability of Algebraic Models Based on Unidirectional Flow to Duct Flow with Lateral Motion," International Journal of Numerical Methods in Fluids, Vol. 1, No. 3, July-September 1981, pp. 225-235.

8. Patankar, S.V. and Spalding, D.B., "Calculation Procedure for Heat Mass and Momentum Transfer in Three-Dimensional Flow," International Journal of Heat and Mass Transfer, Vol. 15, 1972, pp. 1787-1806.

9. Naot, D. and Rodi, W., "Calculation of Secondary Currents in Open Channel Flow," Journal of the Hydraulic Division, Vol. 108, No. HY8, August 1982, pp. 948-968.

10. Naot, D. and Emrani, S., "Numerical Simulation of the Hydrodynamic Behaviour of Fuel Rod with Longitudinal Cooling Fins," Nuclear Engineering and Design, Vol. 73, 1982, pp. 319-329.

11. Naot, D. and Rodi, W., "Interaction of the Turbulent Eddies With Free Surface," Liquid Metal Flow and MHD: AIAA Progress in Astronautics and Aeronautics, Vol. 84, edited by Branover, Lykoudis and Yakhot, AIAA New York, 1983, pp. 98-112.

12. Hossain, M.S. and Rodi, W., "Mathematical Modelling of Vertical Mixing in Stratified Channel Flow," Proceedings of the Second Symposium on Stratified Flows, Trondheim, Norway, 1980.

13. Nikuradze, I. (1933), Boundary Layers Theory, 4th ed., edited by H. Schlichting, McGraw-Hill, New York, 1960, pp. 523.

14. Hinze, J.O., "Secondary Currents in Wall Turbulence," Physics of Fluids, (Suppl.), Vol. 10, Pt. II, 1967, p. S-166.

15. Hinze, J.O., "Experimental Investigation on Secondary Currents in the Turbulent Flow through a Straight Conduit," Applied Scientific Research, Vol. 28, December 1973, pp. 453-465.

16. Grotzbach, G., "Direct Numerical Simulation of Secondary Currents in Turbulent Channel Flow," Structure and Mechanisms of Turbulence II, edited by H. Fiedler, Springer-Verlag, Berlin, 1978.

17. Hua Wang and Nickerson, E.C., "Response of Turbulent Boundary Layer to Lateral Roughness Discontinuities," College of Engineering, Colorado State University, Fort Collins, Colo., Rept. THEMIS-18, 1972.

18. Muller, A. and Studerus, X., "Secondary Flow in Open Channel," Proceedings of the XVIII Congress of the International Association for Hydraulic Research, Italy, 1979, pp. 19-24.

19. Studerus, X., "Sekundarstromungen im offenen Gerinne uber Rauhen Langsstreifen," Institut fur Hydromechanik und Wasserwirtschaft, Edgenossische Technische Hochschle, Zurich, Rept. R19-82, May 1982.

20. Naot, D., "Response of Channel Flow to Roughness Heterogeneity," ASCE Journal of Hydraulic Engineering, Vol. 110, No. 11, November 1984, pp. 1568-1587.

21. Chiu, C.L., Hsiung, D.E., and Lin, H.C., "Three-Dimensional Open Channel Flow," Journal of Hydraulic Division, Vol. 104, No. HY8, August 1978, pp. 1119-1136.

22. Chiu, C.L., Hsiung, D.E., and Lin, R.C.H., "Secondary Currents Turbulence in Open Channels of Various Geometrical Shape," Proceedings of the XVIII Congress of the International Association for Hydraulic Research, Italy, 1979, pp. 47-54.

23. Emrani, S. and Naot, D., "Hydrodynamic Effects of the Gap Between the Fuel Rod Longitudinal Fins and the Tube," Transactions of the Israel Nuclear Society, Vol. 10, Tel Aviv, December 1982, pp. 121-123.

Electrogasdynamic and Kinetic Phenomena in Diffuse Electrical Discharges in Turbulent Gas Flows

Yury L. Khait*
Ben-Gurion University of the Negev, Beer-Sheva, Israel

Abstract

A review is presented of recent developments in the theory of electrogasdynamic and kinetic phenomena in diffuse electrical gas discharges (e.g., coronas) in turbulent flows (turbulent gas diffuse discharges -- TGDD) at pressures from a few hundred torrs to a few atmospheres, which are presently of interest due to their potential for practical applications in lasers, electrogasdynamic energy generators, low-temperature plasma technology, etc. These developments in the TGDD theory suggest interpretations for various interesting experimentally observable effects that were not quite understood before: 1) A sharp increase in electrical power input and currents (of the order of 10^2–10^3 times the nonflow value) in reaching modes without breaking of the discharge stability, 2) rapid nonequilibrium nonstationary gas-to-plasma transitions (GPT) in gas portions passing speedily through the discharge zone, 3) considerable smoothing of discharge spatial and temporal nonuniformities, etc. The TGDD theory is considered on the two levels: 1) the phenomenological one with help of combining of the Kolmgorov-Landau theory of turbulence with semiquantitative models of electrical discharges, and 2) the statistical kinetic one with help of nonstationary statistical kinetics and nonequilibrium statistical physics. Estimates of various important TGDD parameters are presented: characteristic lengths and time intervals, transport coefficients, currents and power inputs, etc.

Paper presented at the Fourth Beer-Sheva Seminar on MHD Flows and Turbulence, Ben-Gurion University of the Negev, Beer-Sheva, Israel, Feb. 27-March 2, 1984. Copyright © 1985 by the American Institute of Aeronautics and Astronautics, Inc. All rights reserved.
*Professor, Department of Physics.

Introduction

Turbulent gas diffuse discharges (TGDD), e.g., coronas, existing in hydrodynamically turbulized gas flows, are of current interest due to their potential for various practical applications to electric discharge convection lasers, low-temperature plasma technology and plasma chemistry, electrogasdynamic energy generators, etc.[1-20] Studies and applications of TGDD phenomena constitute a new interdisciplinary field-turbulent electrogasdynamics, which is associated with a new type of phenomena characterized by interactions between turbulence and electric gas discharge phenomena. But magnetic fields can be neglected in this case due to relatively small currents involved.[2]

The TGDD phenomena are characterized by close interlacing of hydrodynamic, discharge, and kinetic phenomena. Due to this interlacing, the TGDD theory is an interdisciplinary one and involves the theory of turbulence, models of electrical gas discharges, plasma physics, nonequilibrium statistical physics, kinetics, etc. In this work, we shall follow mainly Refs. 2,10,17,19, and 20 in the case of phenomenological TGDD treatment and Refs. 4,10, 12, and 19-22 in the statistical kinetic consideration of the TGDD phenomena. The approaches and results presented in these references lead to a better understanding of observations and suggest explanations for some experimental data that have not been quite understood before. In connection with the theoretical treatment of the TGDD phenomena, the following remarks should be made. The consideration of TGDD's on the phenomenological level is based on combining of the Kolmogorov-Landau theory of turbulence[23] with the semiquantitative theory of electrical discharges.[24,25] Hence, one can hardly expect to solve the problems considered exactly. We shall resort, therefore, to using the approximation analysis and similitude theory to obtain qualitative and semiquantitative results for the main features of the problem and to come to experimentally observable results.[2,4,10,12,17,19-21] While using nonequilibrium statistical physics and nonstationary statistical kinetics, we also bear in mind that they are far from being accomplished. Nevertheless, this approach enables one to consider nonstationary transient phenomena associated with the formation of electron avalanches and plasma in portions of the turbulent gas passing rapidly through the discharge zone during a short residence time $\Delta\tau$ (e.g., $\Delta\tau \approx 10^{-5}-10^{-4}$) in the regions of a strong electric field,[4,10,12,19-22] As a result, one can estimate various parameters of the gas that are changed at high rates and cause

DIFFUSE ELECTRICAL DISCHARGES IN GAS FLOWS 215

dramatic nonstationary alterations in the properties of the gas passing through the discharge. Some of these parameters can be compared directly with observations, others can help to understand and estimate better experimental facts.[2,4,10,12,19-22]

When portions of a neutral gas pass speedily (with velocity v) through a region of characteristic length L_a with a strong electric field of strength E_a, they are affected by this field during the residence time $\Delta\tau_a \approx L_a/v$. In this case, the following phenomena associated with the interaction of gasdynamic and discharge processes occur:

1) Gas portions passing through the electric field E_a during $\Delta\tau_a$ (e.g., during $\Delta\tau_a \approx 10^{-5}$-$10^{-6}$s) experience rapid nonequilibrium nonstationary gas-to-plasma transitions (GPT) and become a partly ionized plasma characterized by a degree of ionization $a(r,t) \ll 1$ (e.g., $a \approx 10^{-4}$-10^{-3}), by a sharp increase in electron (and ion) concentration up to[4,10,12,19-22]

$$n_e(r,t) = a(r,t)N(r,t) = a(r,t)\ P/kT \qquad (1)$$

and by the corresponding sharp decrease in the Debys radius l_d, which satisfies conditions

$$l_d(r,t) = \left[\frac{\bar{\varepsilon}_e(r,t)}{4\pi a(r,t)\ N(r,t)}\right]^{1/2} \ll L_d \qquad (2)$$

where

$$N(r,t) = P(r,t)/kT = \bar{N}(r,t)+\Delta N(r,t); \bar{N}=\bar{P}/kT \text{ and } P=\bar{P}+\Delta P \qquad (3)$$

is the total (and mean) gas particle number density and pressure in the accompanying moving frame (with coordinate r and time t) that suffer random fluctuations ΔN and ΔP near mean values $\bar{N}$ and $\bar{P}$. Equation (2) shows that the ionized gas in the region of the characteristic length L_a can be treated as plasma. The duration $\theta^{(g)}$ of the length L_a depends strongly on E and P and can be shorter than $\Delta\tau_a$ when E/P is not small,[4,10,12,19-22] as shown in Sec. IV.

2) The GPT causes drastic and quick changes of various physical, chemical, electrical, optical, and other properties of the gas that are similar to those taking place in pulse gas discharges of duration $\Delta\tau_a$.[2,4,10,12,19-22] Therefore, the GPT in the turbulent gas in the accompanying frame can be described similarly to processes in pulse gas discharges, as will be shown in Sec. II.[4,10,12,19-22]

3) The GPT causes a sharp and fast increase in the energy of interaction between charged particles (per 1 cm^3)

$$\Delta e_p \sim n_e^2/\Delta r \sim e^2 \, (a \, N)^{4/3} \text{ with } \Delta r \sim n_e^{-1/3} \approx (a \, N)^{-1/3} \quad (4)$$

due to the rapid decrease in distances Δr between charged particles. Here $e=4,8 \cdot 10^{-10}$ CGSE is the electron charge. The energy Δe_p can become of the same order of magnitude or even larger than mechanical energy (per 1 cm^3) of ions of mass M_i associated with the gas flow $\Delta e_v \approx 0.5 \cdot aN \cdot M_i \cdot v_i^2$. The energy $E^2/4\pi$ of the electric field can also become large enough compared to Δe_v. Thus, dimensionless factors

$$\xi p = \frac{\Delta e_p}{\Delta e_v} = \frac{2e^2(aN)^{1/3}}{M_i v^2} \; ; \; \xi E \approx \frac{E^2}{4\pi \cdot \Delta e_v} \; ; \; \xi g = \frac{2 \cdot E^2}{N \cdot M_i \cdot v^2} \quad (5a)$$

can become not too small compared with 1 and even larger than 1. In this case, charged particles can be affected by the field. For instance, at $E \approx 2 \cdot 10^5 - 2 \cdot 10^4$ V/cm, $N \approx 3 \cdot 10^{19}$/cm^3, $a \approx 10^{-4}$, $v \approx 10^4$ cm/s, and $M_i \approx 10^{-22}$ g $\approx 50 M_H$, obtains

$$\xi p \approx 5-0.05; \quad \xi E = 2(10^3 - 10); \quad \xi g = 2(10^{-1} - 10^{-3}) \quad (5b)$$

In the case considered (high $\overline{P}$ and $\overline{N}$), the "asymmetrical flow-charge coupling" takes place: ions are more strongly influenced by the gas flow than the flow is affected by the charged particles. This conclusion results from the consideration of collision integrals in the Boltzmann equation for neutral-neutral ($I_{nn} \sim N^2$), ion-neutral ($I_{in} \sim aN^2 \ll I_{nn}$), and ion-ion ($I_{ii} \sim a^2 N^2 \ll I_{in}$), collisions at $a \ll 1$. Hence, one concludes that the conventional turbulent theory (e.g., the Kolmogorov-Landau theory[23]) can be used approximately to make estimates of parameters in TGDD phenomena at the conditions considered.

The main observed facts to be explained are:
1) The kinetics and parameters of the GPT.[4,10,12,19-22]
2) A sharp increase in power input and currents (of the order of $10^2 - 10^3$ times of the nonflow values) under turbulent conditions, without breaking of the discharge stability.[1,2,17]
3) A sharp enhancement of transport coefficients in the discharge under turbulent conditions.[2,17]

4) A large increase in the spatial and temporal homogeneity of discharge parameters and the corresponding decrease observed in uncontrolled irregularities in the discharge.[2,17]

The approaches discussed in this work enable one to consider the above questions and obtain reasonable agreement with experimental data.

II. Conditions and Geometry of the Problem

Consider a gas flow with controlled turbulence passing through the discharge zone with a large electric field strength E. To be specific, we consider the corona discharge in a hydrodynamically turbulized gas flow with the geometry typified by Figs. 1 and 2.[1,2] Phenomena in corona discharges taking place in turbulent gas flows have not been studied enough in spite of their importance for lasers, electrogasdynamic electric generators, etc. Although various electrode-gas flow geometries are possible,[13,16] we shall concentrate on those shown in Figs. 1 and 2, since other cases can be treated similarly.

Prior to the test section (see A-A' in Fig. 2), the gas passes through a plenum chamber of large cross section with Re<Re**, where Re** is the critical Reynolds number for these particular conditions. The entering flow is purely laminar because any residual upstream turbulence can be purposely removed. Before entering the test section, the

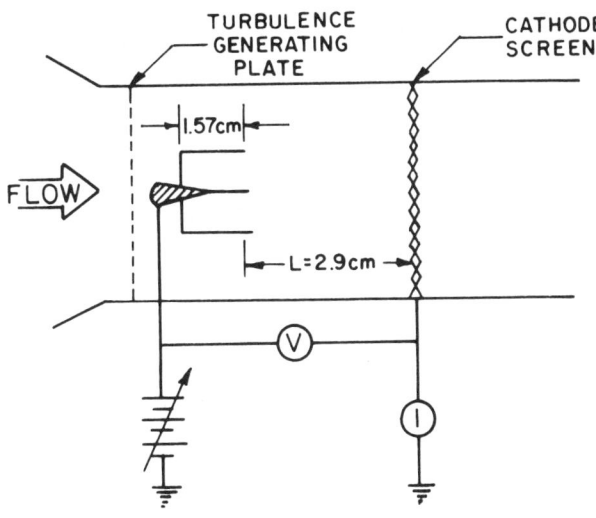

Fig. 1   Experimental setup.

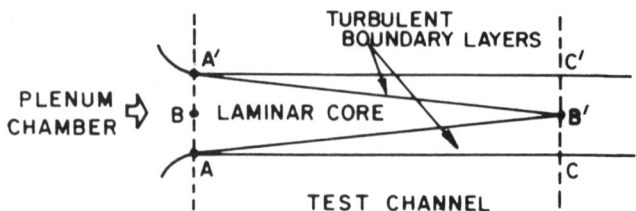

Fig. 2 Developing flow in a channel without the discharge and turbulizing screens.

velocity of the gas increases markedly and Re may become greater than Re**, which leads to the formation of turbulent boundary layers at the test section walls; this comprises the regions A'B'C' and ACB' in Fig. 2. A laminar core exists in the center of the channel (the region A'B'A shown in Fig. 2). All experimental equipment is located within this laminar core. This equipment includes the metallic pins serving as anodes with density b_p of pins per unit cross sectional area. The cathode may consist of several airfoils or, of coarse, wire mesh. The interelectrode distance $L=L_a+L_b$ includes space lengths characterizing internal L_a and external L_b corona regions.

Controlled turbulence is generated in the laminar core by grids and screens located at a distance r_g upstream of the discharge region. The gas with speed v is introduced in the discharge region after passing one or more screen/mesh combinations. In this fashion, various turbulent flow configurations can be generated.[1,2,13,16] The hydrodynamically turbulized gas flow enters in the discharge zone with nonuniform electrical field (see Fig. 3): 1) the averaged electrical field $\bar{E}_a$ in the corona region is very high and nonuniform in space, e.g., $\bar{E}_a \approx 50$ to 150 kV/cm and $\bar{E}_a/P \approx 70-200$ V/cm·Torr at $\bar{P} \approx 760$ Torr; and 2) the averaged electrical field in the external corona zone $\bar{E}_b$ is much smaller and smoother, e.g. $\bar{E}_b \approx 6$ kV/cm and $\bar{E}_b/P \approx 8$ V/cm·Torr.

Consider a small gas portion of linear scale δL_a and volume $\delta V \approx \delta L^3$ that satisfy conditions

$$\delta L \cdot \left|\frac{\partial \bar{E}}{\partial r_o}\right| \ll \bar{E}_a \quad \text{and} \quad \delta L \ll L_a \ll L_b \qquad (6)$$

where $r_o = \{x_o, y_o, z_o\}$ is the coordinate in the laboratory (unmoving) frame. Each of such gas portions contains a very small initial electron number density $n_e^{(o)} = n_e(t_o)$ (formed by cosmic rays, by the natural radioactivity, etc.) when it enters at an instant t_o in the discharge zone with the steady but nonuniform mean electric field $\bar{E}(r_o)$. Different gas portions moving along different trajectories are affected by various electric fields: the closer a trajectory approaches the electrode pins, the stronger $\bar{E}(r_o)$ is. This electric field accelerates the initial electrons and gives a start to the development of the many electron generations that form electron avalanches and the GPT.[4,10,19-22] The GPT kinetics is considered in a later section of this work. Here we want only to stress that the GPT occurs in a small region during a short time interval $\theta^{(8)}$, which can be shorter than $\Delta\tau_a \approx L_a/v$.[4,10,12,19-22] The sharp, fast increase in the electron concentration $n_e(t)$ is shown schematically in Fig. 3. When the gas portion moves ahead and the distance from the electrode pins increase, $\bar{E}(r_o)$ becomes smaller. Then the increase in $n_e(t)$ is slowed down and $n_e(t)$ can start to decrease due to various processes: recombinations, diffusion to the walls, etc. The change of $n_e(t)$ is shown schematically in Fig. 3.

If one considers the corresponding electric field strength $\bar{E}(r;t)=\bar{E}(x,y,z;t)$ acting on the gas portion in the accompanying coordinate system that moves with velocity $v=\bar{v}+\Delta v=\{\bar{v}_k+\Delta v_k\}$ (with k=1,2,3), one finds that $\bar{E}(r,t)$ and $\bar{E}(r,t)/P$ change rapidly with time, as in a pulse discharge.[2,4,10,12,19-22] In the small volume $\delta V \approx \delta L^3$ in which $\bar{E}(r,t)$ satisfies Eq. (6), one can neglect the dependence of $\bar{E}(r,t)$ on r and can approximate this field by a triangular pulse of duration $\Delta\tau$ (see Fig. 4),[10,21] as

$$\bar{E}(r,t) \approx \bar{E}(t) = \begin{array}{l} E_m \dfrac{t-t_o}{t_m-t_o} \quad \text{for } t_o<t<t_m \\[1em] E_m \dfrac{t_o+\Delta\tau-t}{t_o+\Delta\tau-t_m} \quad \text{for } t_m<t<t_m+\Delta\tau \end{array} \quad (7)$$

where t_m is the instant when the mean value $\bar{E}(t)$ of $\bar{E}(t)+\Delta E(t)$ reaches its maximum $E_m=\bar{E}(t_m)$, for a gas portion moving along a given trajectory in the discharge zone. E_m for different trajectories differ from one another. In this case $\bar{E}(t)/P$ is also described by the corresponding triangular approximation. Hence one finds

$$\dot{E} = \frac{d\bar{E}}{dt} = \frac{E_m}{t_m-t_o} \quad \text{at } t_o<t<t_m$$

and (8a)

$$\dot{E} = \frac{E_m}{t_o+\Delta\tau-t_m} \quad \text{at } t_m<t<t_o+\Delta\tau$$

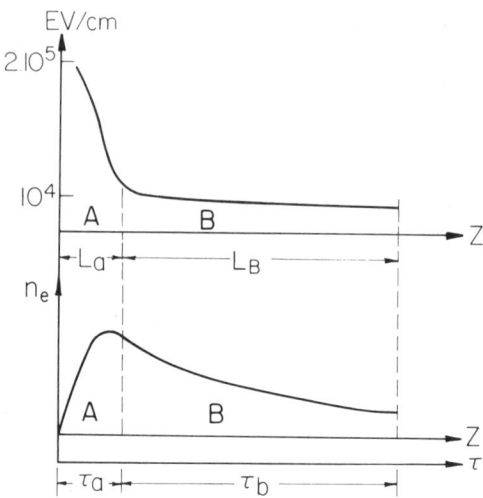

Fig. 3 Diagrams of electric field strength and electron concentration in the gas flow passing through the corona discharge as function of space coordinates Z and time τ (in the accompanying frame). L_a and L_B are the characteristic lengths of the corona A and external B region. τ_a and τ_b are the residence times of moving gas volumes in the regions A and B.

The following numerical examples illustrate orders of magnitudes of the values involved. If $\bar{v} \approx 10^4$ cm/s, $E_m \approx 5\text{-}10 \cdot 10^4$ V/cm, and $L_a \approx 0.3$ cm, one finds that a gas portion moving through the discharge zone is affected by a pulsed electric field characterized by the following parameters:

$$\Delta\tau \approx 2(t_m-t_o) \approx 6 \cdot 10^{-5} s; \quad |d\bar{E}/dt| \approx (1.6\text{-}3.2) \cdot 10^9 \text{ V/cm} \cdot \text{s} \quad (8b)$$

Here we consider a symmetric triangular pulse of $\bar{E}(t)$, but only a slight modification is necessary to treat a more general case. Hence, one can see that in the accompanying frame a small gas portion moving through the discharge zone passes through different states, each of which has a short but finite duration and is characterized by different interrelated properties of the gas (see Fig. 3):

1) The nonionized state of the gas exists at $t<t_o$ before the gas portion enters in the discharge region and starts to be affected by the electric field.

2) The transient state of the formation of ions, electron avalanches, and the GPT in the gas portion occurs at $t_o<t'<t_o+\Delta\tau_a$ during $\Delta\tau_a$ when intensive transient phenomena take place in the gas.

3) The slowing down of the increase in $n_e(t)$ and a partial decay of the plasma formed in the gas portion during

DIFFUSE ELECTRICAL DISCHARGES IN GAS FLOWS   221

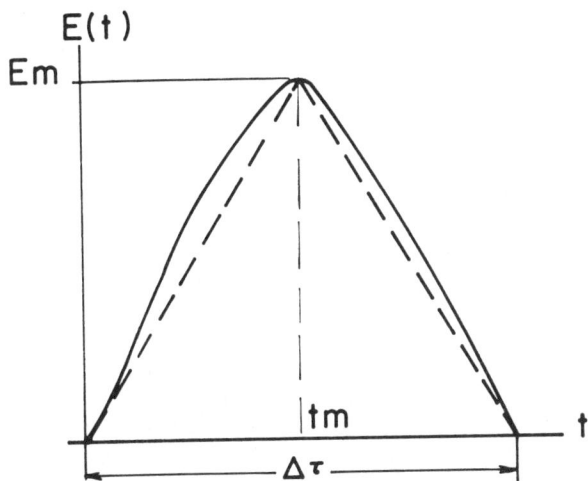

Fig. 4 Diagram of time-dependent pulse-like electric field strength E of duration $\Delta\tau$ in a small gas volume moving through the discharge zone. The dashed line is the triangle approximation of the pulse-like field.

the previous stage take place during $\Delta\tau_b \approx L_b/v$ at $t_0+\Delta\tau_a<t'' <t_0+\Delta\tau_a+\Delta\tau_b$ when the gas portion passes through the external corona region with a weaker electric field $E_b<<E_a$.

4) The final stage of plasma decay and the plasma-to-gas transition (PGT) occurs when the gas portion moves out of the discharge zone at $t^m>t_0+\Delta\tau_a+\Delta\tau_b$.

All these stages are coupled with one another into a "coherent temporal" structure in which parameters of the later stages depend on processes in the previous ones. In the laboratory coordinate system, one finds a spatial non-uniform structure composed of the following interrelated regions: 1) the region of the nonionized gas located upstream of the corona discharge at $z_0<z_0^o$, where z_0^o is the coordinate (accounted along the flow axis) of the beginning of the discharge zone;

2) the corona zone at $z_0^o<z'<z_0^o+L_a$, where a strong electric field E_a exists and intensive ionizations and excitations and the GPTs take place;

3) the external corona region at $z_0^o+L_a<z''<z_0^o+L_a+L_b$ in which the electric field $E_b<<E_a$ and a degree of ionization a_b is lower than that of $a_a>>a_b$ in the corona zone; and

4) the flow region out of the discharge zone at $z''>z_0^o+L_a+L_b$ within which there is no electric field, and where the ionization degree decreases sharply and the PGT takes place. Finer structures of each of these regions can be considered.

It is quite obvious that the aforementioned interrelated stages of moving gas portions in the accompanying coordinate system and the interconnected regions in the laboratory frame are closely related to one another (see Fig. 3). These stages through which the gas portion passes during its motion through the discharge and the related regions in the laboratory system can be described in terms of the fundamental concept of temporal and spatial dissipative structures (DS). The DS concept has been developed and effectively used in physics, hydrodynamics, physical chemistry, etc., by different authors and especially by the Brussels school headed by I. Prigogine.[27-30] Usually, the DS is composed of some number of parts (in particular, in the TGDD case considered here, the DS is composed of the zones of the turbulent flow mentioned above) that are coupled together (e.g., by fluxes of mass, energy, momentum, and electric discharge in the TGDD case) and display a coherent behavior. Recently, the DS concept has been applied by the author to some new fields, namely to: 1) plasma coating and deposition on solid surfaces located in electrical discharges in laminar flows,[31] 2) the kinetic many-body treatment of large energy fluctuations of a small number $N_0 \geq 1$ of particles and rate processes in and on solids (which is closely connected with random instabilities and turbulence in the solid),[32] and 3) sputtering from solids by plasma ions impinging on the surfaces.[31]

Usually the DS formation is associated with some critical values of parameters involved (e.g., temperature differences, electric fields, etc.)[27-32] In the TGDD case, the corresponding DS is associated with the formation of plasma of electronic concentration $n_e^{(g)} = a^{(g)} N$, which satisfies Eq. (2). Such plasma should be formed in every gas portion passing through the discharge zone of characteristic length L_a within which a rather strong electric field E exits. The plasma can be formed in the gas portions passing through the field only if the residence time $\Delta\tau_a \approx L_a/v$ satisfies the condition

$$\Delta\tau_a \approx L_a/v > \theta^{(g)} = g \langle \Delta\theta \rangle_g = b_g \langle \Delta\theta \rangle_g \ln(n_e^{(g)}/n_e^{(o)}) \quad (9)$$

where $\theta^{(g)}$ is the finite duration of the GPT which is connected with the increase in the electron concentration from $n_e^{(o)} = n_e(t_o)$ up to $n_e^{(g)} \gg 1$ and where the ratio $n_e^{(g)}/n^{(o)}$ can be of several orders of magnitude.[4,10,12,19-22] The GPT

DIFFUSE ELECTRICAL DISCHARGES IN GAS FLOWS

appears as a result of the formation of

$$g = b_v \ln(n_e^{(g)}/n_e^{(0)})$$

with

$$b_g = \left[\ln 2 + <\ln(1-\chi/2)>_g\right]^{-1} = \left[<\ln(2-\chi)>_g\right]^{-1} \quad (10)$$

electron generations (EG) each of which has a lifetime of

$$\Delta\theta = \Delta\theta_{og} \exp(B \cdot P/E) \quad \text{with } E = \bar{E} + \Delta E \text{ and } P = \bar{P} + \Delta P \quad (11)$$

of the order of the mean time between the two successive ionizing collisions of the AFE with the gas particles. Therefore, the GPT is a result of the development of electron avalanches due to acceleration of electrons in the strong discharge electric field and ionizations of gas particles. In every ionizing collision, the primary electron of the previous EG produces a new low-energy free electron (that leads to the increase in the electron concentration) and loses a significant part of its energy (see Sec. IV and Fig. 6 below).[4,10,12,19-22] Then these low-energy free electrons are accelerated again in the electric field and produce new ionizations, EGs, etc.[4,10,12,19-22] Here the brackets $<...>_b$ mean averaging over the time interval $\theta^{(g)}$ needed to produce g EGs. Coefficients B and $\Delta\theta_{og}$

$$B = \left[2\delta(\bar{\varepsilon}_e)\right]^{\frac{1}{2}} \frac{\bar{\varepsilon}_e}{kT} \cdot \frac{\sigma(\bar{\varepsilon}_e)}{e} \quad \text{and} \quad \Delta\theta_{og} = \frac{m W_o}{eE[m\delta(\varepsilon)\bar{\varepsilon}_e]^{\frac{1}{2}}}, \approx 1 \quad (12)$$

In Eq. (10) $1 - \chi$ is the mean number of new electrons (per one AFE) formed by ionizing collisions of electrons of the previous EG with gas particles, which is not compensated for by processes (recombination, etc.) leading to the decrease in electron concentration during $\Delta\theta$. Here $0 \leq 1- \int \leq 1$ and $b \rightarrow \infty$ when $\chi \rightarrow 1$. Equations (9-12) together with Eq. (2) determine parameters of the gas flow and the discharge at which the plasma-based DS appears in the TGDD.

In this work we consider gas discharges of relatively low power under the following experimentally observed conditions:[2,17]

1) Main translational energies of heavy particles (atoms $\bar{\varepsilon}_a$ molecules $\bar{\varepsilon}_m$, and ions $\bar{\varepsilon}_i$)

$$\bar{\varepsilon}_y = \frac{M_y \bar{u}_y^2}{2} = \int \varepsilon_y f_y(\varepsilon_y) d\varepsilon_y, \quad y = a, m, i \quad (13)$$

remain close to their values at gas temperature T, but the mean electron translational energies

$$\bar{\varepsilon}_e(r,t) = \frac{m_e \overline{u_e^2}(r,t)}{2} = \int \varepsilon_e f_e(\varepsilon_e;r,t) d\varepsilon_e >> \bar{\varepsilon}_y \gtrsim kT \qquad (14)$$

can be much higher. Here $f_y(\varepsilon_y)$ is the translational energy distribution of a heavy particle of the yth type, which does not differ much from the Maxwell distribution; $f_e(\varepsilon_e;r,t)$ is the electron translational energy distribution determined by the corresponding kinetic equation.[4,12,22] It can differ greatly from the equilibrium distribution and can depend on coordinates r and time t.

2) Electron-electron, ion-ion, and electron-ion collisions can be neglected compared with electron-neutral and ion-neutral collisions due to a low degree of ionization $a(r,t) \ll 1$. At the same time, the electron-ion interaction is to be taken into consideration while calculating turbulent ambipolar diffusion and other parameters associated with turbulence discharge interrelations.

3) Joule heating of the gas is not substantial due to small values of electric currents.

4) The flows are subsonic and the gas is essentially incompressible.

5) Ions (and electrons) are assumed to attain the turbulent motion of the neutral background gas everywhere except at the near-to-wall sheaths. The diffusion of charged particles (i.e., ambipolar and turbulent ambipolar diffusion), as well as the transfer of heat and momentum, can be expressed through transport coefficients of the neutral gas. But such parameters as electrical conductivity should be calculated separately. These assumptions have been discussed in Refs. 3 and 5.

6) The actual processes in TGDD have a three-dimensional character with simultaneous mass, momentum, energy, and charge transfers. We also assume a relatively low field-to-pressure ratio that satisfies the following conditions [4,10,12,19-22]:

$$e\bar{E} \cdot \lambda_{em}(\bar{\varepsilon}_e) \ll W_o \quad \text{or} \quad \bar{E}/\bar{N} \ll (\bar{E}/\bar{N})_{bM} = \frac{W_o}{e} \cdot \sigma_{em}(\bar{\varepsilon}_e) \qquad (15)$$

where $\lambda_{em}(\bar{\varepsilon}_e) = [\sigma_{em}(\bar{\varepsilon}_e) \cdot \bar{N}]^{-1}$ is the minimum AFE mean free path length associated with the maximum total cross section $\sigma_{em}(\bar{\varepsilon}_e)$ in the electron energy range considered $N=P/kT$. W_o is the ionization potential of gas particles. Equation 15 assumes rather high pressures, which are important for applications. For example, for $W_0 \approx 10 \div 20 \, eV$ and $\sigma_{em} \approx 10^{-15}/cm^2$, one finds $(\bar{E}/\bar{N})_{bM} \approx 1-2 \cdot 10^{-14} V \cdot cm^2$, whereas $E/N \approx 3(10^{-16} -10^{-15}) V \cdot cm^2$ and $\bar{E} = 10^4 - 10^5 V/cm$ at $\bar{P} \approx 760$ Torr, $\bar{T} \approx 300$ K. Therefore, Eq. (15) is satisfied. Equation 15 also enables one to develop the nonstationary statistical kinetic theory of formations of electronic avalanches in gas portions moving rapidly through the discharge zone and obtain equations leading to a better understanding of experimental facts and, in particular, to Eqs. (9-12).[4,10,12,19-22]

III. Corona Discharge in Turbulent Gas Flow: Phenomenological Consideration

Many differences exist between corona discharges under nonflow and under hydrodynamically turbulent conditions,[2,10,12,17,19,20] These differences lead directly to various observable consequences, some of which are considered below.

Under nonflow conditions, coronas are very nonuniform in space and time. Typically, the discharge is active on only a few electrode pins. Upon close examination, each discharge looks like a single short-term pulse of current and light. These pulses are connected with avalanches which concentrate along narrow channels with the high local concentration of charged particles. The channel radius and cross section of the avalanches are[25,33]

$$\Delta \rho_0 \approx (6 \int_{\bar{u}_e}^{D_e} \frac{D_e}{\bar{u}_e} dr)^{\frac{1}{2}} \quad \text{and} \quad S_M \approx \pi \rho_0^2 \qquad (16)$$

where D_e is the electron diffusivity equal to

$$D_e = D_{eM} \approx 0.3 \cdot \lambda_e(\bar{\epsilon}_e) \, \bar{u}_e \approx 0.3 \left[\sigma_e(\bar{\epsilon}) N \right]^{-1} \cdot \bar{u}_e \quad \text{at} \quad \rho_0 < 1_d$$

(17a)

and to

$$D_e \approx D_{AM} \approx \frac{D_{eM} \cdot \mu_{iM} + D_{iM} \cdot \mu_{eM}}{\mu_{eM} + \mu_{iM}} \quad \text{at} \quad \Delta \rho_0 > 1_d = (\frac{\bar{\epsilon}}{4\pi e^2 a \, N})^{\frac{1}{2}} \qquad (17b)$$

where μ_{eM} and μ_{iM} are the electron and ion mobilities respectively, 1_d is the Debye Radius [see Eq. (2)], which can depend on the space coordinates and time.

The treatment of turbulent gas discharges is suggested in Refs. 2 and 19 on the phenomenological level with the help of the approximate application of Kolmogorov-Landau theory of locally developed turbulence[23] and is considered below. The statistical kinetic treatment of the related questions suggested in Refs. 2,10,12, and 19-22, is discussed in later sections of this work. We shall consider the turbulence in the discharge space with Re>>Re* everywhere, with a possible exception in small regions near solid surfaces, where specific phenomena associated with the plasma wall interactions can take place.[34-36]

The smallest linear scale Λ_o of turbulence and the highest corresponding frequency are

$$\Lambda_o \approx 1 \left(\frac{Re^*}{Re}\right)^{3/4} \quad \text{and} \quad \Omega_o \approx \frac{\Delta v}{1} \left(\frac{Re}{Re^*}\right)^{3/4} \quad (18)$$

where 1 is the largest scale of the turbulence. Numerical estimates show that the following conditions can take place.[2,17]

$$\Lambda_e(\bar{\varepsilon}) << \Lambda_o << \delta \text{ L} << d_s; \quad \Lambda_o << r_b \approx b_p^{-\frac{1}{2}}; \quad \text{and} \quad \Lambda_o << L_a \quad (19)$$

where $r_b \approx b_p^{-\frac{1}{2}}$ is the mean distance between electrode pins, e.g., $r_b \approx 1$ cm, d_s is the effective linear size of the test tunnel cross section, e.g., $d_s \approx 4$ cm, $\Lambda_o \approx 10^{-2}$ cm, and $\Omega_o \approx 10^6$ s^{-1} at Re/Re* $\approx 4\cdot 10^3$, $\Delta v \approx \bar{v} \approx 10^4$ cm/s, and $1 \approx d_s \approx 4$ cm. Hence one can see that the corona zone volume $L_a\cdot d_s^2 \approx V_a$ and the entire discharge volume $V \approx L\cdot d_s^2 \approx (L_a+L_b)\cdot d_s^2$ contain many small volumes $\delta V \approx \delta L^3$, each of which is in an approximately uniform electric field [see Eq. (6)], has many turbulent degrees of freedom $(\delta L/\Lambda_o)^3 >> 1$, can be described macroscopically since $\Lambda_o >> \lambda_e$, and can be treated approximately by the isotropic-turbulence model.

Components $\bar{E}_j(r,t) = \bar{E}_j(r+t) + \Delta E_j(r,t)$ of the electric field $E(r,t) = \bar{E}(r,t) + \Delta E(r,t)$ change with time according to the following relations, due to the spatial nonuniformity of the electrical field and turbulent random pulsations[10].

$$\frac{dE_j}{dt} = \frac{d\bar{E}_j}{dt} + \frac{d}{dt}(\Delta E_j) = \sum_{k=1}^{3} (\bar{v}_k + \Delta v_k) \cdot \frac{\partial E_j(r_o)}{\partial x_{ok}} ;$$

$$x_{o1} = x_o; \quad x_{o2} = y; \quad x_{o3} = z; \quad j=1,2,3$$

(20)

DIFFUSE ELECTRICAL DISCHARGES IN GAS FLOWS 227

(if $|\partial \cdot \Delta E_j/\partial t|$ are small enough in volume $\delta V \gg \Lambda_0^3$), and

$$E_j(r,t) = \int_{t_0}^{t} (dE_j/dt) \, dt = \bar{E}_j(r,t) + \Delta E_j(r,t) \quad (21)$$

where

$$\partial \bar{E}_j/\partial t = 0; \quad d\bar{E}_j/dt = \sum_{k=1}^{3} \bar{v}_k \left[\partial \bar{E}_j(r_0)/\partial x_{ok} \right] ;$$

and

$$\frac{d}{dt}(\Delta E_j) = \sum_{k=1}^{3} \Delta v_k \left[\partial E_j(r_0)/dx_{ok} \right] \quad (22)$$

where $\Delta E_j(r,t)$ are random oscillations of the field component $E_j(r,t)$ associated with turbulent fluctuations $\Delta v = \{\Delta v_k\}$ of the gas velocity. Hence one can see that the correlation tensor

$$B_{jm}(r_1,t_1;r_2,t_2) = \overline{\Delta E_j(r_1,t_1) \cdot \Delta E_M(r_2,t_2)} \quad (23)$$

of these random fluctuations is closely connected with correlation functions of velocity fluctuations Δv_k.
In the quasi neutral corona zone only high-frequency turbulent fluctuations with frequencies Ω and scales Λ, which satisfy the conditions,[2,17]

$$\Omega_0 < \Omega < \Delta\tau_a^{-1} \approx L_a/\bar{v} \text{ and } \Lambda_0 < \Lambda < L_a \quad (24a)$$

can influence markedly transport coefficients and other properties in the plasma formed in the corona region. Hence one can conclude [if one uses Eq.(18)] that the significant influence of the turbulence on the plasma properties of the corona region is possible only for turbulent flows that satisfy the condition[2,17]

$$Re/Re^* \gg (1/L_a)^{4/3} \quad (24b)$$

This approach leads to the following expression for the turbulent ambipolar diffusion coefficient[2,17]

$$D_{At} \approx \frac{Re}{Re^*} \frac{\bar{\varepsilon}_e}{\bar{\varepsilon}_i} D_{iM} \text{ with } \frac{Re}{Re^*} \gg 1 \text{ and } \frac{\bar{\varepsilon}_e}{\bar{\varepsilon}_i} \gg 1 \quad (25)$$

if one uses Eq. (17b) and the equation $D_t \approx (Re/Re^*) \cdot D_{iM}$ for the turbulent diffusivity in the absence of the discharge. Here, $D_{iM} \approx 0.3\lambda_i \bar{u}_i$ is the ion diffusivity in the absence of the turbulence and the discharge, which is associated with the ion mean free pathlength λ_i, D_{At} is several orders of magnitude larger than D_{iM}. For example, at $Re/Re^* \approx 3 \cdot 10^3$, $\bar{\varepsilon}_e/\bar{\varepsilon}_i \approx 10^2$, and $D_{iM} \approx 0.3$ cm^2/s, one finds $D_{At} \approx 3 \cdot 10^5{}_{iM} \approx 10^5$cm^2/s. On the one hand, this sharp increase in the diffusivity is associated, with the turbulence, and, on the other, with the Coulomb coupling between electrons and ions in the discharge plasma.

The sharp increase in the transport coefficients leads to explanations of main observed facts in the TGDDs: of the large enhancement in the size of the active corona zone, of the increase of the discharge stability and uniformity, of a sharp increase of power input, currents, etc.[2,4,10,17]

The effective length and cross section of the active turbulent corona zone are (see Fig. 5)[2,17]

$$L_{at} \approx 0.5\bar{d} + \Delta L_{Aa} \approx (6\tilde{D}_{At} \cdot \tilde{t}_r)^{\frac{1}{2}} (1+\bar{d}/2\Delta L_{at})^{\frac{1}{2}} \qquad (26)$$

and

$$S_{at} \approx \pi L_{at}^2 \approx 6\pi \tilde{D}_{At} \cdot \tilde{t}_r (1+\bar{d}/2\Delta L_{at}) \approx 6\pi \frac{Re}{Re^*} \cdot \frac{\bar{\varepsilon}_e}{\bar{\varepsilon}_i} \tilde{D}_{iM} \cdot \tilde{t}_r \qquad (27a)$$

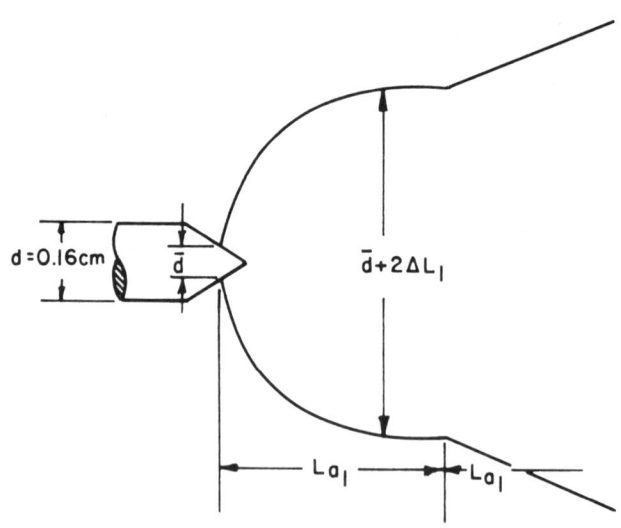

Fig. 5 Detail of a single near-to-electrode pin corona region in the turbulent flow.

DIFFUSE ELECTRICAL DISCHARGES IN GAS FLOWS 229

where $0.5\bar{d} \gg \Delta L_{At}$ is the mean corona space scale under non-flow conditions and t_r is the time interval during which the high electron concentration acquired by the gas in a strong electric field can be kept in gas volumes leaving the corona zone and coming into regions of a weaker electric field due to turbulent fluctuations and the gas flow. In some cases, t_r can be treated as a recombination time. The tilde over the symbols means averaging over the corona region.

Now consider some numerical estimates of ΔL_{Aa} and S_{at} following from Eqs. (25-27a). If one takes $D_{iM} \approx 0.2$ cm^2/s, $\tilde{D}_{At} \approx 0.6 \cdot 10^5$ cm/s, $t_r \approx 10^{-5}$ s, and $0.5\bar{d} \approx 0.2$ cm, one finds

$$\Delta L_{Aa} \approx 2 \text{ cm} \gg 0.5\bar{d} \approx 0.1 \text{ cm}, \quad S_{at} \approx 12 \text{ cm}^2 \approx 10^2 \pi \bar{d}^2 \quad (27b)$$

Therefore, turbulence has the effect of a great spreading of the charged particles from the small region near the pins in all directions and thereby creates a more homogeneous distribution of charged particles in the corona. This spreading of charged particles prevents the formation of streamers and increases the discharge stability. The turbulence fattens the corona regions considerably, as $\tilde{\Delta L}_{Aa} \gg 0.5\bar{d}$ and $S_{at} \gg \pi \bar{d}^2$ (see Fig. 5). When $\Delta L_{Aa} \gtrsim r_p \sim b_p^{-\frac{1}{2}}$ ($r_p \approx 1$ cm), the corona zones of neighboring pins can overlap and create one more or less homogeneous corona discharge common to all pins (one large smoothed corona-like electrode), as observed experimentally. The velocity and flux density of charged particles that participate in the ambipolar turbulent diffusion can be estimated by

$$w_{At} \approx \Delta L_{At}/t_r \approx (6 D_{At}/t_r)^{\frac{1}{2}} \approx (6 \frac{Re}{Re^*} \cdot \frac{\bar{\varepsilon}_e}{\bar{\varepsilon}_i} \cdot \frac{D_{iM}}{t_r})^{\frac{1}{2}} \quad (28)$$

and

$$J_{At} \approx D_{At} \Delta n_e \approx (\frac{Re}{Re^*} \cdot \frac{\bar{\varepsilon}_e}{\bar{\varepsilon}_i} \cdot \frac{D_{iM}}{6 \cdot t_r})^{\frac{1}{2}} \cdot n_e^{(g)} \quad (29)$$

since $\Delta n_e \approx n^{(g)}/\Delta L_{At}$. Hence one can conclude that such large D_{At}, J_{At} and $w_{At} \gg v$ associated with the spreading of electrical charges in various directions (including the direction opposite to the flow one) can lead to preionization of gas portions approaching the discharge zone, since $w_{At} \gg v$, e.g., $w \approx 10^5$ cm/s $\gg v \approx 10^4$ cm/V at $t_r \approx 10^{-5}$ s and $\Delta L_{At} \approx 1$ cm, as estimated above. The preionization, in turn, makes the GPT duration $\Theta^{(g)}$ shorter in Eqs. (9) and (10), since it increases $n^{(o)}$ and decreases $\ln(n^{(g)}/n^{(o)})$.

Summarizing, one can conclude that intensive diffusion and convection in the turbulent gas discharge which spread rapidly charged particles and promote rapid gas mixing lead to some phenomena being in agreement with observations, namely, they can: 1) increase L_{at} and decrease L_{bt}; 2) prevent the formation of large local densities of charged particles and therefore hinder the creation of streamers that, in turn, may enhance the discharge stability (i.e., the interactions of the hydrodynamic instabilities and fluctuations in the turbulent flow with local discharge instabilities cause an increase of the discharge stability and uniformity; 3) create better conditions for more pronounced ionization and excitation and increase the total yield of charged particles; and 4) increase the electric field E_b in the external corona zone (with decrease in its length L_b). These phenomena promote a sharp enhancement of conductivity $R_{at}^{-1} \sim S_{at}$ of the corona zone and an increase of electrical conductivity R_b^{-1} in the external discharge zone. In turn, this leads to large enhancements and power consumption W_t of current I.[2,17]

$$W_t \approx I_t U_t \approx U_t^2/R_t \text{ with } R_t=R_{at}+R_{bt}, \ R_{bt}^{-1}=\Sigma_{bt} \cdot S_{bt} \cdot (L-L_{at})^{-1} \quad (30)$$

$$W_t \approx \Sigma_{bt} \cdot \frac{U_t^2 S_{bt}}{L-L_{at}} (1+\frac{R_{at}}{R_{bt}})^{-1} \sim I_t \text{ with } R_{bt} > R_{at} = \Sigma_{at}^{-1} S_{at}^{-1} L_{at} \quad (31)$$

and

$$I_t \approx 6 \ \Sigma_{bt} \cdot t_r \cdot \frac{\pi U_t \cdot \nu_{pt}}{L-L_{at}} \frac{Re}{Re^*} \frac{\bar{\varepsilon}_e}{\bar{\varepsilon}_i} \frac{\bar{U}_i}{N\sigma_{in}} (1+ \frac{\bar{d}}{2L_{Aa}}+ \frac{L_b}{L_{Aa}})^2$$

$$\times (1+ \frac{R_{at}}{R_{bt}})^{-1} \quad (32)$$

where σ_{in} is cross section of the ion-neutral collisions. The ratios

$$\eta_1 = I_t/I_M \approx \frac{\Sigma_{bt} \cdot S_{bt}}{\Sigma_{bM} \cdot R_{bM}} (\frac{L-L_{aM}}{L-L_{at}}) \frac{\bar{U}_t}{\bar{U}_M} \gg 1$$

and (33)

$$\eta_2 = \frac{W_t}{W_M} = \frac{U_t}{U_M} \eta_1 \gg 1$$

of currents η_1 and energy consumption η_2 under the turbulent I_t and W_t and nonflow I_M and W_M conditions are very large, e.g., $\eta_1 \approx \eta_2 > 10^2$, $2,17$, which corresponds with the experimental values $\eta_{1(obs)} \approx \eta_{2(obs)} \approx 250.^2$

IV. Kinetics of Gas-to-Plasma Transitions in Turbulent Gas Flow Passing through Electric Field at Low Field-to-Pressure Ratios

Consider the moving gas portion of volume $\delta V \approx \delta L^3$ (with $\delta L \gg \lambda_e$) entering the electric field E at the instant t_0 and containing a small number $\delta N^{(o)} = n^{(o)} \cdot \delta V$ of initial electrons (which appear in it due to the preionization discussed in the previous section or due to the action of cosmic rays, natural radioactivity, etc.). These $\delta N_e^{(o)}$ electron forming the zeroth electron generation (zeroth EG) have very low initial mean energy $\bar{\varepsilon}_0^{(o)} = \bar{\varepsilon}_e(t_0) \ll W_0$ and cannot ionize gas particles. When the gas portion moves in the discharge zone with velocity v, the electric field changes with time (in the accompanying frame) and can be approximated by Eq. (7). The $\delta N_e^{(o)}$ initial electrons need time $\Delta\theta^{(o)}$ (see Fig. 6) to be accelerated up to energy $\bar{\varepsilon}^{(o)}(t_0 + \Delta\theta^{(o)}) \geqslant W_0$ and ionize gas particles. Therefore only $t_0 + \Delta\theta^{(o)}$ the AFEs of the zeroth EG can change the electron concentration, according to Refs. 4,10,12, and 19-22.

$$n^{(o)} \longrightarrow n^{(1)} = n^{(o)} + \delta n_{o1} = n^{(o)}(2 - \chi^{(o)}) \text{ with } 0 \leqslant \chi^{(o)} \leqslant 1 \quad (34)$$

and give the start to the first EG with electron concentration $n^{(1)}$ and low electron initial mean energy $\bar{\varepsilon}^{(1)}(t_0 + \Delta\theta^{(o)}) = \bar{\varepsilon}_0^{(1)} \ll W_0$. The $\delta N^{(1)} = n^{(1)} \cdot \delta V$ electrons of the first EG need time $\Delta\theta^{(1)}$ to be accelerated by the electric field up to energy

$$\bar{\varepsilon}_f^{(1)} = \bar{\varepsilon}^{(1)}(t_f^{(1)}) = \bar{\varepsilon}^{(1)}(t_0 + \Delta\theta^{(o)} + \Delta\theta^{(1)}) \geqslant W_0$$

and to give the start to the second EG with electron concentration $n^{(2)} = n^{(1)} + \delta n_{12} = n^{(1)}(2 - \chi^{(1)})$, and so on. Here

$$t_0^{(j)} = t_0 + \sum_{k=0}^{j-1} \Delta\theta^{(k)} \text{ with } j = 1, 2, \ldots$$

is the initial instant of the jth EG. Therefore, the GPT associated with the increase $n^{(o)} \longrightarrow n^{(g)}$ in the electron

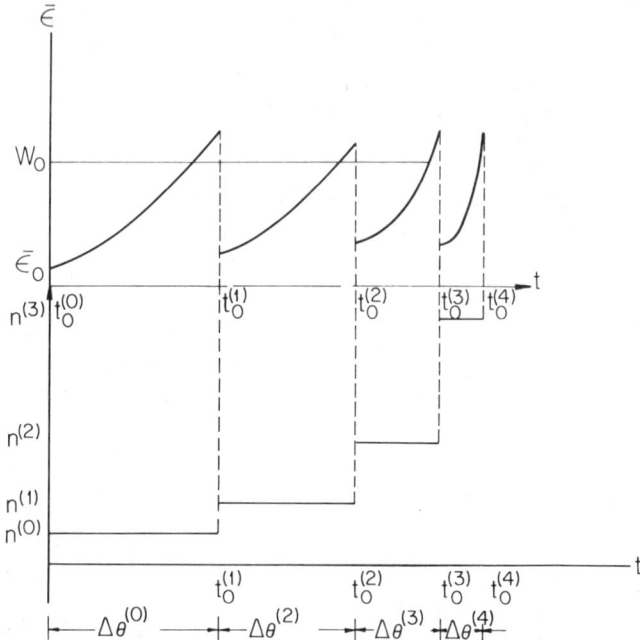

Fig. 6 Diagram of the stepwise kinetic multistage avalanche/formation in a small gas volume moving through the discharge zone which includes: 1) a continuous enhancement of the energy $\bar{\varepsilon}(t)$ of the accelerated average free electrons (AFE) during lifetimes $\Delta\theta^{(j)}$ of single electron generations (EG) between the two successive ionizing collisions at fixed electron concentrations $n^{(j)}$, and 2) stepwise transitions between the successive EGs accompanied by a significant decrease in $\bar{\varepsilon}(t)$ and the increase $\delta n_{j,j+1} = n^{(j+1)}$ in the electron concentration, $j = 0, 1, 2 ...$

concentration is associated with the formation of g EGs (with $g = 1, 2, 3, ...$) where the gth EG has the electron concentration [4,10,12,19-22]

$$n^{(g)} = n^{(o)} + \sum_{k=0}^{g-1} \delta n_{k,k+1} = n^{(o)} \exp\left(\frac{t_o^{(g)} - t_o}{b_g \langle \Delta\theta \rangle_g}\right)$$

$$\text{with } t_o^{(g)} - t_o = \theta^{(g)} = \sum_{k=0}^{g-1} \Delta\theta^{(k)}$$

(35)

where $\theta^{(g)}$ is the duration of the formation of g EGs and of the GPT discussed earlier in connection with Eqs. (9) and (10). Factor b_g determined by Eq. (9) can become large enough when recombinations and other processes causing the decrease in the electron concentration become important, e.g., in the case of high local concentrations of charged particles in coronas in nonflow gases. This slows down the increase in the electron concentration and causes a decrease in $n^{(g)}$ and in the total production of charged particles at given $t^{(g)}-t_0$. Rapid dispersion of charged particles by high-frequency turbulent fluctuations [e.g., with $\Omega_0 \approx 10^6 s$, see Eq. (18)] promotes a decrease in the local electron concentration and b_g and promotes the corresponding increase in the total yield of charged particles.

Equation (35) is in agreement with experimental observations of the electron avalanches[37] and time $\tau_T = b_g <\Delta\theta>_g$ takes a role of the characteristic time (the Townsend time).

A single EG (e.g., the jth) is characterized by the following properties and parameters which determine the development of electron avalanches and the GPTs: 1) the electron concentration $n^{(j)}$ is approximately constant during the jth EG lifetime $\Delta\theta^{(j)}$ (see Fig. 6).[4,10,12,19-22] 2) the time-dependent electron energy distribution $f^{(j)}(\varepsilon^{(j)}, t^{(j)})$ is shifted rapidly toward higher energies during $\Delta\theta^{(j)}$, according to [4,12,22]

$$f^{(j)}(\varepsilon,t) = \int G^{(j)}(\varepsilon^{(j)}, t^{(j)} | \varepsilon_0^{(j)}, t^{(j)}) f^{(j)}(\varepsilon_0^{(j)}, t_0^{(j)}) d\varepsilon_0^{(j)} \quad (36)$$

the Fokker-Plank equation in the energy space

$$\frac{\partial G^{(j)}}{\partial t^{(j)}} = \frac{\partial^2}{\partial \varepsilon^2}(c^{(j)} G^{(j)}) - \frac{\partial}{\partial \varepsilon^{(j)}}(A^{(j)} G^{(j)}) \quad (37)$$

and the stochastic equation of the motion of electrons colliding from time to time with gas particles

$$\frac{d\varepsilon^{(j)}}{dt} = eE^{(j)} u_d - \nu_e^{(j)} \overline{\Delta W}^{(j)} + s^{(j)} \cdot \psi^{(j)}(t) \quad (38)$$

with $u_d \approx m^{-1} \cdot eE^{(j)} \tilde{\tau}_e^{(j)}$

where $E^{(j)}$ is the field in the accompanying coordinate frame [see, e.g., Eq. (7)] during $\Delta\theta^{(j)}$, u_d is the electron drift velocity, $\nu_e \approx \tau_e^{-1} = N \cdot \sigma \cdot u_e$ is the frequency of AFE-molecule collisions,

$$\overline{\Delta W}^{(j)}(\overline{\varepsilon}^{(j)}) = \delta\ (\overline{\varepsilon}^{(j)}) \cdot \overline{\varepsilon}^{(j)}$$

$$(s^{(j)})^2 = C^{(j)} \approx (eE^{(j)})^2 D_{\shortparallel},$$

$D_{\shortparallel}$ is the electron diffusivity along the field, $A^{(j)} = eE^{(j)} u_d - v_e \cdot \overline{\Delta W}^{(j)}$ is the energy the AFE acquires per unit time, $v_d^{(j)}$ is the AFE energy losses per unit time, and $s^{(j)} \psi^{(j)}(t)$ is the stochastic term reflecting the random character of collisions of electrons with gas particles. Hence, one can see that the AFE energy $\overline{\varepsilon}^{(j)}(t^{(j)})$ increases rapidly during $\Delta\theta^{(j)}$, according to the shift $f^{(j)}(\varepsilon^{(j)}, t^{(j)})$. See Fig. 6.

3) The lifetime of the jth EG can be estimated by 4, 10,12,19-22

$$\Delta\theta^{(j)} = \tilde{\alpha}^{(j)} <\tau_e^{(j)}>_{\Delta\theta} = \Delta\theta_{oj} \exp(B^{(j)} P/E^{(j)}) \quad (39)$$

where coefficients $\Delta\theta_{oj}$ and $B^{(j)}$ are given by Eq. (12). Here

$$\tilde{\alpha}^{(j)} = \alpha^{(j)} + \delta\alpha^{(j)} \approx <eW_o(eE^{(j)} u_d^{(j)} - v_e^{(j)} \cdot \overline{\Delta W}^{(j)})^{-1}$$
$$+ \sigma\ (\overline{\varepsilon}^{(j)} \geqslant W_o / \sigma_{in} (\overline{\varepsilon}^{(j)} \geqslant W_o >_{\Delta\theta} \quad (40)$$

is the mean total number of nonionizing collisions of the AFE with gas particles between the two successive ionizing collisions having cross section σ_{in}: $\tilde{\alpha}^{(j)}$ includes $\alpha^{(j)}$ collisions before $\overline{\varepsilon}^{(j)}(t^{(j)})$ reaches magnitude W_o at $t_o^{(j)} + t_1^{(j)}$ and $\delta\alpha^{(j)} \approx \sigma/\sigma_{in}$ nonionizing collisions during $\delta t^{(j)} = t_o^{(j)} + \Delta\theta^{(j)} - t_1^{(j)}$ when $\varepsilon^{(j)}(t^{(j)} > t_o^{(j)} + t_1^{(j)}) \geqslant W_o$, $<\cdots>_{\Delta\theta}$ means averaging over time $\Delta\theta^{(j)}$.

Hence, one can find characteristics of avalanches composed of g EGs. In particular, one can calculate the first Townsend coefficient from Eqs. (11-13)[10,12,21]

$$n_T \approx (b_g <u_d \cdot \Delta\theta^{(j)} >_g)^{-1} \approx F \cdot P \exp\left(-B \frac{P_g}{E_g}\right) \text{with}$$
$$F = eE_b / b_g W_o P_g \quad (41)$$

which is in agreement with the empirical equation for the Townsend coefficient[37] and also leads to a reasonable quantitative agreement with experimental data[37]. Equations (11-13) follow from the consideration of AEFs with random free path times τ and energies $\delta\varepsilon=eE^{(j)}dz$ acquired from the electric field during τ, which are distributed according to [10, 12, 21]

$$\psi(\tau) = \frac{1}{\bar{\tau}(j)} \exp\left(-\int \frac{d\tau}{\bar{\tau}}\right)$$

and
(42)

$$\psi(\tau) = \frac{1}{\bar{\tau}(j)} \exp\left(-\left(\frac{\delta\varepsilon}{\overline{\delta\varepsilon}}\right)^{\frac{1}{2}}\right)$$

where $\delta\varepsilon$ and τ satisfy the conditions

$$\delta e = [\delta\varepsilon - \overline{\Delta W}(\varepsilon)] > 0 \text{ and } \tau > \tilde{\tau}_e = \left[m\delta(\bar{\varepsilon})\bar{\varepsilon}\right]^{\frac{1}{2}}(eE)^{-1} \quad (43)$$

which show that only those electrons which gain energies $\delta\varepsilon$ larger than losses $\delta(\varepsilon)\bar{\varepsilon}=\overline{\Delta W}$ (during $\tau > \tilde{\tau}_e$) can reach energy $\varepsilon \geq W_0$ sufficient for ionizations. Here $\overline{\delta z}=u_d\cdot\tau=\mu E\tau$ is the random way the AFE passes along the field during τ.

$$\overline{\delta\varepsilon} = eE\ \overline{\delta z} = (eE)^2\left[2\bar{\varepsilon}\cdot(N\sigma)^2\right]^{-1} \ll W_0 \quad (44)$$

If one uses the triangular approximation Eq. (7) for $E(t)$ at $t<t_M$, one can obtain

$$n_T(t) = \frac{e\ E_m}{W_o\ b_g}\left(\frac{t-t_o}{t_M-t_o}\right)\exp\left(-\frac{\overline{\Delta T}_d}{t-t_o}\right) \quad (45)$$

where $\overline{\Delta T}_d = B \cdot P \cdot E_M^{-1}$ (t_M-t_o) is a characteristic time interval and $t_M-t_o \approx \Delta L_E/v$ is the time interval the gas portion needs to pass distance ΔL_E in the discharge zone and to reach the point of the maximum electric field E_M. Thus, one can see that the lifetime $\Delta\theta^{(o)}$ of the zeroth EG before the start of the avalanche formation can be rather long, e.g., $\Delta\theta^{(o)} \approx t_M-t_o \approx 3\cdot 10^{-5}$ s (at $\Delta L_E \approx 0.3$ cm and $v \approx 10^4$ cm/s), since at low E electrons cannot gain enough energy to ionize gas particles, and rate $dE/dt \approx E_M v/\Delta L_E$ of the field increase is determined by the gas velocity v and ΔL_E, according to Eqs. (7) and (8) and $t_M-t_o \approx \Delta L_E/v$. Although lifetimes $\Delta\theta^{(j)}$ (with j=1,2,....,g) of other EGs are shorter than $\Delta\theta^{(o)}$, the duration

$$\theta^{(g)} \approx \sum_{i=0}^{g-1} \Delta\theta^{(j)}$$

of the GPT composed of many EGs is not too short compared with the shortest period $\Omega_0^{-1} \approx 10^{-6}$ s of turbulent fluctuations determined by Eq. (18) at large Re/Re*. Hence, one concludes that the turbulent fluctuations have enough time to disperse the charged particles over the large discharge zone before the formation of streamers (that require time $\theta_{st} \gg \theta(g)$) and the discharge breakdown. As a result these phenomena increase the discharge stability.[4,10,12,19-22] Some other important parameters of the GPT can be obtained this way[4,10,12,19-22]:

1) Drift $z(g)$ and diffusion $L(g)$ lengths in avalanches

$$z^{(g)} \approx \theta^{(g)} \langle u_d \rangle \approx b_g \langle \Delta\theta \cdot u_d \rangle_g \ln(n^{(g)}/n^{(o)})$$

$$= b_g \langle \Delta z \rangle_g \ln(n^{(g)}/n^{(o)}) \tag{46}$$

and

$$L^{(g)} \approx \left[g \langle \Delta L^2 \rangle_g\right]^{\frac{1}{2}} = \left[6 b_g \langle D_i \cdot \Delta\theta \rangle_g \ln(n^{(g)}/n^{(o)})\right]^{\frac{1}{2}} \tag{47}$$

characterize the angles of avalanche "wedges" $\psi^{(g)} \approx L^{(g)}/z^{(g)}$. They are in agreement with experimental values $\psi_{obs}^{(g)} \approx (1 \text{ to } 5) \cdot 10^{-2}$ rad.[37]

2) Local time-dependent electron current density J and conductivity Σ rates R of electron impact-induced processes and the Debye radius can be obtained with the help of Eqs. (35) and (2).[4,10,12,19-22]

$$J(t) = e \cdot n(t) \, u_d \approx e \cdot n^{(o)} \mu E(t) \exp\left(\frac{t-t_o}{b_g \langle \Delta\theta \rangle_g}\right) \text{ and } \Sigma(t) = J/E \tag{48}$$

$$R \sim n^{(o)} N \left[\int u_e \cdot \sigma_R(\varepsilon_e) f(\varepsilon,t) \cdot d\varepsilon\right] \cdot \exp\left(\frac{t-t_o}{b_g \langle \Delta\theta \rangle_g}\right) \tag{49}$$

$$l_d \approx l_{od} \exp\left(-\frac{t-t_o}{2 b_g \langle \Delta\theta \rangle_g}\right) \text{ with } l_{od} = (\bar{\varepsilon}/4\pi e^2 n^{(o)})^{\frac{1}{2}} \tag{50}$$

These parameters are exponential functions of time.

3) Effective time-dependent electron statistical weight $\Delta\Gamma_e$, entropy S_e, kinetic energy U_e, kinetic temperature T_e, and "free energy" F_e are determined by

$$\Delta\Gamma_e(t) = \left[n_e(t) \left(\frac{10.8 \pi m \bar{\varepsilon}(t)}{3h^2}\right)^{3/2}\right]^{n_e(t) \cdot \delta V};$$

$$S_e(t) = k \ln \Delta\Gamma_e(t) \sim n(t) \tag{51}$$

$$U_e(t) = n_e(t) \cdot \delta V \, \bar{\varepsilon}_e(t); \quad T_e(t) = (\partial S_e/\partial U_e)^{-1} = 2\bar{\varepsilon}(t)/3k \quad (52)$$

and $F_e(t) = U_e(t) - S_e(t) \cdot T_e(t)$, where S_e, U_e and F_e are proportional $n_e(t)$ and therefore they are exponential functions of time, according to Eq. (35).

Conclusion

The presented consideration of the interaction between turbulence and gas discharge phenomena show that, in spite of great complexity of these phenomena, there exist encouraging possibilities of their theoretical treatment that can help to better understand related experimental facts and practical applications of TGDDs. Only the first steps have been made in this field. In this work, we concentrated mainly on the discussion of transient processes occurring in turbulent flow passing through strong electric fields, since these processes, in our opinion, present a key for understanding of observations and practical applications to lasers, electrogasdynamic energy generators, low-temperature plasma technology, etc.

These studies reveal the unexpected important general fact that the interaction of the two "chaotic" disordered phenomena associated with various instabilities -- turbulent fluctuations and discharge electric and light intermittances and random pulses -- leads to a considerable smoothing, a higher stability, and spatial and temporal uniformity (i.e., to a decrease in the "chaos degree" and the corresponding increase in ordering), at least for observed discharge phenomena. We think that the questions touched upon in this work deserve more attention and hope that the present review will stimulate additional studies in the field.

Acknowledgments

I am very grateful to my wife Nina for her help in the preparation of this manuscript for publication.

References

[1] Biblarz, O., and Nelson, R.E.," Turbulent Effects on an Ambient Pressure Discharge," Journal of Applied Physics, Vol. 45, Feb. 1974, pp. 633-637.

[2] Khait, Yu. L., and Biblarz, O., "Influence of Turbulence on a Diffuse Electrical Gas Discharge under Moderate Pressures," Journal of Applied Physics, Vol. 50, July 1979, pp. 4692-4699.

[3]Shwartz, J. and Lavie, Y., "Effects of Turbulence on a Weakly Ionized Plasma Column,", AIAA Journal, Vol. 13, May 1975, pp. 647-652.

[4]Khait, Yu. L., "Kinetic Aspects and Some Applications of Diffuse Gas Discharges in Turbulent Flows," Journal de Physique (Paris) Colloque, C.9, Supp. 11, Vol. 41., Nov. 1980, pp. 317-323.

[5]Shwartz, J., and Wasserstrom, E., "The Role of Gas Flow and Turbulence in Electric Discharge Lasers," Israel Journal of Technology, May 1975, pp. 122-133.

[6]Nelson, R.E., "Electric Discharge Stabilization by a Highly Turbulent Flow," M.S. Thesis, Naval Postgraduate School, Monterey, Calif., June 1973.

[7]Eckbreth, A.C., and Owen, F.S., "Flow Conditioning in Electric Discharge Convection Lasers," Review of Scientific Instruments, Vol. 43, July 1972, pp. 995-998.

[8]Eckbreth, A.C., and Davis, J.W., "RF Augmentation in CO_2 Closed Cycle D.C. Electric Discharge Convection Lasers," Applied Physics Letters, Vol. 21, July 1972, pp. 25-27.

[9]Reilly, J.R., "High Power Electric Discharge Lasers (EDLs)," AIAA Astronautics and Aeronautics, March 1975, pp. 52-63.

[10]Khait, Yu. L.,"Further Development of the Nonstationary Stochastic Model of the Diffuse Electric Discharges in Turbulent Gas Flows," Gas Flows and Chemical Lasers, Plenum Publishing Co., New York (in press).

[11]Barto, J.L., "Gasdynamic Effects on an Electric Discharge in Air," M.S. Thesis, Naval Postgraduate School, Monterey, Calif., Sept. 1976.

[12]Khait, Yu.L., "Some Nonequilibrium Trends and Concepts in Plasma Chemistry," Invited Paper at International Workshop on Plasma Chemistry in Technology, Ashkelon, Israel, Workshop Abstracts, March 1981, p. 44.

[13]Barto, J.L., "Study of Gasdynamic Effects in Non-Uniform High Pressure Electrical Discharges," Naval Postgraduate School, Monterey, Calif., Rept. 61152 N., RR 000-01-10, Aug. 1980.

[14]Post, H.A., "Subambient Controlled Turbulence Effects on Discharge Stabilization for Laser Applications," M.S. Thesis, Naval Postgraduate School, Monterey, Calif., Sept. 1976.

[15]Khait, Yu.L., "Non-Stationary Heat and Mass Transfer in Plasma Chemical Processes," Doctor of Science Thesis, Dept. Plasma Chemical and Radiation Processes, Moscow Petrochemical Sinthesis Institute, USSR Academy of Sciences, Moscow, 1972 (in Russian).

[16]Biblarz, O., Barto, J.L., and Post, H.A., "Gasdynamic Effects in Diffuse Electrical Discharges in Air," Israel Journal of Technology, Vol. 15, May 1977, pp. 59-69.

[17]Khait, Yu. L., and Biblarz, O., "Some Large Observable Effects of the Influence of Turbulence on Gas Diffuse Discharges," Bulletin of Israel Physics Society, Vol. 24, April 1979, K-5, p. 68.

[18]Davis, C.H., "Aerodynamic Stabilization of an Electrical Discharge for Gas Lasers," Naval Postgraduate School, Monterey, Calif., Sept. 1980.

[19]Khait, Yu. L., "Transport Phenomena in Turbulent Gas Diffuse Discharges and Applications to Plasma Chemistry," Invited Extraprogram Paper at the 5th International Symposium on Plasma Chemistry, Edinburgh, U.K., Aug. 1981.

[20]Khait, Yu. L., "Kinetic Aspects of Plasma Chemical Processes in Diffuse Gas Discharges Affected by Turbulence," Paper Presented at the 4th International Symposium on Plasma Chemistry, Zurich, Sept. 1979.

[21]Khait, Yu. L., "Non-Stationary Kinetics of Transient Electronic Phenomena in Pulse Gas Discharges at Low Field-to-Pressure Ratios," Proceedings of the 6th International Symposium on Plasma Chemistry, Montreal, July 1983, pp. 406-411.

[22]Khait, Yu. L., "A Non-Stationary Stochastic Model of Non-Ionized Gas-to-Plasma Transitions in Short-Term Electric Field Pulses," Abstract, XIVth International Conference on Thermodynamics and Statistical Mechanics, Edmonton, Canada, Aug. 1980.

[23]Landau, L.D., and Lifshitz, E.M., Fluid Mechanics, Addison-Wesley, Reading, Mass., 1959, Chap. III.

[24]Loeb, L.B., Electrical Coronas, University of California Press, Berkeley, 1965.

[25]Brown, S.C., Introduction to Electrical Discharges in Gases, John Wiley & Sons, New York, 1966.

[26]Neumann, J., Theory of Self-Reproducing Automata, University of Illinois Press, Urbana, 1966.

[27]Glandsdorf, P., and Prigogine, I., Thermodynamic Theory of Structures, Stability, and Fluctuations, John Wiley & Sons, New York, 1971.

[28]Nicolis, G., and Prigogine, I., Self-Organization in Non-Equilibrium System, John Wiley & Sons, New YOrk, 1977.

[29]Prigogine, I., From Being to Becoming, Freeman W.H. and Co., San Francisco, 1980.

[30]Haken, H., Sinergetics; An Introduction: Non-Equilibrium Phase Transitions and Self-Organization in Physics, Chemistry and Biology, Springer, Berlin, 1977.

[31] Khait, Yu. L., "Non-Equilibrium Modeling and Dissipative Structures in Solid Material-Plasma Interactions," Proceedings of the 1983 American Material Research Society, Symposium Plasma Processing, Elsevier Science Pub. Co., New York (in press).

[32] Khait, Yu. L., "Kinetic Many-Body Theory of Short-Lived Large Energy Fluctuations of Small Numbers of Particles in Solids and its Applications," Physics Reports, Vol. 99, Oct. 1983, pp. 237-340.

[33] Nasser, E., Fundamentals of Gaseous Ionization and Plasma Electronics, Wiley-Interscience, New York, 1971.

[34] Khait, Yu. L., Inspector, A., and Avni, R., "The Dependence of Coating in Intuctive RF Plasmas on Gas Flow Velocity, Pressure and Power," Thin Solid Films, Vol. 72, Nov. 1980, pp. 249-260. (also Invited Paper at the International Conference on Metallurgical Coating, San Diego, April 1980).

[35] Khait, Yu. L., Carmi, U., and Avni, R., "Mechanism in the Plasma-Substrate Region in a Low Pressure Microwave Plasma," Proceedings of 6th International Symposium on Plasma Chemistry, Montreal, July 1983, pp. 729-736.

[36] Khait, Yu. L., "Collective Non-Equilibrium Models in Plasma-Solid Material Interactions," Invited Paper at the International Workshop on Plasma Chemistry in Technology, Ashkelon, Israel, Workshop Abstract, April 1981, p. 47.

[37] Raether, H., Electron Avalanches and Breakdown in Gases, Butterworths, London, 1964.

Homotopic Structural Invariants in HD and MHD Turbulence

Daniel Weil*

Racah Institute of Physics, The Hebrew University, Jerusalem, Israel

Abstract

Local topological invariants in ideal hydrodynamics and magnetohydrodynamics, well-known consequences of the "freezing" of the vorticity or magnetic flux, lose their importance as dynamical constraints beyond reconnection time-scale, in laboratory non-ideal conditions.

By contrast, the presented three-dimensional topological invariants are global, owe their constancy solely to continuous evolution under some boundary constraints and the absence of null points, and may hold on large diffusion time-scale, in presence of resistivity.

The new conservation laws have as a consequence restricted the combination of the usual local topological structures which may appear during any (even turbulent) evolution. Moreover, they single out unfamiliar toroidal structures, which in simple examples, are solitary ring vortices with swirl in a surrounding laminar flow.

Implications of these results to turbulence analysis are discussed, in particular in the context of the renewed attempts to model turbulent flows as collections of large coherent structures.

Introduction

Topological invariants are known to arise from the "freezing" of magnetic or vorticity flux in fluids that evolve under ideal equations. One celebrated example is

---

Paper presented at the Fourth Beer-Sheva Seminar on MHD Flows and Turbulence, Ben-Gurion University of the Negev, Beer-Sheva, Israel, Feb. 27-March 2, 1984. Copyright © 1985 by the American Institute of Aeronautics and Astronautics, Inc. All rights reserved.
*Research Associate.

the helicity, which, for special cases, has been related to the linking number of closed vortex lines.[1] Their physical relevance in laboratory conditions as effective dynamical constraints of motion is limited by non-ideal effects, such as those associated with resistivity, viscosity, or microturbulence.[2] There exist, however, other topological invariants of a quite different origin that hold for generally less stringent physical conditions.

These are homotopic invariants[3] that are global properties of the magnetic or vorticity field configurations and that owe their constancy to the continuity of the field evolution under two conditions: external constraints that restrict the range of directions allowed at the boundary and non-appearance of zero points in the field. The latter requirement, fulfilled in pure MHD or HD as a well-known consequence of Kelvin's theorem for proper initial conditions, is satisfied in general for a longer lifetime than the non-disconnection of field lines in non-ideal situations. Under some symmetry assumptions, somewhat unfamiliar toroidal structures are singled out where limit cycles for flux lines winding the surface of the toroid, antiparallel to the magnetic axis, constitute an essential feature.[4]

We present here further development of this early research, new conservation laws, and new steady coherent structures, and we discuss some of their implications for turbulence analysis along two modern approaches in this field: on one hand, a growing interest in the invariants of motion on the basis of the conjectured important role they may play on the large-scale properties of turbulent flows as constraints[5]; on the other hand, renewed attempts to model various aspects of turbulent flow as a collection of large coherent flow structures.[6]

The Basic Invariant

When the flow occurs within a simply connected volume (such as channel or a sphere) imbedded in an homogeneous external magnetic field or submitted to uniform rotation, the basic considered homotopic invariant is known in topology as the Hopf invariant and has been introduced in physics in many instances. It has sometimes been assimilated to the total helicity of the flow (see, for example, Ref. 5). But this is true only for special unphysical cases. On the other hand, the Hopf invariant can easily be understood in the context of HD and MHD by analogy with helicity. The conservation of helicity in ideal HD and MHD is essentially related to the fact that the topology of individual vortex or magnetic flux lines and surfaces are preserved. In particular, if the flow includes only two closed vortex lines, the total helicity $\int_V \vec{\omega}\cdot\vec{v}dV$ is proportional to the number ("linking number") of times the two loops link each other. It is preserved, because if these loops move with the fluid, they cannot open during the motion. This is then generalized to a continuous vorticity field. Now, instead of considering a vortex line (along which the vorticity field is tangent at every point), consider a line of constant direction of vorticity (along which the vorticity has same direction at every point): if it does not touch the boundary, it is, in most cases, a closed loop (Fig. 1). Suppose now, that there is no point of zero vorticity in the fluid volume during evolution. Then, two such closed loops, corresponding to two different directions, cannot cross each other: their linking number is conserved. From a theorem of Hopf, it follows that if the field at the boundary of the considered region of the flow has a fixed direction, then this "linking number" is independent of the two chosen directions (provided one does not consider just one, but all closed loops associated with each direction). This number K can be expressed as an integral by describing all these loops as lines of force of a field $\vec{P}: \vec{P} = \sigma(\alpha\beta)\vec{\nabla}\alpha\times\vec{\nabla}\beta$, where α and β parametrize the directions of $\vec{\omega}$. With a proper normalization of P (see Appendix), K is just the total helicity of this field:

$$K = \int_V \vec{P}\cdot\vec{A}\, dV \qquad (1)$$

where $\vec{\nabla}\times\vec{A} \equiv \vec{P}$.

In the assumed conditions, K, the "homotopic charge" of the configuration, is a global constant of motion. An identical discussion can be made for magnetic field of MHD by substituting $\vec{B}$ to $\vec{\omega}$.

The Coherent Structures

Let us first give an example of an axisymmetric magnetic configuration with homotopic charge 1 in a sphere of radius R surrounded by an homogeneous field of strength B_R. An analytical description may be given in radial coordinates (r, θ, γ), with origin at the center of the sphere:

$$B_r = B \cos\theta$$

$$B_\theta = -B \sin\theta \cos w \qquad (2)$$

$$B_\gamma = -B \sin\theta \cos w$$

where w is a function of r that satisfied

$$w(o) = 0 \quad w(r) = 2\pi$$

$$w(o) = o \quad w(r) = 2\pi \quad \frac{dw}{dr} > 0 \qquad (3)$$

This implies that $\vec{n}$ rotates by a complete 2π rotation around any radial direction as it progresses along this direction from the origin to the boundary.

B increases in a spherically symmetric way from the boundary to the origin by

$$B(r) = B_R \exp 2 \int_R^r (1-\cos w) \, r^{-1} \, dr \qquad (4)$$

One can show[4] that such a field is solenoidal with K=1. Roughly, it is a family of nested toroidal surfaces "locked in" by the lines of force originating from the boundary, with no null point (Fig. 2). The x point on the separatrix corresponds to a limit cycle for all lines of force on the surface of the toroid. This magnetic circle has a toroidal component opposite to the magnetic circle at the 0 point (Fig. 3).

Any axisymmetric deformation of the flux surfaces of this simple analytic examples does not modify the value of K. We call such a configuration "Dag" (Hebrew for "fish", appropriate to the characteristic shape of the separatrix.) It turns out that one can derive a formula for the most general axisymmetric configuration with a given K number in terms of the basic structural constituents of the Dag.

Let us assign to an index γ the value +1 for a "0-point" region ("head") and the value -1 for an "x-point" region ("tail") (Fig. 4). Let ε be a second index of module 1/2 and of sign equal to the sign of the poloidal component of the magnetic axis at the x or 0 point.

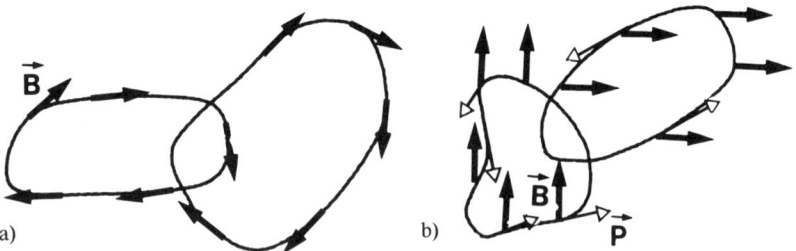

Fig. 1 (a) Linking topology between two B magnetic lines (at the origin of conservation of helicity) vs (b) linking topology between loci of same direction of B (at the origin of conservation of homotopic charge k). These loci are lines of force of field P.

One can show[7]

$$K = \sum_{i=1}^{N} \varepsilon_i \gamma_i \sum_{i=1}^{N} \gamma_i = 0 \qquad (5)$$

that is, any axisymmetric magnetic with N/2 "heads" and N/2 "tails" components has homotopic charge K according to this formula. Inversely, any axisymmetric configuration of charge K necessarily includes these local but extended magnetic structures in a combination imposed by this formula.

For instance, the configuration of Fig. 5 is made of two pairs of "heads" and "tails" with opposite ε; it has

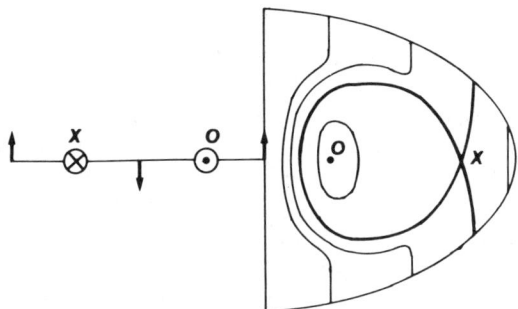

Fig. 2 A typical axisymmetric configuration with K-1 ("Dag"); cross sections of flux surfaces. A: axis of rotational invariance.

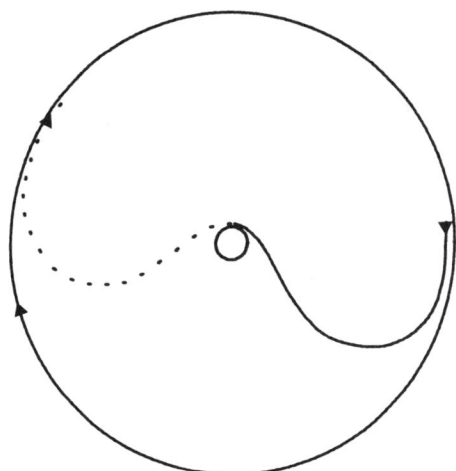

Fig. 3 "Dag" (view from above). The two magnetic circles (O and X of previous figure) have opposite toroidal component. X is a limit cycle for all flux lines on the surface of the outermost toroid.

charge K=0. It is the direct analog to the "breather" of the 1-d Sine-Gordon equation.

Conservation Laws in HD and MHD

In ideal MHD, K is an exact constant of the motion if the magnetic field is imposed in a fixed direction at the boundary (this may happen for instance if the magnetofluid is imbedded in an external homogeneous field) and if, initially, B does not vanish. K is then conserved because "null points" cannot ulteriorly develop as a well-known consequence of the equation:

$$\frac{\partial \vec{B}}{\partial t} = \vec{\nabla} (\vec{\sigma} \vec{B}) \qquad (6)$$

Since an analog equation holds for the vorticity field w in an ideal hydrodynamic, rotational flows remain rotational, and the HD equations have the homotopic constant of motion K [see Eq. (1)]. This strictly holds for an incompressible unforced fluid. For a barotropic fluid in potential fields of external force, $\vec{w}/\rho$ should be substituted for $\vec{w}$. The proper boundary conditions may be satisfied for a rotating fluid. The corresponding coherent structures (Figs. 2-5) in this field are then described in terms of vortex surfaces.

Before we discuss the question whether such Dag structures may exist as steady entities in the ideal case, we wish to emphasize that the full interest of this conservation law appears in non-ideal conditions. This is because, contrary to other known constraints of motions in HD and MHD (besides, perhaps, total helicity), K is conserved independently of the reconnection of the field lines, as long as no "null points" appear. Now, in general, occurrence of reconnection and of null points corresponds to two different time scales. A typical estimate for plasma toroidal configurations[8] is

$$\gamma_{recon} = (\gamma_{dif})^{1/3} (v_A/R)^{2/3} \qquad (7)$$

where v_A is the Alfven velocity and R the major radius of the toroid.

In Tokamak conditions where "tearing modes" have been best studied, when the disruptive reconnection takes place, the magnitude of the magnetic field has changed by only a few percent. Thus, if no competitive process to simple diffusion speeds up the appearance of a zero point, K remains constant well beyond reconnection time.

HOMOTOPIC STRUCTURAL INVARIANTS 247

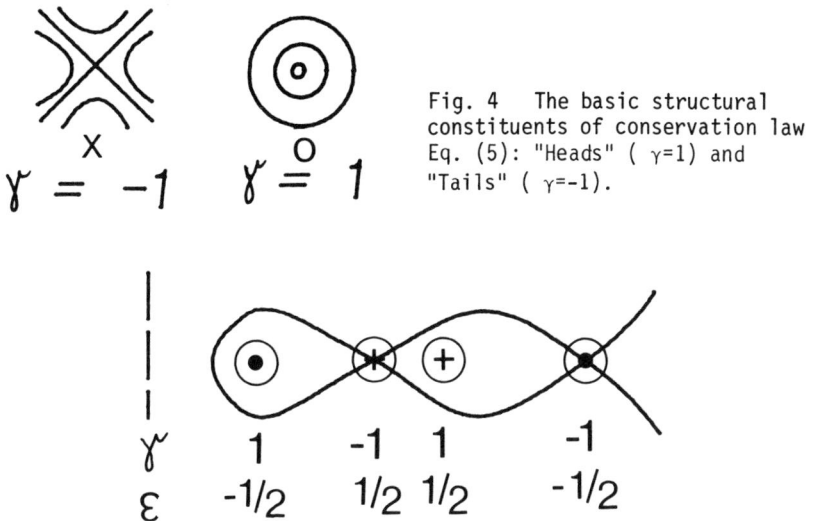

Fig. 4 The basic structural constituents of conservation law Eq. (5): "Heads" ($\gamma=1$) and "Tails" ($\gamma=-1$).

Fig. 5 A "breather"; axisymmetric configuration with K = 0.

Implications for Turbulence

Structural Forbidden Transition

According to Eq. (5), the global invariance of K on a relatively large time scale results in restricting the combination of possible local structures of types "heads" and "tails" (of Fig. 4) that may appear during a turbulent evolution.

The most simple type of structural "allowed transition" is a "pair creation": kinematically, a quasi-homogeneous region (K=0) may evolve to a "breather" configuration (Fig.5) which may then evolve to a pair of separated "Dags" with opposite respective ε ("Dag" and "anti-Dag") (Fig. 6). Note that analog one-dimensional processes have been observed in a chaotic Sine-Gordon molecular chain computer simulation.

Turbulence as a Gas of "Dags"

The possibility of these "creation processes", the relatively long lifetime estimated for the individual basic structures, and the generality of their characteristic features (due to their topological origin) suggest the attempt to analyze turbulence as a gas of such entities along the lines of the precedent models stated in the introduction.

For such an analysis to be quantitative, one needs exact solutions with K=1 of the MHD dynamical equations, which preserve their shape during evolution. An axisymmetric Dag of Fig. 1 cannot be force-free; were this possible, B_γ times the distance to the axis of symmetry would be constant on any magnetic (and current) surface. But the separatrix surface, on one hand, touches the boundary and, therefore, $B_\gamma=0$ and, on the other hand, includes the × point, which cannot have a null toroidal magnetic component unless it is a null point. Thus, $\vec{B}$ parallel to $\vec{J}$ is excluded.

By similar reasoning, the poloidal components of B and J cannot be parallel. They, therefore, yield necessarily a net toroidal component to the magnetic force. Because of axisymmetry, the latter cannot be balanced by a pressure gradient.

In a "Dag", therefore, matter is necessarily in motion. Can this motion be steady? Although we do not have a general theorem at this stage, we can provide a class of examples. In the incompressible case, if the ideal fluid moves everywhere parallel to the magnetic field, the motions are steady provided that

$$v = \rho^{-1/2} B \qquad (8a)$$

$$p + B^2/2 = \text{const} \qquad (8b)$$

In the analytic examples of Eq.(2), Dags are regions of higher magnetic energy density than at their boundary. Condition (8b) can therefore be realized if the imposed boundary total pressure $P_R + B^2/2$ is high enough, and it implies that, for an ideal gas equation of state, these steady Dags have a lower temperature profile than surrounding homogeneous magnetoplasma (we suppose zero thermal conductivity).

These solutions are easily visualized, since the stream lines coincide with the magnetic lines of the Dag (Figs. 2 and 3). They look roughly like cold ring vortices with swirl in a surrounding laminar flow.

Fig. 6   Three stages in a "pair creation".

If Eq.(8a) is not satisfied at the boundary, then one has a traveling steady Dag along the homogeneous lines of force with velocity $V = V_R - \rho^{-1/2} B_R$.

Three conditions are required for these structures to play a significant role in turbulence analysis: stability, lifetime, and "entropy". The fundamental, attractive feature of these structures is that they possess a different origin for their persistence than other known coherent structures: they cannot decay in the "vacuum" (namely, a least energy homogeneous configuration) unless they become singular (B=0 or "shock") at some point, whatever the non-ideal conditions are. As pointed out before, a typical time scale for B to become locally null is the diffusion time, which may be quite long with respect to the time scale of the analysis of the turbulence. However, this time is only an upper limit, and there may be other competitive processes. For instance, consider the steady "Dag" of Eqs.(8) and (4). Its total energy is

$$E = \int_V (\tfrac{1}{2} \rho v^2 + \frac{\rho}{\gamma-1} + \frac{B^2}{2})\, dV$$

$$= E_0 + (2\gamma-3)(\gamma-1)^{-1}\, v\, \frac{B_R^2}{2}\, [1-f(\omega)] \qquad (9)$$

where

$$f(\omega) = 3 \int_0^1 dx\, x^2 \exp 4 \int_1^x (1-\cos\omega)\, x^{1-1}\, dx^1 \qquad (10)$$

E_0 is the "vacuum" energy (corresponding to $\omega \equiv 0$). Each function ω obeying conditions (3) yields a steady equilibrium with a different energy. Suppose that the ideal magnetofluid is initially in such a configuration and that viscosity and resistivity are "turned up" so that all the dynamical constraints[9] are relaxed. Then the fluid can evolve from one such equilibrium configuration to another by a small change δw of w. One can easily check by the analysis of the resulting δE that the energy can be made arbitrarily close to the vacuum. E_0 by constructing and steepening more and more the transition of w from zero to 2π around the circle where $w = \pi$. These "squeezed" configurations will then resemble vortex rings except that the internal structure gives rise to large unstable field gradients. If no constraint stops the shrinking process, they would rapidly disappear through disruption.

A second interesting and open question is the way such homotopic structures interact with one another.

As for the entropy, the only statement that we can make at this stage is that the rough estimate for the probability of occurrence of a pair of Dags and anti-Dag of energy out of vacuum e^2 $(E-E_0)/KT$ (where T is the temperature of the turbulence and where conditions of "easy" field reconnections are supposed)favor the small radia and small section vortex ring types among all such configurations. It is interesting to note that similar structures were postulated by Alfven[10] to be created in pairs in the solar core in a controversed scenario for the origin of sun spots, but those differed at least on two points: they were linear perturbations of the background magnetic field and had trivial homotopic structure.

In conclusion, new conservation laws of topological origin in MHD and HD have been presented as well as steady new solutions. Only further theoretical study will tell whether these may provide useful tools for turbulent analysis in some regimes. Note that in addition a better understanding of the stability properties of these homotopic invariants are most interesting, in the context of plasma confinement. Can these homotopic structural invariants in effect be observed?

Appendix: The Hopf Invariant for (MHD) Pedestrians

Consider a magnetic field $\vec{B}$ that does not vanish anywhere. We can write $B \equiv Bn$ where n is a unit vector in the direction of $\vec{B}$. It can be expressed in terms of two scalar fields θ and ϕ by

$$n_x = \sin\theta \cos\phi \quad n_y = \sin\theta \sin\phi \quad n_z = \cos\theta \quad (A1)$$

θ and ϕ are simply the polar coordinates of a sphere S of radius unity. Statement (A1) is, however, only true if θ is different from zero ($\vec{n}$ not parallel to the z axis, the polar axis of the sphere), in which case ϕ is indefinite. To include these directions, one must patch this parametrization with a second one of this type where $\theta = 0$ corresponds to a different direction.

Consider the vector field

$$\vec{P} = \sigma(\theta,\phi) \vec{\nabla}\theta \times \vec{\nabla}\phi \quad (A2)$$

since $\vec{P} \vec{\nabla} \theta = 0$ and $\vec{P} \vec{\nabla} \phi = 0$, the lines of force of P are loci where θ = constant, ϕ = constant, i.e., $\vec{B}$ has a same direction $\vec{n}_0$. If $\vec{n}$ is analytic, such loci (if $\vec{n}$ is not on the boundary) are closed lines. Let C_0 be such a closed line. Consider the flux of $\vec{P}$ through a surface S_0 bounded by C_o: $F = \int_{S_0} \vec{P} \cdot d\vec{S}$. If σ is the surface density on the sphere of which Θ and ϕ are coordinates (i.e., $\sigma = \cos\Theta$), then $\vec{P} \cdot d\vec{S}$ is just the portion of the surface of the sphere, image of the element dS by the function n. Since C_o corresponds to one direction, the total image surface is bounded by a point. It must, therefore, be an integer number of times, m (which may be zero), the total surface of the sphere, and

$$\int_{S_o} \vec{P} \cdot d\vec{S} = m \qquad (A3)$$

But m is also the number of times any other direction crosses the surface S_o. Thus, Eq. (A3) calls the relation obtained by Moffat[11] for discrete vortex filaments. The P lines may be seen as closed vortex filaments with strength unity. One can pursue the analogy with a vorticity field by noting that, for a continuous evolution of $\vec{n}$ (and $\vec{n}$ analytic), the loops C move also continuously and their motion can be described by a velocity field $\vec{v}$.

F being an integer, it cannot jump and, therefore,

$$\frac{d}{dt} \int_C \vec{P} \cdot d\vec{S} \qquad (A4)$$

It follows that P obeys

$$\frac{\partial \vec{P}}{\partial t} = \vec{\nabla} \times (\vec{\sigma}_\times \vec{P}) \qquad (A5)$$

This equation implies[2] that for every volume bounded by $\vec{P}$ lines, the quantity $K = \int \vec{A} \cdot \vec{P} \, dV$ is an invariant of "motion" ($\vec{\nabla} \times \vec{A} \equiv \vec{P}$). These invariants have, in general, no physical significance, since the P surfaces do not move with the fluid. One important exception is K, the total helicity of P, when the boundary of the volume of the fluid coincides with a P surface, for instance, by imposing $\vec{n} = \vec{n}_o$ on the boundary. Let us show that K is an integer. As long as Θ and ϕ are single valued, $\vec{A}$ can be written:

$$\vec{A} = -\sin\Theta \, \vec{\nabla} \phi \qquad (A6)$$

As noted before, Θ and ϕ can be single-valued except for any two arbitrary directions $\vec{n}_1$ and $\vec{n}_2$ [antiparallel, for the parametrization (A1)] . It follows that in these coordinates (and in this gage) $\vec{A}\vec{P}$ is zero except along the lines $\vec{n} = \vec{n}_1$ and $\vec{n} = \vec{n}_2$. Other lines can be ignored. But for a discrete number of vortex lines with strength 1, Moffat[11] showed that K is just the total linking number between loops C_1 and C_2. Thus, we have recovered a well-known theorem in topology that K is the linking number between any two inverse images of the field $\vec{n}$.

Acknowledgments

I would like to express my thanks for the advice and encouragement I received from D. Finkelstein and M. Kruskal. I would also like to acknowledge very useful conversations with A. Ramani, M.N. Bussac, J. Katz, and G. Laval. This work benefited from the support of the Joseph Broth Foundation.

References

[1] Moffat, H.K., Journal of Fluid Mechanics, Vol. 106, 1981, pp. 27-47.

[2] Taylor, J.B., Physical Review, Letters, Vol. 33, No. 19, 1974, pp. 1139-1141.

[3] Finkelstein, D., and Weil, D., International Journal of Theoretical Physics, Vol. 17, No. 3, 1978, pp. 201-270.

[4] Weil, D., "Magnetohydrodynamic Kinks", Ph.D. Thesis, Yeshiva University, New York, University Microfilm International.

[5] Frenkel, A., Levich, E., and Stilman, L., Physics Letters Section A, Vol. 88a, No. 9, 1982, pp. 461-465.

[6] Saffman, "The Role of Coherent Structures in Modelling Turbulence and Mixing," Lecture Notes in Physics, edited by J. Jimenez, Springer-Verlag, New York, 1981, pp. 1-9.

[7] Weil, D., submitted for publication, 1984.

[8] Furth, H.P., Rosenbluth, M.N., Physics of Fluids, Vol. 6, No. 4, 1963, pp. 459-484.

[9] Kruskal, M.D., and Kulsrud, R.M., Physics of Fluids, Vol. 1, No. 4, 1958, pp. 265-274.

[10] Alfven, H., Cosmical Electrodynamics, Clarendon Oxford, 1950.

[11] Moffat, H.K., Journal of Fluid Mechanics, Vol. 35, Part I, 1969, pp. 117-129.

Chapter III. Two-Phase Flows

Liquid Metal Magneto-Fluid-Mechanic Turbubblence

Paul S. Lykoudis*
Purdue University, West Lafayette, Indiana

Abstract

This paper deals with the experimental results of a two-phase co-current flow inside a plexiglass tube in which a mixture of mercury and nitrogen flow in the presence of a transverse magnetic field. The Reynolds number is kept constant at Re = 100,000. The magnetic field varies between 0 and 1 T. Three series of experiments are conducted for various entrance geometries and void fractions. Measurements are reported of local void fractions obtained through four diametric traverses with the help of a conductivity probe. The motion of some large bubbles on the test section wall is recorded, and their kinematic and geometric behavior is analyzed.

Introduction

Purdue University's Magneto-Fluid-Mechanic Laboratory was founded in 1960 with a grant from the National Science Foundation. Today, the research goals of the laboratory remain the same as those of more than 20 years ago, that is, the conductance of experiments and theoretical work aiming at an understanding of the interaction of liquid metal flows in the presence of electromagnetic fields. Even though the work is basic rather than applied, it is, nevertheless, of importance today, from a technological point of view, to those interested in the following two areas: the design of lithium blankets for proposed fusion reactors and

---

Paper presented at the Fourth Beer-Sheva Seminar on MHD Flows and Turbulence, Ben-Gurion University of the Negev, Beer-Sheva, Israel, Feb. 27-March 2, 1984. Copyright © 1985 by the American Institute of Aeronautics and Astronautics, Inc. All rights reserved.
*School of Nuclear Engineering.

the development of solar liquid metal magneto-hydrodynamic (LMMHD) generators.

Lithium blanket designs introducing two-phase flows have been proposed and, to some extent, researched by several research groups (see, for example, Refs. 1-4). Solar LMMHD generators seem to be under development now in Israel[5] and France.[6] In both of the above cases, the presence of a magnetic field distorts the original flow, and the task is to investigate the changes in local flow structure and, where appropriate, the momentum and heat-transfer rates.

Our understanding of this problem is very limited, even in the ordinary fluid mechanic (OFM) case which, especially for liquid metals, is by no means a solved problem, since most of the literature deals with the overall behavior of the flow rather than local measurements. Credible general theories that can explain and predict such local parameters as void fraction, liquid and gaseous phase geometries, velocity and temperature profiles, turbulence intensities correlations and spectra, slip ratios, pressure drops, and heat-transfer rates are lacking. From this point of view, some of our work at the LMMFM laboratory could be useful to applications where the magnetic field is absent (for example, in sodium-cooled breeder nuclear reactors).

Details of work conducted in our laboratory in the area of two-phase LMMHD can be found in the theses of Wagner,[7] Gherson,[8] and Kirk.[9] For related work in the open literature, see Refs. 10-14.

This paper will not undertake a detailed review of previous work, since all of the work above and, in particular, Refs. 8 and 9 fulfill this purpose. Specific references will be made only to work directly relevant to the present paper, which deals with two-component flow (liquid metal and nitrogen gas) in the bubbly regime, inside circular test sections under isothermal conditions.

When the present work was initiated five years ago, our first goal was twofold: first, investigate to what extent hot-film anemometry and conductivity probes could be developed for local measurements in two-phase flows and, second, study in some detail, extremely simple two-phase flows aiming at giving an insight, always at the local level, of the interaction of the gaseous with the liquid phases in the presence of an externally applied magnetic field.

The first task was essentially accomplished with the work of Gherson[8] and was continued by Kirk,[9] where it was shown that special miniature conical hot-film and conductivity probes could provide reliable data with the help of an on-line minicomputer. This task was accomplished with a stagnant mercury pool experiment consisting of a 125-cm-long Plexiglas cylinder, 10 cm in internal diameter, at the bottom of which nitrogen gas bubbles could be released. All the characteristics of the buoyancy induced turbulence were measured with hot-film and conductivity probes. In order to assess the superposition of a turbulent field, for which we have knowledge of its behavior in the OFM case, with the turbulence generated by a second phase, two experiments were designed. In the first we introduced a grid moving like a piston inside the cylindrical pool as described above with a conical hot-film probe attached behind it and moving with the same speed. In this fashion it was possible to study the superposition of the grid and buoyancy generated turbulence. The second experiment consisted of a small vertical loop (SVL) whose test section was also a Plexiglas cylinder 3.8 cm in diameter and 125 cm long. The flow was in the upward direction. Nitrogen bubbles were released at the entrance. Measurements were made near the exit of the test section with the help of a traversing mechanism capable of azimuthal rotation. The entire test section was immersed inside a horizontal magnetic field whose maximum intensity was about 1 T. Here, again, the study aimed at understanding the nature of the interaction between the wall shear and buoyancy generated turbulence.

Both of the above aspects were investigated in Ref. 8 and reported in Refs. 13 and 14, the main emphasis being the development of the proper instrumentation rather than extensive study of the flows themselves.

The experiments in the SVL have continued (since the publication of Refs. 13 and 14), with the emphasis guided by the puzzles that this work generated rather than the explanations it offered. It is this continuation that will be emphasized in the present paper. However, before we enter into details, the following small special section is inserted to introduce a number of cautionary remarks to this new kind of "superimposed turbulence" and the special problems it creates when looked upon from the point of view of our classical definition of turbulence. The arguments that follow have convinced the author that the word "turbulence" is not appropriate when two phases are present. The new situation needs another word to describe it. I wish to introduce the term "turbubblence," which already appears in the title of the present paper.

Turbubblence

Consider the familiar turbulent field created behind a grid, or by the wall shear inside a pipe or over a solid surface. To these examples, one can add the turbulent field in the wakes behind cylindrical or spherical objects, or turbulence generated by jets. In all these cases, the velocity field is timewise and spacewise continuous even in the presence of so-called "intermittency regions" due to "turbulent bursts," as in the case of sublayers in boundary layers, or superlayers made up of vortices swimming at the interface between the core of the turbulent wake and the flow prevailing away from it.

Consider now the case of a stagnant flowfield in which a single bubble has been injected in the form of a gaseous phase. Because the interface of the bubble and the liquid contains impurities, the normal slip condition is not observed, and, as a result, a shear layer developed there creates a turbulent wake, apparently not dissimilar to our familiar wakes behind solid objects.[15] If measurements are conducted in a frame moving with a bubble, nothing novel is encountered, except perhaps for the wobbliness of the bubble if the bubble happens to be large enough. Now let us add to this bubble another one running just behind it. In this new situation, the velocity field around the second bubble will be modified by the velocity field of the front running bubble. If we can now imagine the liquid phase surrounding a great number of bubbles, the following observations can be made. The signal obtained by a probe stationary in the laboratory will be interrupted every time a bubble passes. This signal will contain portions that belong to turbulent regions as they develop behind the bubble wakes, but equally well to all the other in-between-the-bubbles and the wake regions, where the flow is not turbulent but is either intermittent, laminar, or simply nonsteady owing to the wobbly movement of the bubbles. In such a situation, even if we remove from the velocity signal the portion that belongs to the gaseous phase (essentially what we currently do in such cases, in a more or less arbitrary way, mostly based on intuition and common sense), what still remains is difficult to be characterized as "turbulence" in the classical sense, even though calculations of time and space correlations or spectra could still be made from such a measurement. This new situation, difficult as it may be to analyze, is of course of great technological interest for the study of overall diffusion of momentum, energy, and mass. Overall diffusion coefficients can still be experimentally measured, but it is clear that the classical mechanisms of turbulence we have struggled to understand during the last ten decades or so (in particular, the last five), do not lend themselves readily to an extrapolation that can handle the new situation. A good beginning, having as a goal the study of this new "concoction," is to call it by a different name. We propose the name "turbubblence," a word that has the elements of a good description containing the roots of both bubble and turbulence, with the additional advantage of a very suggestive sound to it.

Rationale and Description of the Experiments

Our small vertical loop (SVL) consists of a Plexiglas test section 3.8 cm in diameter and 125 cm long, as already stated. The flow is in the upward direction provided by a positive displacement pump capable of a Reynolds number of about 100,000. The loop is shown schematically in Fig. 1 and contains among other elements an electromagnetic flow meter, a separator, at its top, and a heat exchanger to remove the heat added by the pump. The test section is inserted inside the poles of an electromagnet 125 cm long,

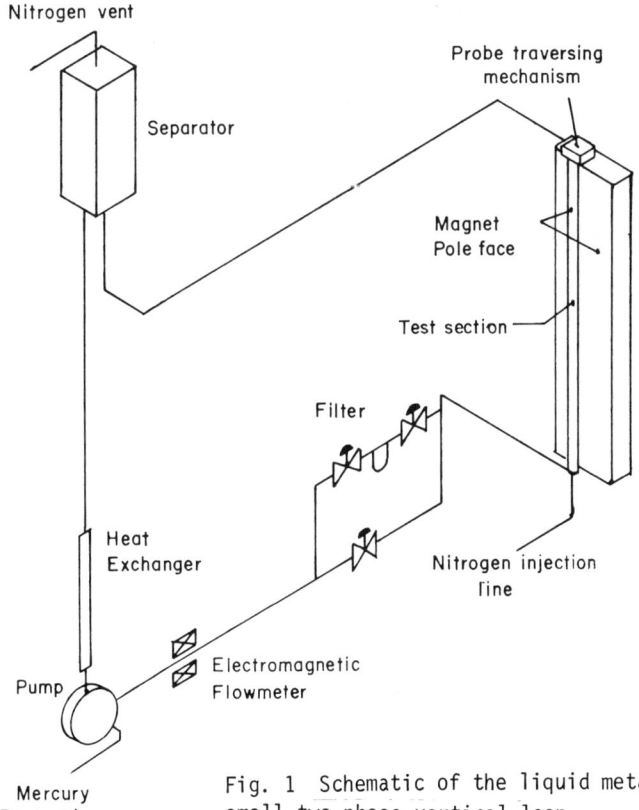

Fig. 1 Schematic of the liquid metal small two-phase vertical loop.

30.5 cm wide, and capable of providing, at a gap of 15 cm, a magnetic field intensity of about 1 T. The nitrogen bubbles are released at the bottom of the test section where the flow enters through a piping tee. A traversing mechanism, very close to the exit of the test section, allows the insertion of a probe to perform local measurements in practically any azimuthal orientation.

Our early work in Refs. 8 and 14 was conducted by releasing bubbles through a number of injectors, protruding through the vertical leg of the piping tee to about the centerline of the incoming flow of mercury from the tee's horizontal leg. At this point, the flow turned 90 deg in the upward direction, entraining the bubbles, which were then intercepted near the exit by the hot-film or conductivity probes. The main result of this experimental effort was the development of the technique for the use of the miniature conical hot-film probes, which, in conjunction with a conductivity probe, could yield in a reliable and repetitious fashion the following quantities: local mean liquid velocities, turbulence intensities, turbulence spectra, void fractions, bubble chord lengths, and bubble velocities. The traverses were conducted along two directions: North-South (N-S) (parallel to the magnetic field) and East-West (E-W).

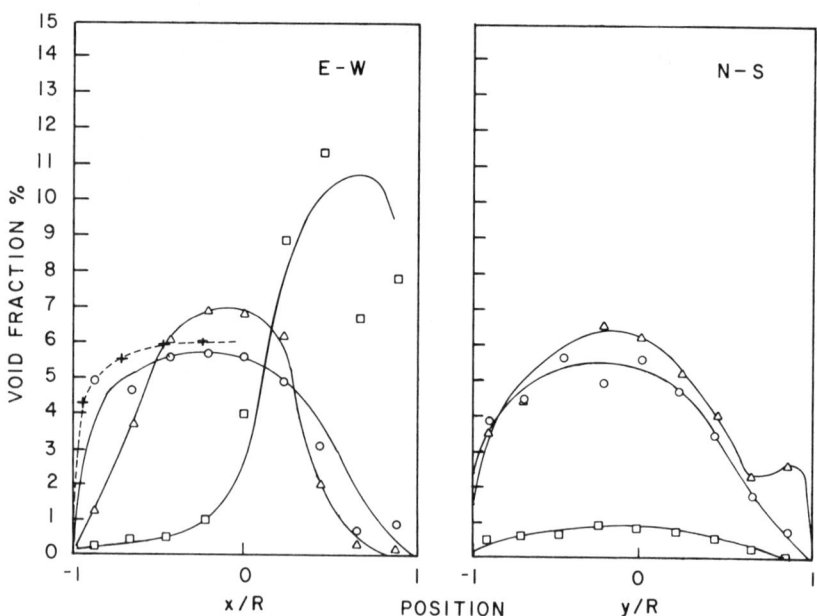

Fig. 2 Profile of the local void fraction (conductivity probe) B (Tesla): (o) 0.0, (□) 0.3, (△) 0.9.

In the OFM case (B = 0), the bubbly flow (average void fractions was of the order of about 5-7%) did not yield any surprises; it was found to be essentially in agreement with previous investigations. The void fraction exhibited a bell-shaped axisymmetric profile of maximum value of about 6%. The mean velocities increased because of the bubble entrainment. The turbulence intensities also increased by the bubble presence. When the magnetic field was increased to the level at which one would expect to see the beginning of "laminarization" for single-phase flow, a degree of intermittency was noticed in the turbulence level pretty much as it was recorded in the single-phase flow work reported in Ref. 17. This occurred for a magnetic field of about 0.3 T. However, the axisymmetry of the profiles of all the quanitities measured ceased to exist, and, instead, a steep maximum of the void fraction appeared in the E-W direction on the west side with a value of almost twice as much as in the case of B = 0, (10% from 6%). The N-S direction was also seen to be depleted of bubbles, the level of void fraction there being around 1%. Analogous increases were noted for the chord length and liquid velocities, as one would normally expect for such a shift in the void distribution. These effects are shown in Figs. 2, 3, and 4. It

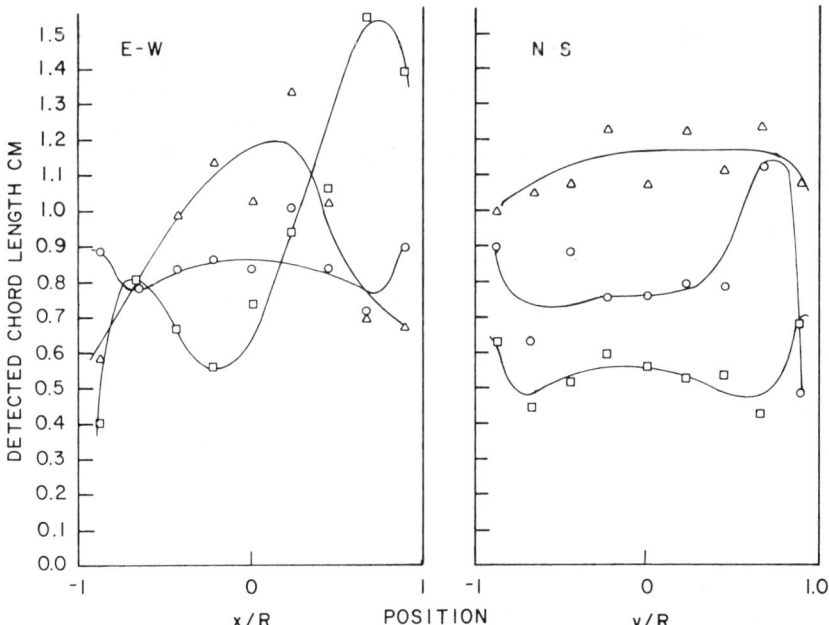

Fig. 3 Profiles of the detected chard length (conductivity probe) B (Tesla): (o) 0.0, (□) 0.3, (∆) 0.9.

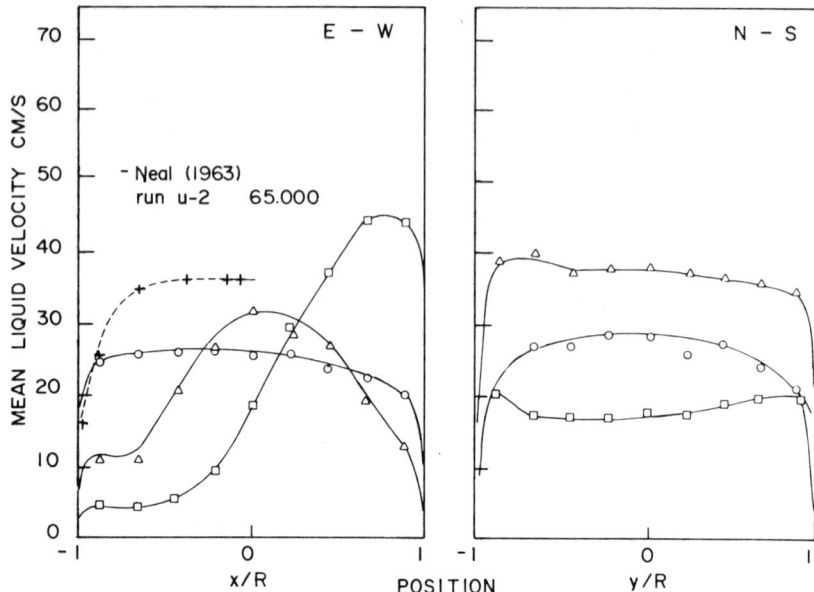

Fig. 4 Profiles of the mean liquid velocity (hot-film) B (Tesla): (○) 0.0, (□) 0.3, (△) 0.9.

was obvious that the magnetic field had acted in such a fashion as to distort the axisymmetry of the flow (expected) and to redistribute the void fraction by promoting bubble coalescence and channeling (also expected), but locating it in a position with clear preference to one side of the wall (not expected). When the field was increased to its maximum value of about 1 T, the void distribution returned to very nearly the one corresponding to the zero magnetic field case, a rather unexpected result. The conclusions drawn from this experiment were that the unexpected location of the void fraction maximum at a position close to the west wall (for B = 0.3) was due to the fact that, at this field, the random dispersion of the entering bubbles was curtailed so that the bubbles, through entrainment, could only follow the path of the entering streamlines, which, because of centrifugal forces at the entrance, would arrive on the west side precisely where the bubbles were seen. In other words, the magnetic field, by damping the turbulence level and the natural wobbliness of the bubbles (as Mori already had reported in his stagnant mercury loop with single bubbles[16]), forced the bubbles to attach themselves to the now laminarized streamlines; and, in doing so, coalescence had increased and so did the void fraction, the bubble size, and the local liquid velocity. When the magnetic field was increased to 0.9 T, the only explanation that could be given as to the new redistribution of the void was that a combination of shear, inertia, and even magnetic pressure forces had torn the large size bubbles apart, and the void maximum could not be sustained anymore.

Bearing in mind the above observations, it was decided that a new series of experiments was necessary to further investigate not only the influence of the entrance geometry, but also the bubble breakup mechanism that caused such a decisive void fraction rearrangement.

In Kirk's M.S. thesis,[9] three new series of experiments were conducted, with the help of a number of modifications of our small vertical loop. A new traversing mechanism was constructed capable of azimuthal rotation. Experiments could be conducted along the N-S, E-W but also NE-SW and NW-SE direction so that isocontour void fraction profiles could be created. The entrance region was also modified.

In the first series of experiments, ten injectors were placed at the inlet as before, but they were positioned at a level much higher than the horizontal axis of the piping tee where the flow had already turned the corner. In the second series of experiments, in an attempt to eliminate the problem of the proximity of the injectors to the entrance tee, a honeycomb was constructed with cocktail straws. Four injectors were inserted through the honeycomb, spaced radially from the center to the wall with the tip of each at the upper edge of the honeycomb. In this series, however, only the central injector was used by releasing one bubble at a time at a rather low frequency of injection. In the third series, we used the same physical arrangement as in the second series, but the bubble release was at the maximum possible rate, aiming at a high void fraction, so that some of the bubbles would attach themselves to the transparent Plexiglas wall of the test section. This allowed us to observe the geometric and kinematic behavior of the bubbles, through still and cinema photography, for different intensity magnetic fields. Here we shall report experiments conducted only with the conductivity probe, which provided us with void fraction contours, bubble velocities, and chord length profiles.

Results

Only the highlights of the results will be given here; details can be found in Kirk's M.S. thesis.[9] In particular the results will be discussed in a series of cross-sectional, constant void fraction contours obtained through a computer program that was capable of synthesizing the data taken in the four traverses, as already explained. In order to make a precise measurement of void fraction, one had to collect data for a sufficiently long time to reach a local time-average value. For high-frequency bubble impact, this time was small, but when the impact frequency was low, as it was in the second series of experiments, the recording time was inordinately high, in some instances well over 240 s for each point. For this reason, the values reported here are good to within about ±10%. It was not uncommon for the value measured at the center to be different from the one of a previous measurement. However, we felt that the accuracy was good enough for our purpose, which was to acquire a qualitative understanding of the void distribution and the mechanism involved for its determination. With this preamble, we now proceed with the results of each one of the three series of experiments.

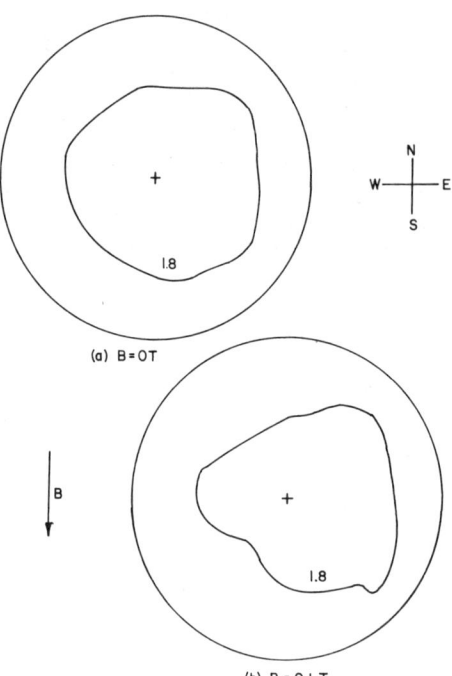

Fig. 5 First series of experiments: cross-sectional contours of constant void fraction, ten injectors.

The First Series

Ten injectors were positioned at the entrance of the test section. Eight of them were evenly spaced every 45 deg forming a ring about two-thirds the radius of the test section. The remaining two were positioned symmetrically close to the center along the N-S direction one-third of the radius away. Figures 5-7 show isocontour void fractions for magnetic field intensities 0, 0.1, 0.2, 0.3, 0.6, and 1.0 T. The recording time for each void measurement was 120 s.

Figure 5a shows the case of zero magnetic field with all ten injectors in operation. The contours, within the accuracy of measurements and the method of the computer interpolation that produced this drawing, are considered to be axisymmetric. Introduction of a weak magnetic field (B = 0.1 T) has not changed this picture significantly, as can be seen in Fig. 5b. In Fig. 6, for B = 0.2 T, we observe a tendency for the bubbles to still maintain a more

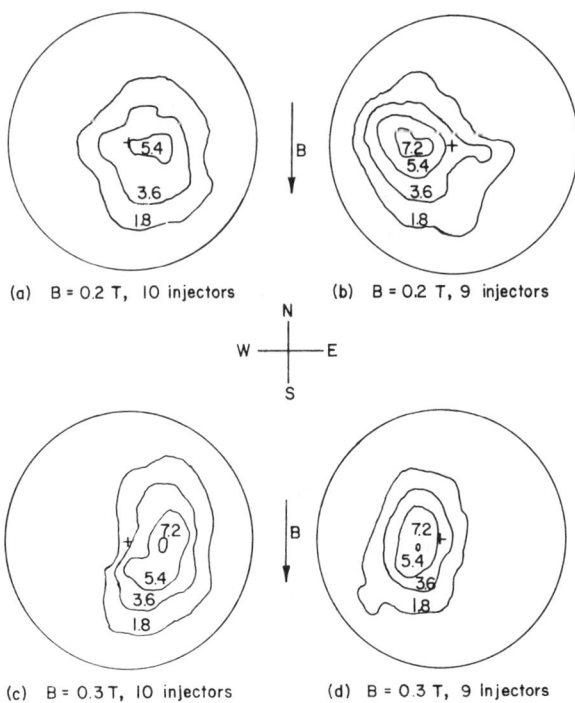

(a) B = 0.2 T, 10 injectors   (b) B = 0.2 T, 9 injectors

(c) B = 0.3 T, 10 injectors   (d) B = 0.3 T, 9 injectors

Fig. 6 First series of experiments: cross-sectional contours of constant void fraction.

or less axisymmetric profile, but the void fraction is depleted at the walls and concentrates at the center with a peak slightly in the eastern direction. We explained this behavior by the fact that the magnetic field had already begun changing the overall velocity field via the ponderomotive force, while, at the same time, both turbulence and bubble wobbliness were suppressed enough for the bubbles to be entrained by the now more rigid path of the liquid streamlines, therefore still carrying with them the memory of the entrance conditions that were dominated by the centrifugal forces that tended to turn the liquid streamlines closer to the eastern wall, precisely as observed. Furthermore, at the point of the maximum void fraction, the chord length had increased from an average of about 0.4 cm for the B = 0 case to the value of 0.5-0.6 cm owing to the greater incidence of coalescence as the bubbles were squeezed into a narrow column. At this point, an effort was made to bias this pattern by providing less flow to the injector located close to the eastern wall. This was done

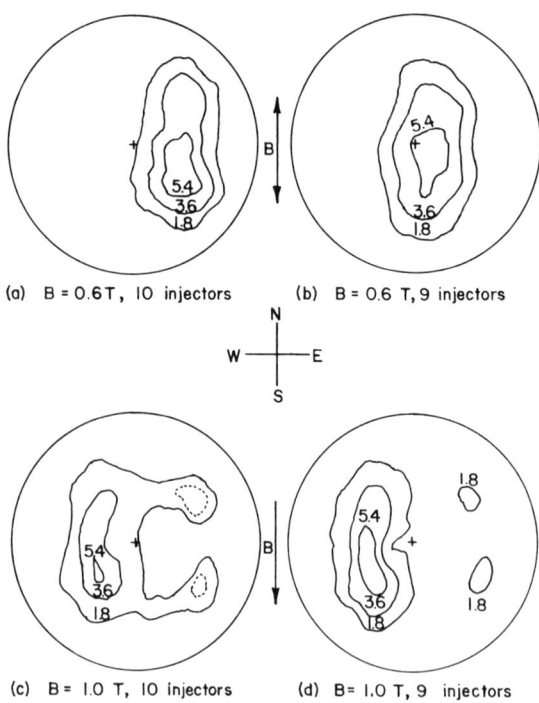

(a) B = 0.6 T, 10 injectors (b) B = 0.6 T, 9 injectors

(c) B = 1.0 T, 10 injectors (d) B = 1.0 T, 9 injectors

Fig. 7 First series of experiments: cross-sectional contours of constant void fraction.

gradually, but the pattern did not change significantly until the moment that it was completely shut off. The void fraction isocontour profiles did then abruptly flip to the western wall, as seen in Fig. 6b. It was clear from this experiment that, notwithstanding our repeated efforts, the central position for the void fraction peak was not a stable one! Figures 6c and 6d tell the same story for B = 0.3 T, a value that corresponds to complete laminarization as we know it for the single liquid phase flow (see Gardner and Lykoudis[17]). Here the void fraction has now shifted distinctly to the eastern wall. The flatness of the contour along the N-S direction can be explained as being the result of the overall accommodation of the nitrogen gas to the eastern side. Here again an attempt to bring the peak to the center by turning off the easternmost injector failed as in the previous case. Figure 6d shows that the maximum flipped to the west side where it stabilized at a pattern very much symmetric with respect to the N-S direction compared to the previous one. At this magnetic field, the chord lengths were measured to be significantly smaller than the ones corresponding to B = 0.2 T. This was a firm manifestation that the combination of shear, inertia, and ponderomotive forces had overcome the surface tension forces which no longer could sustain large size bubbles.

An increase of the magnetic field to B = 0.6 T not only caused the void peak fraction to shift to the east, but at the same time decreased the size of the bubbles by a small amount. Turning off the same easternmost injector brought the void fraction peak close to the center, as can be seen in Fig. 7b, without, however, flipping it on the western side as before. Figure 7c depicts the void isocontour for the case of B = 1 T with all ten injectors in operation. On the basis of the previous findings, we would have expected the void fraction to be positioned in the eastern half of the test section. It is apparent from Fig. 7d that the opposite was observed. It is true that the maximum void fraction did subside to a lower value and that the average chord length was measured to be smaller (0.3 cm) than the one corresponding to B = 0 (0.4 cm). To the extent that the overall gas released remained the same for all magnetic field intensities, the frequency of bubble impact was measured to be higher. Shutting off the easternmost injector further depleted the void on the eastern side. We have no firm explanation for this last case of B = 1 T except to testify that the observations were stubbornly reproducible. It was an attempt to resolve this difficulty that led us to the second series of our experiments.

The Second Series

From the findings of the experiments of the first series, it became apparent that the void distribution geometry was influenced not only by the ponderomotive force action on the liquid, but also by the magnetic pressure forces that are absent along the E-W axis.†

A bubble slightly off the center of the pipe can easily glide in the east or west direction to the east or west corners where not only the magnetic pressure is zero but the $\bar{J} \times \bar{B}$ force is also near zero, since the current there turns around to close in itself because the wall is insulated. In this region, very nearly, $\bar{J}$ is parallel to $\bar{B}$. In order to observe the influence of the $\bar{J} \times \bar{B}$ force and the presence of the slight magnetic pressure, we needed to eliminate the flow from extraneous entrance effects. This was accomplished by the introduction of a honeycomb as described in the previous section. In order to study the force field acting on a single bubble, we decided to release at low frequencies one single bubble at the center of the test section. Four isocontour void fraction profiles were obtained for B = 0, 0.3, 0.6, and 1.0 T. The recording time for each void fraction measurement was 480 s.

Figure 8a, corresponding to B = 0, instead of providing concentric profiles, showed a preference for the bubbles to occupy the S-W portion of the test section. Unexpected as this observation was, it was easily explainable. The Plexiglas test section, due essentially to its age, was warped slightly off from the vertical position leaning on the side that the vertically rising bubbles would land on the conductivity probe exactly where they were observed. As it turned out, this disadvantage was a useful one, since any force distribution that would tend to dislodge them from the position dictated by gravity could be assessed with greater clarity. In fact, at B = 0.3 T (Fig. 8b), the isocontour void fractions are symmetric with respect to the E-W direction, with the maximum slightly to the east. The symmetry still remained at B = 0.6 T, but the contours came closer to the center still elongated somewhat along the N-S direction. At the maximum field of B = 1 T, the peak is unmistakably in the center of the test section, but the

---

†To avoid confusion in this paper we use the word ponderomotive force to be the $\bar{J} \times \bar{B}$ force as computed with Ohm's law assuming that the magnetic field $\bar{B}$ retains its original value since we deal with a very small magnetic Reynolds number. This force acts along the direction of the mean motion of the fluid. Here, by magnetic pressure we mean the pressure acting along the radial direction due to the induced current loops.

isocontour profiles stretch in the E-W direction--no doubt as a result of the decisive influence of the magnetic pressure. The obvious conclusion is that, as the magnetic field increased, the velocity fields of both the liquid and the gaseous phases were dominated by the presence of the ponderomotive and magnetic pressure forces, thus easily erasing the influence of the gravitational force. The bubbles were forced to lock themselves, so to speak, to the streamlines parallel to the generatrix of the test section even though it did not coincide with the vertical direction.

In the experiments of the first series, the flatness of the isocontour void fraction profiles along the E-W direction was not apparent, no doubt because of the much higher number of bubbles released more or less evenly throughout the cross section rather than the center. In the second series experiments, there was no difficulty in bringing the void fraction peak to the center of the cross section. There was no evidence of flip-flop of the isocontours across the N-S axis as had happened in the first series. The only explanation that can be offered is to mention that for the case of ordinary fluid mechanic bubbly flow, two void fraction peaks are known to appear off the center of circular cross sections, a phenomenon still unexplained,

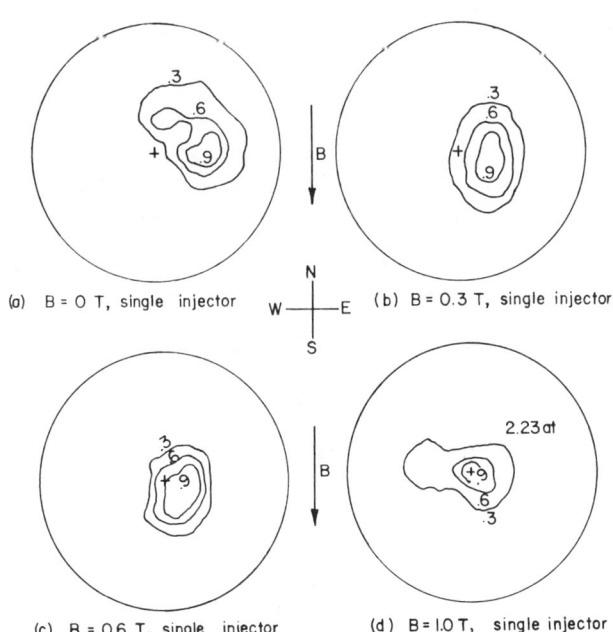

(a) B = 0 T, single injector (b) B = 0.3 T, single injector

(c) B = 0.6 T, single injector (d) B = 1.0 T, single injector

Fig. 8 Second series of experiments: cross-sectional contours of constant void fraction for small α.

notwithstanding the work†† of Ref. 18. The explanation could perhaps lie in the influence of the wall when bubbles at high void fractions, but still within the bubbly regime, swim close to the wall, but no quantitative or qualitative explanation exists at this time. If the void peaks in the OFM case could be explained, perhaps the MFM case could be explained as well.

The Third Series

The experimental arrangement was the same as in the previous series except that a large amount of nitrogen gas was forced out of the centrally located injector. In this fashion, void fraction profiles as high as 10% or more could be secured. The purpose of this series was to force some bubbles to stick to the wall, so that it could be seen through the Plexiglas how the bubbles behave when a magnetic field is on, fully recognizing that the observations dealt this time with three, rather than two phases. Here the recording time for each void fraction measurement was 300 s.

It is instructive to review the isocontour void fraction profiles obtained in this case. The ones at B = 0 in Fig. 9 show an axisymmetric behavior because of the turbulent dispersion of the large number of bubbles and strong interaction among themselves. Because of this, in order to accommodate themselves within the test section, they lose the sense of the nonverticality of the test section. At B = 0.3, the order imposed by the magnetic field is now shown to force the bubbles to move along the liquid streamline paths where gravity would place them, which is closer to the east wall, as explained before. Here the magnetic field is not yet strong enough to obliterate the gravitational force. This, however, clearly happens at B = 0.6 T, where axisymmetric contours are present. In this series, a small increase in bubble size is also observed (referring to their maximum values) from 0.9 cm at B = 0 to 1.3 cm at B = 0.3, to about 1.1 cm at B = 0.6 T and 1 cm at B = 1 T, a sequence of events already familiar from the previous experiments.

The most puzzling observations of this work can be seen in Fig. 10, where the bubble contours have been traced as they are attached to the Plexiglas wall in a sequence of

---

††Drew and Lahey[18] "discover" the peaks of the void fraction by using experimentally obtained turbulent intensity distributions. This is not an explanation of why there are two peaks but rather a quantitative verification that turbulent intensity and void fraction peak go together.

MAGNETO-FLUID-MECHANIC TURBUBBLENCE

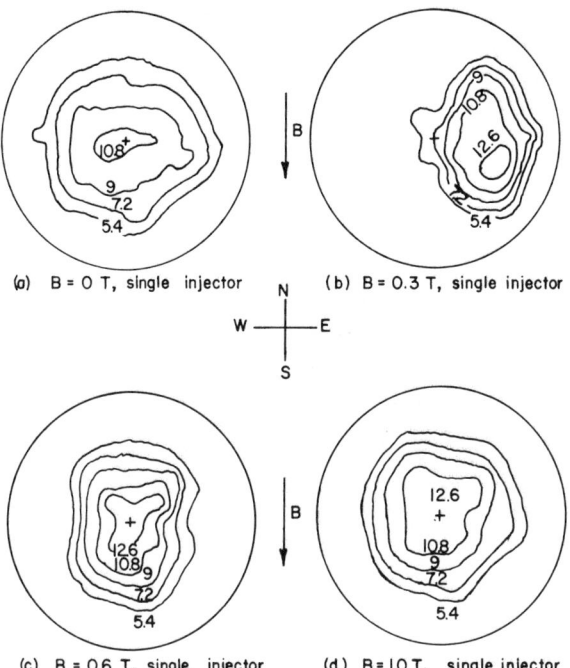

Fig. 9 Third series of experiments: cross-sectional contours of constant void fraction for large α.

time frames 1/30th of a second apart. Because of mercury oxide deposits on the test section wall, the bubbles do not move with a uniform velocity upward, but move more slowly than their average velocity in the core. It was also observed that their speed was smaller for high-magnetic field intensities. To avoid an overcrowded figure, several frames are missing, as depicted by the number labeling each bubble in Fig. 10. The most striking occurrence is that, facing the western side of the test section, two streaks of bubbles are observed at equal distances from the middle moving upward and following precisely the vertical direction. Their shapes are paradoxically triangular. The hypotenuses face the centerline, with their vertical sides being parallel and vertical to the centerline. Occasionally, a bubble on one side will move to the other in the fashion depicted in Fig. 10. During this transition, the triangular shape of the bubble is altered to resume that form as soon as it moves to the other side. These bubbles, as can be seen in Fig. 10, are large, and their vertical dimension to the wall, although not measurable, is certainly thick enough for us to expect it to be exposed to the velocity field of the liquid phase. The bubbles hold their motion and shape as they are submitted to solid-liquid-gas surface tension forces, viscosity, inertia, ponderomotive, and also magnetic pressure forces. It is rather dangerous to venture the suggestion of a mechanism that explains what has been observed and recorded on repeated and extended occasions. It is suspected, however, that an irregular bubble finds one of its sides exposed to a higher flow velocity than the other side, thus being forced to change its shape and position as it glides on the other side. In doing so, it moves against a higher magnetic pressure, as will be explained in a moment. In order to accommodate this pressure, the curvature is flattened, and since the fluid velocity for strong magnetic fields is higher the farther a point is located on the wall from the western (or eastern) side, the bubble is elongated, thus creating a more or less triangular shape. The bubble stabilizes when the shape adjusts itself to the surface tension forces, liquid shear forces, and the magnetic pressure forces,

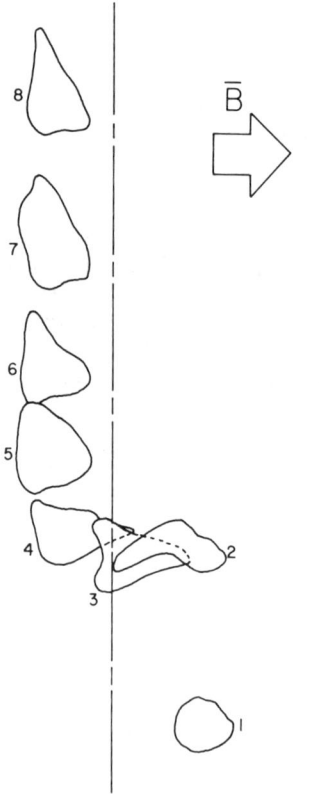

Fig. 10 Bubble tracings from VCR recording; view is from the west wall of the test section.

which are zero at the western (and eastern) side and increase with the square of the cosine of the angle as the angle increases from 0 to 90 deg toward the north (or the south) side locations, where the magnetic pressure assumes its highest value within the distance of the magnetic boundary-layer thickness. The magnetic pressure forces, of course, are very small, but they still could act as the controlling mechanism, just as the small removal of the level instrument from the horizontal position will make the bubble of the level move on to one side or the other. Because of random forces, this "stable shape" may momentarily change, as happens with the bubbles labeled frames 2, 3, and 5 in Fig. 10.

Figure 11 shows a sequence of frames that exhibit, in a rather dramatic fashion, the breakup of a bubble that has appeared for the first time on the center of the test section. In the three frames following its appearance, that is after 3×(1/30) = 1/10 s, the bubble has moved slightly to the left (north side) and suddenly split into five small pieces. The phenomenon is not dissimilar to what can be observed when a few large raindrops hit the dirty windshield of a moving car just after it has begun to rain.

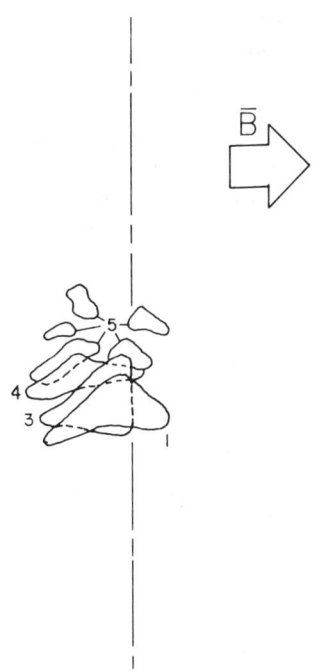

Fig. 11 Bubble tracings from VCR recording; view is from the west wall of the test section.

The drops splash on the glass into several smaller fragments which glue themselves, so to speak, on the windshield, attached by the surface tension forces. Eventually, they move apart slowly and reluctantly in the upward direction under the influence of the wind. Of course, we are not able to observe bubble breakup inside the fluid core but only indirectly by measuring average chord lengths. We could nevertheless gain some insight if we would place a three-dimensional matrix of thin conductivity probes and correlate the signal through a computer. This is not an easy task, and even if accomplished, the amount of information received could not compare with a visual observation (not possible, however, in a medium like mercury).

In the next section we give a brief account of a large vertical loop (LVL) along with its capabilities.

Description of the Large Vertical Loop

Extensive experiments with this facility are in progress as part of the Ph.D. dissertation of David Black.

The large vertical loop is a piping loop that circulates mercury through a vertically aligned test section. A schematic is shown in Fig. 12. The loop operates in a downward direction and is capable of a single- or two-component flow mode. The loop is oriented so that the laboratory magnet can be moved to a position where the test section is between the magnet pole faces. The magnet can also serve either the small or the large vertical loop. The main components of the loop are the pump, the heat exchanger, the electromagnetic flow meter, the control valve, the test section (both heat transfer and isothermal), the traversing mechanism, and, when operating in the two-phase flow mode, gas injector and gas separation tank. All of the components of the loop are made of stainless steel except for the flow meter and the vent piping from the separator, which are made of PVC and plastic. The loop occupies a floor space of 2 m by 4.5 m and has an overall height of 6 m.

The pump is a Gould Model 3196 centrifugal type with a design capacity of about 275 gal/min of mercury at a head of about 100 psi. At the present configuration, the discharge is vertically upward into a 3-in. stainless steel pipe that leads to the water-cooled heat exchanger. The heat exchanger is used to maintain isothermal flow in the test section. The outlet of the heat exchanger leads into the electromagnetic flow meter. The flow meter is made from a piece of 3-in. schedule 80 PVC pipe. The magnetic field is provided by a permanent magnet with a field strength of 1350 G.

The heat-transfer and isothermal circular test section is a 3.12-m-long pipe that has an inside diameter of 3.465 cm. The bottom 1.5 m of the test section is coated with a 0.025-mm-thick layer of Teflon for electrical insulation. The isothermal test section is equipped with pressure taps spaced 12.5 cm along the bottom half.

The heat-transfer section with an inside diameter of 3.62 mm consists of an entering section 111 cm long and a

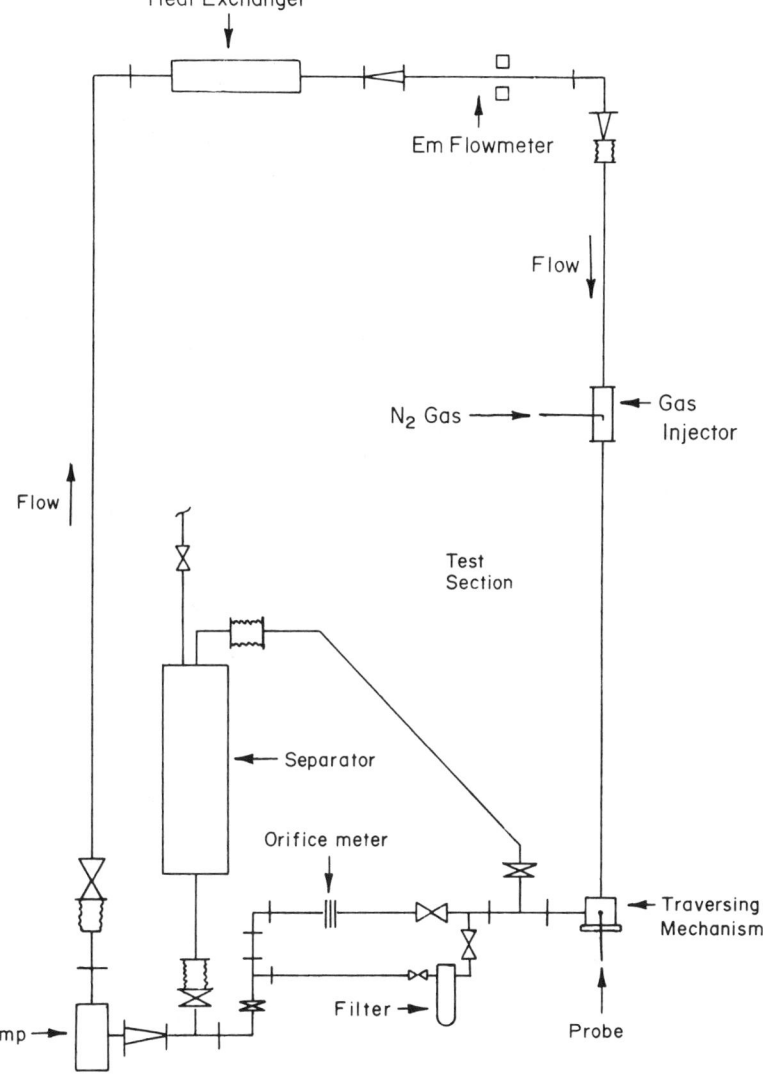

Fig. 12 Schematic of the liquid metal large two-phase vertical loop.

heated section 201 cm in length. When mounted in the loop, the test section will have 32 diam of unheated entrance section after the gas injector, about 20 diam of heated section prior to the magnetic field, and 32 diam of heated section inside the magnetic field. Thirty-eight thermocouples are spot-welded to the exterior pipe in circular grooves at eight axial stations and four azimuthal angles. Local mean temperature profile measurements can be made using a traversing mechanism already in operation. Finally, a rectangular section is also available for isothermal work whose inside dimensions are 38 mm × 190 mm.

The laboratory magnet is a dc electromagnet and provides a uniform field for its entire length of 125 cm. The pole faces are 30.6 cm wide, and at a gap of 6.4 cm the magnet is capable of a field strength of 1.5 T. The electromagnet is water-cooled and consumes 150 kW of power.

Conclusions

We have attempted to clarify the nature of two-phase (or, better, two-component) LMMFM flow by undertaking a series of local experiments inside an electrically insulated pipe. We used a vertical loop with mercury and nitrogen moving concurrently upward. We have shown that hot-film anemometry, with the help of an on-line computer, can give useful information on Turbubblence. These experiments are difficult and demand a high degree of cleanliness for mercury, the probe calibration becoming impossible if a large number of mercury oxide fragments are present. The conductivity probe, because of its nature, is by far more tolerant to such disturbances. The main conclusions reached are the following.

The magnetic field dampens both the dispersion and the wobbliness of the bubbles inasmuch as it also dampens the turbulence level of the liquid phase due to both the shear and buoyancy induced turbulence. Because of this action, the entrance effects, to the extent that they are present, are felt far downstream, causing higher void fraction concentrations and increases in the bubble size.

When the bubble size, owing to coalescence, becomes large enough, the bubbles break up into smaller sizes and redistribute themselves more evenly in the test section. Their size may decrease in some instances below the zero magnetic field value, but the detection bubble frequency increases for a constant overall void fraction.

When bubbles are injected singly from the center of the test section, with a high enough magnetic field, the void fraction peak stabilizes around the center of the test section, but the isocontours exhibit a tendency to spread along the E-W direction, which is the quiet zone from the magnetic pressure point of view. This phenomenon occurs even in the presence of gravitational forces when the test section axis is not aligned with the vertical direction.

On the other hand, when bubbles are injected uniformly from evenly distributed locations in the test section, the competition with the ponderomotive forces, and forces due to bubble interaction with the wall (an interaction not very well understood) seemed to prevent the bubble concentration to peak in the center. If, however, large void fractions are created with bubbles originating with vigorous injections from the pipe center only, axisymmetric void distributions are obtained, since the bubbles reaching the wall cannot interact as vigorously as those that would originate from injection points near the wall.

Bubbles observed on the Plexiglas wall exhibited a triangular, more or less stable shape, and they moved straight upward Indian file at equal distances from the western pipe generatrix. Their hypotenuses were facing the west and occasionally would move from one side to the other by transformation of their shape during this short trip. This unexpected phenomenon was explained by bringing into the picture the known distribution of the magnetic pressure that increases from zero on the east (and west) side to its maximum value close to the walls on the north (and south) locations.

Bubble breakup on the wall was observed and photographed, but it appeared that the breakup was very much similar to a droplet splashing over a surface disintegrating into smaller fragments.

Whereas emphasis in this paper was given to local measurements of void fraction and chord lengths, detailed experimental results of the following additional quantities have also been compiled (see Refs. 8 and 9): mean liquid velocity profiles, bubble velocity profiles, bubble impact frequency profiles, turbulence intensities, and turbulence spectra.

The term Turbubblence was introduced in this paper for the first time in order to describe two-phase liquid and gaseous flows, since the interaction of buoyancy induced turbulence due to the gaseous phase with the turbulence present through wall shear creates a new situation with aspects that cannot be readily understood within the framework of the classical theory of turbulence.

Acknowledgments

The author wishes to thank T. Kirk and D. Black for their careful reading of the manuscript and their discussion. He is also grateful to Prabaddh Riddhagni for his additional help, particularly in reducing and assembling the figures. He also wishes to acknowledge the continuing financial support of the National Science Foundation under Grant MEA 8304743.

References

[1] Fraas, A. P., "Conceptual Design of the Blanket and Shielding Region for a Full Scale Toroidal Fusion Reaction," USAEC Report ORNL-TM-3096, Oak Ridge National Laboratory, Oak Ridge, Tenn., 1973.

[2] Fujii-e, Y. and Suita, T., "IAEA Workshop on Fusion Reactor Design Problem," Paper 14, Culham, England, 1974.

[3] Pendergrass, J. H., Booth, L. A., Peterson, D. R., and Gerstl, S. A., "The Liquid Boiler: A 1500 to 2000 K Fusion Reactor Blanket Concept for Process Heat and/or Electric Power Generation," Report LA-UR-79-1316, Los Alamos National Laboratory, Los Alamos, N. Mex., 1979.

[4] "Fuel Producer, The Technology of Controlled Nuclear Fusion," Proceedings of the Fourth Topical Meeting, Vol. II, King of Prussia, Penn., Oct. 14-17, 1980, pp. 1802-1810.

[5] Branover, H., El-Boher, Sukoriansky, S., and Yakhot, A., "Development of a Low Temperature Liquid-Metal MHD Small Scale Pilot Plant," Proceedings of the 21st Symposium on Engineering Aspects of Magnetohydrodynamics, Report CONF-830634, Argonne National Laboratory.

[6] Thibault, J. P., Joussellin, F., and Alemany A., "Metal Gas MHD Converter Development Plans," Proceedings of the 21st Symposium on Engineering Aspects of Magnetohydrodynamics, Report CONF-830634, Argonne National Laboratory.

[7] Wagner, L. Y., "Single and Two-Phase Liquid Metal Heat Transfer Under the Influence of a Magnetic Field," Ph.D. Thesis, School of Nuclear Engineering, Purdue University, West Lafayeete, Ind., Dec. 1981.

[8] Gherson, P., "Local Measurements in a Liquid-Metal Two-Phase Flow Under the Influence of a Magnetic Field," Ph.D. Thesis, School of Nuclear Engineering, Purdue University, West Lafayette, Ind., Dec. 1981.

[9] Kirk, T., "Co-current Two-Phase Flow in the Presence of a Transverse Magnetic Field," M.S. Thesis, School of Nuclear Engineering, Purdue University, West Lafayette, Ind., May 1984.

[10] Lykoudis, P. S., "Bubble Growth in the Presence of a Magnetic Field," International Journal of Heat and Mass Transfer, Vol. 19, Dec. 1976, pp. 1357-1362.

[11] Wagner, L. Y. and Lykoudis, P. S., "The Effect of Liquid Inertia on Bubble Growth in the Presence of a Magnetic Field," American Institute of Chemical Engineering Symposium Series, Vol. 73, No. 164, 1977, pp. 142-147.

[12] Wagner, L. Y. and Lykoudis, P. S., "Liquid Metal Boiling in a Magnetic Field," Proceedings of the Eighth Symposium in Engineering Problems of Fusion Research, Vol. 4, New York, 1979, pp. 2075-2077.

[13] Gherson, P. and Lykoudis, P. S., "Hot-Film Anemometry in a Two-Phase (Liquid Metal-Gas) Medium," Proceedings of the Seventh Biennial Symposium on Turbulence, University of Missouri-Rolla, Sept. 1981, pp. 27.1-27.9.

[14] Gherson, P. and Lykoudis, P. S., "The Effect of a Traverse Magnetic Field on Liquid-Metal Two-Phase Flow Pattern," Proceedings of the Ninth Symposium on Engineering Problems of Fusion Research," IEEE, New York, 1981, pp. 1410-1413.

[15] Lindt, J. T. and De Groot, R. G. F., "The Drag on a Single Bubble Accompanied by a Periodic Wake," Chemical Engineering Science, Vol. 19, No. 4, 1974, pp. 957-982.

[16] Mori, Y., Hijakata, K., and Kuziyama, I., "Experimental Study of Bubble Motion in Mercury with and without Magnetic Field," Winter Annual Meeting of ASME, Heat Transfer Division, ASME-76/HT-65, Dec. 1976.

[17] Gardner, R. A. and Lykoudis, P. S., "Magneto-Fluid-Mechanic Pipe Flow in a Transverse Magnetic Field. Part I. Isothermal Flow," Journal of Fluid Mechanics, Vol. 47, No. 4, 1971, pp. 737-764.

[18] Drew, D. A. and Lahey, R. T., Jr., "Phase Distribution Mechanism in Turbulent Two-Phase Flow in Channels of Arbitrary Cross Section," Journal of Fluids Engineering, Vol. 103, Dec. 1981, pp. 583-589.

Bubble Growth in a Superheated Liquid Metal in a Uniform Magnetic Field

Paul S. Lykoudis*

Purdue University, West Lafayette, Indiana

Abstract

The problem of bubble growth in a superheated liquid metal in the presence of a uniform magnetic field is examined in two ways. First, an approximate theoretical analysis is introduced by which the liquid region around the bubble is divided into a region in which the flow is orthogonal to the field and into a second conical region in which the growth takes place in a direction parallel to the field and therefore uninhibited by its presence. This procedure enables one to satisfy two separate sets of equations of conservation. This paper shows that, after a characteristic time dictated by the strength of the magnetic field, the bubble grows under a constant pressure difference remaining uncoupled from the energy equation. It is shown that under this condition the equation of conservation of momentum can be nondimensionalized in a universal way, indicating that such variables as pressure and magnetic field do not appear as parameters in the solution, since they are absorbed in the nondimensionalization of the spatial coordinates and time. Second, the problem is also solved numerically with the help of a modified solution algorithm-volume of fluid (SOLA-VOF) numerical code and comparisons are made with theoretical solutions.

Nomenclature

B = magnetic flux density, T
c_p = specific heat at constant pressure, J/kg °C

Paper presented at the Fourth Beer-Sheva Seminar on MHD Flows and Turbulence, Ben-Gurion University of the Negev, Beer-Sheva, Israel, February 27 - March 2, 1984. Copyright © American Institute of Aeronautics and Astronautics, Inc., 1985. All rights reserved.

*School of Nuclear Engineering.

BUBBLE GROWTH IN A UNIFORM MAGNETIC FIELD

E	=	nondimensionalized parameter = $2\,\text{Ja}\,\alpha^{1/2}$, m/s$^{1/2}$
F	=	nondimensionalized parameter = $[(2/3)\gamma\Delta T_s/\rho_1]^{1/2}$, m/s
i_{fg}	=	heat of evaporation, J/kg
Ja	=	Jakob number = $\rho_1 c_{p1} \Delta t_s / \rho_v i_{fg}$ dimensionless
Ja*	=	magnetic Jakob number = $[-1+(1+8\Lambda)^{1/2}]/4$ Λ, dimensionless
k	=	thermal conductivity, W/m °C
K	=	bubble magnetic interaction number = $\sigma_e k B^2 \alpha_1 / \Delta p$, dimensionless
ℓ	=	coordinate in z direction
L	=	characteristic length in longitudinal direction
L*	=	nondimensional length
Nu	=	Nusselt number = kL/k, dimensionless
Nu_0	=	Nusselt number with magnetic field
p	=	pressure, Pa = N/m^2
q″	=	heat flux, W/m^2
r	=	coordinate in r direction, m
R	=	characteristic length in radial direction, m
R*	=	nondimensional radius
T_{sat}	=	liquid saturation temperature
T_∞	=	wall temperature or superheated liquid temperature
ΔT_s	=	liquid superheat = $T_\infty - T_{sat}$, °C
t	=	time, s
t*	=	nondimensional time
u	=	velocity component in liquid for r direction, m/s
w	=	velocity component in liquid for z direction, m/s
α	=	thermal diffusivity, m^2/s
γ	=	Clausius-Clapeyron constant = $\rho_v i_{fg}/T_{sat}$, N/m^2 °C
δ	=	thermal layer thickness = $(\alpha t)^{1/2}$, m
Λ	=	boiling magnetic interaction number = $K\text{Ja}^2$, dimensionless
ρ	=	density, kg/m^3
σ_e	=	liquid electrical conductivity, mho/m
ϕ	=	dimensionless temperature ratio = $(T_v - T_{sat})/(T_w - T_{sat})$
θ	=	angle where $\tan\theta = \dot{R}/\dot{L}$
ψ	=	dimensionless ratio = $R*\dot{L}*/\dot{R}*L*$

Subscripts

c	=	critical value
J	=	Joule
ℓ	=	liquid
ML	=	microlayer

Introduction

Technological interest in the area of boiling liquid metals in the presence of a magnetic field arose more than ten years ago, primarily out of the need to design lithium blankets for magnetically confined fusion reactors. Because the problem of boiling, even under the simplest conditions (such as nucleate pool boiling), still remains only partially understood, the magnetic field provides an additional complexity.

An early attempt to create a theoretical model capable of estimating heat-transfer rates in nucleate boiling in the presence of a magnetic field was undertaken by the author (see Ref. 1) and by Wagner and Lykoudis.[2] The approach taken was to use the method of Forster and Zuber,[3] which was based on the rate of growth of the bubble in a superheated fluid. This problem, in the absence of a magnetic field, has attracted a lot of attention since Lord Rayleigh considered it first and is handled adequately in handbooks and textbooks (see for example, Refs. 4 and 5). From a practical point of view, an extension of the solution of this problem to the magneto-fluid-mechanic (MFM) case would involve the imposition of a uniform magnetic field, which of course would destroy the spherical symmetry of the problem that made its ordinary fluid mechanic (OFM) solution possible. Since in Ref. 1 a physical insight was desired rather than an exact solution, the spherical symmetry was maintained by considering a constant spherical magnetic field, which of course is not physically realizable. However, this work did discover a new nondimensional parameter, Λ, that was capable of describing the temporal growth of the bubble under the influence of the inertia and heat-transfer mechanisms. This parameter was called the boiling magnetic interaction number. It will be seen later that it turns out to be proportional to the square of the magnetic field, the superheat, and a collection of physical properties relating to the fluid and gaseous phases.

To be thorough, mention needs to be made of the work of Wong et al.[6,7] In Ref. 6, the energy conservation equation was not used, and, instead, a principle of "conservation of bubble volume" was involved, an erroneous concept. In Ref. 7, this discrepancy was eliminated, and the conservation equations were integrated with an ad hoc integral method whose accuracy could not be assessed. Heat-transfer rates were evaluated by using the method of Ref. 1, with dimensional numerical results limited to potassium. Since the theoretically obtained values underpredicted the amount by which heat-transfer is diminished upon comparison with the work of Ref. 1, and, since experiments that will be presented later show that the work of Ref. 1 is capable of correlating well heat-transfer data in mercury, the value of the work of Refs. 6 and 7 remains uncertain at best.

The present work is an attempt to extend the theoretical work of Refs. 1 and 2 to the case of a uniform and constant magnetic field, under the hypotheses of small magnetic Reynolds number. The problem is handled in two different ways: theoretically, by seeking analytic solutions for small and very large times (or weak and very strong magnetic fields), and numerically, guided by the results of the theoretical attempt.

Finally, the experimental data already reported by Wagner and Lykoudis[8] will be seen to correlate well with the nondimensional number Λ. The paper will finish with a review of yet another more detailed theoretical attempt to correlate our nucleate boiling experimental data by using the method suggested by Judd and Hwang.[9]

The Approximate Theoretical Model

Consider an initially small spherical bubble that grows under the influence of a superheat ΔT_s in the presence of a uniform and constant magnetic field B. For small times, the bubble will more or less maintain its spherical shape, but as time goes by it will have a tendency to grow more in the direction of the magnetic field (z), eventually assuming the shape of a body of revolution along axis z. Let the characteristic diameter of this shape be equal to 2R along r, and its length 2L along z. At such a time it is not unreasonable to approximate this oblong shape with two surfaces, the one along the coordinate r very long compared to the one along z. This gives the idea of the possibility of dividing the liquid space (at a given time t) into two regions such as shown in Fig. 1. Here, for simplicity, we have approximated the bubble with a cylinder. Furthermore, we make the assumption that the two surfaces of separation are conical with apexes at C and C'.

Let us call θ the angle $\widehat{ACB} = \widehat{A'C'B'}$. Consider now a cylinder D D'E'E resting in the liquid phase. Let the velocity of the surface DD' at a distance r be equal to u, while at the time considered t the velocity of the interface AA' is equal to $dR/dt \equiv \dot{R}$. The corresponding velocities of surfaces DE and D'E' at a distance 2ℓ will be w and $dL/dt \equiv \dot{L}$. In order to insure that the common points A, A', B', and B move at all times together, we need to set the condition

$$\dot{R}/\dot{L} = \tan\theta \tag{1}$$

The angle θ will be, of course, a function of time inasmuch as R and L are unknown functions of time. Clearly, the velocity field at the corners A, A', B', and B are discontinuous and so are they everywhere at the surface of separation. However, the resultant relocities are compatible with the constraint that, being parallel to the dividing surface, they allow no mass to flow from one region to the other. These geometric and kinematic features do allow us to write distinct equations of conservation of mass and momentum for each region. For the region above AA', conservation of mass dictates

$$ur\ell = \dot{R}RL \qquad (2)$$

Solving for u, and using the trigonometric relationship

$$r - R = (\ell - L)\tan\theta$$

we find

$$u = \frac{\dot{R}R}{r} \cdot \frac{L}{\ell} = \frac{\dot{R}R}{r} \cdot \left(\frac{L\tan\theta}{r - R + L\tan\theta}\right) = \frac{\dot{R}RL\tan\theta}{\left[r^2 - r(R - L\tan\theta)\right]} \qquad (3)$$

As shown in Ref. 1, the ponderomotive force per unit volume acting along r is equal to $-\sigma_e u B^2$. It is our intention to find the total forces per unit area contributed by all the volume of the region extending to infinity. After some

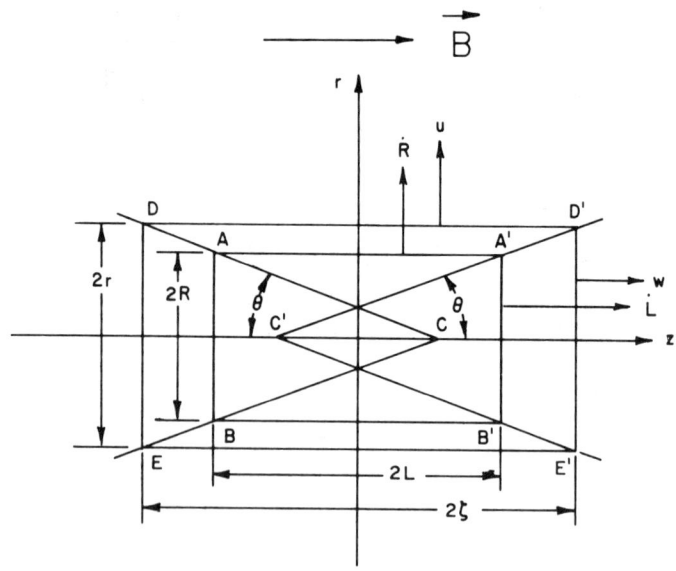

Fig. 1 Approximate bubble geometry under the influence of a horizontal magnet field.

BUBBLE GROWTH IN A UNIFORM MAGNETIC FIELD 285

straightforward integration with the use of Eq. (3), we find

$$-\sigma_e B^2 \int_R^\infty u\, dr = -\sigma_e B^2 \left(\frac{\dot{R}RL \tan\theta}{R - L\tan\theta}\right) \ln\left(\frac{R}{L\tan\theta}\right) \quad (4)$$

Using Eq. (1) in (4), we find

$$\int_R^\infty u\, dr = \frac{R\dot{R}^2}{\dot{L}R - L\dot{R}} \ln\left(\frac{R\dot{L}}{\dot{R}L}\right) \quad (5)$$

Let us define an auxiliary function ψ as follows:

$$\psi \equiv R\dot{L}/\dot{R}L \quad (6)$$

Equation (5) becomes

$$\int_R^\infty u\, dr = \frac{R\dot{R}}{\psi - 1}\ln\psi = R\dot{R}\,\frac{\ln\psi}{\psi - 1} \quad (7)$$

Using the same procedure, we can calculate the pressure and the inertia forces that must counterbalance the ponderomotive force. The result is

$$\frac{\ln\psi}{\psi - 1}\left(\frac{3}{2}\rho\dot{R}^2\right) = p_v - p_\ell - \left(\frac{\ln\psi}{\psi - 1}\right)R\dot{R}\,\sigma_e B^2 \quad (8)$$

We note that Eq. (8) is the same equation we found for the spherical case[1] except for the factor containing the function ψ. The function $\ln\psi/(\psi - 1)$ is expected to vary between the values of 1 and 2, corresponding to a variation of $R \propto L$ for small times, and $R \propto (L)^{1/2}$ for long times. The pressure difference $p_v - p_\ell$ can be readily converted into a temperature difference $(T_v - T_{sat})$ with the help of the Clausium-Clapeyron γ multiplier. Equation (8) becomes

$$\frac{\ln\psi}{\psi - 1}\left(\frac{3}{2}\rho_\ell \dot{R}^2\right) = \gamma(T_v - T_{sat}) - \left(\frac{\ln\psi}{\psi - 1}\right)R\dot{R}\sigma_e B^2 \quad (9)$$

The region along the direction z remains unperturbed from the presence of the magnetic field, since the velocity w is parallel to it. The equation of conservation of momentum is simply†

$$(3/2) \rho_\ell \psi \dot{L}^2 = p_v - p_\ell = (T_v - T_{sat}) \quad (10)$$

The factor ψ is due to the geometric and kinematic limitations imposed in this region.

Finally, we need to write the equation of energy conservation. In words, this principle states that the energy conducted into the bubble is equal to the change of the enthalpy necessary for the transition from the liquid to the vapor phase. This statement in written form is as follows:

$$\left(k \frac{\Delta T}{\Delta r}\right) (\text{Area}) \approx \rho_v i_{fg} \frac{d(\text{Volume})}{dt} \quad (11)$$

Within our approximations, the area around one-half of the bubble is equal to $\pi(2L + R)R$, and the corresponding volume is $\pi R^2 L$. Δr is approximated with the nonsteady thermal skin depth thickness $(\alpha t)^{1/2}$. All of the above reduce Eq. (11) to its final form:

$$\frac{k_\ell(T_\infty - T_v)}{i_{fg}\rho_v(\alpha_\ell t)^{1/2}} = \frac{\dot{R}L(\psi + 2)}{2L + R} \quad (12)$$

The unknowns of the problem are $R(t)$, $L(t)$, and $T_v(t)$, for which we have the three fundamental equations (9), (10), and (12) along with the definition of the auxiliary function ψ given by Eq. (6).

In order to deal only with nondimensional parameters, we introduce, as in Ref. 2, the function Φ:

$$\Phi \equiv \frac{T_v - T_{sat}}{T_\infty - T_{sat}} \quad (13)$$

We also use the customary nondimensional quantities R^*, L^*, and t^* as defined in the nomenclature. The fundamental equation then becomes the momentum equation along r:

$$\dot{R}^{*2} + 4 \Lambda R^* \ddot{R}^* = (\psi - 1)\Phi/\ell n \psi \quad (14)$$

---

†Note that we consistently neglect the acceleration term $R\ddot{R}$ as being small compared with $(3/2)\dot{R}^2$ when computing the inertial forces.

BUBBLE GROWTH IN A UNIFORM MAGNETIC FIELD

momentum equation along z:

$$\psi \dot{L}^{*2} = \Phi \quad (15)$$

energy equation:

$$\frac{1 - \Phi}{2t^{*1/2}} = \frac{\dot{R}^* L^* (\psi + 2)}{2L^* + R^*} \quad (16)$$

auxiliary function ψ:

$$\psi = R^* \dot{L}^* / \dot{R}^* L^* \quad (17)$$

As already discovered in Ref. 1, the boiling magnetic interaction number Λ is given by the relation

$$\Lambda = B^2 \Delta T \left[\frac{\sigma_e \alpha \rho_\ell^2 c_p^2}{\gamma \rho_v^2 i_{fg}^2} \right] \quad (18)$$

Before we solve these equations with the help of a computer, let us examine their behavior for small and large items. When the growing process begins, $T_v = T_\infty$, $\Phi = 1$, and the ponderomotive force is negligible. The system (Eqs. 14-17) has as a solution the familiar result

$$R^* = L^* = t^* \quad (\text{with } \psi = 1)$$

This is the inertial controlled regime, when the bubble is still small and young, beginning to grow with the maximum amount of pressure difference ever to be available to it, namely, $\gamma (T_\infty - T_{sat})$. From there on the temperature inside the bubble T_v will be decreased, but the bubble will still grow, since to its rescue comes the heat-transfer mechanism that delays the decrease of T_v by vaporizing the liquid. When the magnetic field was absent, T_v could reach its terminal value T_{sat} when the bubble had increased to an infinite size. However, with the magnetic field present, the heat-transfer mechanism will have to provide the energy needed to counterbalance the ponderomotive force, and T_v does not approach T_{sat} as in the OFM case.

At this point it is instructive to briefly review the case of the spherical magnetic field for which $\psi = 1$. Equations (14) and (16) are now

$$\dot{R}^{*2} + 4 \Lambda R^* \dot{R}^* = \Phi \quad (19)$$

$$1 - \Phi = 2t^{*2} \dot{R}^* \quad (20)$$

Eliminating Φ, we have

$$\dot{R}^{*2} + 4 \Lambda \dot{R}^* R^* = 1 - 2t^{*1/2}\dot{R}^* \tag{21}$$

For small times, this equation gives the familiar inertia controlled result

$$R^* = t^* \tag{22}$$

For large times, the inertia term $\dot{R}^{*2}$ can be neglected, and the solution is

$$R^* = Ja^* t^{*1/2} \tag{23}$$

where

$$Ja^* \equiv \frac{-1 + (1 + 8\Lambda)^{1/2}}{4\Lambda} \tag{24}$$

Ja* has been called in Ref. 1 a generalized Jakob number and has the value of 1 when $\Lambda = 0$. For times in-between (that is, when both the inertia and heat-transfer mechanisms are present) a numerical solution of Eqs. (19) and (20) has been determined in Ref. 2.

We now wish to determine the point in time when the solution for R(t) ceases to be controlled by inertia and the heat-transfer mechanism begins to be important. Roughly speaking, this should occur at a time t_{CR} when the two asymptotic solutions (22) and (23) yield the same value, that is, when

$$t^*_{CR} = Ja^* t^{*1/2}_{CR} \tag{25}$$

or

$$t^*_{CR} = Ja^{*2} = \left[\frac{-1 + (1 + 8\Lambda)^{1/2}}{4\Lambda}\right]^2 \tag{26}$$

After the time t_{CR}, the bubble size grows parabolically as dictated by Eq. (23), but it is worth mentioning that when this occurs, the temperature (or pressure) distribution Φ has reached an asymptotic value that can easily be obtained by substituting Eq. (23) in (20).

$$\Phi = 1 - 2t^{*1/2}\dot{R}^* = 1 - Ja^* \tag{27}$$

From the above equation, we can estimate the time $t_{C\phi}$ when the pressure reaches this asymptotic value. By making $\dot{R}^* = 1$ (a condition prevailing during the inertia controlled growth), we find

$$t^*_{C\phi} = \frac{Ja^{*2}}{4} = \frac{1}{4}\left[\frac{-1 + (1 + 8\Lambda)^{1/2}}{4\Lambda}\right]^2 \quad (28)$$

In other words, comparing Eqs. (26) with (28), the pressure accommodates itself to its asymptotic value earlier than R^* does. Since $Ja^* < 1$, this means that, in the presence of the magnetic field, the temperature inside the bubble never attains its zero field asymptotic value of T_{sat}, but a value higher than that in order to counterbalance the always present force due to the magnetic field.

At this point, it is instructive to think in terms of physical characteristic times rather than nondimensional ones. Let us recall the definition of t^*:

$$t^*_{CR} \equiv t_{CR}/\frac{E^2}{F} = t_{CR}/6\left(\frac{\alpha_\ell \rho_\ell}{\gamma \Delta T}Ja^2\right) \quad (29)$$

From Eq. (26), when $B = 0$, then $\Lambda = 0$, and $t^*_{CR} = 1$. This means that $t_{CR}/6$ is equal to the value between parentheses in the denominator. We define this value as the Jakob time, that is,

$$t_{CR} \equiv t_{Ja} \equiv \frac{\alpha_\ell \rho_\ell}{\gamma \Delta T}Ja^2 \quad (30)$$

This definition, believed to be given here for the first time, leads us to another physical interpretation of the boiling interaction parameter than the one given in Ref. 1. From Eq. (18), Λ can be rewritten as follows:

$$\Lambda = \left(\frac{\sigma_e B^2}{\rho_\ell}\right)\left(\frac{\alpha_\ell \rho_\ell}{\gamma \Delta T}\right)Ja^2 = \frac{t_{Ja}}{t_J} \quad (31)$$

Where

$$t_J = \rho_\ell/\sigma_e B^2$$

t_J is called the Joulean rollover time, or the time it takes for the magnetic field to suppress the inertia forces. This definition allows us an interpretation for Λ parallel to the one we sometimes give to Reynolds number (viscous dissipation time over flow characteristic time), or the square of the Hartmann number (viscous over Joulean dissipation time), etc. It follows that for weak magnetic fields, t_J will be higher than t_{Ja}, whereas for $\Lambda > 1$, the opposite will be true, making the ponderomotive force the dominant one. In addition, we see from Eq. (26) that for very high values of Λ, $t_{CR}^* \propto \Lambda^{-1}$. Use of definitions (30) and (31) leads to the result that for high magnetic fields $t_{CR} \propto t_J$.

But there is yet another physical interpretation of Λ in terms of other characteristic times to which t_{Ja} can be understood. A small order of magnitude analysis is needed. First, we compare the inertia terms $\rho(R/t)^2$ with the pressure force $\gamma \Delta T$ that yields the characteristic time t_{IC} during which the growth is controlled by inertia. The result is

$$t_{IC} \propto (\gamma \Delta T/\rho_\ell)^{1/2} R$$

We compute the time t_{HC} during which the growth is controlled by heat transfer by comparing the order of magnitude of the conduction term $k\Delta T/(\alpha t)^{1/2}$ and the energy rate of evaporation $\rho_v i_{fg} R/t$. The result is

$$t_{HC} \propto (i_{fg} \rho_v \alpha^{1/2} R/k\Delta T)^2$$

It is now easy to show that Eq. (31) can also take the form

$$\Lambda = t_{IC}^2/t_J t_{HC} \qquad (32)$$

In other words, comparing with Eq. (31) we find that $t_{Ja} = t_{IC}^2/t_{HC}$. Equation (32) states that the square root of Λ is the ratio of the nondissipative time t_{IC} divided by the geometric mean of the two dissipative times t_J and t_{HC}.

We can now formulate a sequence of events, as the bubble grows, in the presence of a magnetic field in terms of these characteristic times. When the bubble is small and starts growing, the phenomenon is inertia controlled. At a time higher than t_{CR}^*, a time which is a mixture of t_{Ja} and t_J, the growth is maintained by the vaporization of the liquid phase through the heat-transfer mechanism while moderated by the presence of the field. The temperature inside the bubble has now dropped from its original high value ($\Phi = 1$). The magnetic field, however, does not allow T_v to reach T_{sat} (that is, Φ cannot go to zero), and, in fact, when $t_{C\Phi}^*$ reaches the value of Eq. (28), the function Φ reaches its asymptotic MFM value equal to $(1 - Ja^*)$. From there on, the growth is controlled by the presence of the magnetic field alone, acting against a constant pressure difference that is maintained by the heat-transfer process.

BUBBLE GROWTH IN A UNIFORM MAGNETIC FIELD

This is an important remark that will be used later in the numerical solution of the problem.

Before we leave these qualitative arguments it will be instructive to examine more carefully the asymptotic behavior of the phenomenon when Λ goes to infinity.

Since we have established that ΔP is a constant, independent of time, but still determined as a function of Λ, we can use an order of magnitude to write the momentum equations (14) and (15) in their natural coordinates as follows:

$$\frac{\sigma_e B^2}{\rho_\ell} \left(\frac{R}{t}\right) \propto \frac{\Delta P}{\rho_\ell R} \tag{33}$$

$$\left(\frac{L}{t}\right)^2 \propto \frac{\Delta P}{\rho_\ell} \tag{34}$$

Solving for R and L and using the time t_J we have

$$R \propto \left(\frac{\Delta P}{\rho_\ell} t_J\right)^{1/2} (t)^{1/2} \tag{35}$$

$$L \propto \left(\frac{\Delta P}{\rho_\ell}\right)^{1/2} t \tag{36}$$

The aspect ratio of the growth is

$$\frac{L}{R} \propto (t/t_J)^{1/2} \tag{37}$$

and depends solely on t_J, the heat-transfer mechanism, since it enters only through t_{Ja} being absent. It should be noted, however, that values of Λ high enough to yield such asymptotic results are only possible in practice for $B \sim 5$ T. The values for Λ for a superheat of 30°C are equal to 60, 580, and 180 for potassium, sodium, and lithium, respectively. Note that for mercury this value is only equal to about 3.0.

The exact solutions of Eqs. (14-16) for high values of Λ ($\psi = 2$) equivalent to Eqs. (35-37) are

$$R^* = \left[\frac{1}{4\ell n 2}\left(\frac{1 - Ja^*}{\Lambda}\right)\right]^{1/2} (t^*)^{1/2} \tag{38}$$

$$L^* = (1 - Ja^*)^{1/2} t^* \tag{39}$$

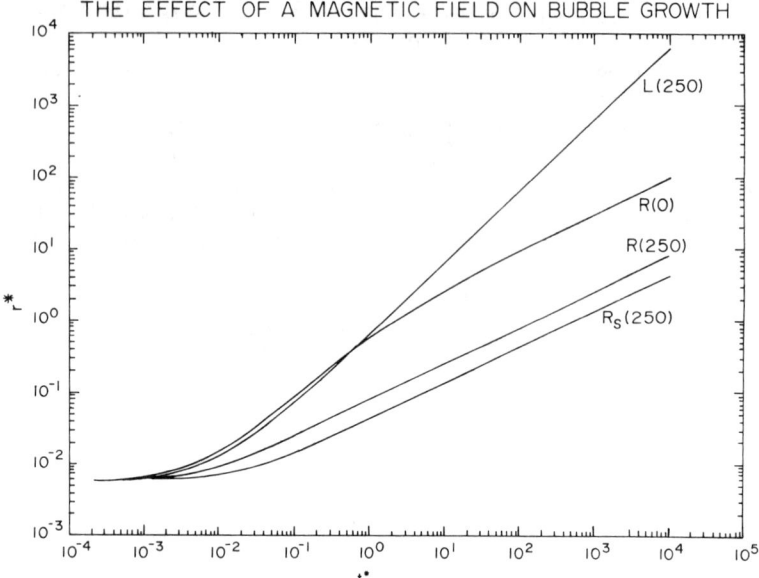

Fig. 2 Theoretical curves showing the effects of a magnetic field on bubble growth.

$$\frac{L^*}{R^*} = (4\ln 2)^{1/2} \left(\frac{t}{t_J}\right)^{1/2} = 1.67 \left(\frac{t}{t_J}\right)^{1/2} \quad (40)$$

The solution of Eqs. (14-17) was found numerically and is shown in Fig. 2. The curve R(0) indicates the classical spherical solution when B = 0. The curve $R_S(250)$ indicates the solution for Λ = 250 for a spherical field. The two curves labeled R(250) and L(250) are the result of the uniform magnetic field case. Note that L maintains a linear temporal behavior, whereas R(250) lies well above the curve $R_S(250)$. This means that the spherical field case was by far more restrictive growthwise than it was originally thought in Ref. 1.

The Numerical Solution

Instead of developing our own numerical code, we searched to find a code suitable for our problem and available in the literature. We found that the Los Alamos solution algorithm-volume of fluid (SOLA-VOF) code was an appropriate one, needing only a few modifications. One of the drawbacks of this code is that it has been developed

BUBBLE GROWTH IN A UNIFORM MAGNETIC FIELD

for cases in which the energy equation is not coupled with the other conservation equations. From the previous section, however, we found that in our case the coupling between the momentum and energy conservation equations was rather weak, especially for high values of Λ, such that, for instance, we will have in the case of a lithium blanket. This means that for ΔP we could use the solution $\Phi = (1 - Ja^*)$, and, since the driving ΔP is constant, it would be possible to solve essentially the Rayleigh problem in the presence of a magnetic field.

The fundamental conservation equations in the liquid region in cylindrical coordinates are as follows: conservation of mass:

$$\frac{\partial u}{\partial r} + \frac{\partial v}{\partial z} + \frac{u}{r} = 0 \qquad (41)$$

momentum in the radial direction:

$$\frac{\partial u}{\partial t} + u\frac{\partial u}{\partial r} + v\frac{\partial u}{\partial z} = -\frac{1}{\rho_\ell}\frac{\partial p}{\partial r} - \frac{\sigma_e u B^2}{\rho_\ell} \qquad (42)$$

momentum in the axial direction:

$$\frac{\partial v}{\partial t} + u\frac{\partial v}{\partial r} + v\frac{\partial v}{\partial z} = -\frac{1}{\rho_\ell}\frac{\partial P}{\partial z} \qquad (43)$$

These equations will be nondimensionalized before integration. The obvious way is to use the familiar nondimensionalization used in the OFM case. If we do so, we should expect the solution to come out in a parametric form in terms of Λ. A more powerful nondimensionalization is available to us now that we do not need to deal directly with the energy equation, since it has been eliminated by providing us with a value for ΔP. This ΔP (a function of Λ) can be used as one of the characteristic quantities for dimensionalization purposes. We need one more quantity and clearly this could be the time $t_J = \rho_\ell/\sigma_e B^2$. We thus construct the following scheme:

$$u_{ref} \equiv (2\Delta P/3\rho_\ell)^{1/2} \qquad (44)$$

$$t_{ref} \equiv \rho_f/\sigma_e B^2 \qquad (45)$$

$$L_{ref} \equiv u_{ref} t_{ref} \qquad (46)$$

Then

$$P_{ref} \equiv \rho_f U_{ref}^2 \tag{47}$$

$$u^+ \equiv u/U_{ref} \tag{48}$$

$$v^+ \equiv v/U_{ref} \tag{49}$$

$$t^+ \equiv t/t_{ref} \tag{50}$$

$$r^+ \equiv r/L_{ref} \tag{51}$$

$$z^+ \equiv z/L_{ref} \tag{52}$$

$$P \equiv (P - P_\infty)/P_{ref} \tag{53}$$

The equations (41-43) become

$$\frac{\partial u^+}{\partial r^+} + \frac{\partial v^+}{\partial z^+} + \frac{u^+}{r^+} = 0 \tag{54}$$

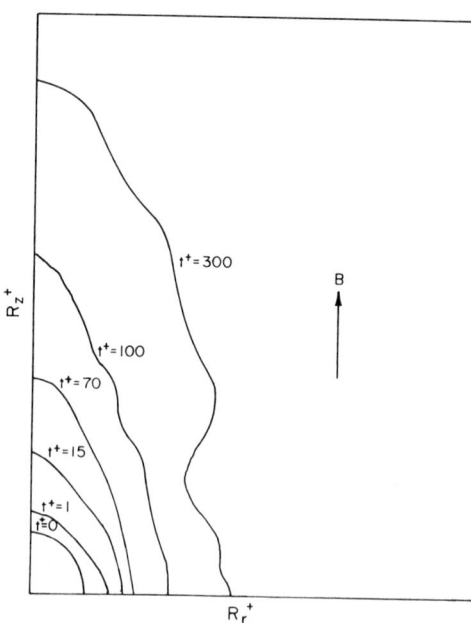

Fig. 3 A sequence of bubble surface shapes (not to scale).

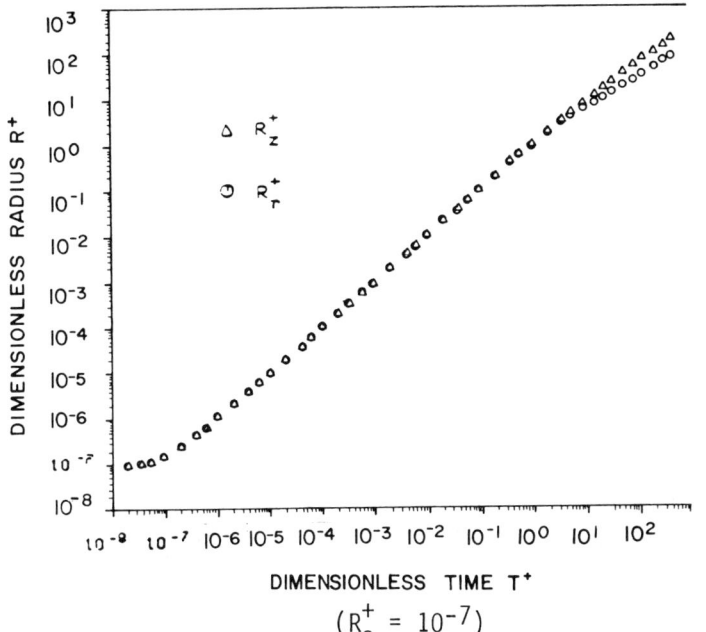

Fig. 4 Nondimensional plot of bubble radii vs time ($R_0^+ = 10^{-7}$).

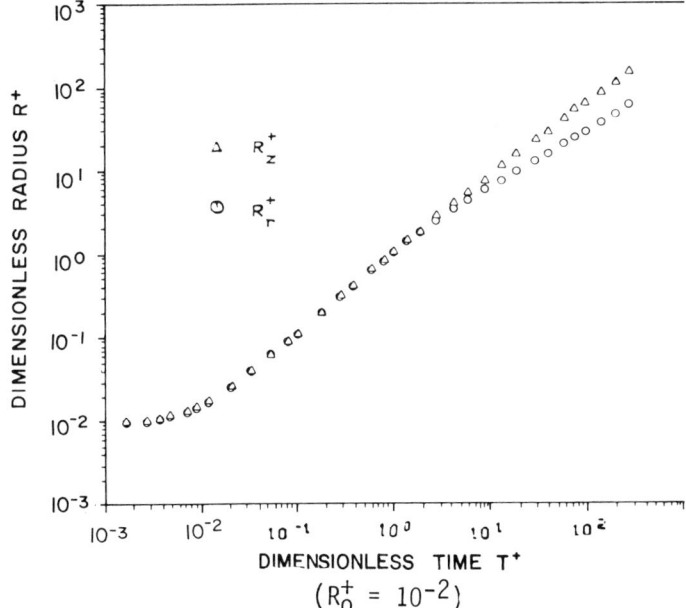

Fig. 5 Nondimensional plot of bubble radii vs time ($R_0^+ = 10^{-2}$).

$$\frac{\partial u^+}{\partial t^+} + u^+ \frac{\partial v^+}{\partial r^+} + v \frac{\partial v^+}{\partial z^+} = -\frac{\partial p^+}{\partial r^+} - u^+ \quad (55)$$

$$\frac{\partial u^+}{\partial t^+} + u \frac{\partial v^+}{\partial r^+} + v^+ \frac{\partial v^+}{\partial z^+} = -\frac{\partial p^+}{\partial z^+} \quad (56)$$

Initial conditions are at $t^+ = 0$ the bubble interface is a sphere of given radius $R_0^+ = R_0/L_{ref}^*$. The initial radius R_0 can be obtained from the familiar condition for $t \simeq 0$:

$$p_v - p_{sat} = 2\sigma/R_0 \quad (57)$$

At the bubble interface, $p^+ = 3/2$ and, far away from the bubble, $u^+ = v^+ = p^+ = 0$.

We can now see that the system (54-56) is a universal set of equations that have no dependence on any parameter, since they have all been absorbed in the nondimensional process.

The SOLA-VOF method numerical code is an Eulerian finite-difference numerical algorithm suitable for two-dimensional transient fluid flows with free boundaries. Details on this code can be found in Refs. 10-13. The extension to the magnetic case was originally undertaken by Efthimiadis, but the final solution as it will be presented here was performed by Fotiadis.[14] The code was first tested for the case B = 0, for which the Rayleigh solution is known, and it was found that it broke down numerically as the value of $t^+ > 300$. It turned out that this failure also occurred in the presence of the magnetic field. Figure 3 shows a sequence of bubble surface shapes for t = 0, 1, 15, 70, 100, and 300. R_z^+ and R_r^+ indicate the characteristic lengths of the bubble in the z and r directions. Figures 4 and 5 show that the initial condition for R_0 stops influencing the solution after one order of time magnitude, indicating that the asymptotic behavior of the system (54-56) does not depend on the original value R_0. Figure 6 shows, in an expanded form, the trend of growth for R_z and R_r given (up to the value of $t^+ \simeq = 300$) by the two approximate relations:

$$R_z \simeq t^{+0.9} \quad \text{and} \quad R_r \simeq t^{+0.7}$$

Figure 7 also shows the solution for the case of a spherical magnetic field but cast in the cross nondimensionalization with a constant ΔP. In this case, the equation

BUBBLE GROWTH IN A UNIFORM MAGNETIC FIELD

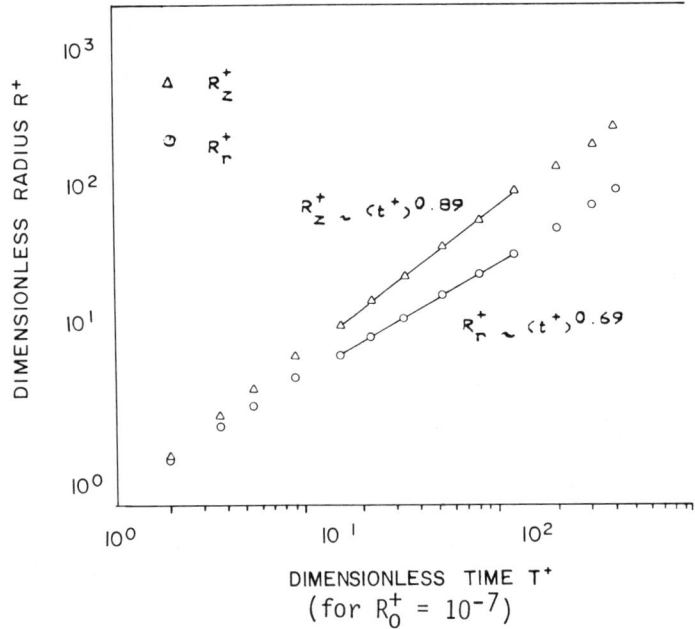

Fig. 6 Nondimensional plot of bubble radii vs time ($R_0^+ = 10^{-7}$).

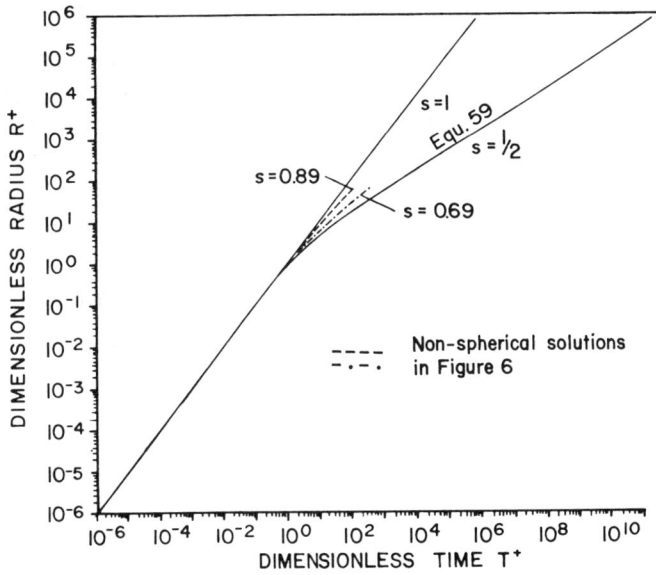

Fig. 7 Nondimensional plot of bubble radii vs time [solution of Eq. (59)].

of motion in the radial direction becomes (including the inertia term R^{+2})

$$\dot{R}^{+2} + (2/3)\ddot{R}^+R^+ - 1 = 0 \qquad (58)$$

The solution is [with $R^+(0) = 0$]

$$t^+ = (3/2) \left[y^2 + y(y^2 + 1)^{1/2} + \ell n(y^2 + 1)^{1/2} + y \right] \qquad (59)$$

with $y = R^+/3$.

We can clearly see that when $t^+ = 1$, $(t = t_J)$ the solution changes from linear to parabolic, that is, $R^{+1/2}$. The reader can verify that this expression coincides with $R^* = Ja^* \times t^{*1/2}$ when account is taken of the two different nondimensionalizations.

From Fig. 7, one can see that the asymptotic solution $R_z^+ = t^+$ and $R_r^+ = (t^+)^{1/2}$ do provide an envelope for the numerical solution of Fig. 11, also shown there.

We are now in the process of modifying the SOLA-VOF code, hoping to be able to advance beyond the value of $t^+ = 300$. We also plan to properly introduce the energy equation in order to justify the use of the asymptotic answer for the pressure distribution given by Eq. (28).

Experimental Work

Our experimental work was conducted at the MFM Laboratory at Purdue University, with mercury pool boiling in the presence of a horizontal magnetic field, at a maximum value of 1.3 T. This work has been already reported in the literature,Eq. 15 and only the highlights will be given here. The boiler consisted of a vertical stainless steel 316 tube (60 mm i.d. x 910 mm long) with a horizontal heater surface at the bottom and a water cooled condenser region at the top. The maximum power input was 1000 W. Boiling surface temperature measurements were made with a platinum resistance thermometer extended over the entire length of the surface. The liquid bulk temperature was measured with a traversing resistance thermometer and several fixed thermocouples. Data were provided for saturated pressures of 0.0066 and 0.1 MPa at magnetic field strengths of 0.0, 0.4, 0.8, and 1.26 T. Figures 8 and 9 show the boiling curves for the cases of pressures corresponding to 0.1 and 0.0066 MPa.

The work of Forster and Zuber[3] suggests a Nusselt number based on a Reynolds number that is calculated with the radius of the bubble and its growth rate as the character-

BUBBLE GROWTH IN A UNIFORM MAGNETIC FIELD 299

istic length and velocity. For the heat controlled growth we found R^* to be equal to $Ja^* \times t^{*-1/2}$. The Reynolds number then turns out to be proportional to $Ja^{*2} \times Ja^2$ rather than Ja^2 as in the OFM case. For the Prandtl number variation, the one-third power was suggested. Therefore, we have the relationship

$$Nu_B/Nu_0 = (Ja^{*2})^m \qquad (60)$$

The power m is determined from the boiling curve q" versus ΔT_S. From our OFM data, we determined $q'' \sim \Delta T_S^{3.2}$. This value corresponds to a value of m = 1.1. (In Ref. 1, we cite the value m = 0.62 as suggested for the data of Ref. 3.)

Since for large fields, Ja^* falls off with $\Lambda^{-1/2}$, it is clear from Eq. (60) that Nu falls off with the power Λ^{-m}. In Ref. 15, we fabricated an empirical relationship guided by this asymptotic behavior, but, at the same time, we introduced a pressure dependence. The suggested equations from Ref. 15 were as follows:

$$Nu_B = \left[\frac{10^{-2}(P)^{1/2}}{1 + 17.5 \, P^{1.25}(\Lambda)^{1/2}} \right]^{2.44} \left(\frac{Ja}{2Pr} \right)^{1.22} Pr^{0.33} \qquad (61)$$

or, in the working form,

$$q'' = 1.92 \left[\frac{10^{-2}(P)^{1/2}}{1 + 17.5 \, P^{1.25}(\Lambda)^{1/2}} \right]^{2.44}$$

$$\times \left[\frac{c_{p\ell}^{1.05} \rho_\ell^{1.69} k_\ell^{1.39}}{\mu_\ell^{0.89} \rho_v^{0.69} i_{fg}^{0.69} T_{sat}^{0.75} \sigma^{0.5}} \right] \Delta T_s^{3.2} \qquad (62)$$

The solid lines in Figs. 8 and 9 show the prediction (62) to be in good agreement with the data.

It is worth noting that, pressure variation apart and the determination of the power of the boiling curve (which even in the OFM case are constants determined empirically for every experimental situation), the simple theory of Ref. 1 needed only one additional empirical constant [essential-

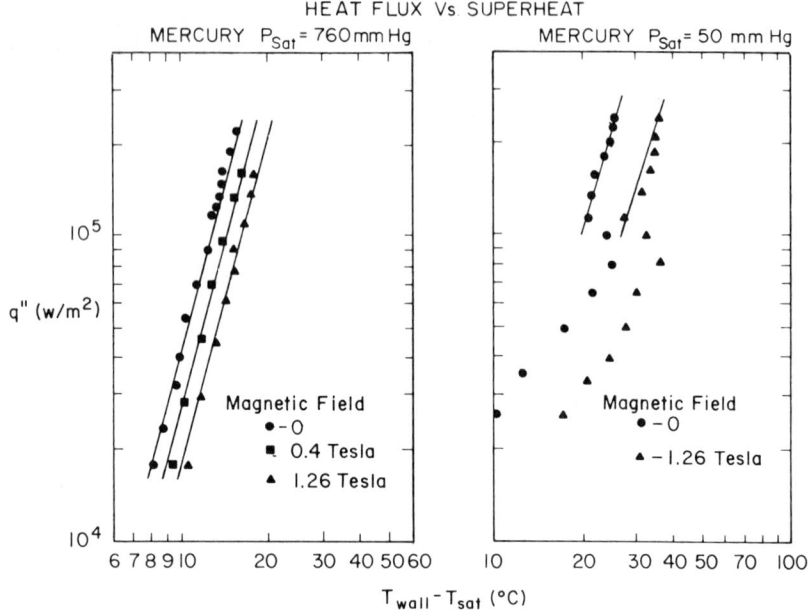

Fig. 8 Heat flux vs superheat (P_{sat} = 760 mm Hg).

Fig. 9 Heat flux vs superheat (P_{sat} = 50 mm Hg).

ly the coefficient $17.5P^{1.25}$ in Eq. (61)] to correlate all of the experimental data in terms of the boiling magnetic interaction parameter Λ.

In a paper now under preparation by Wagner and Lykoudis,[16] a more detailed analytical model has been developed in which the total nucleate boiling heat transfer q_T'' computed by calculating separately the contributions from each mode of heat transfer, namely q_{TC}'' transient heat conduction through the liquid, q_{ML}'' microlayer evaporation, and q_{NC}'' natural convection, as suggested first by Graham and Hendricks[17] and later followed up experimentally by Judd and Hwang.[9] For completeness, we show in Fig. 10 the total heat flux q_T'' versus ΔT_s for the two extreme cases of B = 0 and B = 1.26 T along with the separate contributions due to q_{TC}'', q_{ML}'', and q_{NC}''. The q_T'' experimental data coincide with the theoretically obtained curves, but the data are not shown here in order to avoid confusion.

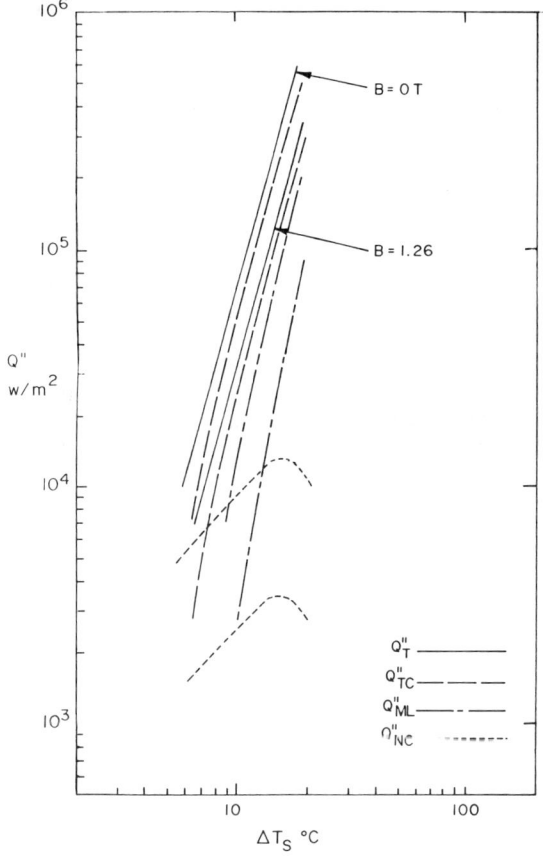

Fig. 10 Comprehensive model heat flux components.

Conclusions

This paper dealt with the problem of the growth of an initially small spherical bubble in the presence of a horizontal uniform and constant magnetic field. Because of the complexity of the problem, no closed-form solution could be obtained. An approximation, which divided the liquid domain around the bubble into two regions, provided separate solutions for the growth of the bubble both vertically and horizontally with respect to the direction of the magnetic field. In this work, the Jakob characteristic time was introduced, a time above which the heat-transfer mechanism becomes stronger than the inertia mechanism. The boiling magnetic interaction parameter Λ was then identified as being the ratio of the Jakob time divided by the Joulean rollover time $\rho_L/\sigma_e B^2$.

It was determined that after a critical time $t_{C\phi}$ (Λ) the vapor temperature inside the bubble settled to a value higher than T_{sat}, a number that is set by the strength of the magnetic field. For high values of Λ, the aspect ratio L/R of the oblong bubble grows parabolically with t and depends solely on t_J and not on the heat-transfer mechanism. In the limit, it was found that the growth perpendicular to the horizontal field is timewise parabolic and linear along the horizon.

The limiting value ΔP hinted at a new nondimensionalization to be used in the exact differential equations of motion by bringing them to a universal form free from any nondimensional parameters such as Λ. The equations were integrated with the help of a modified SOLA-VOF code, but the solution could only be obtained up to a value t^+ less than 300. R and L varied at the higher times with $t^{0.9}$ and $t^{0.7}$, but not yet quite as t and $t^{1/2}$ as the asymptotic solution of the approximate theory predicted. The values and trends however were fairly close.

Experimentally obtained boiling curves for mercury in pool boiling were reviewed and correlated with the help of the theoretical obtained results. Notwithstanding the complexity of the problem, it appeared that the parameter Λ was capable of correlating all of the data taken for values of magnetic intensities and ambient pressures acceptably.

Acknowledgments

The author wishes to thank L. Wagner for useful discussions and Riddhagni for his careful reading of the manuscript and his discussions. He also wishes to acknowledge the continuing financial support of the National Science Foundation under Grant MEA 8304743.

References

[1] Lykoudis, P. S., "Bubble Growth in the Presence of a Magnetic Field," International Journal of Heat and Mass Transfer, Vol. 19, Dec. 1976, pp. 1357-1362.

[2] Wagner, L. Y. and Lykoudis, P. S., "The Effect of Liquid Inertia on Bubble Growth in the Presence of a Magnetic Field," American Institute of Chemical Engineers Symposium Series, Vol. 73, No. 164, 1977, pp. 142-147.

[3] Forster, K. and Zuber, N., "Dynamics of Vapor Bubbles and Boiling Heat Transfer," American Institute of Chemical Engineers Journal, Vol. 1, No. 4, 1976, pp. 425-429.

BUBBLE GROWTH IN A UNIFORM MAGNETIC FIELD

[4] Rohsenow, W. M. and Hartnett, J. P. (ed.), Handbook of Heat Transfer, Section 13, McGraw-Hill, New York, N.Y., 1973, pp. 12-14.

[5] Collier, J. G., Convective Boiling and Condensation, McGraw-Hill, New York, N.Y., 1980, pp. 118-121.

[6] Wong, C. P. C., Vliet, G. C., and Schmidt, P. S., "Analytical and Experimental Studies of Bubble Growth in Superheated Liquid Metals Under a Uniform Magnetic Field," American Institute of Chemical Engineers Symposium Series, Vol. 73, No. 164, 1977, pp. 148-154.

[7] Wong, C. P. C., Vliet, G. C., and Schmidt, P. S., "Magnetic Field Effects on Bubble Growth in Boiling Liquid Metals," Journal of Heat Transfer, Vol. 100, No. 3, 1978, pp. 466-472.

[8] Wagner, L. Y. and Lykoudis, P. S., "Mercury Pool Boiling Under the Influence of a Horizontal Magnetic Field," International Journal of Heat and Mass Transfer, Vol. 24, No. 4, 1981, pp. 635-643.

[9] Judd, R. L. and Hwang, K. S., "A Comprehensive Model for Nucleate Pool Boiling Heat Transfer Including Microlayer Evaporation," Journal of Heat Transfer, Vol. 98, No. 4, 1976, pp. 623-629.

[10] Hirt, C. W., Nichols, B. D., and Romero, N. C., "SOLA-A Numerical Solution Algorithm for Transient Fluid Flows," Report LA-5852, Los Alamos National Laboratory, Los Alamos, N.M., 1975.

[11] Hotchkiss, R. S., "Simulation of Tank Draining Phenomena with the NASA SOLA-VOF Code," Reports LA-8163-MS and UC-32, Los Alamos National Laboratory, Los Alamos, N.M., 1979.

[12] Nichols, B. D., Hirt, C. W., and Hotchkiss, R. S., "SOLA-VOF: A Solution Algorithm for Transient Fluid Flow with Multiple Boundaries," Reports LA-8355, UC-32, and UC-34, Los Alamos National Laboratory, Los Alamos, N.M., 1980.

[13] Hirt, C. W. and Nichols, B. D., "Volume of Fluid Method for the Dynamics of Free Boundaries," Journal of Computational Physics, Vol. 39, No. 1, 1981, pp. 201-225.

[14] Fotiadis, I., "Isobaric Bubble Growth Inside a Liquid Metal in the Presence of a Uniform Magnetic Field," Unpublished MS Project Report, School of Nuclear Engineering, Purdue University, West Lafayette, Indiana, December 1983.

[15] Wagner, L. Y. and Lykoudis, P. S., "Liquid Metal Boiling in a Magnetic Field," Proceedings of the Eighth Symposium on Engineering Problems of Fusion Research, Vol. 4, IEEE, New York, N.Y., 1979, pp. 2075-2077.

[16] Wagner, L. Y. and Lykoudis, P. S., "A Theoretical Model for Liquid Metal Nucleate Boiling Heat Transfer Under the Influence of a Magnetic Field," Submitted to the International Journal of Heat and Mass Transfer, Nov. 1983.

[17] Graham, R. W. and Hendricks, R. C., "Initiation of Cooling Due to Bubble Growth on a Heating Surface," NASA TN-D-2990, 1964.

Analysis of Two-Phase MHD Flow in Converging-Diverging Ducts

Shin-ichi Kamiyama*

Institute of High-Speed Mechanics, Tohoku University, Sendai, Japan

Abstract

To clarify the effect of a magnetic field on two-phase bubbly flow characteristics in converging-diverging ducts, an analytical study of one-dimensional two-phase MHD flow is developed for a low-quality case, taking into account slip and pulsation of gas bubbles, where the gas-phase momentum equation is replaced by the equation of relative translational motion of a single gas bubble. Numerical calculation shows that the effect of bubble volume change on bubble slip and pressure distribution is not negligible compared with the acceleration of liquid phase in the equation of bubble translational motion. The application of a magnetic field causes the position of the critical flow condition (i.e., a Mach number of 1) to move down the throat. Comparison of pressure distributions between the separated and homogeneous flow models shows that an assumption of isothermal and homogeneous bubbly flow is reasonably acceptable in the low-quality case. Furthermore, a simple analysis is developed to show the effect of the loading factor on a critical flow condition.

Introduction

Much interest has recently been shown in liquid metal MHD flows in relation to the development of liquid metal MHD power generation systems, MHD devices in nuclear power, metallurgical applications, and also the possibility of liquid metal cooling systems in thermonuclear fusion reac-

---

Paper presented at the Fourth Beer-Sheva Seminar on MHD Flows and Turbulence, Ben-Gurion University of the Negev, Beer-Sheva, Israel, February 27 - March 2, 1984. Copyright © American Institute of Aeronautics and Astronautics, Inc., 1985. All rights reserved.

*Professor.

tors. However, compared with the fruitful results of the basic studies on single-phase liquid metal flows, little attention has been directed toward two-phase MHD flows.
Although several papers $^{1-8}$ have dealt with pressure drop and void distributions in two-phase MHD flows with relatively large void fraction, it seems necessary to obtain a deeper insight into the basic flow characteristics of high-speed two-phase MHD flows for the development of the above-mentioned MHD devices.

On the other hand, cavitation occurrence or boiling phenomena in liquid metals are also serious problems in the design of high-speed MHD flow devices. When we consider cavitation occurrence in a single-phase liquid metal MHD flow, it is necessary to regard the flow as two-phase because of the presence of micro-gas-bubbles as cavitation nuclei even if the quality is very low. Concerning this problem, we have recently proposed an analytical method for predicting cavitation threshold based on a two-phase flow analogy^{9,10} for ordinary liquid flow with no applied field. Further, this approach was extended to liquid metal MHD flows,^{11,12} with the assumption of a small void fraction.

Therefore, the clarification of two-phase MHD flow characteristics is very important in connection with not only the development of two-phase MHD flow devices, but also for the prediction of cavitation occurrence in such flows.

In the previous analyses,$^{9-12}$ isothermal behavior of the gas (or vapor) phase was assumed because of the relatively large heat capacity of the liquid phase, especially for small void fractions.

To confirm the applicability of such assumptions, we have recently conducted an analysis of two-phase MHD flows

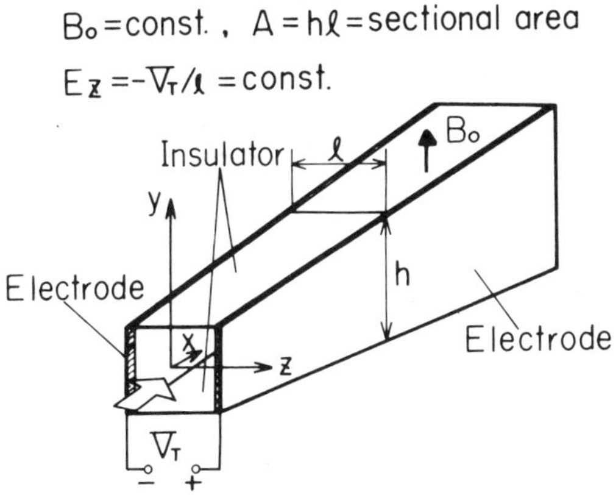

B_o = const., $A = h\ell$ = sectional area

$E_{\bar{z}} = -V_T/\ell$ = const.

Fig. 1 Diverging portion of MHD duct.

using a separated flow model,[13] taking into account slip and expansion of gas bubbles.

In the present paper, a more detailed two-phase MHD flow analysis is carried out using again the separated flow model to show the effects of virtual mass of expanding bubbles on slip and pressure distribution. Two-phase flow characteristics, particularly the choked flow condition, are analyzed and compared with the earlier approximate analysis, where simple isothermal and homogeneous flow has been assumed. Moreover, the effect of the electromagnetic loading factor on choking is clarified by a simple analytical flow model.

Analysis

Basic Equations

Let us consider a quasi-one-dimensional two-phase MHD flow in a converging-diverging duct. The configuration of the diverging portion from the throat is shown in Fig. 1.

It is assumed that the gas phase is in thermal equilibrium with the liquid phase and consists of many tiny bubbles of the same equivalent radius R. The mass flow rate is constant. The effects of aggregation or breakup of bubbles or of evaporation or condensation are thus neglected. The magnetic flux density B_0 and electric field $E_z = V_T/\ell$, where V_T is the potential difference and ℓ the channel width, are assumed to be constant.

The basic equations governing such two-phase MHD flows can then be written as follows[13]: The continuity for the gas phase is

$$\frac{d(\rho_g u_g \alpha A)}{dx} = 0 \qquad (1)$$

and for the liquid phase,

$$\frac{d[\rho_\ell (1-\alpha) u_\ell A]}{dx} = 0 \qquad (2)$$

The combined momentum equation is, if viscosity effects are not taken into account,

$$\rho_g \alpha u_g \frac{du_g}{dx} + \rho_\ell (1-\alpha) u_\ell \frac{du_\ell}{dx} = -\frac{dp_\ell}{dx} - \sigma_T B_0 (E_z + u_\ell B_0) \qquad (3)$$

The combined energy equation is

$$\rho_\ell u_\ell^2 (1 - \alpha)\frac{du_\ell}{dx} + \rho_g u_g^2 \alpha \frac{du_g}{dx} + \left[\rho_\ell(1 - \alpha)u_\ell c_{p\ell} + \rho_g c_{pg}\alpha u_g\right]\frac{dT}{dx}$$
$$= -\sigma_T E_z(E_z + u B_0) + Q \qquad (4)$$

where A is the cross-sectional area of the duct, c is the specific heat, Q the heat added to fluid per unit volume from duct wall, T the temperature, u the velocity, α the void fraction, ρ the mass density, and σ the electrical conductivity. The suffixes g, ℓ, and T denote the gas, liquid, and two-phase mixtures, respectively.

To analyze the effects of slip and radial expansion of the bubbles, the motion of a single gas bubble is here considered to represent the momentum equation of the gas phase, i.e., the equation of the relative translational motion of a single gas bubble is as given by Johnson-Hsieh,[14] taking into account Auton's modification,[15]

$$\frac{4}{3}\pi R^3 \rho_g \frac{D_g u_g}{Dt} = -\frac{4}{3}\pi R^3 \frac{dp_\ell}{dx} - F_d - F_{vm} \qquad (5)$$

where

$$F_d = \frac{1}{2}\rho_\ell c_d (u_g - u_\ell)|u_g - u_\ell|\pi R^2$$

is the drag force;

$$F_{vm} = \beta \rho_\ell a_{vm}$$

is the virtual mass force; and

$$a_{vm} = \frac{4}{3}\pi R^3 \left(u_g \frac{du_g}{dx} - u_\ell \frac{du_\ell}{dx} + \frac{3}{R}(u_g - u_\ell)u_g \frac{dR}{dx}\right) \qquad (6)$$

where β = 1/2 is the virtual mass coefficient.
Equation (5) shows that inertial force on a bubble with virtual mass is balanced by pressure and drag forces.
If we neglect the small terms for $\rho_g \ll \rho_\ell$, Eq. (5) reduces to

$$\tfrac{1}{2}\rho_\ell \left(u_g \frac{d(u_g - u_\ell)}{dx} + (u_g - u_\ell)\frac{du_\ell}{dx} + 3(u_g - u_\ell)u_g \frac{dR}{dx}\right)$$

(continued)

$$= - \tfrac{1}{2}\rho_\ell (u_g - u_\ell)|u_g - u_\ell|\frac{3c_d}{4R} - \frac{dp_\ell}{dx} \qquad (7)$$

The drag coefficient c_d of a spherical bubble is determined, using Haberman's empirical equation,[14] as

$$c_d R_b/24 = 1 + 0.197 R_b^{0.63} + 2.6\times 10^{-4} R_b^{1.38}$$

$$R_b = 2R(u_g - u_\ell)/\nu \qquad (8)$$

The equivalent bubble radius R is assumed given by the equation for an isolated oscillating bubble in an infinite liquid,[16] i.e.,

$$R\frac{d^2 R}{dt^2} + \frac{3}{2}\left(\frac{dR}{dt}\right)^2 = \frac{1}{\rho_\ell}\left(p_g - p_\ell - \frac{2\gamma}{R} - \frac{4\mu}{R}\frac{dR}{dt}\right) \qquad (9)$$

where t is the time, γ the surface tension, and μ the fluid viscosity.

Using the variable transformation $d/dt = u_g d/dx$, Eq. (9) becomes

$$Ru_g^2\frac{d^2 R}{dx^2} + Ru_g\frac{du_g}{dx}\frac{dR}{dx} + \frac{3}{2}u_g^2\left(\frac{dR}{dx}\right)^2$$

$$= \frac{1}{\rho_\ell}\left(p_g - p_\ell - \frac{2\gamma}{R} - \frac{4\mu}{R}u_g\frac{dR}{dx}\right) \qquad (10)$$

Equations (1-5) and (10) are the basic equations for a bubbly flow, taking into account slip and bubble pulsation. The equation of state $p_g = f(\rho_g, T)$ is also needed.

An analytical model considering the effects of single bubble oscillation on ordinary two-phase flow was first proposed by van Wijingaarden.[17] In his analysis, a single homogeneous flow model was considered without MHD effect. Morioka et al.[18-19] also analyzed a choking phenomenon in ordinary two-phase bubbly flow by applying Eq. (9).

Calculating Procedure

For a small void fraction ($\alpha = 0.2$), Maxwell's equation is pertinent for the effective electrical conductivity of two-phase mixtures.[20] Then,

$$\sigma_T = \sigma_\ell \frac{2(1-\alpha)}{(2+\alpha)} \tag{11}$$

Assuming the gas component follows the ideal gas law,

$$p_g = \rho_g R_g T \tag{12}$$

where R_g is the gas constant. Also, the mixture quality χ is related to the void fraction α as

$$\chi = \frac{\alpha \rho_g S}{\rho_\ell (1-\alpha) + \alpha \rho_g S} \tag{13}$$

where $S = u_g/u_\ell$ is the slip ratio.
If the initial bubble radius and void fraction are specified at a reference state such as STP (standard temperature and pressure), the following relation is also utilized:

$$\left(\frac{R}{R_{ref}}\right)^3 = \left(\frac{\alpha}{1-\alpha}\right)\left(\frac{1-\alpha}{\alpha}\right)_{ref} \tag{14}$$

The unknown quantities u_g, u_ℓ, T, χ, p_ℓ, p_g, R, and α are solved from Eqs. (1-5), (10), (12), and (13) with the initial condition given at the upstream position.

Numerical Calculation and Discussion

Comparison Between Adiabatic and Isothermal Assumption Results

The basic equations were numerically solved by the Runge-Kutta-Gill method for a given duct area variation such as

$$A^*(x^*) = 1 + \varepsilon \sqrt{x^{*2} + 100H^2} - 10\varepsilon H \tag{15}$$

This results in

$$\left(\frac{dA^*}{dx^*}\right)_{x^*=0} = 0$$

$$A^* = 1 + \varepsilon x^* \text{ for large } X^* \tag{16}$$

where $A^* = A/A_t$; $x^* = x/h_t$ (the suffix t refers to the throat); and ε is a nondimensional parameter expressing duct expansion and H the distance interval employed in the numerical calculation (equal to 10^{-4} for nonpulsating pressure distribution and 10^{-5} for pulsating pressure distribution).

Mercury at 473 K (200°C) is the working fluid used for the numerical calculations. Throat pressure is 200 kPa for the initial value.

The loading factor K of the electric circuit is defined at the throat as

$$K_t = - E_z/u_t B_o \qquad (17)$$

Figure 2 shows pressure and temperature distributions along the diverging passage after passing through the throat, using the duct expansion parameter ε to calculate an adiabatic flow condition (Q = 0) in a homogeneous flow model ($u_g = u_\ell$). The pressure distribution for the isothermal case (T = const) and the single-phase flow (α = 0) are also shown for comparison. The open circles in Fig. 2 show the sonic flow condition (i.e., Mach number = 1), which means that choking conditions occur far from the throat owing to the magnetic field effect. It is clear from Fig. 2 that temperature increase due to Joule heating is negligibly small, and the approximate isothermal analysis gives good description of the two-phase flow characteristics, for small duct expansion ratio and small void fraction. As explained in a previous paper,[12] the applied magnetic field suppresses a pulsating pressure distribution near a critical flow[18] and causes monotonous pressure decrease along the flow passage in diverging ducts, as shown in Fig. 2.

Effect of Slip

Figures 3 and 4 show the effect of the virtual mass term in Eq. (7) on pressure distribution and slip ratio, respectively. The full line is the original Johnson-Hsieh formula where the second term in the parenthesis of the left-hand side of Eq. (7) is omitted, and the broken line shows the modification by Auton where all terms in Eq. (7) are considered. It is clear from these figures that the bubble expansion term, i.e., the last term of the left-hand side of Eq. (7), influences the flow characteristics, but the effect of the liquid acceleration term, i.e., $(u_g - u_\ell) u_g du_\ell/dx$, is negligibly small in this situation.

The behavior of the slip ratio is also shown in Fig. 4 changing the initial slip ratio from 1.5 to 0.5. The slip ratio approaches a certain value (1.1) at the downstream portion irrespective of the initial values.

Figure 5 shows a comparison of pressure distribution between separated and homogeneous flow models. It is clear from Fig. 5 that two-phase flow characteristics such as the pressure distribution and choking condition are well modeled by simple homogeneous flow model if the behavior of bubble slip is not considered.

Effect of Loading Factor

As the effect of slip on the pressure distribution is small in the present analysis (see Fig. 5), the homogeneous two-phase flow model (i.e., $u_\ell = u_g = u$) is considered adequate for simple analysis.

Equations (3) and (4) can then be written as

$$\rho_T u \frac{du}{dx} = -\frac{dp}{dx} - \sigma_T B_o (E_z + u B_o)$$

$$\rho_T u c_{pT} \frac{dT}{dx} = -\rho_T u^2 \frac{du}{dx} + \sigma_T E_z (E_z + u B_o) + Q$$

(18)

where

$$\rho_T = \rho_\ell (1 - \alpha) + \rho_g \alpha \quad \text{(apparent density)}$$

(continued)

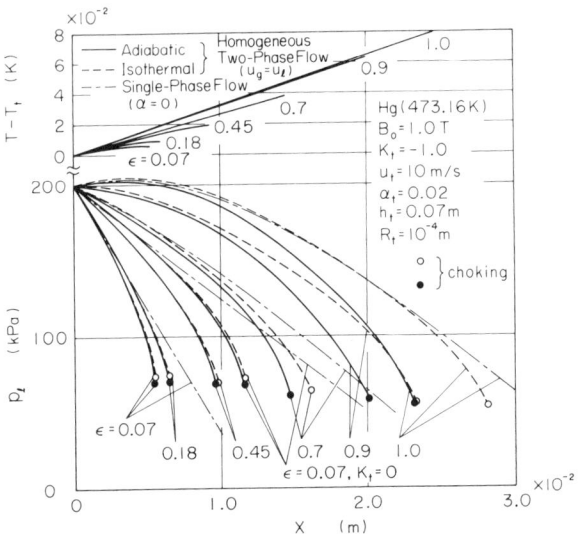

Fig. 2 Pressure and temperature distributions.

$$c_{pT} = \rho_\ell(1-\alpha)c_{p\ell} + \rho_g\alpha c_{pg} \quad \text{(specific heat)} \quad (19)$$

Combining these equations, we obtain

$$\rho_T u\frac{du}{dx} = -\frac{dp}{dx} - \frac{B_o}{E_z}\left[\rho_T u\left(c_{pT}\frac{dT}{dx} + u\frac{du}{dx}\right) - Q\right] \quad (20)$$

Neglecting the heat addition from the duct wall ($Q = 0$) and further assuming $dT/dx = 0$ for the small temperature variation, Eq. (20) is reduced to

$$u\frac{du}{dx}\left(1 + \frac{u}{K_t u_t}\right) = -\frac{1}{\rho_T}\frac{dp}{dx} \simeq -\frac{1}{\rho_\ell(1-\alpha)}\frac{dp}{dx} \quad (21)$$

Also, from the ideal gas law for isothermal behavior,

$$p\rho_g^{-1} = \text{const} \quad \text{or} \quad pR^3 = \text{const} \quad (22)$$

Substituting Eq. (22) into Eq. (21), the following simplified relation, i.e., the modified Bernoulli equation, is obtained:

$$\rho_\ell\left(\frac{u^2}{2} + \frac{u^3}{3K_t u_t}\right) + p + b \ln p = \text{const} \quad (23)$$

where

$$b = \left(\frac{\alpha}{1-\alpha}p\right)_{ref} \quad (24)$$

The pressure coefficient C_p is then expressed as

$$C_p = \frac{2(p - p_t)}{\rho_\ell u_t^2} = 1 - u^{*2} + \frac{2}{3K_t}(1 - u^{*3}) - \frac{2b}{\rho_\ell u_t^2}\ln p^* \quad (25)$$

where $u^* = u/u_t$ and $p^* = p/p_t$.

The effect of the loading factor on pressure distribution can be derived from Eq. (25). However, this is not valid near the short-circuit condition ($K_t = 0$) because of the divergence of the second term in Eq. (25) due to approximate derivation.

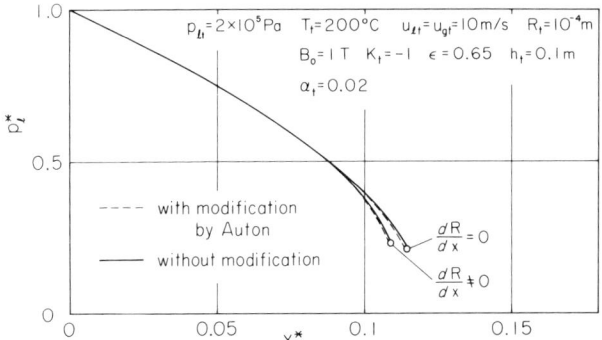

Fig. 3 Effect of virtual mass on pressure distribution.

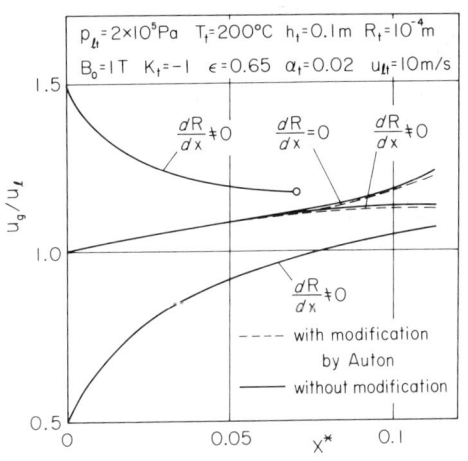

Fig. 4 Effects of initial slip ratio and virtual mass on slip ratio.

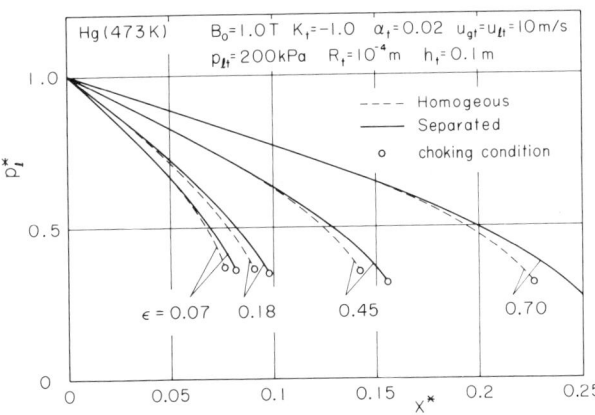

Fig. 5 Comparison of pressure distribution between separated and homogeneous flow models.

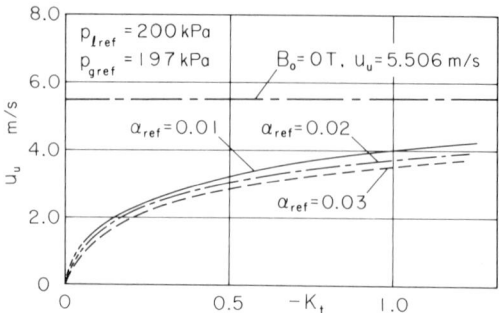

Fig. 6 Effect of loading factor on choked flow condition.

Figure 6 shows the effect of loading factor K_t on the critical upstream velocity u_u for choking, if the duct shape is described for the simple analysis as $A*(x*) = 1 + 0.1(\cos 2x* - 1)$. It is clear from Fig. 6 that the choking flow velocity is reduced with application of magnetic field, depending on loading factor and void fraction.

Conclusion

Numerical calculations of two-phase MHD flow are presented, taking into account radial expansion of bubbles and slip. The assumption of isothermal behavior of the gas phase and homogeneous bubble flow (i.e., no slip) is reasonably acceptable in the case of small void fraction. Further, the analysis shows that the effect of a magnetic field on two-phase flow reduces the critical (choking) flow velocity and, thus, the flow rate.

Acknowledgments

This work was mainly conducted at Cambridge University in Great Britain and at Michigan University in the United States. The author wishes to express his thanks to J. A. Shercliff, F. G. Hammitt, J. C. R. Hunt, and T. R. Auton for their useful discussions on the work.

References

[1] Thome, R. J., "Effect of a Transverse Magnetic Field on Vertical Two-Phase Flow Through a Rectangular Channel," Argonne National Laboratory Report 6854, Argonne, Ill., March 1964.

[2] Petrick, M., "Two-Phase Flow Liquid Metal MHD Generator," Proceedings of Bat-Sheva Seminar on MHD Flows and Turbulence, Ben-Gurion University, Israel, March 1975, pp. 125-145.

[3]Owen, R. C., Hunt, J. C. R., and Collier, J. G., "Magnetohydrodynamic Pressure Drop in Ducted Two-Phase Flows," International Journal of Multiphase Flow, Vol. 3, July 1977, pp. 23-33.

[4]Michiyoshi, I., Funakawa, H., Kuramoto, C., et al., "Local Properties of Vertical Mercury-Argon Two-Phase Flow in a Circular Tube under Transverse Magnetic Field," International Journal of Multiphase Flow, Vol. 3, August 1977, pp. 445-457.

[5]Lindgren, E. R., Kurzweg, U. H., Elkins, R. E., et al., "Two-Phase Hartmann Flows in the MHD Generator Configuration," University of Florida, Gainsville, Fla., Annual Report NR 099-412, 1977.

[6]Saito, M., Inoue, S., and Fujiie, Y., "Gas Liquid Slip Ratio and MHD Pressure Drop in Two-Phase Liquid Metal Flow in Strong Magnetic Field," Journal of Nuclear Science and Technology, Vol. 15, July 1978, pp. 476-489.

[7]Saito, M., Nagae, H. Inoue, S., et al., "Redistribution of Gaseous Phase of Liquid Metal Two-Phase Flow in a Strong Magnetic Field," Journal of Nuclear Science and Technology, Vol. 15, Oct. 1978, pp. 729-735.

[8]Dunn, P. F., "Single Phase and Two Phase Magnetohydrodynamic Flow," International Journal of Heat and Mass Transfer, Vol. 23, March 1980, pp. 379-385.

[9]Kamiyama, S. and Yamasaki, T., "One Predicting Method of Gaseous Cavitation Occurrence in Water and Sodium," Bulletin of Japanese Society of Mechanical Engineers, Vol. 23, Nov. 1980.

[10]Kamiyama, S. and Yamasaki, T., "Prediction of Gaseous Cavitation Occurring in Various Liquids Based on Two-Phase Flow Analogy," Transactions of the ASME, Journal of Fluid Engineering, Vol. 103, Dec. 1981, pp. 551-556.

[11]Yamasaki, T. and Kamiyama, S., "Analytical Study on High Speed Liquid Metal MHD Flow," Transactions of the Japanese Society of Mechanical Engineers, Vol. 48, April 1982, pp. 715-721 (in Japanese).

[12]Kamiyama, S. and Yamasaki, T., and Watai, T., "Effect of Magnetic Field on Two-Phase Flow Phenomena and Cavitation Occurrence," ASME Cavitation and Poly-Phase Flow Forum, St. Louis, Mo., June 1982, pp. 24-26.

[13]Yamasaki, T. Kamiyama, S., and Hammitt, F. G., "Effect of Magnetic Field on Two-Phase Liquid Metal Flow," University of Michigan, Ann Arbor, Mich., Report 014571-53-I, July 1982.

[14]Johnson, V. E. and Hsieh, T., "The Influence of Trajectories of Gas Nuclei on Cavitation Inception," Sixth Symposium on Naval Hydrodynamics, Washington, D.C., Sept. 1966, pp. 163-182.

[15] Auton, T. R., "The Force on an Expanding Spherical Bubble in a Non-uniform Flow of a Perfect Fluid," private communication (to be published).

[16] Hammitt, F. G., "Cavitation and Multiphase Flow Phenomena," first edition, McGraw-Hill Book Co., New York, 1980.

[17] Wijingaarden, L. van, "On the Equations of Motion for Mixtures of Liquid and Gas Bubbles," Journal of Fluid Mechanics, Vol. 33, Sept. 1968, pp. 465-474.

[18] Morioka, S. and Matsui, G., "Choking Phenomena in Nozzle Flows of Bubbly Liquid," Journal of the Physical Society of Japan, Vol. 42, June 1977, pp. 2014-2022.

[19] Morioka, S. and Yoshinaga, T., "Transonic Nozzle Flows of Dispersive Compressible Fluids," Physics of Fluids, Vol. 23, April 1980, pp. 689-694.

[20] Tanatsugu, N., Fujiie, Y., and Suita, T., "Electrical Conductivity of Liquid Metal Two-Phase Mixture in Bubbly and Slug Flow Regime," Journal of Nuclear Science and Technology, Vol. 9, Dec. 1972, pp. 753-755.

Stability of Two-Phase Liquid Metal MHD Channel Flow

Shigeki Morioka* and Toshinori Toma†
University of Tsukuba, Sakura, Ibaraki, Japan

Abstract

Two-phase liquid metal flow in an MHD channel is distinguished by inner-instability mechanism and stratified structure. The influence of these features on the development of a disturbance and on the stability of the flow is investigated on the basis of the extended van Wijngaarden's bubbly liquid model and the homogeneous two-phase flow model. These features bring about a growing wave that does not occur in single-phase liquid metal, although it is not yet developed to a critical situation.

Introduction

The flow of two-phase liquid metal without an external field behaves as ordinary gas-liquid two-phase flow. The dispersion relation of such a bubbly liquid flow has six modes of wave, in which a temporarily and spatially growing wave associated with the slip of bubbles and a spatially growing wave associated with the volume oscillation of bubbles are involved. These waves have remarkable features in the main stream condition of their appearance, the spectrum of the growing wave, and their relation to the behavior of bubbles.[1,2]

In the flow in an MHD channel having no external road, the electrostatic field appears against the induced electromotive force, and a current-free uniform flow is possi-

---

Paper presented at the Fourth Beer-Sheva Seminar on MHD Flows and Turbulence, Ben-Gurion University of the Negev, Beer-Sheva, Israel, Feb. 27-March 2, 1984. Copyright © 1985 by the American Institute of Aeronautics and Astronautics, Inc. All rights reserved.
*Professor, Institute of Engineering Mechanics.
†Senior Scientist, Institute of Engineering Mechanics.

ble. The dispersion relation of this flow has a term multiplied by the interaction parameter in addition to that in ordinary bubbly liquid flow, although the degree of the equation does not change, and any new mode of wave does not appear. The growth rate of the unstable wave associated with the slip of bubbles is reduced by the MHD effect for moderate void fraction, but it is promoted for very small and large void fractions. This result suggests that the turbulence does not remain so small as in single-phase flow of liquid metal with a transverse magnetic field[3] because of the inner-instability mechanism peculiar to the two-phase flow. On the other hand, the instability associated with the volume oscillation of bubbles is supressed a little for higher frequency, but, for lower frequency, the wave that has been stable in the absence of a magnetic field becomes unstable, and the growth rate increases with increasing interaction parameter. However, if this instability brings about the breakup of bubbles, the MHD effect might be useful for atomization of bubbles.

In the MHD channel with an external road, there are the pressure gradients parallel and perpendicular to the main stream owing to the external and induced magnetic fields, and they change the distribution of void fraction and set up a stratified structure in the channel flow. Then, we are interested in knowing the influence of such a stratified structure on the amplification or damping of a disturbance. In this paper, we shall investigate the effect of the stratified structure in the main stream direction on the propagating disturbance on the basis of a homogeneous two-phase flow model. We can find a solution for a plane monochromatic wave in which the amplitude is proportional to the average value for large void fraction, although it does not grow temporarily and spatially. On the other hand, for a small void fraction, we cannot find such a solution, but we can find that the relative strength of a disturbance along the downstream characteristics from the numerical solutions by the method of characteristics. The growth rate increases proportionally to the interaction parameter.

Dispersion Relation

We consider the flow in an MHD channel without an external road. We assume that the magnetic Reynolds number is so small that the deformation of the magnetic field can be neglected and that the bubbles are sufficiently large so that the drag and surface tension can be neglected. Then, the basic equations describing the one-dimensional motion of bubbly liquid perpendicular to the external field (extending the van Wijngaarden's bubbly liquid model[4] to include MHD effect) are as follows:

$$\frac{\partial}{\partial t}(1 - \alpha) + \frac{\partial}{\partial x}(1 - \alpha)u = 0 \qquad (1)$$

$$\rho_L(1 - \alpha)\left(\frac{\partial u}{\partial t} + u\frac{\partial u}{\partial x}\right) = -\frac{\partial p}{\partial x} + jB \qquad (2)$$

$$j = \sigma_L(1 - \alpha)(E + uB) \qquad (3)$$

$$\frac{\partial}{\partial t}\rho_G\alpha + \frac{\partial}{\partial x}\rho_G\alpha v = 0 \qquad (4)$$

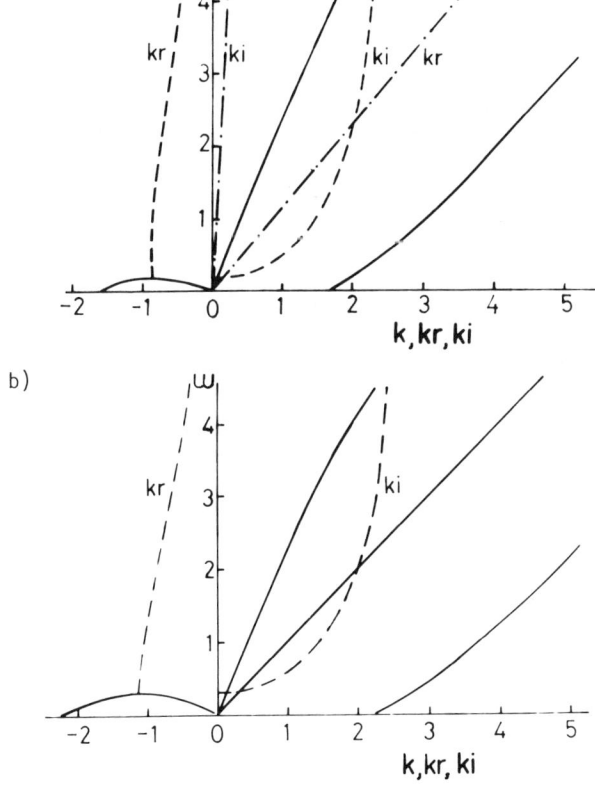

Fig. 1 Numerical solution of dispersion relation (9) for $N = 0$, $\delta = 0.05$, $m^2 = 6$, $\alpha_0 = 0.1127$: a) $\mu = 1.2$ (with slip of bubbles, b) $\mu = 1.0$ (without slip of bubbles). The imaginary part of conjugate complex roots is plotted only in the first quadrant.

$$\frac{1}{2} \rho_L \frac{d}{dt} \left[R^3 (v - u) \right] = -R^3 \frac{\partial p}{\partial x} \quad (5)$$

$$p_G - p = \rho_L R \frac{d^2 R}{dt^2} + \frac{3}{2} \rho_L \left(\frac{dR}{dt} \right)^2 \quad (6)$$

$$\frac{dp_G}{p_G} = \frac{d\rho_G}{\rho_G} = -3 \frac{dR}{R} \quad (7)$$

$$B = \text{const} \quad E = \text{const} \quad (8)$$

where ρ_L and ρ_G are the densities of liquid and gas, respectively; u and p the velocity and pressure in two-phase fluid, respectively; v and p_G the velocity and pressure of bubbles, respectively; α the void fraction; R the radius of bubble; σ_L the electric conductivity of liquid metal; and $d/dt = \partial/\partial t + v\partial/\partial x$ the time derivative riding the bubble.

We consider a uniform flow in which there is a difference between the velocity of bubbles and that of two-phase fluid. Such a uniform state is denoted with the subscript 0 ($p = p_0$, ...). Then we consider the perturbations from such a uniform state and denote them in the nondimensional form $p = p_0 (1 + \varepsilon p')$, ... The length and time are also made nondimensional ($x = L\xi$, $t = L\tau/u_0$). Substituting these expressions into Eqs. (1-8), neglecting the higher-order terms of ε, and assuming a plane monochromatic wave $[p' = \hat{p} \exp \{i(k\xi - \omega\tau)\}, ...]$, we obtain the following dispersion relation:

$$\left(\frac{\delta}{3} m^2 (\omega - \mu k)^2 - 1 \right) k^2$$

$$\times \left(\frac{1}{\alpha_0} (\omega - \mu k)^2 + \frac{(\omega - \mu k)(\omega - k)}{1 - \alpha_0} + 2(\omega - k)(\omega - k + iN) \right)$$

$$+ m^2 [\omega - (2\mu - 1)k](\omega - \mu k)(\omega - k)(\omega - k + iN) = 0 \quad (9)$$

where the five parameters have been introduced; that is, α_0 is the void fraction, $\mu = v_0/u_0$ the slip ratio, $\delta = R_0^2/L^2$ the cross-sectional area ratio of bubble to channel, $m^2 = \rho_L u_0^2/p_0$, and $N = \sigma_L B^2 L/\rho_L u_0$ is the interaction parameter. In the absence of the magnetic field ($N = 0$), Eq. (9) reduces to that in ordinary bubbly liquid flow. This equation is the sixth order with respect to k regardless of N, and it shows that there are six modes of wave.

Bubbly Liquid Metal Flow Without a Magnetic Field

Figure 1 shows the numerical solution of the dispersion relation (9) for $N = 0$. Case a ($\mu = 1.2$) has a relative velocity between the bubbles and the surrounding fluid, and case b ($\mu = 1.0$) has no relative velocity between them. The other parameters are assumed to be $\delta = 0.05$, $m^2 = 6$, and $\alpha_0 = 0.1127$ for both cases.

One of the conjugate complex roots ($k_r > 0$, $k_i < 0$) that can be found over the whole range of angular frequency for $\mu = 1.2$ (with slip of bubbles) presents a temporarily and spatially growing wave. The phase and group velocities of this wave coincide with the flow velocity u_0 as the slip of bubbles vanishes, and it may be considered as an entropy wave mode. Since such an instability does not occur in a dusty gas flow, the cause may be attributed to the virtual mass force acting on the bubbles. From the asymptotic expression for small slip ($|\mu - 1| \ll 1$), the temporal and spatial growth rates can be expressed as

$$\begin{bmatrix} \omega_i \\ k_i \end{bmatrix} \sim (\mu - 1) \, X_i \, (\alpha_0) \begin{bmatrix} u_0 k \\ \omega/u_0 \end{bmatrix} \quad (10)$$

where

$$X_i = \frac{\alpha_0(1 - \alpha_0)}{1 + 2\alpha_0(1 - \alpha_0)} \, \text{Im} \left(\frac{1}{4(1-\alpha_0)^2} - \frac{2}{\alpha_0} \right)^{\frac{1}{2}} \quad (11)$$

and Im denotes the imaginary part of the square root.

On the other hand, one of the conjugate complex roots ($k_r < 0$, $k_i > 0$) that appears at the angular frequency above a critical value ω_c presents a spatially growing wave. For $\mu = 1.0$, the critical angular frequency can be expressed by

$$\omega_c = \omega_0 (1 - M^{2/3})^{3/2} \quad (12)$$

where ω_0 is the proper angular frequency for the volume oscillation of a spherical bubble, $M = u_0/c$ is the Mach number in the bubbly liquid flow, and ω_0 and the speed of sound c are expressed as

$$\omega_0^2 = \frac{3p_0}{\rho_L u_0^2} \qquad c^2 = \frac{[1 + 2\alpha_0(1 - \alpha_0)]p_0}{\alpha_0(1 - \alpha_0)\rho_L} \quad (13)$$

The critical value ω_c decreases as u_0 increases. As u_0 vanishes, the four roots, except for the double root $\omega = k$, are reduced to

$$\omega = \pm ck(1 + c^2k^2/\omega_0^2)^{-1/2} \pm \omega_0 \qquad (14)$$

and these show that they have been the hybrid modes of the sound wave and the bubble oscillation. The feature of this wave is in the phase and group velocities. As seen in Fig. 1, the phase velocity points upstream, while the group velocity points downstream and is very large. The growth rate is large except in the neighborhood of the critical frequency, and it has a value near ω_0/c. In an experiment on the nozzle flow of a bubbly liquid performed by using a nitrogen-water two-phase loop, it has been found that this kind of instability is related to the breakup of bubbles, and then the turbulence in the main stream is appreciably reduced.[2]

Bubbly Liquid Metal Flow With a Magnetic Field

For a finite N, the asymptotic expressions for the growth rates of the unstable wave associated with the slip of bubbles can be expressed as

$$\begin{bmatrix} \omega_i \\ k_i \end{bmatrix} \sim (\mu - 1) Y_i (\alpha_0, n) \begin{bmatrix} u_0 k \\ \omega/u_0 \end{bmatrix} \qquad (15)$$

where

$$Y_i = \frac{\alpha_0(1 - \alpha_0)}{1 + 2\alpha_0(1 - \alpha_0)}$$

$$\times \left[-n + \mathrm{Im} \left(\frac{1}{4(1-\alpha_0)^2} - \frac{2}{\alpha_0} - n^2 - i\frac{(2-\alpha_0)n}{\alpha_0(1-\alpha_0)} \right)^{1/2} \right] \qquad (16)$$

and

$$n = N/(\mu - 1)\omega \qquad (17)$$

Figure 2 shows the variation of Y_i as n increases for the various values of the void fraction α_0. The growth rate tends to reduce with increasing n for moderate void fractions as shown by the curves for α_0 = 0.1, 0.3, and 0.5, but it is not so large. Since an increase in n corresponds to a decrease in $|\mu - 1|$ and ω in addition to an increase in N, we can say that the growth rate is considerably reduced with a decrease in the slip of bubbles and in the frequency. On the other hand, for very small and large void fraction, the growth rate tends to increase with increasing n, as shown by the curves for α_0 = 0.01 and 0.7. These results suggest that, for two-phase liquid metal flow, we could not expect a remarkable reduction of the turbulence by a transverse magnetic field, as seen in single-phase liquid metal, because of the inner-instability mechanism peculiar to the two-phase flow.

Numerical solutions for the unstable wave associated with the volume oscillation of bubbles are shown in Fig. 3 for δ = 0.05, m^2 = 6, α_0 = 0.1127, and N = 0, 0.2, and 0.4. The complex roots are no longer conjugate for finite N, and

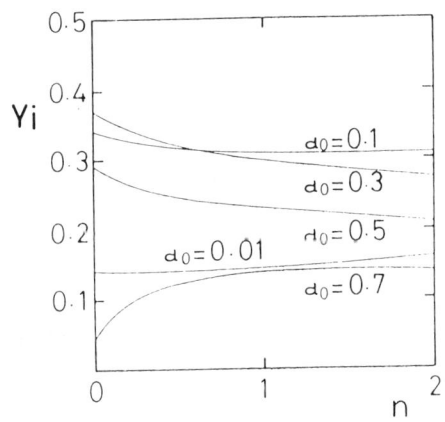

Fig. 2 Y_i as a function of n for α_0 = 0.01, 0.1, 0.3, 0.5, and 0.7.

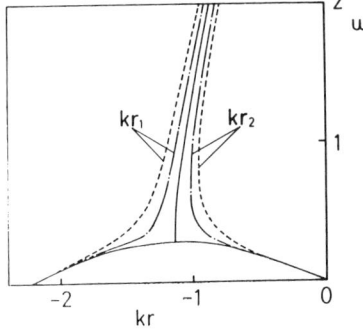
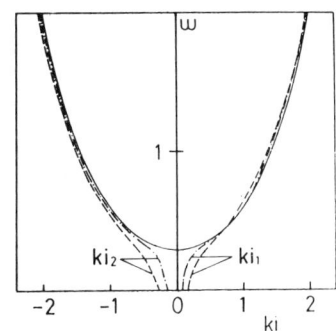

Fig. 3 Numerical solution of dispersion relation (9) for μ = 1.0, δ = 0.05, m^2 = 6, α_0 = 0.1127, and N = 0 (——), 0.2 (— · —), 0.4 (----).

They extend to all the range of frequency. The subscripts 1 and 2 in the real and imaginary parts of the wave number in Fig. 3 show the corresponding parts of two complex roots. The remarkable change in this mode due to the MHD effect is that the lower-frequency waves, which are stable in the absence of a magnetic field, are unstabilized and their growth rate increases with an increasing interaction parameter, while the growth rate decreases a little for the higher-freqency waves. For the lower-frequency waves, the group velocity is not so large but still points downstream. However, if this instability is related to the breakup of bubbles, as in ordinary bubbly liquid flow, the MHD effect might be useful for atomization of bubbles and for reduction of turbulence.

Stability of Homogeneous Two-Phase Flow in an MHD Channel With an External Road

We assume a homogeneous two-phase flow. This flow may be considered as an ideal bubbly liquid flow consisting of infinitesimal but innumerable bubbles and providing a finite void fraction. We consider the flow (x direction) of the homogeneous two-phase liquid metal in an MHD channel having a constant height (y direction) and a variable width (z direction). A uniform magnetic field is applied to the z direction. The equations describing the quasi-one-dimensional motion are as follows:

$$\frac{\partial}{\partial t}(1 - \alpha)A + \frac{\partial}{\partial x}(1 - \alpha)Au = 0 \qquad (18)$$

$$\rho_L(1 - \alpha)\left(\frac{\partial u}{\partial t} + u\frac{\partial u}{\partial x}\right) = -\frac{\partial p}{\partial x} + jB \qquad (19)$$

$$j = \sigma_L(1 - \alpha)(E + uB) \qquad (20)$$

$$\alpha_p/(1 - \alpha) = \text{const} \qquad (21)$$

where A is the channel cross-sectional area. Equation (21) can be obtained by eliminating ρ_G between the no-slip condition $\rho_G \alpha/\rho_L(1 - \alpha)$ = constant and the equation of isothermal variation of state p/ρ_G = constant. The variations of the void fraction, the pressure, and the channel cross-sectional area in the steady flow with a constant velocity can be found by dropping the time derivatives in the above equations and putting u = constant:

$$1 - \frac{\alpha_0^*}{\alpha_0} + \alpha_0^* \ln\left(\frac{\alpha_0}{\alpha_0^*} \frac{1 - \alpha_0^*}{1 - \alpha_0}\right) = \frac{x}{a} \qquad (22)$$

$$\frac{p_0}{p_0^*} = \frac{\alpha_0^*}{\alpha_0} \frac{1 - \alpha_0}{1 - \alpha_0^*} \qquad (23)$$

$$\frac{A}{A^*} = \frac{1 - \alpha_0^*}{1 - \alpha_0} \qquad (24)$$

Where the superscript asterisk denotes their values at a reference point, say, the entrance of channel, and

$$a = \frac{L}{(1 - K)(1 - \alpha_0^*)m^2 N} \qquad (25)$$

is a characteristic length and $K = E/u_0 B$ is the road factor. The asymptotic expressions of Eqs. (22-24) for large and small void fractions can be represented as

$$\frac{p_0}{p_0^*} \sim \frac{1 - \alpha_0}{1 - \alpha_0^*} = \frac{A^*}{A} \sim \exp(-\frac{x}{a}) \quad \text{for} \quad \alpha_0 \sim 1 \qquad (26)$$

$$\frac{p_0}{p_0^*} \sim \frac{\alpha_0^*}{\alpha_0} \sim 1 - \frac{x}{a} \quad \frac{A}{A^*} \sim 1 \quad \text{for} \quad \alpha_0 \sim 0 \qquad (27)$$

Thus, we find that the Lorentz force has a similar effect on the gravity force, and it sets up a stratified structure in the homogeneous two-phase flow in the MHD channel.

Now, we consider a perturbation from such a steady flow $[p = p_0^*\{p_0(x) + \varepsilon p'\}, u = u_0(1 + \varepsilon u')]$. The length and time are made nondimensional as $x = a\zeta$ and $t = a\eta/u_0$, unlike in the preceding sections. Substituting these expressions into Eqs. (18-21) and neglecting the higher-order terms with respect to ε, and further eliminating α' and j', we have

$$\frac{\partial p'}{\partial \eta} + \frac{\partial p'}{\partial \zeta} + \frac{P_0}{\alpha_0}\frac{\partial u'}{\partial \zeta} = -\frac{\alpha_0^2}{\alpha_0^*}p' \qquad (28)$$

$$\frac{\partial u'}{\partial \eta} + \frac{\partial u'}{\partial \zeta} + \frac{1}{(1-\alpha_0)m^2}\frac{\partial p'}{\partial \zeta} = -\frac{1}{(1-\alpha_0^*)m^2}\left(\frac{\alpha_0}{P_0}p' + \frac{1}{1-K}u'\right) \qquad (29)$$

For large void fraction, we can find a solution which is a plane monochromatic wave whose amplitude is proportional to the average value:

$$\frac{p'}{P_0(\zeta)\hat{p}} = \frac{u'}{\hat{u}} = \exp[i(k\zeta - \omega\eta)] \qquad (30)$$

$$\alpha_0^*(1-\alpha_0^*)m^2(\omega-k)^2 + \frac{i\alpha_0^*}{1-K}(\omega-k) - k^2 = 0 \qquad (31)$$

The dispersion relation shows that there is no unstable wave. Thus, we can say the homogeneous two-phase flow in the MHD channel is stable for the large void fraction. Since the void fraction always increases and the pressure always decreases along the stream in the MHD channel, the flow sooner or later comes into the low-pressure region with the large void fraction. Therefore, if a disturbance might grow in the restricted upstream region of the small void fraction, it could not further develop in the downstream region.

In fact, for the small void fraction, the stratified structure is not as simple as for the large void fraction, and we cannot find a solution for the plane monochromatic wave having a simple dependence of the amplitude on the main stream. However, Eqs. (28) and (29) can be solved numerically by the method of characteristics. Figure 4 shows the variation of the relative strength of a pulse disturbance along the two characteristics for $\alpha_0 = 0.01$, $K = 0.05$, and $m^2 = 25$ and 230. One of the characteristics goes downstream and the other upstream for $m^2 = 25$ (subsonic flow), but both of the characteristics go downstream for $m^2 = 230$ (supersonic flow). As seen in Fig. 4, the relative strength of the disturbance increases as it propagates along the downstream characteristics, while it decreases along the upstream characteristics. If we consider the results by the

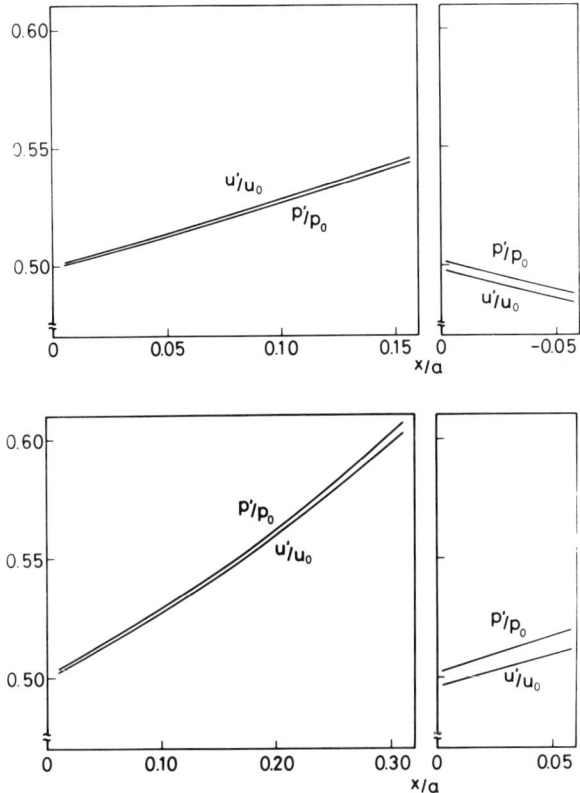

Fig. 4 Development of a pulse disturbance along the characteristics for $\alpha_0^* = 0.01$, $K = 0.5$: a) $m^2 = 25$ (subsonic flow), b) $m^2 = 230$ (supersonic flow).

original dimensional coordinate, that is,

$$x = \frac{L\zeta}{(1-k)(1-\alpha_0^*)m^2 N} \quad (32)$$

we find that the growth rate increases in proportion to the interaction parameter.

References

[1] Morioka, S. and Matsui, G., "Mechanism and Effect of Dispersion and Dissipation in Nozzle Flows of Bubbly Liquid," *Proceedings of the First Asian Congress of Fluid Mechanics*, Bangalore, India, Indian Institute of Science, December, 1980, A 54.

[2] Matsui, G. and Morioka, S., "Two-Phase Instability in Nozzle Flows of Bubbly Liquid," Proceedings of the Second International Meeting on Nuclear Reactor Thermal-Hydraulics, Santa Barbara, California, ANS/ASME/AICLE, January, 1983, Vol. 1, pp. 161-169.

[3] Lock, R. C., "The Stability of the Flow of an Electrically Conducting Fluid Between Parallel Planes Under a Transverse Magnetic Field," Proceedings of the Royal Society of London, Series A, Vol. 233, No. 1192, 1955, pp. 105-125.

[4] van Wijngaarden, L., "One-Dimensional Flow of Liquids Containing Small Gas Bubbles," Annual Review of Fluid Mechanics, Vol. 4, 1972, pp. 369-396.

An Analytical Model for Bubbly Flow

M. Mond* and S. Sukoriansky†

Ben-Gurion University of the Negev, Beer-Sheva, Israel

Abstract

The equations governing the one-dimensional flow in vertical pipes of compressible bubbles immersed in fluid are derived. Mass and momentum conservation for each phase are invoked in order to obtain those equations. The effect of the multitude of bubbles on the drag coefficient for a single bubble was modeled with a single parameter. A single value of that parameter was found to fit various different flow conditions. In the case of gas bubbles in liquid, the bubbles' inertia can be neglected. The equations are numerically solved for that particular case and the effects of various initial conditions are observed and discussed.

I. Introduction

Bubbly flow plays an important role in a wide variety of technological applications such as nuclear reactors[1] and recent concepts of magnetohydrodynamics (MHD) generators[2,3]. The decisive role of the upward motion of the bubbles in existing devices and planned pilot plants makes it of crucial importance to devise an analytic model of bubbly flow, which lends itself easily to numerical computations, in order to predict the effects of various initial on the bubbly flow, which determines to a large extent the efficiency of the energy production process. In this paper, we take the first step toward such a general model by treating a one-dimensional bubbly flow.

Paper presented at the Fourth Beer-Sheva Seminar on MHD Flows and Turbulence, Ben-Gurion University of the Negev, Beer-Sheva, Israel, Feb. 27-March 2, 1984. Copyright © 1985 by the American Institute of Aeronautics and Astronautics, Inc. All rights reserved.

*Senior Lecturer.
†Lecturer.

The bubbles are described by a density distribution function that depends on space, time, and radius. In the case of a uniform radius distribution, the density distribution reduces to the bubbles' number density function. The expansion inertia of the bubbles is also taken into account, thus giving rise to volume oscillations. The interaction of the bubbles with its surrounding fluid is via the drag forces. The drag force on a single bubble is modeled in such a way as to take into account the influence of the rest of the bubbles.

In Sec. II the conservation laws governing the motion of the bubbles as well as that of the fluid are used in order to derive the equations of motion. Section III lists the equations of state governing the radius and pressure of the bubbles as well as the fluid density. In Sec. IV a particular case (gas bubbles in liquid) is solved numerically and the effect of various initial conditions are discussed.

The present model can be easily extended to include two-dimensional effects and can be used as a basis to a study of the linear stability of bubbly flows.

II. One-Dimensional Bubbly Flow

In this section we derive the equations governing the motion of multitude of bubbles flowing in moving incompressible continuous medium (fluid). We start by defining the density distribution $n(x,R,t)$ such that $ndxdRdt$ is the number of bubbles located between x and $(x + dx)$ along the x axis have radii between R and $(R + dR)$ at the time between t and $(t + dt)$. Note that if the bubbles have all the same radius, n becomes the bubbles number density. If $N(x,R,t)$ bubbles are created (or annihilated) per unit time at x and t with radius R, the conservation of the number of bubbles reads

$$A \frac{\partial n}{\partial t} + \frac{\partial}{\partial x}(A n u) + \frac{\partial}{\partial R}(A n \dot{R}) = \dot{N} A \qquad (1)$$

where $\dot{R}(x,R,t)$ is the rate of change of the radius of the bubbles, $u(x,R,t)$ their velocity and $A(x)$ the pipe's cross section. The right-hand side of Eq. (1) is responsible for such processes as coalescence, break-up or any source or sink of bubbles.

Next, the mass of the flowing bubbles is conserved up to a mass source or sink $\dot{M}(x,R,t)$. Thus,

$$A \frac{\partial}{\partial t}(m_b n) + \frac{\partial}{\partial x}(m_b n A u) + \frac{\partial}{\partial R}(m_b n A \dot{R}) = \dot{M} A \qquad (2)$$

where m_b is the mass of bubble with radius R. It should be noted that in the case of no mass and number sources or sinks ($\dot{M}=\dot{N}=0$) and uniform radius distribution, Eqs. (1) and (2) are dependent.

Next, we consider the momentum of the bubbles. There are mainly three sources of momentum change. The first is supplied by gravity, while the two others are given by the fluid in the form of both buoyancy forces and drag forces. Thus, the force balance and equation for the bubbles reads

$$A \frac{\partial}{\partial t}(m_b n u) + \frac{\partial}{\partial x}(m_b n u^2 A) + \frac{\partial}{\partial R}(m_b n u A \dot{R}) = F_B - F_{gb} + n F_d A + \dot{M} u_m A \quad (3)$$

where F_B is the buoyancy force given by the pressure gradient dP/dx,

$$F_B = -A_b \frac{dP}{dx} \quad (4)$$

where A_b is the fraction of the area seen by the bubbles. F_{gb} is the gravitational force and is given by

$$F_{gb} = n m_b g A \quad (5)$$

F_d is the drag force exerted on the bubbles by the fluid and is given by

$$F_d = \frac{1}{2} C_d \rho_\ell \pi R^2 (u - v)|u-v| \quad (6)$$

where ρ_ℓ is the material density of the fluid and v(x,t) its velocity. The drag coefficient C_D will be discussed in the next section. The last term on the right-hand side of Eq. (3) is the change of momentum caused by a source or sink $\dot{M}$ that causes a mass of bubbles to enter or leave the system respectively at velocity u_m. For instance, in the case of evaporation, $\dot{M}$ is the evaporation rate and u_m is u.

The next equation would describe the energy transfer mechanisms into and out of the bubbles. However, the flow is assumed to be isothermal; hence, no energy equation is needed. Even though this assumption of isothermal process can easily be relaxed in order to include energy transfer mechanisms, it provides quite an accurate account of most of the processes taking place at the MHD power generators at the Ben-Gurion facilities where the flow was observed to be isothermal due to the large heat capacity of the fluid. Note, however, that in some cases, as in the processes that involve evaporation or condensation, this assumption is not consistent with Eq. (2) and an energy equation has to be used in order to properly account for the latent heats.

We turn now to the equations which describe the motion of the fluid. The conservation of the fluid mass is expressed in the familiar form as

$$A \frac{\partial \overline{\rho_\ell}}{\partial t} + \frac{\partial}{\partial x}(A\overline{\rho_\ell} \, v) = 0 \tag{7}$$

where $\overline{\rho_\ell}$ is defined as

$$\overline{\rho_\ell} = \lim_{\Delta V \to 0} \frac{\Delta mf}{\Delta V} \tag{8}$$

where ΔV is a volume element containing fluid of mass Δmf as well as $n\Delta \, v$ bubbles.

The fluid's momentum is changed by the gravitational force as well as by the pressure gradient and the drag forces. Thus, the force balance equation for the fluid is given by

$$A \frac{\partial}{\partial t}(\overline{\rho_\ell} \, v) + \frac{\partial}{\partial x}(A\overline{\rho_\ell} \, v^2) = -A_\ell \frac{dP}{dx} - nF_d A - \overline{\rho_\ell} \, g \, A \tag{9}$$

where A_ℓ is the fraction of the area occupied by the fluid.

As an alternative to Eq. (9) a force balance equation can be derived for the "combined" fluid. In this case, the drag force acting on the bubbles and on the fluid cancel each other and we obtain

$$A \frac{\partial}{\partial t}(\overline{\rho v}) + \frac{\partial}{\partial x}(A\overline{\rho v^2}) = -A \frac{dP}{dx} - F_g \tag{10}$$

where the average momentum and momentum flux are

$$\overline{\rho v} \equiv \overline{\rho_\ell} \, v + \int m_b \, n \, u \, d R$$

$$\overline{\rho v^2} \equiv \overline{\rho_\ell} \, v^2 + \int m_b \, n \, u^2 \, d R \tag{11}$$

Equations (1-3), and (7) and (9) [or (10)] provide a set of differential equations that describe the bubbly flow. However, the number of unknowns is bigger than the number of equations and some constitutive relations between the various variables have to be given. Those relations are called equations of state and are discussed in the next section.

III. Equations of State

Even though the fluid is taken to be incompressible, its density changes due to the presence of the bubbles that are discreet and compressible. Thus, the fluid density is given by

$$\overline{\rho_\ell} = \rho_\ell (1-\alpha) \tag{12}$$

where α is commonly known as the void fraction and is given by

$$\alpha = \int n \frac{4\pi}{3} R^3 \, dR \tag{13}$$

On the other hand, the bubbles, being compressible, require a connection between their pressure and material density. This relation is generally provided by an expression of the form

$$P_g = f(\rho_g, T_g) \tag{14}$$

where p_g, ρ_g, and T_g are the pressure, material density, and temperature inside the bubble, respectively. For our calculations in the next section, we use the ideal gas relation, which is

$$P_g = \rho_g R_g T_g \tag{15}$$

where R_g is the universal gas constant. After obtaining the pressure inside the bubble, a relation is needed that relates it to the surrounding fluid pressure. For that relation, we use the dynamical equation given by Wijngaarden,[5] which after neglecting surface tension is given by

$$\frac{P_g - P}{\rho_\ell} = R \frac{d^2 R}{dt^2} + \frac{3}{2} \left(\frac{dR}{dT}\right)^2 \tag{16}$$

where d/dt is the convective derivative along the bubbles' trajectory and in steady state is given by

$$\frac{d}{dt} = u \frac{d}{dx} \tag{17}$$

Note that for large expansion Euler number (Eu: $2P/\rho_\ell u_{ex}$, u_{ex} is the expansion velocity), the inertial effects of the expansion of the bubbles are negligible and the pressures outside and inside the bubble equalize.

The last equation needed is an expression for the drag coefficient. We assume that the drag force acting on a single bubble is modified by the presence of the rest of the bubbles. We model this modification as a power law of $(1-\alpha)$. (There is no modification when there are no other bubbles and no drag force without fluid.) Thus, the modified drag coefficient is given by

$$C_D = C_D^o (1-\alpha)^q \qquad (18)$$

where C_D^o is the drag coefficient without any other bubbles around and q is a positive power to be determined empirically. Also C^o_D is determined semiempirically and different best fit expressions exist for different regimes of Reynolds numbers as well as for different types of bubbles. Now, Eqs. (1-3), (7) and (9), together with Eqs. (12-18) provide a complete set of differential equations governing the bubbly flow.

IV. Gas Bubbles in Liquid: Numerical Solution

In this section, we solve numerically the equations formulated in the previous two sections for a particular case of gas bubbles in liquid medium. The solution is carried out under the following assumptions:

1) Steady state

$$\frac{d}{dt} = \begin{cases} u \frac{d}{dx} & \text{bubbles} \\ v \frac{d}{dx} & \text{liquid} \end{cases}$$

2) $\rho_g \ll \rho_\ell$ ($nm_b \ll \rho_\ell$)

3) There are no mass or number sources or sinks.

4) Uniform radius distribution

Under assumption 2, the momentum equation for the bubbles is reduced to the expression of zero total force acting on a bubble (the bubble's inertia is neglected), thus becoming an equation of state for the velocity of the bubbles,

$$u = u(C_D, R, \alpha, v) \qquad (19)$$

Under assumptions 1, 3, and 4, the rest of the conservation equations become

ANALYTICAL MODEL FOR BUBBLY FLOW   335

$$\alpha u \rho_g A = G = \text{const} \tag{20}$$

$$(1-\alpha) v \rho_\ell A = Q = \text{const} \tag{21}$$

$$\frac{dP}{dx} = \rho_\ell g(1-\alpha) - \frac{1}{A}\frac{d}{dx}(A\rho_\ell(1-\alpha)v^2 + A\rho_g \alpha u^2) \tag{22}$$

$$Ru\frac{2d^2R}{dx^2} + R\frac{du}{dx}\frac{dR}{dx} + \frac{3}{2}u^2\left(\frac{dR}{dx}\right)^2 = \frac{P_g - P}{\rho_\ell} \tag{23}$$

Thus, Eqs. (19–23) and (15), together with the conservation of a bubble's mass ($R^3\rho_g$=const), provide seven equations for the seven unknowns (ρ_g, u, v, P, P_g, α, and R). After a simple transformation, it turns out that one needs to solve only two ordinary differential equations for two unknowns, while the rest of the quantities are obtained algebraically. A fourth-order Runge-Kutta method was used to solve the two ordinary differential equations.

In the range of Reynolds numbers similar to that of the Ben-Gurion experiments, the drag coefficient is almost constant and is given by

$$C_D^o \simeq 0.09$$

The available experimental data enabling us to check and calibrate the numerical solution is scarce and not well defined.

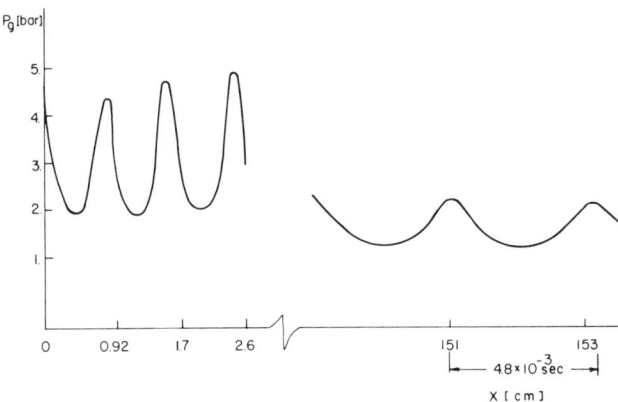

Fig. 1 Pressure inside the bubbles as a function of the coordinate x along the pipe (X=0 is the bottom of the pipe). In all figures in this paper, the following values of the initial conditions were used: $u_L(0)$ = 1 m/s, $\alpha(0)$ = 0.2, P(0) = 3 bar, R(0) = 0.3 cm. In each figure one of the initial parameters was varied while the others remained constant.

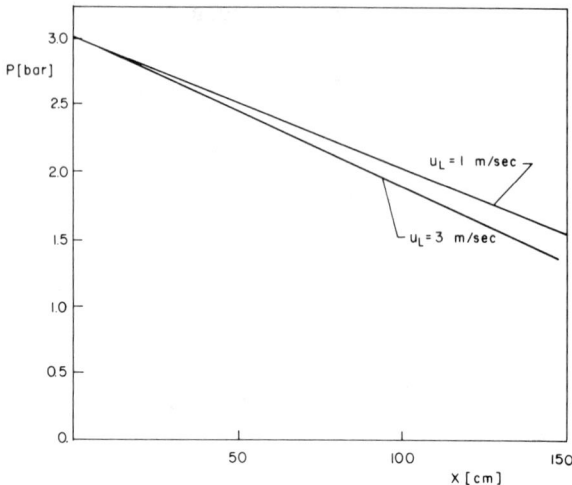

Fig. 2 Fluid pressure as a function of the coordinate x along the pipe for two different initial (x-0) fluid velocities.

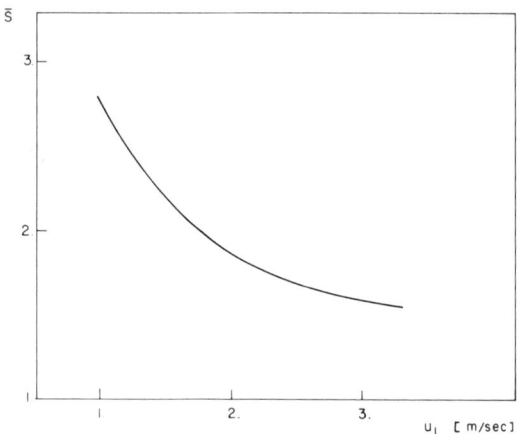

Fig. 3 Average slip ratio as a function of the initial liquid velocity.

However, the power q was picked in such a way that the numerical results yield similar results to both those based on Smissaert's semiempirical laws,[7] as well as to some of the preliminary measurements performed at the Ben-Gurion facilities. It seems that one value of q fits those two different data, a fact that raises hope for a universal value. This value was found to be

$$q \simeq 4$$

ANALYTICAL MODEL FOR BUBBLY FLOW 337

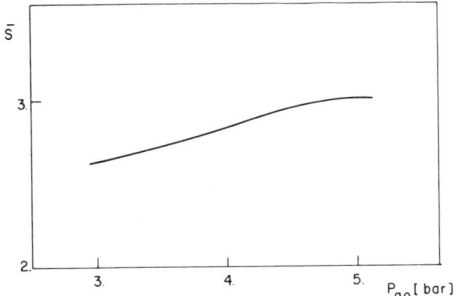

Fig. 4 Average slip ratio as a function of the initial pressure inside the bubbles.

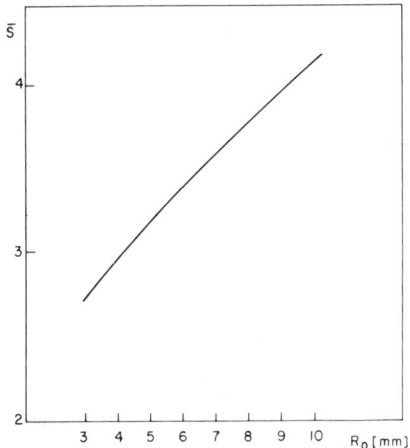

Fig. 5 Average slip ratio as a function of the initial bubbles' radius.

The gas used in the calculation is Freon R113, while the fluid is mercury.

The numerical solutions were carried out under a variety of different initial conditions in order to check their influence on the average quantities of the flow. Figure 1 demonstrates a typical distribution of the pressure inside the bubbles along the pipe. At the bottom of the pipe ($x \simeq 0$), thi difference between the fluid and the bubbles' pressure is considerable, thus giving rise to non-linear effects in Eq. (23). These effects can be noticed through the change of the oscillation period near the bottom part of the pipe. Further upward, the outside and inside pressures become closer and the oscillations become linear with

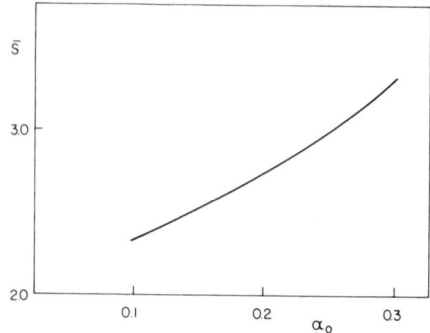

Fig. 6 Average slip ratio as a function of the initial void fraction.

a constant period given by

$$T = 4.8 \times 10^{-3} s$$

This result agrees with the expression for the local linear volume oscillations given by

$$T = 2\pi \, R_o \, \sqrt{\rho_\ell/3P} \qquad (24)$$

Figure 2 shows the liquid pressure drop for two different initial liquid velocities. It is clear that the pressure drop is not very sensitive to the initial liquid velocity.

Figures 3-6 demonstrate the effects of various initial conditions on the average slip ratio S. The average slip ratio is defined as

$$\bar{S} = \frac{1}{L} \int_0^L \frac{u(x)}{v(x)} \, dx \qquad (25)$$

where L is the pipe's length. The slip ratio is a crucial quantity in MHD power generation devices like the one at Ben-Gurion University because it is closely related to the device's efficiency.

Figure 3 indicates that the average slip ratio decreases as the initial liquid velocity is increased. This is due to the fact that the velocity difference between the phases depends weakly on the initial liquid velocity. On the other hand, we can see from Fig. 4 that the average slip increases as the initial bubbles pressure increases. This is due to the fact that for higher pressures the bubble expand faster to higher radii. The effect of varying the initial liquid velocity is much more pronounced.

ANALYTICAL MODEL FOR BUBBLY FLOW

Figures 5 and 6 show the increase of the average slip ratio as a result of increasing the number density of the injected bubbles, as well as its initial radius separately. There is a clear increase in the average slip ratio as both variables are increased separately.

V. Conclusions

Equations were formulated which described the one-dimensional flow bubbles in fluid. The effect of the multitude of the bubbles on a single one was taken into account through a modified drag coefficient. The modification depends on one parameter and it was found that one value of this parameter fit a variety of flow conditions.

A particular case of gas bubbles in liquid was solved numerically and the effects of various initial on the average slip ratio were observed and discussed.

References

[1] Flinn, W.S. and Petrick, M., "Performance and Potential of Natural Circulation Boiling Reactors," Paper presented at ASME Nuclear Engineering and Science Confernece, March 1958.

[2] El-Boher, A, Branover, H., and Petrick, M., "The Etgar Liquid Metal MHD Project," AIAA Progress in Astronautics and Aeronautics: appearing elsewhere in this volume.

[3] Branover, H., El-Boher, A., Lesin, S., and Marcus, B., "Tests of ER-4 Mercury Steam Power Generation Research Facility," AIAA Progress in Astronautics and Aeronautics: appearing elsewhere in this volume.

[4] Thibault, J.P., Jousselin, F., Laborde, R., Alemany, A., and Werkoff, F., "Tin-Water Faraday Generation," AIAA Progress in Astronautics and Aeronautics; appearing elsewhere in this volume.

[5] Van Wijngaarden, L., Journal of Fluid Mechanics, Vol. 33, 1969, p. 465.

[6] Clift, R., Grace, J.R., and Weber, M.E., "Bubbles, drops and Particles," Academic Press, New York, 1978.

[7] Smissaert, G.E., "Two-Component Two-Phase Flow Parameters for Low Circulation Rates," Argonne National Laboratory, Argonne, IL, Rept. ANL-6755, July 1963.

Computer Modeling for Single-Phase Reacting Flow Patterns

E-D. Cristea,* D. Mihai,† and N. Lemnean‡
Icsitee, Bucharest, Romania

Abstract

The idea that an intellectual experiment performed on an electronic computer corroborates a practical experiment today represents an economical way of predetermining combustion processes. This work approaches the aerodynamic and mixture of an MHD cylinder-shaped combustion chamber in which two jets concentrical and coaxial to it expand, mix and burn. The mathematical model of natural gas burning under the simplified conditions of an axial-symmetrical chamber covers three partial models: the flow model, the burning chemical reactions model, and possibly the heat-transfer model, which require the solving of an equation system governing the turbulent, axial, or vortex steady-state flows. The number of equations being less than the number of unknowns requires the introduction of a "closing scheme" using the "k-W" turbulence model, the system being solved by digital technique with a finite-difference method. Special attention has been paid to setting the boundary conditions (free and solid), with special emphasis on the input boundary and solid walls. The computer program, as issued, permits table and graph chart printing of a greater number of calculation arrays but, under the conditions of an input data set corresponding to an MHD combustion chamber of 5 MW_T input heat power, the arrays discussed here are only the ones related to the stream lines, mixing, isotherm isolines for CH_4, and isolines for axial and radial velocity, namely, those capable of giving information concerning the aerodynamics and mixing inside the combustion chamber.

Paper presented at the Fourth Beer-Sheva Seminar on MHD Flows and Turbulence, Ben-Gurion University of the Negev, Beer-Sheva, Israel, Feb. 27-March 2, 1984. Copyright © 1985 by the American Institute of Aeronautics and Astronautics, Inc. All rights reserved.
*Principal Research Scientist, Head of MHD Laboratory.
†Principal Research Scientist, Automation Laboratory.
‡Scientific Manager, Research Department.

COMPUTER MODELING FOR FLOW PATTERNS 341

Nomenclature

$A_\phi, B_\phi, C_\phi, D_\phi$	=	coefficients of balance general equation
a_{ij}	=	tensor of deformation speeds, m/s
C_1, C_2, C_3, C_D	=	coefficients of turbulence "k-W" model
D_{jl}	=	diffusion coefficient for mixture of components j and l, m^2/s
D_j^T	=	thermodiffusion constant of component j, $kg_j/m \cdot s \ln K$
$\vec{f}$	=	body force, N/kg
h	=	specific total enthalpy, kJ/kg
h_j^g	=	negative of component j formation enthalpy, kJ/kg
k	=	pulsatory specific kinetic energy, m^2/s^2
m_j	=	mass concentration of chemical species component j, kg_j/kg_Σ
p	=	static pressure N/m^2
R_k	=	reaction speed, kg/m^3
T	=	absolute temperature, K
T_{ij}	=	tensor of viscous tension, N/m^2
t	=	time, s
v	=	velocity vector, m/s
v_z, v_r, v_o	=	axial, radial, and tangential components of velocity vector, m/s
W	=	square of frequency of turbulent pulsations, $1/s^2$
δ_{ij}	=	unity tensor, Kronecker symbol
η	=	dynamic viscosity, $kg/m \cdot s$
ν_{jk}	=	stoichiometric coefficients
ρ	=	density, kg/m^3
ϕ	=	dependent variable
ω	=	vorticity, 1/s
ψ	=	stream function, kg/s
$\vec{\Pi}_v$	=	tensor of diffusion flux density of specific impulse, $m/s/m^2$
$\vec{\Pi}_j$	=	diffusion flux density of component j, kg_j/m^2

Abbreviations

mol	=	molecular
rad	=	radiation
eff	=	effective

Introduction

Over the last two decades, the study of fuel burning has developed an application of mathematical modeling that predetermines the actual physical processes in industrial flames, an economical solution of the equation systems describing the complex combustion process.

The study of aerodynamics and mixing in steam boiler or industrial furnace fireboxes based on a procedure of predetermination by means of a digital technique and finite-difference method, respectively, was initiated by Spalding and the research team of the Imperial College of London[1] for the conditions of steady-state turbulent flows in axial-symmetrical shapes.

Pay et al[2] successfully tested this predetermination procedure in cement kiln furnaces and steam boilers, and Richter[3] applied it both in the case of natural gas burning (one-phase flow) and in pulverized coal burning (two-phase flow).

In 1983, Cristea et al[4] and Ustimenko and Krol[5] reported on the use of such a predetermination procedure in an MHD combustion chamber for one-phase flow.

Cristea[6] applied this predetermination procedure for isothermal vortex turbulent flow in an extended flow domain, divergent quarl-cylindrical furnace.

Although this procedure of predetermining the performances of an axial-symmetrical combustion chamber is based on the fundamental laws of aerodynamics, thermodynamics, and heat transfer, it requires a great number of tests and controls in order to reach results of correct physical significance.

For these reasons, its application in the overintensified burning conditions characteristic of combustion chambers incurs several difficulties, among which, but not the least of which, is the shortage of experimental data published in the specialized literature that could permit comparison with the theoretical results.

The PACAAS computer program has brought some improvements in the technique of writing the mathematical model: the determination of vortex exchange coefficients by "closing schemes" with two equations ("k-W" model) and the utilization of "wall functions" in determining the boundary conditions on solid walls; as well as in the programming technique itself: the application of a nonuniform grid on the integration domain, thus permitting the improvement of the calculation accuracy for the areas characterized by high gradients, and the utilization of memory dynamic assignment.

Physical Remarks

The general physical processes developing in combustion chambers, in the case of the use of natural gas, the entering conditions are considered, the fuel and the oxident enriched are independently introduced transmitting their properties by convection. Such jets get mixed with the recirculated hot combustion products. This mixture is macroscopical in the case of turbulent flow (turbulent diffusion phenomena) and microscopically controlled by the molecular diffusion, both resulting in the heating of the mixture. Heating of such a mixture is also performed by the heat transmitted by radiation from the combustion gases and walls. When the temperature reaches the combustion kindling point, strongly exothermal sharp reactions occur. Heat is transmitted instantly by radiation and, thereafter, the remaining is carried by the combustion products by convection. Some of the combustion products flux leaving the combustion chamber is recirculated to the flame basis, thus enabling the ignition and stability thereof.

The heat transfer from the combustion gas flux to the walls is made by radiation, convection, and conductibility (in the boundary layer attached to the wall).

Inside the combustion chamber, the thermochemical processes are closely related to the flow processes and the heat- and mass-transfer mechanisms. The complete mathematical model must consider such mechanisms, allowing the predetermination of combustion chamber performances.

It must be pointed out that the geometrical dimensions of a combustion chamber operating on natural gas are not determined by the combustion process; they are determined by the evaporation time of the ionizing seed introduced to increase plasma electroconductibility.

Predetermination of the flow model installed in the combustion chamber space allows a decision to be made concerning entry point of the ionizing seed and, in the case of its entering as liquid, it can give some indications about the technical characteristics of the type of atomizer to be used.

The partial mathematical model of combustion reactions is considered so much simplified for the following reasons:

1) The elementary combustion reactions are written easily only for simple gas combustibles such as hydrogen and methane.

2) The complete description of the combustion process by means of elementary chemical reactions requires a great number of components.

3) There are difficulties in determining the reactions speed of a gas combustible when the flow in the combustion chamber occurs in a turbulent regime.

That is why the "gross reactions" were used, neglecting the intermediate stages and drawing-up the balance only for the initial and final components.

Another simplifying hypothesis is that the combustion reactions occur at infinite reaction speed.

The model used considers simple combustion reactions, neglecting the turbulence, so that the average reaction speeds are only a function of simple field sizes (concentrations, average temperatures).

As to the partial mathematical model for heat transfer, it must be noted that the equations governing the heat exchange by radiation are integrodifferential, while the equations related to the heat transfer by convection and diffusion are the type of differential equations with partial derivatives.

Mathematical Model Working-Out

A general form of balance equations for a general characteristic ϕ of the fluid, as expressed with the Euler description of motion, is written as follows:

$$\frac{\partial(\rho\phi)}{\partial t} + \text{Div}\ (\rho\phi\vec{v} + \dot{\Pi}_\phi) - S(\phi) = 0 \qquad (1)$$

The balance equations for total mass and total impulse result from Eq. (1) for an unsteady-state, laminar flow regime of compressible viscous fluid.

The continuity equation is described by the following expression:

$$\frac{\partial \rho}{\partial t} + \text{Div}\ (\rho\vec{v}) = 0 \qquad (2)$$

The Navier-Stokes equation is expressed by the following relation:

$$\frac{\partial(\rho\vec{v})}{\partial t} + \text{Div}\ (\rho\vec{v}\vec{v} - T_{ij_v}) + \text{grad}\ p - \rho\vec{f} = 0 \qquad (3)$$

The molecular diffusion is a complex phenomenon representing a macroscopical transport of substance due to the molecular thermal agitation, and the internal friction can be expressed as a diffusion of specific impulse:

$$T_{ij} = 2\eta \left[a_{ij} - (1/3 \text{div}\ \vec{v})\delta_{ij} \right] - p\delta_{ij} \cong T_{ij_v} - p\delta_{ij} \qquad (4)$$

The tensor of diffusion flux density of specific impulse is

$$\vec{\Pi_v} = T_{ij_v} = -2\eta \left[a_{ij} - 1/3 \operatorname{div} \vec{v}) \delta_{ij} \right] \quad (5)$$

where

$$a_{ij} = 1/2 \left(\frac{\partial v_i}{\partial z_j} - \frac{\partial v_j}{\partial z_i} \right) \quad i,j = 1,2,3$$

The balance equation of chemical species j for dynamic reactions is

$$\frac{\partial(\rho m_j)}{\partial t} + \operatorname{div}(\rho m_j \vec{v} + \vec{\dot{\pi}}_j) - \sum_{k=1}^{n} \nu_{jk} R_k = 0 \quad (6)$$

The mass flux density for two components is

$$\vec{\dot{\Pi}}_j = -\rho D_{j1} \operatorname{grad} m_j + D_j^T \operatorname{grad}(\ln T) \quad (7)$$

in which the first term represents Fick's law and the second one is the Soret effect (thermodiffusion).

The balance equation for enthalpy is

$$\frac{\partial(\rho h_g)}{\partial t} + \operatorname{Div}(\rho h_g \vec{v} + \vec{\dot{\Pi}}_2) - \operatorname{Div}(\rho \vec{v} + \vec{v} T_{ij_v})$$

$$- \frac{\partial p}{\partial t} - \rho(\vec{v} \vec{f}) = 0 \quad (8)$$

The density of total heat flux consists of

$$\vec{\dot{\Pi}}_2 = \vec{\dot{\Pi}}_{2mol} + \vec{\dot{\Pi}}_{2rad}$$

where the molecular heat flux density is

$$\vec{\dot{\Pi}}_{2mol} = -\lambda \operatorname{grad} T + \sum_{j=1}^{p} h_j \vec{\dot{\Pi}}_j \quad (9)$$

covers Fourier's law in its first term and, in the second term, the negative enthalpy of forming up the components j.

$\vec{\Pi}_{2rad}$ was expressed by means of a model with two radial fluxes.

The processes of mass and heat transfer developed in turbulent flow regime; the equations shown above are transcribed for the turbulent movement, where the moment value of a size has a temporal average component and a pulsatory component.

The balance equations for the average value $\vec{v}$, $\bar{p}$, $\bar{m}_j$, and $\bar{h}_g$ of the dependent variables are drawn from such rewriting, and they represent the essence of turbulent flow. They have in their formulation, in addition to the writing for moment values, correlation terms representing the transport increased by convection in a turbulent flow of the impulse, chemical species, and flow as compared with the laminar flow.

This equation system is solved by introducing some "closing schemes" that allow the calculation of the correlation terms. The k-W turbulence model was adopted, namely, a closing scheme of two equations: the turbulent kinetic energy equation and the square of turbulent pulsations average frequency equation introduced by Spalding. Thus, a system is obtained composed of equations expressing continuity, impulse, chemical species, enthalpy, pulsatory specific kinetic energy, and square of turbulent pulsations average frequency, as well as supplementary equations expressing turbulent viscousness, vortex angular velocity, total enthalpy, and equation of state.

The basic principles of the procedure are well known and, consequently, they will be only briefly described.

The formulation of the equations system can be made in two different manners: "velocity pressure" (v_z, v_r, p) or "vortex angular velocity stream function" (ω_ϕ, ψ).

The second system (ω_ϕ, Ψ), recommended by Gosman[1], was adapted so that at the first stage the pressure can be eliminated from the balance equations.

For bidimensional flows, the stream function Ψ (z,r) can be defined by

$$v_z = 1/\rho r \left(\frac{\partial \Psi}{\partial r}\right)$$
$$v_r = 1/\rho r \left(\frac{\partial \Psi}{\partial z}\right)$$
(10)

and pressure, which is not an independent variable, can be eliminated by introducing the vortex angular velocity $\vec{\omega}$ as a new dependent variable:

COMPUTER MODELING FOR FLOW PATTERNS

$$\vec{\omega} = \text{curl } \vec{v} = \partial v_r/\partial z - \partial v_z/\partial r \tag{11}$$

The rewriting of balance equations was made based on such considerations, thus obtaining an equation system made of the following equations: stream function Ψ, vortex angular velocity, as related, ω_ϕ/r, specific impulse rv_ϕ, chemical species m_i, enthalpy h_g, pulsatory specific kinetic energy k; square of pulsations average frequency W, the general form of which is

$$A_\phi \left[\frac{\partial}{\partial z} (\phi \frac{\partial \Psi}{\partial r}) - \frac{\partial}{\partial r}(\phi \frac{\partial \Psi}{\partial z}) \right] - \frac{\partial}{\partial z} \left[rB_\phi \frac{\partial}{\partial z} (C_\phi \Psi) \right]$$

$$- \frac{\partial}{\partial r} \left[rB_\phi \frac{\partial}{\partial r} (C_\phi \Psi) \right] + rD_\phi = 0 \tag{12}$$

The influence of turbulence is expressed by an effective viscosity that is, for the k-W mode, given by

$$\eta_{eff} = \rho k/W^{1/2} \tag{13}$$

Thus, the simultaneous solving of k and W equations is necessary for obtaining the effective viscosity distribution.

The diffusive fluxes of the k and W equations are given by

$$\dot{\Pi}_{\phi,z} = \eta_{eff}/\sigma_{eff} (\frac{\partial \phi}{\partial z})$$

$$\dot{\Pi}_{\phi,r} = -\eta_{eff}/\sigma_{eff} (\frac{\partial \phi}{\partial r}) \tag{14}$$

They are necessary for the specification of six functions (C_1, C_2, C_3, C_D, σ_k, σ_W) for using the k-W model, which generally depend on Reynolds number but may be taken as constants for high Reynolds Numbers[1-6].

Technique for Solving the Equation System

Due to the elliptical, bidimensional, axial-symmetrical character of the flow, the equation system was solved by using the digital method based on the finite-difference technique.

The approximation of the functions searched in the grid nodes or therearound was made on control volumes of grid.

The general differential equation of transport was written for each variable, and such equations, together with the boundary conditions, form a system of (n+m) nonlinear algebra equations, according to the grid nodes, with (n+m) unknowns.

The convergency matter was settled by the combined use of methods of "underrelaxation", estimation of gross values, implicit treating of boundary conditions, and calculation of grid pitch.

The nonuniform grid used was rectangular type of 20x20 lines, such lines being parallel to the directions z and r, with a denser packing in the vicinity of the burner "mouth".

The boundary conditions were indicated by the values of variables on the domain boundary, the gradients of variables on the domain boundary, and a combination of the two methods.

The flowfield input section used two methods of expression: the distribution of variables as established by empirical formulas and the distribution of variables as determined by experimental measurements, respectively.

The expression of boundary conditions on solid walls used the "wall functions" method, ensuring an approximate introduction of the influence of a turbulent boundary layer attached to the wall upon the impulse and energy transport, which is satisfactory for the study of flow models in combustion chamber.

Computer Program

The computer program PACAAS was written in Fortran IV language and rolled on a computer IBM 360/340.

The analysis of the computer program flow chart showing the organization of the main program, subroutines, and functions reveals the particular attention paid to the introduction and preparation of input data performed with BLOK DATA, DATINT, and PREDIN, which provide much increased mobility and flexibility in approaching various practical cases (Fig. 1).

The program prints the solutions on the mathematical model in the form of local properties, giving the following distributions: stream function; angular velocity component; velocity vector components; pulsatory specific kinetic energy; square of average frequency of turbulent pulsations; effective viscosity; mixture distribution; CH_4O_2 distributions; and temperature distribution.

COMPUTER MODELING FOR FLOW PATTERNS 349

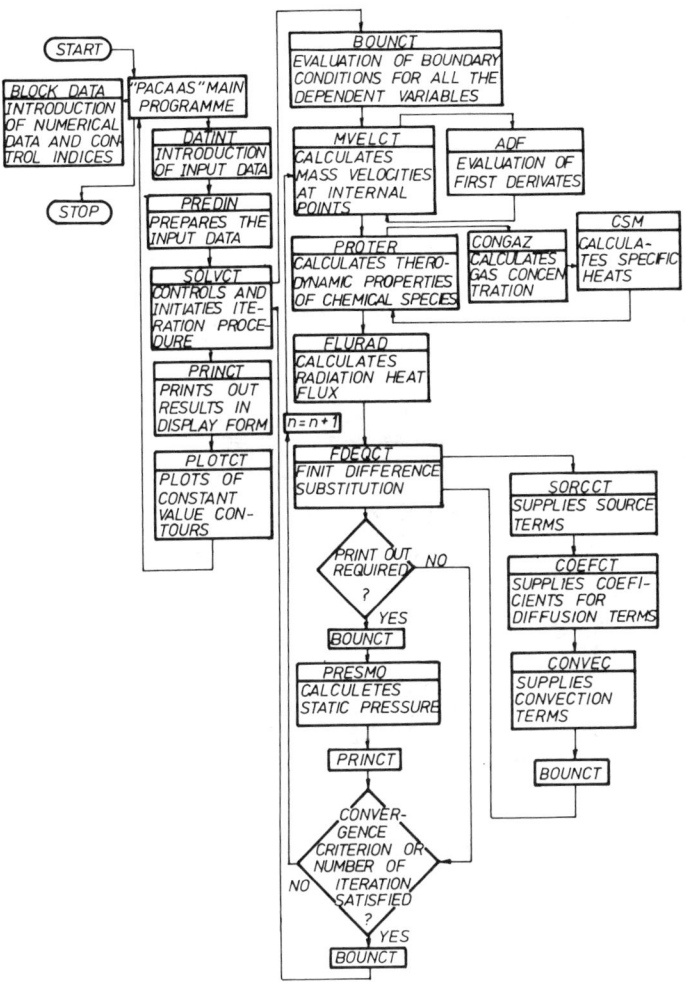

Fig. 1 Flow diagram of the "PACAAS" computer program.

Computer Model Applications

The computer model was applied to a design combustion chamber operating on natural gas, with a total mass flow rate of combustion products of up to 1.5 kg/s. An example of an input data set follows: thermal input (fuel), 5.3 MW_T; elementary analysis of natural gas, CH_4=99.2%, CO_2=0.1%, air=0.7%; net calorific value of natural gas, 33739 kJ/kg; oxidizer preheat temperature, 1823 K; oxygen concentration in oxidizer, 40%; pressure, 0.11 MPa; dimensions; 0.400 m in diameter and 1.0 m long. The analysis of the results obtained implies the interpretation of distributions of isolines traced for some variables.

Figure 2 shows a model of the dimensionless stream lines $\Psi/\Psi_p + \Psi_s$, which gives a clear image of the flow model as installed in combustion chamber. Notice the selection of expressing manner for which $\Psi/\Psi_p + \Psi_s = 1$ on the symmetry axis and $\Psi/\Psi_p + \Psi_s = 0$ at the top limit of the secondary circuit (oxidizer).

In the upper part of the combustion chamber, a large peripheral recirculation zone is installed, and the stream line $\Psi/\Psi_p + \Psi_s = 0$ points the axial flow area itself, with beneficient results upon stability.

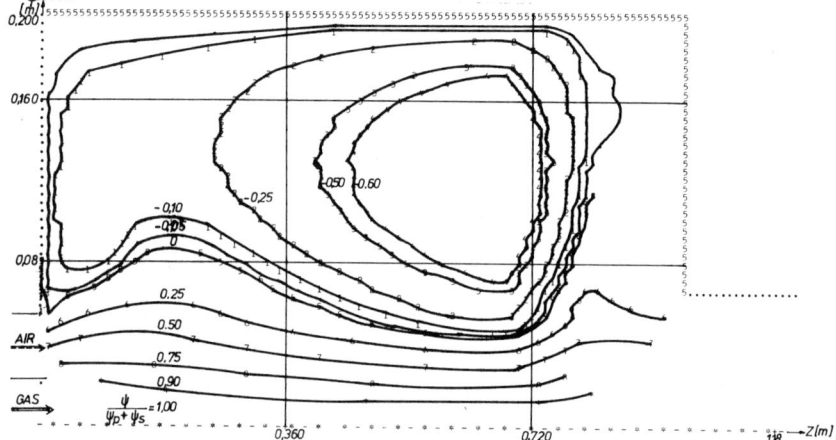

Fig. 2 Stream-line patterns.

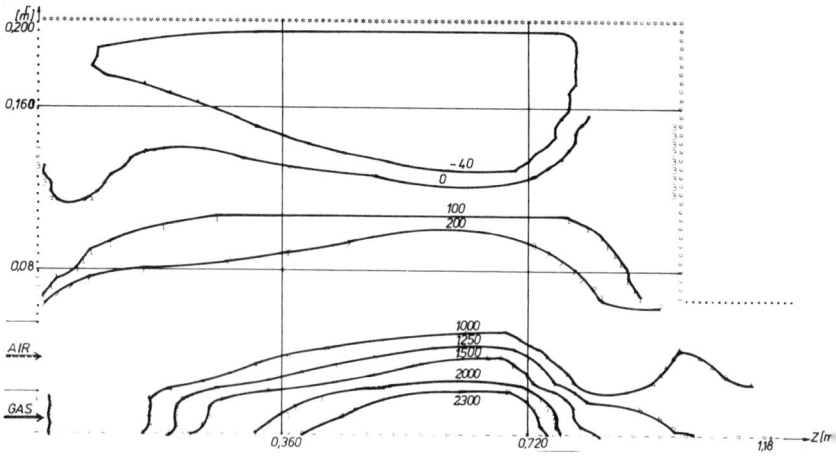

Fig. 3 Axial velocity isolines.

COMPUTER MODELING FOR FLOW PATTERNS   351

The streamline model can also reveal the penetration way and place of the ionizing additive, whose residence in combustion chamber should be the longest possible. This matter leads to some remarks on choosing the type of atomizer when a K_2CO_3 solution (50% K_2CO_3 + 50% H_2O) is used and the same is radially located in the front area of combustion chamber:

1) When pressure mechanical spray atomizers (approximately 35-40 bars) are used, an atomized liquid jet of a very high dynamic impulse is obtained, but the basic characteristics of this jet are lower. The liquid jet would penetrate doubly close in an extremely short running route in

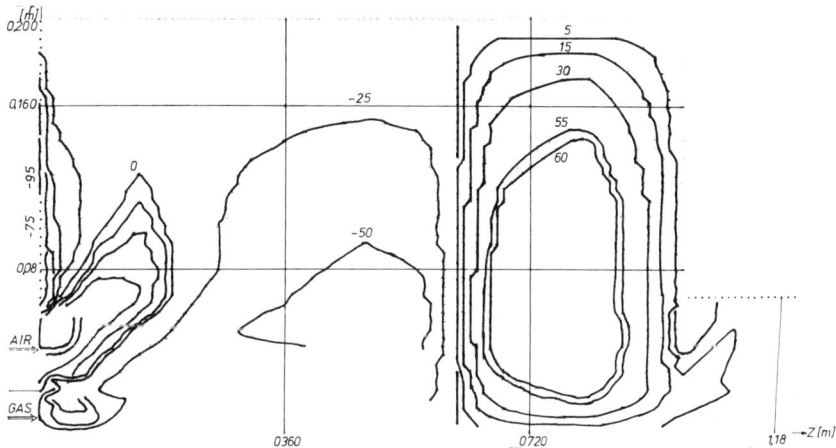

Fig. 4   Radial velocity isolines.

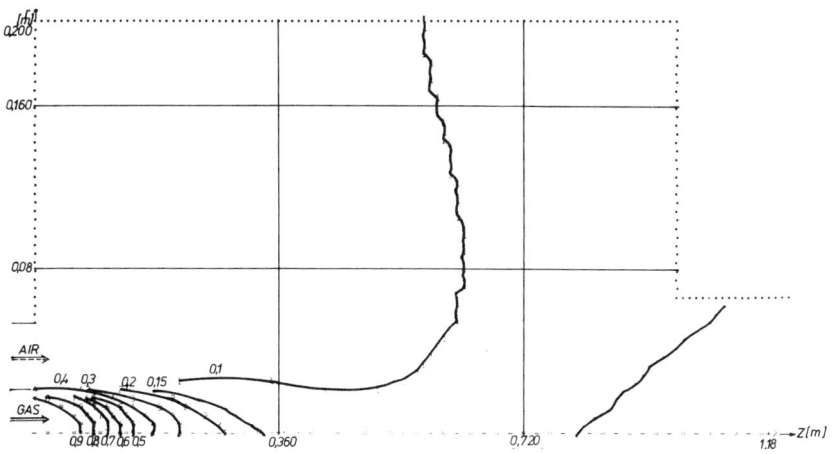

Fig. 5   Mixture fraction isolines.

the combustion chamber and hence has a short residence therein:

2) When atomizers with auxiliary agent atomizing are used, there is the advantage of higher basic characteristics of jet (particularly atomizing fineness), but the dynamic impulse is lower. That leads to penetration being reduced, not exceeding the peripheral recirculation zone, which has a favorable effect by triplication of the running route inside the combustion chamber.

All these aspects make us consider and propose the use, in the case of installing axial flow models, of some ultrasoics-type atomizers, particularly with the Levavasseur generator. This type of generator operating at pressures lower than Hartmann type, has a better acoustic rated output[7].

The acoustic field favorably influences the uniformization of the combustion products-drops mixture, creating a continuous relative movement between drops and combustion products even when they are very fine, greatly favoring the operation process thereof.

Figures 3 and 4 show the isoline distributions for axial and radial velocities, marking, for axial velocities, a decrease in the value of isolines in the radial direction, the maximum being near the symmetry axis. A very developed area of negative axial velocities located in the upper part of combustion chamber can also be noticed.

Figures 5 and 6 trace the mixing distributions and CH_4, respectively, marking the process completion at a distance of approximately 0.360 m from the burner "mouth". This theoretically shows that burning ends in the first half of

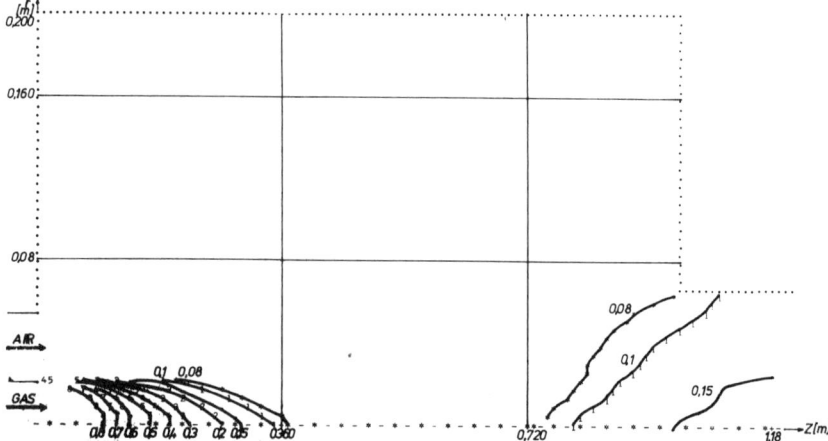

Fig. 6 CH_4 isolines.

combustion chamber and that the length thereof will not be determined by the burning process but by the evaporation process of the ionizing additive.

Conclusion

This work was mainly intended to demonstrate the possibility of predetermining the flow and mixture model in an MHD combustion chamber on natural gases using the computing procedure based on finite-difference method.

The use of a "closing scheme" with two equations ("k-W" model) as well as the use of "wall functions" for the boundary conditions on solid wall proved their validity, the results obtained being of correct physical significance.

Moreover, the application of a nonuniform grid and a memory dynamic assignment actually pointed out ways to improve the computer program.

The stream lines patterns permit the study of aerodynamics inside the combustion chamber and give some information about the place of introducing the ionizing additive in liquid solution as well as the type of atomizer that must be used. The mixing isoline model demonstrated that mixing ends a short distance from the burner "mouth," and the isolines tracing for CH_4 shows that burning ends at about half-length of the combustion chamber, which means that the length thereof is actually determined by the residence time of the ionizing additive necessary for its evaporation process.

A final concluding remark is that the flow model type installed in the combustion chamber, as well as the inlet place of the ionizing additive, are elements influencing the residence time.

References

[1] Gosman, A.D., et al, *Heat and Mass Transfer in Recirculating Flows*, Academic Press, London, 1969, pp.226-323.

[2] Pay, B.R., Richter, T.M., and Lowes, T.M., "Flow and Mixing in Confined Axial Flows," *Journal of the Institute of Fuel*, Vol. 48, NO. 397, Dec. 1975, pp. 185-196.

[3] Richter, W., "Mathematische Modelle Technischer Flamme," (in German), Ph.D. Thesis, University of Stuttgart, F.R.G., pp. 252-262.

[4] Cristea, E.D., Mihai, D., and Lemnean, N., "Computer Modeling of Natural Gas Burning with Application on MHD Combustion Chamber," Eighth International Conference on MHD Electrical Power Generation, Moscow, Sept. 12-18, 1983, Vol. 3, pp. 53-57.

[5] Ustimenko, B.P., and Krol, V.O., "Computer Modeling of Aerodynamic and Burning inside MHD Generator Combustion Chamber," (in Russian), Eighth International Conference on MHD Electrical Power Generation, Moscow, Sept. 12-18, 1983, Vol. 3, pp.58-61.

[6] Cirstea, E.D., "Contributions to Research on Auxiliary Fluid Spray Burners for Industrial Furnaces," (in Romanian), Ph.D. Thesis, "Traian Vuia", Polytechnic Institute Timisoara, Romania, 1984, pp. 61-87.

[7] Lemnean, N., Cristea, E.D., and Jianu, C., Burning Installations on Liquid Fuels, (in Romanian), Technical Publisher, Bucharest, Romania, 1982, pp. 289-298.

Two-Phase Flow Measurement Using a Modified Laser Doppler Anemometry System

Y. Levy* and Y.M. Timnat†

Technion, Israel Institute of Technology, Haifa, Israel

Abstract

Complete theoretical predictions of two-phase flow structures are not possible at present, and existing numerical programs require preliminary knowledge of initial conditions, such as velocity distribution of particles of various size ranges. A system for measuring simultaneously the size and velocity of discrete particles is described. The principle of the size measurement technique is based on the almost linear relation between the diameter of spherical particles and the maximum amplitude (pedestal value) of the signal, as detected by the photomultiplier of a laser Doppler anemometry (LDA) system. The system, which allows simultaneous two-phase flow measurements, consists of a conventional Doppler frequency processor (frequency counter or frequency tracker) for the velocity measurements and a specially built peak detector unit to record the pedestal amplitude. This information is interfaced to a minicomputer for data analysis. A calibration curve of the size-amplitude relation is obtained using a recently developed calibration technique that records simultaneously the frequency (and thus the velocity), the pedestal amplitude, and the time of flight of particles crossing the LDA control volume. The accuracy of the measurement system is still to be determined. Initial calibration curves indicated repeatability limits in size measurements of better than 25%, while the effect of the direction of light detection is still to be evaluated. The accuracy of velocity measurements for the whole size range of particles is generally better than 4%.

Paper presented at the Fourth Beer-Sheva Seminar on MHD Flows and Turbulence, Ben-Gurion University of the Negev, Beer-Sheva, Israel, Feb. 27-March 2, 1984. Copyright © 1985 by the American Institute of Aeronautics and Astronautics, Inc. All rights reserved.
*Lecturer, Department of Aeronautical Engineering.
†Professor, Department of Aeronautical Engineering.

Introduction

Numerous engineering applications incorporate two-phase flow systems (gas plus solid, or gas plus liquid), for example, fuel injection systems, fluidized beds, rocket propulsion, PF (pulverized fuel) combustion, and filtration processes. Complete theoretical predictions of the two-phase flow structure are not possible at present, and existing numerical programs require preliminary knowledge of initial conditions[1], such as the velocities distribution of the gas and of particles of various size ranges and of some empirically based drag coefficients. Obtaining these initial conditions, as well as confirming the selected drag coefficients, requires a technique for simultaneous size and velocity measurement of discrete particles.

Laser Doppler anemometry (LDA) is a relatively new technique[2], that is mainly used for single-phase flow measurement (liquid or gas). The fluid flow velocity is obtained by detecting velocities of micron-sized tracking particles artificially seeded in the flow. A similar but modified technique can be applied to two-phase flows. When measuring in two-phase flows, signals are obtained simultaneously from both phases. Particles from the entire size range are detected, of which the low range serves as the tracking particles and represents the continuum phase. A typical signal detected from a relatively large particle (of the order of the control volume diameter) is shown in Fig. 1. The important signal characteristics are also illustrated in the figure. The velocity of the detected particle is linearly related to the Doppler frequency; the size of the particle can be related to the signal amplitude. The type of relation between the particle diameter and the amplitude of its signal depend upon the particle size and can be practically classified in three size ranges[3]:

1) The low range ($0.1\mu m \lesssim d_p \lesssim 10\mu m$), where the intensity of scattered light and thus the signal amplitude can be described according to the Mie theory.[4]

2) Particles of the high range are of the order of, or bigger than, the diameter of the beam at the control volume ($200\mu m \leq d_p$). Within this group the diameter of a spherical and smooth particle is measured using a technique by which the two incident laser beams are reflected and/or refracted at the particle surface and form an imaged control volume at the photodetector or pinhole[5,6]. The signal obtained is related to the diameter of the particle as well as its velocity. In the case of large nonspherical particles, there is not yet a technique for direct size measurements. An alternative technique[7,8] can be used in cases where the

particles are of known size and only discrimination between the two phases is required.

3) The intermediate size range ($10\,\mu m \leq d_p \leq 200\,\mu m$) is of immediate interest to the authors. Two types of signal characteristics can be used to give information about the particle diameter, the visibility of the signal, and the pedestal amplitude. The visibility of the signal V is defined as

$$V = \frac{I_{max} - I_{min}}{I_{max} + I_{min}}$$

I_{max} and I_{min} are defined in Fig. 1. The advantage of using visibility values as a measure of the particle size is due to the fact that it is a nondimensional characteristic of the signal, independent of the absolute signal intensity,[6,9] and requires no calibration. Its main disadvantage is the

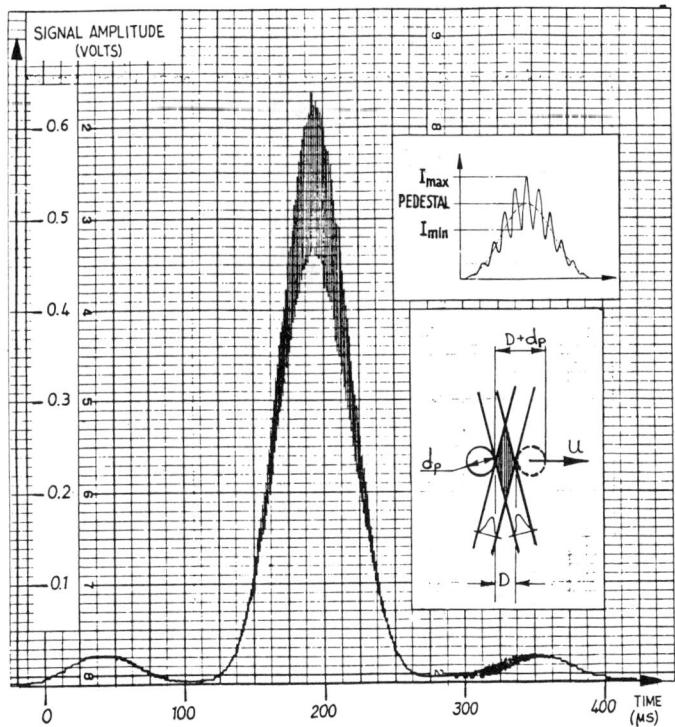

Fig. 1 Typical plot of a signal detected from a large liquid droplet.

relatively low dynamic particles size range where an unambiguous one-to-one relation can be obtained.

The pedestal amplitude is defined as the maximum amplitude of the low-frequency component of the signal as detected by the photomultiplier (see Fig. 1). It was confirmed to have a linear relation to the particle diameter over a relatively wide size range[10-13]. It is easier to measure (only one amplitude information is required), but it necessitates performing calibration tests for each experimental configuration. A new technique to obtain the calibration curve is described in Ref. 10, and a typical calibration curve is given in Fig. 2. The coefficient of this relation depends on the particle's index of refraction, the geometry

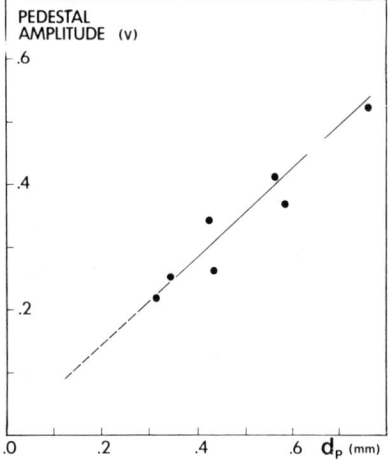

Fig. 2 Variation of pedestal amplitude with particle diameter (Ref. 10).

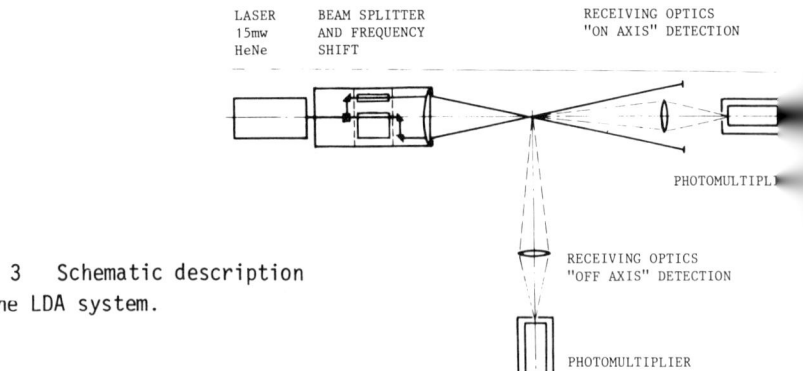

Fig. 3 Schematic description of the LDA system.

of the LDA system, the supply voltage to the photomultiplier tube, the extent of dust on the optical windows, and on various amplifier's gains. The larger dynamic range and the simpler method required for automatic data acquisition, due to the measurement of a single amplitude parameter, has led to the authors' preference of the pedestal technique for size measurements.

The laser light intensity inside the control volume has a spatial distribution with a maximum at the center. Particles of similar characteristics crossing the control volume by different paths will generate signals of different amplitudes. The calibration curves are obtained by detecting signals from particles crossing exactly at the center of the control volume, and the same should be done during the flow measurement. A common technique used to ensure this condition is to perform "off-axis" detection, where the photomultiplier is aligned at some angle to the optical axis, with the best accuracy obtained at 90 deg. LDA systems with off-axis detection are more difficult to assemble in cases of enclosed flow systems such as channel flows. The complication is mainly due to the optical windows required. The present study was performed using forward scatter (on-axis) detection. This arrangement introduced some statistical errors in the size measurement. The signal processing system, which was developed and is described later, processes the signal from the photomultiplier. It allows us to obtain more accurate results when off-axis detection or an alternative technique is used.

Description of the Technique

The experimental system can be divided into three main sections: 1) the optical system, 2) the signal processing system, and 3) the data acquisition and data processing.

The Optical System

A forward scattered LDA system is described in Fig. 3, where the option for off-axis detection not used in the present study, is also indicated. The system includes a 15-mW HeNe laser (Spectra Physics No. 124) directed to a DISA modular optic unit (model 55X). The unit includes a beam splitter with 60-mm beam separation and a 50-MHz Bragg cell for frequency shift. The focal length of the focusing lens is 300 mm.

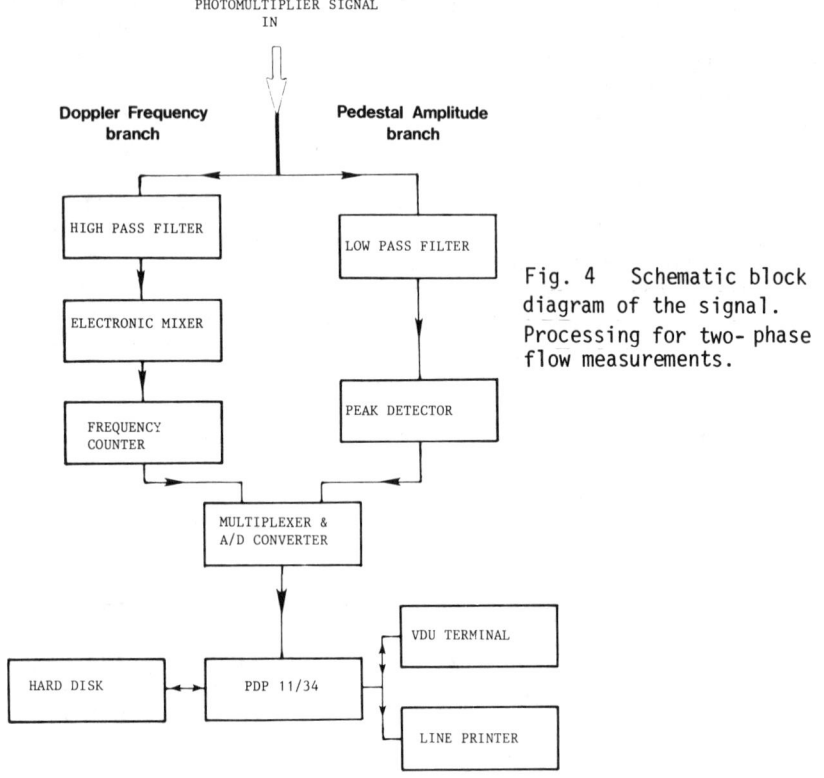

Fig. 4 Schematic block diagram of the signal. Processing for two-phase flow measurements.

The Signal Processing System

A schematic block diagram of the signal processing system is given in Fig. 4. The electric signal generated by the photomultiplier is connected to two parallel branches for simultaneous measurements of the Doppler frequency and the pedestal amplitude.

The Doppler Branch

The signal from the photomultiplier is connected to a Doppler frequency processor (a frequency counter DISA model 55L 90a) through a high-pass filter and an electronic mixer (DISA model 55 N 10). The high-pass filter acts as a buffer between the photomultiplier and the mixer. The filter prevents reverse influence of the mixer on detected signals, which may cause significant deterioration of their shape, precluding the possibility of obtaining reliable pedestal amplitude measurements. The filter (TTE Inc. USA, Model

TWO-PHASE FLOW MEASUREMENT 361

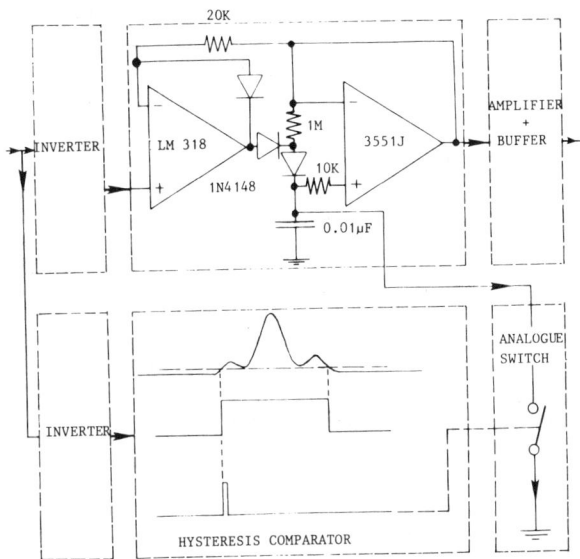

Fig. 5 The peak detector.

H67E) used is a 1-MHz high-pass filter with a cut-off frequency of 100 MHz. Any buffer amplifier of appropriate frequency response could serve for this purpose. The electronic mixer combines the signal from the photomultiplier with an internally selected frequency to form an effective frequency shift between 0.01 and 9 MHz. The frequency counter converts the frequency to a linearly related analog voltage. In order to obtain the pedestal amplitude, the signal from the photomultiplier is connected to a low-pass filter (self-built, 1 MHz) that removes the Doppler frequency oscillation, revealing the pedestal amplitude of the signal. The signal is then directed to the specially built peak detector. Owing to the uniqueness of the detector, its main components are shown in Fig. 5. In the detector, a capacitor is charged to the level of the input signal. The charging is performed whenever the signal is higher than the capacitor voltage. Leakage from the capacitor is prevented: controlled and quick discharge of the capacitor is achieved by a solid-state analog switch. The switch is operated by the front edge of a trigger pulse, generated from the low-pass filtered signal. The timing diagram of the peak detector operation is shown in Fig. 6. It can be seen that the detector maintains the pedestal amplitude until a new signal appears. It is then reset to zero and follows the increase of the new signal up to its peak. It should be noted that all signals with amplitude higher than the trigger level are detected by the peak detector. This is not the case with the frequency counter, where an internal signal validation test is performed. It follows that the number of particles measured by the peak detector is much larger than for the counter.

Data Acquisition and Data Processing

The output results of the frequency counter and the peak detector are in the form of a linear related analog voltage. Both instruments are connected to an analog to digital (A/D) converter with a maximum sampling rate of 100 kHz and 12 bits resolution. Both signals are sampled sequentially by a multiplexer. The data are stored temporarily in the minicomputer (PDP 11/34) buffer memory. They are then processed and stored on the hard disk.

The sequential sampling, which is performed at a preselected and fixed rate, records in the computer memory pairs of integers corresponding to the Doppler frequency and the pedestal amplitude. A schematic description of the sampling sequence is given in Fig. 7. The velocity and the pedestal amplitude information are obtained by performing a special validation test on all of the recorded values. The

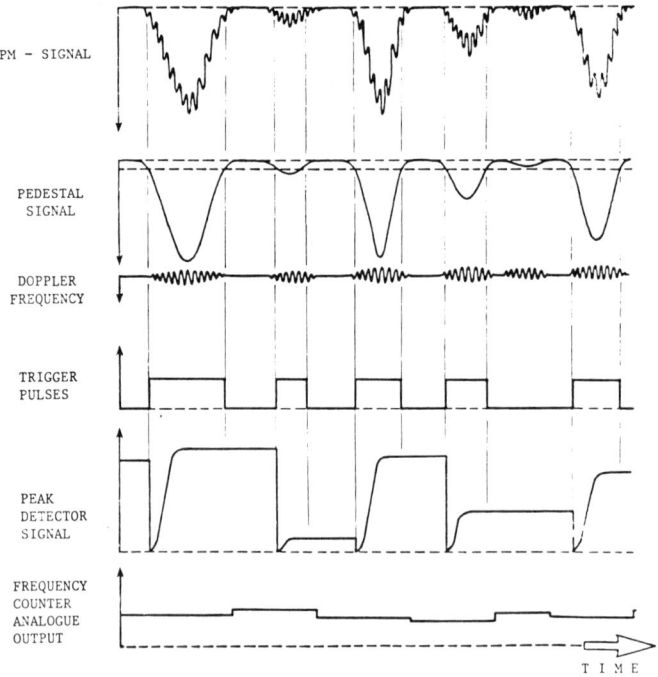

Fig. 6 Timing diagram of the signal processing.

TWO-PHASE FLOW MEASUREMENT 363

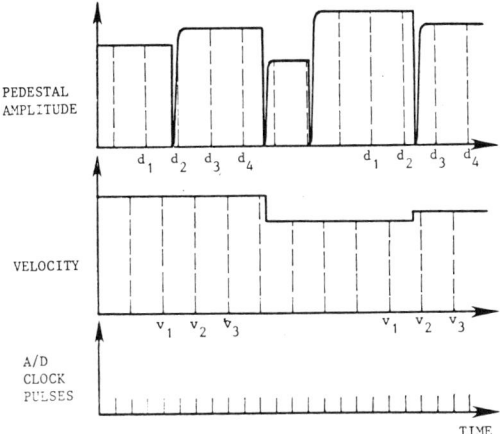

Fig. 7 Sampling sequence.

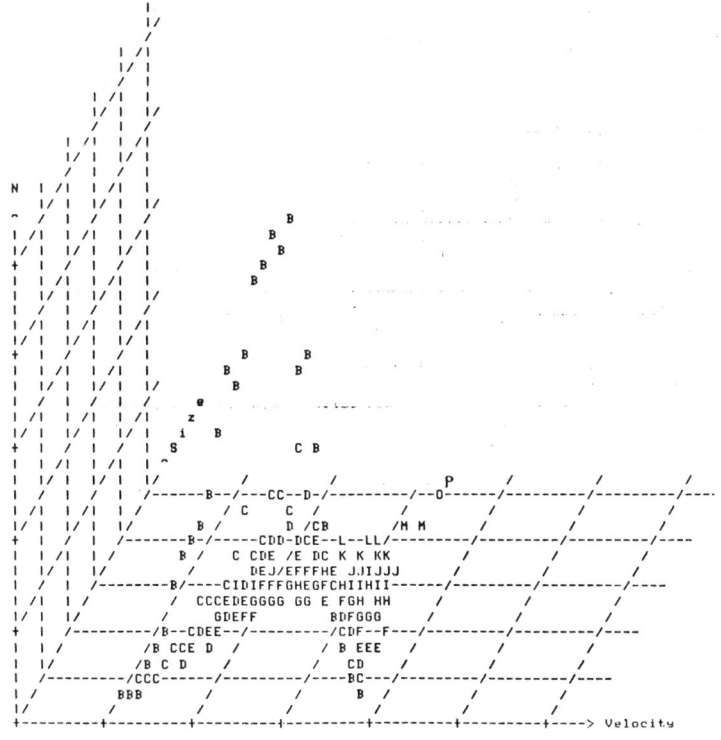

Fig. 8 A typical velocity and pedestal amplitude histogram as obtained from the data analysis routine.

criteria of the validation tests were selected to ensure that the measured velocity and the associated pedestal amplitude will belong to the same particle. The validation criterion is

$$\begin{bmatrix} |V_2/V_1 - 1| > \varepsilon \\ \\ |V_3/V_2 - 1| < \varepsilon \end{bmatrix} \text{ and } \begin{bmatrix} \begin{pmatrix} |d_2/d_1 - 1| > \varepsilon \\ \\ |d_3/d_2 - 1| < \varepsilon \end{pmatrix} \text{ or } \begin{pmatrix} |d_3/d_2 - 1| > \varepsilon \\ \\ |d_4/d_3 - 1| < \varepsilon \end{pmatrix} \end{bmatrix}$$

where V_1, V_2, V_3 (velocity values) and d_1, d_2, d_3, d_4 (pedestal amplitude values) are shown in Fig. 7. ε is an accuracy criterion factor depending on the noise level and has a typical value of 1%. If the validation test is passed, then V_3 and d_3 represent the velocity and pedestal amplitude of a validated signal. This pair of information together with other validated data are used to perform ensemble averaging according to groups of similar pedestal amplitudes. For each group, the velocity mean and rms values are calculated and a velocity histogram is displayed. A typical two-dimensional histogram is shown in Fig. 8. Its principal axes are **number** of particles **N, velocity,** and pedestal amplitude groups with the letter A corresponding to the lowest value and P to the highest (the value of the letters increases along the pedestal amplitude axis).

Results and Discussion

In order to test the performance of the system, a set of cold flow measurements was performed in a dump combustor (see Fig. 9). The Reynolds number was about 300,000. They

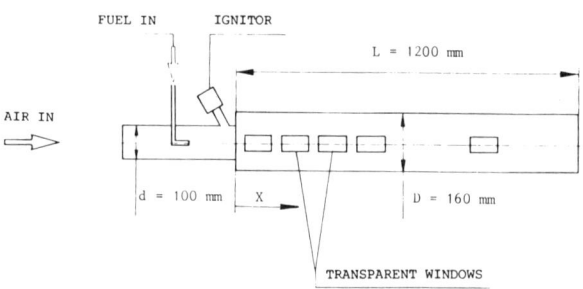

Fig. 9   The dump combustor.

were taken along a radial profile at a distance of 250 mm from the area enlargement. Kerosene was sprayed axisymmetrically. At each location, 10,000 measurements of velocity and pedestal amplitude were recorded within about five minutes. The results are given in Figs. 10-12. Figure 10 illustrates the velocity distribution along a radial profile. The mean and rms values are given for gas of single-phase flow (without droplets) and for the droplets in a two-phase flow. In the two-phase flow, the velocity values were averaged for all droplets. The figure also shows local velocity distributions for different particle size ranges. Figure 11 demonstrates in more detail the local variation of velocity histograms for different particle size groups. The velocities, particle sizes and dimensions are normalized. Figure 12 shows the particle size distribution at the center of the combustor.

It is difficult to estimate quantitatively the accuracy of the measurement system. Qualitative analysis confirms the tendencies of the results. The accuracy of the velocity

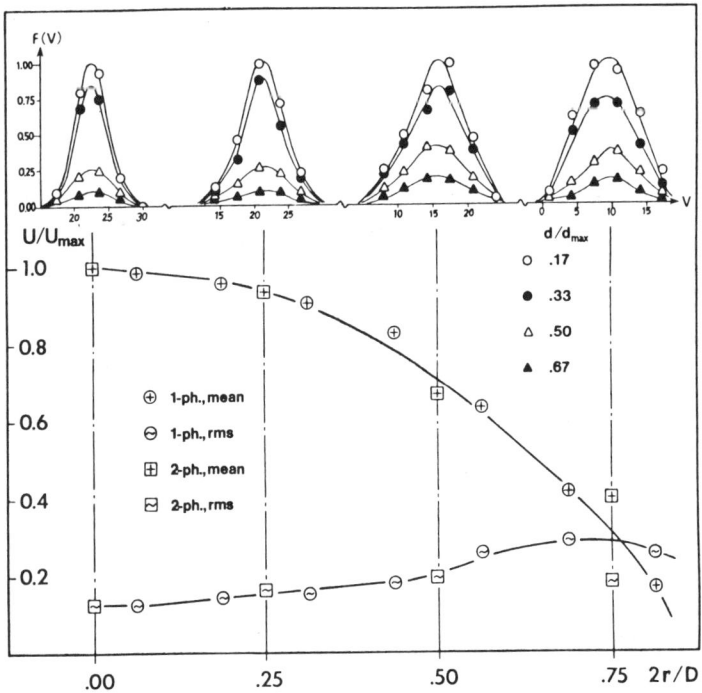

Fig. 10 Velocity distribution along a radial profile of the dump combustor (X = 250 mm).

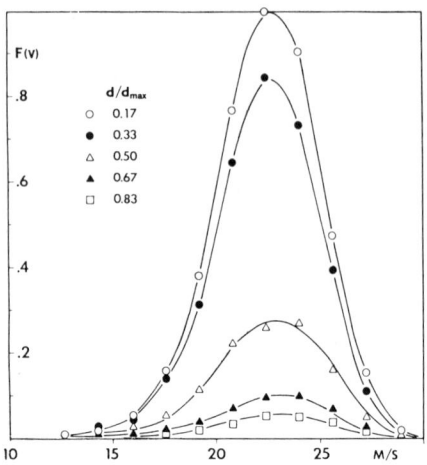

Fig. 11 Variation of velocity histograms with particle size at the centerline of the combustor (X = 250 mm).

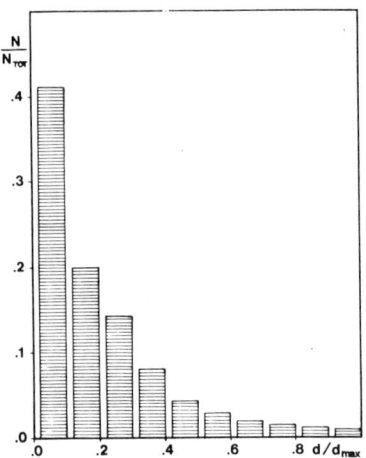

Fig. 12 Particle size distribution at the centerline of the combustor (X = 250 mm).

measurements is determined by the optical alignment, by the signal-to-noise ratio, and by the frequency counter settings. The error is expected to be less than ±2%. The accuracy of the peak detector can be deduced from its repeatability. The initial calibration curve (see Fig. 2) shows repeatability limits of better than 25%.

This preliminary set of results demonstrates the proper operation of the measuring system. However, significant improvements are still possible. In the optical system development should be directed toward an alternative technique to the off-axis detection.

The calibration technique, as described in Ref. 10, should also be improved to demonstrate with higher repeatability the linear relation between particle size and its pedestal amplitude.

References

[1] Tambour, Y., "A Sectional Model for Evaporation and Combustion of Sprays of Liquid Fuels," Israel Journal of Technology, Vol. 18, 1980, Issue No. 1-2, pp. 47-56.

[2] Durst, F., Melling, A., & Whitelaw, J.H., Principles and Practice of Laser-Doppler-Anemometry, 2nd ed., Academic Press, London, 1981.

[3] Levy, Y., "Two-Phase Flow in the Freeboard of a Fluidized Bed," Ph.D. Thesis, University of London, 1981.

[4] Van de Hulst, H.C., Light Scattering by Small Particles, Wiley, New York, 1957.

[5] Durst, F & Zare, M., "Laser-Doppler Measurements in Two-Phase Flows," University of Karlsruhe, F.R.G., Sonderforschungsbereich 80, Report SFB 80/TM/63, July 1975.

[6] Bachalo, W.D., "Methods for Measuring the Size and Velocity of Spheres by Dual-Beam Light Scatter Interferometry," Applied Optics, Vol. 19, No. 3, 1980, pp. 363-370.

[7] Levy, Y. & Lockwood, F.C., "Laser Doppler Measurements of Flow in Freeboard of a Fluidized Bed," AICHE Journal, Vol. 29, No. 6, 1984, pp. 889-895.

[8] Levy, Y. & Lockwood, F.C., "Velocity Measurements in Particle Laden Turbulent Free Jet," Combustion and Flame, Vol. 40, March 1981, pp. 333-339.

[9] Farmer, W.M., "Measurement of Particle Size, Number Density and Velocity using Laser Interferometer," Applied Optics, Vol. 11, No.11, 1972, pp. 2603-2612.

[10] Levy, Y. & Timnat, Y.M., "Study of Sudden Expansion Combustion for Propulsion Applications," Paper No. IAF-82-372 presented at the XXXIV IAF Congress, Budapest, Oct. 9-15, 1983.

[11] Yule, A.J., Ereaut, P.R., & Ungut, A., "Droplet Sizes and Velocities in Vaporizing Sprays," Combustion and Flame, Vol 54, December 1983, pp. 15-22.

[12] Yule, A.J., Chigier, N.A., Atakan, S., & Ungut, A., "Particle Size and Velocity Measurement by Laser Anemometry," AIAA Paper 77-214, 15th Aerospace Science Meeting, Los Angeles, Calif., Jan. 1977.

[13] Mizutani, Y., Kodama, H., & Miyasaka, K., "Doppler Mie Combination Technique for Determination of Size Velocity Correlation of Spray Droplets," Combustion and Flame, Vol. 44, January 1982, pp. 85-95.

Chapter IV. MHD Power Generation and Application to Fission and Fusion Reactors

Liquid Metal MHD Power Generation—
Its Evolution and Status

Michael Petrick*
Argonne National Laboratory, Argonne, Illinois
and
Herman Branover†
Ben-Gurion University of the Negev, Beer-Sheva, Israel

Abstract

The paper deals with the history of the development of liquid metal magnetohydrodynamic power conversion systems from the inception of this concept in the late 1950s up to the present time. Different cycles and systems are reviewed and analyzed in chronological sequence. The last sections of the paper are dedicated to a new liquid metal MHD concept called OMACON (optimized magnetohydrodynamic conversion system) that has been developed during the last few years. Preliminary results of application studies of this concept are presented and indicate essential advantages regarding attainable efficiencies.

Introduction

During the past 30 years, liquid metal MHD power generation has been passing through an evolutionary cycle that has produced both major shifts in conceptualization and progress. The major conceptual changes that have evolved have been driven primarily by potential applications and by the need to improve thermodynamic performance in order to compete with alternative technologies.

The concept of liquid metal magnetohydrodynamic energy conversion evolved from the search for an alternate working fluid for an MHD cycle that possessed a high electrical conductivity, would not have the complexities of a plasma flow,

---

Paper presented at the Fourth Beer-Sheva Seminar on MHD Flows and Turbulence, Ben-Gurion University of the Negev, Beer-Sheva, Israel, Feb. 27-March 2, 1984. This paper is declared a work of the U.S. Government and therefore is in the public domain.
*Director, Fossil Energy Programs.
†Professor.

and could operate in a heat engine cycle in a lower temperature range.

In contrast to the turboelectric Rankine cycle, which employs a pure vapor phase, the liquid metal MHD cycle must be designed to transfer efficiently the thermal energy of the thermodynamic working fluid (vapor phase) to kinetic energy or pressure head of the liquid (the electrodynamic working fluid), which then passes through the MHD generator. Thus, in reality, liquid metal MHD power cycles operate with two phases (gas to liquid) whose quality (gas to liquid mass flow ratio M_g/M_ℓ) generally falls in the range of 10^{-2} - 10^{-4}. Since a number of flow regimes can exist in this quality range, a considerable variety of systems is possible -- at least in principle. Much of the work to date has been involved in classifying these various possibilities and exploring their characteristics in a preliminary way.

It is, therefore, interesting and informative to review the various concepts that have been proposed and studied in order to document the evolutionary pathway and, most importantly, to understand how we have come to be where we are now. The data and descriptions presented are derived from both published and unpublished information obtained from personal contacts; information on the Soviet work, in particular, has come from the latter sources.

Initial Liquid Metal MHD Cycles and Systems

The liquid metal MHD energy conversion concept was originally developed in the late 1950s and early 1960s as one of the most promising methods for generating electric power in space, using nuclear reactors as the heat source. A number of concepts evolved that offered the prospect of elimination of moving mechanical parts, high performance, specific power, and efficiency; and in several systems the ability to produce ac power directly. The system concepts proposed were tailored specifically to function within parameter ranges dictated by the space application and the heat source under development. Each is described briefly below; also discussed are the results that affected the evolution of the concept.

The Two-Phase Two-Component Separator Cycle.

Proposed by Elliott[1], this is generally acknowledged to be the original liquid metal MHD concept; the cycle schematically illustrated in Fig. 1 was intensively studied.

In operation, the two-phase mixture generated in the mixer by combining the condensate from the vapor loop and

the liquid from the reactor passes through a two-phase
nozzle, where a high-kinetic-energy mixture is formed. The
liquid is then separated in the separator and passed through
the MHD generator where its kinetic energy is reduced and
electrical energy extracted. The fluid then passes by diffuser back to the reactor where heat is again added. The
vapor is taken from the separator, passed to the condensor,
and pumped back into the mixer.

Extensive analysis and component testing have been completed and the performance of this system can be specified
with some degree of confidence. The analysis and tests on
two component mixtures have shown that the efficiency of the
nozzle, where the conversion from thermal to kinetic energy
takes place, can approach 90%. Efficiency is defined as the
ratio of the actual kinetic energy of the liquid emerging
from the nozzle to the isentropic homogeneous kinetic energy.
The major problems in the separator cycle are encountered
after the thermal-to-kinetic energy conversion takes place,
when one attempts to collect and divert the high-velocity
liquid droplets into the MHD generator. Separator efficiencies (i.e., the velocity at separator exit to the velocity
at the nozzle exit) that appear realistic lie between 0.8
and 0.9. This yields kinetic energy losses attributed to

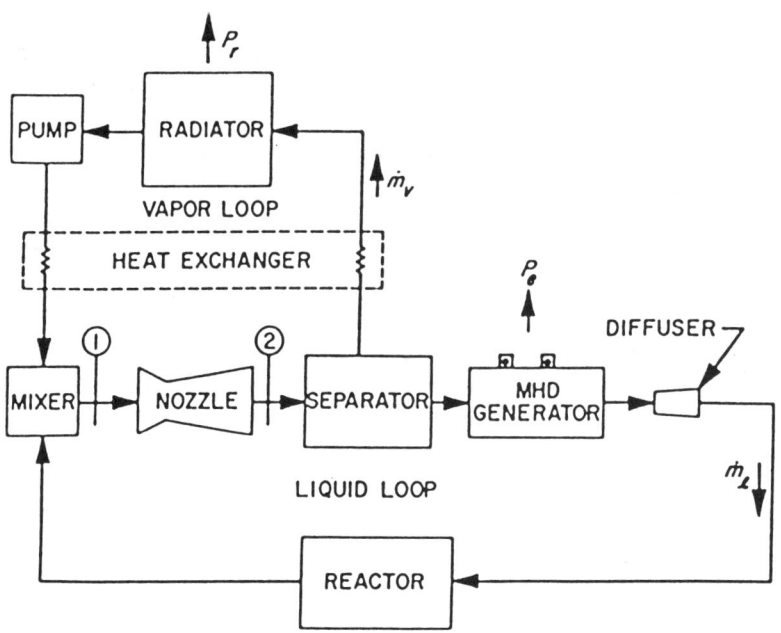

Fig. 1   Liquid MHD power conversion cycle.

friction on the surface of the separator of approximately 20-36%. Although the separation of the phases is excellent, it is not necessarily complete. Generally, some vapor (1-2% by weight) is entrained with the liquid as it is discharged from the separator surface. The result is a supersonic two-phase mixture, since the sonic velocity at this void fraction range can be less than 100 ft/s. A convergent-divergent diffuser must then be used to raise the pressure of the fluid to the prescribed inlet condition to the generator. The efficiency of the supersonic two-phase diffuser (ratio of measured to isentropic pressure rise) has been found to vary 20-70% over the void fraction range of 0.71-0.45; again, this represents a substantial energy loss. Thus, in the separator cycle, a large fraction of the fluid energy (up to 50%) can be lost prior to entering the MHD generator. Additional viscous losses in the generator and downstream diffuser reduce performance even further.

A total 300 kW integrated system using NaK-N_2 to simulate performance at parameter ranges of interest was constructed (Fig 2) and tested (Fig. 3). Measured performance compared favorably with predicted; the ratio of actual system efficiency to Carnot efficiency that appeared achievable was ~0.20-0.30.

One- and Two-Component Two-Phase Condensing Cycles

In these cycles, the vapor is condensed without being separated from the liquid prior to entering the generator.

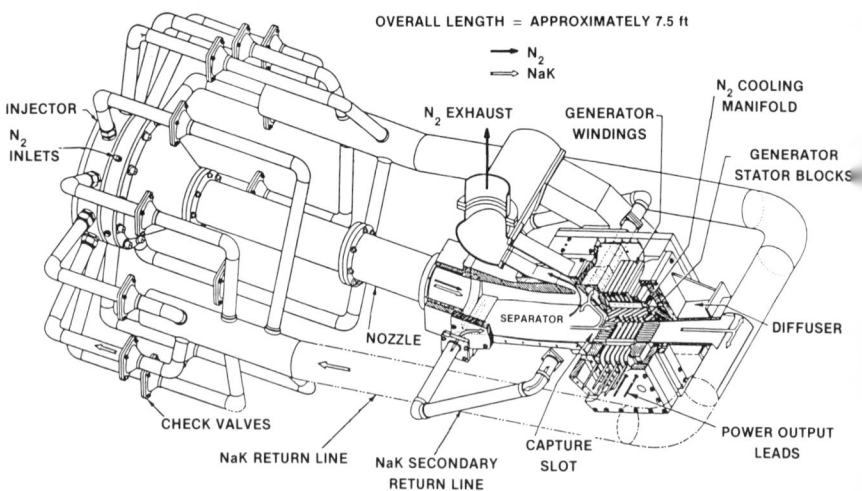

Fig. 2   Experimental liquid metal MHD power system.

LIQUID METAL MHD POWER GENERATION

Fig. 3 Closed-loop system installed in hydraulic laboratory for water-nitrogen tests.

Condensation of the vapor is achieved by injecting a subcooled liquid at the nozzle exit. Again, the objective was to achieve a compact (high-specific-power) system that could produce ac power directly. The condensing injector cycle proposed by Brown and Lee,[2] schematically illustrated in Fig. 4, consists of a vapor loop (heat input loop) and a liquid loop (heat rejection loop). The vapor is generated in the heat source and passes into the condensing injector where it is mixed with the liquid stream emerging from the waste heat exchanger. In the condensing injector, the vapor is condensed and a high-stagnation head liquid, in principle, is generated. The liquid passes through the MHD generator, where electric energy is extracted at the expense of the stagnation pressure head, and then is separated into two streams -- one passes into the vapor loop and hence to the reactor where it is vaporized and the other is sent to the liquid loop where heat is rejected in a heat exchanger.

The performance of this cycle is determined essentially by the performance of the condensing injector itself. The injector is not a new device, having been developed as a boiler feed water pump. Brown and Levy[3] analyzed the mixing, condensing, and pressure recovery in such a device and performed experiments with steam and water, which agreed reaso-

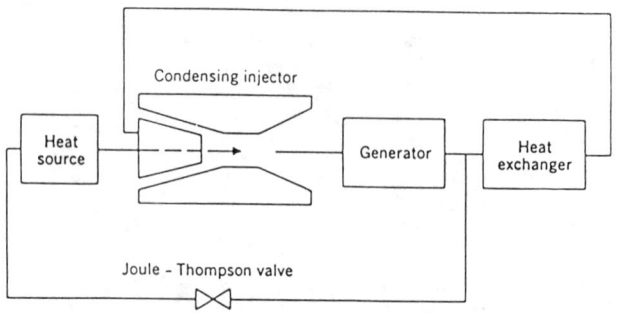

Fig. 4 Schematic diagram of condensing injector liquid metal MHD power cycle proposed by Jackson and Brown.

nably well with theory in limited parameter ranges. Grolmes et al[4] also conducted extensive theoretical and experimental studies of the injector; their results agreed with Brown and Levy and also showed that the device had an inherent low internal efficiency. He concluded that a cycle based on this concept would have very poor performance.

Several complete condensing injector power systems were built in the Soviet Union operating with potassium at 900°C in the late 1960s. At the Kryhizhanovsky Power Institute a 100 kW (thermal input) facility was placed into operation in 1968[5] that operated for 1000 h including 300 h at 900°C. The facility was used to test the condensing injector under conditions of an actual power cycle. Self-circulation operation with the injector was achieved and the net overall efficiency was ~1-1.5% (the ratio of actual to Carnot was very low, <0.10). The results were in line with the predictions of Grolmes et al. A similar 300 kW potassium power system was also built and tested at the High Temperature Institute. The performance achieved has not been disclosed.

The condensing cycle proposed by Prem and Parkins[6], while conceptually differing somewhat in the operation of the "drift tube" condensing device, faces the same internal and overall system efficiency limitations.

Intermittently Vaporizing Liquid Metal Cycle
===

This concept was studied[7,8] as a space power system. In this cycle (see Fig. 5), electricity is generated by the intermittent system of slugs of liquid metal through a magnetic field driven by high-pressure metal vapor from a heat source. While the concept had a relatively high maximum potential ratio of actual/Carnot efficiency (45-75%), the experimental work conducted identified serious problems

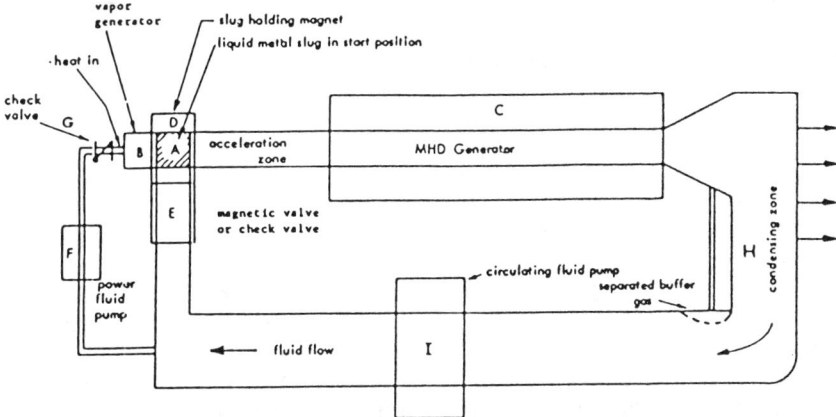

Fig. 5 Schematic diagram of intermittently vaporizing liquid-metal MHD power system concept.

in cycle rates, slug stability, liquid kinetic energy recovery, and generator shorting problems that seriously diminished the performance potential.

Second-Generation Liquid Metal MHD Cycles and System Concepts

In the early and middle 1960s, several factors emerged that strongly impacted the evolution of liquid metal MHD energy conversion. First and foremost was the fact that, with the U.S. emphasis on space programs declining, missions requiring high-power nuclear systems all but disappeared. The result was that if the liquid metal MHD concept were to survive, it would have to compete with alternative technologies for various terrestrial applications. The systems would have to compete at lower temperatures and offer at least the same thermodynamic performance. To achieve these objectives, several approaches were pursued.

Two-Phase Generator Cycle Concept

This concept was proposed by Petrick and Lee[9] to circumvent the basic loss mechanisms inherent in the separator and condensing cycles that had been studied. The cycle concept, illustrated in Fig. 6 was based on the fact that a two-phase mixture is a compressible fluid and thus an effective thermodynamic working fluid that could be expanded directly through an MHD generator. In operation the gas or vapor and liquid metal working fluids are mixed to form a two-phase mixture that then expands directly through the generator, where electrical power is extracted. As the mixture leaves the generator, it is further expanded to increase its kinetic energy and is then sent to the separator. There the liquid metal and gas are separated, with the liquid returning via a diffuser through the heat source to the mixer. The gas (or vapor) working fluid is then handled as in a normal cycle, either Brayton or Rankine. It is passed through the regenerative heat exchanger (if required) to the heat sink and is then compressed (or pumped) and sent back to the mixer via the heat source, if desired. The gas is the thermodynamic working fluid and the liquid metal, which remains in a closed loop, is the electrodynamic working fluid. The liquid metal, pumped through the generator by the gaseous working fluid, interacts with the magnetic field, thus inducing a voltage across the generator.

As originally conceived, the cycle had the following features, which were then, and still are, thought to be very important: 1) the velocity in the generator and the separator can be set independently, thus controlling losses more easily; 2) as the two-phase mixture expands, heat is transferred from the liquid metal to the gaseous working fluid -- the generator thus acts as an infinite reheat turbine; 3) various working fluids can be chosen to match the temperature range of interest; and 4) by choice of working fluid, the cycle can operate as either a Brayton or Rankine cycle and yield attractive efficiencies operating in either the topping mode or as a pure MHD conversion cycle.

In order to achieve the attractive performance potential of the system, the performance values of nonstandard components must reach certain values commonly used in system performance studies; these are listed in Table 1 along with minimum acceptable values.

Table 1 Performance values of two-phase generator cycle

	Goal	Min Acceptable
Mixer: pressure drop psi	5	-
Generator		
Voltage, V	(Depends on power and application)	
Exit void fraction	0.95	~0.85
Efficiency	0.8	~0.7
Nozzle Efficiency	0.9	~0.8
Separator		
Liquid void fraction at exit	0	~0.5
Liquid kinetic energy loss	0.1	~0.3
Diffuser Efficiency	0.9	~0.8

LIQUID METAL MHD POWER GENERATION

The probable performance values of the separator, nozzle, and diffuser system have been developed from extensive test programs conducted by the Jet Propulsion Laboratory, Argonne (ANL), Ben-Gurion University, and others.[10-18] Expected performance values were cited in the previous section that addressed the separator cycle. The unique component in the cycle is the two-phase flow generator. It inherently has a high efficiency if the loss mechanisms not basic to its operation (end, viscous and shunt electrical, slip) can be controlled. (Internal ohmic loss is basic, and is controlled by adjusting the ratio of internal resistance to load resistance.) Extensive studies of generator performance that focused on defining the individual losses and characterizing two-phase flows in a magnetic field were conducted at ANL utilizing an ambient temperature $NaK-N_2$ loop (Fig. 7) and a high-temperature (1000°F) $Na-N_2$ loop (Fig. 8).

End losses (i.e., ohmic losses due to current reversal in the generator end regions as a result of spatially decreasing magnetic fields) are a crucial factor because they

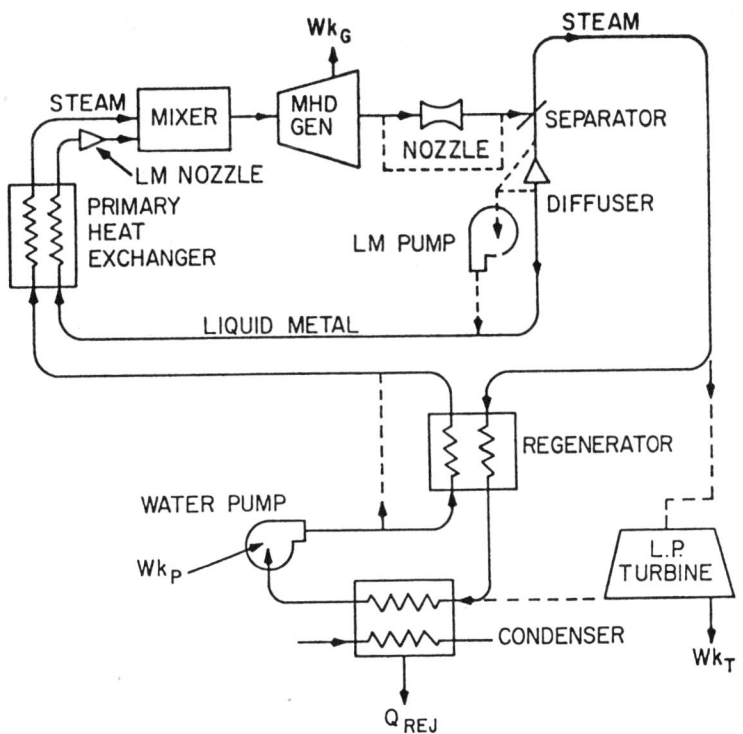

Fig. 6 Schematic of liquid metal MHD Rankine cycle, two-phase generator concept.

Fig. 7 General view of the ambient temperature NaK-N_2 loop at Argonne National Laboratory.

set a lower limit to the generator's length and an upper limit to the generator's voltage. Early ANL work established the use of insulating vanes to minimize end currents and losses[19] and generator experiments demonstrated that the desired increase in efficiency with vanes (decrease in losses) was attained[20]. The concept to use multiple generators connected electrically in series and fluid dynamically in parallel to partially uncouple the end losses and the voltage limitation[21] was studied and shown to be a viable technology for reaching acceptable voltage levels. A model developed at Purdue University allows the calculation of the end loss for an arbitrary arrangement of insulating vanes (number, lengths, locations)[22].

Viscous losses (due to wall shear) are small because the electromagnetic forces are so much larger than all other forces. However, wall shear means that there is a pure-liquid layer adjacent to the wall with a low velocity. Current reversal occurs in this layer, the effect of which is magnified because the liquid conductivity is higher than the two-phase core flow conductivity. Analysis has shown

LIQUID METAL MHD POWER GENERATION

Fig. 8 General view of high-temperature (1000°F) Na-N_2 loop at Argonne National Laboratory.

the effect of this loss on generator efficiency to be very small for practical generator parameters.[23]

Slip, where the gas velocity is higher than the liquid velocity, reduces the efficiency of the generator and the cycle. The slip ratio (gas-to-liquid velocity ratio) was large in early generator tests, but has decreased as the electromagnetic interaction has increased. The most recent data from the ambient temperature generator experiments[24] show that at higher liquid flow rates (velocities), the slip ratio approaches unity -- as desired. Figure 9 clearly shows the sharp decrease in slip ratio with the increasing liquid flow rate (velocity). Data from the high temperature tests of an open-circuited generator show that the slip ratio also decreases as the temperature is increased.[25]

Generator efficiency has increased with experience (time) and the most recent results show substantially higher efficiencies at the high void fractions of interest for power systems[24] (see Fig. 10). The power density (.i.e., the electrical power output per unit volume) is comparable to, or above that anticipated for, commercial generators, thus minimizing the chances of encountering unanticipated problems in scaling to larger generators. Efficiencies in excess of 0.60 were obtained with a small generator (~ 20 kWe) that had no provision (such as vanes) to minimize end

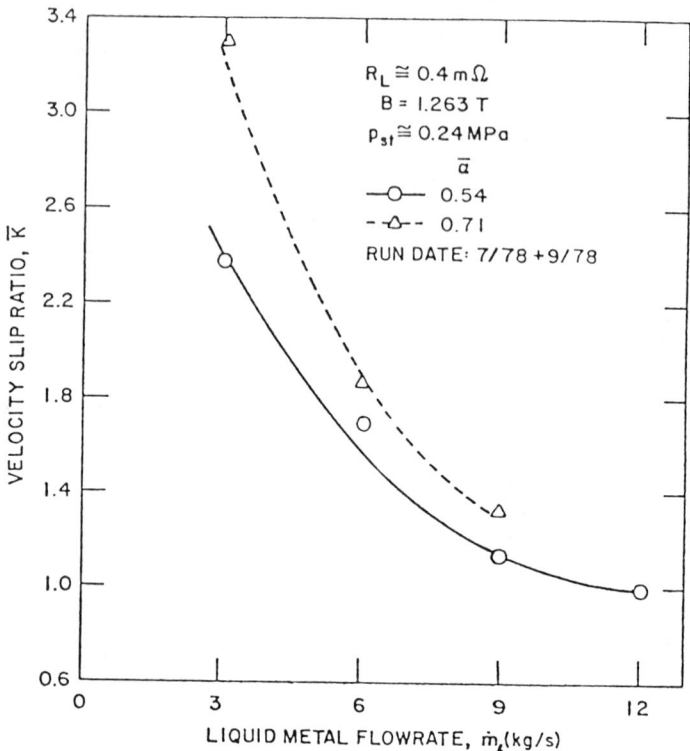

Fig. 9 Measured slip ratio as a function of liquid mass flow rate.[26]

losses. Thus, these results present very encouraging evidence that larger generators can be built to meet the efficiency goal of 0.8 and that even small generators, if designed initially for high efficiency, should meet or exceed the minimum acceptable value of 0.7.

The two-phase flow generator concept was also studied in the Soviet Union. A schematic diagram of the system[26] that was apparently built is shown in Fig. 11. Nominal design conditions disclosed for the 1 MW (t) system are a maximum temperature of 300-400°C and a maximum pressure of 50-60 atm. Working fluid combinations considered included H_2O and CO_2 with Sn, Ga, and In. Nominal design power is 100 kWe with a possible maximum output of 200 kWe, which if achieved produced an overall efficiency of 20% ($\sim \eta_{act}/\eta_{Carnot}$ of 0.40). Performance data on this system have not been reported in the open literature.

The two-phase flow generator cycle emerged as the lone concept that appeared to have the potential for commercial application. During the 1970s, the two-phase flow generator

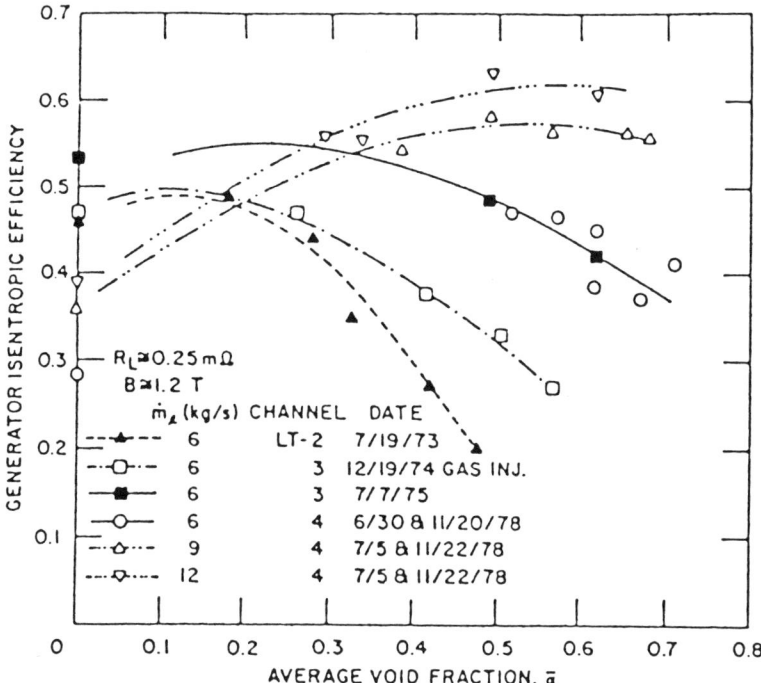

Fig. 10 ANL measured generator efficiencies for LT-2, LT-3, and LT-4, R_L - 0.25 mΩ.[26]

liquid metal MHD cycle was studied in sufficient depth to clearly delineate the critical technical issues that must be resolved in order to achieve its maximum performance potential. Reference 30 provides an excellent summary of the data. The technical problems are:

1) Self-circulation of liquid metal by thermal energy via a two-phase nozzle separator diffuser system. To achieve maximum performance of the cycle, the nozzle, separator, and diffuser components must operate at relatively high efficiencies, ~0.8 for each component. Even under these conditions, 50% of the energy of the fluid is lost in the self-circulation mode. The experimental data tend to indicate that lower performance levels are likely for each component in an integrated system -- especially in the two-phase nozzle and diffuser, resulting in a probable 60% loss of kinetic energy of the liquid in a practical system. This represents a significant loss that puts a cap on overall performance.

2) Generator performance is directly affected by two-phase flow parameters, especially slip (relative phase velocity). As slip increases, the amount of gas required to

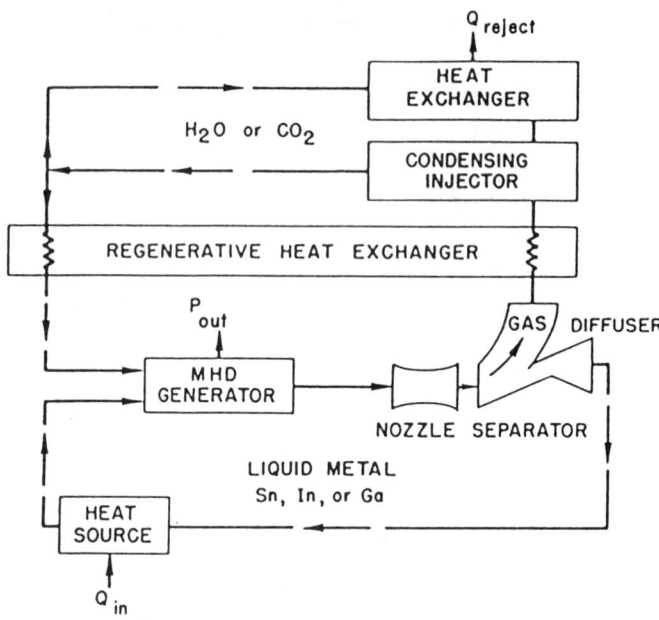

Fig. 11 Schematic of 1 MW liquid metal MHD system under development at ENIN-II.

generate the same power output increases proportionally, reducing cycle performance. Also, an exit void fraction limit exists on generator performance beyond which a flow transition occurs that destroys liquid continuity and, hence, electrical conductivity. This then sets the mass flow ratio of liquid and gas and also impacts attainable performance. A third major factor is that the channel must be contoured to accommodate the changing flow parameters in order to maintain as uniform a velocity as possible. The latter condition is required to be able to load the generator so as to achieve a high electrical efficiency. These problems make it difficult, but not impossible, to achieve a high efficiency (0.80) generator.

3) Separator performance. One hundred percent separation of phases is extremely difficult to achieve with the flat-plate separators. As a result, even if 98% separation efficiency is obtained, it is still necessary to devise measures for returning the liquid carryover and the residual liquid from the separator to the flow loop. While not impacting performance significantly, this complicates the design and operation of the system and exacerbates the control problem.

4) Startup and shutdown of the system. The self-cir-

culating system requires that special subsystems be built in to achieve startup and shutdown. This again adds to system complexity in the intake and circulation system and to the hardware, but does not impact performance.

5) Inversion from dc to ac. The generator produces a low-voltage/high-amperage current. The generator electrical characteristics are set primarily by cycle conditions. Little leeway exists to adjust the voltage in a single-loop, self-circulating system. In turn, the complexity of the self-circulating system discourages utilizing multiple electrically isolated loops in which the generator can be coupled electrically in order to build up voltage for more efficient inversion.

6) Two-phase flow electrical conductivity. In general, the two-phase flow in the generator as dictated by cycle and generator design conditions is such that the electrical conductivity is reduced by about an order of magnitude. This, in turn, requires that more costly, higher magnetic field volumes must be utilized -- a potential economic penalty.

7) Off-design performance. In the self-circulating system, highly efficient off-design performance is difficult to maintain because the generator and nozzle geometrical contours are set for the design point. As system performance parameters are varied from the design point, the performance of those two components deteriorate and thus overall system performance degrades.

8) Production of ac power. The two-phase flow in the generator generally precludes the generation of ac power directly in the MHD generator because it causes a low power factor.

While all the above problems can be addressed through proper design and engineering, as evidenced in the ETGAR-1 (Fig. 13) and ETGAR-2 designs[31] to produce an attractive conversion system (in performance), the inevitable result is that system complexity is increased and economic viability is threatened. This reality has very recently led to the development of a new liquid metal MHD system concept, which represents the latest refinement and approach toward developing a viable system that is commercially viable.

To achieve the higher performance needed for terrestrial application, proponents of the separator and condensing cycles attempted to employ staging to boost the thermodynamic performance of these systems. When a two-phase mixture expands through a nozzle, between cyclic pressure bounds, the fluid velocity produced is considerably higher than can be utilized efficiently. In an attempt to improve the overall efficiency of the thermal-to-kinetic energy

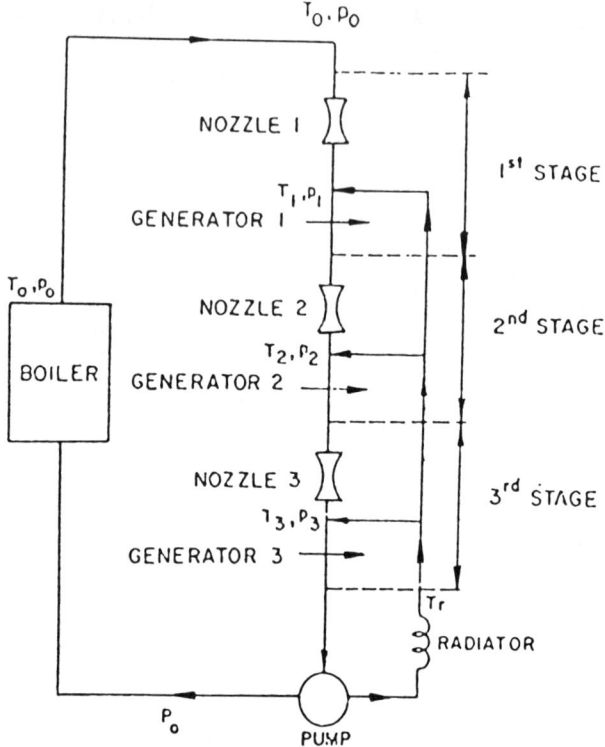

Fig. 12 Multistage cycle with two-phase MHD generators.

transformation, the expansion is made to take place in stages as shown in Fig. 12. After the working fluid expands in the first nozzle, the vapor content is reduced and the fluid stream is slowed down by the injection of a subcooled stream prior to entering the next nozzle where additional expansion and fluid acceleration take place. This process repeats itself until the fluid expands to the system back pressure, thus producing a large mass of fluid at a relatively low liquid velocity. The basic multistage condensing cycle process described can be modified in a number of ways to conform to the objectives of the various liquid metal cycles. Also, electrical energy can be extracted between steps through the use of multiple generators. With this type of cycle, regenerative heating and reheats can be introduced to theoretically produce an attractive efficiency.

The application of staging, as well as other techniques to minimize losses during momentum and mass transfer (e.g., the use of a higher heat capacity coolant[27] and regenerative nozzles[28]) to the condensing and separator cycles indicate that the maximum efficiency potential of a topping

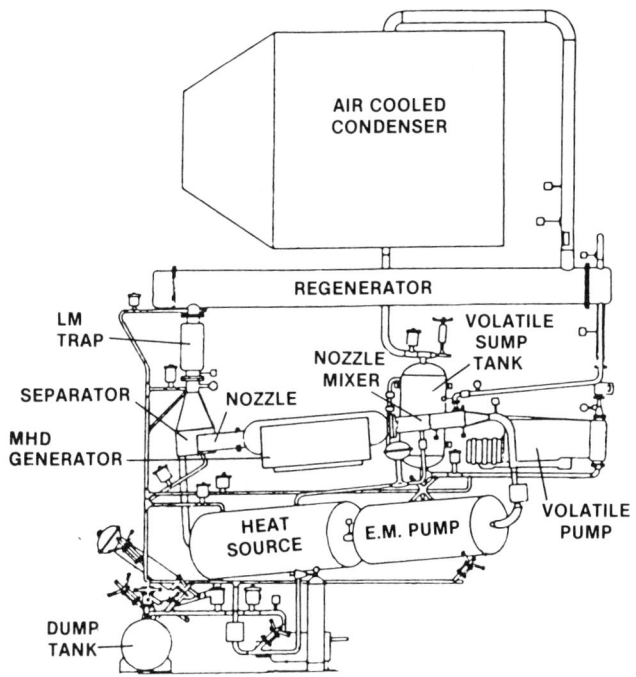

Fig. 13 Layout of ETGAR-1 - a NaK-Neo-Hexane two-phase MHD generator system designed at Ben Gurion University.

cycle is ~15%. Such a cycle would necessarily need to operate above 1200°F (up to 1600°F) and employ liquid metal working fluid pairs for which there is little relevant technology. The modest increase achievable in a real multi-staged system (accounting for losses) in the temperature range indicated, plus the lack of a data base, provided little incentive to continue development of the separator and condensing cycles for commercial application. Some work continues and significant progress is being made on the condensing injector cycle[29] in the Soviet Union utilizing a variable throat injector. The reality, however, is that these cycles do not appear to have sufficient performance potential for extensive commercial exploitation. They may still be of interest for special applications.

Optimized Magnetohydrodynamic Conversion System

The optimized MHD conversion system (OMACON) has been developed to circumvent and/or eliminate the technical issues of the two-phase flow generator concept cited above. The concept is schematically illustrated in Fig. 14; it is a

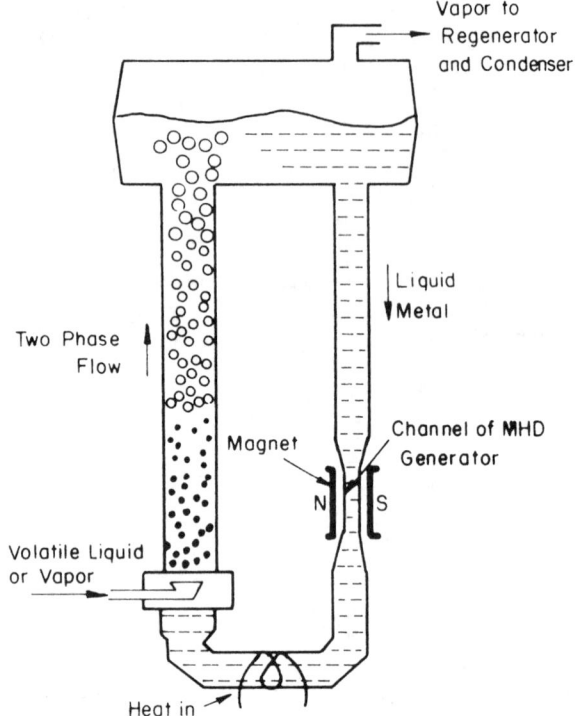

Fig. 14 Schematic of optimized magnetohydrodynamic conversion system (OMACON).

very simple direct-conversion, natural circulation power system. The basic system consists of two pipes (an upcomer and downcomer) connected at the bottom with a crossover pipe and with a simple tank (the separator) joining them at the top. A mixer is located at the bottom of the upcomer and an MHD generator, from which electrical power is extracted, is located in the downcomer or lower crossover pipe.

Operation of the system is as follows. A vapor or gas (the thermodynamic working fluid) is introduced into the mixer at the bottom of the upcomer at an appropriate temperature and pressure. A two-phase fluid of lower density is created. As the two-phase fluid flows to the separator, the gaseous phase undergoes an expansion from the high pressure in the mixer to the low pressure in the separator, accelerating the fluid and lowering its density. The gaseous phase (working fluid) is disengaged in the separator plenum by buoyancy (gravity) forces, thus producing a single-phase flow return into the downcomer.

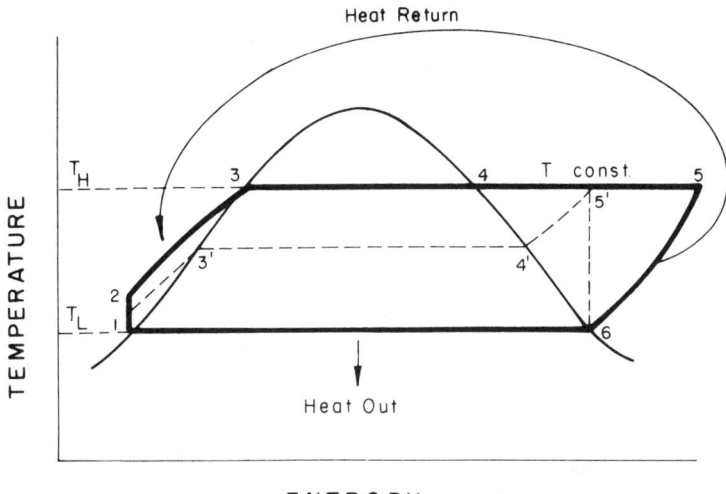

Fig. 15 Thermodynamic cycle performed by an OMACON system (solid lines) as compared with a turbine cycle (dotted lines).

The pressure differential that exists between upcomer and downcomer due to the density difference causes the liquid metal (the electrodynamic fluid) to circulate in the system; as the single-phase liquid metal passes through the MHD generator, an electrical potential is generated and power is extracted. The flow rate in the loop adjusts to balance the density differential between the upcomer and downcomer with the frictional and acceleration flow losses and the power extracted from the MHD generator. Therefore, the system converts pressure head (potential energy) to electrical energy via the MHD generator.

Major advantages of the system are the simplicity of design and the absence of moving mechanical parts. An analysis of the thermodynamic cycle indicates even more important advantages (see Fig. 15). Since the vapor in the upcomer is completely surrounded by hot liquid metal, its expansion is almost isothermal. Heat is added to the vapor continuously during its expansion, causing the thermodynamic cycle to become what can be defined as an infinite reheat cycle. Most of the sensible heat contained in the vapor, after it is separated from the liquid metal, is returned to the system by regeneration. The system can use wet vapor and does not require initial superheating; therefore, the volatile fluid boils at the highest available temperature, which is particularly important when low-temperature heat sources are utilized.

Staging and Scale-Up

The height H of the basic OMACON system needed to obtain maximum performance is determined by T_{max} and the vapor pressure of the working fluid (liquid) such that

$$P_2 = V.P.|_{T_{max}}$$

and

$$H = \frac{V.P.|_{T_{max}} - P_{condenser}}{\bar{\alpha}\,\rho_\ell}$$ (here friction and acceleration losses are neglected)

where $P_{condenser}$ = V.P. at $T_{condenser}$, ρ_ℓ = density of the liquid, and $\bar{\alpha}$ = the mean void fraction created by injection of the volatile working fluid.

In order to control the height to reasonable levels, develop higher voltages for inversion, control α in the upcomer over narrow ranges to insure bubble flow and utilize efficient donut-type magnets, a multistage system is desired as depicted in Fig. 16. Each module functions in the same manner as described for the basic system with the difference that the volatile fluid is passed from one module to the other and the MHD generators are electrically connected in series. The number of modules utilized is set by consideration of the optimum heights, the practical height of a single module, the voltage desired, and the cost of the liquid inventory. With this approach, important technical objectives that can be achieved are:

1) Maintain reasonable height for individual natural circulation units while still developing a high pressure P_t for obtaining the maximum efficiency.

2) Control and optimize the void fraction in each loop to achieve a maximum for each stage and reduce the void fraction levels in the final stage. This also allows one to keep the void fraction in the range where bubbly flow is likely to exist, thus minimizing the slip velocity.

3) Provide for modular construction, which in turn allows one to electrically isolate the loop and develop higher voltage and thus to increase the efficiency of and reduce the cost of the inverter.

4) The volatile fluid flows through each loop whose pressure levels are automatically controlled by the solid-liquid levels. Startup, system control, and shutdown are highly simplified.

LIQUID METAL MHD POWER GENERATION

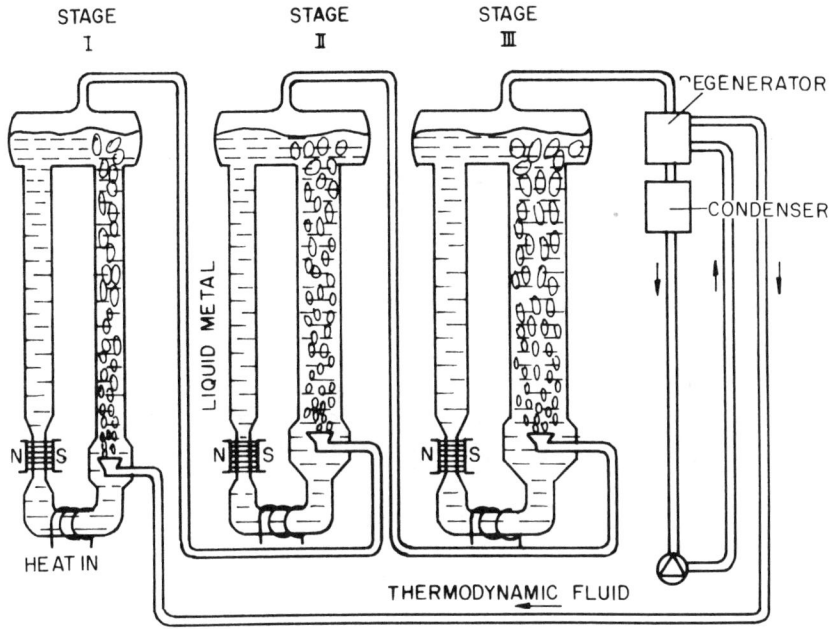

Fig. 16 Multistage OMACON system.

Scale-up of the OMACON power system appears to be relatively straightforward since the system as described is easily modularized and there is strong technical incentive to proceed in that direction. It is anticipated that standardized modules of various power levels will evolve for various applications that can be shop-fabricated and site-assembled. Utilizing modules, staging can be readily accomplished and efficient dc-ac inversion can be obtained by connecting the MHD generators in series.

One more improvement in performance can be achieved by switching the thermodynamic fluids when expansion of the vapor becomes too large. This happens particularly with steam under pressures lower than 1 bar and here it can be advantageous to condense the steam after separation in the n-1 stage and transfer the heat to a different fluid with a lower boiling point. This fluid can then be used in the n stage (fig. 17).

Advantages of the OMACON System

The OMACON system eliminates or minimizes many of the technical problems cited earlier. The specific benefits that accrue from this concept are:
1) Eliminate the critical, high-loss components in the

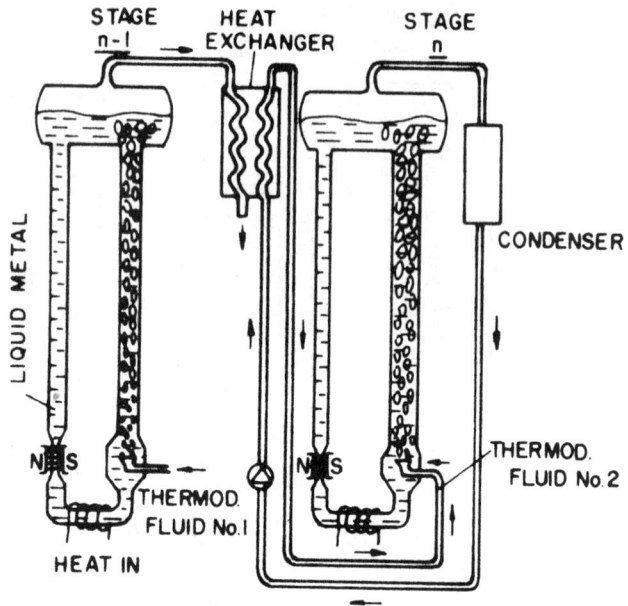

Fig. 17 Multistage OMACON system with a different fluid in the last pressure stage.

two-phase flow generator MHD system, namely, the two-phase nozzle, separator, and diffuser.

2) Facilitate ultra-simple startup, operation, and shutdown.

3) Eliminate two-phase flow limitation of MHD generator performance and hence overall system performance.

4) An 80-85% efficient single-phase MHD generator can be achieved and designed with confidence from the data base that exists today.

5) The single-phase flow through the MHD generator facilitates the use of low-field, low-cost magnets.

6) The high electrical conductivity single-phase flow through the generator makes possible the use of an ac induction generator to produce ac power directly -- eliminating the inverter and replacing the magnet with a conventional type of coil system found in electrical motors.

7) The overall system can be designed to have low kinetic energy (frictional) losses, thus maximizing performance potential.

8) Performance of the system is not a function of void fraction. One can obtain the same overall efficiency at both low and high void fractions, in contrast to the two-phase flow cycle.

9) Since the fluid passing through the generator is single-phase and at constant velocity, a high electrical efficiency generator design can be utilized.

10) Electrical parameter fluctuations caused by two-phase flow instabilities in the generator are eliminated.

11) The system can be operated as a Rankine or Brayton cycle and be readily adapted to various temperature ranges applications) by proper selection of the working fluids.

Unknowns and Tradoffs in OMACON System Design

There is <u>only one</u> major technical unknown in the OMACON system concept: the relative velocity of the gas to liquid in the upcomer (slip ratio). However, this unknown is the only one that requires additional research and development. The slip can be controlled by use of surfactants and by maintaining low void fractions, that is, keeping the flow in the bubble regime. The former needs to be researched further, whereas the latter approach can be taken directly in the design of the system. As discussed, a unique feature of the OMACON system is that it can operate efficiently at low values of void fraction or at $\bar{\alpha}$: As $\bar{\alpha}$ decreases, however, the recirculation flow rate must increase. In turn, this leads to larger liquid inventories and system size, which is an economic drawback. Tradeoff studies must be made to set the proper void fraction levels consistent with minimizing slip and maintaining realistic (economical) liquid inventories and system dimensions.

Initial calculations for OMACON systems have been performed on the basis of previous studies related to natural circulation two-phase flow systems[32] and of experiments with two-phase mercury-nitrogen flows[33].

A number of slip studies have been undertaken at Ben-Gurion University in connection with the ETGAR project developed jointly by Ben-Gurion and Argonne National Laboratory. These studies were performed under conditions closest to those existing in systems built according to the above described OMACON concept. Both mercury-Freon 113[34] and mercury-nitrogen[35] has been investigated. Results obtained in the mercury-Freon 113 facility with two fast-closing valves coincided very well with Smissaert's correlation[33]. The mercury-nitrogen experiments are still in progress and are reported in more detail in Ref.35.

A complete mercury-steam facility producing 0.45 kWe called ER-4 has been assembled at Ben-Gurion University and is described in detail in Ref. 36. This facility has as one of its major goals the accumulation of more data about slip and is provided with necessary measurement equipment for this purpose.

Finally, the ETGAR-3 semi-industrial pilot plant being built at Ben-Gurion University is going to provide evidence about the actual performance as compared with the present predictions[37].

Preliminary Results of Application Studies

Application of the ETGAR-3 system has already been studied (using the computer code developed at Ben-Gurion for the OMACON system) for a number of heat sources and for different system sizes. A summary of the cases studied is given in Table 2. For cases with cycle temperatures below 800 K, the OMACON system performing a Rankine cycle is quite attractive, as can be seen from Fig. 18, where overall efficiencies for a staged 2 MW_e liquid metal system and a dual MHD/turbine 5 MW_e system are presented. It is supposed that this system works with steam and lead (or lead alloys). In more detail, the basic assumptions for this calculation are given in Table 3. One of the calculated cases relates to a

Table 2 Application studies of ETGAR 3 system

Type of heat source	Highest temperature in cycle, K	Size (power output level), KW_e
Waste industrial heat	350-700	100-15,000
Geothermal	350-550	1,000-5,000
Fuel combustion	600-1,000	1,000-250,000
Solar	350-1,100	100-30,000

Table 3 Basic parameters for calculating Rankine cycle efficiencies

Pressure in the mixer of first stage (highest pressure in cycle)	Equal to pressure at saturation temperature but not higher than 90 bar
Riser mean void fraction	0.4
Minimal output voltage of a single MHD Generator electrical efficiency	6 V
Regeneration heat exchange effectiveness	90%
Riser slip ratio	Calculated according to empirical correlation
Liquid metal	Lead (lead alloys for lower temperature)
Thermodynamic fluid	Steam

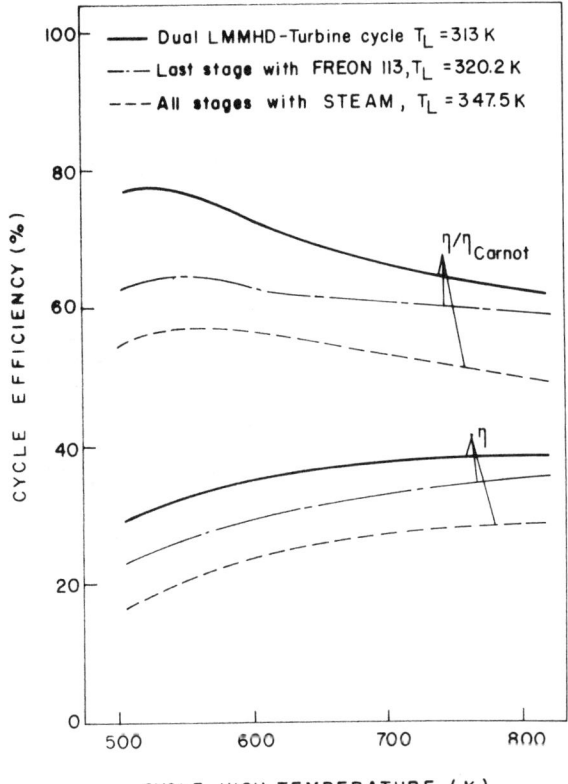

Fig. 18 Overall efficiencies for multistage Rankine cycle liquid metal MHD system with 2MW$_e$ total electrical power output and for a dual liquid metal MHD-turbine system with 5MW$_e$ total electrical power output (for other parameters - see Table 2).

system where, in the n-1 stage, steam is condensed at higher pressure and heat is transferred to the Freon-113 that issued as the thermodynamic fluid in the n stage. This eliminates the otherwise high expansion ratio in the last stage. For this case, at temperatures of about 550-600 K, the efficiency is higher than 70% of Carnot efficiency, which should be regarded as very positive for a small-sized system.

It should be mentioned that in many cases hybrid concepts are desirable. For a number of applications, a combined (dual) liquid metal MHD and turbine (steam, organic, or gas) cycle turns out to be most advantageous. In the case with steam, for instance, the liquid metal MHD system can utilize saturated steam at the highest possible temperature. While producing electrical power in the liquid metal MHD system, the steam expands at constant temperature (as explained above) and is superheated. This superheated steam then enters a low-pressure steam turbine. For a proper division of expansion rates between the liquid metal MHD system and the turbine, extremely high efficiencies are attainable, especially if feed water preheating is used (Fig. 19). Here overall efficiencies in some cases exceed 70% of Carnot efficiency. This version can be attractive for central power plants and for dispersed smaller plants using either fuel, solar, or other renewable heat sources.

In relation to future applications, the following comment should be made regarding power conditioning. Detailed calculations show that a 4-8 V dc output can be achieved from a single channel. Thus, by using multistage systems and/or splitting the channels of each stage, it is feasible to increase the total output voltage to at least 20-40 V and in some cases much more. At this voltage level dc to ac inversion is already straightforward either with solid state or with homopolar inverters. Efficiency of the Westinghouse homopolar inverters is higher than 93-94% and cost of inversion less than \$120/kW even for dc inputs of less than 10 V.

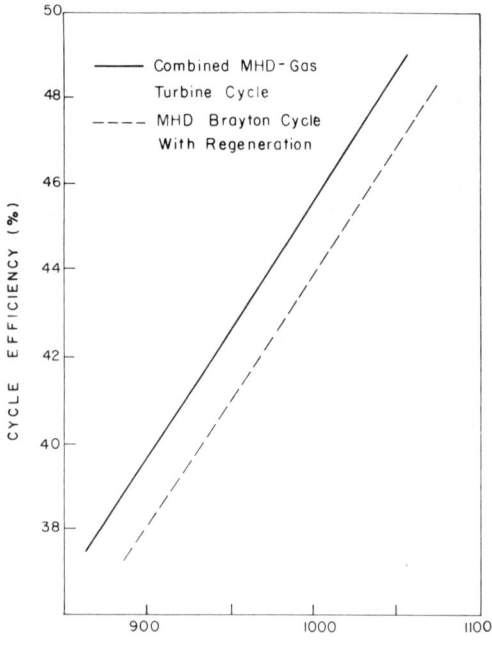

Fig. 19 Overall efficiencies for multistage Brayton cycle liquid metal MHD system (for other parameters - see Table 3).

In all cases, when the OMACON concept is used, the required magnetic flux density is very moderate, usually between 0.4-0.8 T. Thus, a conventional electromagnet can be easily used with no need for any special development and power consumption by the magnet is a reasonably small fraction of the total electrical power output.

Acknowledgments

The authors wish to express their gratitude to Solmecs Corporation for supporting this work. The submitted manuscript has been authored by a contractor of the U.S. Government under Contract W-31-109-ENG-38.

References

[1] Elliott, D., "Two Fluid Magnetohydrodynamic Cycle for Nuclear-Electric Power Conversion," ARS Journal, June 1962.

[2] Brown, G.A. and Lee, K.S., "A Liquid Metal MHD Power Generation Cycle Using a Condensing Ejector," Proceedings of International Symposium on Magnetohydrodynamic Electrical Power Generation, Paris, Vol. 4, July, 1964.

[3] Brown, G.A. and Levy, E.K., "Liquid Metal MHD Power Generation with Condensing Ejector Cycles," Proceedings of International Symposium on Electricity from MHD, Salzburg, July 1966, Vol. 2.

[4] Grolmes, M.A., Petrick, M., and Jerger, E.W., "Condensing Injector Experiments and Analysis of Performance with Supersonic Inlet Vapor," Proceedings of International Symposium on Electricity from MHD, Warsaw, Vol. III, July 1968.

[5] Aladiev, I.T., Teplov, S.V., and Telmach, I.M., "Experimental Potassium MHD Installation of 1 KW," Fifth International Conference on Magnetohydrodynamic Electric Power Generation, Munich, Vol. III, April 1971.

[6] Prem, L.L. and Parkins, W.E., "A New Method of MHD Power Conversion Employing a Fluid Metal," Proceedings of International Symposium on Magnetohydrodynamic Electrical Power Generation," Paris, Vol. 4, July 1964.

[7] Bjerklie, J.W., and Powell, J.R. Jr., "A Liquid Metal MHD Power Generation Scheme Using Intermittent Vaporization," Proceedings of Symposium on Electricity from MHD, Vol. III, July 1968, Warsaw.

[8] Bazeev, E.T., Bratishchev, Yu. A., Odnorozhenko, I.G., and Schegolev, G.M., "Piston Acceleration of a Liquid and the Separation of Two Phase Piston Flows Applied to Liquid Metal Installations," Sixth International Conference on Magnetohydrodynamic Electrical Power Generation, Washington, Vol. III, June 1975.

[9]Petrick, M. and Lee, K.Y., "Performance Characteristics of a Liquid Metal MHD Generator," Proceedings of International Symposium on Magnetohydrodynamic Electrical Power Generation, Paris, Vol. 4, July 1964.

[10]Liquid Metal Magnetohydrodynamics Technology Transfer Study, Vol. II, App. C, Jet Propulsion Laboratory, Pasadena, Calif., Rept. 1200-59, May 1973.

[11]Elliott, D., Cerini, D., Hays, L., O'Connor, D., and Weinberg, E., "Liquid MHD Power Conversion," Space Programs Summary, Vol. IV, Jet Propulsion Laboratory, Pasadena, Calif., Rept. 37-28, Aug. 1964.

[12]Lenzo, C.S., Dauzvardis, P.V., and Hantman, R.G., "An Experimental Investigation of Rotating Drum Separators for Liquid Metal MHD Applications," 17th Symposium on Engineering Aspects of MHD, Stanford, Calif., 1978.

[13]Cerini, D.J. and Hays. L.G., "Power Production from Geothermal Brine with the Rotary Separator Turbine," Proceedings of the 15th Intersociety Energy Conversion Engineering Conference, AIAA, New York, August 1980.

[14]Cerini, D.J., "Circulation of Liquids for MHD Power Generation," Proceedings of International Symposium on Electricity from MHD, Warsaw, Vol. III, 1968.

[15]Fabris, G., Chow, J.C.F., and Dunn, P.R., "On Formation of a Homogeneous Two-Phase Foam Flow," Journal of Power Engineering, Vol. 102, 1980.

[16]Elliott, D.L. and Weinberg, E., "Acceleration of Liquids in Two-Phase Nozzles," Jet Propulsion Laboratory, Pasadena, Calif., Rept. TR32-987, July 1968.

[17]Branover, H., Yakhot, A., and El-Boher, A., "Solar-Powered Liquid Metal MHD Generators and Some Peculiarities of the Performance of Two-Phase Generators," Proceedings of the Seventh International Conference on MHD Electrical Power Generation, Cambridge, Mass., Vol. I, 1980.

[18]Klem, E., "Preliminary Results Obtained from Operation of Two-Phase Nozzles with Potassium, Freon and Water," 5th International Conference on MHD Electrical Power Generation, Munich, Vol. IV, April 1971.

[19]Moszynski, J.R., "Reduction of Electrical End Losses in MHD Generator Channels by Insulating Vanes," Argonne National Laboratory, Argonne, ILL, Rept. ANL-7188, Sept. 1967.

[20]Petrick, M. and Roberts, J.J., "Analytical and Experimental Studies of Liquid Metal Faraday Generators," Proceedings of the Symposium on Magnetohydrodynamic Electric Power Generation, Warsaw, paper SM-107/20, 1968.

[21]Hsu, C., Petrick, M., and Pierson, E.S., "A Study of Factors Pertinent to the Development of an Efficient High Power Two-Phase Liquid Metal MHD Generator System," Proceedings of the Fifth International Conference on MHD Electrical Power Generation, Munich, 1971.

[22]Gherson, P., Lykoudis, P.S., and Lynch, R.E., "Analytical Study of End Effects in Liquid Metal MHD Generators," 7th International Conference on MHD Electrical Power Generation, Boston, Mass, June 1980.

[23]Lykoudis, P.S., "Liquid Metal MHD Generators with Shunt Layer in MHD Flows and Turbulence - II," Proceedings of the Second Bat-Sheva Seminar on MHD Flows and Turbulence," Beer Sheva, Israel, 1978.

[24]Fabris, G., Pierson, E.S., Pollack, I., Dauzvardis, P.F., and Ellis, W., "High-Power-Density Liquid Metal MHD Generator Results," Proceedings of the 18th Symposium on Engineering Aspects of MHD, Butte, Mont., 1979.

[25]Dunn, P.F., Pierson, E.S., Staffon, J.D., Pollack, I., and Dauzvardis, P.V., "High-Temperature Liquid Metal MHD Generator Experiments," Proceedings of the 18th Symposium on Engineering Aspects of MHD, Butte, Mont., 1979.

[26]Personal Communications.

[27]Basov, S.N., Kirillov,P.L., Subbotin, V.I., and Turchin, N.M., "Injection of Liquid Metals in MHD Converters," Symposium on Electricity from MHD, Vienna, Vol. II, 1966.

[28]Radebold, R., Lang, H., Schulz, T., Weh, H., Klein, E., and Wagner, K.H.,"Energy Conversion with Liquid Metal Working Fluids in MHD-Staustrahlrohr," Symposium on Electricity from MHD, Vienna, Vol. II, 1966.

[29]Personal Communications.

[30]Pei, R.Y. and Purnell, S.W., Proceedings of the Rand Corporation Conference on Liquid-Metal MHD Power Generation, Rand Corp. Rept. R-2290-DOE, Sept. 1977.

[31]Branover, H., El-Boher, A., Sukoriansky, S., Yakhot, A., Petrick, M., Pierson, E., and Smith, I., "Development of a Low Temperature Liquid Metal MHD Small Scale Pilot Plant," Proceedings of the 21st Symposium on Engineering Aspects of Magnetohydrodynamics, Argonne National Laboratory, Argonne, ILL, Conf.-830634, June 1983.

[32]Lottes, P.A., Petrick, M. et al, "Experimental Studies of Natural Circulation Boiling and their Application to Boiling Reactor Performance," Proceedings of 2nd United Nations Conference on the Peaceful Uses of Atomic Energy, Geneva, Vol. 7, Sept. 1978.

[33]Smissaert, E., "Two-Component Two-Phase Flow Parameters for Low Circulation Rates," Argonne National Laboratory, Argonne, ILL, ANL-6755, July 1963.

[34]Branover, H., El-Boher, A., and Yakhot, A., "Testing of a Low-Temperature Liquid Metal MHD Power System," Energy Conversion and Management, Vol. 22, 1982, pp. 163-169.

[35]Unger, Y., Kiel, J.H.A., Zuckerman, B., and Branover, H., "Two-Phase Liquid Metal-Gas Flows with Medium and High Void Fractions," Proceedings of the 4th Beer Sheva Seminar on MHD Flows and Turbulence, Feb-27 - March 2, 1984, Beer Sheva, Israel.

[36]Branover, H., El-Boher, A., Lesin, S., and Marcus, B., "Test of ER-4 Mercury-Steam Power Generation Research Facility," AIAA Progress in Astronautics and Aeronautics, ed. H. Branover, 1985, New York.

[37]El-Boher, A., Branover, H., and Petrick, M., "The ETGAR Liquid Metal MHD Project," published elsewhere in this volume.

Tin-Water Faraday Generator

J.P. Thibault,* F. Joussellin,† R. Laborde,† and A. Alemany‡
Institut de Mecanique de Grenoble, Saint Martin d'Heres, France
and
F. Werkoff‡
Centre d'Etudes Nucleaires de Grenoble, France

Abstract

The main part of this paper deals with the evaluation of the flow of a two-phase conducting mixture submitted to a constant magnetic field (two-phase MHD flow). This theoretical work leads us to a six-equation, time-dependent, one-dimensional model, the solution for which is evaluated by a numerical code called FARADEX. The first tests presented fit the imperfectly suited transfer laws, but they are in good enough agreement with the predicted results to show that the code is numerically operational. In the second part, the experimental tin-water loop, EPEE, built at the Institut de Mecanique de Grenoble, is described. The purpose of EPEE is to test the performances and reliability of metal-gas FARADAY converters in the mean temperatures range (500-600 K) and to collect experimental data that will be used to improve the accuracy of the transfer laws.

Introduction

Since the 1960's, liquid metal MHD (LMMHD) conversion has been the subject of extensive research around the world.[1-3] In France, unfortunately, the development of MHD conversion was stopped in 1969, essentially because of technological difficulties with plasma MHD. In 1980, following the visit of Branover[4],the revival of FARADAY conversion was

---
Paper presented at the Fourth Beer-Sheva Seminar on MHD Flows and Turbulence, Ben-Gurion University of the Negev, Beer-Sheva, Israel, Feb. 27-March 2, 1984. Copyright © 1985 by the American Institute of Aeronautics and Astronautics, Inc. All rights reserved.
*Research Associate.
†Research Engineer.
‡Research Head.

promoted by the Centre National de la Recherche Scientifique (CNRS) at the request of the Institut de Mecanique de Grenoble (IMG), with the aim of developing an alternative solution to convert heat into electricity in the temperature range of 500-600 K. In 1982, the Centre National d'Etudes Spatiales (CNES) asked us to estimate the possibility of using a metal-gas Faraday converter for space applications in the temperature range of 1200-1400 K (Ref. 5).

Modelization of a Metal-Gas Faraday Generator

Flow Model

The essential difference between single-phase (liquid metal) and two-phase (metal-gas) MHD is the existence of interfaces. The momentum and energy transfers between the two phases lead to nonequilibrium phenomena that are taken into account by transfer laws.[6] In order to represent the effect of electromagnetic forces on two-phase flows in a conducting generator, we use a time-dependent, one-dimensional, two-fluid model. This particular case, of more general tri-dimensional models, is sufficient to describe the flow in our reference case: almost homogeneous flow, weakly diverging channel, and no mass transfer between the two components of the flow[7] (see Fig. 1). Electromagnetic forces are dominating (Hartmann number M > 100), flow equations are separated from electromagnetic equations (Reynolds Magnetic $R_m \angle 1$), and end effects are neglected. It follows the six phasic conservation equations (suscripts ℓ and v represent respectively liquid and vapor).

Mass conservation

$$\frac{\partial}{\partial t} \left[S(1-\alpha) \rho_\ell \right] + \frac{\partial}{\partial z} \left[S(1-\alpha) \rho_\ell V_\ell \right] = 0 \qquad (1)$$

$$\frac{\partial}{\partial t} (S \alpha \rho_v) + \frac{\partial}{\partial z} (S\alpha \rho_v V_v) = 0 \qquad (2)$$

Momentum conservation

$$S(1-\alpha) \rho_\ell (\frac{\partial V_\ell}{\partial t} + V_\ell \frac{\partial V_\ell}{\partial z}) + S(1-\alpha) \frac{\partial P}{\partial z} = -Fmt + Fem \qquad (3)$$

$$S\alpha\rho_v (\frac{\partial V_v}{\partial t} + \frac{\partial V_v}{\partial z}) + S\alpha \frac{\partial P}{\partial z} = Fmt \qquad (4)$$

Thermal energy conservation

$$S(1-\alpha)\rho_\ell \left(\frac{\partial h_\ell}{\partial t} + V_\ell \frac{\partial h_\ell}{\partial z}\right) - S(1-\alpha)\left(\frac{\partial P}{\partial t} + V_\ell \frac{\partial P}{\partial z}\right) =$$

$$- Q_{2P} + Q_{jo} \qquad (5)$$

$$S\alpha \rho_v \left(\frac{\partial h_v}{\partial t} + V_v \frac{\partial h_v}{\partial z}\right) - S\alpha \left(\frac{\partial P}{\partial t} + V_v \frac{\partial P}{\partial z}\right) = Q_{2P} \qquad (6)$$

In addition, in order to characterize the two components of the flow, we use the state equations

$$\rho_{\ell,v} = \rho_{\ell,v}(P, h_{\ell,v}) \qquad (7)$$

$$T_{\ell,v} = T_{\ell,v}(P, h_{\ell,v}) \qquad (8)$$

where the z axis is the flow direction; α is the instantaneous volumetric gas fraction; P the pressure; $h_{\ell,v}$ the enthalpies; $T_{\ell,v}$ the temperatures; $V_{\ell,v}$ the velocities; and $\rho_{\ell,v}$ the densities. All these quantities are functions of time and z position; they are the average values in the cross-section area S which is calculated with a homogeneous model[7].

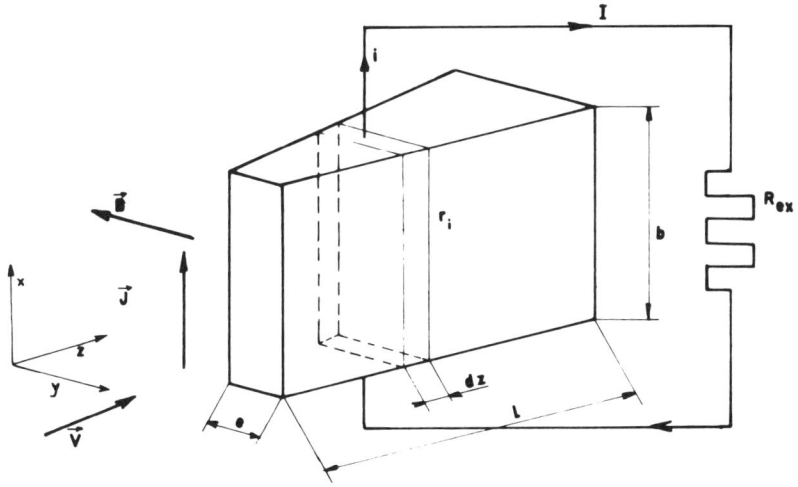

Fig. 1 Geometry of the problem.

Because the vapor is electrically insulating, the electromagnetic terms (Fem and Q_{jo}) concern only the liquid. We make the following assumptions: no heat transfer at the wall, no pressure drop at the interfaces, only a liquid film wetting the wall.

In Eqs. (3) and (4), Fmt is the interfacial momentum transfer force per unit of length. It takes into account two effects:

$$Fmt = Fvm + Fd \qquad (9)$$

where Fvm, the virtual mass unitary force, is the product of the virtual mass of liquid swept along by a vapor bubble and the relative interfacial acceleration:

$$Fvm = S\ \beta\alpha(1-\alpha) \left[\alpha\ \rho_v + (1-\alpha)\ \rho_\ell\right] \qquad (10)$$
$$(\frac{\partial V_\ell}{\partial t} + V_\ell \frac{\partial V_\ell}{\partial z} - \frac{\partial V_v}{\partial t} - V_v \frac{\partial V_v}{\partial z})$$

The drag unitary force Fd is the product of the interfacial slip $(V_\ell - V_v)$ and the local momentum transfer coefficient K:

$$Fd = K\ (V_\ell - V_v) \qquad (11)$$

The K model corresponds either to a complicated expression for slug-annual dispersed flows or to a simpler theoretical model for spherical bubbles in an infinite medium [see Eq. (12)].[8]

$$K = 3$$
$$K = \frac{3}{8} S\ \alpha(1-\alpha)\ \frac{1}{R_b^2} \left(12\ \mu_\ell + \frac{1}{2} R_b\ \rho_\ell |V_v - V_\ell|\right) \qquad (12)$$

where R_b is the bubble radius and μ_ℓ the liquid dynamic viscosity.

This expression, actually used for both zero and nonzero electromagnetic forces, corresponds to a theoretical model for spherical bubbles in an infinite medium [8]. It does not point out the effect of the high pressure gradient due to the electromagnetic forces, on the bubble shape. In addition the interfacial slip increases with the magnetic field because the electromagnetic forces act only on the liquid metal.[9] Unfortunately we are still unprovided with a K model as a function of the magnetic field.

The electromagnetic aspect of the problem is assumed to be simple, because the velocity $\vec{V}$, magnetic field strength $\vec{B}$, and current density $\vec{j}$ vectors form an orthogonal trihedron (see Fig. 1). We determine the current density $\vec{j}$ with the Ohm's law :

$$\vec{j} = \sigma_{2p} (\vec{E} + \vec{V} \times \vec{B}) \qquad (13)$$

Where the external electric field $(\vec{E})$ is on x axis, and σ_{2p} is the apparent conductivity of the two phase mixture. We use optionally two expressions, as function of σ_1 the electrical conductivity of the liquid and α the gas fraction.

$$\sigma_{2p} = \sigma_1 \frac{2(1-\alpha)}{2+\alpha} \qquad (14a)$$

$$\sigma_{2p} = \sigma_1 \exp(-3.8\,\alpha) \qquad (14b)$$

The first one (14a) is a theoretical expression obtained from Maxwell's equation with the gas bubbles assumed to be zero conductivity solid spheres. The second one (14b) is an empirical expression proposed by Petrik and Lee.[10] Finally, the $\vec{j} \times \vec{B}$ force (Fem) is on z axis :

$$Fem = \sigma_{2p} S B (V_\ell B - E) \qquad (15)$$

Owing to the fact that the Hartmann number M (M^2 equals the electromagnetic forces divided by the viscous forces) is greater than 100, we neglect the viscous losses at the wall.

In Eqs. (5) and (6), Q_{2p} is the interfacial thermal power per unit of length :

$$Q_{2p} = C (T_\ell - T_v) \qquad (16)$$

Considering the complexity of heat transfers in two-phase flow, for the first numerical tests, we make the assumption of a quasi-zero temperature difference between the two phases, which is consistent with experimental data[11]. This lead us to a high heat transfer coefficient C.

Q_{jo} is the unitary Joule effect thermal power due to the current circulation in the liquid metal :

$$Q_{jo} = S \sigma_{2p} (V_\ell B - E)^2 \qquad (17)$$

In order to solve the set of equations (1-6), we develop a numerical code FARADEX, which uses some studies done at the Commissariat a l'Energie Atomique (CEA) on the field of safety problems in pressurized water reactors. For the space discretization, a ICE-type method[12] is used with staggered mesh and donor cell principles. For the time discretization, the scheme is fully implicit [13-14]. This leads to a linear system of tridiagonal (6 x 6) blocks. However, the advantage of this method is that we do not have to satisfy any Courant-Hilbert condition on the time step[15].

First Results

The first numerical tests have been done with the liquid tin + water steam pair corresponding to our experimental facility EPEE (see Sec. : The Tin-Water Loop). We have simulated both zero and non-zero electromagnetic forces. In the last case, we apply a constant magnetic field $B = 1.2$ T and an external electric field $E = 22$ V:m which correspond to a load factor of about $\phi = 0.75$ ($\phi = \frac{E}{V\,B}$). At the inlet cross section, the void fraction is $\alpha_0 = 0.4$, the liquid-vapor velocities are around 25 m/s and the enthalpies correspond to a temperature : $T_{l,v} \# 523$ K.

In both cases, steady state conditions are reached after a time of about 1 second. The variation of the cross section (see Fig. 2), which is a datum, is the one calculated by our previous program VICOPE[16] (constant velocity, variable area, one dimensional, time independent, two phase equivalent one fluid). The cross section is globally expanding in order to compensate, at a quasi constant velocity, the volumic expansion due to the high pressure gradient which ba-

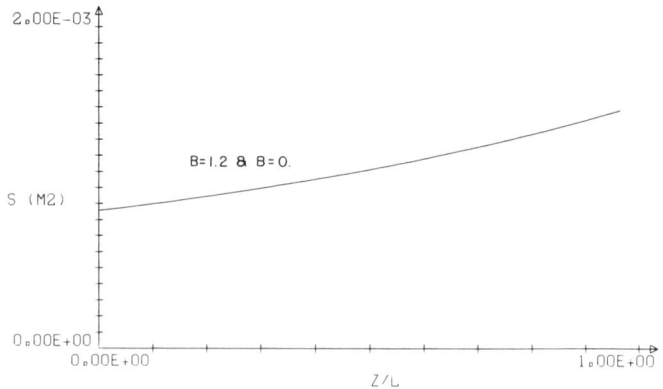

Fig. 2 Variation of cross section with fractional distance from channel inlet.

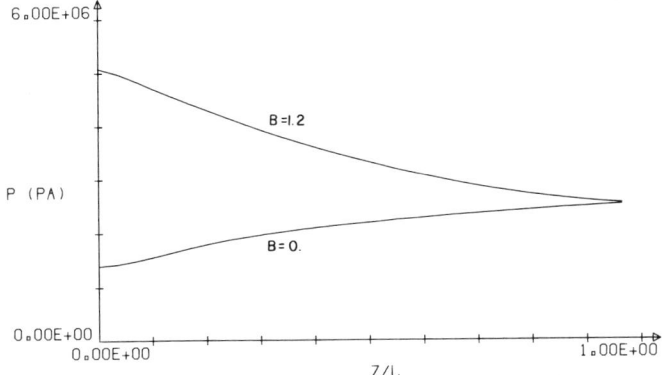

Fig. 3 Variation of pressure with fractional distance from channel inlet.

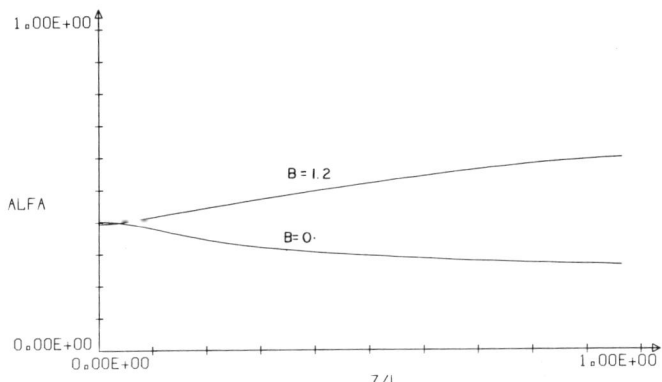

Fig. 4 Variation of gas fraction with fractional distance from channel inlet.

lances the electromagnetic forces (see Fig. 3). In the same duct, with zero magnetic field, the problem is similar to a two phase diffuser, like that we can observe an evident inversion in the pressure gradient. The void fraction repartition are also strongly affected by the very different pressure behaviours. It follows that the void fraction, at the outlet cross section, is higher with electromagnetic forces than without (see Fig. 4).

The computation done with FARADEX in a duct dimensioned by VICOPE, leads to quasi constant velocities with non-zero magnetic field. It shows that FARADEX is working correctly (see Fig. 5). The difference is due to the non-constant gas to liquid velocity slip ratio $G = V_v / V_l$ (see

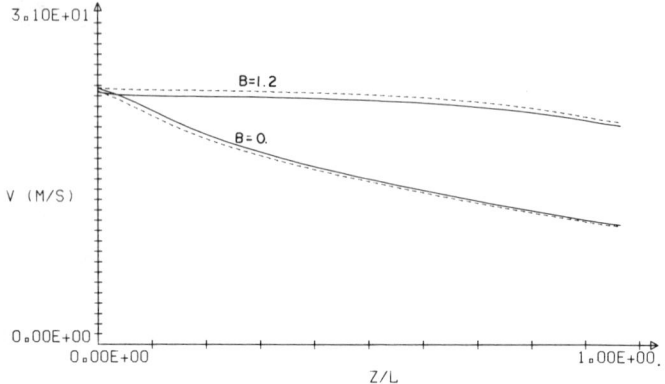

Fig. 5 Variation of liquid and gas velocities with fractional distance from channel inlet.

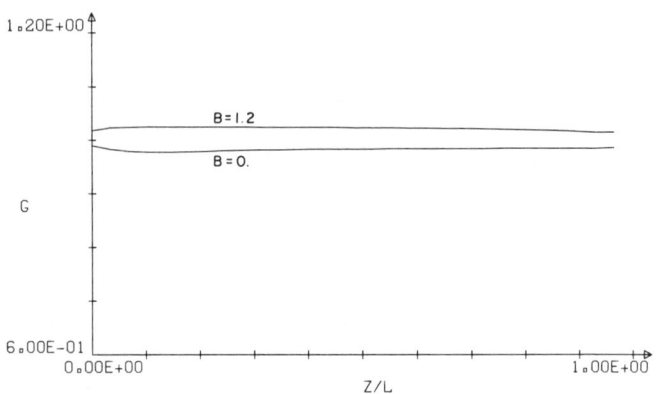

Fig. 6 Variation of gas to liquid velocity slip ratio with fractional distance from channel inlet.

Fig. 6). Unfortunately the K model (see equation 12) used is a zero magnetic field theoretical two-phase flow model, which gives only a qualitative value of the slip ratio.

The comparison of non-zero and zero magnetic field velocities graphs, shows an inversion between liquid velocity (solid line) and gas velocity (dashed line) (see Fig. 5). In the first case, the momentum provided by the expansion of the gas is the driving force, this explains that the gas velocity is superior to the liquid one. In the second case, due to the higher specific weight and consequently higher momentum of the liquid, the gas is firstly slow down in the diffuser, the increasing slip velocity produces an increasing interfacial momentum transfer force which slows down the liquid.

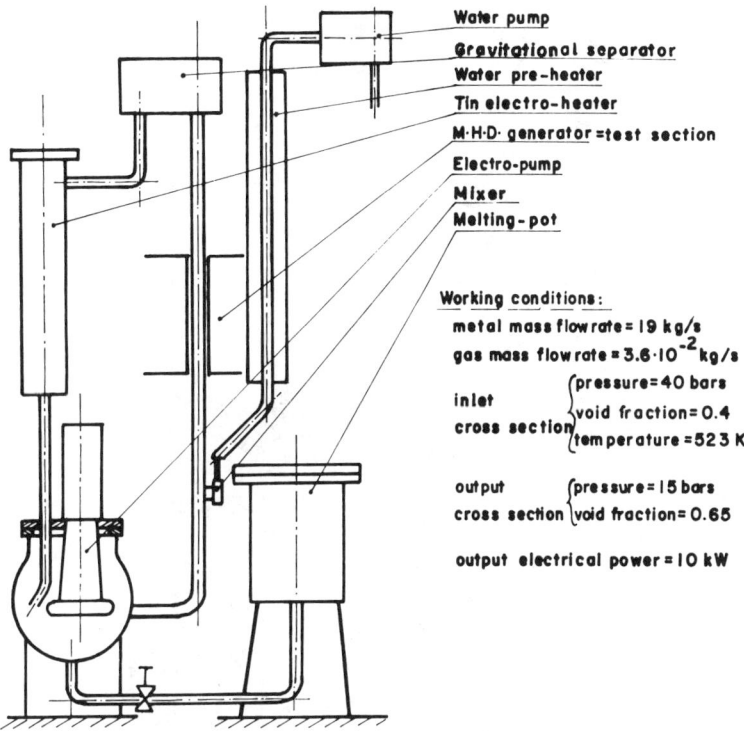

Fig. 7   The experimental tin-water loop: EPEE.

All the figures presented correspond to the expression (14b) for the apparent electrical conductivity. This leads to pessimistic results which could be improved with (14a) expression. This one would be probably more realistic especially in a two-phase flow exploiting a surfactant additive to creat homogeneous foam [17].

The Tin-Water Loop : EPEE

Description

The modular conception of the tin-water experimental loop (see Figs. 7 and 8), built at the IMG, allows us to adjust all parameters of each component of the two-phase flow. All the elements of the loop are made of stainless steel ; they are preheated with electrical resistance in order to suppress thermal shocks ; and they can be dismantled and modified.

The working fluid is a two-phase metal-gas conducting flow : liquid tin plus water steam, which seems to be the

Fig. 8   The experimental facility.

more suitable pair for the following working conditions : heater temperature, 500-600 K ; cooler temperature, 300 - 350 K. The rated electric power is about 10 kW, with a direct current electromagnet of 1.2 T.

The MHD channel has an increasing rectangular cross section of about 7 cm^2 and 1 m length. Water will be injected, in the flow, by means of a porous stainless steel tube. The flow rate of injection is controlled by a volumetric water pump and a valve. After electricity production, the two phases are separated in a gravitational separator. This choice introduces large kinetic energy losses but it makes experiments simpler. Obviously, in industrial development, a centrifugal or inclined plate separator shoud be used, and thus the centrifugal pump could be suppressed. An electroheater, the heat source of the loop, makes restitution of heat consumed by the MHD expansion. The furnace, utilized as a melting pot and as a stocking pot, communicates with the loop by a siphon and a valve.

Experimental Program

The major purposes of EPEE are firstly to demonstrate the feasibility of industrial tin-water 50-150 kW generators and, secondly, to collect experimental data that will be extrapolated to any other working conditions by numerical simulation (see Sec.: Modelization of a Metal-Gas Faraday Generator).

The building of the loop is now finished. We have tested the thermal and pressure resistance with a neutral gas and pressurized hot water. We are now testing with liquid tin. The next steps will be to realize stable and homogeneous two-phase flow and, at last, to produce electricity.

Concluding Remarks

We have presented the status of Faraday conversion in Grenoble in the mean temperature range. With FARADEX, we now have an operational code adapted to the two-phase flows in the presence of a magnetic field ; and with EPEE, a tin-water experimental facility.

Our next purpose is, first, to couple FARADEX with an electrical two-dimensional model in order to take into account the edge effects and to try to represent the supersonic nozzle flow (without magnetic field).

We hope to realize tin-water emulsions in the presence of a magnetic field into EPEE during 1985.

Acknowledgments

We wish to thank J.C. Rousseau. His great knowledge and experience of two-phase flow modelization, as well as his comprehensive interest, were very useful to us in the choice of a numerical method.

References

[1] Bidard R., "Les Générateurs Electriques à Métaux Liquides (Liquid-Metal Electric Generators)", Rev. Energie Nucl., Vol. 14, No 4, 1972, pp. 3-20.

[2] Petrick M. and Roberts J., "Performance of Space-Power Liquid Metal MHD Cycles Utilizing a Two-Phase Flow Generator", J. Spacecraft, Vol. 4, No. 8, 1967, pp. 967-973.

[3] Elliott D.G., Cerini D.J. and Weinberg E., "Liquid-Metal MHD Power Conversion", Space Power Syst. Eng., Vol. 16, 1966, pp. 1275-1298.

[4] Branover H., "On the Feasability of Solar Energy Conversion into Electricity by Means of MHD Generators", Institut de Mécanique de Grenoble Report, Grenoble, France, 1979.

[5] Thibault J.P., Joussellin F., Alemany A. and Dupas A., "The possible Use of Metal-Gas MHD Energy Converter in Space", Acta Astronautica, Vol. 10, No. 8, 1983, pp. 587-589.

[6] Bergles A.E., Collier J.G., Delhaye J.M. et al, "Basic Equations for Two-Phase Flow Modeling", Two-Phase Flow and Heat Transfer in the Power and Process Industries, Hemisphere Publishing Corp., New York, 1981, pp. 40-97.

[7] Thibault J.P., "Générateur de Faraday à Métal Liquide (Liquid-Metal Faraday Generator)", Thèse Docteur-Ingénieur, INPG, Grenoble, France, 1983.

[8] SOO S.L., "Fluid Dynamics of Multiphase Systems", Blaisedell Publishing, Lexington, Mass., 1967.

[9] Hantman R.G., Cole R.L., Fabris G. and Cutting J.C., "Performance of a Constant-Velocity Variable-Area Two-Phase Liquid-Metal MHD Generator", Proc. of the 6th Int. Conf. on MHD Power Generation, Wastington, 1975, pp. 295-307.

[10] Tanatugu N. Fujü-E Y. and Suita T., "Electrical Conductivity of Liquid-Metal Two-Phase Mixture in Bubbly and Slug Flow Regime", J. of Nucl. Sci. and Techn., Vol. 9, No. 12, 1972, pp. 753-755.

[11] Fabris G. and Pierson E.S., "The Role of Interfacial Heat and Mechanical Energy Transfers in a Liquid-Metal MHD Generator", ASME Publication 78-WWA/HT-33, New-York, 1978.

[12] Harlow F.H., and Amsden A., "Numerical Calculation of Multiphase Fluid Flow", J. of Comp. Phys., Vol. 17, 1975, pp. 19-52.

[13] Rousseau J.C., "Le module de Base du Code Cathare Développements Physiques et Performances (Base-Module of Cathare System Physical Developments and Performances)", La Houille Blanche, No. 314, 1984, pp. 209-215.

[14] Dube D.A., "Development of a Fully Implicit Two-Fluid, Thermal-Hydraulic Model for Boiling Water Reactor Transient Analysis", Ph. D. Thesis, MIT, Cambridge, Mass., 1980.

[15] Richtmyer R.D. and Morton K.W., "Fluid Dynamics in One Space Variable", Difference Methods for Initial-Value Problems, Interscience Publishers J. Wiley and Sons, New York, 1967, pp. 288-350.

[16] Thibault J.P., Joussellin F. and Alemany A., "Conversion MHD Métal-Gaz. Aspects Théoriques et Perspectives de Développements (Metal-Gas MHD Conversion. Theorical Aspects and Development Prospects)", 6ème Congrès Français de Mécanique, Résumés des Communications, Ass. Univ. de Mec. IMTA, Paris, 1983, pp. 6.13-6.16.

[17] Fabris G., "Mixers and Surfactants", Proc. of the Rand Corp. on LMMHD Power Generation, Rand Santa Monica, Calif., 1977, pp. 95-120.

The ETGAR Liquid Metal MHD Project

Arik El-Boher* and Herman Branover†
Ben-Gurion University, Beer-Sheva, Israel
and
Michael Petrick‡
Argonne National Laboratory, Argonne, Illinois

Abstract

This paper presents the ETGAR Liquid Metal MHD Program at Ben-Gurion University. An 8kW$_e$ LMMHD pilot plant is currently under construction and should be completed and tested during 1984. This pilot plant, called ETGAR-3, is a part of the ETGAR Project elaborated under a contract from Solmecs Corporation and supported by the Ministry of Trade and Industry of the State of Israel. Argonne National Laboratory takes permanent and active part in this development. The present paper reflects the iterative process of parametric studies as well as the conceptual and detailed design that led to the finally accepted version of the pilot plant, ETGAR-3. The previously designed versions, ETGAR-1 and ETGAR-2, which have not been chosen for construction at the present time, are also presented in detail.

Introduction

The MHD Laboratory at Ben-Gurion University (BGU) in Beer Sheva, Israel, has been developing LMMHD power systems for direct conversion of heat into electricity since 1977. Initially, this development was intended mainly for utilization of low-grade heat, e.g., solar, geothermal, and waste industrial heat, in relatively small units, up to several

---

Paper presented at the Fourth Beer-Sheva Seminar on MHD Flows and Turbulence, Ben-Gurion University of the Negev, Beer-Sheva, Israel, Feb. 27-March 2, 1984. Copyright © 1985 by the American Institute of Aeronautics and Astronautics, Inc. All rights reserved.
*Project Manager, ETGAR-3.
†Professor, Department of Mechanical Engineering.
‡Director, Fossil Energy.

hundred kilowatts. Later, the scope of the program became wider, and fuel burning plants of 1-10 MW_e size were also considered. In the latter case, much attention was given to cogeneration cases with simultaneous production of electricity and process steam. Low-quality fuels and biomass cases were considered as well. Finally, several megawatt systems with either central solar tower or parabolic linear focus solar collectors were studied.

Since 1980 the LMMHD program has been subsidized, organized, and promoted by Solmecs Corporation, a British-Israeli investment company, which acquired the patents and know-how rights from BGU and committed itself to bring the program through to full commercialization. Solmecs brought additional participants into the program, among them, Argonne National Laboratory (U.S.A.), The City University of London, Kvaerner Engineering (Great Britain), Oxford Instruments, Liquid Metal Laboratory of Nottingham University, FRY Company of the Cookson Group (all Great Britain), and others. These institutions and companies deal with different engineering problems related to the program, study chemical compatibility of liquid metals, and develop conventional and superconductive magnets, etc.

At the present time, the program has reached the stage of assembly of the first semi-industrial, fully engineered pilot plant, ETGAR-3, which is expected to be operational during 1984.

The purpose of this pilot plant is manifold, the major goals being the following:
1) To demonstrate the viability of the concept on a semi-industrial scale.
2) To verify the calculated performance and thus enable reliable predictions for larger-scale units.
3) To study and develop engineering aspects of the system.
4) To establish a practical basis for cost calculations
5) To study off-design performance characteristics.
6) To perform further long-term chemical compatibility studies for the working fluids and confinement materials.
7) To continue two-phase flow studies in different components of the system.

Development of the Program and Its Present Status

The basic concept of an LMMHD energy conversion system, which was accepted in the early stages of the BGU program, was a concept developed earlier by Argonne National Laboratory.[1,2] It is presented in Fig. 1. Here all the heat, which will be converted into electricity, is added to the liquid metal. A volatile liquid is injected into the hot metal and evaporated through direct contact heat exchange. The expanding vapor propels the two-phase liquid metal-vapor mixture across the magnetic field. The vapor is permitted to expand in the generator's channel to the highest void fraction value at which the liquid metal phase is still continuous. After the liquid metal phase breaks into separate droplets, the vapor expands further in a supersonic nozzle, accelerating the liquid metal droplets. The separator is designed to preserve most of the kinetic energy of the liquid metal. This kinetic energy is then recovered into pressure in a diffuser. A simpler but usually less efficient alternative to the nozzle-separator-diffuser system is a liquid metal pump (usually an MHD pump).

The system described performs a modified Rankine-type thermodynamic cycle with the unique property of almost isothermal expansion of the working fluid (Fig. 2). Providing that sensitive heat is regenerated (or alternatively used for process purposes), this cycle is highly efficient. In different variations of this concept, heat can be added to both liquid metal and volatile liquid, or just to the latter.

Between 1978 and 1983 the following studies were performed[3-7]:

Phenomena Studies

1) Boiling of volatile liquid droplets in direct contact with another hot liquid.

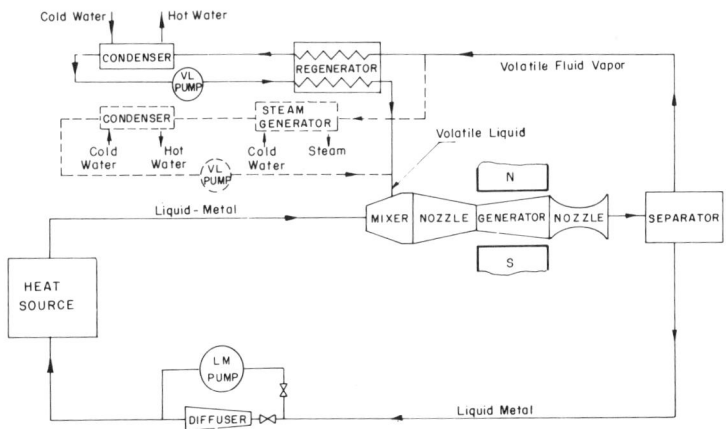

Fig. 1 Basic concept of an LMMHD energy conversion system with two-phase flow generator.

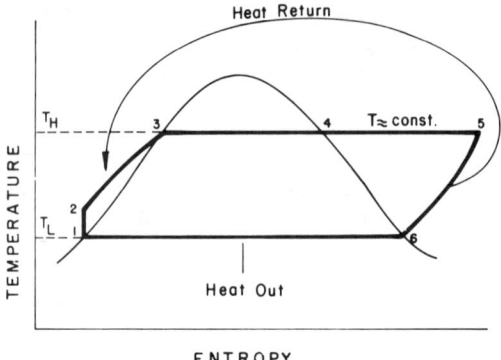

Fig. 2 Typical thermodynamic cycle performed by an LMMHD system (the cycle shown in this figure corresponds to steam as thermodynamic).

2) Influence of liquid droplet initial size on homogeneity of two-phase flow.
3) Two-phase vapor-liquid flow at high void fraction values.
4) Separation of two-phase vapor-liquid flows with maximal preservation of kinetic energy of the liquid.
5) Influence of surface active additives on size and stability of bubbles.

Component Studies

1) Solar collector (flat plate and parabolic with linear focus and spherical dishes) performance with liquid metal circulation through the absorber.
2) Optimization of mixers (design and experimental tests).
3) Optimization of separators (design and experimental tests).
Generator channels have not been tested at BGU since results of an extensive study of this problem performed by Argonne National Laboratory could be used.

Material Studies

1) Compatibility of NaK and neo-hexane up to 200°C.
2) Compatibility of mercury and freon-113 up to 100°C.
3) Compatibility of other fluids (proprietary to Solmecs) up to 300°C.

ETGAR LMMHD PROJECT

Table 1 Temperature ranges of various heat sources

Type of heat source	Highest temperature, K	Size power output level, kW_e
Waste industrial heat	350-700	100-15000
Geothermal	350-550	1000-5000
Fuel combustion	600-1000	1000-150,000
Solar	350-1100	100-3000

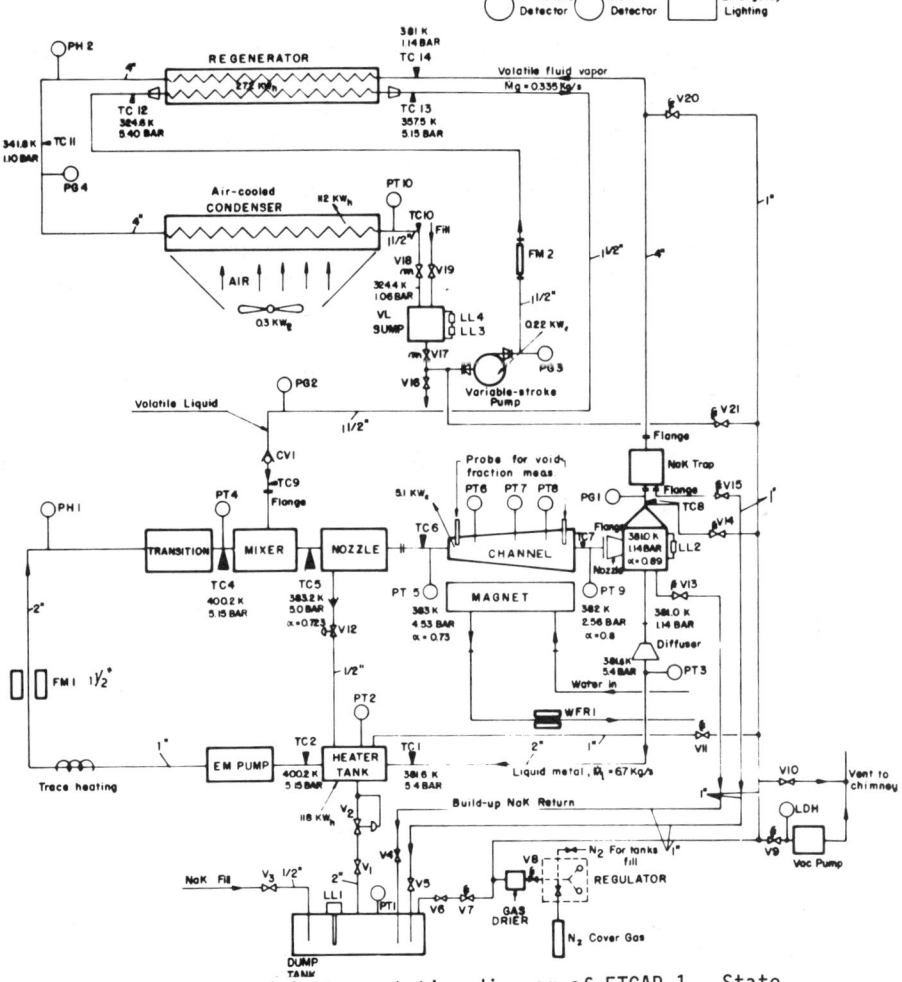

Fig. 3 Piping and instrumentation diagram of ETGAR-1. State points of the cycle are indicated.

Parametric Studies

A computer code for system performance calculation[8], design, and optimization has been developed. A great variety of applications with different working fluids have been studied by use of this code. The main cases considered and temperature and size ranges are given in Table 1.

For most of the cases considered, the calculations showed a very favorable conversion efficiency, superceding the efficiencies that are characteristic for conventional energy conversion methods. System cost analysis performed by Atkins and Partners[9], a leading London engineering firm, demonstrated cost per kW_e installed up to 40% lower than for turbine systems.

In 1981, the first small, but complete, system was assembled. This system, working with mercury and freon-113 at 90°C, produced just about 20 kW of power, but it performed according to predictions and worked accumulatively for more than 3000 hours without failures.

In 1982 the BGU program reached the stage of building a 5-10 kW semi-industrial pilot plant[10]. The design was done in close cooperation with Argonne National Laboratory. The design went through an iterative process in which several system concepts were developed up to working drawings. Finally, the ETGAR-3 version was accepted for construction, while the ETGAR-1 and ETGAR-2 designs are being kept for possible future applications.

ETGAR-1

Concept

ETGAR-1 is based on the two-phase flow MHD generator concept with a supersonic nozzle-separator-diffuser pressure recovery system. An MHD pump is used only for start-up of the system. Working fluids are NaK and neo-hexane. The basic schematic of the system can be understood from the piping and instrumentation P & I diagram presented in Fig. 3.

System Analysis and Established Performance Parameters

The energy conversion process is based on a modified Rankine cycle. The working fluid is a liquid metal and volatile fluid mixture that flows as a homogeneous fluid during the power-producing process in the MHD channel but which separates into its constituent components after expansion is completed. A variety of flow paths can be chosen for the heating process involving mixing before heating, separate heating of the two fluids before mixing, or heating of the liquid metal only, followed by direct contact heat transfer to the volatile fluid in the mixing process. The system layout for direct contact heating of the volatile fluid is shown in Fig. 4, where the principle components are indicated.

For all types of heating process, adiabatic expansion of the fluid mixture in the MHD channel results in a small temperature drop because of the large thermal capacity of the liquid metal relative to the volatile fluid. Thus the classical Rankine cycle for a pure volatile fluid shown in Fig. 5a is modified to that shown in Fig. 5b, and this has been verified by exact calculation. If the volatile fluid is desuperheated in a regenerative heat exchanger, the resulting cycle comes closer to the Carnot ideal that the Rankine cycle, and consequently the system cycle efficiency is improved. The liquid metal content of the mixture must be kept to a minimum compatible with adequate electrical conductivity in order to keep the parasitic losses due to feed pump work as small as possible. This implies the specification of the void fraction at the MHD channel exit, and hence, for exact calculations involving the matching of the cycle to a specified heat source, iterative calculations are required to obtain the required fluid duct inlet conditions. This, together with the complex calculations required for obtaining other thermodynamic property values of the working fluid at the principal state points in the cycle, makes the study of such systems by hand calculation very difficult. A large computer program was therefore prepared to analyze the entire system. This was developed from a computer program for Rankine cycle system analysis that evolved over a period of 10 years of studies in power recovery from low-grade heat sources at The City University, London. The program has been prepared for different types of heat sources: exhaust gases of combustion products, diesel engine exhaust gases, gas turbine exhaust gases, waste hot gases, geothermal water, direct solar insolation, hot condensate, etc. Three types of cooling systems were considered: water cooling from a river or sea where the coolant is not, in turn, cooled and recirculated; water cooling recirculated through a cooling tower; and air cooling systems.

Two types of the LMMHD system were considered:

Solar Heated System. Here the collectors were assumed to heat the liquid metal directly, which in turn transfers the heat to the volatile fluid. Since efficiency is of paramount importance, in order to reduce the cost of the collectors, a conventional water-cooled system together with a cooling tower was included.

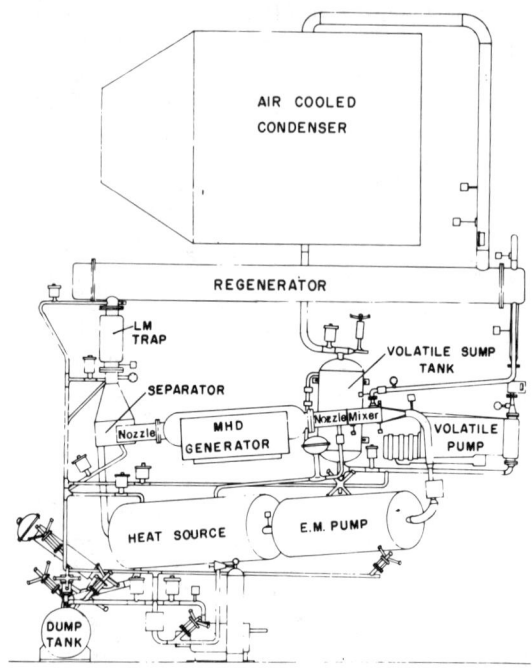

Fig. 4 Layout of ETGAR-1.

Waste Heat and Geothermal Systems. An isothermal heat source such as from wet steam and condensing vapors is essentially identical to that of an infinite heat source such as the sun, and hence the results of the solar studies are directly comparable, differing only in the collector and boiler efficiencies. Where the heat source is a liquid or gas stream, the ideal cycle is not that of Carnot but of a series of infinitesimal Carnot cycles, each at successively lower heat reception temperature. For this case a hot gas stream at 300°C cooling to 170°C was considered as the source. This corresponds to typical values for oil-fired boilers or large marine diesel exhaust systems where the sulphur content of the fuel prevents further cooling of the combustion products because of acid dew point corrosion due to the formation of sulphuric acid in the exhaust gases. For this case, both direct contact heating of the volatile fluid by the metal and separate heating of the volatile fluid and the liquid metal were considered because the latter permits a higher cycle temperature while still withdrawing the maximum possible heat from the combustion products. Water cooling with a cooling tower was the only case considered, although more appropriately for marine applications, for which steam powerplants are used currently, the cooling tower should be omitted.

The basis of the cycle analysis computer code is a suite of subprograms for the estimation of thermodynamic properties of volatile fluids and liquid metals. The volatile subprograms are interchangeable and can be based on either the Martin-Hou vapor equation of state together with suitable correlation for the liquid phase or the Lee-Kesler equations of state, which are valid for all fluid phases but more restricted in the range of fluids for which they are suitable. The liquid metal subprograms are based on curve-fitting equations derived from density and specific heat data, the assumption of liquid metal incompressibility, and the use of generalized thermodynamic property relationships. These subprograms are called from a further set of subprograms to estimate mixture properties and property changes associated with reversible and irreversible adiabatic processes and other processes involving heat transfer. The program for cycle analysis uses these sets of subprograms to calculate the principal state points in the cycle and hence to obtain overall cycle efficiency.

Mass flow rates can then be calculated for any specified power output. The cycle analysis program only requires the input of the desired power output, the channel inlet pressure and temperature together with basic data of the working fluids. Special features of the cycle analysis program can be summarized as follows:

1) The permitted void fraction can be specified either for the MHD channel inlet or the exit.

2) Analysis is possible for all types of fluid mixing paths.

3) The amount of regenerative desuperheating of the vapor leaving the MHD channel can be specified.

4) The cycle analysis is not limited to saturated vapor conditions at the expander inlet but includes superheated and supercritical cycles, as shown in Figs. 5c and 5d.

5) The analysis includes pressure loss factors for the estimation of pressure drops in the heat exchangers and component efficiencies for nonisentropic flow in the nozzles, MHD channel, diffusers, and pumps.

6) The program includes the matching of the thermodynamic cycle to the heat source and sink so that overall conversion efficiencies from solar, geothermal, condensate, and single-phase heat sources can be estimated together with water or air cooling.

7) More advanced versions of the program are now under development to optimize the pressure and temperature choice

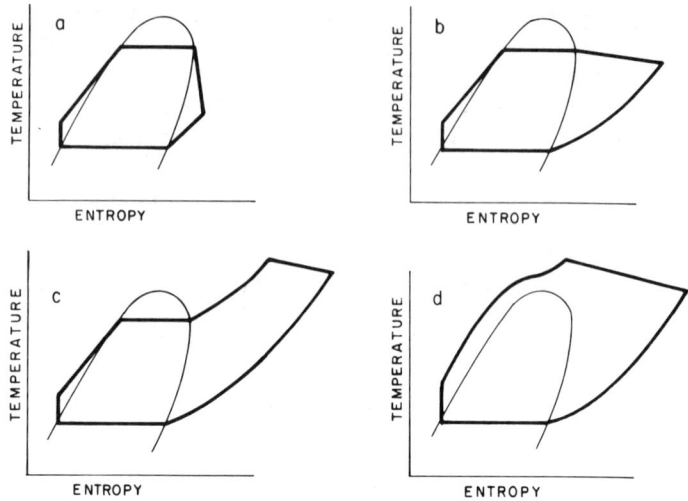

Fig. 5 Classical (a) and modified (b,c,d) organic Rankine cycles.

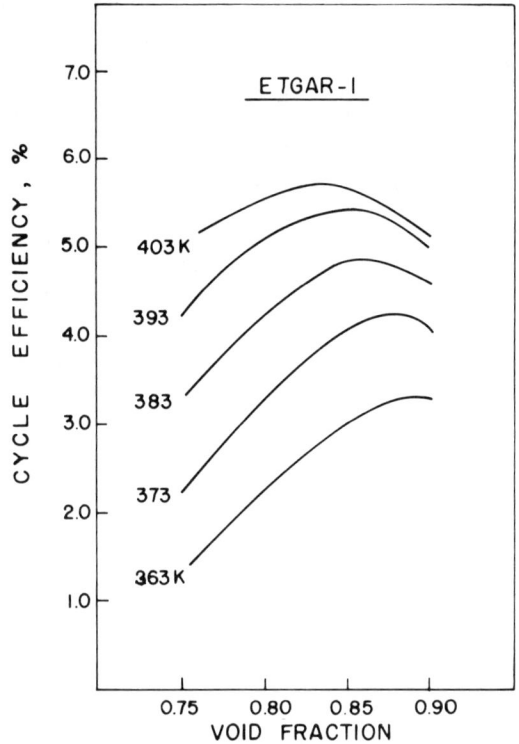

Fig. 6 Calculated cycle efficiencies vs void fraction at channel exit for ETGAR-1 system.

at the MHD duct inlet to obtain the maximum efficiency of conversion of available heat to power. This is especially important for heating media in which there is a large temperature drop associated with the heat-transfer process.

8) The overall system efficiency includes allocation of parasitic losses associated with the circulation of the heating and cooling media through the heat exchangers. The thermodynamic property estimation techniques are quite accurate and, as shown in Refs. 4 and 5, errors in computed values should not be greater than ±1.5%. The overall accuracy of the analysis is therefore mainly dependent on the reliability of the assumed values for the component efficiencies where the greatest uncertainties are in the two-phase nozzle and MHD channel performance.

Volatile fluids (VF) considered during these studies include most common refrigerants, light hydrocarbons, and a variety of fluids with heavier molecular weight and with suitable thermodynamic and thermal stability properties. Liquid metals (LM) considered in the analyses were sodium and sodium-potassium eutectics.

Typical sets of calculated efficiencies and power output curves are presented in Figs. 6-8. In establishing design parameters, not only efficiencies, but also level of thermal input, output voltage, sensitivity to variation of working conditions, etc., have been taken into account. The main parameters established for ETGAR-1 are summarized in Table 2.

It should be noted that, because of safety precautions, it was required that in the whole system, including the condenser, a positive gage pressure exists, and this determined the high sink temperature (51.2ºC) and essentially reduced the cycle efficiency.

The liquid metal MHD computer code of Argonne National Laboratory has also been used for typical cases considered during system analysis. The results confirmed fairly the calculations made at Ben-Gurion University.

ETGAR-2

Concept

ETGAR-2 is a natural circulation system in which a vapor-lift principle is used for recirculation of the liquid metal. There is no acceleration nozzle, and separation is achieved in a gravitational separator. Working fluids are mercury and refrigerant-113.

The concept can be understood from the P & I diagram (Fig. 9).

Table 2 ETGAR-1 design parameters

General parameters:

Liquid metal	Na + K (22%+78%)
Volatile fluid	Neo-hexane
High temperature in cycle	383 K
Low temperature in cycle	324.2 K
Maximum liquid metal temperature	400 K
Heat source pressure	5.4 Bar
Mixer pressure	5.15 Bar
Generator inlet pressure	4.53 Bar
Generator outlet pressure	2.56 Bar
Condenser pressure	1.06 Bar
Thermal input	118 kW
Thermal output	112 kW
Regenerated heat	27.2 kW
LM mass flow rate	6.76 kg/s
VF mass flow rate	0.335 kg/s
Entry nozzle efficiency	0.9
Acceleration nozzle efficiency	0.60
Separator efficiency	0.70
Diffuser efficiency	0.70
Overall cycle efficiency	3.67%
Net power output	4.32 kW

Generator parameters:

Electrode length	0.800 m
Electrode space	0.100 m
Channel inlet width	0.015 m
Channel outlet width	0.020 m
Magnetic field	1.2 T
Inlet void fraction	0.73
Outlet void fraction	0.80
Inlet velocity	20.0 m/s
Inlet mach number	0.4
Load resistance	1.12 Ω
Output current	2188 A
Output voltage	2.33 V
Output power	5.1 kW
Conversion efficiency	0.65

System Analysis and Established Performance Parameters

The computer code described above has been adjusted to calculate the thermodynamic cycle and components for the systems of the ETGAR-2 type. Gravity forces and performance of the vapor-lift have been taken into consideration.

Several typical examples of calculation results are presented in Figs. 10 and 11. The main parameters established for ETGAR-2 are summarized in Table 3.

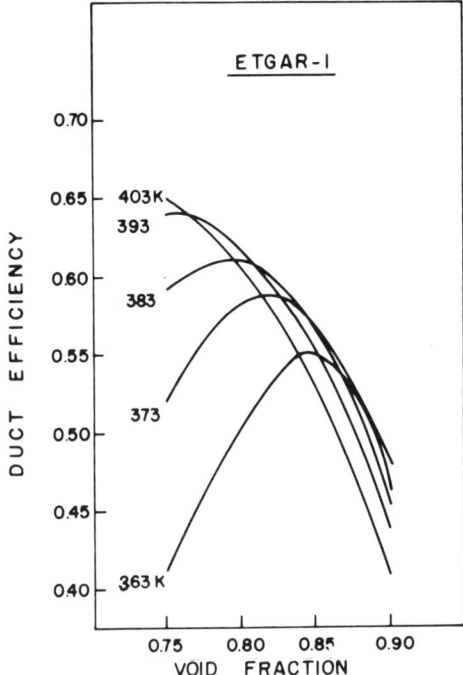

Fig. 7 Calculated MHD duct efficiency vs exit void fraction for ETGAR-1 system.

ETGAR-3

Concept

ETGAR-3, schematically illustrated in Fig. 12, is a very simple direct conversion natural circulation power system. The basic system consists of two pipes (an upcomer and downcomer) connected at the bottom with a crossover pipe and with a simple tank, the separator, joining them at the top. A mixer is located at the bottom of the upcomer, and an MHD generator, from which electrical power is extracted, is located in the downcomer or lower crossover pipe.

Operation of the system is as follows: A vapor or gas (the thermodynamic working fluid) is introduced into the mixer at the bottom of the upcomer at an appropriate temperature and pressure. A two-phase fluid of lower density is created. As the two-phase fluid flows to the separator, the gaseous phase undergoes an expansion from the high pressure of the mixer to the low pressure in the separator, accelerating the fluid and lowering its density. The gaseous phase (working fluid) is disengaged in the separator plenum

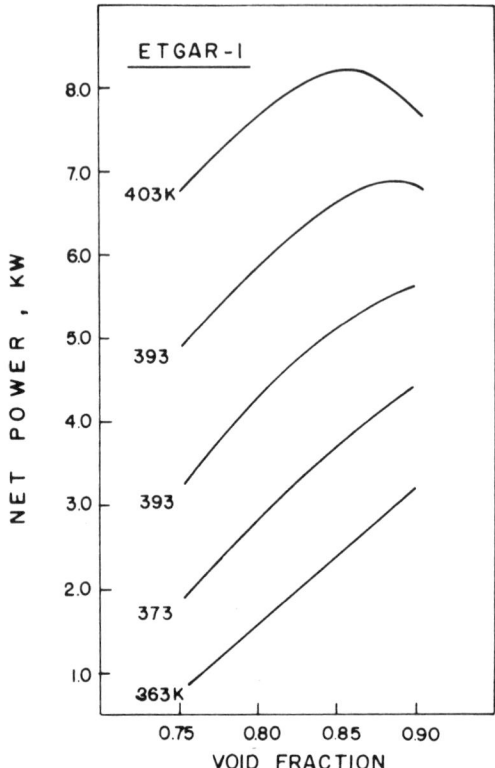

Fig. 8 Calculated net power vs channel exit void fraction for ETGAR-1 system.

by buoyancy (gravity) forces, thus producing a single-phase flow return into the downcomer. The pressure differential causes the liquid metal (the electrodynamic fluid) to circulate in the system; as the single-phase liquid metal passes through the MHD generator, an electrical potential is generated and power is extracted. The flow rate in the loop adjusts to balance the density differential between the upcomer and downcomer with the frictional and acceleration flow losses and the power extracted from the MHD generator. The ETGAR-3 system, therefore, converts pressure head (potential energy) to electrical energy via the single-phase flow constant velocity MHD generator.

The accepted ETGAR-3 concept minimizes many of the technical problems that are characteristic of LMMHD systems with two-phase generators. The specific benefits that accrue from this concept follow:

1) Elimination of the critical, high-loss components in the two-phase flow generator MHD system, namely, the two-phase nozzle, separator, and diffuser.

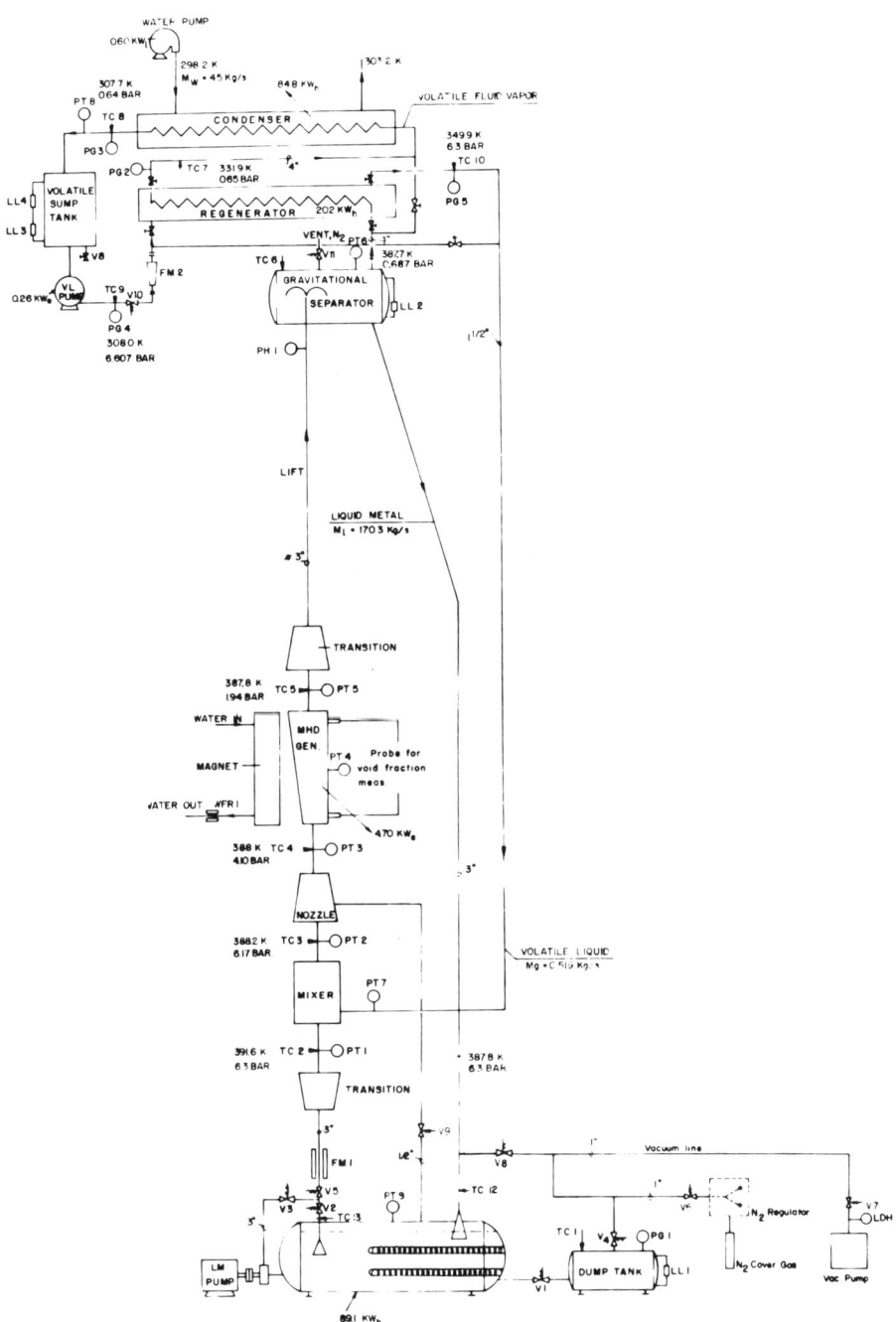

Fig. 9 Piping and instrumentation diagram of ETGAR-2. State points of the cycle are indicated.

Table 3   ETGAR-2 design parameters

General parameters:

Liquid metal	Mercury
Volatile fluid	Refrigerant-113
High temperature in cycle	388 K
Low temperature in cycle	307.7 K
Maximum liquid metal temperature	391.6 K
Heat source pressure	6.3 Bar
Mixer pressure	6.17 Bar
Generator inlet pressure	4.10 Bar
Generator outlet pressure	1.94 Bar
Condenser pressure	0.64 Bar
Thermal input	89.1 kW
Thermal output	84.8 kW
Regenerated heat	20.2 kW
LM mass flow rate	170.3 kg/s
VF mass flow rate	0.519 kg/s
Overall cycle efficiency	4.04%
Net power output	3.6 kW

Generator parameters:

Electrode length	0.400 m
Electrode space	0.125 m
Channel inlet width	0.031 m
Channel outlet width	0.050 m
Magnetic field	1.8 T
Inlet void fraction	0.522
Outlet void fraction	0.70
Inlet velocity	6.70 m/s
Inlet mach number	0.60
Load resistance	0.394 Ω
Output current	3760 A
Output voltage	1.25 V
Output power	4.70 kW
Conversion efficiency	0.66

2) Facilitation of ultrasimple startup, operation, and shutdown.

3) Elimination of the two-phase flow limitation of MH generator performance and, hence, overall system performanc

4) An 80% to 82% single-phase flow MHD generator can be achieved and designed with confidence from the data base that exists today.

5) The single-phase flow through the MHD generator facilitates the use of low-field, low-cost magnets.

6) The high electrical conductivity single-phase flow through the generator makes possible the use of an ac induction generator to produce ac power directly -- eliminating

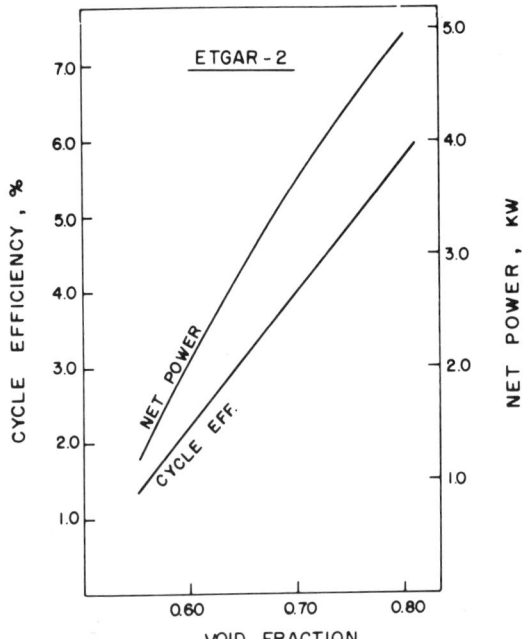

Fig. 10 Calculated cycle efficiency vs void fraction at channel exit for ETGAR-2 system.

the costly inverter and replacing the magnet with a conventional type of coil system found in electrical motors.

7) The overall system can be designed to have low kinetic energy (frictional) losses, thus maximizing performance potential.

8) Performance of the system depends little on void fraction value, and, most importantly, one can obtain the same overall efficiency at low void fractions as well as high void fractions in contrast with the two-phase generator cycle (where for low void fraction values the power needed for pumping the liquid metal becomes unacceptably high).

9) Electrical parameter fluctuations in the generator are eliminated.

10) The system can be operated as a Rankine or Brayton cycle and can be readily adapted to various temperature ranges (applications) by proper selection of working fluids.

System Analysis and Established Performance Parameters

The ETGAR-3 system, as described, is a natural circulation system in which fluid flow is derived through a density

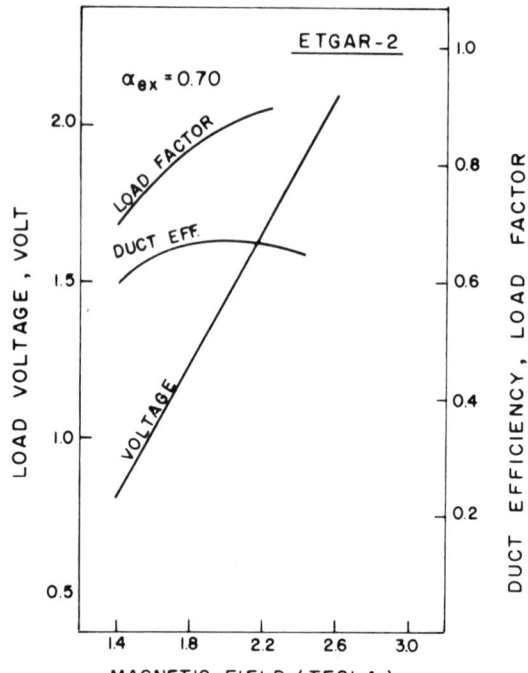

Fig. 11 Calculated voltage, MHD duct efficiency, and load factor vs magnetic flux density for ETGAR-2 system.

potential existing between the vertical segments of the system. The analysis of such a natural circulation system is complicated by the interrelationship existing between the various sytem parameters, such as power void fraction, weight fraction, circulation velocity, heat input, and output. A thermodynamic-hydrodynamic computer code was developed for calculating the ETGAR-3 performance in order to carry out the parametric and sensitivity studies for developing a conceptual design of upscaling the system. The basis of the cycle analysis computer code is a suite of subprogram for estimation of thermodynamic properties of volatile fluids and liquid metal and principal state points, hydrodynamic for calculating friction and acceleration losses in the loop, and magnetohydrodynamic equations for calculating power output and generator performance. The computer code simultaneously solves a set of differential equations in the riser for acceleration, two-phase friction losses using different models, and slip ratio using different models for two-phase flow in vertical pipes.

The cycle analysis program requires operating temperatures, system geometry, magnetic field, load factor, and

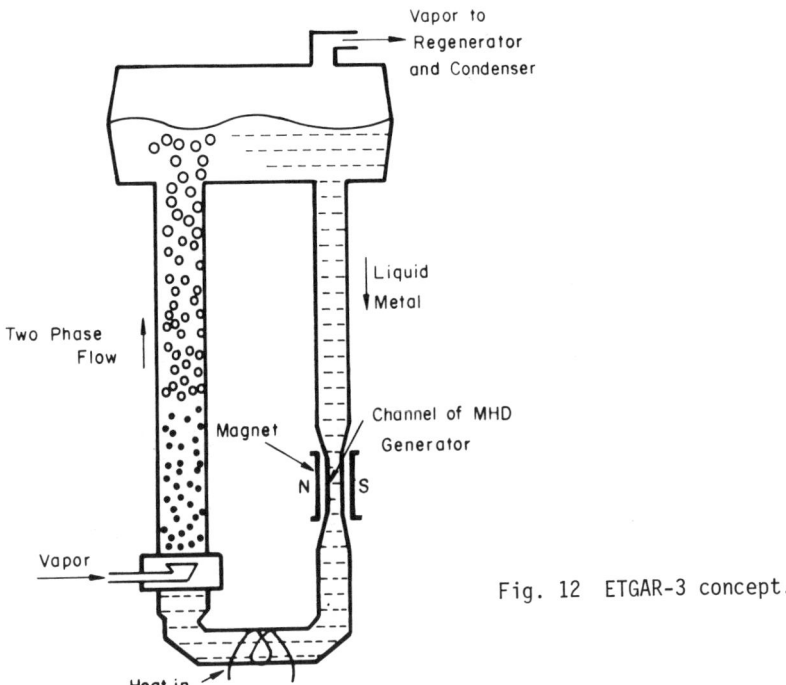

Fig. 12 ETGAR-3 concept.

average void fraction in the riser and plots net power output, output voltage, output current, mass flow rates, void fraction, and slip ratio as functions of height of the riser, heat input, heat output, generator efficiency, and overall efficiency.

In establishing design parameters, not only efficiencies, but also level of thermal input, output voltage, sensitivity to variation of working conditions, etc., have been taken into account. The main parameters established for ETGAR-3 are summarized in Table 4.

ETGAR-3 has been designed as a complete power system both to demonstrate the system's unique features and performance potential and to function as a flexible test facility to provide the basic engineering information needed to proceed with confidence in scale-up and application studies. Thus, it represents the intermediate stage between laboratory scale experiments and an industrial scale demonstration project (Fig. 13).

ETGAR-3 has been heavily instrumented to facilitate the conduct of a comprehensive test program that will achieve the goals outlined above. The instrumentation has been designed to produce data that can be used to evaluate the

Table 4 ETGAR-3 system parameters

Electrical power output	8 kW
High temperature in cycle	$423°K$
Low temperature in cycle	$338°K$
Overall efficiency	8.2%
Mixer pressure	4.9 Bar
Thermal input	97.5 kW
LM mass flow rate	435.2 kg/s
Steam mass flow rate	0.0303 kg/s
Average void fraction in upcomer	0.4
Average slip ratio in upcomer	1.42
Effective height of the systam	7.5 m
Upcomer diameter	20.32 cm
Downcomer diameter	20.32 cm
MHD generator channel width	6 cm
MHD generation electrode spacing	15 cm
Magnetic field	0.73 T
LM	Proprietary
VL	Proprietary

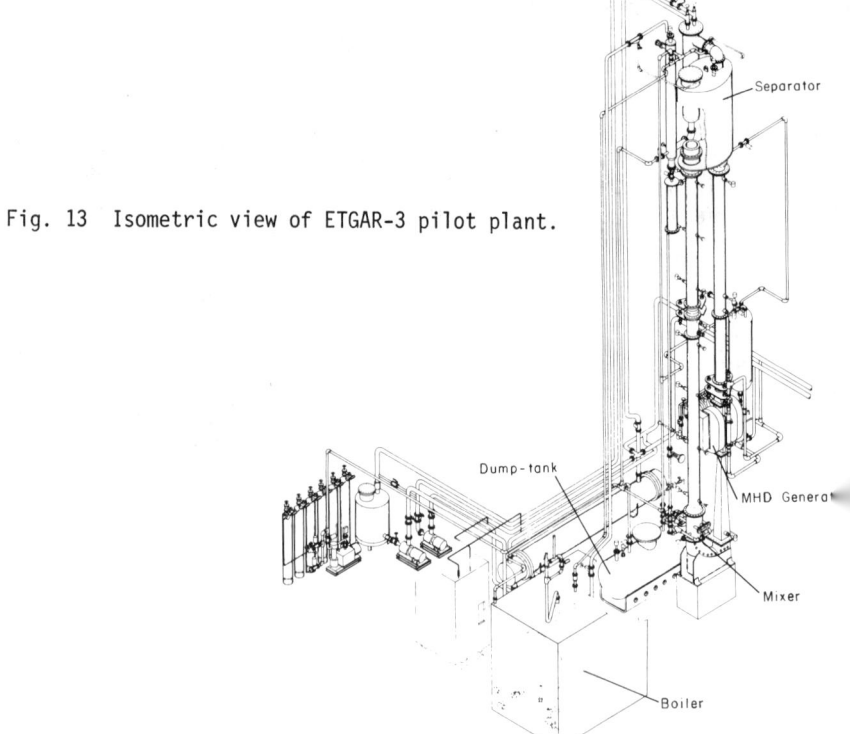

Fig. 13 Isometric view of ETGAR-3 pilot plant.

performance of every major system component as well as the overall system. Temperature and pressure measurements will be made before and after every component and at other strategic positions specified by the detailed computer simulation analysis in order to provide data needed for analytical and design validation. Special emphasis is being placed on developing those data that are needed to resolve the only major technical issue in the system, namely, the variation of slip ratio in the upcomer as a function of mixture quality, flow rate, pressure, and temperature.

In addition to temperature and pressure measurements, vapor and liquid flow rates, heat flows (in and out), and power output will be measured. Also, detailed electrical measurements on the generator will be made to facilitate validation of generator design models. Long duration testing will be conducted to demonstrate material compatibility. In addition, transient testing will be conducted to provide data for evaluation of load following characteristics and potential system failure modes.

In parallel with the construction and assembly of the ETGAR-3 pilot plant, an extensive supporting research program has been elaborated. This program includes the following:

1) Two-phase vapor-liquid metal studies (main problem: slip values).
2) Chemical compatibility studies (performed by Nottingham University).
3) Enhancing of bubble stability.
4) Studies of electrode contact resistance and its possible reduction (performed by Nottingham University).
5) Study and design of superconductive magnets for LMMHD application (performed by Oxford Instruments).
6) Expansion of the computer code including costs and optimization routines (performed by City University, London).
7) dc to ac current invertor adjustment to LMMHD systems.
8) Separator studies have been performed with water-air flows.

For design verification, a complete smaller size system has been assembled. Components of this system, called ER-4, can be easily modified and exchanged. ER-4 works with steam and mercury at 165°C and produces 0.45 kW_e at the design point.

Acknowledgments

This work was subsidized by Solmecs Corporation, Jerusalem, and by the Ministry of Trade and Industry of the

State of Israel. The authors are grateful to Zvi Raveh of Bateman Engineering Ltd. for his most valuable participation in the design and building of ETGAR-3.

References

1. Petrick, M., "MHD Generators with Two-Phase Liquid-Metal Flows", Electricity from MHD, International Atomic Energy Agency, Vienna, Vol. II, (1966), p. 889.

2. Petrick, M., et al., "Experimental Two-Phase Liquid Metal Magnetohydrodynamic Generator Program", Argonne National Laboratory, Argonne, IL., ANL/MHD-78-2, 1978.

3. Branover, H., Borda, I., El-Boher, A., Leitner, A., "On the Possible Use of MHD Generators in Solar Energy Systems", MHD Flows and Turbulence, Proceedings of the Second Bat-Sheva International Seminar, Beer Sheva, Israel Universities Press, 1979.

4. Pierson, E.S., Branover, H., Fabris, G., and Reed, C.G. "Solar-Powered Liquid Metal MHD Power Systems", Transactions of the ASME, Mechanical Engineering, Vol. 102, 1980, 79-WA/SOL-22, p. 32.

5. Branover, H., El-Boher, A., Yakhot, A., "Testing of a Complete Closed-Cycle Two-Phase Liquid Metal MHD Power System", 19th Symposium, Engineering Aspects of Magnetohydrodynamics, Tullahoma, Tenn., 1981.

6. Yakhot, A. and Branover, H., "An Analytical Model of a Two-Phase Liquid Metal Magnetohydrodynamic Generator", The Physics of Fluids, Vol. 25, No. 3, 1982, pp. 446-451.

7. Branover, H., El-Boher, A., Sukoriansky, S., et al., "Development of a Low Temperature Liquid Metal MHD Small Scale Pilot Plant", 21st Symposium on Engineering Aspects of Magnetohydrodynamics, Argonne National Laboratory, Argonne, Ill., 1983, pp. 7.1.1.-7.1.13.

8. Smith, I.K., Branover, H., and Yakhot, A., "Performance Studies of Two-Phase Low Temperature Liquid-Metal MHD Systems," 20th Symposium Engineering Aspects of Magnetohydrodynamics, University of California Irvine, California, 1982.

9. "Estimated Costs of LMMHD and Turbine Schemes", Report 51413/EIP/RFM/AS, W. S. Atkins and Partners, Engineering Consultants, London, 1981.

10. Branover, H., El-Boher, A., Sukoriansky, S., et al., "ETGAR Liquid Metal MHD Program at Ben-Gurion University", 22nd Symposium, Engineering Aspects of Magnetohydrodynamics, Mississippi State University, Starkville, Miss., 1984.

Investigation of a Lithium-Caesium Faraday Converter

F. Joussellin,* R. Laborde,* A. Alemany,† and J.P. Thibault‡
Institut de Mecanique de Grenoble, Saint Martin d'Heres, France
and
F. Werkoff†
Centre d'Etudes Nucleaires de Grenoble, France

Abstract

The Faraday conversion is mainly studied at low-temperature levels. Another application is revisited here for the temperature range of 750 to 1500 K. Here, the Lithium (liquid)-Caesium (vapor) pair is chosen. This paper deals with the evaluation of the converter dimensions, mass and global efficiency.

Introduction

Faraday conversion, also called L.M.M.H.D. (liquid metal magnetohydrodynamic), could be used with various thermal sources. Among them, the possibility of using a nuclear reactor has been considered for a long time[1]. About 20 years ago a study on space application was done, but during the sixties the available boosters did not allow the launching of such reactor-converters into space, so the project was stopped. A few years ago another concept was proposed using the coupling of a tin-water loop with a terrestrial fast breeder reactor[2].

Today the situation, for the space application, is quite different due to the success of both U.S. shuttle and the European Ariane launchers: thus the production of elec-

---

Paper presented at the Fourth Beer-Sheva Seminar on MHD Flows and Turbulence, Ben-Gurion University of the Negev, Beer-Sheva, Israel, Feb. 27-March 2, 1984. Copyright © 1985 by the American Institute of Aeronautics and Astronautics, Inc. All rights reserved.
*Research Engineer.
†Research Head.
‡Research Associate.

tricity in space from a nuclear reactor is becoming realistic. A project of the National Center of Spacial Research (C.N.E.S. Centre National d'Etudes Spatiales) is studying the interest in an orbital transfer vehicle using electrical propulsion. As a matter of fact, outside the atmosphere it is possible to take advantage of the high specific impulse of an electrical thruster (ten times greater than that of a chemical one). For this application, calculations on the ARIANE project show that the carrying capacity is increased compared to a classical third stage.

Only a thrust of few Newtons is needed to transfer satellites between low and geosynchronous orbits in a few weeks. This requires a large electrical power source (100 to 500 kW). The cost of such a generator is more than high enough to prove the interest in a reusable orbital transfer vehicle called ERATO (Fig. 1). The heat source could be a small liquid metal fast reactor (L.M.F.R.) of about 1 to 1.5 MW of thermal power which would operate at a high temperature (1100 to 1500 K)[3]. The thrust would be obtained by using a set of electrical thrusters (ionic or M.P.D.). It would need 100 to 300 KWe delivered at 1000 to 1500 V dc.

For the energy conversion between these two elements, four possibilities are studied jointly by three partners : CNES, the French Atomic Energy Commission (C.E.A.) and the National Scientific Research Center (C.N.R.S.). The main constraint is the global system weight. The study frame for the conversion system is the following :

1) the hot temperature : T_h = 1373 K,
2) the cold temperature : T_c = 773 to 900 K,
3) the electrical power range : P_{el} = 100 to 300 kW ;
   U = 1500 V dc ; failure protection by several units.

Objectives

This paper is a preliminary global calculation of a lithium-caesium Faraday converter. It differs from the U.S. concept by the fact that in the loop the separation occurs after the generator[1]. We have chosen a constant velocity flow into the generator, and we compare a pure liquid generator to one using a two-phase flow.

We have developed a code based on global balance equations which evaluates one by one size and mass of each component of the loop. Loss evaluations are based on experiments [4,5,6].

The choice of two components of the mixture which are immiscible and chemically compatible at 1373 K, leads to the Lithium (liquid) and Caesium (thermodynamic condensable fluid) pair[7]. The Caesium enables the converter to work at

LITHIUM-CAESIUM FARADAY CONVERTER

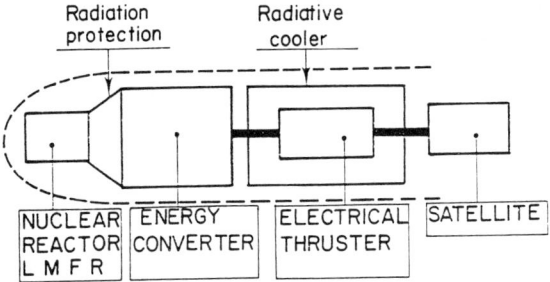

Fig. 1 Orbital transfer vehicle : ERATO.

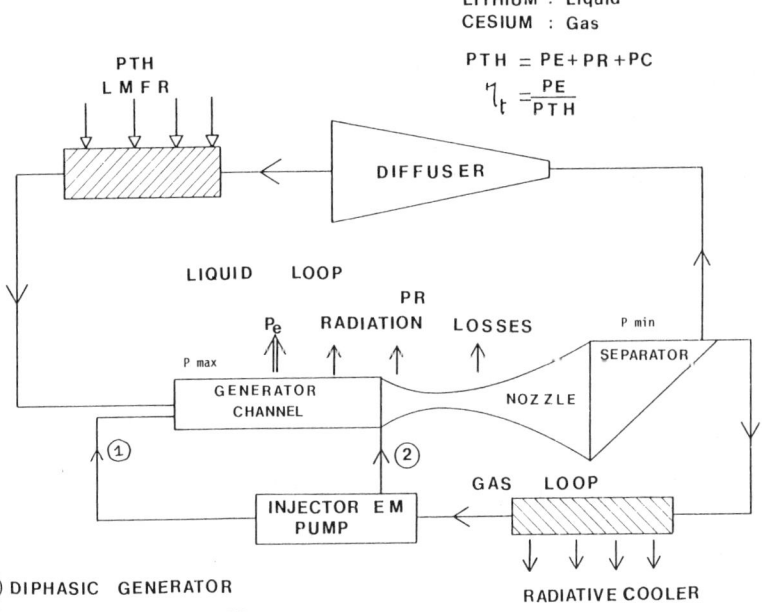

① DIPHASIC GENERATOR
② MONOPHASIC GENERATOR

Fig. 2 Converter loop.

low pressures (between 0.14 and 16 bars). The Lithium has the advantage of weighting half than water.

Faraday Loop Characteristics

The components of the loop are divided into four parts (Fig. 2) : 1) Generator channel : Liquid metal mechanical energy is transformed into ac or dc power depending on the nature of the applied magnetic field (static or travelling). 2) Diphasic motor : The liquid metal is accelerated through a supersonic nozzle by the gas expansion. The gas, using thermal energy contained in the metal, creates mechanical energy. To separate the two phases in weightlessness, an inclined plate type separator is used to preserve the liquid mechanical energy. 3) Liquid loop : Only the liquid flows into the heater which restores the thermal energy used during gas expansion or lost through radiation. A diffuser is needed to return to the pressure and velocity of the flow at the generator inlet section ; thus the liquid loop is closed. 4) Vapor loop : The radiator has a maximal surface of 90 m^2 to cool and condensate the vapor. The two extremum temperatures of the cycle determine two vapor pressure: the injection pressures $P(T_H)$ and the separation pressure $P(T_C)$ as one can see on the thermodynamic cycle of the gas (Fig. 3).

Vapor and Liquid Mass Flow Rate Calculations

The vapor flow rate Q_g calculation needs an iterative process : at first we choose a starting value of this flow rate which defines the gas volume and gives the kinetic energy for the entire system, and all the component losses included in the vapor flow rate calculation (Fig. 4).

The liquid mass flow rate Q_L is deduced from an energy balance between the input and output of the generator. At constant velocity, the pressure drop is determined by thermodynamic considerations. The liquid to vapor flow rate ratio obtained is about 100. Diphasic flow modelization depends on the void fraction. The apparent density and dynamic viscosity of the two-phase flow are modeled with the classical proportional laws of the homogeneous model[8]. The apparent electrical conductivity of the two-phase flow is modeled with an empirical expression proposed by Petrick and Lee[9].

Generator Families

The first family is determined by the flow nature :
1) The monophasic type, where only liquid metal flows into the generator, has the advantage of reducing the inductor mass. The gas injection takes place at the generator output; For this family, the pressures are determined as follows :

$$\left[\begin{array}{l} P_{gene-output} = P(T_H) \\ \\ P_{min} = P(T_c) \end{array} \right. \quad (1)$$

LITHIUM-CAESIUM FARADAY CONVERTER 439

P_{max} is choosen as a compromise between technological problems due to the high pressure and generator efficiency. 2) The diphasic type, where both liquid and vapor flow into the generator, corresponds to the following conditions :

$$\left[\begin{array}{l} P_{max} = P(T_H) \\ P_{min} = P(T_c) \end{array} \right. \quad (2)$$

The second family is determined by the nature of the electrical current : 1) A conducting M.H.D. converter using a constant magnetic field collects the induced current with electrodes. It produces high dc current and low voltage. So, 1500 V dc requires a heavy power transformation system, which leads us to consider only the following inductor : 2) An induction M.H.D. converter acts like an inverted induction pump[10]. A travelling magnetic field is generated by a field inductor in which induced alternating current is collected by direct magnetic coupling, which means that there is no direct contact between fluid and load. The produced sinusoïdal field moves at synchronous velocity V_c, and the fluid with velocity V_g : if V_g is greater than V_c, mechanical energy is converted into electricity, the inductor consumes a great amount of reactive power which requires a large power supply. This induction converter is lighter than the conducting converter even in the case of monophasic flow.

Moreover for a three phase inductor, to obtain 1500 V dc, only 600 V ac are needed from each phase.

All electrical parameters in an asynchronous generator depend on the slip s which gives the working mode. The power delivered by the converter, P_{el}, is defined by thrusters power, P_{up}, Caesium pump power, D_{pp}, and electrical system losses D_{ptr}. The hydraulic power, P_h, needed to obtain, P_{el}, depends on Joule losses in the coils, D_{pi}, and in the flow, D_{pv}, and iron losses, D_{pf} : this gives the electrical efficiency, η_{el}.

$$s = (V_g - V_c) / V_c \quad (3)$$

$$P_{el} = P_{up} + D_{pp} + D_{ptr} \quad (4)$$

$$P_h = P_{el} + D_{pi} + D_{pv} + D_{pf} \quad (5)$$

$$\eta_{el} = P_{el} / P_h \quad (6)$$

By using an equivalent electrical scheme (Fig. 5), these machines are modeled by phase. The inductor winding is represented by an inductance (L). The inductor working mode needs to consume reactive power delivered by capacitors the values of which (C) are determined from the inductance-frequency product. This system is self-exciting, the frequency corresponds to the resonnance frequency of the L - C circuit[6].

The magnetic Reynolds number, R_m, represents the ratio between the active power P_{el} absorbed by the load and the reactive power Q. In addition, it is the ratio of inducting B and induced b magnetic field :

$$R_m = \mu_o \sigma V_c / k \qquad (7)$$

$$P_{el} / Q \# b/B \# s R_m \qquad (8)$$

where k is the wave number and μ_o is the liquid magnetic permeability.

Generator friction losses depend on the Hartmann number[4]. In our case, it is very large : M = 200 to 500

$$M = B e \sqrt{\sigma_{ap} / 4 n_{ap}} \qquad (9)$$

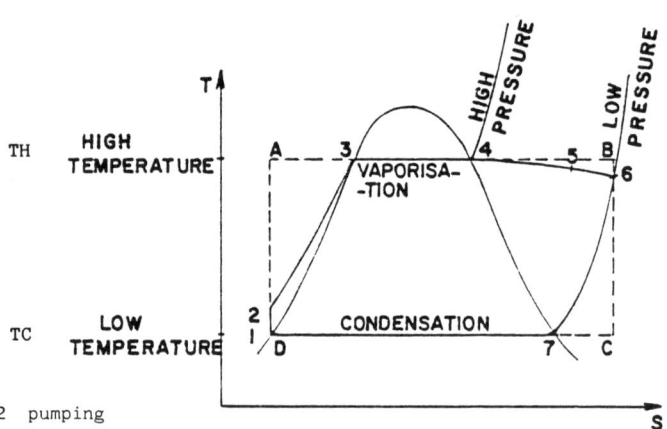

1 - 2 pumping
2 - 3 heating
3 - 4 vaporisation
4 injection point
4 - 6 vapor expansion
6 - 7 cooling
7 - 1 condensation

Fig. 3 Theoretical thermodynamic Caesium cycle.

where B, e, σ, η are respectively the magnetic field (B∿1T) the air gap (e ∿ 1.5 cm), the apparent conductivity and dynamic viscosity.

The gas to liquid slip ratio g is taken as one because of the high working velocity.

$$g = V_{liq} / V_{gas} \# 1 \qquad (10)$$

Nozzle

Gas expansion developed in the accelerator gives the liquid mechanical energy for all or part of electricity generation depending on the monophasic or diphasic nature of the converter ; it is also used to compensate all the losses. The nozzle is composed of a convergent and a supersonic divergent duct. Because of the relatively unknown nozzle characteristics, we have taken a variable efficiency between 50% and 90% waiting experimental data.

Throat parameters are determined by the integration of the momentum equation in the accelerator between input and throat. If the pressure at the throat P_t is known, the sonic velocity can be approximated as the theorical diphasic sonic velocity, a_d :

$$a_d^2 = P_t (1 - \beta P_t) / \rho \alpha (1 - \alpha) \qquad (11)$$

$$\rho_g P = (1 - \beta P) RT \qquad (12)$$

where α is the void fraction, β is a virial coefficient of the gas equation and R is the gas constant.

Separator

The separator proposed ia an inclined plate type that collects the high void fraction flow leaving the nozzle and delivers the liquid to the diffuser as a thin film at a high velocity[12]. The specific mass ratio between liquid and gas is about 2500 and the void fraction is higher than 90% at the separator input ; the two-phase flow is composed of droplets dipersed in gas. The droplets have higher inertia than the gas which follows the wall contours, so that they are projected on the inclined plate where they form a film. The viscous losses come mainly from the liquid. The losses can be calculated by a model of a turbulent boundary layer (for a Reynolds number Re of about 10^7 and for a fixed angle of divergence).

At the separator exit, the liquid rate in the vapor loop could be 2%, but we consider a negligible vapor rate in the liquid loop.

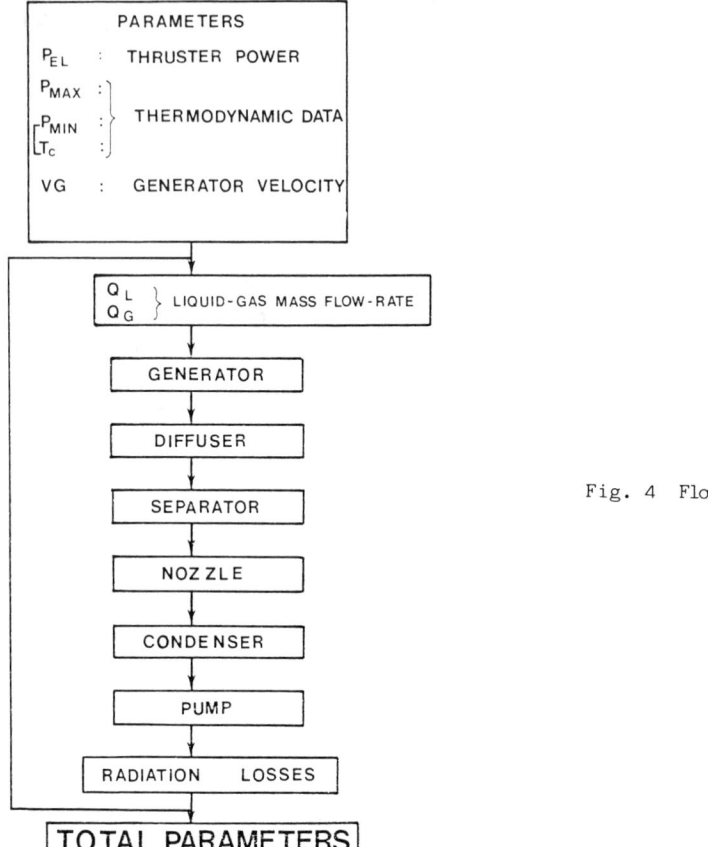

Fig. 4 Flow-chart.

Fig. 5 Induction machine.

Liquid Loop

After the separation, the liquid flow is at low pressure (< 1 bar) and high velocity (100 m/s). The purpose of the diffuser is to transfrom these conditions to those of the generator input, lower velocity (30 to 50 m/s) and higher pressure (16 or 30 bars), by a divergent section. The optimum efficiency of the diffuser is obtained by a compromise between the divergence angle and the friction losses [13].

Vapor Loop

After the vapor has been cooled into a radiator from 1373 to 573 K, the vapor is liquified in weightlessness by a condenser similar to the one of heat pipes[14]. Several tubemodules in Niobium enable the droplets to be pumped by capillarity forces. For a particular case based on a vapor flow rate Q_{go} of 0,8 kg/s, the condensor weight is 62 kg.

An electromagnetic pump is used to raise the Caesium liquid pressure to the injection pressure 16 bars. For a Caesium mass flow rate Q_{go}, the mechanical power is about 700 W. Even for a low pump efficiency of 10%, the electrical power is reasonable. In that case the pump mass is 130 kg.

Power and losses are taken to be proportional to the Caesium flow rate Q_g. The total thermal power lost Q_c at the cold source is then:

$$Q_c = 480 \; Q_g / Q_{go} \quad (kw) \tag{13}$$

Total Parameters

The global converter efficiency is the ratio between the electric power delivered to the thrusters P_{el} and the thermal power given by the nuclear source P_{th} which represents the addition of all the output powers. All friction losses contribute to heat the mixture :

$$P_{th} = P_{el} + Q_c + H_r \tag{14}$$

where H_r are radiation losses calculated by the Stephan-Boltzmann law with an emissivity ε of a polished metal :

$$\varepsilon = 0.05$$

The global efficiency is then :

$$\eta_G = P_{el} / P_{th} \qquad (15)$$

The total mass M_t takes into account all loop components : channels, inductor, capacitors, liquid and vapor mass. We have not included the radiator and the heat exchanger.

Results Presentation

Table 1 (named Reference Case on the figures), shows that the monophasic converter presents the advantages of being lighter and more efficient than the diphasic converter for a lower velocity in the generator channel. The variation studies have been limited to the monophasic converter which proposes a better known flow in the generator channel.

Variation Studies for Mass Optimisation of a Monophasic Generator

The four following parameters have been varied P_{up} is the electrical power delivered to the thrusters, P_{max} is the high pressure at the generator input, P_{min} is the low pressure at the nozzle exit, V_g is the generator velocity.

Generator velocity variation (V_g). For a quasi-constant efficiency, raising the velocity from 30 to 50 m/s leads to an increase in mass 1.1 to 1.4 t (cf. Table I - monophasic). For a similar electrical efficiency, a high velocity leads to a longer inductor : $L_g = p\ V_g/F\ (1 + s)$; where F is the frequency.

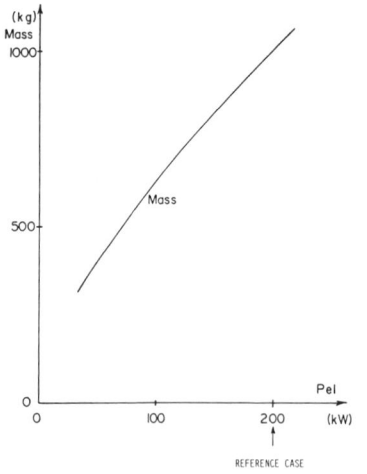

Fig. 6 Variation of the total mass with the electrical output power.

Table 1 Comparison of monophasic and diphasic converters

PARAMETERS	MONOPHASIC		DIPHASIC
P_{max}, bars	35		16
$P_{gene-outlet}$, bars	16		4
P_{min}, bars	0.52		0.52
T_h, K	1373		1373
T_c, K	880		880
V_G, m/s	50	30	50
Q_G, kg/s	1.3	1.2	1.2
Q_L, kg/s	109	63	72
P_{el}, kW	217	215	213
η_{el}	0.91	0.86	0.76
F, Hz	100	100	70
p	8	8	4
L_G, m	1.6	1	1.2
Δ HGenerator, kW	64	16	34
Δ HNozzle, kW	43	40	27
Δ HSeparator, kW	60	62	34
Δ HDiffuser, kW	31	27	10
Mass, kg	1,400	1,100	1,500
η_g	0.21	0.21	0.21
P_{th}, MW	1	0.936	0.920

Electrical power variation (P_{up}). For a nearly-constant efficiency, the mass M_t variation for a converter divided into several units leads to the following results: $M_t = 1.10^3$ kg for producing 1 x 200 kW and $M_t = 1.610^3$ kg for producing 4 x 50 kW. So, four units are 60% heavier than one element (Fig. 6).

High pressure variation (P_{max}). For a high pressure of 30 bars, there is an optimum of efficiency 21% though the mass optimum is at 35 bars (Fig. 7).

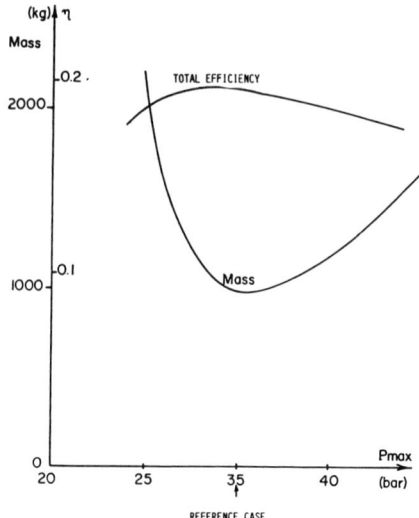

Fig. 7 Variation of the total mass and efficiency with the maximum of pressure.

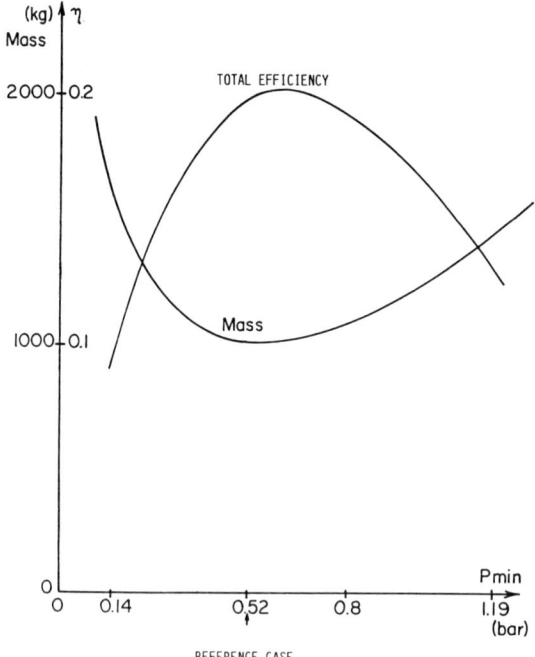

Fig. 8 Variation of the total mass and efficiency with the minimum of pressure.

Low pressure variation (P_{min}, T_c). These graphs show that the best low pressure is about 0.52 bar which corresponds to 880 K. The total mass M_t and efficiency are respectively 1.1 t and 21% (Fig. 8) :

Inductor mass :	710 kg
Capacitors mass :	100 kg
Duct and Lithium mass :	100 kg
Caesium pump mass :	200 kg
M_t :	1 100 kg

Conclusions

These numerical results show the possibility of finding a minimum mass that satisfies the constraints. The specific converter mass is 5 kg/kWe including the inductor, the channel, the Lithium and Caesium mass, the Caesium e.m. pump and the condenser.

To improve the accuracy of this basic study, we are building two experimental setups : 1) The first one concerns studies on injection, separation and interfacial slip in a water-air loop composed of nozzle and an inclined plate separator. 2) The second one concerns the transformation of mechanical to electrical power in an induction generator with a mercury loop.

References

[1] Elliott D.G., Cerini D.J., and Weinberg E., "Liquid-Metal Power Conversion", Space Power Syst. Eng., Vol. 16, 1966, pp. 1275-1298.

[2] Chow S., "Application of Liquid Metal Magnetohydrodynamic Generator to Liquid Metal Fast Breeder Reactors", Georgia Institute of Technology Thesis, Geor., USA, 1981.

[3] Repairoux A., "Des Réacteurs Nucléaires dans l'Espace", La Recherche, France, Vol. 14, no 146, Juillet-Août 1983, p. 977.

[4] Pierson E.S., "Boundary-Layer Analysis of Turbulent MHD Channel Flows", Quaterly Progress Report, No. 78, M.I.T., Cambridge, Mass., USA, 1965, pp. 152-159

[5] Fujü E.Y., "Experimental Studies on Liquid-Metal Two-Phase Flow in Travelling and dc Magnetic Field", Third Beer-Sheva

International Seminar on MHD and Turbulence, Ben-Gurion University of the Negev, Beer-Sheva, Israël, March 23-27, 1981.

[6] Glukhikh V.A. and Kirillov I.P., "Experimental Investigations of an Asynchronous Self-Excited Liquid-Metal MHD Generator", Magnetohydrodynamics, Plennum Publishing Coporation, New-York, USA, Vol. 2, No. 4, 1966, pp. 64-68.

[7] Petrick M. and Roberts J., "Performance of Space-Power Liquid-Metal MHD Cycles Utilizing a Two-Phase Flow Generator", J. Spacecraft, Vol. 4, No. 8, 1967, pp. 967-973.

[8] Bergles A.E., Collier J.G., Delhaye J.M. et al, "Basic Equations for Two-Phase Flow Modeling", Two-Phase Flow and Heat Transfer in the Power and Process Industries, Hemisphere Publishing Corp., New-York, 1981, pp. 40-97.

[9] Tanatugu N., Fujü-E Y. and Suita T., "Electrical Conductivity of Liquid-Metal Two-Phase Mixture in Bubbly and Slug Flow Regime", J. of Nucl. Sci. and Techn., Vol. 9, No. 12, 1972, pp. 753-755.

[10] Pierson E.S. and Jackson W., "MHD Induction Machine with Laminar Fluid Flow", Quaterly Progress Report, No. 77, M.I.T., Cambridge, Mass. USA, 1965, pp. 218-232.

[11] Fabris G., "Mixers and Surfactants", Proc. of the Rand Corp. Conf. on LMMHD Power Generation, Rand Santa Monica, Calif., 1977, pp. 95-120.

[12] Cerini D.J., "Circualtion of Liquid for MHD Power Generation", Conf. on Electricity from MHD, IAEA, Vienna, Austria, Vol. 3, 1968.

[13] Idel'cik I.E., Memento des Pertes de Charge, ed. Eyrolles (Press), Paris, France.

[14] Dunn P.D. and Reay D.A., Heat-Pipes, 3rd ed., Pergamon Press, New-York, USA, 1982.

The Feasibility of Remote Power Generation based on LMMHD and Biomass Energy

D.G. Malcolm*
D.G. Malcolm & Associates, Inc., Saskatoon, Canada

Abstract

The paper introduces the concept of using indigenous biomass fuel as the energy source to drive a two-phase LMMHD electrical power generator.

If forest surplus or wood residues are used as fuel, LMMHD could develop into a highly competitive technology, replacing wood gasifiers and/or diesel-generator sets for remote site, off-grid electrical power generation.

If low pressure steam is used as the volatile second phase, a two-phase LMMHD generator can be used to provide both electrical power and space heating for remote communities.

Introduction

The purpose of this paper is to present the concept that liquid-metal magnetohydrodynamics (LMMHD) electrical generation technology be seriously considered for some industrial applications and for remote site off-grid electrical power using indigenous biomass fuel. Because such fuel represents renewable energy, remote industrial plants and communities could, in many cases, be self-sufficient in energy at costs well below that of a common petroleum-fuelled internal combustion engine generator unit.

The concept of two-phase LMMHD power generation was first discussed in depth by Petrick.[1] Petrick pointed out that single phase LMMHD has a disadvantage in that a liquid metal is basically incompressible and cannot be used to convert thermal energy directly to mechanical and electri-

---
Paper presented at the Fourth Beer-Sheva Seminar on MHD Flows and Turbulence, Ben-Gurion University of the Negev, Beer-Sheva, Israel, Feb. 27-March 2, 1984. Copyright © 1985 by the American Institute of Aeronautics and Astronautics, Inc. All rights reserved.
*Managing Director.

cal energy. The use of a compressible vapor or gas phase in a LMMHD power generation cycle allows expansion and the transfer of mechanical energy to the liquid phase.

Basic performance studies have been carried out for two-phase low-temperature LMMHD power generation systems by Smith et al.[2] Also, various industrial applications for LMMHD generators have been discussed in an unpublished paper by Smith.[3]

In his unpublished paper, Smith[3] reported a preliminary study of the suitability of LMMHD power generation systems for utilizing thermal energy from three sources: high-temperature combustion products, solar energy, and waste heat (including geothermal sources). He concluded that an overall fuel-to-electrical conversion efficiency of 30% was realistic using high-temperature combustion products, and suggested railway locomotives and heat pumps for heating purposes as two major applications. For low-grade thermal energy, Smith found the LMMHD system to be generally much more efficient than organic Rankine systems and steam power systems.

It appears that two-phase LMMHD power generation is approaching practical commercial reality under the ETGAR development program of El-Boher et al.[4] The key problems do not, in fact, concern biomass utilization as an energy source, but rather the practicality of the LMMHD technology for electrical power generation.

In remote communities and industrial operations, the key requirements are electrical power and thermal energy for either space heating or process uses. The use of steam as the volatile phase in the LMMHD scheme, therefore, appears to have significant advantages from a practical point of view.

Biomass Availability

The consideration of biomass inventories is beyond the scope of this introduction. Biomass inventories are being compiled and updated in many countries of the world. In Canada, work in this area has been carried by a number of companies and institutions (see, for example, Procos[5] and Cunningham[6]). In the United States, work on inventories has been carried out by such researchers as Clark.[7]

In the Caribbean area, where sugar production is carried out from sugar cane, there is often considerably more bagasse available than is used for process heat (see Ref. 8). There are also considerable quantities of coconut shells and waste palm tree biomass that can be utilized.

Throughout the world, peat has come to be valued as a useful biomass energy resource, as discussed by Matthews[9] and Punwani.[10]

Overend[11] has given some thought to current world usage of forest and agricultural lands. He estimates that on the terrestrial surface there is some $40 \times 10^6 km^2$ of forest area, producing about 1.15×10^{12} kg/yr of forest biomass growth. In the agricultural field, he estimates some $15 \times 10^6 km^2$ of cultivated area, producing about 1.5×10^{12} kg/yr of agricultural biomass.

Sorenson[12] has estimated that the actual energy flow of organic matter from the biosphere to human society is 1.2×10^9 kW, of which half is food. The other 0.6×10^9 kW is mostly wood, of which a small fraction is burned directly, with the rest going to wood products and production losses. If these numbers are taken as representative, with 15,000 kJ/kg taken as the energy density of dry wood, then over three times more forest biomass is produced than is normally consumed by society.

An example of the vast forest reserves in Canada is indicated by a privately financed project in northern Saskatchewan, which is presently awaiting government approval, to build a 50-MW powerplant based on wood.[13] The wood to be used is primarily poplar, which produces several tons per hectare annually of new growth.

Energy Utilization Efficiency: Biomass vs Petroleum

Biomass utilization efficiency has been discussed by Overend[11] and by Juneja et al.[14] Generally speaking, biomass utilization is feasible wherever security of energy supply is an overriding factor or in remote areas where electricity from a supply grid may not be available.

Juneja et al. point out that the most cost-effective option is the direct combustion of biomass to produce heat or steam. It must be remembered in an analysis of operating costs that wood-fired boilers do not usually have thermal efficiencies higher than 50%, whereas petroleum-fueled boilers may have a thermal efficiency as high as 80%. Also, whereas crude oil typically has an energy density of 42,000 kJ/kg, dry wood has an energy density of about 15,000-19,000 kJ/kg.

In the case of forest biomass, the feedstock preparation and handling systems and the boiler systems are now a matter of commercially proven, traditional technology. Various industrial and government groups in Canada, for example, have studied the technical problems in some depth.[15] Also, the author has observed the operation of a wood boiler system in the state of Florida providing process steam to a fuel ethanol plant where the entire wood handling and boiler system operated continuously for one year on wood chips as fuel.

Two-Phase LMMHD Power Generation Scenarios Using Biomass Energy

Possibilities for viable LMMHD power generation systems exist, both for some industrial applications and for supplying remote site off-grid communities.

In the Caribbean region, there are cases where sugar factories use bagasse-fired boilers to generate high-pressure steam. This steam is fed to a turbine generator unit, with the exhaust steam being used for process uses. This industrial application would be one candidate for two-phase LMMHD power generation.

Other industrial applications would be power generation for saw mills and other northern industries.

Remote communities that are not connected to the power grid usually use diesel electric generating sets for the production of electrical power. Since wood-biomass-fueled two-phase LMMHD systems have the potential of being more cost-effective than this conventional technology, they have considerable development potential.

Another important practical aspect of LMMHD is the simplicity of design and the absence of moving mechanical parts. This means that maintenance of LMMHD systems should be minimal. This consideration is especially important at remote locations.

Northern communities often require heating as well as electrical power. If steam generation is used in the LMMHD powerplant, the exhaust low-pressure steam can be used for community heating schemes. If steam is not used as the volatile phase in the two-phase LMMHD system, heat exchange apparatus can be used to transfer the thermal energy from the steam to the volatile phase.

The Market For Biomass-Fueled LMMHD Systems

It has been estimated that there are 43 remote communities in the province of Ontario, Canada, that are off-grid. In all of Canada, there are several hundred remote communities. Therefore, in Canada alone there is a substantial market for remote site power generation systems that are more cost-effective than diesel electric units. These small communities tend to require electrical power in quantities ranging from 100 to 500 kW.

As an example, a Canadian firm is supplying a wood gasifier fueled by wood chips from a local forest products company,[16] to replace the diesel powerplant for a 400-kW diesel generator in the community of Ramsey, Ontario. The supplier of the equipment is looking at possibilities for an installation in at least one other Ontario community. The gas produced fuels an internal combustion engine, replacing the expensive diesel fuel.

Wood gasifiers are not noted for reliability or low maintenance under continuous operation. The use of gasifiers and other renewable energy options have been reviewed for ten remote communities in Ontario by Robillard[17] and by Robillard and Love.[18]

The intense worldwide activity in industrial energy conservation and in power generation for remote communities assures a ready market for two-phase LMMHD technology. Because biomass handling and steam generation technology is standard technology, the problem becomes one of the commercial viability of LMMHD systems, such as the ETGAR system of El-Boher et al.[4] at Ben-Gurion University.

On a worldwide basis, Lawand et al.[19] have carried out a detailed survey of village renewable energy projects for the United Nations Conference on New and Renewable Sources of Energy in Nairobi, Kenya, in August 1981. Their work was still continuing in this area in August 1982. Among the various sources of renewable energy studied by these researchers are photovoltaics, wind, and small hydro. There is no doubt that, worldwide, there would be a market for 100 to 400-kW two-phase LMMHD power generation systems.

References

1. Petrick, M., "Two-Phase Flow Liquid Metal MHD Generator", Proceedings of the Bat-Sheva International Seminar on MHD-Flows and Turbulence, Ben-Gurion University of the Negev, Beer-Sheva, Israel, 1975.

2. Smith, I.K., Branover, H., and Yakhot, A., "Performance Studies of Two-Phase Low-Temperature Liquid-Metal MHD Systems," Proceedings of the 20th Symposium on Engineering Aspects of Magnetohydrodynamics, University of California, Irvine, Calif., 1982, pp. 14.1.1 - 14.1.5.

3. Smith, I.K., "Prospects for Branover LMMHD Generators," Department of Mechanical Engineering, The City University, London, England, 1981 (unpublished).

4. El-Boher, A., Branover, H., and Petrick, M., "The ETGAR Liquid Metal MHD Project," Proceedings of the Fourth Beer-Sheva Seminar on MHD-Flows and Turbulence, Ben-Gurion University of the Negev, Beer-Sheva, Israel, 1984.

5. Procos, D., "Land-Use Settings for the Local Use of Energy Derived from Biomass Wastes," Proceedings of the Fourth Bioenergy R&D Seminar, National Research Council, Winnipeg, Canada, 1982, pp. 97-102.

6. Cunningham, R.A., "Computerized Mapping of National Forest Inventory," Proceedings of the Fourth Bioenergy R&D Seminar, National Research Council, Winnipeg, Canada, 1982, pp. 81-88.

7. Clarke, A., "Biomass Distribution of Coastal Plain Hardwood Stands," Proceedings of the Symposium on Energy from Biomass and Wastes, Vol. VII, Institute of Gas Technology, Lake Buena Vista, FLA, 1983, pp. 101-115.

8. Alexander, A.G., "The Energy Cane Alternative to Sugar Planting," Proceedings of the Symposium on Energy from Biomass and Wastes, Vol. VII, Institute of Gas Technology, Lake Buena Vista, FLA, 1983, pp. 185-203.

9. Matthews, R.D., "The Peat Resource -- Overview of Quantities and Locations Worldwide," Proceedings of the Symposium on Peat as an Energy Alternative, Vol. II, Institute of Gas Technology, Arlington, VA, 1981, pp. 25-53.

10. Punwani, D.V., "Peat as an Energy Alternative: 1981 Update", Proceedings of the Symposium on Peat as an Energy Alternative, Vol. II, Institute of Gas Technology, Arlington, VA, 1981, pp.1-27.

11. Overend, R.P., "Biomass Conversion Technologies," Proceedings of the ISES Congress, Solar World Forum, Brighton, England, 1981.

12. Sorenson, B., Renewable Energy, Academic Press, New York, 1979.

13. "Generator Project Proposed for North," Saskatoon Star Phoenix, February 10, 1984.

14. Juneja, S.C., Pnevmaticos, S.M. and Charron, J.M., "The Effect of Operating and Financial Variables on the Economics of Biomass Combustion Systems," Proceedings of the Symposium on Energy from Biomass and Wastes, Vol. VII, Institute of Gas Technology, Lake Buena Vista, FLA, 1983, pp. 333-366.

15. "Status of Biomass/Energy Conversion Technology: ENFOR Program Review 1982," Forintek Canada Corp., Ottawa, Ontario, March 1982.

16. "Wood Gas Unit to Replace Logging Camp's Diesel," Renewable Energy News, Nov. 1983, p. C3.

17. Robillard, P., "Northern Community Energy Assessments: Final Report," Ministry of Energy, Ontario, Canada, 1982.

18. Robillard, P. and Love, P. "Assessment of Alternative Electricity Generation Alternatives: Phase II Final Report," Ministry of Energy, Ontario, Canada, 1982.

19. Lawand, T.A., Alward, R., and Barrett, V.H., "A Worldwide Survey of Village Projects Using Renewable Energy Sources," Proceedings of the Energex '82, Solar Energy Society of Canada, University of Regina, Regina, Canada, 1982, pp. 260-270.

Interaction of Hall Currents and Turbulent Boundary Layers in Closed-Cycle MHD Experiments

Willem F.H. Merck* and Johannes G.A. Arts†

Eindhoven University of Technology, The Netherlands

Abstract

This work was performed to obtain better understanding of the phenomenon of static pressure rise in the downstream half of an experimental noble gas MHD generator during power extraction. A loss mechanism model, based upon a resistor network, was used to estimate the Hall currents in the electrode wall boundary layers. A two-dimensional computing method based upon Patankar's scheme was used to solve the gasdynamic boundary-layer equations for flat electrode walls with superimposed current-potential distribution. The theoretical results show boundary-layer separation at high magnetic interaction levels. Increases in both load currents and Hall-currents lead to the upstream shift of the separation point. The boundary-layer separation is preceded by a slight increase in the static pressure. The experiments show the pressure increase at much lower levels of power extraction. This discrepancy is probably caused by the semicylindrical electrodes protruding into the supersonic flow and creating a situation much more favorable for boundary-layer separation.

I. Introduction

The aim of the work reported here is to find a more thorough explanation of a phenomenon observed in the Eindhoven Blow Down Facility (EBDF), which is the sharp pressure

---

Paper presented at the Fourth Beer-Sheva Seminar on MHD Flows and Turbulence, Ben-Gurion University of the Negev, Beer-Sheva, Israel, February 27 - March 2, 1984. Copyright © American Institute of Aeronautics and Astronautics, Inc., 1985. All rights reserved.

*Scientist, Department of Electrical Engineering, Direct Energy Conversion.

†Former Assistant Scientist, Department of Electrical Engineering, Direct Energy Conversion.

rise in the downstream half of the ACs (alternate current) MHD generator at strong magnetic interaction (Fig. 1) and the subsequent deterioration of the generator performance. In the early 1970s, this specific phenomenon was also observed at MIT by the Kerrebrock group without satisfactory explanation.[1,2]

The hypothesis that the Hall current in the electrode wall boundary layer (EBL) plays an important role has now been worked out in more details by the authors.

Brederlow[3] and Hellebrekers[4] have independently observed that the currents in argon-alkali plasmas flow through streamer-like discharge structures (Fig. 2).

By means of two-dimensional time-dependent calculations in periodic generator segments, Hara et al.[5] have theoretically proved the creation of concentrated discharge structures, moving downstream. Recent work by Flinsenberg and Uhlenbusch[6,7] gives more physical understanding of the interaction of streamers with the gasdynamic flow and proves that streamers act like impermeable cylinders moving downstream with velocities slightly smaller than the velocity of the surrounding gas.

Furthermore, it has been observed that the discharges are bent downstream, electrically connecting two or more electrode pairs in axial direction and thus producing strong axial current components in the boundary-layer (BL) region (Fig. 2).

At present, however, the picture of streamer-creation, streamer-bulk interaction and streamer-BL interaction is not

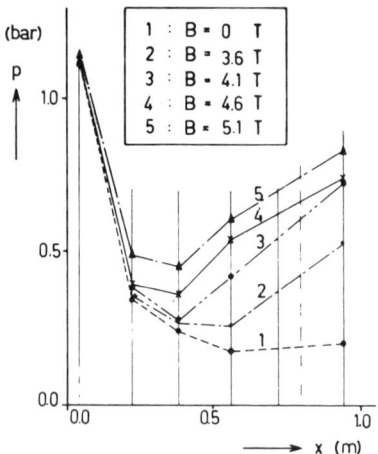

Fig. 1 Static pressure distribution in the MHD generator at different magnetic inductions during RUN 303.

HALL CURRENTS AND TURBULENT BOUNDARY LAYERS

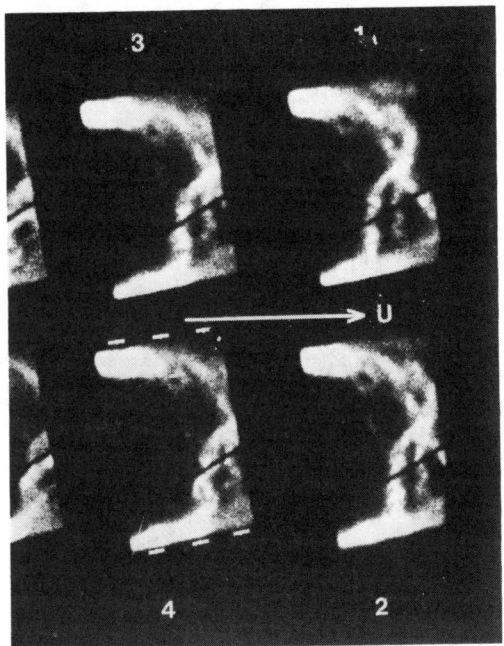

Fig. 2 Image converter picture of streamers in the Eindhoven shocktube facility at B = T; exposure time 2.5 µs.

omplete. To complete this picture, the addition of experimental data is necessary. In the next sections it will be shown how the measured load currents and voltages and the Hall voltage can be used to calculate the Hall currents in the BL and how the BL development can be calculated successively.

II. Loss Mechanism Model

A model to find approximate values for diverse loss mechanisms in MHD generators, using experimental data, was introduced by Hoffman[8] and further elaborated by Houben.[9] The base of this model is a resistor network replacing the diverse possible current paths in the segments of an MHD generator (see Fig. 3). The following phenomena are taken into account: 1) insulator and electrode wall leakages R_{yw} and R_{xw}, respectively; 2) insulator and electrode boundary-layer losses R_{yis} and R_{xel}, respectively; 3) bulk resistance losses R_{iy}; and 4) cathode and anode voltage drops, expressed through R_k and R_a, respectively.

Assuming a uniform core flow, Ohm's law, the simplified electron energy equation and the Saha equations can be defined. Using Ohm's law and Kirchhoff's law, we can elimi-

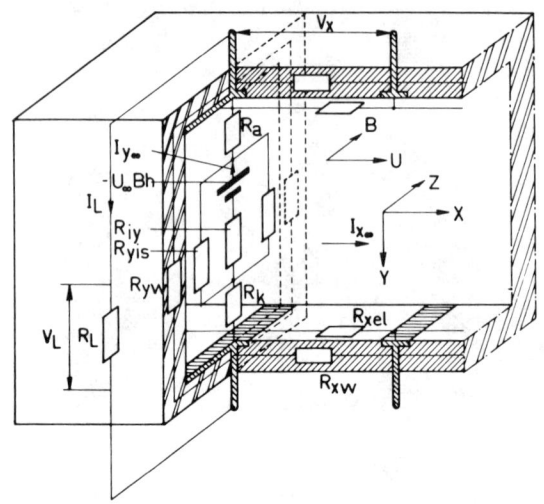

Fig. 3 Equivalent resistor network for an MHD generator segment.

nate $I_{x\infty}$, yielding an effective core resistance per segment:

$$R_{y\infty} = \left(1 + \frac{\beta_\infty^2}{\rho_x(h^2/\ell^2) + 1}\right) R_{iy} \qquad (1)$$

where

$$R_{iy} = \frac{\ell}{\sigma_\infty} \frac{h}{\ell b} \qquad (2)$$

ℓ, h, and b are the segment dimensions in the x, y, and z direction, respectively, and

$$\rho_x = \frac{R_{xe\ell} R_{xw}}{R_{xe\ell} + R_{xw}} \frac{\ell}{2 R_{iy}} \qquad (3)$$

By eliminating $I_{x\infty}$, we obtain the reduced resistor network shown in Fig. 4, on which the following definitions and expressions are based:

$$\rho_{is} = \frac{R_{yis}}{R_{yis} + 2 R_{y\infty}} \qquad (4)$$

HALL CURRENTS AND TURBULENT BOUNDARY LAYERS

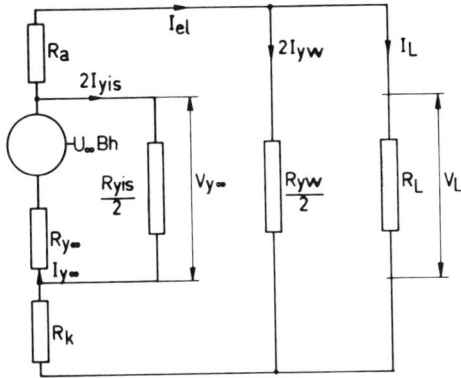

Fig. 4 Equivalent resistor network of a segment reduced to the y component.

$$\rho_y = \frac{R_a + R_k}{R_{y\infty}} + \rho_{is} \quad (5)$$

$$\rho_w = \frac{R_{yw}}{R_{yw} + 2\rho_y R_{y\infty}} \quad (6)$$

$$K_L = \frac{R_L}{R_L + \rho_w \rho_y R_{y\infty}} \quad (7)$$

Defining

$$I_s = -u_\infty Bh/R_{iy} \quad (8)$$

we obtain three equations:

$$V_L/-u_\infty Bh = K_L \rho_w \rho_{is} \quad (9)$$

$$\frac{I_L}{I_s} = \rho_{is} \frac{1 - K_L}{\rho_y} \left(1 + \frac{\beta_\infty^2}{\rho_x(h^2/\ell^2) + 1} \right)^{-1} \quad (10)$$

$$\frac{V_x}{-u_\infty Bh} = \frac{-\beta_\infty \rho_x(h/\ell)\rho_{is}^2}{1 + \beta_\infty^2 + \rho_x(h^2/\ell^2)} \left(\frac{1 - \rho_w K_L}{\rho_y} + \frac{1 - \rho_{is}}{\rho_{is}^2} \right) \quad (11)$$

Equations (9-11) show the measured data I_L and V_x (Fig. 5) and $V_L = I_L R_L$. u_∞ can be found from preliminary quasi-one-dimensional calculations. The wall temperatures measured in the EBDF experiments make clear that wall leakage currents are negligibly small, so $\rho_w = 1$ and $R_{xw} = \infty$.

The experiments also give data for the electrode voltage drop (ΔV), so we obtain

$$\Delta V = I_{e\ell}(R_a + R_k) \qquad (12)$$

Taking further into account that the current y component is found just within the streamer body, a limiting solution for $I_{yis} = 0$ should be found.

From the electron gas equation and Solbes' theory on instabilities[10] a first approximation of the effective electrical conductivity $\sigma_{eff,\infty}$ and the critical Hall parameter ∞_{cr} can be found for the core flow.

The value of $\sigma_{eff,\infty}$ thus found appears to be an overestimation. In the calculations, the effective Hall parameter β_{eff} is now taken equal to β_{cr}, whereas the value of $\sigma_{eff,\infty}$ is decreased step by step until the limiting solution with $I_{yis} = 0$ is found. In this way ρ_x, ρ_y, ρ_{is}, and $\sigma_{eff,\infty}$ are found, and the Hall current through the electrode boundary layer is found as

$$I_{xe\ell} = V_x/R_{xe\ell} \qquad (13)$$

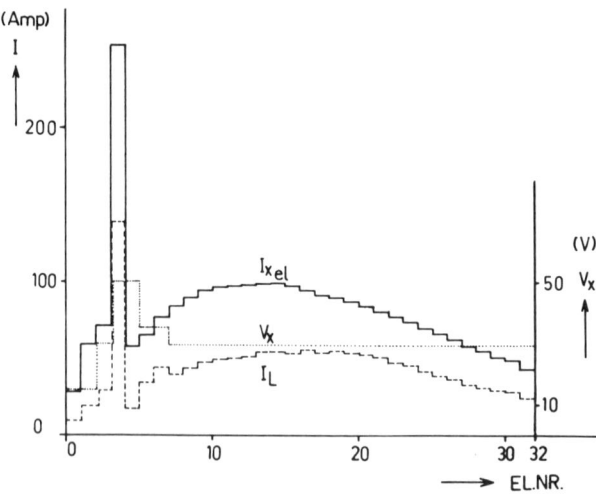

Fig. 5 Measured load current I_L and Hall voltage V_x and calculated Hall current I distribution for RUN 303 at 40 s, B = 5.1 T.

Figures 5 and 6 reflect the experimental and calculated data. It should be noted that in most segments the BL Hall current $I_{xe\ell}$ is about twice the load current I_L.

In the specific experiment RUN 303 at 40.00 s running time, the general gasdynamic and electromagnetic data are: stagnation pressure p_s = 7.4 bars; stagnation temperature T_s = 1900 K; mass flow $\dot{m}$ = 5.1 kg/s; inlet Mach number M_o = 1.6; maximum induction B_{max} = 5.1 T; load resistance R_L = 6Ω; and total electrical power P_e = 362 kW. The experiment was performed so as to obtain preionization by means of shorting the first four electrode pairs, which are shown in Figs. 5 and 6, where at the fourth electrode a huge load current and Hall current are found together with a sharp peak of the bulk conductivity $\sigma_{eff,\infty}$.

III. Boundary-Layer Model

The model is based upon a segmented electrode wall with flat electrodes (Fig. 7). The BL equations include the continuity, momentum, and stagnation enthalpy equations for the heavy particle gas using the following assumptions: steady, turbulent, two-dimensional flow, $\delta/\delta t = \delta/\delta z = 0$;

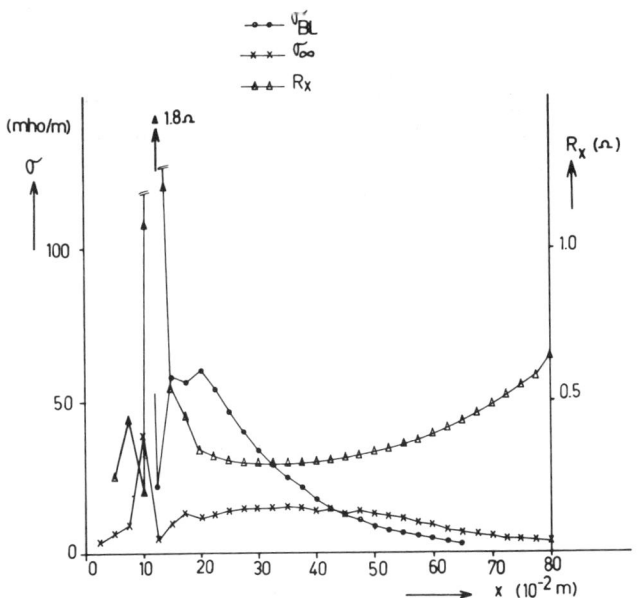

Fig. 6 Calculated electric conductivity of core σ_∞ and electrode wall boundary layer σ_{BL} and BL resistance R_x for RUN 303 at 40 s.

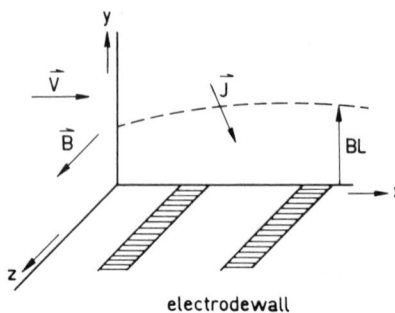

Fig. 7 Coordinate system for the MHD boundary-layer problem.

constant pressure over a cross section; small magnetic Reynolds number, so $B = B_z$ = constant; radiation losses neglected; and superimposed current-potential distribution. The latter means that the electron gas equations are not involved.

The method to solve this set of gasdynamic equations is extensively described by Patankar and Spalding,[11] whereas Pian and Merck[12] and Arts and Merck[13] give the extension of this method to incorporate the electromagnetic interaction terms, yielding a computing program analogous to STAN 5 (Ref. 14) that has been designed for open-cycle MHD flow.

The following terms were added to the regular gasdynamic equations for turbulent BL flow: x-momentum equation, source term $+j_y B_z$; stagnation enthalpy equation, source term $+j_x E_x + j_y E_y$. The common boundary conditions are used: impermeable flat wall with nonslip condition $u(x,o) = v(x,o) = 0$; fixed wall temperature $T(x,o) = T_w(x)$; and, at the outer boundary, the core flow equations apply.

A. Pressure Gradient

To calculate the pressure gradient in confined flows, which is a part of the complete solution of boundary-layer and bulk flow, Patankar and Spalding developed an elegant method. At any step, the pressure gradient is estimated first. Then, the forward step calculations are performed, yielding solutions for the gasdynamic quantities. The criterion is the area occupied by the flow A_f, found through $\iint \rho u dA = \dot{m}$, and the actual area provided by the duct A_d. If they do not match, then the pressure gradient is corrected in a sense that the difference $A_d - A_f$ is counteracted. Arts and Merck[13] found the following expression after integration over a cross section:

$$\frac{dp}{dx} = \frac{1}{1 - M_s^2} \left(\frac{\rho_s u_s I_2}{A} \frac{dA}{dx} + \frac{u_s W}{T_s C_p} I_2 (I_{B1} - F_w u_\infty) \right.$$

$$\left. - \frac{u_s W}{T_s C_p} (I_E + \Phi) + I_{B2} - F_w \right) \quad (14)$$

where

$$I_{B1} = \frac{\ell}{h} \int_0^h jBu \, dy \quad I_E = \frac{\ell}{h} \int_0^h \vec{j} \cdot \vec{E} \, dy$$

$$I_{B2} = \frac{\ell}{h} \int_0^h jB \, dy \quad F_w = \frac{\ell}{A} \int \tau_s \, ds \quad (15)$$

$$\Phi = (1/A) \int q_w \, ds$$

and q_w is the heat flux at the wall; W is the molecular weight of the gas; and ρ_s, u_s, T_s, and M_s are averaged values over the cross section.

B. Couette Layer

A disadvantage of the Patankar-Spalding scheme[11] using normalized stream function coordinates, is the coarse grid close to the wall, especially the first step, where the axial velocity u is small. This problem is overcome by defining the first step as a Couette layer. Two nondimensional differential equations have been derived that have to be integrated across the Couette layer, using the Patankar-Spalding notation, where the index + indicates the nondimensionality:

$$\frac{du_+}{dy_+} = \frac{1 + p_+ y_+}{\mu_+} + \frac{m_+ u_+}{\mu_+}$$

$$\frac{d\Phi_+}{dy_+} = \frac{1 + m_+ \Phi_+ - 0_+ y_+}{u_+/Pr_{eff}} + \frac{1 - Pr_{eff}}{2} W \frac{du_+^2}{dy_+} \quad (16)$$

where the electric power term $\vec{j}\vec{B}$ is incorporated in O_+, and the Lorentz force j_yB_z in $p_+\Phi_+$ is the dimensionless enthalpy.

The dimensionless viscosity

$$\mu_+ = 1 + K^2 y_+^2 \left[1 - \exp\left(\frac{-y_+ \tau_+^{\frac{1}{2}}}{A_+}\right)\right]^2 \frac{du_+}{dy_+} \quad (17)$$

contains the Van Driest damping term.

Finally, the effective Prandtl number Pr_{eff} contains the refined expressions for the turbulent Prandtl number given by Crawford and Kays[14]:

$$Pr_{eff} = \frac{1 + \mu_t/\mu}{1/Pr + \mu_t/\mu Pr_t} \quad (18)$$

where

$$Pr_t = \{\frac{\alpha^2}{2} + \alpha c Pe_t - (cPe_t)^2 \left[1 - \exp\left(\frac{-\alpha}{cPe_t}\right)\right]\}^{-1} \quad (19)$$

and

$$c = 0.2 \quad \alpha = \left(\frac{1}{0.86}\right)^{\frac{1}{2}} \quad Pe_t = \frac{\mu_t}{\mu} Pr$$

C. Current-Potential Distribution

As mentioned before, the current-potential distribution is superimposed. Known methods to solve the Maxwell equations in order to find the current-potential distribution give no account for the streamers that have been observed and yield far too optimistic values for effective electric conductivity and Hall parameter. Also, the power production predicted this way is much higher than that achieved in the experiments.

So, to calculate the development of the electrode boundary layer, in this case at the cathode wall, it is advisable to take the measured electric quantities into account. These are the load resistance and current R_L and I_L, respectively; the Hall voltage V_x; and the total electrode voltage drop ΔV. From the loss mechanism model, we obtain values for the Hall current along the electrode wall $I_{xe\ell}$, the electrode wall resistance $R_{xe\ell}$, the effective bulk conductivity $\sigma_{eff,\infty}$, and the effective Hall parameter β_{eff}.

Figure 7 represents the current-potential distribution used.[12] The assumptions within the BL are $j_y = j_y(x)$ and $E_x = 0$ over the conductor; $j_y = 0$ and $E_x = E_x(x)$ over the insulator; and $j_{xe\ell}$ = constant and fills the whole boundary layer.

For the source term in the enthalpy equation, we find that over the conductor,

$$\vec{j} \cdot \vec{E} = \frac{j_y^2(1 + \beta^2_{eff})}{\sigma_{e\ell}} + uBj_y \quad (20)$$

and over the insulator

$$\vec{j} \cdot \vec{E} = j_{xe\ell} E_{xe\ell} \quad (21)$$

where

$$\sigma_{e\ell} = (a + d)/R_{xe\ell} b\delta$$

and

$$j_{xe\ell} = I_{xe\ell}/b\delta$$

The source term in the momentum equation reads

$$j_y B_z \quad (22)$$

Over the conductor, $j_y(x)$ is expressed by[12]

$$j_y = C_1 j_{y\ max} \left[\exp(\alpha x/a) - 1 \right] \sin(\pi x/a) \quad (23)$$

with C_1 and α chosen so as to fulfill

$$I_L = b \int_0^a j_y\ dx \quad (24)$$

The electric field over the insulator is divided in two regions in order to avoid discontinuities and numerical instabilities (Fig. 8). For $0 < x_{ins} < e$,

$$E_{xe\ell}(x) = E_{x\ max}\ (x_{ins}/e) \quad (25)$$

For $e < x_{ins} < d$,

$$E_{xe\ell}(x) = E_{x\,max} \frac{d - x_{ins}}{d - e}$$

Here, $E_{x\,max}$ follows from the measured Hall voltage $E_{x\,max} = -2\,V_x/d$, and e is taken to be $0.05d$.

With this approach we have calculated the boundary-layer development within the experiment for the case of flat electrode walls.

IV. Results

Some of the calculation results are shown in Figs. 9-14. At first, we did not obtain the expected pressure rise and

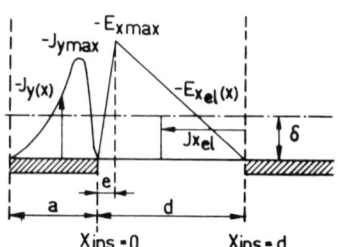

Fig. 8 Current density j_y and electric field $E_{xe\ell}$ distribution within the boundary layer.

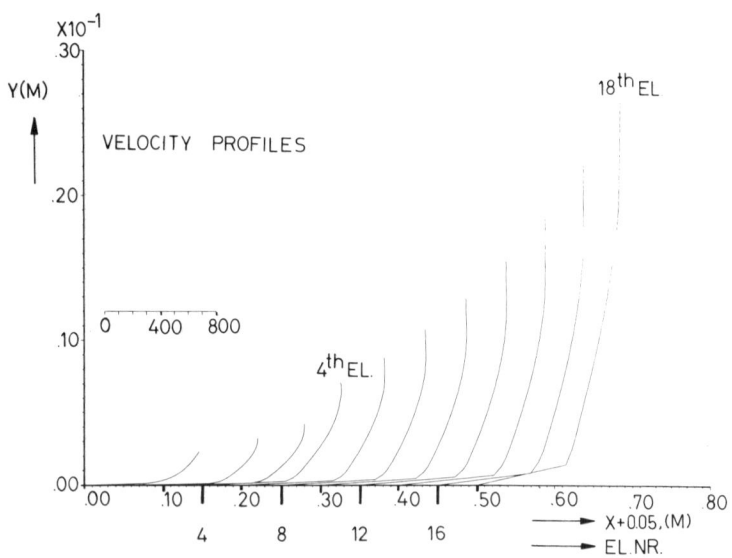

Fig. 9 Velocity profiles in the electrode wall boundary layer at different x locations for the situation of measured load currents of RUN 303, 40 s (Fig. 5) multiplied by $C_L = 1.5$.

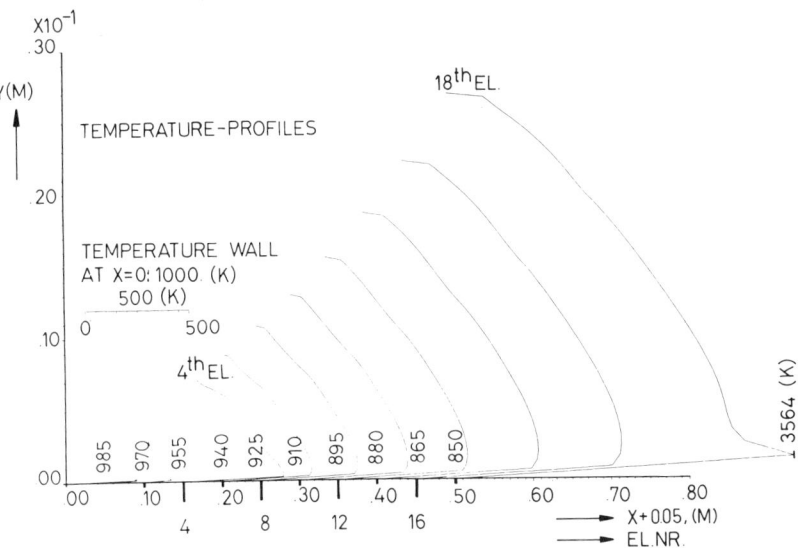

Fig. 10 Temperature profiles in the electrode wall boundary layer at different x locations for the conditions of Fig. 9.

BL separation in the generator downstream half with the aforementioned calculation method at the given magnetic interaction.

Increasing all load currents by a factor $C_L > 1$ led to the wanted effect of decreasing values of the location of the separation point x_s for increasing values of C_L, as shown by Fig. 15.

Figures 9-14 reflect the situation where $C_L = 1.5$. During the experiments, a high current peak (>200 A) was observed at the fourth electrode pair due to short-circuiting the first four electrode pairs. The influence of this current peak is found in most curves: big changes in the velocity and temperature profiles (Figs. 9 and 10); dips in the axial profiles of the core flow quantities p_s, u_∞ (Fig. 11), and M_∞ (Fig. 14) and the friction coefficient C_f (Fig. 13); jumps in the axial profiles of core flow quantities T_∞ and p (Fig. 11) and BL quantities δ (Fig. 12), q_w and shapefactor d_1/d_2 (Fig. 13).

Figure 13 clearly demonstrates the influence of the current peaks at each electrode on the friction coefficient C_f and the wall heat flux q_w. We see that C_f is sharply decreasing and q_w is sharply increasing at the current maxima because of the high local values of the Lorentz force and the ohmic heating.

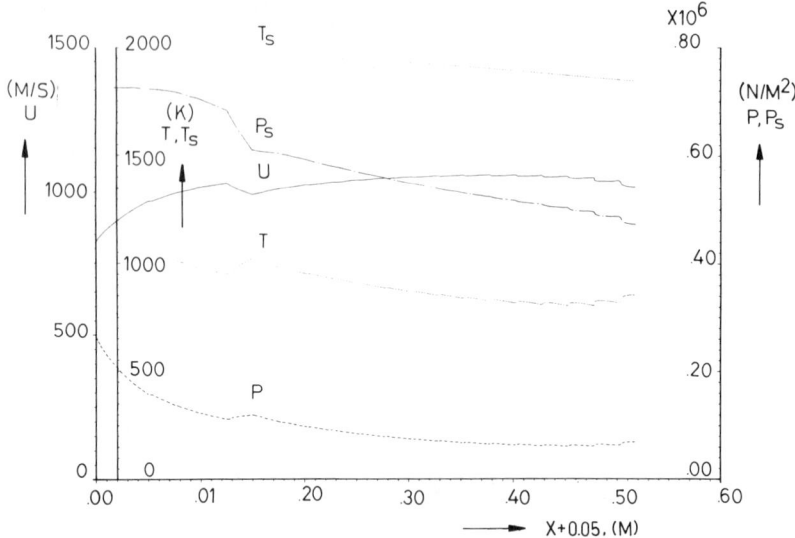

Fig. 11 Core flow distribution of velocity u, pressure p, stagnation pressure p_s, temperature T, and stagnation temperature T_s for the conditions of Fig. 9.

Notice the big changes in C_f and q_w close to the separation point. The separation criterion was taken to be C_f = 0. The BL thickness δ and the displacement thickness d_1 increase step by step at each current peak.

The accumulation of heat caused by the currents through the BL is clearly demonstrated by the successive temperature profiles in Fig. 10.

During the calculations, the effect of the Hall current was simulated by multiplying the Hall current values obtained in Sec. II by a factor C_H between 0 and 2. Figure 16 shows that the heat accumulation increases with increasing Hall currents.

At the location x = 0.425 m, which is the separation point for C_H = 2, the temperature profiles for different values of C_H ranging from 0.02 to 2 are compared, while C_L = 1.5. The effect of increasing Hall currents upon the BL is twofold: firstly, the gas temperature within the BL is enhanced due to the increased ohmic dissipation; secondly, the thickness of the BL is increased, which causes a stronger reduction of the effective cross section and, thus, while the Mach number is larger than 1, a higher core flow temperature and pressure will be found. Table 1 reflects this performance. The first three columns show the increase of BL temperature and thicknesses with increasing Hall currents, whereas the last five columns show the reaction of the in-

HALL CURRENTS AND TURBULENT BOUNDARY LAYERS 469

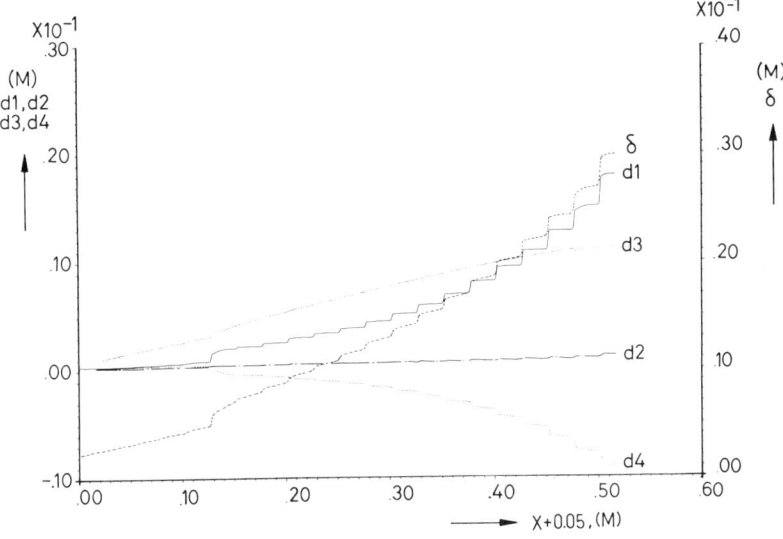

Fig. 12 Development of boundary-layer thickness δ, displacement thickness d_1, momentum thickness d_2, energy thickness d_3, and enthalpy thickness d_4 in the MHD generator for conditions of Fig. 9.

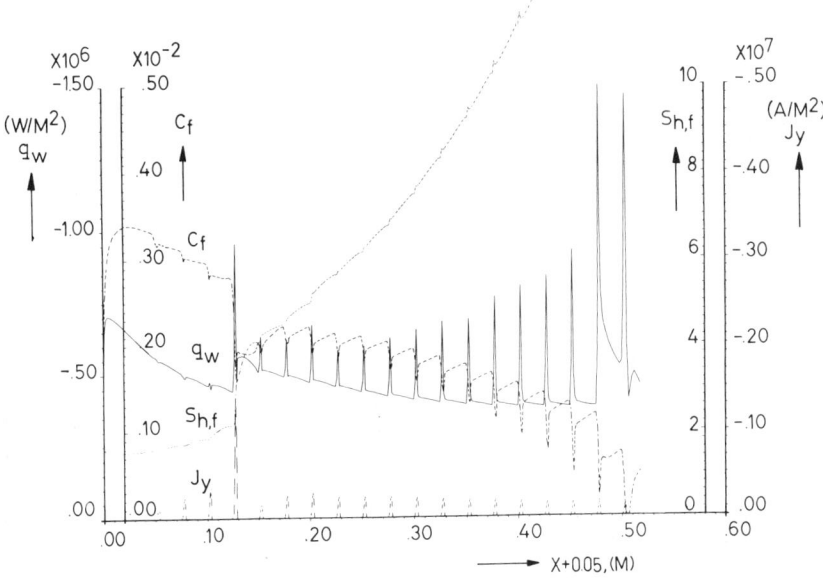

Fig. 13 Distribution of friction coefficient C_f, wall heat flux q_w, shape factor d_1/d_2, and current density j_y along the duct for conditions of Fig. 9.

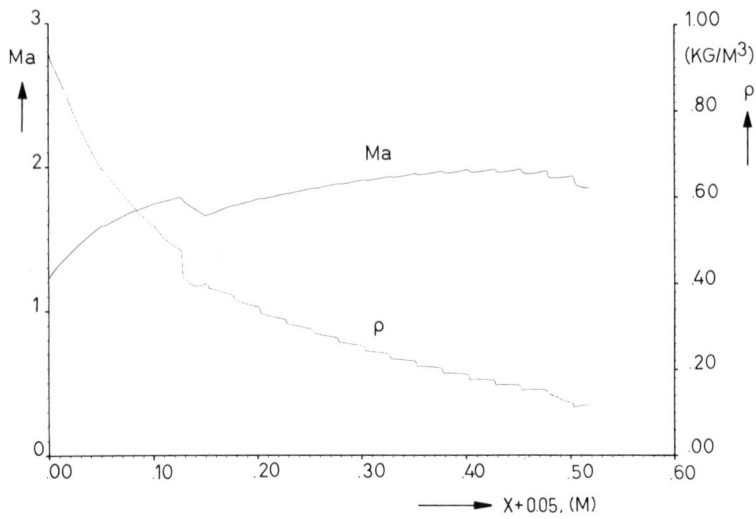

Fig. 14 Axial distribution of core flow Mach number Ma and mass density at the edge of the Couette layer ρ for conditions of Fig. 9.

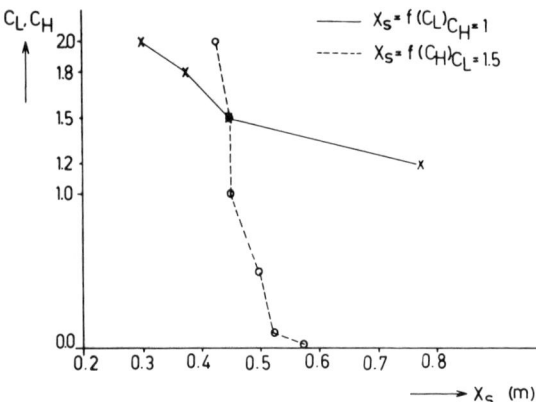

Fig. 15 Location of the boundary-layer separation point x_S in dependency of the multiplication factors C_L for the load currents and C_H for the Hall currents.

creasing displacement thickness d_ℓ upon the core flow quantities. The pressure p, temperature T_∞, and mass density ρ_∞ show an increase, and the velocity v_∞ and the Mach number M_∞ show a decrease in their numeric values, which is in agreement with general physical insight in supersonic flow. It may be expected from this discussion that the separation point will move upstream with increasing Hall currents. This effect is shown in Fig. 15, where we also see that the effect of increasing Faraday currents is much more pronounced.

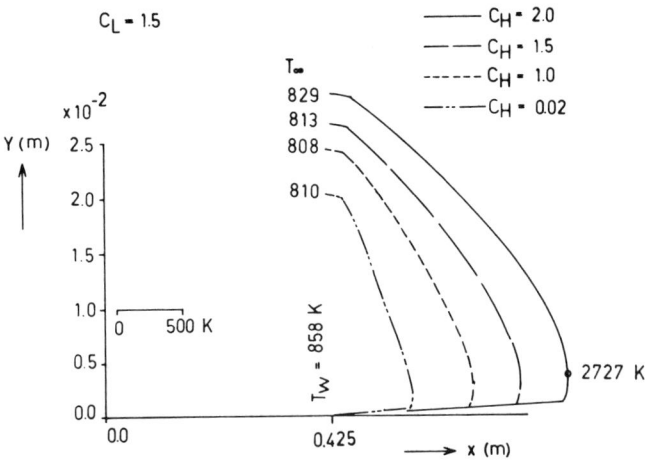

Fig. 16 Impact of boundary-layer Hall current, expressed by the factor C_H, upon the temperature profiles within the boundary layer.

Table 1 Core flow and boundary layer quantities for different values of the Hall current multiplication factor C_H (for $C_L = 1.5$)

C_H	T_{max}, K	δ, 10^{-2} m	d_1, 10^{-2} m	p, 10^{-5} Nm^{-2}	T_∞, K	ρ_∞, kg m^{-3}	v_∞, ms^{-1}	M_∞, 1
0.02	1472	2.00	0.81	0.568	810	0.336	1060	2.00
0.1	1509	2.03	0.85	0.571	810	0.339	1059	1.999
0.5	1711	2.21	1.04	0.588	808	0.350	1055	1.99
1.0	1996	2.43	1.23	0.615	808	0.366	1047	1.98
1.5	2329	2.67	1.54	0.652	813	0.386	1037	1.95
2.0	2727	2.93	1.83	0.709	829	0.411	1020	1.90

V. Comparison with Experiments

The main data of the EBDF are: mass flow 5 kg/s; stagnation pressure 7 bars; stagnation temperature 2000 K; thermal input power 5 MW; magnetic induction 5T; argon gas with 1 - 0/00 cesium seed. For more details, also see Blom et al.[15] The specific data for the experiment RUN 303 at 40.00 s are listed in Sec. II.

The dimensions of the generator duct are h = 0.154 m, b = 0.050 m at the inlet; h = 0.154 m, b = 0.180 m at the

outlet; ℓ_t = 0.800 m is the overall length; and ℓ = 0.025 m
is the segment length. Electrodes are semicylindrical with
diameter 0.004 m and protrude into the flow by 0.002 m.
   From the measured axial pressure profiles in Fig. 1, we
see that the tendency of pressure increase already exists at
rather low interaction levels (curve 2 with B = 3.6 T) and
that this pressure increase progresses upstream at higher
interaction levels, as shown by curves 3, 4, and 5.
   The pressure increase starts somewhere behind x = 0.38
m. From the calculations, no pressure increase or BL separation is found even at the highest interaction level with
induction B = 5.1 T. Increasing the load currents with a
factor C_L = 1.5, which means increasing the power level by a
factor 2.25, leades to BL separation at x = 0.45 m and a
pressure minimum at x = 0.40 m (Figs. 11 and 13). So, the
effects measured in the experiments are found only by calculations at much higher interaction levels.
   The disagreement may be explained by the difference in
electrode wall structure. The BL model is based upon a flat
electrode wall, whereas the actual experimental situation is
an electrode wall with semicylindrical electrodes protruding
2 mm into the supersonic flow. This will introduce pressure
waves at each electrode where the BL flow has to pass
through, thus undergoing frequent pressure rises and drops.
It is clear that this situation is much more favorable for
the onset of BL separation than the flat electrode wall. If
BL separation occurs, then the decrease in the effective
flow area will cause a considerable increase in pressure and
might lead to the creation of oblique shocks, with consequently stronger pressure rise.
   On the other hand, the influence of the streamers, acting like rigid bodies that move downstream with velocities
smaller than the core flow velocity, upon the electrode wall
BL is not clear at this moment. We expect that this problem can be solved only by a three-dimensional time-dependent
approach.

VI. Conclusions

   From the theoretical and experimental considerations
mentioned in the preceding sections some conclusions can be
drawn:
   1) The developed computing method is able to predict
the occurrence of BL separation on a flat electrode wall,
provided the current potential distribution is given.
   2) Increasing the electromagnetic interaction causes
the separation point to move upstream in the channel.
   3) The primary effect of the Hall currents in the BL
is to increase the BL temperature and BL thickness.

4) Increasing the Hall currents in the BL will shift the location of the separation point upstream.
5) The phenomena of BL separation is preceded by a slight increase of the static pressure.
6) Obstacles on the electrode wall increase the chance that BL separation occurs.
7) For an exact determination of the separation point, a more thorough understanding of the streamer-BL interaction is necessary.
8) In the EBDF MHD generator, BL separation occurs at rather low levels of power extraction (at 3.6 T).
9) The EBDF MHD generator will not achieve the high power level of 1 MW_e indicated by quasi-one-dimensional calculations owing to premature BL separation.

Acknowledgments

The authors would like to thank Professor Rietjens, chairman of the Division Direct Energy Conversion, for the opportunity to complete this work.

References

[1] Reilly, J. P., "Open- and Short-Circuit Experiments with a Nonequilibrium MHD Generator Using Both Cold and Hot Insulator Walls," Energy Conversion, Vol. 10, No. 1, 1970, pp. 13-23.

[2] Decher, R., Hoffman, M., Kerrebrock, J., "Behaviour of a Large Nonequilibrium MHD Generator," AIAA Journal, Vol. 9, No. 3, 1971, pp. 357-364.

[3] Brederlow, G., Witte, N. J., Zinko, H., "Investigation of the Discharge Structure in a Noble Gas Alkali MHD Generator Plasma: Part I," AIAA Journal, Vol. 11, No. 8, 1973, pp. 1065-1072.

[4] Hellebrekers, W. M., "Instability Analysis in a Nonequilibrium MHD Generator," Ph.D. Thesis, Eindhoven University of Technology, The Netherlands, 1980.

[5] Hara, T., et al., "Numerical Analysis of the Nature of the Streamers in Noble Gas MHD Generators," Proceedings of the Eighth International Conference on MHD Electrical Power Generation, Vol. 4, Moscow, Sept. 12-18, 1983, pp. 136-139.

[6] Flinsenberg, H. J., "Fossil Fuel Fired Closed Cycle MHD Power Generating Experiments," Ph.D. Thesis, Eindhoven University of Technology, The Netherlands, 1983.

[7] Flinsenberg, H. J. and Uhlenbusch, J., "Streamer Dynamics in MHD Generators," Paper presented at the Fourth Beer-Sheva Seminar on MHD Flows and Turbulence, Beer-Sheva, Israel, Feb. 27-March 2, 1984.

[8] Hoffman, M. A., "Nonequilibrium MHD Generator Losses Due to Wall and Insulator Boundary Layer Leakages," Proceedings of the IEEE, Vol. 56, 1968, pp. 1511-1519.

[9] Houben, J. W. M. A., "Loss Mechanisms in an MHD Generator," Ph.D. Thesis, Eindhoven University of Technology, The Netherlands, 1973.

[10] Solbes, A., "Instabilities in Nonequilibrium MHD Plasmas, a Review," AIAA Paper 70-40, AIAA Eighth Aerospace Sciences Meeting, New York, Jan. 19-21, 1970.

[11] Patankar, S. V. and Spalding, D. B., Heat and Mass Transfer in Boundary Layers: A General Calculation Procedure, Intertext Books, London, 1970.

[12] Pian, C. C. P. and Merck, W. F. H., "Boundary-Layer Separation from the Electrode Wall of an MHD Generator," AIAA Journal, Vol. 14, No. 11, 1976, pp. 1585-1588.

[13] Arts, J. G. A. and Merck, W. F. H., "Two-Dimensional MHD Boundary Layers in Argon-Cesium Plasmas," Department of Electrical Engineering, Eindhoven University of Technology, The Netherlands, EUT Report 83-E-139, 1983.

[14] Crawford, M. E. and Kays, W. M., "STAN 5: A Program for Numerical Computation of Two-Dimensional Internal and External Boundary Layer Flows," NASA CR-2742, 1976.

[15] Blom, J. H., et al., "Design of the 5 MW Thermal Blow Down Experiment," Proceedings of the 17th Symposium on the Engineering Aspects of Magnetohydrodynamics, Stanford, Calif., March 27-29, 1978, edited by C. H. Kruger, pp. H.4.1-H.4.6.

Streamer Dynamics in MHD Generators

H.J. Flinsenberg* and J. Uhlenbusch†

Eindhoven University of Technology, The Netherlands

Abstract

Experiments with the Eindhoven closed-cycle MHD facilities show a pronounced nonhomogenous current carrying discharge structure in the MHD generator. These very constricted regimes of elevated temperature, called streamers, with a dimension r_σ, act as arc discharges exposed to mutual orthoganol external magnetic field and gas flow. The flowfield inside and outside the streamer can be calculated from the equation of continuity and the momentum balance, where friction and $\vec{j} \times \vec{B}$ forces are taken into account. By introducing a stream function ψ, the flowfield can be described by a linear fourth-order partial differential equation. It can be derived that a backflow inside the streamer is established if the Hartmann number appropriate to the problem exceeds a critical value. Two stagnation points occur in the flowfield, defining a characteristic aerodynamic radius r_{aer} of the streamer. The aerodynamic radius strongly depends on the Hartmann number, and, for practical MHD generator applications, it is found that $r_{aer}/r_\sigma > 1$. At first, the flowfield inside and outside a single streamer located inside an infinitely extended channel is considered. The theory is extended to a situation of several streamers present in a channel of finite width. This is done by applying the principle of superposition. The results of our calculations are compared with experimental results obtained in the blowdown experiment.

Paper presented at the Fourth Beer-Sheva Seminar on MHD Flows and Turbulence, Ben-Gurion University of the Negev, Beer-Sheva, Israel, Feb. 27-March 2, 1984. Copyright © 1985 by the American Institute of Aeronautics and Astronautics, Inc. All rights reserved.

*Scientist in the Group Direct Energy Conversion; presently at Philips Datasystems, Apeldoorn, The Netherlands.

†Extraordinary Professor in the Group Direct Energy Conversion, Department of Electrical Engineering; Professor in the Physikalisches Institut II of the University of Düsseldorf, Düsseldorf, Germany.

Introduction

From experiments with Faraday-type MHD generators using an Ar-Cs mixture as working gas, it is seen that the plasma is strongly nonuniform. This behavior is observed under power generating conditions in a shock tube experiment[1] as well as in a blowdown facility.[2] Photographic exposures of the discharge show arc-like structures, called streamers, that move with nearly the velocity of the gas downstream the channel.

Some data of these streamers are briefly reviewed here. The streamers are burning in an argon atmosphere at a pressure of about 1 bar with a Cs additive of 1 - 2 0/00. From experiments, it follows that the seed is fully ionized.[2] Coupled with this, the current density within the streamer is on the order of $j = 10^5$-10^6 A/m^2, the electron temperature is about 4500 K, the gas temperature is 2000 K (1000 K outside the streamer), and the electron density reaches the value of 2×10^{21} m^{-3}. Typical streamer velocities are $u_s \approx 900$ m/s, where the gas velocity reaches $u_\infty \approx 1000$ m/s. An external magnetic field $\vec{B}$ on the order of 5 T is applied. Balancing the Lorentzian force and the drag force acting on a streamer, moving with the velocity ($u_\infty - u_s$) relative to the fluid, gives the typical aerodynamic streamer dimension. A value of the order of 5×10^{-3} m results for a streamer current of about 5 A. The current carrying part of the streamer has a size of roughly 10^{-3} m following from the j value mentioned.

In practice, many of these streamers combine to a larger structure with typical size of the current-carrying part of about 10^{-2} m. Those "superstreamers" have a current of about 50 A.

The streamer problem strongly resembles the situation of free burning arcs under the influence of mutually orthogonal transverse flow and magnetic fields. This problem is studied extensively in the literature.[3,4] Up to now, this work has been restricted to local thermodynamic equilibrium (LTE) conditions and comparatively low $\vec{B}$ fields.

In the following, the physical situation is described using the monofluid approach. Assuming a radial symmetry of the temperature profile within the streamer, formulas for the flowfield inside and outside the streamer are given. As an important result, accurate streamer dimensions are derived from these calculations. The measured relative velocity between fluid and streamer is confirmed by these calculations.

The Basic Equations

Figure 1 shows the geometrical arrangement of a streamer in an MHD channel transformed to rest at the position $x = y = 0$. Throughout this paper, radial symmetry of the temperature, density, and thus of the conductivity profile of the streamer is assumed, whereas the viscosity η remains constant over the channel.

The flowfield in a laboratory frame, which is fixed to the channel wall, is given by $\vec{u}$. In this system, a streamer moves with a constant velocity $u_s \vec{e}_x$; thus, in the coordinate systems of Fig. 1, one obtains the velocity field

$$\vec{v} = \vec{u} - u_s \vec{e}_x \qquad (1)$$

In order to take into account the finite channel height L, we introduce a profile of the flowfield far away from the streamer in the laboratory frame

$$\vec{u}_\infty = \left[1 - \frac{z^2}{(L/2)^2} \right]^p u_\infty \vec{e}_x \qquad (2)$$

Seen from the streamer, the flowfield obeys the formula

$$\vec{v}_\infty(z) = v_\infty \left\{ \frac{u_\infty}{v_\infty} \left[\left(1 - \frac{z^2}{(L/2)^2} \right)^p - \frac{u_s}{v_\infty} \right] \vec{e}_x \right\} \qquad (3)$$

where

$$v_\infty = u_\infty - u_s \qquad (4)$$

We adopt this behavior of $\vec{v}_\infty$ for all points of the flow-

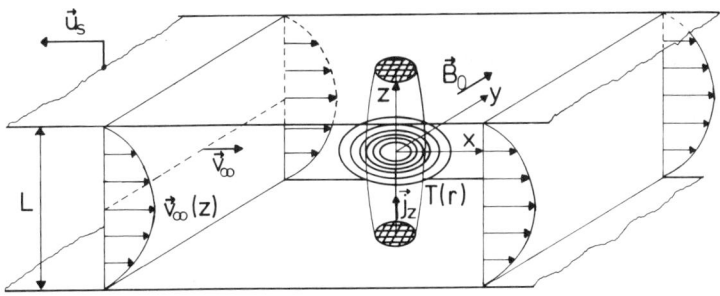

Fig. 1 Streamer in the flowfield of an MHD channel of height L and infinite width.

field and finally get

$$\vec{v}(x, y, z) = \left[\frac{u_\infty}{v_\infty}\left(1 - \frac{z^2}{(L/2)^2}\right)^p - \frac{u_s}{v_\infty}\right]\vec{v}_0(x,y)$$

$$= g(z)\,\vec{v}_0(x,y) \tag{5}$$

with

$$g(z) = 1 - \frac{u_\infty}{v_\infty}\left[1 - \left(1 - \frac{z^2}{(L/2)^2}\right)^p\right] \tag{6}$$

where

$$g(0) = 1 \quad g'(0) = 0 \quad g''(0) = -\frac{2p}{(L/2)^2}\frac{u_\infty}{v_\infty} \pm 0 \tag{7}$$

v is derived from the equation of continuity

$$\mathrm{div}(\rho_m \vec{v}) = 0 \tag{8}$$

and the momentum balance, which reads neglecting the inertial term and neglecting induced $\vec{B}$ fields,

$$0 \simeq -\mathrm{grad}\,p + \vec{j}\times\vec{B}_0 + \eta\Delta\vec{v} + (\eta/3)\mathrm{grad}\,\mathrm{div}\,\vec{v} \tag{9}$$

Instead of solving the coupled energy balance, the radial temperature profiles of electrons and the heavy particles are given in accordance with the experimental results. The electrical current density in the streamer then follows from Ohm's law:

$$\vec{j} = j_z\vec{e}_z = \sigma(r)E_z\vec{e}_z \tag{10}$$

where E_z is the field strength at the position of the streamer and σ is the electrical conductivity.

Equation (8) can be solved by introducing a stream function ψ via

$$\rho_m\vec{v}(x, y, z) = \mathrm{rot}(\psi\vec{e}_z) \tag{11}$$

Analogous to Eq. (5), one writes in cylindrical coordinates

$$\psi(x, y, z) = g(z)\psi_0(x, y) \tag{12}$$

STREAMER DYNAMICS IN MHD GENERATORS

Defining the vector quantity

$$\vec{W} = (\text{grad } \psi_0) \rho_m^{-1} \qquad (13)$$

one also has

$$\vec{v} = g(z) \vec{W} \times \vec{e}_z \qquad (14)$$

After some vector operations, it follows that for the plane $z = 0$,

$$(\text{rot } \vec{v})_{z=0} = -g(z) \text{ div} \vec{W} \qquad (15)$$

$$(\text{rot rot rot } \vec{v})_{z=0} = g''(o) \text{div} \vec{W} - \vec{e}_z \cdot \text{rot rot}(\vec{e}_z \text{div} \vec{W}) \qquad (16)$$

Now the curl of Eq. (9) can be reduced to the vector relation

$$\text{rot } \vec{j} \times \vec{B}_0 = (\vec{B}_0 \nabla) \vec{j}$$
$$= -\eta \text{ rot rot } \vec{e}_z \text{ div } \vec{W} + g''(0)\eta \cdot \vec{e}_z \text{div } \vec{W} \qquad (17)$$

Introducing

$$\chi = \text{div } \vec{W} \qquad (18)$$

the following set of equations describing the flowfield is derived from Eqs. (17) and (18):

$$\Delta_\chi + g''(0) \chi = \frac{B_{yo}}{\eta} \frac{\partial j_z}{\partial R} \sin \phi \qquad (19)$$

$$\chi = \text{div} \left[(\text{grad } \psi_0)/\rho_m \right] \qquad (20)$$

In cylindrical coordinates, the flow components at $z = 0$ read

$$V_r = \frac{1}{\rho_m(r)} \frac{1}{r} \frac{\partial \psi_0}{\partial \phi} \qquad V_\phi = -\frac{1}{\rho_m(r)} \frac{\partial \psi_0}{\partial r} \qquad (21)$$

For completeness, the Cartesian components are given as

$$V_x = \frac{1}{\rho_m(r)} \left(\cos \phi \, \frac{1}{r} \frac{\partial \psi_0}{\partial \phi} + \sin \phi \, \frac{\partial \psi_0}{\partial r} \right) \qquad (22)$$

$$V_y = \frac{1}{\rho_m(r)} \left(\sin \phi \, \frac{1}{r} \frac{\partial \psi_0}{\partial \phi} - \cos \phi \, \frac{\partial \psi_0}{\partial r} \right) \qquad (23)$$

The boundary conditions follow from Eqs. (22) and (23) when the limit $r \to \infty$ is performed:

$$\lim_{r\to\infty} V_x = V_\infty = \lim_{r\to\infty} \frac{1}{\rho_m(r)} \left(\cos\phi \frac{1}{r} \frac{\partial\psi_0}{\partial\phi} + \sin\phi \frac{\partial\psi_0}{\partial r} \right) \quad (24)$$

$$0 = \lim_{r\to\infty} \frac{1}{\rho_m(r)} \left(\sin\phi \frac{1}{r} \frac{\partial\psi_0}{\partial\phi} - \cos\phi \frac{\partial\psi_0}{\partial r} \right) \quad (25)$$

These boundary conditions are fulfilled for the asymptotic behavior

$$\lim_{r\to\infty} \psi_0 = \rho_{m\infty} V_\infty r \sin\phi \quad (26)$$

Thus, we try for entire regime

$$\psi_0(r,\phi) = \rho_{m\infty} V_\infty r f(r) \sin\phi \quad (27)$$

and try for Eq. (20) in a similar way:

$$\chi(r,\phi) = (V_\infty/r_\sigma^2) r h(r) \sin\phi \quad (28)$$

The functions f and h are dimensionless quantities. We further introduce the dimensionless variables

$$\alpha^2 = -g''(0) r_\sigma^2 = \frac{2p}{(L/2)^2} \frac{u_\infty}{V_\infty^2} r_\sigma^2$$

$$r = r_\sigma \rho \qquad \rho_m = \rho_{m\infty} \hat{\rho}_m(\rho) \quad (29)$$

$$\sigma = \sigma_0 \hat{\sigma} \qquad \beta^2 = \frac{2 B_{y0} E_{z0} \sigma_0 r_\sigma^2}{\eta V_\infty} \quad (30)$$

where β^2 is two times the Hartmann number squared. σ_0 and E_{z0} refer to the streamer center, and $\rho_{m\infty}$ is the density far away from the streamer.

Now the ordinary differential equations for f and h read

$$\frac{1}{\rho} \frac{d}{d\rho} \rho \frac{d}{d\rho}(\rho h) - \frac{h}{\rho} - \alpha^2 \rho h = \frac{\beta^2}{2} \frac{d\hat{\sigma}}{d\rho} \quad (31)$$

$$\frac{1}{\rho}\frac{d}{d\rho}\frac{\rho}{\hat{\rho}_m(\rho)}(\rho f) - \frac{f}{\rho\,\hat{\rho}_m} = \rho\,h \qquad (32)$$

which must be solved using the boundary conditions

$$f;\ \frac{df}{d\rho}\quad h;\ \frac{dh}{d\rho}\quad \text{regular at } \rho = 0 \qquad (33)$$

$$f \to 1 \text{ and } h \to 0 \quad \text{for } \rho \to \infty$$

Solution of the Basic Equations for a Channel of Infinite Width

Instead of solving Eqs. (31) and (32) using a digital computer, we look for an analytical solution that follows after some simplifications. As shown in Fig. 2, a so-called "channel model" is introduced, writing

$$\hat{\sigma}(\rho) = (1-\rho^2) \quad 0 \leq \rho \leq 1$$
$$= 0 \qquad 1 \leq \rho < \infty \qquad (34)$$

$$\rho_m(\rho) = \rho_{mi}\quad 0 < \rho < 1$$
$$= 1 \quad 1 < \rho < \infty \qquad (35)$$

This simplification directly leads to the following split-

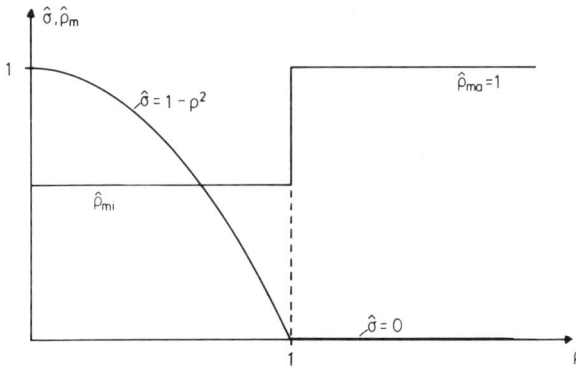

Fig. 2 Channel model for a streamer.

ting of Eqs. (31) and (32).

$$\frac{1}{\rho}\frac{d}{d\rho}\rho\frac{d}{d\rho}(\rho h_i) - \frac{h_i}{\rho} - \alpha^2 \rho h_i = -\beta^2 \quad 0 \leq \rho \leq 1$$

$$\frac{1}{\rho}\frac{d}{d\rho}\rho\frac{d}{d\rho}(\rho f_i) - \frac{f_i}{\rho} = \hat{\rho}_{mi} \rho h_i \quad 0 \leq \rho \leq 1$$

(36)

$$\frac{1}{\rho}\frac{d}{d\rho}\rho\frac{d}{d\rho}(\rho h_a) - \frac{h_a}{\rho} - \alpha^2 \rho h_a = 0 \quad 1 < \rho < \infty$$

$$\frac{1}{\rho}\frac{d}{d\rho}\rho\frac{d(\rho f_a)}{d\rho} - \frac{fa}{\rho} = \rho h_a \quad 1 < \rho < \infty$$

(37)

We postulate that, besides the boundary conditions (33), the functions f and h and their derivatives must be steady at $\rho = 1$.

The solutions of Eqs. (36) and (37) including the correct boundary and jump conditions, can be expressed in terms of the modified Bessel functions I_n and K_n of the order n as

$$h_i(\rho) = \left(\beta^2 \frac{1}{\alpha^2} - K_2(\alpha) \right) \frac{I_1(\alpha\rho)}{\alpha\rho}$$

$$f_i(\rho) = 1 + \hat{\rho}_{mi}\frac{\beta^2}{\alpha^2}\left(\frac{\rho^2}{8} - K_2(\alpha) \frac{I_1(\alpha\rho)}{\alpha\rho} - \frac{1}{\hat{\rho}_{mi}} \frac{I_2(\alpha) K_0(\alpha)}{2} \right.$$

$$\left. - \frac{1}{4} + \frac{K_2(\alpha) I_0(\alpha)}{2} \right) \quad \text{for} \quad 0 \leq \rho \leq 1 \quad (38)$$

and

$$h_a(\rho) = \beta^2 I_2(\alpha) \frac{K_1(\alpha\rho)}{\alpha\rho}$$

$$f_a(\rho) = 1 + \frac{\beta^2}{\alpha^2}\left(\frac{K_1(\alpha\rho)I_2(\alpha)}{\alpha\rho} - \frac{1}{2\rho^2} \right.$$

$$\left. \times \frac{\rho_{mi}}{4} + (1 - \hat{\rho}_{mi}) K_2(\alpha)I_2(\alpha) \right) \quad \text{for} \quad 1 < \rho < \infty \quad (39)$$

To evaluate these functions, a value $\hat{\rho}_{mi} \approx 0.5$ is assumed in the following, which is quite adequate to MHD streamer conditions.

Figures 3 and 4 show the calculated functions $H = h/\beta^2$ and $F = (f-1)/\beta^2$, which are independent of β. As the curves in Fig. 4 demonstrate, one can find β values for which, at a radial position ρ_0, the condition $F(\rho_0) = -1/\beta^2$ is fulfilled. At this position, $f(\rho_0) = 0$. For a given α, which follows according to Eq. (29) from the MHD channel geometry, the streamer dimensions, and the flow conditions, such β values only exist if they fulfill the condition $\beta > \beta_c$.

For $\beta = \beta_c$, the value $f(\rho_0) = 0$ is reached at the position $\rho_0 = 0$. Using Eq. (38), the critical value β_c follows from

$$\frac{1}{\beta_c^2} = \frac{\hat{\rho}_{mi}}{\alpha^2} \left(\frac{K_2(\alpha)}{2} + \frac{1}{\hat{\rho}_{mi}} \frac{I_2(\alpha) K_0(\alpha)}{2} + \frac{1}{4} - \frac{K_2(\alpha) I_0(\alpha)}{2} \right) \quad (40)$$

For $\beta < \beta_c$, no solution $f(\rho_0) = 0$ can be found.

As can be seen from Eqs. (21) and (27), a vanishing $f(\rho_0)$ also means zero radial velocity components on a circle with radius ρ_0. Thus, the incoming gas does not penetrate into the streamer bulk if $\beta > \beta_c$.

The streamer acts as a solid body of aerodynamic radius r_{aer}, where

$$\rho_0 \, r_\sigma = r_{aer} \quad (41)$$

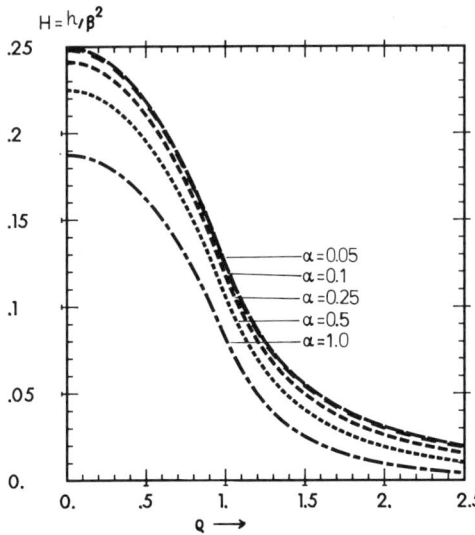

Fig. 3 Radial profile of the auxiliary function H; parameter is α.

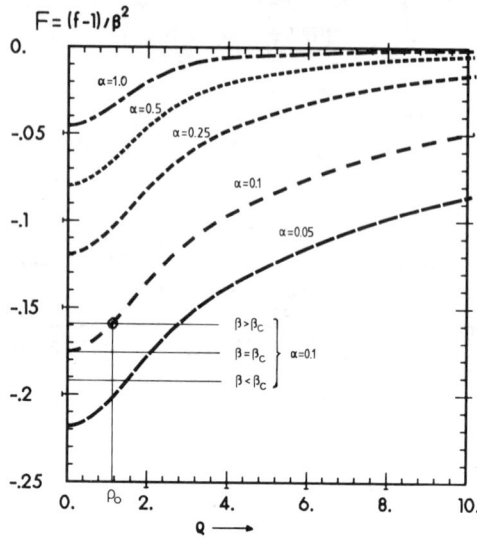

Fig. 4 Radial profile of the auxiliary function F; parameter is α.

Figure 5 gives the aerodynamic radius as function of β/β_c and α as parameter, showing the growth of the streamer dimensions with increasing Hartmann number. On the other hand, for $\beta < \beta_c$, the incoming gas flows through the hot low-density streamer. The streamlines are curved in such a way that the mass flow through the streamer is reduced.

In the case $\beta > \beta_c$, the flow recirculates around the streamer in the regime $\rho > \rho_0$. For $\rho < \rho_0$, the function f and also the stream function change sign; thus, the flow direction in the center of the streamer is opposite to the external flow, producing stagnation points at S_1 and S_2 (see Fig. 6). For reasons of continuity, a double vortex field is generated, the driving force of which is ultimately the jB force.

To obtain more convenient formulas, the general solutions were developed with respect to the parameter α for the condition that $\alpha\rho \ll 1$.

Now Eqs. (38) and (39) read, with $\gamma = 1.781$,

$$h_i(\rho) = \frac{\beta^2}{4}\left(1 - \frac{\rho^2}{2}\right) \qquad 0 < \rho < 1$$

$$f_i(\rho) = 1 + \hat{\rho}_{mi}\frac{\beta^2}{32}\left(\rho^2 - \frac{\rho^4}{6} - \frac{3}{2} - \frac{2}{\hat{\rho}_{mi}}\ln\frac{2}{\gamma\alpha}\right) \qquad 0 < \rho < 1$$

(continued)

STREAMER DYNAMICS IN MHD GENERATORS

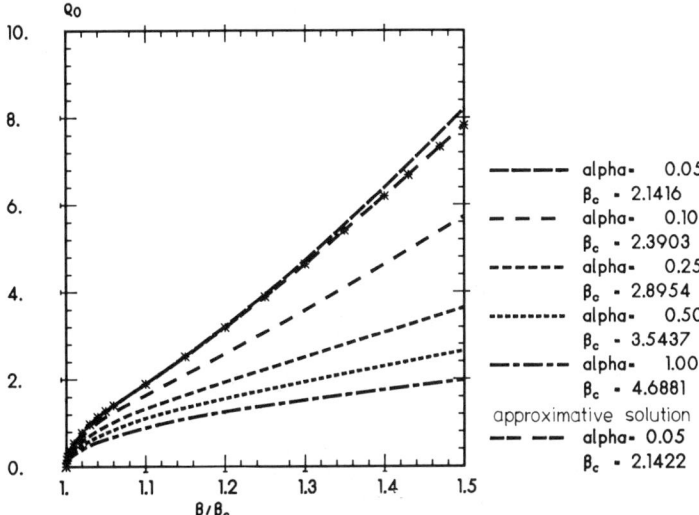

Fig. 5 Normalized aerodynamic radius ρ vs normalized Hartmann number for different α factors.

$$h_a(\rho) = \frac{\beta^2}{8} \frac{1}{\rho^2} \quad 1 < \rho < \rho_M \quad (42)$$

$$f_a(\rho) = 1 + \frac{\beta^2}{32}\left[-1 -2 \ln \frac{2}{\gamma\alpha\rho} + \frac{1}{\rho^2}\left(1 - \frac{2}{3}\hat{\rho}_{mi}\right)\right] \quad 1 < \rho < \rho_M$$

where we also postulate $\alpha \rho_M \ll 1$.

From these approximations, setting $f_i(0) = 0$, we directly conclude

$$\beta_c \approx 1\left/\sqrt{\left(\frac{3}{64}\hat{\rho}_{mi} + \frac{1}{16} \ln \frac{2}{\gamma\alpha}\right)}\right. \quad (43)$$

and the aerodynamic radius in the regime $\rho_0 < 1$ follows from the condition $f_i(\rho_{0i}) = 0$:

$$\rho_{oi} = \sqrt{3}\sqrt{1 - \sqrt{1 - \frac{64}{3\hat{\rho}_{mi}}\left(\frac{1}{\beta_c^2} - \frac{1}{\beta^2}\right)}} \quad (44)$$

which simplifies to

$$\rho_{oi} \approx 8\sqrt{\frac{1}{2\hat{\rho}_{mi}}\left(\frac{1}{\beta_c^2} - \frac{1}{\beta^2}\right)} \quad (45)$$

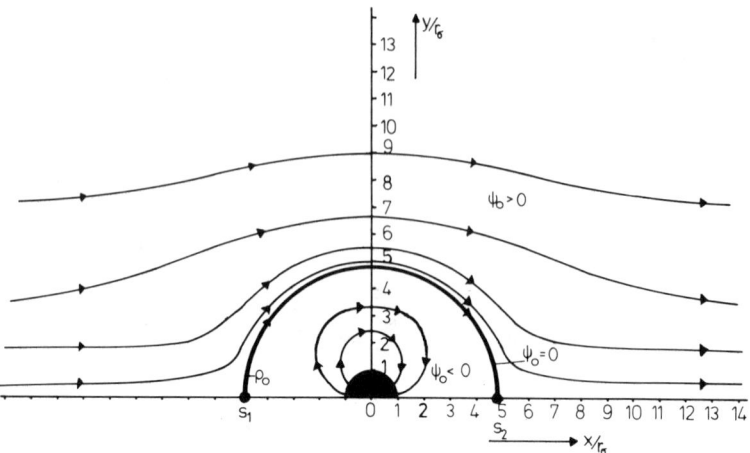

Fig. 6 Flowfield inside and outside the streamer with stagnation points S_1 and S_2. Shown are lines of constant ψ_o.

near threshold $\beta = \beta_c$.

Appropriate formulas for $\rho_o > 1$ can be derived from the relation $f_a(\rho_{oa}) = 0$. The transcendental equation that determines ρ_{oa} reads

$$\ln \rho_{oa}^2 + \frac{1}{\rho_{oa}^2}\left(1 - \frac{2}{3}\hat{\rho}_{mi}\right) = 32\left(\frac{1}{\beta_c^2} - \frac{1}{\beta^2}\right) + 1 - \frac{3}{2}\hat{\rho}_{mi}$$

$$= 1 + 2\ln\frac{2}{\gamma\alpha} - \frac{32}{\beta^2} \quad (46)$$

To show the validity range of these formulas, they are compared in Fig. 5 with the exact solutions.

The numerical evaluation shows that, under realistic conditions in MHD channels (B = 5 T; u_s = 900 m/s; v_∞ = 100 m/s; r_σ = 10^{-2} m; η = $10^{-4} J_s/m^3$; L = 0.15 m; $\hat{\rho}_{mi}$ = 0.5; p = 1/7; α = 0.225; σ_o = 100 $\Omega^{-1}m^{-1}$; E_{zo} = 500 V/m; β_c = 2.84; β = 71), the aerodynamic radius reaches values that are in contradiction to the experience. Thus, for practical purposes, this streamer theory, valid for an infinitely extended channel width, must be modified, as is done in the next section.

Solution of the Basic Equations for a Channel of Finite Width

The aerodynamic radius r_{aer} is a strong function of the Hartmann number, and, as shown before, for realistic

STREAMER DYNAMICS IN MHD GENERATORS 487

conditions r_{aer} reaches too large values if only one streamer in an infinitely extended channel is considered. Two effects may reduce the streamer size: 1) The channel geometry has a finite width in the y direction with a spatial distance D of the walls, and 2) more than one streamer is present in the channel (see Fig. 7).

To fulfill the boundary condition valid for streamlines that are parallel to the channel walls, 2N virtual streamers are introduced in the positive and negative y direction at a constant distance D. Further, it is assumed that 2M + 1 subsequent streamers in the positive and negative x direction are present in the channel with distance E in between each other. All streamers are assumed to have the same velocity and are transformed at rest in the same way as described in the section entitled The Basic Equations. In the preceding section, a solution of the basic equations is presented that describes the flowfield belonging to one streamer in a channel of infinite width whose position is

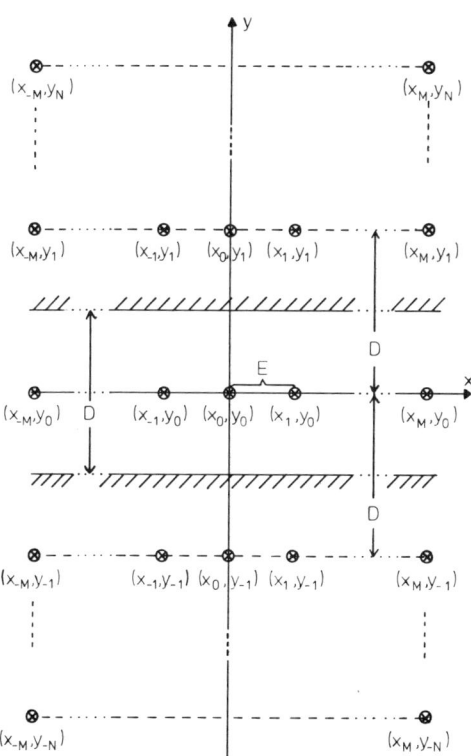

Fig. 7 Arrangement of streamers in a channel of width D.

at $x = 0$, $y = 0$. To perform the superposition of $2M + 1$ streamers in the x direction and $2N + 1$ streamers in the y direction, consider a streamer that is shifted to a position x_m, y_n, where (m,n) are integers with $-M < m < M$ and $-N < n < N$, respectively (see Fig. 7). The mutual arrangements of the streamers are such that the current carrying parts do not overlap; in other words, $D, E \gg r_\sigma$. It is easy to show a solution of Eqs. (19) and (20) in a region, where $\rho_m = \rho_{m\infty} = $ const and $j_z = 0$, that means a solution of

$$\Delta \chi + g''(0) \chi = 0 \qquad (47)$$

$$\Delta \psi_0 - \rho_{m\infty} \chi = 0 \qquad (48)$$

can be written as

$$\psi_0(x,y,x_m,y_n) = \rho_{m\infty} v_\infty (y - y_n) f_a(\rho_{mn}) \qquad (49)$$

Here, $f_a(\rho_{mn})$ follows from Eq. (39), where

$$\rho^2 = \rho_{mn}^2 = \frac{(x-x_m)^2}{r_\sigma^2} + \frac{(y-y_n)^2}{r_\sigma^2} \qquad (50)$$

Because Eqs. (47) and (48) are linear and homogeneous, a superposition of solutions like Eq. (49) also satisfies the differential equations (47) and (48). Thus, one expects that a very general solution results as a superposition of several individual streamers:

$$\psi_0 = \rho_{m\infty} v_\infty \sum_{m=-M}^{m=+M} \sum_{n=-N}^{n=+N} C_{mn} \cdot (y-y_n) \cdot f_a(\rho_{mn}) \qquad (51)$$

where the C_{mn} are constants.

To procede we choose a regular arrangement of the streamers as indicated in Fig. 7, where the streamers are located at

$$x_m = mE \qquad y_n = nD$$
$$-M \leq m \leq M \qquad -N \leq n \leq N \qquad (52)$$

Additionally, all streamers are weighted in the same way,

so one figures

$$C_{mn} = \frac{1}{(2M+1)} \cdot \frac{1}{(2N+1)} \qquad (53)$$

Figure 8 qualitatively shows the flowfield around a set of three streamers. As is evident, the "virtual" streamers indicated by n = 1 and n = -1 produce, after superimposing them on the flowfield of the streamer at $(x,y) = 0$, the following two effects:
1) The backflow velocity $v_x(0,0)$ at $(x,y) = 0$ is reduced; as a result, a small aerodynamic radius follows.
2) The streamlines at $y = \pm D/2$ are more parallel to the wall after superposition.

In order to quantify the flowfield inside the streamer center, the backflow velocity at $(x,y) = 0$ for one single streamer is considered, which reads, using Eqs. (22), (27), (38), and (40),

$$v_x(0,0) = (v_\infty/\hat{\rho}_{mi})(1 - \beta^2/\beta_c^2) \qquad (54)$$

Assuming the distance D is so large that $D/\rho_0 r_\sigma \gg 1$, the superposition of 2N additional "virtual" streamers in the y

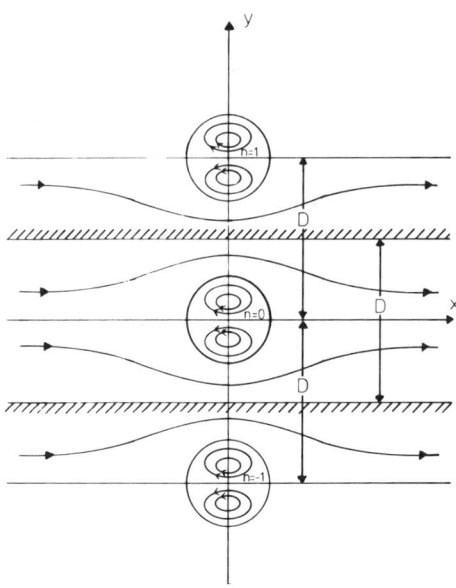

Fig. 8 Influence of two virtual streamers (n = ±1) on the flowfield.

direction produces the backflow velocity at $(x,y) = 0$

$$v_x^N(0,0) = \left[\frac{v_\infty}{\hat{\rho}_{mi}}\left(1 - \frac{\beta^2}{\beta^2}\right) + 2N\, v_\infty\right] (2N + 1) \quad (55)$$

A comparison of Eqs. (54) and (55) directly leads to

$$v_x^N(0,0) = \frac{v_\infty}{\hat{\rho}_{mi}} \frac{2\hat{\rho}_{mi} + 1}{2N + 1}\left(1 - \frac{\beta^2}{\beta_{cN}^2}\right) \quad (56)$$

where

$$\beta_{cN}^2 = \beta_c^2(1 + 2N\, \hat{\rho}_{mi}) \quad (57)$$

Thus, the superposition of 2N streamers enhances β_c to $\beta_{cN} \approx \sqrt{1+2N\hat{\rho}_{mi}}\,\beta_c$. In other words, one needs a much larger β to initiate a backflow in case of N streamers than in case of a single streamer.

A similar result can be derived if 2M additional streamers in the x direction at a distance E are considered, where, again, $E/\rho_0 r_\sigma \gg 1$ is postulated. In case of stream-

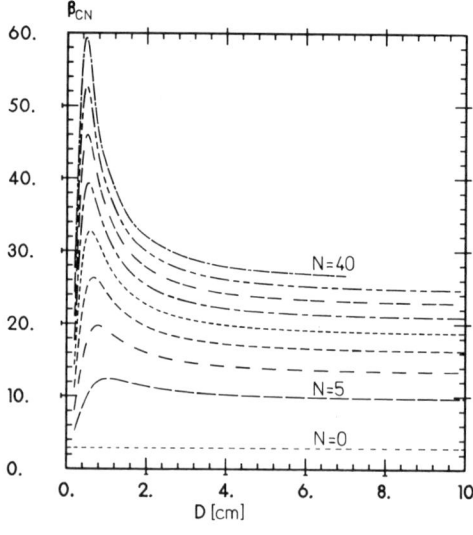

Fig. 9 Critical Hartmann number vs channel width with the number of virtual streamers in the ±y direction as a parameter.

STREAMER DYNAMICS IN MHD GENERATORS

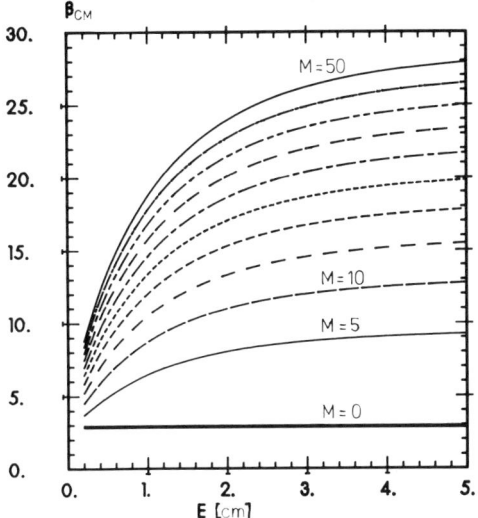

Fig. 10 Critical Hartmann number vs distance between the streamers with the number of superimposed streamers in the ±x direction as a parameter.

er superposition in the x and y direction, one gets

$$\beta_{cNM}^2 \approx \beta_c^2 \{1 + [(2N+1)(2M+1)-1]\hat{\rho}_{mi}\} \quad (58)$$

The situation is more complex if the condition $D, E \gg \rho_0 r_\sigma$ is no longer valid. Then the resulting velocity vector component added at the position $(x,y) = 0$ has no longer the value $[2N/(2N+1)] v_\infty$, as derived in Eq. (55). A numerical solution performed by determining the zeros of Eq. (51) at $x = 0$ leads to β_{cNM} data that are quite different from the result of Eq. (58), as can be seen from Figs. 9 and 10. Here, β_{cN} is plotted as a function of the channel height D, and β_{cM} as function of the streamer distance E. It can be easily checked that the asymptotic data for large E and D are in agreement with formula (58). The increase in β_{cN} with decreasing D (see Fig. 9) is due to the fact that the velocity component one has to add at the position $(x,y) = 0$ becomes larger than $[2N/(2N+1)] v_\infty$ if the conditions $D/\rho_0 r_\sigma \gtrsim 1$ is valid. In the case of $D/\rho_0 r_\sigma \lesssim 1$, the velocity component one has to superimpose decreases again and β_{cN} decreases, too. The situation is quite different when the E dependence of β_{cM} is studied (see Fig. 10). Here, β_{cM} is monotonously decreasing with decreasing E because of the

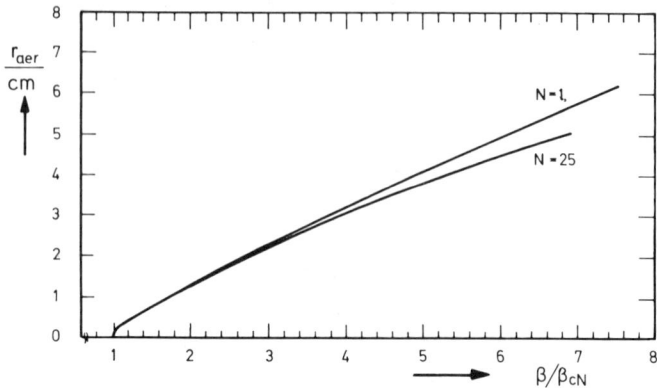

Fig. 11 Aerodynamic radius vs normalized Hartmann number with the number of virtual streamers (in the ±y direction) as a parameter.

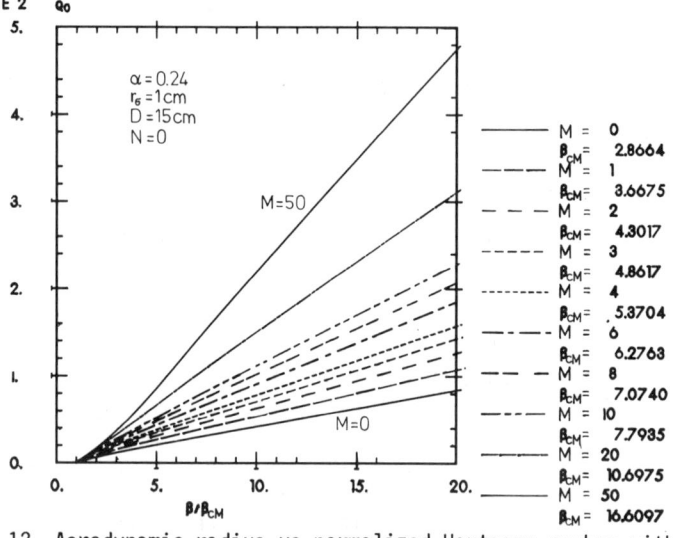

Fig. 12 Aerodynamic radius vs normalized Hartmann number with the number of superimposed streamers (in the ±x direction) as a parameter.

monotonous reductions in velocity if approaching the stagnation points at r_{aer}.

The increase in β_c that occurs if streamers are superimposed in a way as described before is accompanied by a decrease of the aerodynamic radius r_{aer}. Because the flow line belonging to $\psi_0 = 0$ is now no longer of circular shape, the y value at $x = 0$ is chosen to characterize the aerodynamic radius, as mentioned before. As an example, Fig. 11 shows the zeros of Eq. (51) (for $x = 0$) as a function of β/β_{cN}. The result is, at least for $\beta/\beta_{cN} < 10$, not very

STREAMER DYNAMICS IN MHD GENERATORS 493

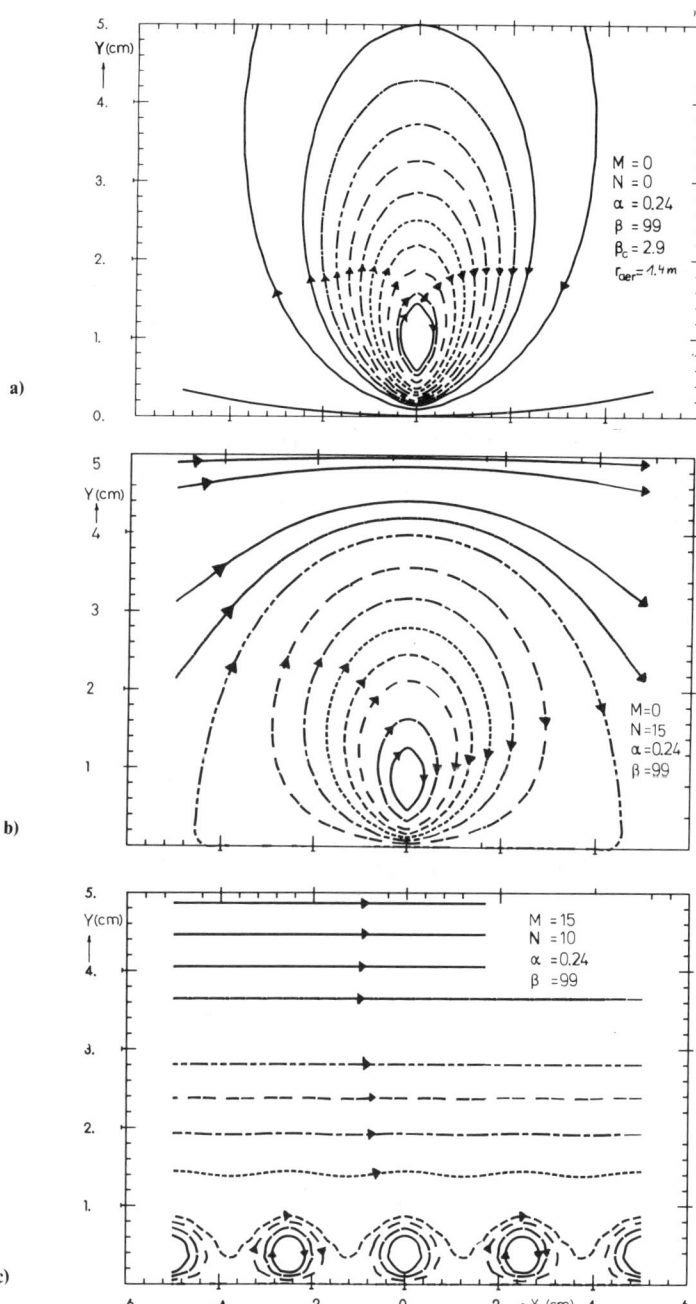

Fig. 13a) Flowfield of one streamer in an MHD channel of infinite width. b) Flowfield of one streamer after superposition of 15 virtual streamers in the ±y direction. c) Flowfield of 15 streamers in the ±x direction and 10 virtual streamers in the ±y direction.

sensitive with respect to the number of superimposed streamers in the y direction if the normalized variable β/β_{cN} is chosen.

On the other hand, the resulting aerodynamic radius is a much stronger function of the number of streamers superimposed in the x direction, as can be seen from Fig. 12. Notice, however, that a larger β/β_c interval is shown. Both figures postulate a reduction of the aerodynamic radius if for a given β value the number of streamers is increased. With increasing M or N, the ratio β/β_{cN} decreases, leading to a smaller value r_{aer} (see also Fig. 13).

The superposition of streamers as described here produces, as mentioned, the effect that the flow lines in the vicinity of the channel wall are smoothly aligned with the wall contour. This can be seen best from the flow pattern. Figure 13a shows the situation for one streamer alone, where $\alpha = 0.24$, $\beta = 99$ was assumed. The aerodynamic radius reaches the value $r_{aer} = 1.4$ m and exceeds the channel width (D = 0.1 m). Now 30 "virtual" streamers in the $\pm y$ direction are superimposed leading to the flowfield in Fig. 13b. The aerodynamic radius is reduced to ≈ 0.04 m. Notice that the flow lines are now nearly parallel to the wall. In Fig. 13c, finally 40 streamers in the x direction (M = 20) and 20 streamers in the y direction are chosen. The aerodynamic radius is now about 0.01 m. This plot characterizes very well the situation in a realistic MHD experiment.[5]

As a closing remark, it must be stated that in the vicinity of the wall our solution does not describe the physical solution very well, because in reality the fluid is at rest near the wall (in the laboratory frame). Thus, a boundary layer is developed that must be analyzed by other methods.[6]

Conclusions

The nonhomogeneous discharge type with a strongly constricted current density distribution in closed-cycle MHD generators in connection with the external $\vec{B}$ field leads to $\vec{j}\vec{B}$ forces that can produce a backflow in the streamer center. The results of an appropriate analysis are only in accordance with the experience if the flowfield of several interacting streamers is studied.

Acknowledgment

The authors wish to thank A. Müller for writing the computer programs.

References

[1] Sens, A. F. C., et al., "Investigation on the Gasdynamical Effects of a Nonuniform Supersonic Flow with Streamers in a Noble Gas MHD Generator," Symposium on Engineering Aspects of MHD, Irvine, CA, 1982.

[2] Flinsenberg, H. J., Balemans, W. J. M., Rietjens, L. H. T., "Power Extraction Experiments with the Eindhoven MHD Blow-down Facility," Proceedings of the Eighth International Conference on MHD, Vol. 1, 1983, pp. 80-87.

[3] Uhlenbusch, J., "Miscellaneous Arc Devices," Physics, Vol. 82c, March - April 1976, pp. 61-85.

[4] Jones, G. R. and Fang, M. T. C., "The Physics of High Power Arcs," Reports on Progress in Physics, Vol. 43, August - September 1980, pp. 1415-1465.

[5] Flinsenberg, H. J., "Fossil Fuel Fired Closed Cycle MHD Power Generating Experiments," Ph.D. Thesis, Eindhoven University of Technology, The Netherlands, 1983.

[6] Massee, P., "Gasdynamic Performance in Relation to the Power Extraction of an MHD Generator," Ph.D. Thesis, Eindhoven University of Technology, The Netherlands, 1983.

Magneto-Fluid-Dynamic Issues for Fusion First-Wall and Blanket Systems

B. Picologlou,* C.B. Reed,* R. Nygren,† and J. Roberts‡
Argonne National Laboratory, Argonne, Illinois

Abstract

Magnetohydrodynamic considerations are shown to be of paramount importance in self-cooled liquid metal blankets for fusion reactors. Coupling of the stress and MHD pressure drop through the conduit wall thickness is such that self-cooled liquid metal blankets without electrical insulators may not be feasible, unless acceptable designs that address satisfactorily MHD-related constraints can be developed. Such an acceptable design, based on current understanding and available analyses and/or experimental data on high Hartmann number, high interaction parameter MHD, is presented. The confidence in the success of the design is directly related to uncertainties on the validity of assumptions necessary to carry out the MHD calculations. Thus, the need for developing analytical tools, supported by experiment, is identified, and a program for addressing this need is formulated. The experimental facility and the first tests are outlined, and the broad goals of companion analytical work are discussed.

Introduction

In fusion reactors fueled with deuterium (D) and tritium (T), the hot plasma will produce alpha particles (He) and energetic neutrons (n) as products of the fusion reaction:

$$D + T \rightarrow He\ (3.4\ MeV) + n\ (14.1\ MeV)$$

Paper presented at the Fourth Beer-Sheva Seminar on MHD Flows and Turbulence, Ben-Gurion University of the Negev, Beer-Sheva, Israel, Feb. 27-March 2, 1984. Copyright © 1985 by the American Institute of Aeronautics and Astronautics, Inc. All rights reserved.
*Engineering Division.
†Manager, Blanket Technology.
‡Associate Laboratory Director, Energy and Environmental Technology.

FIRST-WALL AND BLANKET SYSTEMS

Eighty percent of the energy from the reaction is carried away from the plasma by the energetic neutrons, and most of this energy is deposited in a component called the blanket, which contains lithium-bearing material that breeds tritium for fuel through nuclear transmutation reactions

$$^6Li + n \rightarrow He + T + n, \quad ^7Li + n \rightarrow He + T + 2n$$

The major components that surround the plasma in a magnetically confined fusion reactor are shown schematically in Fig. 1.

The first wall and its superstructure, such as a limiter or divertor, provide the physical boundary that confines the edges of the plasma. Electromagnetic radiation from the plasma heats the surface of the first wall. Parts of the superstructure receive intense surface heating and bombardment by energetic particles from the plasma.

In most conceptual designs for power-producing fusion reactors, process heat is extracted from both the first wall and the blanket. Therefore, the coolant(s) for these components must reach temperatures useful for power production. About 320°C is generally held to be a lower bound for supply to steam cycles.

Both solid and liquid lithium-bearing materials have been utilized in reactor designs. Solid breeders are ceramics such as Li_2O, $LiAlO_2$, and Li_8ZrO_6. The liquid metals Li and ^{17}Li-^{83}Pb are prime candidates for lithium metal blankets. The seemingly small fraction of lithium in

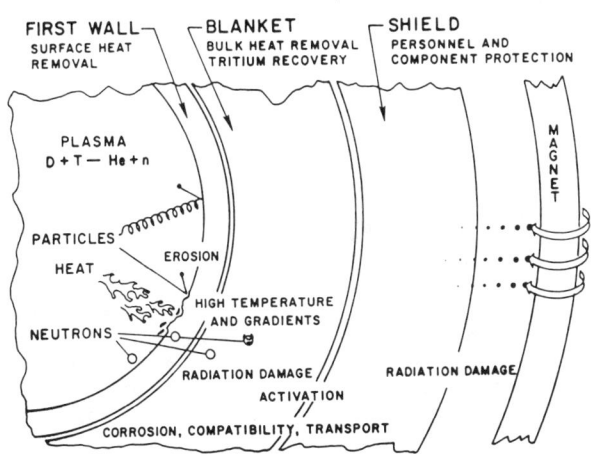

Fig. 1 First wall, blanket, and shield of a fusion reactor.

$17Li$-$83Pb$ produces sufficient tritium when enriched in ^6Li, which has a larger cross section than ^7Li, because of neutron multiplication by the lead.

The shield protects personnel and sensitive components, such as the magnets, from neutron radiation, and the magnets provide the magnetic confinement. The maximum strength of the magnetic field in locations where liquid metal might be circulated in a blanket is about 4 to 8 T depending on the type of reactor.

The major effort in the U.S. Department of Energy's (Fusion) Magnetic Confinement Program focuses on two types of reactors: tokamaks and mirrors. ("Inertial confinement" depends on lasers or particle beams and does not use magnets to confine the plasma.) Figs. 2 and 3 show the basic configurations of tokamak and mirror reactors.

The issues in designing liquid metal blankets for tokamaks and for mirrors are similar, but the following three features of tokamaks make the engineering of their liquid metal blankets somewhat more difficult: 1) higher magnetic fields, 2) longer flow lengths in the high field region, and 3) greater surface heating at the first wall.

Because of the inherent radial gradient in the magnetic field of a tokamak, the confining fields are somewhat higher than those of a mirror to achieve similar parameters in the plasma. The peak field in the blanket of a tokamak fusion reactor will probably be in the range of 6 to 8 T, whereas the comparable value for mirror reactors is about 4 T.

The flow paths in mirrors are simpler because of access for manifolds. In the central cell of a mirror reactor, the

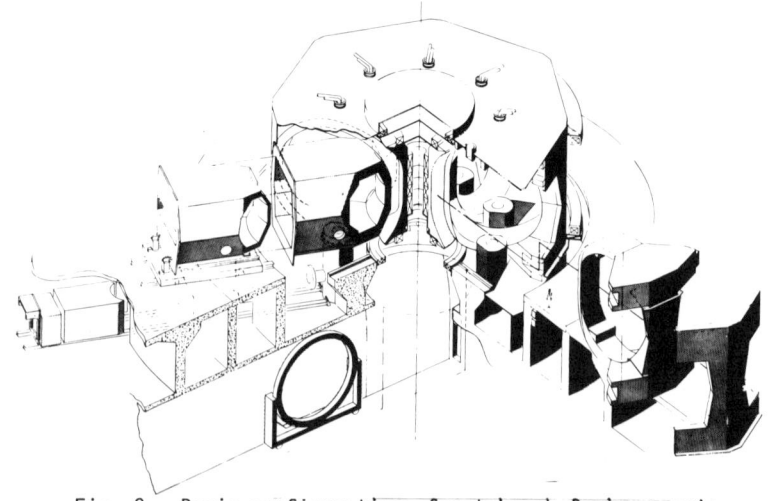

Fig. 2   Basic configuration of a tokamak fusion reactor.

FIRST-WALL AND BLANKET SYSTEMS

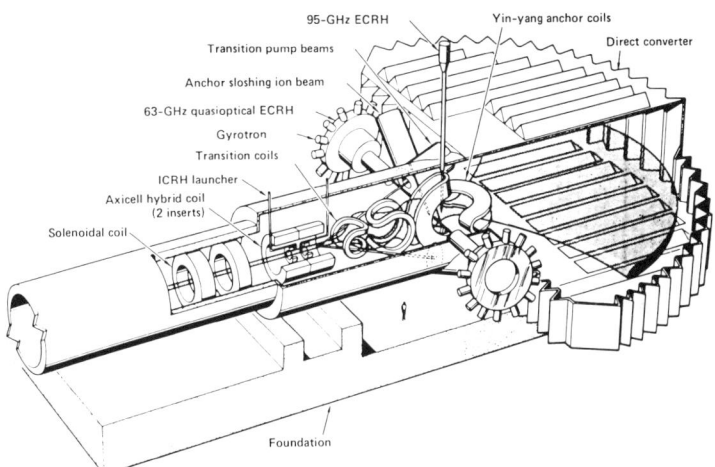

Fig. 3   Basic configuration of a mirror fusion reactor.

magnets are evenly spaced, with access for manifolds to the blanket both from above and below and, if need be, from the sides. Near the center of a tokamak, the magnets butt together, and all manifolding must be accomplished on the outboard side of the blanket. Consequently, the flow paths in the high field region are comparatively longer for tokamaks than for mirrors.

The third feature that makes the engineering of liquid metal blankets more difficult in tokamaks is the demanding cooling requirement of the first wall. The anticipated surface heat flux on the first wall of tokamak reactors is in the range of 20-100 W/cm^2. This becomes problematic for liquid metal blankets because the magnetic field suppresses turbulent heat transfer. With heat penetrating the (liquid metal) coolant essentially by simple conduction, the implied rise in temperature of the fluid layers adjacent to the first wall would result in excessively high temperatures in the structure for any simple "straight through" flow path. Our analyses to date, discussed later, indicate that this temperature limit is set by corrosion limits, which are sensitive to the temperature of the liquid metal in contact with the first wall, rather than by mechanical properties in the region of peak temperature on the plasma side of the first wall.

The structural temperatures would be reduced if the flow rates were greater, but higher velocities would increase the coolant pressure owing to MHD effects and would place greater stress on the structure. Increasing the thickness of the structure to reduce stress is an ineffective design solution because the increased electrical conductance of the wall exacerbates the MHD problem. Also, increasing the thicknesses of structural members is undesirable because the neutronic performance of the blanket is degraded. A design solution for liquid metal blankets in tokamaks has been identified in the Blanket Comparison and Selection Study, a fusion design study conducted in the U.S. during 1983 and 1984. The next section of this paper describes the preliminary analyses performed for this design effort.

The M-shaped velocity profiles induced by three-dimensional effects at high magnetic interaction parameters and Hartmann numbers are extremely important with respect to heat transfer and corrosion rates in blanket designs for fusion reactors. The beneficial or adverse impacts on the design of this phenomenon may be significant but have not yet been evaluated. Developing appropriate calculational capability, supported by adequate experimental confirmation, to describe pressure and velocity fields at high Hartmann numbers and magnetic interaction parameters, $M \sim N \sim 10^5$, is a prerequisite for realistic engineering of liquid metal blankets and is the central goal of an experimental and analytical task recently implemented in the U.S. The latter portion of this paper will briefly describe the activities planned in this program.

Blanket Comparison and Selection Study

MHD Considerations for Self-Cooled Liquid Metal Blankets

The two-year Blanket Comparison and Selection Study (BCSS) was initiated by the U.S. Department of Energy/Office of Fusion Energy in October 1982. The study is being carried out by a multidisciplinary team led by Argonne National Laboratory, with personnel from national laboratories, industry, and universities with objectives to: 1) define a small number (three) of blanket design concepts that should be the focus of the blanket R & D program; 2) identify and assign priorities to the critical issues for the leading blanket concepts; and 3) provide the technical input necessary to develop a blanket R & D program plan.

The viability of blanket concepts based on the use of liquid metals, either lithium or ^{17}Li-^{83}Pb, as the tritium breeding material are being evaluated in the BCSS. The blanket concepts considered for both tokamak and mirror reactors represent two classes: 1) self-cooled systems in which the liquid metal serves as both breeder and coolant, and 2) separately cooled liquid metal blanket concepts.

FIRST-WALL AND BLANKET SYSTEMS

In the following, only liquid metal self-cooled blankets for tokamak reactors will be discussed in detail. The development of liquid metal cooled tandem mirror designs will also be completed within the framework of the BCSS.

The main feature of self-cooled blanket concepts relates to the use of the same liquid metal as both tritium breeder and coolant. This feature greatly simplifies both materials and design considerations since the blanket requires only a structure and a breeder-coolant. Coolant-breeder compatibility/reactivity is not a factor, and structure compatibility considerations are less restrictive. Heat-transfer requirements are also reduced because most of the nuclear heating is deposited directly in the breeder-coolant. The liquid metals have good heat-transfer characteristics with high thermal conductivities and heat capacities, which are beneficial for normal and transient operation. Lithium and Li-Pb both provide relatively high tritium breeding capability, and tritium recovery with relatively low tritium inventory is feasible.

Under the fusion reactor parameters, common to all blanket designs (e.g., magnetic field, dimensions, neutron and surface heat fluxes, etc.), dictated by the guidelines of the BCSS, the self-cooled liquid metal blanket designs have been driven primarily by magnetohydrodynamic considerations. To understand the reason for this, consider a typical coolant conduit, as may be found in the design of an inboard liquid metal cooled blanket. Such a conduit, with a cross-sectional area 2a × 2b and length ℓ, is shown in Fig. 4 (the arguments to follow are also valid for a circu-

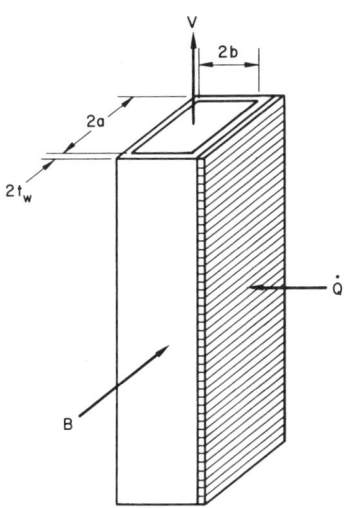

Fig. 4 Typical coolant conduit in the inboard blanket of a tokamak reactor.

lar conduit). The conduit of wall thickness t_w is normal to a magnetic flux density B and carries liquid metal coolant at an average velocity V. Heat is supplied to the coolant at a rate of $\dot{Q}$ per unit surface area of one of the lateral walls of the conduit.

For the conditions of high Hartmann number ($M \sim 10^5$) and thin conducting walls $\phi = O(10^{-2} - 10^{-3})$, valid in the fusion blanket environment, the pressure drop along the conduit is given by

$$\Delta p = \ell \sigma V B^2 \phi \tag{1}$$

where ϕ is the wall conductivity ratio $\sigma_w t_w / \sigma a$.

The mixed mean temperature rise along the conduit is given by

$$\Delta T = \frac{\dot{Q} \ell\, 2a}{\dot{m}\, C_p} \tag{2}$$

where the mass flow rate $\dot{m}$ is given by $\dot{m} = \rho V A = 4ab\rho V$. Combining the above equations one obtains

$$\Delta p = \frac{\dot{Q} \ell^2 B^2}{\rho C_p \Delta T} \frac{\sigma_w t_w}{a} \frac{1}{2b} \tag{3}$$

If the pressure at the conduit outlet is of the same order of magnitude as that of the pressure exterior to the conduit, the conduit walls near the inlet will have to sustain a pressure difference given by Eq. (3). As a result, in the walls of width 2b, a tensile stress $S = \Delta p\, a/t_w$ will develop. By virtue of Eq. (3), this tensile stress will be equal to

$$S = \frac{\dot{Q} \ell^2 B^2 \sigma_w}{\rho C_p \Delta T\, 2b} \tag{4}$$

For a conduit in the inboard blanket, material, geometric, and functional constraints dictate that $B \sim 7T$, $\ell \sim 6m$, $\dot{Q} \sim 6MW/m^2$, $\sigma_w \sim 10^6 S/m$, $\Delta T \sim 50°C$, $2b \sim 0.5m$, and $\rho C_p \sim 2MJ/m^3K$. With these values of the parameters, the stress is about 200 MPa. In actuality, the stress will be even higher because the limit on interface temperature at the first wall dictates a higher velocity than that resulting from a specified mixed-mean temperature rise of $50°C$. The stress of 200 MPa is very close to the allowable primary stress in the wall material S_m. Had S, as computed from Eq. (4), been an order of magnitude lower than S_m, MHD considerations, although still important, would not be critical to the design. Had S been an order of magnitude higher than S_m, liquid metal blankets without insulators would be impossible to design. As it stands, the fact that S is so close to the allowable stress means that MHD considerations are of paramount importance, and designs that can adequately cool the first wall without significant rise in pressure drop appear to be feasible but still require development.

As can be seen from Eq. (4), the stress is independent of the wall thickness t_w. The common practice of increasing the wall thickness in order to reduce the stress can not be applied here. An increase in t_w will result in a proportional increase in Δp, leaving the stress in the wall the same.

There are two possible solutions to the situation. The first is to develop a design with tapered walls so that the walls will be thicker near the inlet and thinner near the outlet. If the tapering is gradual, the overall Δp will be proportional to the average wall thickness and the stress at the inlet can be reduced. The design should also incorporate first-wall coolant channels that are parallel to the magnetic field so that the high velocities needed for cooling the first wall can be achieved without MHD pressure drop penalty.

The second solution is to decouple the wall stress from the MHD pressure drop by developing a sandwich construction, in which the pressure is carried by a heavier wall, electrically insulated from the wall in contact with the liquid metal. The obvious solution of insulating the conduit wall so that the pressure drop will be reduced by a factor of M_ϕ (at least 10^2 for the fusion blanket conditions) requires the development of appropriate insulating coatings compatible with liquid metals at elevated temperatures.

The possibilities of sandwich construction or insulating coatings are currently being evaluated and appear to be promising, but only an acceptable reference design for a self-cooled liquid metal blanket without the use of such insulators will be reported here. The design and the details of the analysis are reported in Ref. 1. Here, only the salient features of the design and the methods of computing the MHD pressure drops are given.

Reference Design Concept for Self-Cooled Liquid Metal Blankets

Figure 5 shows the schematic of a toroidal/poloidal flow blanket. It is composed of slightly slanted poloidal manifolds and relatively small toroidal channels. Each manifold supplies a number of toroidal channels. The toroidal channels are exposed to both the surface heat flux and the relatively high nuclear heating rate, while the poloidal manifold is heated mainly by nuclear heating. The poloidal manifold is protected by the toroidal channels both thermally from the surface heat flux and structurally from radiation damage. The mean velocity in the poloidal manifold can be kept at relatively low values to reduce the MHD pressure drop through the manifold. Since the pressure drop through the manifold is the single largest pressure drop of the entire blanket, its reduction will reduce significantly the overall pressure drop of the blanket.

A second advantage of the toroidal/poloidal flow blanket is that the walls of the poloidal manifold can take higher stress (primary and thermal) levels that can be the first wall, since the former is not exposed to the surface heat flux and receives less radiation dosage than the latter. The first wall is cooled by the coolant flowing through the small toroidal channels that are parallel to the toroidal magnetic field. The flow in the toroidal channels is perpendicular to the poloidal field, which is much smaller than the toroidal field (~0.5T). Thus, the velocity in the toroidal channels can be increased considerably over that in the poloidal manifold without increasing significantly the overall pressure drop through the blanket. It is this relatively high velocity in the toroidal channels (compared

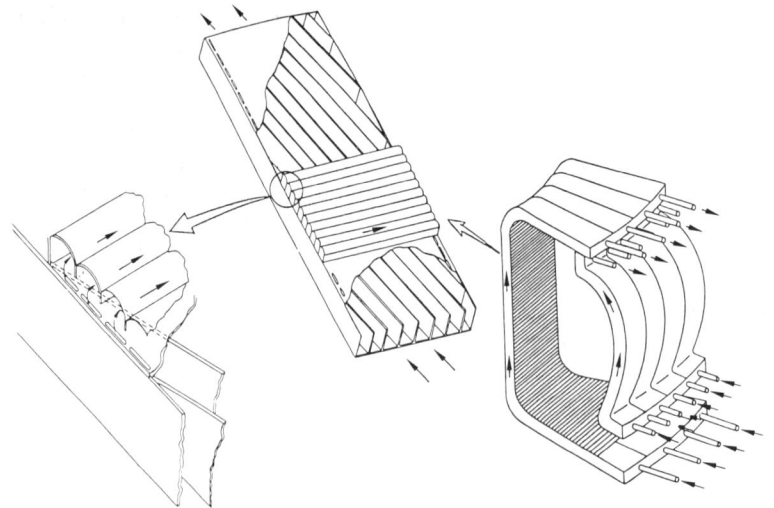

Fig. 5   Poloidal/toroidal reference design schematic.

with the velocity in the manifold) and the relatively short length of the toroidal channels (compared with the full length of the manifold in the poloidal direction) that can reduce the maximum interface temperature to an acceptable level, provided that the flow is evenly distributed in the toroidal channels. There is, of course, some additional pressure drops associated with the bends. The important concept is to provide adequate and efficient cooling of the first wall region while minimizing these additional pressure drops.

The design of both the inboard and outboard blankets involves poloidal manifolds and toroidal first-wall channels. The layout of the blankets is shown to scale in Fig. 6. Coolant enters the magnetic field at a radial location of 12.2 m (point 1, not shown in Fig. 6) through a conduit of uniform cross-sectional area of 0.43m × 0.7m. The effective cross-sectional area changes at the point 2 to 0.43m × 0.66m because of the presence of two 2-in.-thick reflector plates placed near the outer wall of the conduit. Starting at point 3, the channels become inclined by 18 deg, and their manifold cross-sectional area for flow becomes 0.405m×0.6m. As a result of this inclination, the poloidal manifolds terminate at the end walls of the sector, and the flow is forced through appropriately designed openings to the toroidal channels. The toroidal channels run the entire width

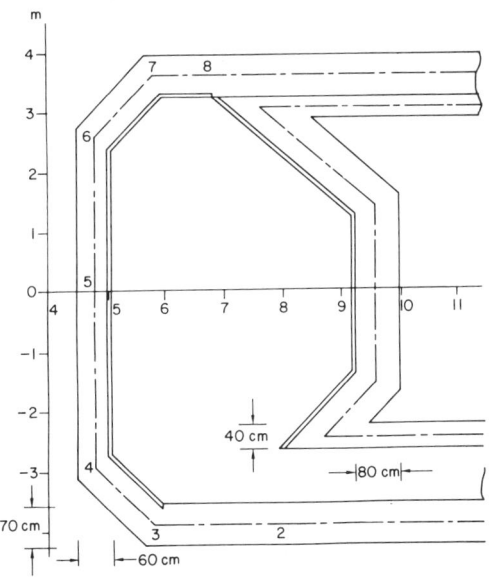

Fig. 6   Reference design blanket layout.

of the sector and communicate through similar openings with another poloidal manifold that takes the coolant out of the reactor. Geometric considerations dictate that there are seven manifolds/sector for the inboard blanket and 49 toroidal channels/manifold. The manifold cross section changes again at point 4 to 0.405m×0.5m and remains constant thereafter until it reaches point 6, where it becomes again 0.405m×0.6m. Beyond point 8, the cross-sectional area remains constant at 0.43m×0.7m.

For the purposes of MHD analysis, only the wall thickness normal to the magnetic field is assumed to be of consequence. This assumption is identical to assuming perfectly conducting sidewalls. In reality, the finite conductance of the sidewalls will result in a smaller pressure drop than that calculated here. The common wall thickness is 8 mm from point 1 to point 3; 6 mm from point 3 to point 4; changes linearly from 6 to 2 mm from point 4 to point 6; and remains constant at 2 mm until the exit from the magnetic field at point 9 (not shown in the figure). In the proposed design, the channels have common walls; in this case, the wall thickness that enters in the calculation of ϕ is equal to one-half of the common wall thickness. Fig.7 shows the details and dimensions of the cross-sectional area of the poloidal manifolds and the toroidal channels. A development of the inboard blanket from point 3 to point 8 is shown in Fig.8.

To compute the pressure distribution along the flow path, the average velocity distribution must be known. The

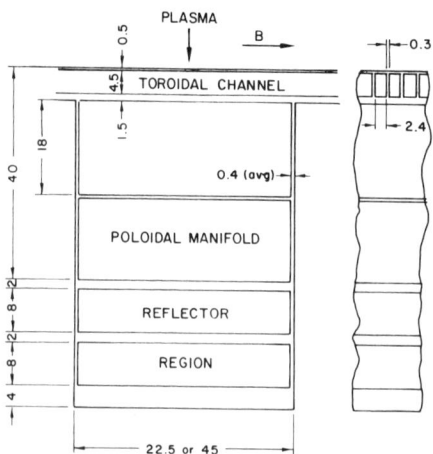

Fig. 7 Cross-sectional views of the toroidal/poloidal blanket. All dimensions in centimeters.

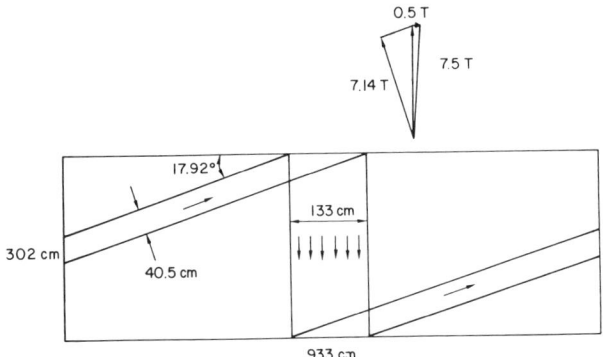

Fig. 8 Development of the inboard region of one sector of the poloidal/toroidal design concept.

velocity, of course, depends on the cross-sectional areas and the mass flow rate of the coolant, which, in turn, depends on the energy deposition rate in the blanket and the specified mixed mean temperature rise. For a specified energy deposition rate of 27 MW/manifold and the specified mixed mean temperature rise of 150°C, the mass flow rate/manifold is 42.3 kg/s. The corresponding average coolant velocities in the inlet and outlet conduits and poloidal manifolds are given in Table 1. The velocity in the toroidal channels, as determined by geometric considerations, is 1.60 m/s.

MHD Calculations

The coolant flows through a series of interconnected channel segments of uniform cross section, most of which are at right angles to the toroidal magnetic field. The Hartmann number, interaction parameter, and wall conductivity ratio for these conduits are of the following order of magnitude, whereas the magnetic Reynolds number is of the order of 10^{-3}.

$$M = O(10^5), \quad N = O(10^5), \quad \Phi = O(10^{-2}\text{-}10^{-3})$$

As a result of the large value of M, the fluid flow and current density are uniform throughout most of the flow cross section, with variations confined in thin layers near the walls. The large value of N makes inertial effects negligible; turbulence is also suppressed everywhere, with the possible exception of fluctuations that may exist in thin shear layers under some special circumstances.

Finally, the condition $\phi \ll 1$ combined with $M\phi \gg 1$ defines the "thin wall" regime, in which the induced currents and, thus, the pressure drops are controlled by the wall resistance. Under such conditions, the pressure gradient for a uniform channel segment in a uniform transverse magnetic field is given by[2]

$$\frac{dp}{dx} = \sigma VB^2 \frac{\phi}{1 + \phi} \qquad (5)$$

For those cases where the magnetic field, the wall thickness, the channel dimensions and geometry, a and b, change gradually in the flow direction, the equation for dp/dx is still approximately valid locally, and it can be integrated to yield the overall pressure drop. Obviously, the more gradual the aforementioned variations, the more reliable the calculation of the overall pressure drop. Unfortunately, at present, accurate criteria for establishing whether a variation is sufficiently gradual do not exist, especially for non circular conduits. Nevertheless, it is the consensus of the MHD community[3], that, although deviations from the fully developed character of the flow can have a strong effect on velocity profiles, the effects on pressure drop are expected to be moderate.

Three-dimensional perturbations result from changing B field (direction or magnitude), changing cross-sectional area dimensions, changing wall thickness, bends, manifolds, etc. Such variations set up axial electric field gradients

Table 1 Corresponding average coolant velocities in the inlet and outlet conditions and poloidal manifolds

Location	Average flow velocity, m/s	Coolant pressure, MPa
Inlet	0.28	3.28
1	0.28	3.25
2	0.28/0.30	2.90
3	0.30/0.35	2.54
4	0.35/0.42	2.18
6	0.42/0.35	0.91
7	0.35/0.28	0.79
8	0.28	0.74/0.33
9	0.28	0.21
Outlet	0.28	0.20

that, in turn, set up circulating currents within the
liquid. These currents interact with the magnetic field
and result in pressure drop over and above that predicted
by fully developed flow theory.

Analyses that have been carried out to date predict
almost stagnant regions at the center of the conduit with
most of the flow rate carried in layers adjacent to the
walls (see, for example, Refs. 4 and 5). Also, theory
predicts significantly different behavior for rectangular
and for circular conduits, with the rectangular ones being
more prone to the adverse three-dimensional effects.

Unlike the case for a straight duct normal to a uniform magnetic field, analysis of three-dimensional effects
is highly case-specific, involved, and, for most cases, not
amendable to solution with currently available analytical
tools. Nevertheless, fair estimates of the pressure drop
associated with a number of three-dimensional effects exist
and are supported by limited experimental data[6], albeit at
much lower values of M and N than those prevailing in the
blanket.

It was assumed in the analysis of the MHD pressure drop
in the fusion blanket that the variation of magnetic field
strength is sufficiently gradual, so that the associated
three-dimensional effects are minimal. The other remaining
three-dimensional effects are associated with the abrupt
change of the magnetic field at the inlet and outlet regions,
the effects associated with abrupt changes in wall thickness,
and those associated with conduit bends, either in a plane
normal to B or, as is the case of the reference design
manifolds, from a direction normal to B to a direction
parallel to B.

The pressure drop associated with abrupt changes in B
(or, equivalently, with abrupt changes in wall thickness or
cross-sectional areas) has been analyzed for thin-wall circular ducts[4]. This pressure drop is found to be equal to

$$\Delta p = C \sigma VaB^2 \sqrt{\phi}$$

with the coefficient C depending on the magnitude of the
discontinuity. The peak value of C was computed to be 0.16.
A conservative value of 0.2 is adopted in the analysis of
the blanket. The pressure drop for a bend in a plane normal
to B can also be estimated by the same equation.

The pressure drop for a bend from a direction normal
to B to one parallel to B can be computed from the following
formula:

$$\Delta p = 0.5 \, \sigma VaB^2 N^{-1/3}$$

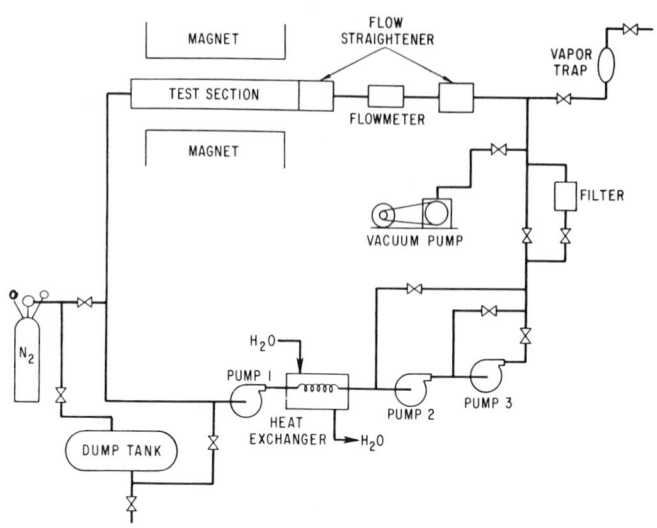

Fig. 9 Schematic of facility for LMMHD experiments.

Such a correlation is based on theoretical considerations, and the coefficient of 0.5 has been established through experimental work at ϕ = 0.155 and N<90 (Ref. 6). If the theoretical foundation of the above correlation is valid, the formula is applicable to values of N prevailing in the fusion reactor (10^5). The coefficient of 0.5 refers to the value of ϕ of 0.155. Although the scaling of the numerical coefficient with ϕ has not been established, it will undoubtedly decrease with decreasing ϕ. Therefore, for the fusion reactor, with its lower value for ϕ, the pressure drop will be smaller than that given by the above formula. Nevertheless, a conservative value of 0.5 has been adopted in the analysis.

The analysis of the reference design performed using the aforementioned formulas resulted in the pressure distribution along the flow path, shown in Table 1, for an outlet pressure of 0.2 MPa.

It was assumed in the pressure distribution shown in Table 1 that the poloidal to toroidal bend was made at point 8, since this distribution will stress the thin walls of the blanket the most. The two values for P_8 correspond to positions upstream and downstream of the toroidal channels.

This pressure distribution results in a maximum primary stress of 170 MPa at the common manifold wall at the location 3. This stress is comfortably lower (by 24%) than the allowable material stress of 210 MPa.

FIRST-WALL AND BLANKET SYSTEMS   511

Liquid Metal MHD Program

The MHD analysis performed in the BCSS was based on the best, if not the only available, information suitable for engineering calculations. Early in the study, it was recognized that such information is limited, and a number of assumptions had to be made in order to carry out the design process. The assumptions were based on engineering judgment, and an attempt to justify them was made whenever possible.

There is little doubt that the theory for fully developed flow in circular and rectangular ducts is well understood and, although experimental confirmation of the theory has been limited to $M \sim 10^3$, the theory is expected to be valid at $M \sim 10^5$. All of the uncertainties in the MHD analysis are centered around three-dimensional effects. The fact remains that three-dimensional effects have been analyzed only for a handful of cases, and experimental confirmation of the theory is virtually nonexistent at a parameter range approaching that of the fusion blanket. Although some data have been obtained at $N = 0 \times 10^2$ and $M \sim 10^2$ (Ref. 6), no velocity profiles were obtained to confirm the very high velocity wall jets predicted by the theory. The only three-dimensional data that include velocity profiles have been obtained at $N \sim 1$ and $M \sim 10^2$, and in nonconducting ducts[7].

Because of the limited information on which the MHD analysis for the blanket was based, and the small margin between the computed and the allowable stresses, MHD was identified as a critical feasibility area for liquid metal cooled blankets. This conclusion was well anticipated, and preliminary plans for a fusion liquid metal MHD program were initiated in early 1983. In August 1983, the Fusion Power Program of Argonne National Laboratory organized a two-day workshop on liquid metal MHD in fusion to arrive at a consensus on the specific direction that the MHD program should take in order to provide the analyses and data needed for a realistic design of a liquid metal blanket.

The participants, presentations, and detailed conclusion of the workshop may be found in Ref. 2. Here, only the general conclusions are repeated. The conclusions were circulated to the workshop attendees, and their concurrence has been obtained.

1) Both new analyses and new data on MHD effects are needed in regions of high Hartmann number M, and magnetic interaction parameter N (M, N > 1000).

2) There are important differences between circular and rectangular ducts that affect scaling parameters in the situations of interest for fusion blankets.

3) Large unanticipated increases in pressure drops from unknown/uncertainties are unlikely in blanket designs with few bends and orifices. Even where three-dimensional effects are present and flow is not fully developed, pressure drop estimates for straight pipes based on somewhat inexact assumptions will probably be valid (within 20% to 30%).

4) In contrast, the flow velocity profiles, which are very important in analysis of heat transfer and corrosion, may be quite dependent on the exactness of MHD analysis and fulfillment of bounding criteria for theory, e.g., inertialess regime. Measurements of velocity profiles in experiments will be important for strengthening our understanding of the phenomena.

5) Analysis will be needed to plan and to interpret experiments. The most productive approach in the near future will be to use scaling laws based on inertialess flow theory and to verify experimentally that the theory can provide useful physical predictions well outside the theoretical limits of its validity.

6) A reasonable approach to liquid metal MHD (LMMHD) research for fusion is to begin with tests that are technically aggressive, i.e., test conditions sufficient to produce magnetic interaction parameters in the range of 1000 and above, and to use existing facilities where possible (as opposed to the construction at this time of a major new facility to study LMMHD).

Based on the recommendations of the workshop and the clarification of issues in the BCSS, Argonne National Laboratory (ANL) has formulated an experimental program on MHD and is sponsoring research on companion analyses. The experimental program will utilize an existing ANL NaK loop and an existing 2.5-T iron core magnet. A schematic of the major loop components is shown in Fig. 9. Major components include a dump tank, pumps, heat exchangers, filter, flow meters, a test section, valving, and a magnet. The NaK-78, which will be used as the working fluid in these experiments, represents an optimum tradeoff between handling considerations and favorable MHD fluid properties. Since velocity measurements are an important part of the experimental program, a room temperature working fluid was considered mandatory.

The maximum loop flow rate is 18 kg/s (300 gallons per minute), and the maximum test section inlet pressure is 2.0 MPa (275 psig). The magnet, borrowed from the high-energy physics program at Fermi National Accelerator Laboratory, is an iron core, water-cooled device with a field strength of

FIRST-WALL AND BLANKET SYSTEMS

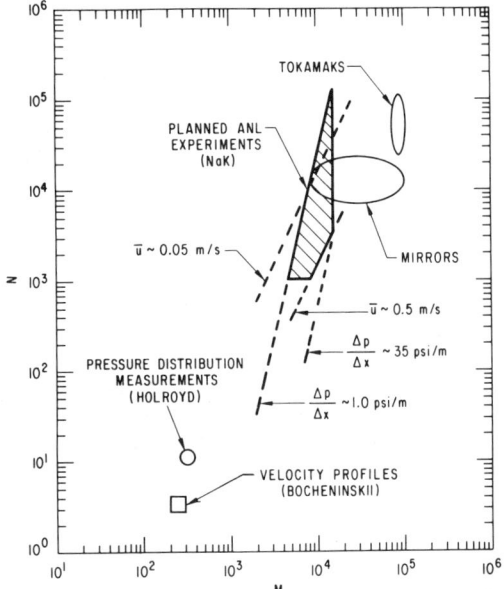

Fig. 10 Hartmann number and interaction parameter ranges for proposed experiments.

2.5 T and weighing 5.7×10^4 kg (62.5 tons). The pole faces are parallel (rather than tapered as for open-cycle MHD magnets) with dimensions approximately 76 cm wide by 1.83 m long. The air gap is 20 cm, nominally. The magnet was designed as a beam line bending magnet and has a highly uniform magnetic field.

The experimental parameter range that can be covered by this loop magnet combination is shown in Fig. 10. This figure was developed assuming a test section of square cross section having a half-width of 0.1 m and a conductivity ratio $\phi = 0.03$. Planned measurements include pressure gradients, voltage distributions, current and velocity profiles. State-of-the-art, as well as advanced instrumentation techniques, will be used.

The figure also shows in an M-N space the region of interest for fusion reactors, existing three-dimensional data, as well as the region for the planned ANL experiments. The region of the planned experiments is determined by the magnetic flux density of the magnet, the size of the conduit, the available capacity in pressure and flow rate of the NaK loop, and the limits of 1 psi/m for the pressure gradient and 0.05 m/s for velocity. The latter limits are close to values for which reliable pressure gradient and velocity measurement can be made.

The measurement of velocity profiles is important in these studies. Existing analyses predict very-high-velocity jets adjacent to the sidewalls. The existence of such jets has not been confirmed experimentally. As was noted earlier, Holroyd's data[6] did not include velocity profile measurements and Bocheninskii's data[7] were obtained in conduits with insulated walls. Both Holroyd's and Bocheninskii's results were obtained in circular conduits whose behavior in three-dimensional effects differ significantly from that of rectangular conduits.

The first phase of the experimental program proposed by ANL will include, but will not necessarily be limited to, detailed measurements on the following configurations.

1) A uniform rectangular conduit in a uniform transverse magnetic field. In the course of the experiment, three-dimensional effects in rapidly varying magnetic flux density will also be obtained at the ends of the magnet pole face.

2) A uniform rectangular conduit in a slowly varying transverse magnetic field.

3) A rectangular conduit with a 180-deg bend on a plane normal to the magnetic flux density vector.

Subsequent possible experiments may include a rectangular duct in a transverse magnetic field not parallel to either of the conduit sides, investigation of three-dimensional effects in a duct with thin walls normal to B but thick sidewalls (as is the case for the reference design), pressure drop and flow distribution through manifolds, and other configurations not amendable to analysis, etc.

The companion analytical effort is directed towards obtaining solutions for the configurations and the conditions of the test articles. The analysis will be based on the intertialess assumptions, although that assumption may not be theoretically valid in all cases. In fact, as mentioned earlier, one of the goals of the experimental program is to establish the limits of the inertialess theory for the cases of interest. An effort will also be made to develop seminumerical techniques for obtaining solutions for three-dimensional effects, with the hope that such an approach will yield results of engineering value in a routine fashion.

Acknowledgments

Work on the development of the reference design was done within the framework of the Blanket Companion and Selection Study. Development of the design is the result of the collective effort of D. Smith, Y. Cha, S. Majumdar, and B. Picologlou.

FIRST-WALL AND BLANKET SYSTEMS   515

References

1. "Blanket Companion and Selection Study," Argonne National Laboratory, Argonne, Ill., ANL/FPP-83-1, October 1983.

2. "Proceedings of Workshop on Liquid Metal MHD in Fusion," Argonne National Laboratory, Argonne, Ill., Aug. 1983.

3. Chang, C.C. and Lundgren, T.S., "Duct Flow in Magnetohydrodynamics," ZAMP, 12, 1961, pp. 100-114.

4. Holroyd, R.J. and Walker, J.S., "A Theoretical Study of the Effects of Wall Conductivity Non-Uniform Magnetic Fields and Variable Area Ducts on Liquid Metal Flows at High Hartmann Number," Journal of Fluid Mechanics, Vol. 96, 1978, pp. 471-495.

5. Ludford, G.S.S. and Walker, J.S., "Current Status of MHD Duct Flow," Proceedings of the Second Beer Sheva International Seminar, Beer-Sheva, Israel, Israel Universities Press, March 1978, pp. 83-95.

6. Holroyd, R.J., "An Experimental Study of the Effects of Wall Conductivity, Nonuniform Magnetic Fields and Variable Area Ducts on Liquid Metal Flows at High Hartmann Number: Part 2. Ducts with Conducting Walls," Journal of Fluid Mechanics, Vol. 96, 1980, pp. 355-374.

7. Bocheninskii, V.P., Tanaev, A.V., and Yakovlev, V.V., "Experimental Investigation of the Flow of an Electrically Conducting Liquid in Curved Tubes of Circular Cross Section in an Intense Magnetic Field," Magnitnaya Gidrodinamika, No. 4, pp.1977, pp. 61-65.

Experiments on a Large Thin-Wall Duct

C.C. Alexion* and A.R. Keeton†

Westinghouse Electric Corporation, Pittsburgh, Pennsylvania

Abstract

A thin-wall duct has been tested at the Westinghouse Research and Development Center in a 60-liter/s (∼1000-gal/min) NaK loop. In a continuing National Science Foundation (NSF) study, entrance and exit effects on liquid metal flows at high interaction parameters have been examined experimentally using the 1.8-m-long test duct and an internal instrumentation probe. Currents up to 16 kA and magnetic flux densities up to 0.3 T have been applied to the liquid metal, resulting in pressure rises up to 36 kPa (5.2 lb/in^2). Pumping efficiencies up to 66% have been attained and differ by less than 9% from theoretically predicted values. These results show qualitative agreement with results from the insulating-wall duct tested in a previous NSF study. Magnetic Reynolds number effects have also been observed, on the same order of magnitude as in the insulating-wall duct. Interaction parameters have been modest ($\lesssim$10), but further tests will bring this value to well over 100.

Introduction

The National Science Foundation (NSF) program, "High Interaction Parameter Magnetofluidynamic (MFD) Studies in Liquid Metal Flows," has combined a rigorous theoretical investigation into entrance/exit effects in real MFD duct flows with an extensive laboratory test program. The goal of this program is to bring together the ultrahigh interac-

---

Paper presented at the Fourth Beer-Sheva Seminar on MHD Flows and Turbulence, Ben-Gurion University of the Negev, Beer-Sheva, Israel, Feb. 27-March 2, 1984. Any opinions, findings, conclusions or recommendations expressed in this publication are those of the authors and do not necessarily reflect the views of the National Science Foundation. Copyright © by the American Institute of Aeronautics and Astronautics, Inc., 1985. All rights reserved.

*Senior Engineer, Electrotechnology Department, Research and Development Center.

†Senior Engineer, Advanced Chemical Research Department, Research and Development Center.

tion parameter analytical solutions and the moderately high numerical solutions with some experimental support between them. Analytical solutions to the problem of MFD flows in finite-length devices with finite-length magnetic fields have been addressed by investigators at the University of Illinois,[1,2] and numerical solutions have been developed by Carnegie-Mellon University,[3] while the experimental program has been carried out at Westinghouse.[4]

In the initial two-year grant, a thin-wall rectangular duct operating in an electromagnetic pump mode was tested in a 60-liter/s NaK loop. This duct had an active region of uniform magnetic and electric fields 23.6 cm (9.3 in.) long and a fluid cross section 7.6 by 8.9 cm (3.0 by 3.5 in.). The side walls of this duct were later insulated with several layers of epoxy, and a successful test program followed. Efficiencies up to 60% were attained, and the overall performance of the duct agreed well with simple analytical theories.[5]

In the current grant period, a thin-wall conducting duct has been tested in the same loop system under similar conditions and having the same physical dimensions. Internal probe tests have also been conducted (as in the previous program), with the goal of detecting fine differences between the insulating and thin-wall designs. These results are presented and compared with theoretical values, together with a discussion of some results from the earlier insulating-wall duct studies.

Test System

Loop Piping and Components

The 60-liter/s experimental facility consists of a NaK loop constructed largely from a 16.1-cm-diam (6-in.-diam) stainless steel pipe. A dump tank and expansion tank are provided, together with an argon pressurization system and vacuum system. Fluid circulation is obtained primarily as a result of the $\vec{I} \times \vec{B}$ forces provided by the interaction of the applied current with the applied magnetic field in the test duct. However, the loop is also connected, via closable valves and through a 8.3-cm-diam (3-in.-diam) stainless steel pipe, to a mechanical pump capable of circulating up to 19 liter/s (300 gal/min) of NaK. Flow in the main circuit may be controlled by the two disk valves--one motor-driven, one hand-operated--which permit the flow to be throttled down. A calibrated turbine flowmeter measures the main flow rate in the large piping. At flow rates below 19 liter/s, diversion of the flow through the 8.3-cm-diam (3-in.-diam) pipe section permits a high-precision electromagnetic flowmeter to be used. Also in this leg is a NaK-to-air heat exchanger capable of dissipating 16 kW at 65°C. A schematic diagram of the loop system is shown in Fig. 1.

Magnet

The magnetic field was generated by a dc-excited C-shaped electromagnet. It was originally a toroidal-coil iron core design with four circular arc segments and four 3.8-cm (1.5-in.) air gaps. For the NSF programs, it was reconfigured into two semicircular arcs with a straight segment joining them (see Fig. 2). Cobalt-steel pole pieces were bolted to each segment on the open side to give a 14-cm (5.5-in.) air gap, each pole face measuring 15.2 by 23.8 cm (6.0 by 9.3 in.). A maximum field of 0.30 T was attained at the centerline of the air gap.

Current Supply

The current source for the test section was a dc homopolar generator having Ga-In liquid metal slip rings built

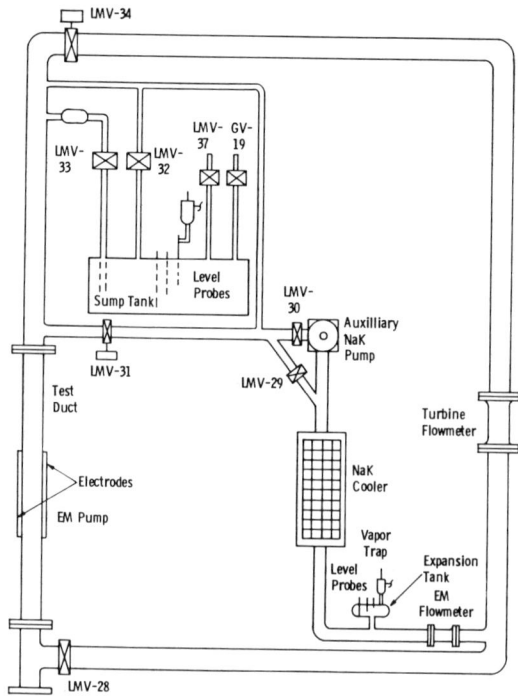

Fig. 1 Schematic diagram of the 60 liter/s loop system with the 1.8-m test section installed.

EXPERIMENTS ON A LARGE THIN-WALL DUCT 519

Fig. 2 Direct-current toroidal-coil electromagnet during construction.

by the General Electric Company Ltd. in England (Fig. 3). A 50-hp drive motor provided power to generate up to 18 kA at approximately 1.5 v. Aluminum bus-bars connected the generator to braided aluminum flexible connectors. These connectors, which were bolted to the outside electrode and the return conductors, allowed the components to move without stress during thermal expansion and contraction.

Internal Instrumentation Probe

The internal instrumentation probe is composed of three stainless steel tubes measuring 3.2, 6.4, and 12.7 mm o.d., respectively. The 3.2-mm tube extends acially through the 6.4-mm tube, and they both extend through the 12.7-mm tube. The combination of the 3.2-mm tube and the 6.4-mm tube forms a pitot tube for measuring NaK velocity (Fig. 4). An epoxy seal at the end of the 6.4-mm tube separates the total and static pressures. These pressures are sensed by a differential pressure transducer and then converted to velocity by a computer program that analyzes all of the data from the test section.

The sensing end of the probe is offset 31.8 mm (1.25 in.) from the main axis of the probe tubing so that measure-

Fig. 3 General Electric Company homopolar generator and control panel.

ments can be made from the NaK stream centerline to the electrode face by rotating the probe. Two Hall probes are located 90 deg apart to measure orthogonal components of the magnetic field in the liquid metal. Axial movement of the probe is made by loosening a Swagelok®fitting at the flange and pushing or pulling the probe to the desired location. Scribe marks are located axially and circumferentially for precise positioning of the probe. The double seal within the interseal gas cavity is designed to prevent external NaK leakage.

Data Acquisition

Data from the test system are collected by a Fluke 2240C Data Logger®(Fig. 5), which has a 12-channel/s scanning capability. The functions of the Data Logger are controlled remotely by a PDP-11 microcomupter program. This program is interactive and utilizes a CRT terminal and keyboard for inputting data. Data reduction and analysis are

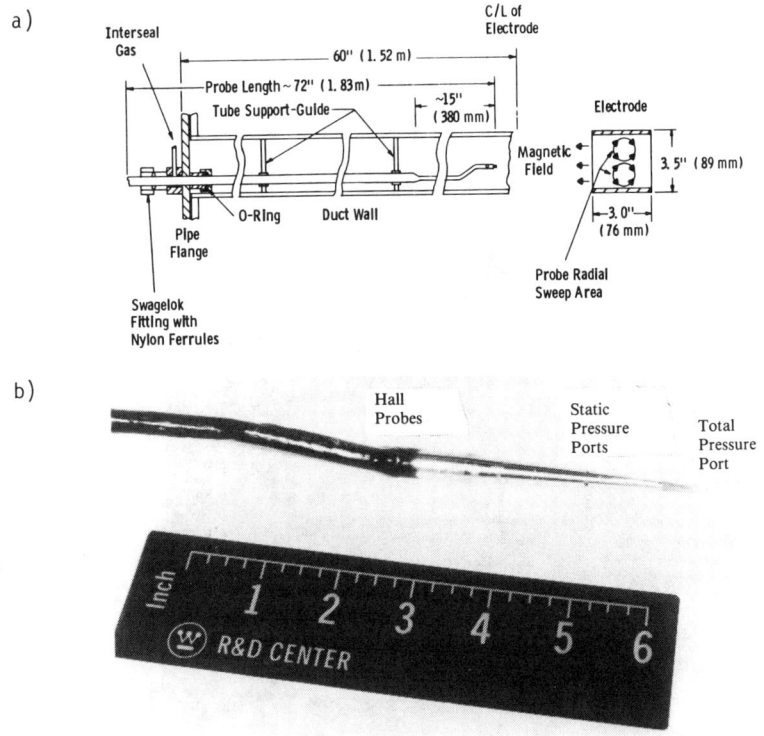

Fig. 4a) Internal instrumentation probe configuration, b) Sensor end of the modified internal instrumentation probe.

also performed by the PDP-11 in a separate computer program. Special graphics packages on the Apple minicomputer are also available through a PDP-Apple hardware link, but this will not be completed until the latter part of the project.

Test Duct

Mechanical Design

The thin-wall NSF duct is constructed of 0.5-mm-thick Inconel 600 sheets bent and spot-welded into a 7.6 by 8.9 cm rectangular box 182.9 cm (72.0 in.) long. Two rectangular windows were cut into the shorter sides, opposite one another, for the electrode faces. Rectangular steel flanges were fabricated and positioned above these openings. Eighteen holes (3.2 mm in diameter) were drilled on each side of both electrode windows--14 of them along the centerline of the wall--to locate the pressure stations. Then, the entire assembly was wrapped in fiberglass and machined flat at the

Fig. 5 Fluke 2240C Data Logger® and Apple II minicomputer in the ELMAG control room.

electrode flanges and end flanges. As a safety precaution, all areas of the duct outside of the active region were covered by U-shaped steel plates. Areas not covered by these plates were coated with potting epoxy, both for its sealing ability at the surface and for fire protection. Swagelok® fittings and circle-seal NaK valves were then fitted to the protruding stainless steel tubes to allow connections to the transducer lines. An overall view of the duct at this stage of fabrication is shown in Fig. 6.

Electrodes

The two OFHC copper electrodes were inserted in their windows with face gaskets and tightened down on the flanges with stainless steel studs. This procedure did not seal the electrodes adequately, so they were potted in place and sealed with epoxy. The electrode faces are flush with the inside surface of the duct walls to minimize flow disturbance, but they are not directly in contact with the walls.

Fig. 6 View of the thin-wall test duct with fiberglass structural shell showing electrode (in foreground) before installation.

The electrodes measure 7.6 by 23.6 cm long (3.0 by 9.3 in.), which is the same as the earlier duct design.[4]

Instrumentation

Pressure transducers are connected via NaK-filled sensing lines to 27 stainless steel tubes penetrating the duct side wall. In the present program, these penetrations are located on both sides of the duct, but the transducers only sense the pressures on one side of the duct--the one outside of the magnet. Twenty-three of these transducers sense pressures along the centerline of the duct side wall, while the other four are offset--two at each end of the active region. Nine of these 23 locations are within the electrode.

Electric potentials are measured at these same locations by wires spot-welded to Swagelok® fittings. There are 27 potential taps on the "outside" wall (where the pressures are measured), but only 20 on the "inside" wall, including five in the electrode. These electric potential measurements provide some indication of the integrity of the thin liner and the amount of current fringing into the end regions. They will be discussed with the test results in the next section. Table 1 lists the axial locations of the pressure stations and potential taps.

Test Results

Pressure Profiles

The duct side-wall pressures are calculated after the calibrated voltage signals of the pressure transducers are converted to pressures by a computer program. As mentioned earlier, there are nine stations within the electrode and 14 outside of the active region. The shape of the pressure profile is mostly dependent on the flow rate of NaK in the duct and is relatively insensitive to current or magnetic field. Figure 7 shows two profiles taken along the duct

Table 1 Locations of pressure stations and voltage taps.

Location, cm (ref. to centerline)	Pressure stations	Voltage taps Outside	Inside
-67.3 (inlet)	P1	V28	V55
-27.3	P4	V31	V59
-24.8	P5	V32	V60
-22.2	P6	V33	V61
-19.7	P7	V34	V62
-17.1	P8	V35	V63
-14.6	P9	V36	V64
-11.4	P10	V37	V65
- 8.6	P11	V38	...
- 5.7	P12	V39	...
- 2.9	P13	V40	V66
0	P14	V41	V67
2.9	P15	V42	V68
5.7	P16	V43	...
8.6	P17	V44	...
11.4	P18	V45	V69
14.6	P19	V46	V70
17.1	P20	V47	V71
19.7	P21	V48	V72
22.2	P22	V49	V73
24.8	P23	V50	V74
27.3	P24	V51	V75
62.2	...	...	V77
67.3 (outlet)	P27	V54	V78

side wall. One profile corresponds to a dead-head, i.e., no flow condition, while the other represents a full flow condition. On the latter curve, note the depressions at the inlet and outlet of the active region. These are readily explained by the current fringing almost parallel to the wall from the edges of the electrodes into the end regions of lower back electromotive force (EMF) thus creating a body force perpendicular to the wall. This force points away from the wall at the inlet and toward the wall at the outlet of the active region.

The centerline pressure profiles cannot be obtained at a single instant in time. Instead, they must be constructed one point at a time for each axial position of the internal probe. These pressure readings are then adjusted to account for overall pressure changes over the entire duct, for example, due to current or field fluctuations between test intervals. There is little effect of flow on the shape of these profiles, however, as shown in Fig. 8.

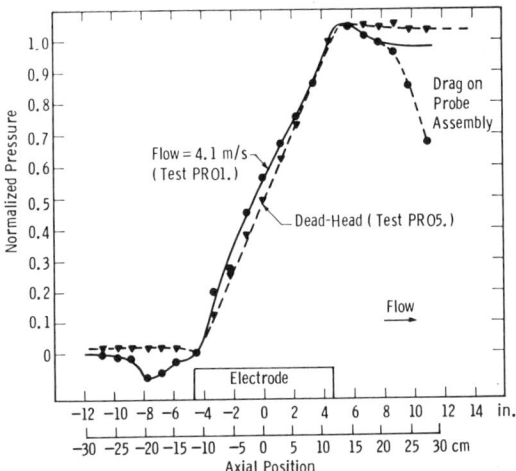

Fig. 7 Side-wall pressure profiles for dead-head and flow test (28 liter/s), with I = 12 kA and B = 0.20 T. Reference pressure rises are 29.0 kPa (dead-head) and 23.7 kPa (flow).

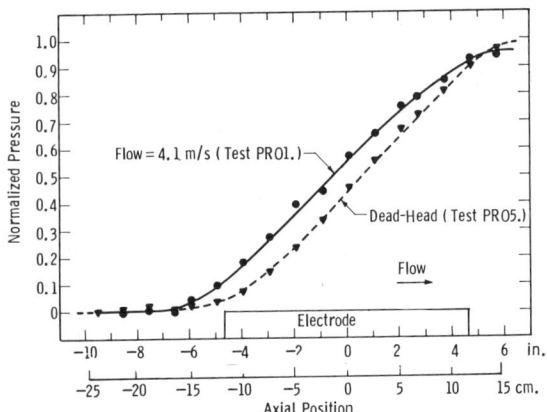

Fig. 8 Centerline pressure profiles for dead-head and flow test (28 liter/s), with I = 12 kA and B = 0.20 T. Reference pressure rises are 20.2 kPa (dead-head) and 18.0 kPa (flow).

Magnetic Field Profiles

The magnetic field profiles are constructed from the normal component of magnetic flux density for each of 17 axial locations along the duct centerline. As was the case with the pressure profiles, the magnetic field profiles are adjusted to account for minor changes ($\approx 0.5\%$) in the overall field due to excitation current fluctuations between test intervals. These effects are much smaller than the pressure fluctuations, however, which are on the order of 0.3 kPa, or about 2%.

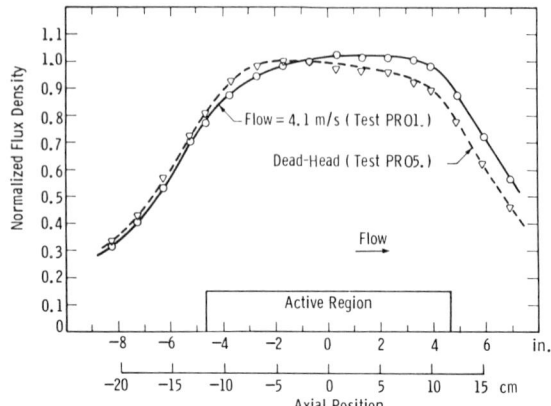

Fig. 9 Magnetic field profiles (at centerline) for dead-head and flow test (28 liter/s), with I = 12 kA and B = 0.20 T.

For the condition of zero flow in the duct, i.e., no self-induced field, the profile looks just as expected. The field is nearly uniform over approximately 80% of the active region, dropping slightly (~20%) near the edges. Then, the field decays nearly exponentially outside of the active region, with a decay constant of roughly 10 m^{-1} (3.05 ft^{-1}), which corresponds to an "e-folding" distance of approximately 10 cm (3.9 in.). This is shown in Fig. 9 along with the profile taken during a typical flow test. The flow velocity in that case was 4.1 m/s (volumetric flow of 28 liter/s), which corresponds to a magnetic Reynold's number of 0.6. This is on the order of a modest perturbation to the applied field (since $R_m \approx 1$), as discussed by Turner.[6]

It can be seen that the high-flow magnetic field profile (Fig. 9) has been augmented near the trailing edge of the active region and diminished near the leading edge. This tilts the profile such that it favors the exit end, so one would expect greater electromagnetic forces at the outlet end. However, looking back at Fig. 8, we find that slightly more than 50% of the total head generated comes from the inlet half of the duct. The current density (which is proportional to the difference between the terminal voltage and the back-EMF) is much greater over the inlet half of the duct, so the product of current density and magnetic flux density (which equals the pressure gradient) is also greater over the inlet half of the duct.

Efficiency

The duct geometry and magnetic field make this test program ideal for the quasi-one-dimensional analysis of

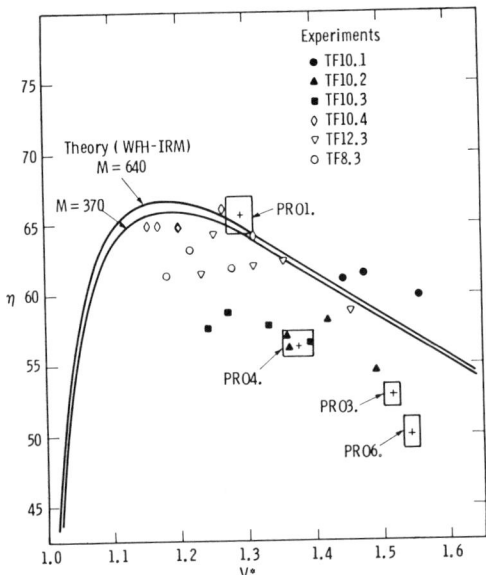

Fig. 10 Efficiency (η) comparison to theory as a function of reduced voltage (V*) for the thin-wall duct experiments.

Hughes and McNab.[5] Under the assumptions of negligible self-induced fields (which are true under most flow conditions) and compensated electrode buswork, the one-dimensional approach has been demonstrated an effective modeling tool, especially at duct centerline.[4] Therefore, the test results from this program will also be compared to that theory.

The first group of tests that were performed were throttled-flow tests. Here, the current and field were held constant while the flow rate was varied by partially closing the motorized main flow valve. The pumping efficiencies§ were calculated and plotted as a function of the reduced voltage V* (see Fig. 10).†† Efficiencies of other tests, using the internal instrumentation probe, were also plotted in this manner (Fig. 10). The theoretical efficiency curves, plotted for Hartmann numbers of 370 and 640, would encompass all of the test points, ideally. The agreement is good ($\leq 10\%$), even though there is appreciable data scatter. This scatter is primarily due to the calculation of the pressure rise along the test section.

---

§Pumping efficiency is defined as the hydraulic power output from the duct (flow rate times pressure rise) divided by the electrical power input to the duct (current times voltage).

††Reduced voltage is defined as the applied voltage divided by the calculated back-EMF generated by the liquid metal interacting with the magnetic field.

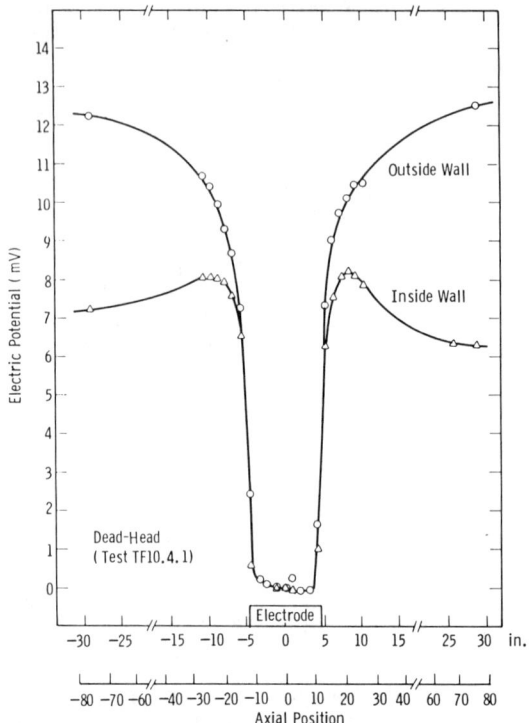

Fig. 11 Axial potential profiles (at side wall) for dead-head test, with I = 10 kA and B = 0.29 T. Reference potentials are 19 mV (outside wall) and zero (inside wall). (Note: Outside-wall profile has been inverted.)

Potential Profiles

The axial variation in electric potential has been plotted in Fig. 11 for a dead-head test and in Fig. 12 for a flow test. Notice how the potential rises smoothly in Fig. 11 for the outside wall but not for the inside wall. The hump on either side of the electrode indicates a 1- to 2-mV driving potential from the ends of the duct to the fringe region approximately 10 cm from the ends of the electrodes. This may be caused by fluid convection cells driven by non-uniform J B forces near the electrode edges. Fluid motion along the walls would depress the potential profile along one wall near the electrode edges and raise the profile along the other wall. However, the absolute value of the outside-wall potential has been plotted, so both profiles should appear to be raised. The fact that only one profile shows this "hump" indicates some asymmetry across the duct, possibly in the current distribution.

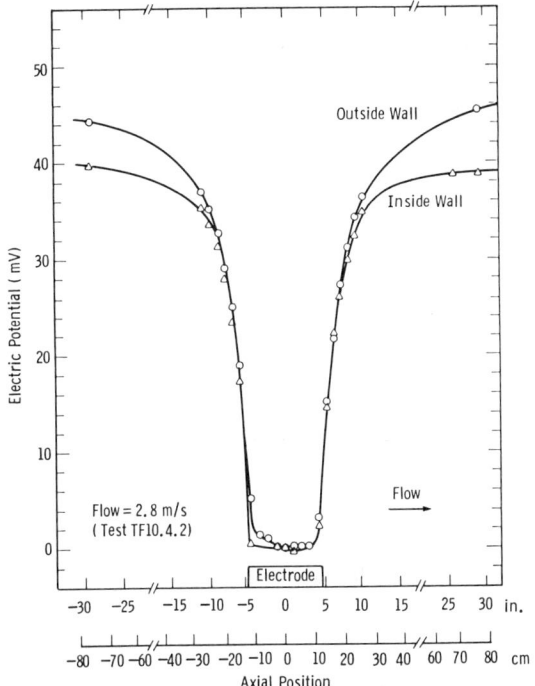

Fig. 12 Axial potential profiles (at side wall) for flow test (19 liter/s), with $I = 10$ kA and $B = 0.29$ T. Reference potentials are 83 mV (outside wall) and zero (inside wall). (Note: Outside-wall profile has been inverted.)

This effect is not present in Fig. 12 (the flow test) for either of the two profiles. Here the back-EMF generated by the flow is very strong along the electrodes, where the magnetic flux density is highest, and there is uniform fluid motion (to first order). As a result, there is a more uniform decrease in electric potential throughout the end regions.

Conclusions

The results presented in the last section were, for the most part, predictable by the quasi-one-dimensional theory. For this part of the test program, the interaction parameters were rather small (≤ 10) and magnetic Reynold's numbers very small (≤ 1). In the next phase of the program, we have scheduled high interaction parameter tests (≥ 100) with even smaller magnetic Reynold's numbers ($<<1$) because the flow velocities will be quite small (≤ 10 cm/s). However, the Reynold's numbers will be on the order of 10^4, so we can

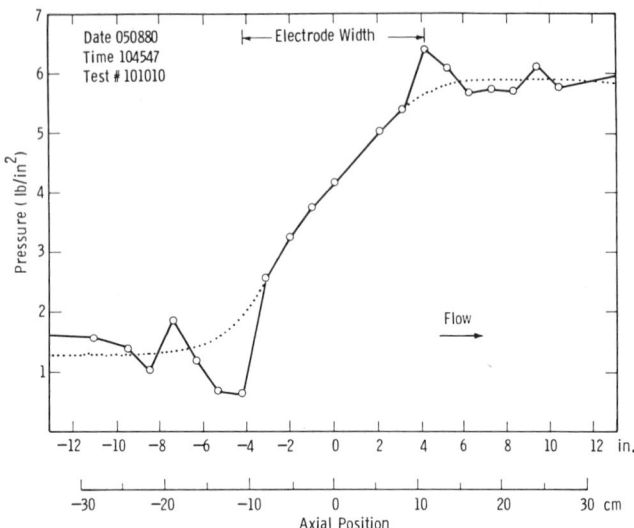

Fig. 13 Side-wall pressure profile for a typical flow test using the insulating-wall duct. Dotted line shows expected centerline profile.

still expect turbulent velocity profiles. The one-dimensional theory assumes slug profiles, which are a close approximation to turbulent profiles (except for the losses).

In reflecting back to the first NSF program, there were a few items that were not observed this time around. First, the "wiggles" in the side-wall pressure profiles (see Fig. 13) were not present with this duct. Since the two ducts were also constructed in different manners, we have concluded that the wiggles were probably not an MFD effect but, rather, a hydrodynamic effect due to surface roughness at the perimeters of the penetrations for the pressure stations.

Also, the efficiencies obtained in the thin-wall duct were systematically lower than those from the insulating-wall duct (Fig. 14), as predicted by the theory. With the changes made to the internal probe, we found it difficult to use the velocity data derived from the pitot tube measurements. The extreme length of the pitot tube part of the probe (~5 cm) made the static pressure of the "static" port different from the static pressure at the "dynamic" port, so velocity could not be calculated directly. Although it was possible to subtract the pressure difference due to electromagnetic forces <u>after</u> the fact, we learned that too much random error (from two measurements instead of one) had been introduced. It is recommended that nonintrusive local velocity measurement techniques be developed for submersion i

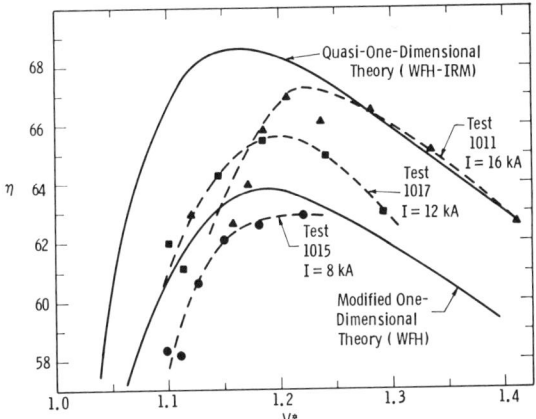

Fig. 14 Efficiency (η) comparison to theory as a function of reduced voltage (V*) for the insulating-wall duct experiments.

alkali metals (e.g., NaK, lithium, and sodium) to examine velocity distributions inside ducts and pipes.

Acknowledgments

The authors wish to thank I. R. McNab for his direction throughout the program and R. C. Palumbo for his help with the experiments. The work reported here was supported by National Science Foundation Grant CPE-8108952, R. E. Rostenbach, Program Director. Many thanks are also given to R. D. Nathenson for presenting the paper in our absence.

References

[1] Walker, J. S., "Magnetohydrodynamics Flows in Rectangular Ducts with Thin Conducting Walls, Part I: Constant-Area and Variable-Area Ducts with Strong Uniform Magnetic Fields," Journal de Mecanique, Vol. 20, 1981, pp. 79-112.

[2] Walker, J. S., "Magnetohydrodynamic Flows in Rectangular Ducts with Thin Conducting Walls, Part II: Pumps with Finite Length Electrodes and with Strong Uniform Magnetic Fields," (in preparation).

[3] Hughes, W. F. and Winowich, N. S., "A Finite-Element Analysis of Two-Dimensional MHD Flow," Liquid-Metal Flows and Magnetohydrodynamics: AIAA Progress in Astronautics and Aeronautics, Vol. 84, AIAA, New York, 1983, pp. 313-322.

[4]McNab, I. R., Alexion, C. C., Keeton, A. R., and Ciarelli, P. A., "High Interaction Parameter Studies in a Large NaK Loop," Liquid-Metal Flows and Magnetohydrodynamics: AIAA Progress in Astronautics and Aeronautics, Vol. 84, AIAA, New York, 1983, pp. 263-265.

[5]Hughes, W. F. and McNab, I. R., "A Quasi-One-Dimensional Analysis of an Electromagnetic Pump Including End Effect," Liquid-Metal Flows and Magnetohydrodynamics: AIAA Progress in Astronautics and Aeronautics, Vol. 84, AIAA, New York, 1983, pp. 287-312.

[6]Turner, R. B., "Aspects of Magnetohydrodynamics Duct Flow at High Magnetic Reynolds Number," Ph.D. dissertation, University of Warwick, England.

Demonstration of Flow Couplers for the LMFBR

R.D. Nathenson,* C.C. Alexion,† and A.R. Keeton‡
Westinghouse Electric Corporation, Pittsburgh, Pennsylvania
and
O.E. Gray III§
Electric Power Research Institute, Naperville, Illinois

Abstract

Operation of an experimental "proof-of-principle" high-efficiency flow coupler has been successfully demonstrated. The flow coupler is a specialized direct-current electromagnetic device that has application as the primary sodium pump in a large liquid metal fast breeder reactor. Operating like a fluid transformer, the flow coupler draws the hydraulic power to drive the primary sodium flow from the intermediate sodium heat-transfer loop. It has long been recognized that electromagetic pumps offer significant advantages in the difficult reactor environment because of their inherent simplicity and lack of moving parts. One of the main problems in utilization of a large dc electromagnetic pump has been the need for large electrical conductors to supply high currents at low voltage to the device. A flow coupler eliminates the need for the external conductors because the large currents are generated locally in the device itself. Only a small current supply is required to produce the needed magnetic field. This paper describes experimental operation of a "proof-of-principle" flow coupler. Theoretical and experimental performance are compared and are found to be in close agreement.

Paper presented at the Fourth Beer-Sheva Seminar on MHD Flows and Turbulence, Ben-Gurion University of the Negev, Beer-Sheva, Israel, Feb. 27-March 2, 1984. Copyright © American Institute of Aeronautics and Astronautics, Inc., 1985. All rights reserved.
*Principal Engineer.
†Engineer.
‡Senior Engineer.
§Project Manager, Nuclear Power Division, Consolidated Management Office for the LMFBR.

I. Introduction

Early in the development of liquid metal fast breeder reactors (LMFBR), it was recognized that the liquid metal could be pumped by electromagnetic forces. Electromagnetic (EM) pumps offer significant advantages in this difficult reactor environment due to their inherent simplicity and lack of moving parts. A number of direct-current and alternating-current EM pumps have been operated successfully in such an environment: For example, at EBR-II.

A specialized dc electromagnetic pump known as a flow coupler has particular application as the primary sodium pump in an LMFBR. In the flow coupler, as shown in Figs. 1 and 2, the primary liquid sodium is pumped by a dc electromagnetic pump. The necessary direct current is supplied locally by a companion liquid metal dc generator in the intermediate sodium loop. The power input necessary to drive the liquid sodium through the generator is provided by the intermediate loop pump. In this manner the two sodium flows are electromagnetically "coupled" together through the generator/pump sections. The local generation of current enables the use of lower voltages and higher currents than would be possible with a conventional dc pump and allows the transfer of energy to be made at a high efficiency.

In 1983, the basic principle of the flow coupler was successfully demonstrated. Experimental and analytical predictions of the flow coupler performance agree well for the full range of operating parameters. In the sections that follow, first a description of the theory behind the flow coupler is given. Next the "proof-of-principle" experiment is described. Finally, the application of the flow coupler to the LMFBR is discussed.

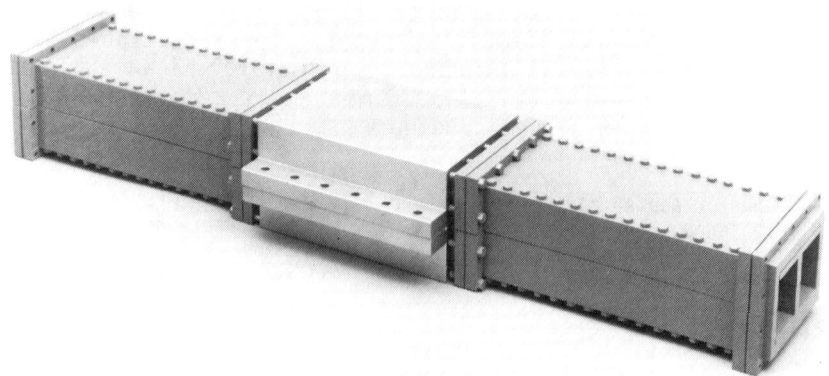

Fig. 1 Model of flow coupler (assembled).

FLOW COUPLERS FOR THE LMFBR

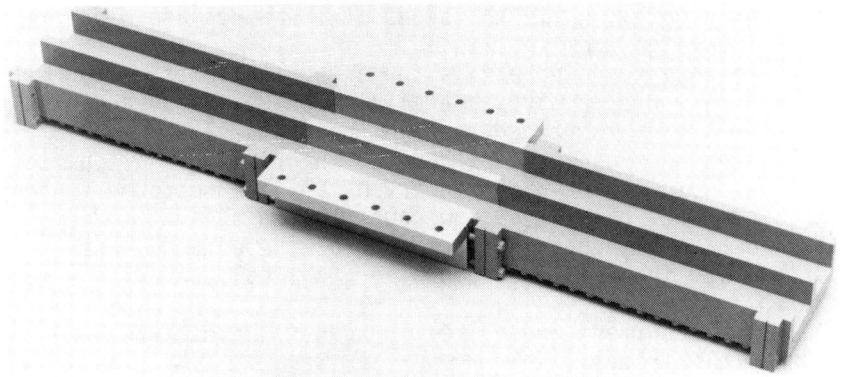

Fig. 2 Model of flow coupler (cutaway).

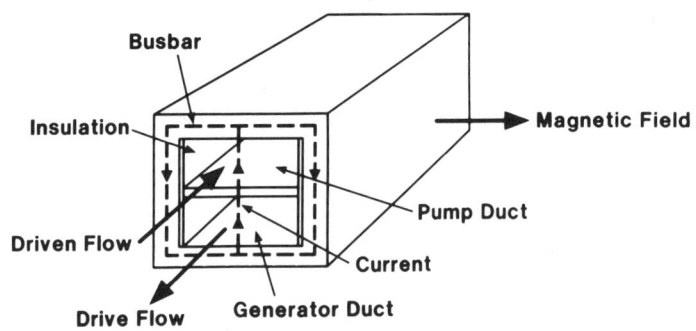

Fig. 3 Basic geometry of a typical flow coupler.

II. Flow Coupler Theory

Figure 3 shows the basic geometry of a typical flow coupler, which comprises a direct current generator duct and a pump duct. The two ducts are separated by a center bar (which is a common electrode to both ducts) and surrounded by a return busbar. Liquid sodium in the intermediate loop (the drive flow) is forced through the active region, where it interacts with the perpendicular magnetic field to generate a current in the liquid metal. The current acts in a direction mutually perpendicular to the magnetic field and the drive flow and produces a pressure drop in the generator duct.
    The current also conducts up through the center bar into the pump duct. Here it interacts with the magnetic field to produce a pressure rise on the primary loop liquid metal in the pump duct. The current is returned to the generator duct by the parallel paths of the busbar. Notice that the top and bottom of each duct are electrodes in contact with

the liquid metal, while the sides of the ducts are electrically insulated from the liquid metal.

The design shown here is for a counterflow coupler. It is also possible to have a parallel-flow design by changing the geometry of the two ducts and the busbars.

Before discussing flow coupler theory, it is useful to consider briefly the basic theory of an ideal dc pump (which is, of course, half of a flow coupler) and then to modify the "ideal" pump theory to account for the effects that can be observed in a real dc pump. For an ideal pump we assume the following:

1) The current goes directly across the duct between the two electrodes.

2) A magnetic field is directed straight across the duct, perpendicular to the current.

3) The liquid metal is contained by insulating walls except at the electrodes.

4) Liquid metal has a uniform slug flow profile.

5) No interfacial resistances exist, and the busbars/electrodes are perfectly conducting (zero resistance).

Under these assumptions we can derive simple expressions for the pump pressure rise and flow rate:

$$P = IB_0/h$$

where I is the applied current, B_0 is the applied field, and h is the duct height in the field direction.

$$Q = V_0 h/B_0$$

where V_0 is the counter-electromagnetic flow developed in the pump.

However, in a real dc pump we find the following:

1) Current fringes from the ends of the electrodes into the liquid metal in the end regions (i.e., regions before and after the "active" region containing the main current and field).

2) The magnetic field also fringes into the end regions of the liquid metal and interacts with the current there.

3) Current leakage occurs in the duct walls, which are metallic (and thus, conductors).

4) Current leakage also occurs in thin boundary layers (called Hartmann layers) in a path parallel to the wall leakage.

5) Resistive losses occur in the busbars and at the interfaces (liquid metal to busbar).

FLOW COUPLERS FOR THE LMFBR

Since these effects are not included in the ideal pump theory, a more detailed theory must be developed to account for them.

A quasi-one-dimensional theory that accurately models the effects in a real pump has been developed.[1-3] It derives analytical, closed-form solutions of the parameters-- pressure rise, flow rate, and efficiency--that are of interest to us. These three parameters can be expressed as functions of the applied current and voltage, plus the following six parameters: 1) flux density, 2) field fringing, 3) fluid viscosity, 4) fluid electrical conductivity, 5) busbar and interface resistance, and 6) wall conductance (includes wall thickness and electrical conductivity).

The theory is also applicable to a generator because the interactions in the generator are the same as those in the pump. From the pump and generator theories, we can construct a flow coupler theory, using the link of common total current between them.

It is convenient to express the pressures and flows in the coupler in terms of ratios in the following form:

$$p = \Delta P_p / \Delta P_g$$

$$q = Q_p / Q_g$$

where p is the pressure ratio, ΔP_p the pump pressure rise, ΔP_g the generator pressure drop, q the flow ratio, Q_p the pump flow rate, and Q_g the generator flow rate.

The hydraulic efficiency η may be defined as the hydraulic output power of the pump divided by the hydraulic input power to the generator:

$$\eta = Q_p \Delta P_p / Q_g \Delta P_g = pq$$

Thus, the efficiency is a measure of the flow coupler's ability to transform flow power from one duct to another. Efficiency can also be expressed as the product of the flow ratio q and the pressure ratio p.

Figure 4 shows the general operating characteristics, i.e., p and η vs q, for a flow coupler. For small values of the flow ratio q, the efficiency increases almost linearly with the flow ratio. At larger values of q, the efficiency reaches a maximum, then drops to zero. The pressure ratio remains nearly constant for small values of q, then drops to zero in the same manner. Since the efficiency equals the flow ratio times the pressure ratio, it is clear why the ef-

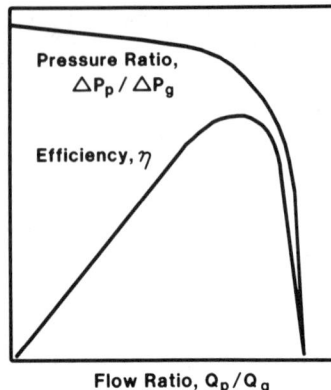

Fig. 4 Performance curves.

ficiency, as reflected in the shape of the pressure ratio curve, also peaks and then falls to zero.

In a given flow coupler design, efficiency is maximized for a certain value of the flow ratio. Thus, it is logical to ask what factors influence the flow ratio and, hence, the efficiency. To maximize efficiency, the following steps are desirable:

1) Minimize wall conductance by making thin containment walls and using high-resistivity alloys to minimize current leakage.

2) Minimize bus resistance by using low-resistance busbars and minimizing the number of interfaces in the return circuit (especially the liquid metal to busbar interface).

3) Optimize field fringing by choosing the best magnetic field profile in the end regions to give the best current/field interactions.

4) Maximize the Hartmann number to result in a thin boundary layer and to minimize current leakage.

These parametric variations are plotted in Fig. 5. Figure 5a shows efficiency vs the wall conductance parameter, which is the ratio of wall conductance to fluid conductance. For small values of the wall conductance, the efficiency drops nearly linearly from its maximum value, then tapers off as the parameter increases. The effect of busbar resistance (as shown in Fig. 5b) is similar, but less pronounced. The busbar resistance parameter is the ratio of the busbar and interface resistance to the fluid resistance in the active region. As shown in the graph, increasing the busbar resistance results in a decrease in efficiency. Thus both wall conductance and busbar resistance should be minimized to maximize efficiency.

Figure 5c shows efficiency vs the magnetic fringing parameter, which is the ratio of magnetic field fringing

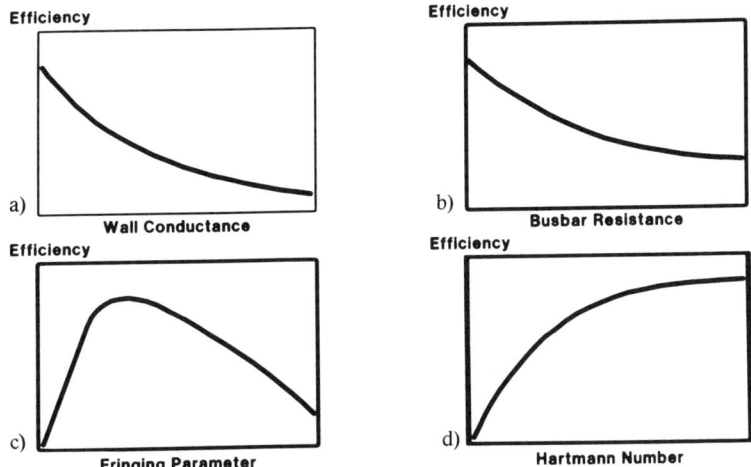

Fig. 5 Parametric variations.

length (into the end regions) to the active length. Efficiency is maximized for a particular value of the fringing parameter; that value depends on the specific flow coupler design.

Figure 5d shows efficiency vs Hartmann number, which is a magnetofluidynamic parameter representing the ratio of magnetic forces to viscous forces on the liquid metal. For low values of this ratio, the efficiency rises nearly linearly from zero; then it levels off as the Hartmann number is increased further. Thus, there is an optimum value for the Hartmann number beyond which the efficiency is not significantly improved.

In practice, this flow coupler theory can be adapted to design flow coupler systems of various sizes and for specific applications. Once the desired performance of the flow coupler has been determined in terms of flow rates and pressures, it is possible to model the magnetofluidynamics of the two flows for a given set of physical duct dimensions and magnetic field values. The next task is to optimize the efficiency of the flow coupler using the several dimensionless groups (for example, Hartmann number) cited earlier to determine the best design. Note that since the flow coupler theory is written in terms of dimensionless groups, many different flow couplers of various sizes, but dynamically similar characteristics, can be confidently designed based upon experimental confirmation from scale flow coupler test systems. As described in the following section, a proof-of-principle experimental program has been carried out to validate this theory.

III. Proof-of-Principle Experiment

To provide a proof-of-principle demonstration of the flow coupler theory, a test program was carried out using a 250-gal/min NaK test system.[4] The results of the tests agree very well with theory.

Description of the Flow Coupler Test Module

To demonstrate the flow coupler theory described in Sec. II, a flow coupler module was designed to fit into an existing liquid metal facility.

A photograph of the actual flow coupler is shown in Fig. 6. Note the copper busbars in the center as shown earlier in the cutaway view (Fig. 2); the channels in the copper are for locating sensing wires to measure the electrical current in the busbars. The two flow ducts can be seen at the end where the interface is made to the liquid metal loop piping. The tube fittings along each side are pressure ports; there are four at the entrance and exit of each duct. The flow coupler is constructed of NEMA grade G-11 glass epoxy and tough pitch copper and is completely symmetrical. The seals are made with BUNA-N rubber gaskets or O-rings.

The machanical features of the demonstration flow coupler module are listed in Table 1. The overall length of the unit is just under 4 ft, with the active length, the area of uniform magnetic flux, a little over 9 in. The internal dimensions of the two ducts are 2x2½ in.

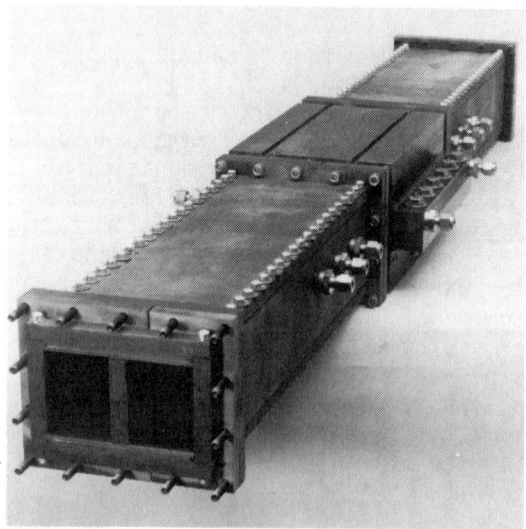

Fig. 6 Experimental flow computer module.

FLOW COUPLERS FOR THE LMFBR

Table 1 Flow coupler mechanical specifications

Overall length	47.5 in.
Active length	9.25 in.
Overall height	5.0 in.
Flow Duct size:	
Generator	2.0 in. wide x 2.5 in. high
Pump	2.0 in. wide x 2.5 in. high
Electrode material	ETP copper
Insulation material	NEMA G-11 glass epoxy sheet
Design pressure	50 psig
Design temperature	160°F

Flow Coupler Test Systems

The NaK loop test facility with the flow coupler installed is shown in Fig. 7. NaK was chosen as the test fluid instead of sodium, first because it is more convenient and less costly to work with, and second because the physical properties of NaK at room temperature are very close to those of sodium at elevated temperatures. The facility features a computer-linked data acquisition system that allows real time display of important information such as flow and pressure ratios, efficiency, and Hartmann number.

The C-shaped magnet is clearly shown in the photograph; the magnet pole pieces are partially visible at the location of the flow coupler. The large 6-in. piping is part of the original 1000-gal/min test system used for magnetofluidynamic studies, but was not used for the flow coupler test. The smaller 3-in. pipe is part of the flow coupler test loop itself. Not shown, behind the magnet, is a 250-gal/min mechanical pump used for circulating the liquid metal through the generator duct of the flow coupler.

A schematic of the NaK loop system, with the flow coupler at its center, is shown in Fig. 8. The mechanical pump for this test system was rated at 250 gal/min at 40 psi head. In the drive (generator) loop, NaK flows from the pump, through an air cooler to prevent extensive heat build-up, through a flow meter, through the generator channel of the flow coupler, and back around to the pump inlet.

A counterflow of NaK is pumped from the flow coupler pump duct, through an electromagnetic flow meter, through a remotely controlled back pressure valve, and back to the flow coupler inlet. As previously stated, four pressure transducers are located at the inlet and exit of each flow coupler duct. Sump and expansion tanks are used to control

Fig. 7 Flow coupler test system.

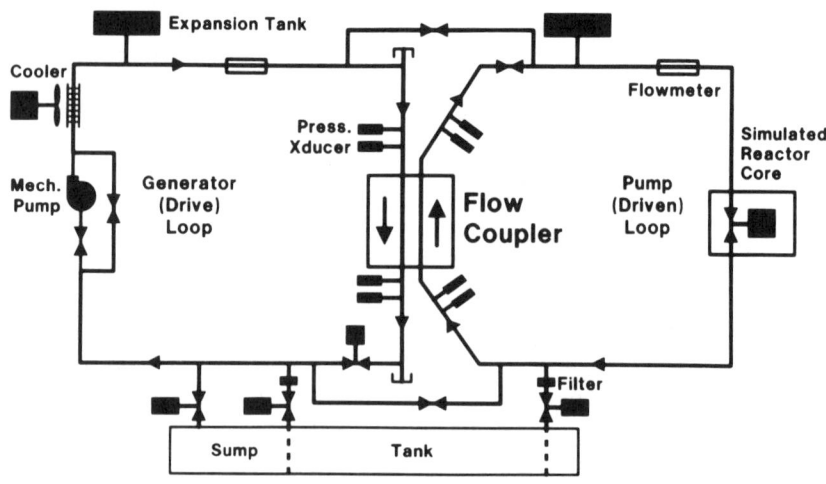

Fig. 8 Flow coupler demonstration experiment.

the NaK levels in the two loops. The expansion tanks allow the gas overpressure to be controlled independently in each loop and provide a means for pressure transducer calibration.

Flow Coupler Test Program

The experimental program was designed with three general objectives in mind:
1) To provide a proof-of-principle demonstration of the flow coupler.
2) To characterize the flow coupler performance as a function of a) generator (drive) flow rate, b) pump (driven) flow rate, and c) magnetic flux density.
3) To compare our experimental results with our theoretical predictions.

Figure 9 shows the experimental results in terms of pressure profiles. In this experiment, the NaK flow rate in the generator was from left to right in the figure, while that in the pump was from right to left. The copper electrode is shown along with the location of the pressure sensing ports. The error bars at each location are the actual variations of pressure over a series of data readings. Note that the pressures before and after the active region are relatively uniform for both the generator and the pump. The difference in pressure level between the generator duct and the pump duct is independently controlled by the gas overpressure on the expansion tanks.

Figure 10 shows the results of a deadhead test where the pump (driven) flow rate was held at zero while the generator (drive) flow rate was increased in increments from 0 to 250 gal/min. The plots show the total current flowing in

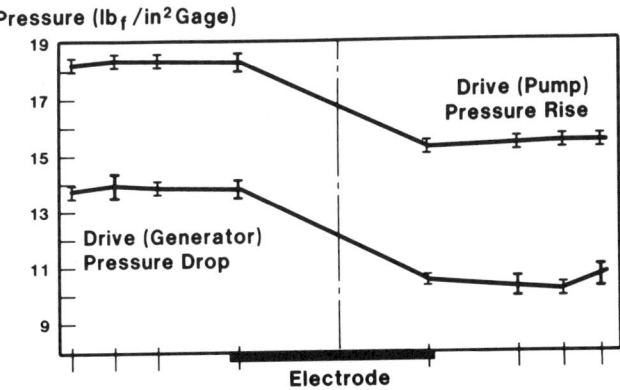

Fig. 9 Pressure profile.

the busbars and the developed pressure heads in the pump and generator ducts. As expected for the deadhead condition all of the plots are straight lines.

Figure 11 shows the test results as a function of driven flow rate. The drive flow rate was held at a constant 150 gal/min, while the driven flow rate was increased steadily from 0 to 110 gal/min. The generator pressure drop, pump pressure rise, and efficiency are plotted as a function of the driven (pump) flow rate. The solid lines indicate the theoretical predictions, and the points with error bars give the experimental data; there is excellent agreement between the two. The efficiency curve shows a maximum of ~ 60% for both predicted and measured values.

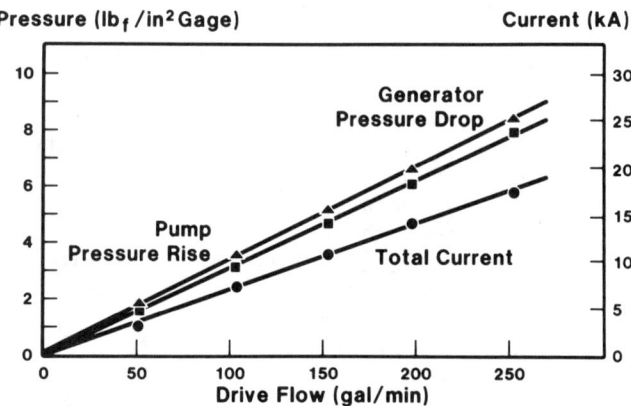

Fig. 10 Deadhead test results.

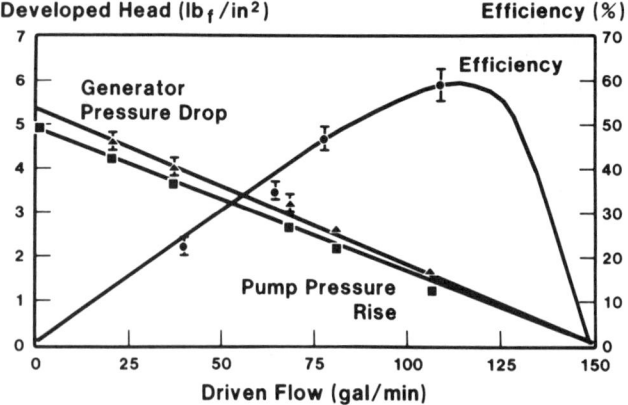

Fig. 11 Flow test results.

Figure 12 shows the coupling effect of the magnetic field. The flux density was increased from 0 to 0.27 T, and the dirve (generator) and driven (pump) flow rates were plotted as a function of flux density. At low magnetic fields, very little flow is induced in the driven loop. As the magnetic field increases, the flow in the driven loop increases accordingly. The drive loop flow decreases slightly as the load on the mechanical pump due to the power transferred to the driven loop increases.

IV. Application of the Flow Coupler to the LMFBR

In an LMFBR using a flow coupler, liquid sodium would be pumped by a conventional centrifugal pump through the intermediate sodium loop, which also passes through one side of the flow coupler. Low-power dc electromagnets and/or permanent magnets would be used to create a magnetic field in the coupler. In much the same way as a transformer, the flow of intermediate (drive) sodium in one side of the coupler would cause the primary (driven) sodium to flow in the other side of the device, which is connected to the primary heat-transfer loop.

In the system illustrated in Fig. 13, the flow coupler is located in the cold leg of both loops. This setup could be varied, depending upon the design of the plant, in a number of ways. For example, the flow coupler could be located in either the hot or cold leg of either of the heat-transfer loops.

The flow coupler concept offers a number of advantages for the LMFBR:

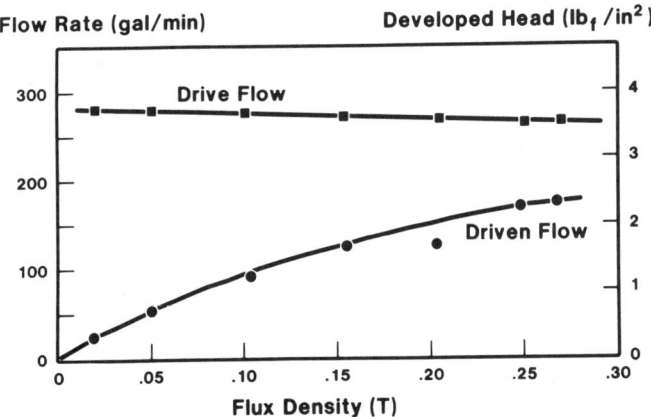

Fig. 12 Magnetic field test results.

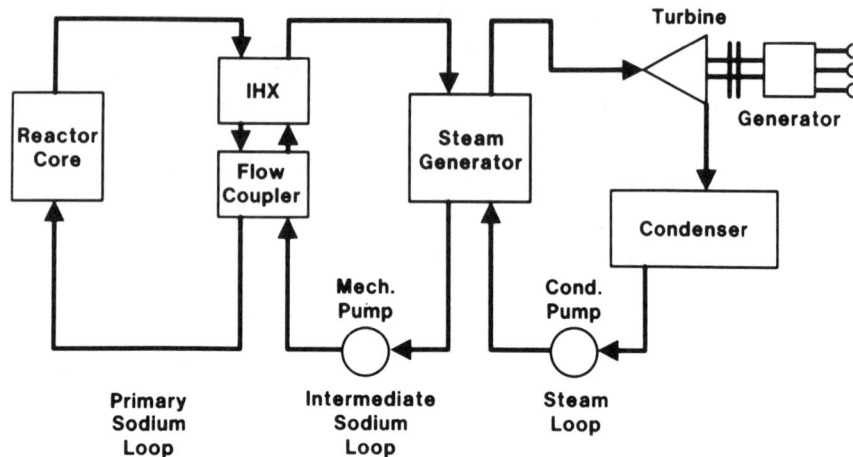

Fig. 13 Reactor system using a flow coupler.

1) In contrast with other dc or ac pump concepts, high-power electrical busbars do not have to enter and be electrically isolated within the liquid sodium pool.

2) The flow coupler ensures that a flow in one loop, intermediate or primary, will cause a flow to occur in the other loop, even as a result of natural convection. Thus, the primary coolant flow rate can be directly controlled by the flow rate in the intermediate coolant loop, and the integrity of the heat exchanger may be improved during thermal upset events.

3) Operational reliability is enhanced by the inherent simplicity and lack of moving parts in the flow coupler. Problems associated with vibration or thermal distortion in areas of closely toleranced moving parts, such as bearings or seals, are eliminated; and the cavitation problems associated with a rotating impeller do not exist.

4) The relative simplicity and easy operation of the direct current flow coupler should enable a relatively low-cost component to be developed.

5) Finally, the smaller flow coupler volume, compared with that of a mechanical pump and motor, permits an improved reactor layout. This includes the possibility that the flow coupler could be integrated directly with the intermediate heat exchanger. Thus, significant savings in containment size and cost can be effected.

V. Summary

Summarizing the results of flow coupler development to date, we find that:

1) A 250-gal/min flow coupler has been designed, built, and installed in an existing NaK loop.

2) This test program has demonstrated conclusively the basic principle of the flow coupler.

3) Good agreement has been achieved between theoretical predictions and experimental results, over a full range of operating parameters.

4) These experimental results lay the foundation for the design and testing of large high-temperature sodium pumps.

Acknowledgments

The authors wish to acknowledge the many suggestions and constructive advice of D. C. Gibbs and R. Balent, who have contributed to the success of this project. We also wish to thank A. F. Berringer, D. G. Boles, R. A. Evans, R. C. Palumbo, E. J. Wagner, and M. E. Walko for contributing their time and skills in the construction and test of the flow coupler test module and loop system. This program was sponsored by the Electric Power Research Institute, Consolidated Management Office for the LMFBR.

References

[1] McNab, I. R. and Alexion, C. C., "DC EM Pumps and Flow Couplers for LMFBRs," Electric Power Research Institute, Naperville, Ill., EPRI Report NP-1656, Jan. 1981.

[2] McNab, I. R., Alexion, C. C., and Winkleblack, R. K., "High Efficiency DC Electromagnetic Pumps and Flow Couplers for Pool-Type LMFBRs," Liquid Metal Flows and Magnetohydrodynamics: AIAA Progress in Astronautics and Aeronautics, Vol. 84, AIAA, editors H. Branover, A. Yakot, and P. S. Lykoudis, New York, 1983, pp. 266-286.

[3] Hughes, W. F. and Alexion, C. C., "A Theoretical Analysis of the DC Electromagnetic Flow Coupler," Nuclear Engineering and Design, Vol. 74, March 1982, pp. 367-376.

[4] Nathenson, R. D., Alexion, C. C., and Keeton, A. R., "Demonstration of Flow Couplers for the LMFBR," Westinghouse Electric Corporation for Electric Power Research Institute Consolidated Management Office for the LMFBR, Naperville, Ill., Final Report, Dec. 1983.

Disk Generator Performance Prospects

H.K. Messerle*
School of Electrical Engineering, University of Sydney, Australia

Abstract

There has been revived interest in the study of MHD disk generators in recent years. Two types of generator designs are feasible: one with the flow outward and the other with the flow inward. They have been shown to be comparable in performance with each other and competitive with linear generator designs. Experimental and theoretical studies on the disk generator are reviewed and the principal factors determining the disk's performance are outlined. The gas flow is subsonic for the inflow design and supersonic for the outflow design, and each design has advantages and disadvantages. Calculations have been made, predicting generator performance as load conditions are varied, for base-load inflow and outflow designs and the results are presented. The induced azimuthal gas flow (swirl) has a significant effect on off-design point performance.

Introduction

The MHD disk generator represents one of a number of geometries that can be used to generate electric power. Most attention in the past two decades has focused on the linear geometry using Faraday, diagonal, or Hall connection for the multitude of electrode pairs required. In recent times, the disk geometry has been receiving increasing attention.

Paper presented at the Fourth Beer-Sheva Seminar on MHD Flows and Turbulence, Ben-Gurion University of the Negev, Beer-Sheva, Israel, Feb. 27-March 2, 1984. Copyright © 1985 by the American Institute of Aeronautics and Astronautics, Inc. All rights reserved.
*School of Electrical Engineering.

The important features of a disk generator are its circular geometry, leading to an inherent simplicity of design; its compactness; and, in particular, the simplification of its electrical structure. These features should lead to potential economic advantages.

In a disk generator, the gas flows either radially outward or inward, and an axial magnetic field is applied across the flow region. The flow passes between two insulated circular disks, and the induced rotating current closes upon itself. The electrical energy is extracted by making use of the resulting radially induced voltage and power takeoff can be achieved by one or more pairs of ring electrodes. One of these pairs of electrodes is placed at the inner entry radius, and the others are placed farther outward along the circular disks, with the final pair at the end of the effective interaction region before the plasma enters the circular diffuser in an outflow geometry. We then talk of single-pair or segmented ring electrode arrangements (see Figs. 1a-c).

In a linear generator, the induced current is extracted by a multitude of electrodes along the channel. This electrode segmentation, involving hundreds of pairs in a large generator, is essential for satisfactory operation whether we use Faraday or diagonal connection, and the electrodes must be electrically isolated. To reduce the complexity of the external electrical circuitry, complex schemes for electrode consolidation have been developed. At the same time, the longevity and integrity of these finely segmented electrodes represent a major design problem, and a limitation is the axial Hall voltage that can be allowed to develop.

In contrast, the disk generator walls are made up of insulating material, and the number of ring electrodes required is small. Hence, the voltage gradient developed along the disk walls can be raised from 3 to 4 kV/m in a linear generator to 12 to 18 kV/m in the disk generator. This means that the length of a channel can be considerably reduced. In addition, the provision of a circular superconducting magnet should lead to a considerable reduction in cost compared with the saddle-shaped magnet in a linear system. In the following sections, performance characteristics and limitations of the disk geometry are discussed with emphasis on open-cycle systems.

Background

Theoretical and experimental studies of disk generators go back to the early days of MHD development.[1-3]

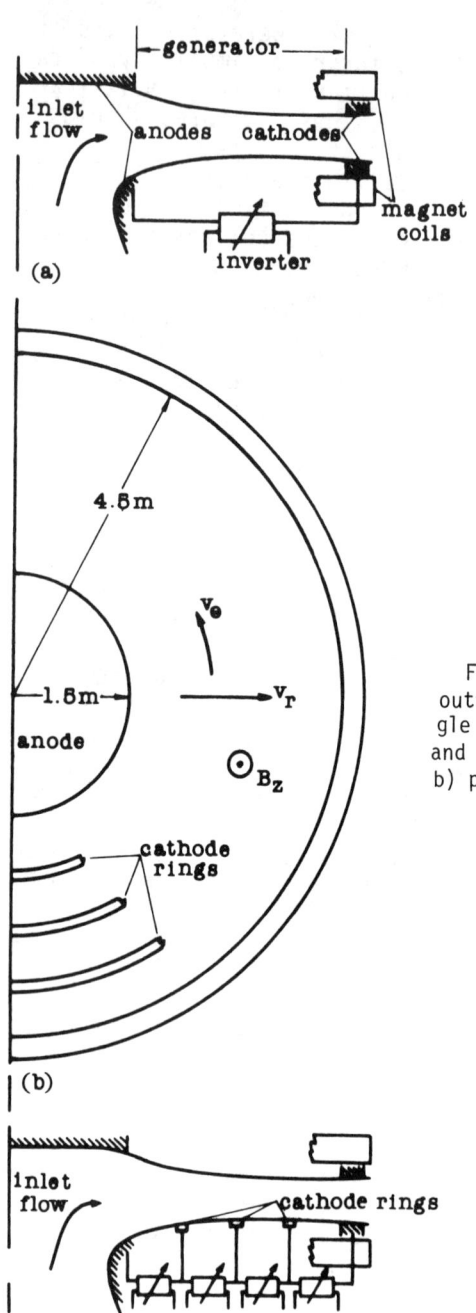

Fig. 1 Schematic diagram for outflow disk geometry showing a single electrode pair: a) side view; and showing segmented electrodes: b) plan view, c) side view.

During 1964-65, Klepeis and Rosa[1] used an inert gas blowdown facility for the first disk geometry experiments. Cesium-seeded argon was used, and the operation typically lasted for 10 s, the gas flowing outward. The electric output was about 5 W, owing to the small diameter of about 7.5 cm and weak magnetic field of about 0.6 T. Later, Louis[2] described the use of a shock tube involving a 30-cm-dia disk. Using B = 2.8 T, a significant MHD interaction became possible. A power output of over 700 kW was achieved corresponding to about 10% enthalpy extraction. This early work was carried out without added counterswirl, and output could be raised by 50% with the introduction of swirl at the inlet.[4] At the same time, a larger shock tunnel driving a disk generator with a 1.30-m outside diameter produced 11 MW with an enthalpy extraction of 17% using cesium-seeded argon.[6] Since the flow was supersonic, only a small magnetic field was needed (B = 1 T). Experiments with a simulated coal combustion mixture of H_2, CO_2, and N_2 using B = 4.5 T to compensate for a smaller electrical conductivity achieved similar results.

Disk Geometry

The schematic diagram of Fig. 1 illustrates the basic features of the outflow generator. In an inflow generator, the flow is reversed. After leaving the combustion chamber, the gas expands radially outward (or inward) through a magnetic field B parallel to the cylindrical axis. A unique feature of the disk geometry is that the Faraday current j_θ short-circuits on itself without crossing a material boundary; $\underline{j_\theta}$ is induced azimuthally in the direction of $\bar{v}_r \times \bar{B}$, where $\bar{v}_r$ is the radial velocity.

The Hall effect causes electrons to drift toward the anode, which is at the inner radius for the outflow generator and at the outer radius for the inflow generator, and a Hall current $\bar{j}_r$ develops when the generator is under load. The two Lorentz force components are $\bar{F}_r = \bar{j}_\theta \times \bar{B}$, the major radial retarding or compressive force, and $\bar{F}_\theta = \bar{j}_r \times \bar{B}$, a tangential force that induces a swirl velocity v_θ. This induced swirl degrades the generator performance, since a current, proportional to the $\bar{v}_\theta \times \bar{B}$, is developed that opposes the output current $\bar{j}_r$. When swirl is added at the generator inlet in the opposite direction to the induced $\bar{v}_\theta$, this effect is reversed and the output current and performance are enhanced.

The relations for radial current j_r and azimuthal current j_θ are

$$j_r = \sigma E + \sigma v_\theta B + \beta j_\theta \qquad (1)$$

$$j_\theta = -\sigma v_r B + \beta j_r$$

where σ is conductivity, and β is the Hall factor. E is the radial voltage gradient, and its integral along the radius is the terminal voltage.

The open-circuit voltage gradient for zero radial current ($j_r = 0$) is given by

$$E_{oc} = -(S + \beta) v_r B$$

and

$$j_{\theta oc} = -\sigma v_r B$$

where $S = v_\theta/v_r$ is the swirl factor. E_{oc} is the maximum voltage, and, considering values of B approaching 10 T, the Hall voltage can reach extremely high values.

At short circuit, when $E = 0$ we get

$$j_{rsc} = \sigma(\beta + S) v_r B (1 + \beta^2) \qquad (1a)$$

$$j_{\theta sc} = \sigma(\beta S - 1) v_r B / (1 + \beta^2) \qquad (2a)$$

It is important to note that the azimuthal current is zero for $S = 1/\beta$. This swirl condition eliminates the Hall contribution to j_r and corresponds with maximum power extraction, as shown later. Negative swirl reduces performance.

The local power extracted is given by

$$W = -E j_r = \frac{\sigma v_r^2 B^2 (S + \beta)^2}{1 + \beta^2} K_h (1 - K_h) \qquad (3)$$

where we define the load factor

$$K_h = \frac{-E}{E_{oc}} = \frac{E}{(S + \beta) v_r B} \qquad (4)$$

as the ratio of the electric field to that at open circuit. The power for a specific β and S maximizes for

$$W_{max} = \frac{\sigma u^2 B^2}{4} \frac{1}{1 + S^2} \frac{(\beta + S)^2}{1 + \beta^2} \qquad (5)$$

where

$$u^2 = v_r^2 + v_\theta^2 = v_r^2(1 + S^2)$$

corresponding to $K_h = 1/2$. Plotting W_{max} for different values of β as function of S in Fig. 2, we find it reaches a maximum for $S = 1/\beta$. An important conclusion is that large swirl may be good for low Hall factors; however, for the large β encountered in full-scale disk generators, the swirl should be relatively small. Normal ranges of β are $1 < \beta < 10$, and for swirl S, $-\frac{1}{4} < S < 2$.

Enthalpy Extraction Rate

The power output is strongly affected by the Mach number in the disk. The energy equation is

$$\rho v_r \frac{d}{dr}\left(h + \frac{u^2}{2}\right) = W \tag{6}$$

For $h_0 = h + u^2/2$,

$$\frac{dh_0}{dr} = \frac{W}{\rho v_r} = \frac{\sigma v_r}{\rho} B^2 \frac{(S + \beta)^2}{1 + \beta^2} K_h(1 - K_h) < \frac{\sigma v_r}{\rho} B^2 \frac{1 + \beta^2}{4B^2} \tag{7}$$

Thus, the enthalpy extraction depends on $\sigma v_r/\rho$, and this factor normally reaches a maximum for Mach numbers in the range $1.4 < M < 1.9$.

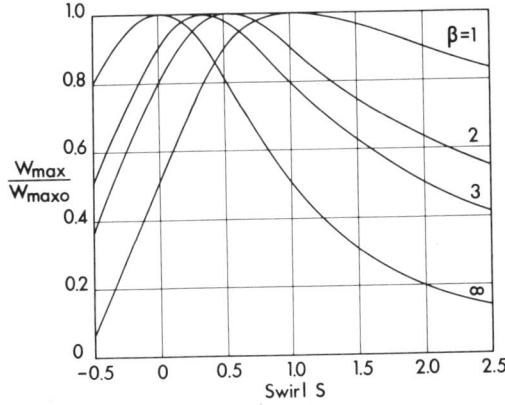

Fig. 2 Power ratio W_{max}/W_{maxo} as a function of swirl S for several values of β; $W_{maxo} = \sigma v^2 B^2/4$.

Effective Length

The channel length or distance between inner and outer ring electrodes depends on interaction effectiveness. An indication of the effect of local parameters can be gained by using relation (6) above. We can write

$$\frac{dh_0}{h_0} = \frac{-dr}{L} \quad \text{or} \quad L \sim \frac{uh_0}{W} \qquad (8)$$

where L is the effective enthalpy extraction length. Thus, from Eq. (5),

$$L > \frac{\rho h_0}{\sigma u B^2} \frac{4(1 + S^2)(1 + \beta^2)}{(S + \beta)^2} \qquad (9)$$

For maximum power extraction when $S = 1/\beta$, this becomes

$$L > 4(\rho h_0/\sigma u B^2) \qquad (10)$$

The effective length depends on the factor $\sigma u/\rho$ and is optimized in the range indicated above for supersonic flow. At subsonic velocities, with u fairly fixed, we must aim for high temperature to get a large σ and we must have a large B.

Local Electrical Efficiency

The conversion effectiveness is measured by the local electrical efficiency:

$$\eta_e = \frac{\overline{E \cdot J_r}}{\overline{u \cdot (J \times B)}} = \frac{(\beta + S)^2 K_h (1 - K_h)}{(1 + S^2) + (\beta^2 - S^2) K_h} \qquad (11)$$

This is a maximum when

$$K_{hmax} = \frac{S_1}{\beta_1 + S_1} \quad \eta_{emax} = \left(\frac{\beta + S}{\beta_1 + S_1}\right)^2 \qquad (12)$$

where

$$S_1^2 = 1 + S^2 \quad \beta_1^2 = 1 + \beta^2$$

As shown in Fig. 3, the maximum efficiency increases with increasing β and moves towards higher load factors, i.e., to higher voltages in the range of $K_h = 0.7$ to 0.9. The swirl has a pronounced effect once it drops below zero.

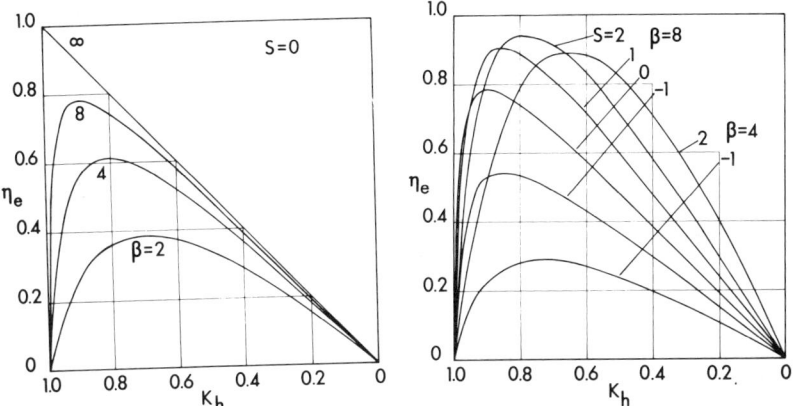

Fig. 3 Local electrical efficiency η_e as a function of load factor K_h: a) $S = 0$ and $\beta = 2, 4, 8, \infty$; b) $S = -1, 0, 1, 2$ for $\beta = 8$ and $S = -1, 2$ for $\beta = 4$.

Overall Disk Performance

The discussion in the last section has already indicated some critical requirements for satisfactory operation as a power generator. A careful analysis of the overall performance is still necessary, and a number of specific studies have been presented in the literature that discuss performance optimization. The question of performance when changing load away from design conditions has not been dealt with in such detail and poses severe problems. To start with, we must distinguish between open-cycle and closed-cycle systems. Early experiments dealt with noble gases in pulsed shock driven facilities.[1-6] Experimental data on combustion driven generators are more limited, but most theoretical analyses have considered the feasibility of large-scale disk facilities using fossil fuels.[7-10] Hence, most of the discussion here will also deal with open-cycle studies.

Studies generally have led to the conclusion that outflow generators operate optimally with supersonic flow in the impulse mode, and inflow generators with subsonic flow in the reaction mode at constant Mach number. Calculations carried out by various groups and authors indicate that both approaches may lead to similar overall enthalpy extraction efficiencies and performance.[7-10] It is possible to extract up to 19% of enthalpy with a total inlet pressure of about 6 atm. This corresponds closely with similar results for linear generators operating with an inlet pressure of 6 atm.[11] Higher conversion efficiencies have been reported for linear generators operating at higher pressures. This includes a most recent analysis by Seikel.[11] A similar improvement in performance could be expected for disk generators.

In the analysis of a disk generator there are a number of specific limitations imposed by the medium used and the geometry and materials for insulating walls and electrodes. In a simple one-dimensional analysis, overall estimates of performance are feasible. The upper gas temperature limit for combustion systems is of the order of 2800 K; in a closed-cycle noble gas system it is 2000 K. For supersonic flow, the inlet Mach number is optimal at about $M = 1.8$, and, in this case, the Hall voltage must be limited to prevent wall breakdown. Experimentally it has been shown that 18 kVm^{-1} may be sustained on ceramic walls. The Hall voltage is usually limited to about 12 kVm^{-1} and this compares to a limit of 3 kVm^{-1} in linear systems. A high wall voltage gradient reduces the radial length of the disk channel, and thus the outer diameter can be kept to less than about 9 m even for a 6000-MW thermal design.[10]

Obviously, the permissible wall gradient depends on the wall material and its temperature, and the latter becomes critical because of the severe conditions imposed by the flow boundary layers. The disk geometry creates a very large wall surface area to flow volume ratio because the channel height between the disk walls is generally only a fraction of the radius. For example, in a 3100-MW thermal input disk with radial outflow geometry, the channel height would vary from about 0.5 m at the 1.2-m inlet radius to 0.3 m at the 4.2-m outlet radius.[10]

Recent analyses have emphasized the critical role of boundary layers. Different methods to allow for layers are used. One approach[7] is to decide on a prespecified layer profile that adapts to temperature conditions. In a more detailed approach,[12] turbulent boundary layers can be allowed to develop using a mixing length model. Here a very basic difference in the behavior of the subsonic inflow and supersonic outflow designs emerges. For the supersonic outflow, viscous dissipation in the boundary regions can cause a temperature overshoot. The plasma in the layers is hotter, and its electrical conductivity rises. This allows a large azimuthal current to flow that decelerates the plasma and causes rapid boundary-layer growth. The thickening boundary layer tends to block the flow toward the narrowing exit, and oblique shocks will start to develop in these layers with a final overall shock transition. Deterioration in performance associated with this effect is more marked in the case of rough walls. A considerable performance improvement can be achieved by keeping the

DISK GENERATOR

walls intentionally cool to prevent boundary-layer temperature overshoot.[12] There is uncertainty about disk generator behavior after the shock, and no experimental data are available for disk operation in the presence of shocks.

For the subsonic inflow design, the cool boundary layers imply a reduced interaction, and this results in a flow acceleration and velocity overshoot in the boundary layers. Again, this can lead to shock formation near the channel exit. However, in the case of the inflow, the boundary layers are less significant, since the channel height increases rapidly approaching the inner radius exit.

For the inflow geometry, the velocity overshoot in the boundary layers means that the diffuser should be more efficient. The diffuser pressure recovery significantly affects the enthalpy extraction. There is still uncertainty about diffuser performance with the complicated swirling flow, but it is suggested[13] that the pressure recovery coefficient for the inflow design may be up to 0.8 owing to reduced likelihood of diffuser inlet blockage because of boundary-layer steepening and velocity overshoot. For the outflow disk, the diffuser is usually assumed less efficient and the recovery coefficient is taken to be about 0.5.

Another important factor affecting generator performance, especially that at the combustor and disk inlet is the heat loss to the walls. The combustor losses are relatively small in the outflow design. For the inflow generator, a multiple combustor system arranged around the outer disk diameter may involve substantial losses. Also the hottest part of the flow is exposed to a large circumferential scroll and channel inlet area. Here, again, uneven flow entry and heat loss information is lacking and experimental study is needed. Some information is being established at the disk facility at Stanford University, which, at present, is the only operational facility for the study of open-cycle performance with fossil fuel.

The disk generator is basically a constant current device. The swirl introduces another degree of freedom that makes operation extremely sensitive to load current variations. The open-circuit voltage gradient is approximately β times as large as that for a Faraday generator. However, the channel length is reduced because of the use of higher sustainable voltage gradients; hence, the load current will not be that much smaller than the corresponding Faraday current for the same power. The Hall factor varies from 5 to 10, going from channel inlet to outlet for supersonic operation, and from 1.5 to 8 for subsonic operation for commercial size 300- to 1000-MW designs.

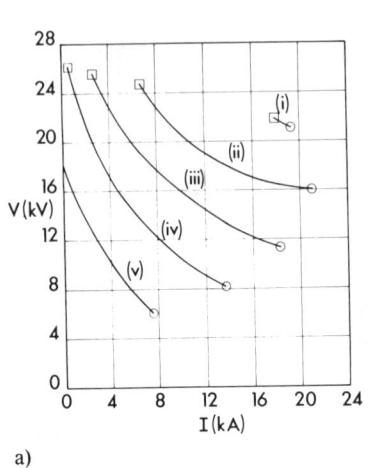 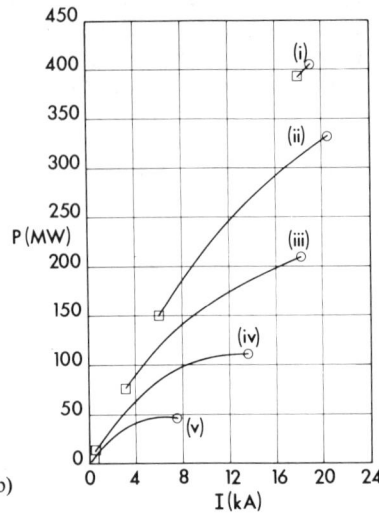

Fig. 4 Typical inflow disk generator characteristics: a) Voltage against current for mass flows of (i) 513 kgs^{-1}, (ii) 456 kgs^{-1}, (iii) 382 kgs^{-1}, (iv) 303 kgs^{-1}, and (v) 216 kgs^{-1}. b) Power output for same values of mass flow. □ exit M → 1.0; ○ inlet M → 1.0. Channel profile was chosen to maintain M = 0.9 for optimum power design case on curve i.

Performance Calculations

The behavior of disk generators subjected to off-design loading was investigated for the two cases: 1) an inflow generator operating in the reaction mode with the Mach number equal to 0.9 at the design point and 2) an outflow generator with supersonic flow where the design criterion was a constant E-field of 12 kV/m along the channel. The operating conditions for the disks were those used by Nakamura[7] for the inflow and Teare[14] for the outflow. The respective thermal inputs chosen were 2000 MW for the former and 3100 MW for the latter.

In Fig. 4, the voltage and power output vs. load current are given for the inflow generator with various mass flows. The ends of the curves indicate the loadings for which the exit Mach number is 1. The exit swirl is positive at the end of the curve corresponding to the lower current and negative at the other end. It can be seen that the subsonic operating regime at the design mass flow exists over a very small range of load current. It should be pointed out, however, that the design investigated here was not optimized to operate under a wide range of loading conditions, and other designs may improve this aspect of per-

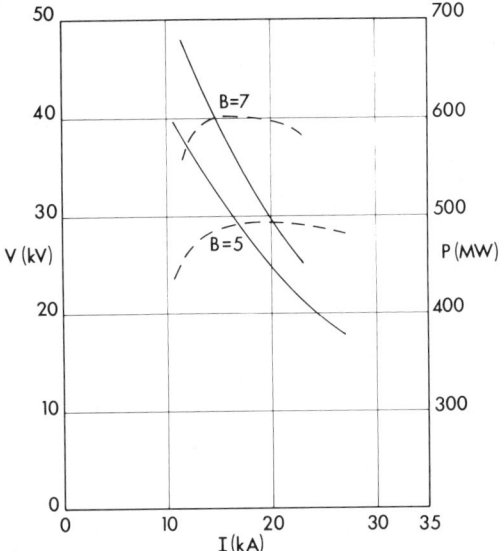

Fig. 5 Typical outflow disk generator characteristics for a mass flow of 764 kgs^{-1}. The voltage V (solid curves) and power P (dashed curves) are shown as functions of the load current I for the magnetic fields B indicated.

formance. This is currently being investigated. The choice of the inlet swirl is an important consideration here.

In Fig. 5, the voltage and power output vs load current are given for the supersonic outflow generator at the design mass flow but with two different magnetic fields. Previous calculations[14] suggested that this mode of operation is unstable and that perturbations of the load current from the design point caused large variations in the electrical and flow variables. The placement of multiple electrodes at various radii along the channel was suggested as a means of controlling the channel for off-design load currents. The present study determined, however, that the method of computation was responsible for the apparent instability and that operation is feasible over a wide range of load currents even with a single electrode pair. These preliminary calculations indicate that the performances of the outflow generator is more satisfactory for base-load operation than was previously thought, and further work is being done in this area.

Conclusion

Analyses and some restricted experiments in the last

few years indicate that a disk generator might be a feasible and simple alternative to the linear generator. Experiments have been carried out with shock tube driven disks, and these have demonstrated that significantly large enthalpy extraction is feasible. Performance studies indicate that both outflow and inflow geometries may offer alternative solutions to the linear generators for baseload application. They indicate that outflow generators should operate in the supersonic regime, and inflow generators in the subsonic regime. Performance is affected by the development of a swirl motion, which can be offset by providing counterswirl at the inlet.

In recent studies, outflow and inflow geometries have been shown to each have specific advantages. The outflow has lower thermal wall losses, whereas the inflow offers operation with high swirl without the need for guide vanes, and the counterswirl is preserved along the flow path in the channel by conservation of angular momentum. The inflow can operate more effectively at subsonic speed, whereas the outflow geometry requires supersonic speed. This means, however, that the outflow does not need the high magnetic fields, which leads to cheaper field structures.

There is also the possibility of operating a disk in a closed-cycle as well as open-cycle system. Again, both offer specific advantages, and a great deal of work is still required to assess and compare the performance of the outflow and inflow disk geometries under various operating conditions.

Acknowledgments

The author would like to express his thanks to S. W. Simpson, S. Marty, and M. Rados for their assistance in preparing this paper.

References

[1] Klepeis, J. E. and Rosa, R. J., "Experimental Studies of Strong Hall Effects of V x B Induced Ionization," Proceedings of Fifth Symposium on Engineering Aspects of Magnetohydrodynamics, MIT, Cambridge, Mass., April 1964, pp. 60-65, and Proceedings of Sixth Symposium on Engineering Aspects of Magnetohydrodynamics, Pittsburgh, Pennsylvania, April 1965, pp. 26-30.

[2] Louis, J. F., "Studies on an Inert Gas Disk Hall Generator Driven in a Shock Tunnel," Proceedings of Eighth Symposium on Engineering Aspects of Magnetohydrodynamics, Stanford, California, March 1967, pp. 75-88.

[3]Louis, J. F., "Disk Generator," AIAA Journal, Vol. 6, Sept. 1968, pp. 1674-1678.

[4]Loubsky, W. J., Hruby, V. and Louis, J. F., "Detailed Studies in a Disk Generator with Inlet Swirl Driven by Argon," Proceedings of 15th Symposium on Engineering Aspects of Magnetohydrodynamics, Philadelphia, Pa., May 1976, pp. VI.4.1-4.5.

[5]Klepeis, J. E. and Hruby, V., "The Disk Generator Applied to Open Cycle MHD Power Generation," Proceedings of 16th Symposium on Engineering Aspects of Magnetohydrodynamics, Pittsburgh, Pa., 1977, pp. I.5.31-5.37.

[6]Klepeis, J. E., "Open-Cycle Disk Generator Studies," Proceedings of 17th Symposium on Engineering Aspects of Magnetohydrodynamics, Stanford, Calif., March 1978, pp. B.1.1-1.5.

[7]Nakamura, T. and Jenkins, M. K., "Performance of Disk Generators for Open-Cycle MHD Power Generation," Proceedings of 18th Symposium on Engineering Aspects of Magnetohydrodynamics, Butte, Mont., June 1979, pp. B.3.1-3.11.

[8]Roseman, D., Nakamura, T., and Eustis, R., "Current Distribution and Non Uniformities in a Combustion Driven MHD Disk Generator," Proceedings of 20th Symposium on Engineering Aspects of Magnetohydrodynamics, Irvine, Calif., June 1982, pp. 8.3.1-8.3.5.

[9]Nakamura, T., Lear, W. E., and Fang, Y., "Results of Combustion Driven Inflow Disk Generator Experiments," Proceedings of 20th Symposium on Engineering Aspects of Magnetohydrodynamics, Irvine, Calif., June 1982, pp. 6.2.2-6.2.5.

[10]Simpson, S. W., Marty, S. M., Rankin, R. R., and Messerle, H. K., "Disk Generator Project at Sydney University," Proceedings of 20th Symposium on Engineering Aspects of Magnetohydrodynamics, Irvine, Calif., June 1982, pp. 64.1-64.5.

[11]Seikel, G. R., "Power Systems Integration," Specialist Meeting on Coal Fired MHD Power Generation, Institute of Engineers, Australia National Conference Publication 82/13, Sydney, Australia, Nov. 1981, pp. 7.1.1-7.1.9.

[12]Nakamura, T., "An Integral Method Analysis of the Disk Generator Boundary Layer," Proceedings of 20th Symposium on Engineering Aspects of Magnetohydrodynamics, Irvine, Calif., June 1982, pp. 6.6.1-6.6.5.

[13]Nakamura, T., Lear, W. E., and Eustis, R. H., "Feasibility Study of the Inflow Disk Generator for Open-Cycle MHD Power Generation," Proceedings of 19th Symposium on Engineering Aspects of Magnetohydrodynamics, Tullahoma, Tenn., June 1981, pp. 3.1.1-3.1.14.

[14]Teare, J. D., Loubsky, W. J., Lytle, J. K., and Louis, J. F., "Optimization of Disk Generator Performance for Open-Cycle MHD Power Generation," Proceedings of 7th International Conference on MHD Electrical Power Generation, Cambridge, Mass., June 1980, Vol. II, pp. 644-652.

High-Temperature Liquid Metal MHD Solar Thermal Systems

E.S. Pierson* and W.D. Jackson**
HMJ Corporation, Washington, D.C.
and
G. Berry,† M. Petrick,‡ and C. Dennis§
Argonne National Laboratory, Argonne, Illinois

Abstract

The performance potential for the Brayton-cycle two-phase generator, liquid metal magnetohydrodynamic energy conversion concept (LMMHD) coupled to a high-temperature solar receiver is evaluated. Three configurations are considered -- simple LMMHD, LMMHD combined with a gas turbine in the same gas loop, and LMMHD coupled with a steam bottoming cycle. Two liquid metals are considered -- sodium for solar receiver temperatures up to 922 K and lithium for higher temperatures. The sensitivity of the cycle efficiency to both design parameters and component efficiencies is demonstrated, and the results compared with a conventional steam power reference system design. It is shown that, for sodium cycles, efficiencies up to 40% may be obtained with high-performance components at a heat source temperature compatible with established containment materials. With lithium, efficiencies up to 50% are possible by raising the heat source by 100-150 K. Materials engineering is required in the lithium case to devise containment systems.

I. Introduction

Three basic considerations can be cited to support the claim that liquid metal MHD (LMMHD) conversion systems

---

Paper presented at the Fourth Beer-Sheva Seminar on MHD Flows and Turbulence, Ben-Gurion University of the Negev, Beer-Sheva, Israel, Feb. 27-March 2, 1984. Copyright © 1985 by the American Institute of Aeronautics and Astronautics, Inc. All rights reserved.

* Professor and Head, Department of Engineering, Purdue University Calumet, Hammond, Indiana.
** President, HMJ Corporation.
† Section Manager, Systems Analysis and Modeling.
‡ Director, Fossil Energy Programs.
§ Engineer.

couple effectively with high-temperature solar central receivers or power towers because:
1) Liquid metals have the most favorable heat-transfer characteristics for solar receivers.[1] The temperature difference between the receiver tubes and the coolant is low, and the risk of coolant boiling is minimal. Since liquid metals have low vapor pressures at elevated temperatures, the receiver can operate at lower pressures with thinner tubes than with other fluids.
2) A LMMHD conversion system permits the use of a liquid metal as the receiver coolant without introducing the complexity of a liquid metal/steam heat exchanger and does not incur the losses associated with the liquid return pump.
3) The operating temperature of an LMMHD energy-conversion system is limited only by available containment materials. The material temperature limits are different from other energy-conversion systems because a) there are no solid moving parts, and b) the temperature differences in the liquid metal loop are so small that material transport is minimal. Thus, the full temperature potential of a solar receiver system can be utilized to maximize efficiency or reduce collector area and, hence, cost and environmental impact.

Other features of LMMHD that relate to its application to solar uses are:
1) The system is modular with a cycle efficiency that has low dependence on module size, so that part-load operation may be accomplished by switching out an appropriate number of modules without impacting system efficiency. In addition, the modules can be factory-assembled for minimum cost.
2) The almost-constant-temperature operation means that containment material lifetimes should be longer. Material transfer or transport from the hotter to the cooler parts of the flow system due to differences in chemical activity or solubility is established as a major limiting factor on lifetime. In LMMHD systems, temperature differences are typically 10 to 30 K, compared with the 245 K temperature difference in the sodium for the Carrisa Plain preliminary design study.

The application of LMMHD systems to solar heat sources, and in particular for high-temperature solar power towers, has been considered by Pierson et al.[2,3] and Pierson and Herman.[4] All three works present performance data, and the last one also includes cost estimates.

LMMHD cycles for lower-temperature applications, including solar, are considered in Refs. 2 and by Branover et al.[5,6]

The objective of the present study is to define the efficiency potential of the high-temperature LMMHD cycle for solar applications. A revised and improved computer model is used for the calculations.[7] First, the performance of the reference system, a steam power plant coupled to a sodium-cooled solar receiver, is described in Sec. II and the solar receiver model is treated in Sec. III. The cycle configurations chosen for study are described in Sec. IV. The results using sodium and lithium as the liquid metal are presented in Secs. V and VI respectively. The conclusions of the study appear in Sec. VII. A fuller treatment of the topics of the paper has been prepared by Pierson et al.[8]

II. Selection of Reference System

The preliminary design study[9] for the Carrisa Plain Solar Central Receiver Power Plant Project currently being undertaken by Rockwell International Corporation with U.S. Department of Energy support was chosen as the reference or basis of comparison for the solar thermal LMMHD system. It has a sodium-cooled central receiver coupled to a conventional Rankine cycle with main steam conditions of 99.7 atm (1,450 psig) and 811 K (1000°F). The steam plant uses superheat and feedwater heaters, but no reheat because of its small size. The net power output is 30 MWe, the thermal efficiency of the steam plant is 37.5%, and the net plant efficiency (net electrical output divided by heat into the liquid metal) is 33%.

In comparing an LMMHD plant with the reference plant it is important to note that the LMMHD plant provides the circulation power for the receiver coolant directly, while in the reference case the pumping power is provided by separate pumps and accounts for most of the difference between the thermal efficiency and the net efficiency. Also, the LMMHD cycle efficiency has to be adjusted for any changes in solar receiver efficiency (see Sec. III), and allowance has to be made for the efficiency of the power conditioning equipment. Pending further work on this latter topic, an efficiency of 95% has been assumed. The correct comparison is then accomplished by multiplying the LMMHD efficiency by 0.95 and comparing the result with the net reference plant efficiency of 33%, i.e., the LMMHD efficiency should exceed 35%.

III. Receiver Model

A receiver model has been developed for analyzing the performance of a sodium-cooled external receiver for a wide variety of operating conditions. This model was used to

HIGH-TEMPERATURE LMMHD SOLAR THERMAL SYSTEMS 565

calculate heat and pressure losses, and the results indicate that at the same sodium outlet temperature (838.7 K) and for the same peak solar energy input as for the Carrisa Plain design, the receiver efficiency decreases from 91.6% for the reference system to 89.4% for the LMMHD system. The reduced efficiency for LMMHD is caused primarily by the higher radiation loss resulting from the higher average sodium temperature in the receiver tubes. (Reflection loss is set at 5%, and convection losses are much smaller than radiation loss for these temperatures). For the reference design the sodium inlet temperature is 594 K, while for the LMMHD system it is 10 to 30 K cooler than the sodium outlet temperature.

One of the primary advantages of the LMMHD system over the steam cycle is the ability of effectively utilizing higher temperatures. The calculated receiver efficiency as a function of sodium exit temperature for a sodium inlet temperature 10 K lower than the exit temperature decreases as the sodium temperature is increased. The optimum LMMHD solar system could require an alternative concept, such as a cavity receiver with significantly reduced radiation loss.

IV. LMMHD Cycles Considered

Three versions of the LMMHD Brayton cycle (Fig. 1) were considered: 1) a simple LMMHD cycle where all of the output power comes from the MHD generator; 2) LMMHD-gas turbine dual cycle with a gas turbine in the gas stream after the gas-liquid separator to extract additional energy from the gas stream; and 3) a LMMHD-steam turbine binary cycle where the gas from the separator supplies heat to a steam bottoming plant.

Two liquid metal working fluids were considered -- sodium and lithium. For the lower receiver temperatures, sodium is used because sodium technology is better established. For temperatures above approximately 950 K, lithium is superior because of its lower vapor pressure. Helium was selected as the gas for both liquids.

The impact on cycle efficiency of the major parameters - temperature, pressure, cycle configuration, generator inlet and exit void fractions, regenerator effectiveness, and pipe lengths - are considered. The effects of the component efficiencies -- e.g. diffuser or nozzle efficiencies -- are presented in the sensitivity studies. The effects of exceeding or not meeting base component efficiency rates are presented in the form of gain or reduction, respectively, of the overall cycle efficiency. The component parameters used for the calculations are listed in Table 1. The tur-

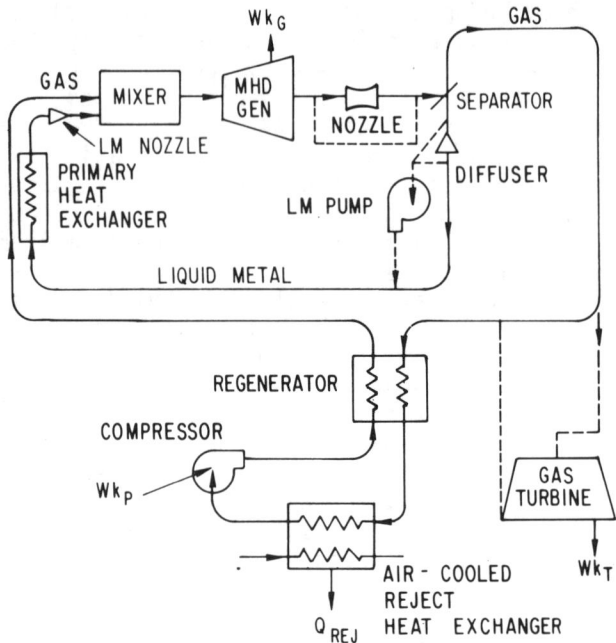

Fig. 1 Schematic of the LMMHD system analyzed.

bine and compressor efficiencies were taken from the Energy Conversion Alternatives Study[10]. The two-phase nozzle and diffuser efficiencies, the separator kinetic energy loss and the generator efficiency were based on an extensive review of available data conducted by the Argonne National Laboratory.[11]

The specified temperature is the exit temperature of the coolant from the solar receiver. The liquid metal is heated in the solar receiver and the gas is then heated by direct contact with the liquid metal in the two-phase mixer. This results in a simpler cycle because there is no separate gas heater, a large and expensive component, but reduces the cycle efficiency slightly because the temperature at the generator inlet is slightly lower, and this is the temperature which controls the thermodynamic cycle efficiency. (In some cases the temperature decrease is not small and this is one factor which determines the optimum operating conditions. The specified pressure is the pressure at the mixer inlet. This pressure differs from the solar receiver pressure by the velocity heat (from 5 m/s in the receiver and pipes to 30.5 m/s at the mixer inlet), the gravity head due to the height difference and pressure losses. The receiver pressure will normally be lower.

Table 1 Component Performance Parameters

Pressure drop (indicated pressures at component exit)

$\Delta p_{mixer} = 0.33$ atm $\qquad \Delta p_{reg\ hot} = 0.015 p_{sep}$

$\Delta p_{LMHX} = 0.67$ atm $\qquad \Delta p_{reg\ cold} = 0.015 p_{RejHX}$

$\Delta p_{sep} = 0$ $\qquad\qquad\quad \Delta p_{rejHX} = 0.025 p_{reg\ hot}$

Efficiencies

Compressor	0.873	Liquid metal nozzle	0.95
LMMHD generator	0.8	Two-phase nozzle	0.9
Turbine	0.933	Diffuser	0.8

Others

Five compressor stages with four interstage coolers
Ambient temperature 297.2 K, pinch point 11.1 K
Generator velocity 30.5 m/s
Separator loss, 15% of inlet kinetic energy

For all of the data presented in Secs. V and VI, the base conditions are used unless stated otherwise. These are a system pressure of 50 atm, a heat source temperature of 922 K, an exit void fraction of 0.85, a regenerator effectiveness of 0.95, and the component parameters specified in Table 1.

V. Sodium Cycles

The simple LMMHD cycle is considered in Sec. VA, the LMMHD-gas turbine combined cycle is dealt with in Sec. VB, and the LMMHD-steam turbine binary cycle is the topic of Sec. VC.

A. Simple LMMHD Cycle

The efficiency of the simple LMMHD Brayton cycle as a function of (generator) inlet void fraction and heat-source temperature is shown in Fig. 2 for the base conditions. The efficiency for the same parameters except an effectiveness of 0.90 has a peak value for each temperature that is approximately two percentage points lower and occurs at a slightly lower inlet void fraction. The regenerator effectiveness plays a significant role because of the almost-constant-temperature expansion in the generator and the resulting high gas temperature after the separator.

The efficiency increases rapidly with increasing temperature up to approximately 978 K at a rate of approximately

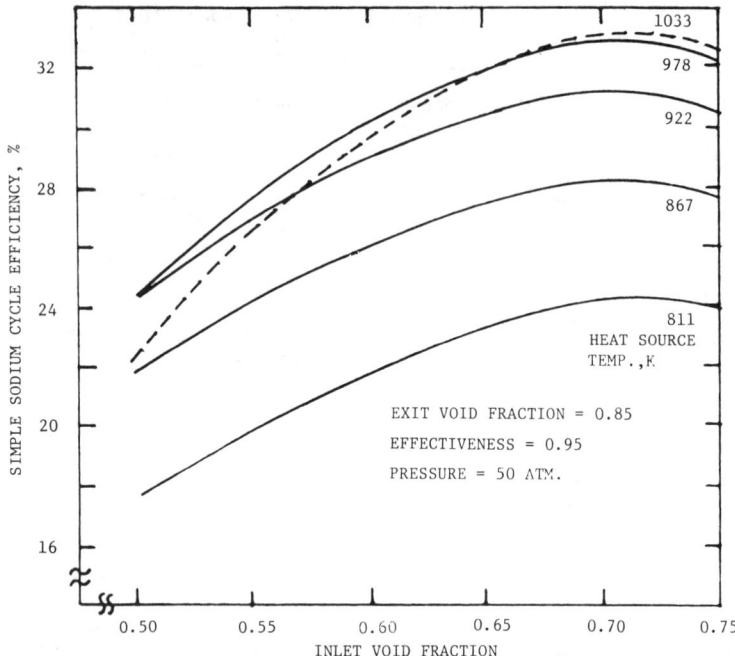

Fig. 2 Sodium cycle efficiency vs inlet void fraction as a function of temperature for the simple LMMHD cycle.

four points per 55 K (1000°F). This demonstrates the significant advantages for this cycle at the higher heat-source temperatures. For still higher temperatures, the efficiency does not increase further and may even decrease because of the sodium vapor carried over with the gas leaving the separator. This vapor will condense as the gas is cooled and thus will impact the design of the downstream components. The curves cross in Fig. 2 because the vapor carry-over depends on both temperature and pressure and, as the generator exit pressure is decreased (corresponding to lower inlet void fraction for the fixed exit void fraction), the separator pressure is reduced and the amount of vapor carry-over increased. At higher temperatures, a fluid with a lower vapor pressure is preferable, as illustrated for lithium in Sec. VI.

The effect of increasing the generator exit void fraction to 0.90 is to increase the efficiencies by about 3.5 percentage points because there is less circulating liquid metal (and associated losses). At 922 K, the efficiency of over 34% could be attractive because of the simplicity of the cycle. The feasibility of such high void fractions is discussed in Sec. VB.

HIGH-TEMPERATURE LMMHD SOLAR THERMAL SYSTEMS 569

The dependence of the efficiency on mixer inlet pressure is shown in Fig. 3. The efficiency increases significantly as the pressure is increased for low pressure, but peaks at about 100 atm. For higher pressures (not shown because they are too high for the receiver), the efficiency decreases. The increase occurs, because as the pressure increases the gas density increases, resulting in less liquid metal flow (and associated circulation losses in the nozzle, separator, and diffuser) needed for a given power output. As the pressure is increased further, the liquid metal flow rate becomes sufficiently small so that the temperature drop in the mixer as the gas is heated and the temperature drops in the LMMHD generator and two-phase nozzle increase enough to lower the cycle thermodynamic efficiency. The optimum pressure depends on the cycle parameters and on the constraints imposed in coupling to the solar receiver. For this study, 50 atm was chosen as the base because the efficiency was close to the optimum value and the pressure was well within the receiver limits.

The simple LMMHD Brayton cycle is a single cycle and, as extensive thermodynamic experience confirms, does not attain the required thermodynamic performance within the sodium temperature range when compared with conventional Rankine cycle turbomachinery. It may, however, be an

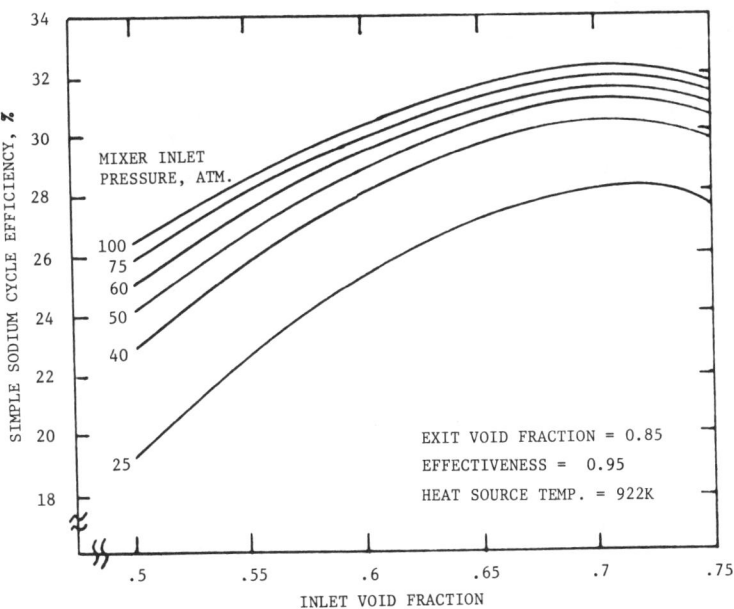

Fig. 3 Sodium cycle efficiency vs inlet void fraction as a function of pressure for the simple LMMHD cycle.

attractive first approach to demonstrate the potential of the LMMHD cycle and its coupling to a solar central receiver.

B. LMMHD-Gas Turbine Cycle

The LMMHD-gas turbine or combined cycle has a substantially higher efficiency, uses the gas turbine to provide most or all of the compressor drive power and requires a significantly smaller regenerator. The LMMHD combined cycle is superior to the conventional gas turbine cycle because 1) the improved coupling to the solar receiver provided by the liquid metal results in a higher conversion cycle top temperature, and 2) the almost-constant-temperature expansion results in a higher cycle efficiency and eliminates any need for a reheat heat exchanger.

The efficiency of the combined cycle as a function of temperature and turbine pressure ratio for generator pressure ratios of 0.327 and 0.704 is shown in Figs. 4 and 5 respectively. (The standard generator pressure ratios used and the corresponding inlet void fractions for an exit void fraction of 0.85 are listed in Table 2). As in the case of the pure LMMHD cycle, the efficiency increases with tempera-

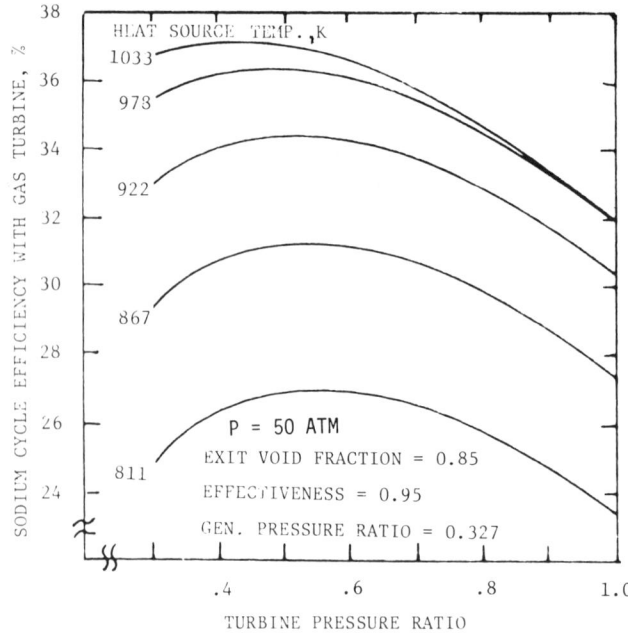

Fig. 4 Sodium cycle efficiency vs turbine pressure ratio as a function of temperature for 0.327 generator pressure ratio.

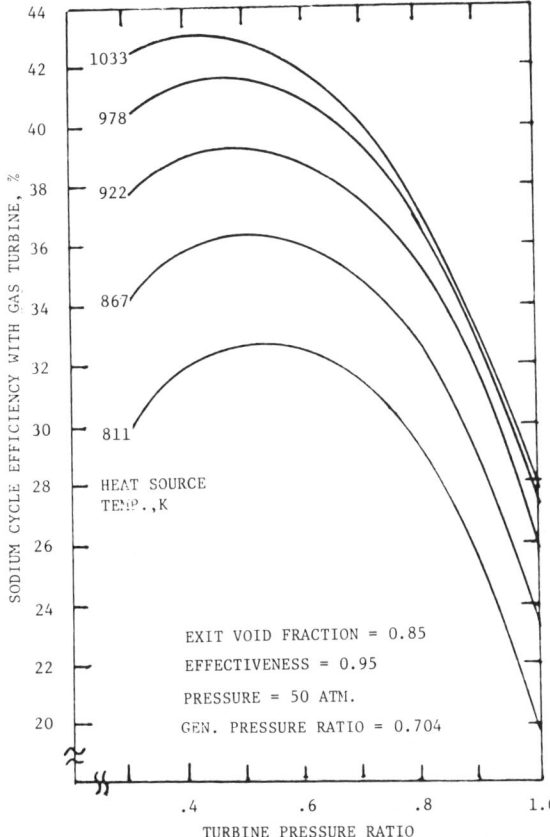

Fig. 5 Sodium cycle efficiency vs turbine pressure ratio as a function of temperature for 0.704 generator pressure ratio.

ture, but the impact of vapor carry-over on the cycle efficiency is reduced because part of the power comes from the turbine.

The effect of pressure is similar to that for the simple LMMHD cycle. As the turbine pressure ratio is decreased from unity to 0.2 for a 0.410 generator pressure ratio, the percent of the total gross power from the gas turbine increases from zero to approximately 55% and the temperature difference across the turbine increases from zero to approximately 400 K. The former means that the effect of pressure is reduced for lower turbine pressure ratios because more of the total power comes from the turbine and the turbine efficiency is independent of pressure. The latter means that the energy transferred in the regenerator at a turbine pressure ratio of 0.2 is approximately 20% of

Table 2 Sodium Cycle Generator Pressure Ratios
and Corresponding Inlet Void Fractions
for 0.85 Exit Void Fraction

Generator Pressure Ratio	Inlet Void Fraction
0.264	0.60
0.327	0.65
0.410	0.70
0.528	0.75
0.704	0.80

the value at unity, and the impact of the effectiveness on performance is reduced.

The attainable cycle efficiency is strongly influenced by the void fraction limits. The maximum attainable generator inlet (mixer exit) void fraction is set by the void fraction at which a homogeneous two-phase flow can be established. The inlet void fraction limit is normally considered to be approximately 0.6-0.65, although higher values may be attainable with careful design at high velocities and the magnetic field may assist for the LMMHD case.

In the LMMHD cycle the gas enters the mixer substantially colder than the liquid, and thus the void fraction at which mixing occurs is lower than the generator inlet void fraction, which is always calculated in this report for equal gas and liquid temperatures. For example, the highest efficiency for a generator pressure ratio of 0.410 occurs at a turbine pressure ratio of 0.5 (Fig. 6). The generator inlet void fraction is 0.70, but the mixer inlet void fraction is only 0.635 because the gas enters the mixer 232 K colder than it leaves the mixer.

The generator exit condition is set by the void fraction to which the two-phase flow can be expanded in the generator while maintaining homogeneity and electrical conductivity. Values to 0.85 have been demonstrated in an LMMHD generator, and higher values may be attained with the improved flow obtained in the latest reported experiments.[12] The recent data indicate that a two-phase flow can be very uniform with 1) a low slip or ratio of gas to liquid velocities of approximately unity, and 2) two-phase to pure liquid electrical conductivity ratio close to the Maxwell equation. The previous 0.85 limit was established with a non-uniform two-phase flow, i.e., higher slip and lower electrical conductivity, but this flow is considered to have a lower void fraction limit than the uniform flows now attainable. The use of surfactants may further increase

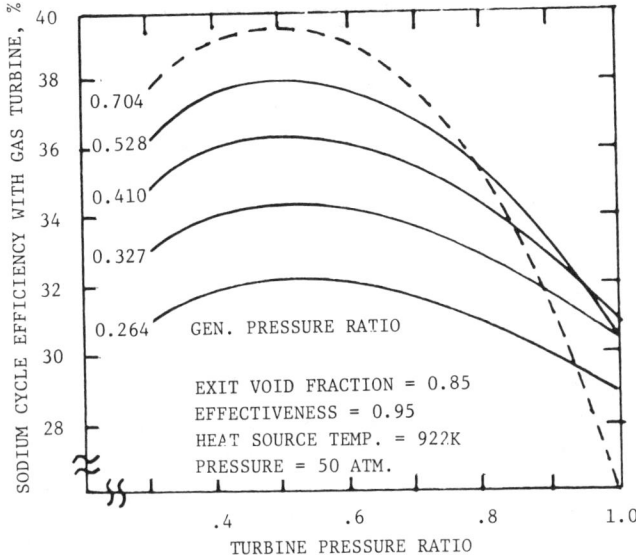

Fig. 6 Sodium cycle efficiency vs turbine pressure ratio as a function of generator pressure ratio.

the maximum attainable values for both inlet and exit void fractions.

The effect of the generator exit void fraction on cycle efficiency is illustrated by Figs. 6 and 7 as a function of the turbine pressure ratio and the generator pressure ratio (or inlet void fraction). Note that if the turbine contribution is small, the best efficiency is for the smaller generator pressure ratios as for the simple LMMHD cycle, whereas if the turbine power is large, the best efficiency is for the larger generator pressure ratios. Clearly, since mixer inlet void fractions of at least 0.65 are within the current state of the art, efficiencies of approximately 36% or more are attainable.

Increasing the exit void fraction from 0.85 to 0.90 increases the efficiency, with the increase being larger for the smaller generator pressure ratios. Decreasing the effectiveness from 0.95 to 0.90 decreases the efficiency by approximately two points.

The increase and subsequent decrease in efficiency with increasing exit void fraction is clearly shown in Fig. 8 for a generator pressure ratio of 0.327. The impact of changing the effectiveness to 0.90 for 0.85 exit void fraction is also shown.

The efficiency limits set by an upper limit on the inlet void fraction are demonstrated in Fig. 9 for an in-

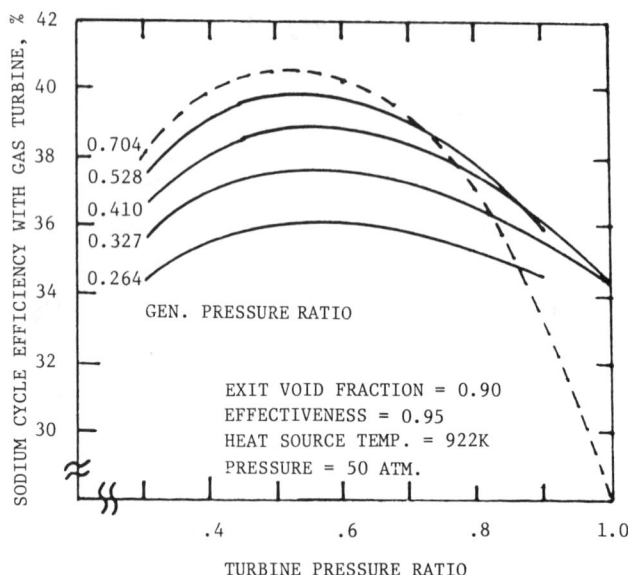

Fig. 7 Sodium cycle efficiency vs turbine pressure ratio as a function of generator pressure ratio for 0.90 exit void fraction.

let void fraction (without taking account of the effect of the colder gas entering the mixer) of 0.65. Significant gains will result from increasing this limit, e.g., at an inlet void fraction of 0.70, the efficiency increases by two percentage points.

One significant advantage of the LMMHD cycle coupled to a solar receiver is that the liquid pumping power is supplied directly from the cycle by the two-phase nozzle and, thus, has a lower impact on overall plant efficiency and cost. All of the preceding data use an assumed pipe length of 10 m. enough to connect all of the components together but insufficient to reach a solar receiver located on a tower. The velocity in the pipe is 5 m/s, while the generator velocity is 30.5 m/s. The impact of changing the pipe length, shown in Fig. 10, is small. Increasing the pipe length up to 500 m will decrease the efficiency by less than one point for all three generator pressure ratios shown. (The pipe loss can be decreased further by reducing the pipe velocity at a cost of a larger liquid inventory and slightly increased diffuser losses.) The effect of increasing the pressure drop across the receiver from 0.67 to 3.0 atm (Fig. 10) is to decrease the efficiency by slightly more than one point for a 0.410 generator pressure ratio.

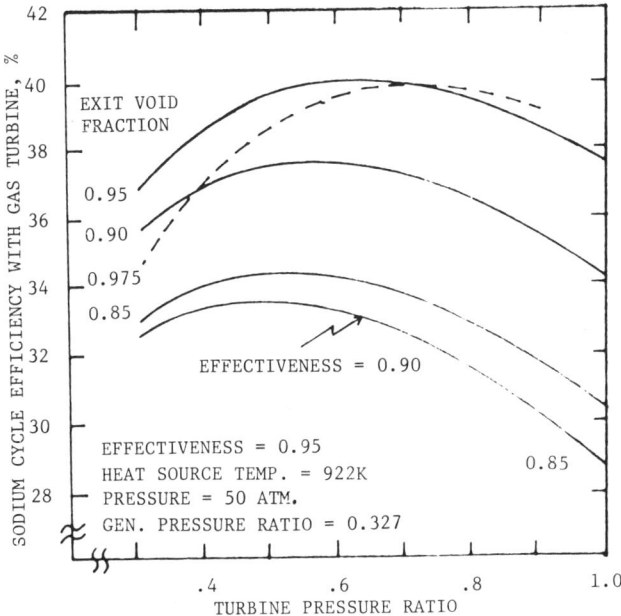

Fig. 8 Sodium cycle efficiency vs turbine pressure ratio as a function of generator pressure ratio for 0.327 inlet void fraction.

The assumed component efficiencies have a significant impact on the cycle efficiency. The most uncertain values are the two-phase nozzle efficiency, the kinetic energy loss in the separator and the (two-phase) diffuser efficiency. The base values used for the preceding curves are 0.90, 0.15, and 0.80, based on the Argonne National Laboratory technology assessment already referenced.[11] The impact of the component parameters depends on the system parameters, as shown in Tables 3-5 for 0.5 turbine pressure ratio (the value for maximum system efficiency.) Table 3 shows the effect of the generator pressure ratio and the regenerator effectiveness. As the generator pressure ratio decreases, the pressure drop across the generator increases, the mass flow rate of liquid metal increases, more "pumping" work is required from the nozzle-separator-diffuser combination, and the component losses have a greater impact on the system efficiency. The influence of the effectiveness decreases as the generator pressure ratio decreases because there is more liquid metal flow and, thus, less temperature drop in heating the gas. The effect of compressor efficiency, the number of compressor interstage coolers, and the turbine efficiency as a function of the same parameters is shown in Table 4 for 0.410 generator

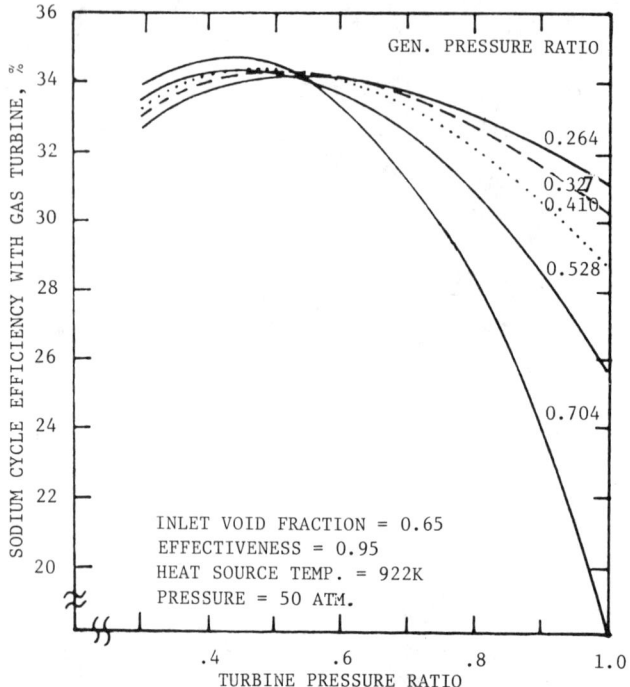

Fig. 9 Sodium cycle efficiency vs turbine pressure ratio as a function of generator pressure ratio for 0.65 inlet void fraction.

pressure ratio. Note in particular that decreasing the number of intercoolers from four to two decreases the efficiency by approximately three points. Increasing exit void fraction, Table 5, decreases the sensitivity to the nozzle, separator, and diffuser losses because of the decreased liquid metal flow rate, but increases the sensitivity to the effectiveness.

C. LMMHD-Steam Cycle

Analysis of a combined cycle with a steam plant provides insight into the Brayton/Rankine cycle tradeoffs in the temperature range of interest. Accordingly, a model for a steam bottoming plant was developed that calculates the turbine power and steam thermal plant efficiency as a function of the gas temperature entering the boiler, the gas temperature leaving the boiler, and the gas mass flow rate. The feedwater heater duty is adjusted to correspond to the specified gas exit temperature -- lower gas exit temperatures yield more steam plant output power but at a lower efficiency. Superheating of the steam is included, but not

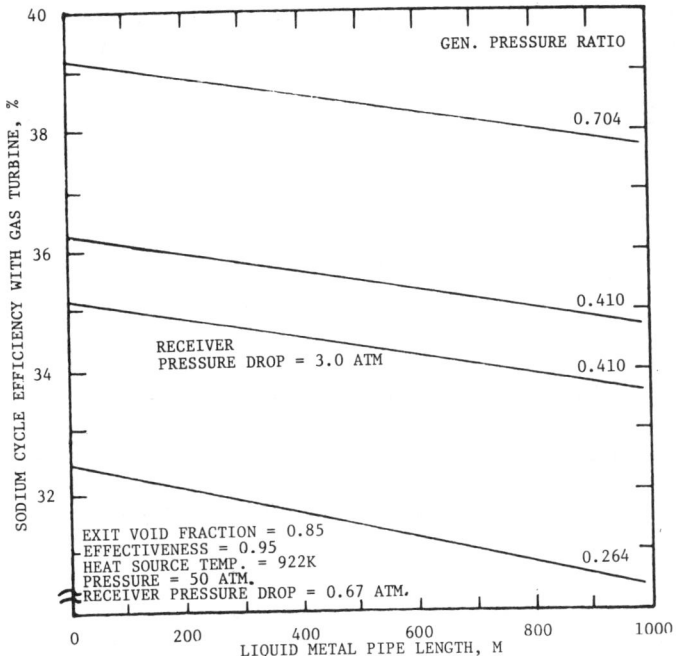

Fig. 10 Sodium cycle efficiency vs liquid metal pipe length.

reheating because, like Carrisa Plain, the plant is too small to make reheating economical with conventional ASME turbines. The steam conditions exiting the boiler are 99.7 atm and 811 K, as in the reference (Carrisa Plain) plant.

For the LMMHD-steam cycle the gas leaving the LMMHD separator enters the boiler, and the gas leaving the boiler flows to the regenerator, the reject heat exchanger if there is no regenerator, or the compressor if both the regenerator and the reject heat exchanger are not used. The best LMMHD results are obtained if both the regenerator and the reject heat exchanger are eliminated, the boiler feedwater is heated by cooling the helium to 335 K, and there are no compressor interstage coolers. This results in a steam plant without feedwater heaters. While the steam cycle efficiency is reduced, the best performance is obtained from the combined cycle. This situation arises because inclusion of the reject heat exchanger results in much of the energy in the helium not being transferred to the steam, with the net result that the work output and efficiency are reduced. If the regenerator is eliminated, then the interstage coolers are detrimental because, without them, the compressor work input heats the gas, decreases the heat input, and increases the two-phase fluid temperature at the generator inlet.

Table 3 Sodium LMMHD-Gas Turbine Cycle Efficiencies
for System Parameters and 0.5 Turbine Pressure Ratio

General Pressure Ratio Regenerator Effectiveness			0.704		0.528	0.410	0.327	0.264	
			0.90	0.95	0.95	0.95	0.95	0.90	0.95
Noz. Effy	Sep. Loss	Diff. Effy							
0.90	0.10	0.90	39.1%	41.1%	40.4%	39.4%	38.1%	35.9%	36.6%
0.85	0.15	0.85	37.7	39.6	38.2	36.5	34.6	31.8	32.4
0.80	0.20	0.80	35.7	37.5	35.0	32.5	29.8	26.3	26.8
0.90	0.15	0.80	37.6	39.5	38.1	36.3	34.4	31.6	32.2

Table 4 Sodium LMMHD-Gas Turbine Cycle Efficiencies
for System Parameters, 0.410 Generator Pressure
Ratio, and 0.5 Turbine Pressure Ratio

Compressor Effectiveness			0.873	0.873	0.900	0.873
No. Coolers			4	2	4	4
Turbine Efficiency			0.933	0.933	0.933	0.900
Noz. Effy	Sep. Loss	Diff. Effy				
0.90	0.10	0.90	39.4%	36.3%	41.1%	38.6%
0.85	0.15	0.85	36.5	33.1	38.2	35.6
0.80	0.20	0.80	32.5	28.6	34.3	31.6
0.90	0.15	0.80	36.3	32.9	38.1	35.5

Table 5 Sodium LMMHD-Gas Turbine Cycle Efficiencies
for System Parameters, 0.410 Generator Pressure
Ratio, and 0.5 Turbine Pressure Ratio

Exit Void Fraction Regenerator Effectiveness			0.85		0.90		0.95	
			0.90	0.95	0.90	0.95	0.90	0.95
Noz. Effy	Sep. Loss	Diff. Effy						
0.90	0.10	0.90	38.3%	39.4%	39.5%	40.7%	39.6%	41.0%
0.85	0.15	0.85	35.4	36.5	37.7	38.9	38.8	40.1
0.80	0.20	0.80	31.6	32.5	35.4	36.5	37.7	39.0
0.90	0.15	0.80	35.3	36.3	37.6	38.8	38.8	40.1

HIGH-TEMPERATURE LMMHD SOLAR THERMAL SYSTEMS

The power contributions from the LMMHD generator and steam turbine, and the power to the compressor per kilogram per second of gas flow are listed in Table 6 for several sets of inlet and exit void fractions. The LMMHD power clearly makes a significant contribution to the overall plant output, but at the lowest inlet void fraction the LMMHD power is less than the compressor input power, and the total efficiency is less than that for the steam plant alone. Clearly, care must be taken in the choice of operating conditions or there may be no efficiency benefit from the LMMHD cycle. The best case has an efficiency of 38.5%, which is significantly above the target efficiency.

The sensitivity of the LMMHD-steam binary cycle to component (nozzle, separator, diffuser) efficiencies is shown in Table 7. The impact of varying these parameters together is larger than the decrease due to each alone. Clearly, for this system, high average component efficiencies results in a very attractive system efficiency, while it is of marginal interest if the average efficiency is low.

The selection of a steam bottoming cycle introduces cost and operational considerations that can only be resol-

Table 6 Sodium Cycle LMMHD-Steam Powers and Efficiencies[a]

Inlet Void Fraction	Exit Void Fraction	Total Efficiency %	Steam Efficiency %	LMMHD Power MW	LMMHD Power MW	Compressor Power MW
0.60	0.85	27.5	33.2	2.78	1.002	2.96
0.65	0.85	34.5	33.2	2.25	0.996	2.15
0.70	0.85	37.7	33.2	1.73	0.998	1.50
0.75	0.85	38.5	33.2	1.20	0.979	0.94

[a] 922 K, 50 atm, no interstage coolers, power in MW per kg/s of gas flow.

Table 7 Sodium Cycle LMMHD-Steam Efficiency as a Function of Component Parameters[a]

Component	Base Case	Nozzle		Separator		Diffuser	
Efficiency or Loss	...	0.85	0.80	0.10	0.20	0.90	0.85
Cycle Efficiency, %	37.7	36.7	35.5	38.7	36.6	39.6	38.8

[a] 922 K, 50 atm, no interstage coolers, 0.70 and 0.85 inlet and exit void fractions.

ved through further analysis. It is important to note, however, that the heat exchange is gas-steam, a significantly easier condition to handle than the sodium-steam heat exchanger of the reference plant.

VI. Lithium Cycles

Lithium yields higher cycle efficiencies than sodium in part because it has better properties (lower density, higher specific heat capacity), but primarily because it has a lower vapor pressure and thus can be used at temperatures up to approximately 1260 K (1800°F). The disadvantage with lithium is the limited availability of containment materials compatible with both lithium and air at temperatures above approximately 900 K. Pending a materials evaluation, it may be observed that there are coating methods for lithium-containing pipes, or sodium could be used in the solar receiver and the lithium energy-conversion system jacketed.

The general efficiency behavior for the lithium cycle as a function of the various parameters -- temperature, pressure, void fractions, and component efficiencies -- is similar to that for sodium except that the efficiency is higher. Thus, the lithium data presented below are not as detailed as for sodium. Both the pure LMMHD cycle and the LMMHD-gas turbine combined cycle are covered. The LMMHD-steam binary cycle with lithium is not considered because the limits imposed by the steam plant efficiency do not allow the overall cycle to take full advantage of the higher gas inlet temperature.

The base condition for all of the lithium data is the same as for the sodium data except that the temperature is 1033 K.

A. Simple LMMHD Cycle

The efficiency as a function of (generator) inlet void fraction and heat-source temperature is shown in Fig. 11 for the base condition. For an effectiveness of 0.90, the efficiency is approximately 2.5 percentage points lower. Note that the efficiencies at 922 K with lithium are two percentage points higher than the corresponding sodium value Even higher efficiencies are attainable if increasing the void fractions proves to be feasible, but, for lithium, even the simple LMMHD cycle at quite reasonable void fractions exceeds the target efficiency of 35%.

B. LMMHD-Gas Turbine Cycle

The dependence of efficiency on temperature for a generator pressure ratio of 0.327 is illustrated in Fig. 12.

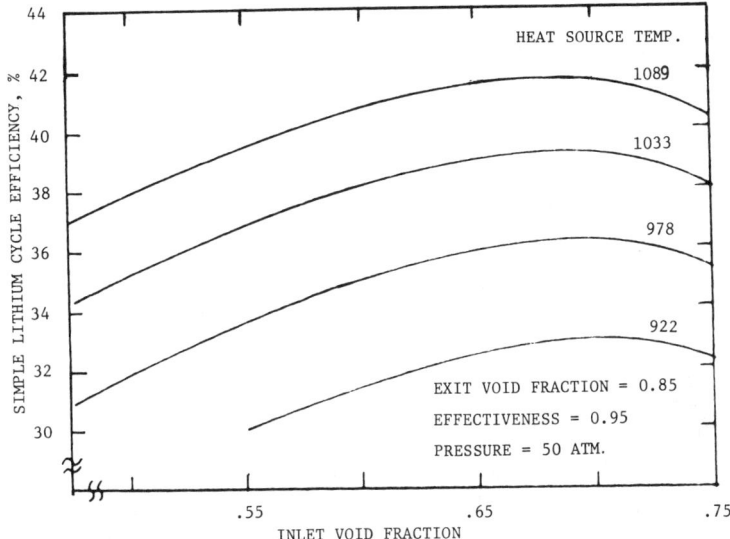

Fig. 11 Lithium cycle efficiency vs inlet void fraction as a function of temperature for the simple LMMHD cycle.

Efficiencies well in excess of 40% are easily attained at 1033 K and inlet void fractions of 0.65 or less. At 922 K, the efficiencies are again approximately two percentage points higher than the corresponding sodium values. The efficiency increases by 3 to 3.5 points for each 55 K (100°F) increase in temperature.

The effect of generator pressure ratio (inlet void fraction) for the base condition and an exit void fraction of 0.85 is summarized in Fig. 13. Even with a conservative input void fraction of 0.60, allowance for power conditioning losses and with the lower-solar receiver efficiency of Sec. III, the efficiency is 10% higher than the reference case and the cycle is simpler. In evaluating these results, note that at 0.5 turbine pressure ratio and a range of 0.264 to 0.704 generator pressure ratio, the LMMHD generator provides a corresponding range from 72% to 37% of the gross output power before compressor power and processing losses.

The efficiency for higher exit void fractions increases as with sodium and the efficiency spread with generator pressure ratio decreases. The efficiency constraint imposed by an inlet void fraction limit is similar to the sodium case. The peak efficiency is almost independent of generator pressure ratio. Efficiencies in excess of 42% are attainable at 0.65 inlet void fraction, and 44% at 0.70 inlet void fraction. The impact of the liquid metal pipe length is

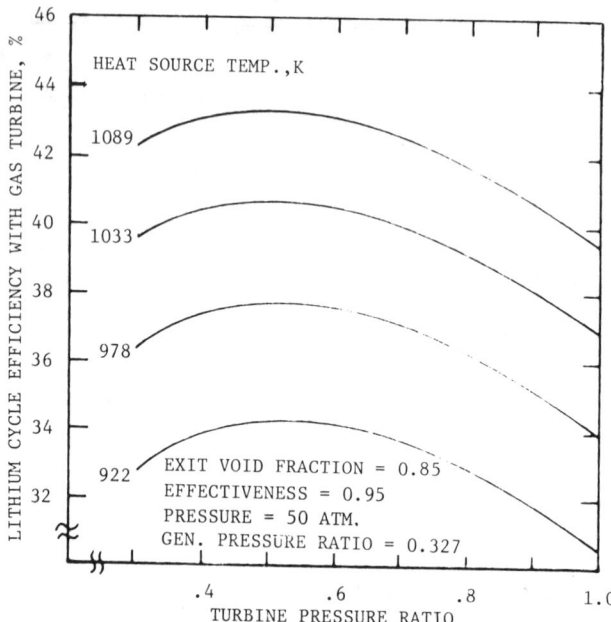

Fig. 12 Lithium cycle efficiency vs turbine pressure ratio as a function of temperature for 0.327 generator pressure ratio.

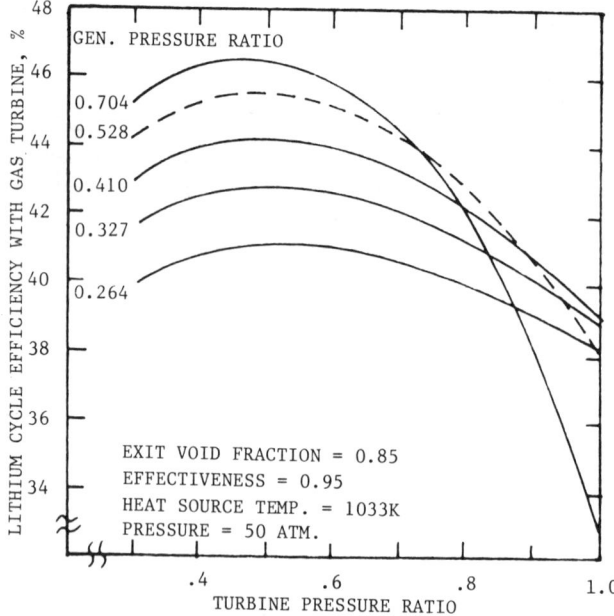

Fig. 13 Lithium cycle efficiency vs turbine pressure ratio as a function of generator pressure ratio.

HIGH-TEMPERATURE LMMHD SOLAR THERMAL SYSTEMS 583

also similar to that for sodium except that the sensitivity is smaller.
The sensitivity of the system efficiency to the component efficiencies is shown in Tables 8 and 9. These tables for lithium are for the same conditions as Tables 3-5 for sodium except for the higher temperature. The sensitivity of the lithium system is rather lower than for the sodium system, and this is a further advantage of lithium.

The lithium LMMHD cycle at temperatures of 922 K and above is clearly potentially more efficient than the reference case. The practicality of the lithium system needs to be resolved in a future design-oriented study with appropriate attention to materials engineering issues.

Table 8 Lithium LMMHD-Gas Turbine Cycle Efficiencies for System Parameters and 0.5 Turbine Pressure Ratio

Generator Pressure Ratio Regenerator Effectiveness	0.704 0.90	0.95	0.528 0.95	0.410 0.95	0.327 0.95	0.264 0.90	0.95
Noz. Effy	Sep. Loss	Diff. Effy					
0.90	0.10	0.90	45.2% 47.8%	47.5%	46.8%	45.9%	43.8% 44.8%
0.85	0.15	0.85	44.1 46.6	45.6	44.3	42.9	40.4 41.3
0.80	0.20	0.80	42.4 44.8	42.9	40.9	38.8	35.8 36.6
0.90	0.15	0.80	44.0 46.5	45.5	44.2	42.8	40.2 41.2

Table 9 Sodium LMMHD-Gas Turbine Cycle Efficiencies for System Parameters, 0.410 Generator Pressure Ratio, and 0.5 Turbine Pressure Ratio

Compressor Efficiency	0.873	0.873	0.900	0.873
No. Coolers	4	2	4	4
Turbine Efficiency	0.933	0.933	0.933	0.900

Noz. Effy	Sep. Loss	Diff. Effy				
0.90	0.10	0.90	46.8%	44.1%	48.3%	46.1%
0.85	0.15	0.85	44.3	41.3	45.8	43.5
0.80	0.20	0.80	40.9	37.5	42.6	40.1
0.90	0.15	0.80	44.2	41.2	45.8	43.4

VII. Conclusions

The preceding results have demonstrated that the LMMHD system 1) is compatible with solar central receivers employing a liquid metal as a heat-transfer medium; 2) eliminates the sodium-steam heat exchanger and return pump; and 3) offers significant increases in the system efficiency over conventional turbomachinery. With sodium at 922 K, efficiencies to almost 40% were calculated, while with lithium at 1033 K (1400°F), efficiencies to 46.5% were calculated, and efficiencies up to 50% are possible by further increase of the heat-source temperature. One reason for the higher plant efficiency is that the LMMHD cycles provides the pumping power for the receiver coolant directly. From a materials viewpoint, the sodium cycle can utilize the extensive high-temperature system development undertaken for the space and nuclear programs, while a materials engineering effort on containment structures is evidently required in the case of lithium.

The simple LMMHD cycle uses only a two-phase LMMHD generator to produce the output power, and this arrangement does not yield a thermodynamically competitive system at 922 K. At higher temperatures, especially with lithium, the efficiencies are significantly above the reference system, 39% being obtained at 1033 K. The thermal efficiency is increased to the values given above if a gas turbine or a boiler/steam turbine is added in the case of sodium. For lithium, the high efficiencies are attained with gas turbine combined cycles.

The study reported here has addressed the thermodynamic aspects of a solar LMMHD system and identified the performance potential but has left cost and oeprational issues to future work. It has also identified those areas where engineering efforts are required both to establish actual component performance limits and to develop an engineering basis for the design and construction of a demonstration system operationally acceptable to a utility power system.

Acknowledgments

This work was supported in part by the U.S. Department of Energy, San Francisco Operations Office under Contract DE-AC03-835F11943 and, in part, by the HMJ Corporation.

References

[1] Hildebrand, A.F. and Dasgupta, S. "Survey of Power Tower Technology," _Journal of Solar Energy Engineering_, Vol. 102, May 1980, pp. 91-104.

[2] Pierson, E.S., Branover, H., Fabris, G., and Reed, C.B., "Solar Powered Liquid Metal MHD Power Systems," Mechanical Engineering, Vol. 102, Oct. 1980, pp.32-37.

[3] Pierson, E.S., Cohen, D., and Grammel, S.J., "Liquid Metal MHD for Solar and Coal," Proceedings of the Seventh International Conference on MHD Electrical Power Generation, Boston, Mass., 1980, pp. 150-157.

[4] Pierson, E.S. and Herman, H., "Solar Powered Liquid Metal MHD Performance and Cost Studies," Liquid Metal Flows and Magnetohydrodynamics: AIAA Progress in Aeronautics and Astronautics, Vol. 84, edited by H. Branover, P.S. Lykoudis, and A. Yakhot, AIAA, New York, 1983, pp. 138-159.

[5] Branover, H., El-Boher, A., Sukoriansky, S., and Yakhot, A., "Development of a Low Temperature Liquid Metal MHD Small Scale Pilot Plant," Proceedings of the 21st Symposium on the Engineering Aspects of MHD, Argonne National Laboratory, Argonne, IL, June 1983.

[6] Branover, H., El-Boher, A., Lessin, S., and Marcus, B., "Test of ER-4 Mercury-Steam Power Generation Research Facility," Proceedings of the Fourth Beer Sheva Seminar, Beer Sheva, Israel, Feb. 1984.

[7] Geyer, H.K. and Pierson, E.S., "Solar Liquid Metal MHD Performance Predictions," Proceedings of the 20th Symposium on the Engineering Aspects of MHD, University of California, Irvine, Calif, June 1982.

[8] Pierson, E.S. et al, "Performance of Solar Thermal Systems with Liquid Metal MHD Conversion," HMJ Corporation, Washington, D.C., Rept. No. 4MHDSA-DOE-P/84/R1, June 15, 1984.

[9] "Preliminary Design of the Carrisa Plain Solar Central Receiver Power Plant, Volume 1, Executive Summary." DOE Report ESG-DOE 13404, Dec. 1983, prepared by Rockwell International Corporation under Contract DE-SFC03-82SF11674.

[10] Energy Conversion Alternatives Study, ECAS, General Electric Phase II Final Report, Vol. II, Pt. 2, pp. 1.3-23, NASA-CRI34949, Dec. 1976.

[11] "Liquid Metal MHD Systems: Concepts, Applications and Status," Attch. 7, App. D., "Status of LMMHD Technology," Argonne National Laboratory Rept, Feb. 1981 (available from M. Petrick, ANL).

[12] Fabris, G. et al, "High-Power-Density Liquid Metal MHD Generator Results," Proceedings of the 18th Symposium on the Engineering Aspects of MHD, Butte, Mont., 1979.

Chapter V. Metallurgical Applications

Metallurgical Applications of MHD

Marcel Garnier

Groupement d'Interet Scientifique, Madylam, Saint Martin d'Heres, France

Abstract

Some examples of techniques that involved MHD phenomena are presented. Before significant progress was made in the knowledge of MHD, especially concerning alternating magnetic fields, they show that these phenomena have long been supported. In the last few years, a conviction appeared that MHD could offer efficient solutions to metallurgical problems; at the same time, a rational approach to typical mechanisms used was developing. Electromagnetic stirring in continuous casting is a good illustration of a successful application of MHD as a result of this. Some scientific problems, connected with this example and with other general ones, are presented and are to be solved to enable a wider development of MHD in metallurgy. Some new possible applications are given to prove that MHD can give rise to very innovative metallurgical technologies.

Introduction

Even though MHD appeared very early in metallurgy (at the beginning of this century), it would be incorrect to claim that MHD phenomena were used to solve metallurgical problems. Initially the application's goal was merely to take advantage of an electric current flowing through a metal to dissipate energy by Joule effect in order to melt the metal, or to give rise to an electrolysis, as in the elaboration technique for aluminum. In all the applications, MHD phenomena were present. Their main effect was to induce an organized or unorganized flow in the metal, but the phenomena were secondary and generally supported.

Thus, in induction furnaces, electromagnetic stirring had positive effects on the correct bath thermal homogeni-

---

Paper presented at the Fourth Beer-Sheva Seminar on MHD Flows and Turbulence, Ben-Gurion University of the Negev, Beer-Sheva, Israel, Feb. 27-March 2, 1984. Copyright © 1985 by the American Institute of Aeronautics and Astronautics, Inc. All rights reserved.

zation, and negative effects, for example, on the mechanical erosion of the refractory walls and the resulting pollution of the molten metal. Today, they are still designed as fusion systems offering an optimization of the thermal efficiency, but not as metallurgical systems considering metallurgical quality and elaborating conditions of the metal. The same occurs with channel furnaces in which, for many years, only thermal free convection was responsible for transferring the Joule energy necessary for melting the metal load. Just recently have we acquired the first rational approaches of the mechanisms involved that lead to modifying the channels geometry in order to induce rational electromagnetic forces able to stir the metal, to improve heat transfers, and reduce melting time. The example of aluminum electrolysis cells is also typical; an important part of energy consumed by the cells (30%) is used only to compensate ohmic losses in the alumina bath located between the anode and the previously elaborated metal. Increasing the horizontal size of the cells and reducing the thickness of the alumina layer can lead to an important reduction of ohmic losses, and therefore to an increase of the process efficiency. The present limits of these two parameters are imposed by the motion of the liquid metal, which is not well-known and consequently uncontrolled. Also, in this particular case, MHD phenomena, although they are secondary, represent certain drawbacks, and thus are limiting factors in the development of the technique.

Recent studies, stemming from both a phenomenological approach of electromagnetic effects and a global and local modeling of the cells, lead to a better understanding of the geometric influence of liquid aluminum channels and of the position of the electric conductors feeding the cells[1-3]. The definition of certain rules to respect in the design of the plants is the result. Among the classical techniques, remelting processes for elaborating ESR and VAR are to be mentioned. For a long time, they have been victims of the incorrectly controlled constraint of MHD effects. They were accepted when favorable, as with the stirring in the ESR process, and they were unfortunately accepted when unfavorable, as with the free surface motion in the VAR process responsible for short circuits. During the past few years, decisive progress in both knowledge and comprehension of electromagnetic stirring[4-6] has rendered possible a significant improvement for these two techniques. Some modifications have been introduced to the design of the devices, such as the addition of a stirring inductor around the ingot mould of the VAR process: the electromagnetic stirring induced promotes a rapid removal of the free surface and improves the quality of the elaborated product.

A very interesting and characteristic application of MHD is electromagnetic stirring in continuous casting and electromagnetic stirring in ladle metallurgy; this involves an important progress because of a precise analysis of metallurgical effects produced. In this paper, this application will be chosen as an example to demonstrate how MHD phenomena can bring forth important improvements in metallurgy. The present experimental, physical, and mathematical approaches of the mechanism involved are analyzed. The scientific problems to be solved are discussed in connection with this particular application and other general ones. Then certain examples of very recent applications are given. These last applications are opposed to the previous ones whose technique existed before a precise knowledge of the phenomena. In comparison, for the recent applications, the exploitation of physical mechanisms of MHD guides the conception of the devices.[7,8]

An Example of Successful Metallurgical Application of MHD: Electromagnetic Stirring

The Technique

It is interesting to note that the first attempt to electromagnetically stir a volume of liquid metal during its solidification, in order to modify the final metallurgical structure, occurred very early in 1917. Following this experiment, induction stirring has become a widely used technique in continuous casting of billets, blooms, or slabs and in ladle metallurgy. The acquired metallurgical benefits have encouraged a worldwide research effort, and important progress has been made in the last ten years. An excellent review, which gives an updated picture of the state of the art, was published by Birat and Chone in 1982 (Ref. 9).

Today, inductive stirring is the best example of MHD metallurgical application, and this is proved by the impressive number of solutions arrived at to stir molten metals, which are currently in operation on many stands worldwide. Two types of electromagnetic stirrers have been developed in continuous casting, the first for billets and blooms, and the second, the later one, for slabs: in mold and beneath the mold in the secondary cooling zone. Different kinds of inductors are used: cylindrical inductors generating rotating or traveling magnetic fields; linear flat inductors parallel to the strand axis in the case of billets and blooms; or horizontal ones along the broad side of the slabs, generating traveling magnetic fields. A combination of rotating and traveling magnetic fields produced by the same inductor placed around the mold has also been used to give rise to helicoidal stirring. These inductors are generally provided with alternating currents of low frequency (1 → 20 Hz) in order to overcome the magnetic shield effect caused by the copper mold or the thick solid metal shell. Other original stirrers have been developed, for example, the purely magnetic stirring device using permanent samarium; Cobalt magnets[10]; the stirrer partially using the copper mold as a single turn coil provided with higher-intensity currents[11]; the conductive stirrer combining the magnetic field produced by permanent magnets and the current injected into the metal through the solid skin with the help of brushes[12]. Precise description of these stirrers are presented in many publications[13-18]. Some examples of electromagnetic stirring techniques used for slabs are given in Fig. 1.

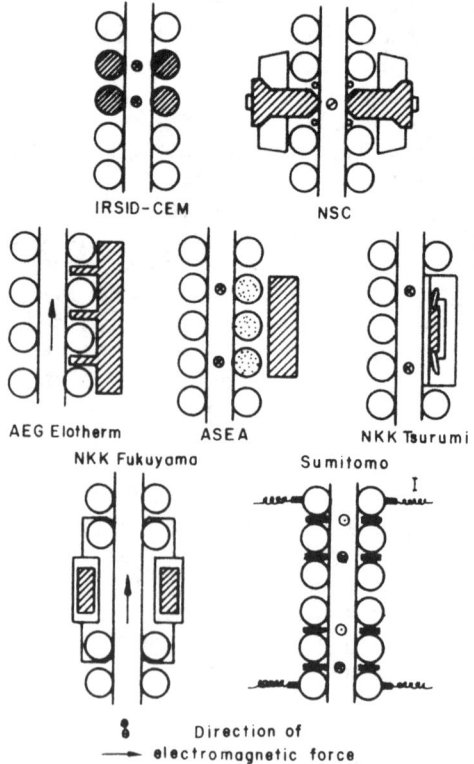

Fig. 1 Electromagnetic Stirring Techniques used for slabs.

Developments concerning induction stirring in ladle metallurgy, which began approximately 25 years ago, continue today. The same techniques are used[19]: cylindrical coils generating rotating magnetic fields, which result in imposing a main rotational motion in the ladle with secondary vertical recirculating flows: or vertical flat coils generating traveling magnetic fields, upward or downward, which induce vertical flows with one or several main vortices and secondary horizontal recirculating flows[20]. Fig. 2 summarizes some induction stirrers and presents schematic views of the resulting flow patterns.

Metallurgical Effects Produced by Electromagnetic Stirring

During the past four years, a great deal of literature has discussed the various benefits of electromagnetic

Fig. 2 Stirring patterns in a ladle.

stirring in continuous casting and in ladle metallurgy[21,22]
For continuous casting, the principal advantages are as
follows:

1) Improvement in the outer surface quality by a
spectacular decrease of slag traps (Fig. 3) with rotary
stirring in the mold while pouring out in an open stream,
and an extremely small number of pinholes.

2) Improvement in subsurface quality, especially with
linear upward stirring in the mold (Fig. 4) due to the fact
that the inclusions must flow with the stirred molten metal
towards the free surface, where they are then trapped by
the slag.

3) Beneficial effects concerning internal quality by
stirring beneath the mold, or inside the mold, separately
or together. The electromagnetic stirring tends to eliminate
the formation of bridges and smooths the solidification
fronts. In the secondary cooling zone, the time of superheat and the latent heat removal is reduced, and the temperature homogenization is improved because of better heat
transfers by convection and remelting of the broken-up
dendrites formed at the front of solidification. This
promotes equiaxed zone and leads to a more uniform distribution of non-metallic inclusions along the axis. However,
in stirring of slabs, with horizontal linear inductors, the
presence of high velocities and of small streamline
curvature radii near the narrow face of the slab can lead
to the formation of "white bands" (negative segregation).

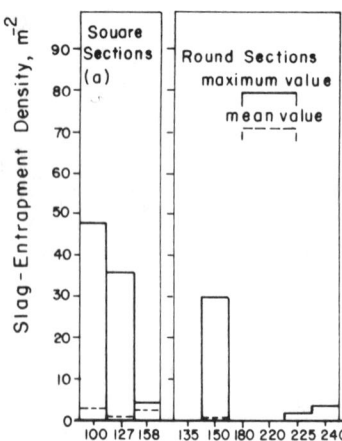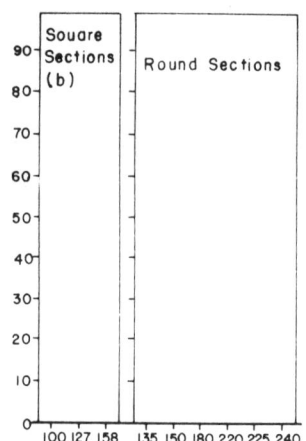

Fig. 3   Improvement in outer surface quality obtained with in-mold rotary stirring.

METALLURGICAL APPLICATIONS OF MHD

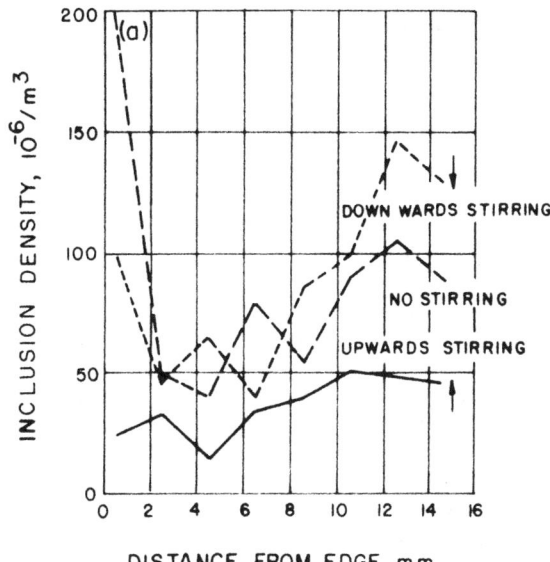

DISTANCE FROM EDGE, mm

Fig. 4 Improvement in subsurface quality with in-mold linear stirring.

To counteract this negative effect, alternate stirring[23] has a very good influence.

Also important are the advantages of electromagnetic stirring in ladle metallurgy through the acceleration of chemical reaction and homogenization of the molten metal bath[24]:

1) The stirring velocity induced in the metal near the interfaces increases the speed of heterogeneous reactions, for example, slag-bath reactions, and it reduces the time of transport of elements reacting to the interfaces. In some cases, the limiting factor may be the time of the chemical reactions.

2) The turbulence created in the bath accelerates the homogenization kinetics of the melt when reactive elements or alloys are added.

3) The removal time of inclusions, for example, particles of Al_2O_3 formed during the operations of deoxydization, can also be reduced by induction stirring.

The kinetics mechanism of inclusion removal is very complex in an electromagnetically stirred bath. Indeed, the magnetic field existing in the molten metal acts differently on the particles depending on their external shape and their electrical conductivity[25]. Moreover, the flow pattern and the turbulence characteristics might lead

to significant opposed effect: either they can promote
their growth and consequently their removal, or, on the
contrary, they lead to a dispersion and homogenization and
therefore lead to a negative uniform distribution of non-
metallic particles in the bath. In addition, inclusions
can escape entrapment at the interfaces if the stirring
velocity is too high. Particles can also be washed away
from the removal zones (for example, free surface or slag-
metal interface) if the liquid metal velocity is not low
enough.

This last limit, concerning the advantages of electromagnetic stirring, together with the possible formation of
white bands in continuous steel casting, demonstrates the
difficulty in determining the characteristics of a "good"
electromagnetic stirrer: What constitutes a "good" stirrer,
and what are the "correct" stirring conditions for a given
geometry of products (billet, bloom, slab) or ladle, and
for given metals or alloys with various compositions,
considering the precise result to be obtained? For continuous casting, Birat[9] gives very interesting and useful
guidelines for tailoring a stirring system with given
metallurgical objectives and continuous casters. However,
some question marks remain, for there is not much experience
dealing with particular linear systems or helicoidal
stirrers, or with new stirrers such as conductive or
alternate systems. Therefore, much work is still needed
to enable the designing of a stirring system. Recently,
interesting results have been obtained with two parallel
and complementary approaches: the physical approach and
the mathematical modeling.

Physical Approach and Mathematical Modeling
of MHD Phenomena

Physical Approach

The physical approach was at the origin of most of the
progress made in the development of continuous casting.
One should notice that IRSID plays a very important part in
the research based on metallurgical analysis of stirred
products and correlations with the parameters of the stirrers and of the characteristics of the molten metal.

The basic information involving the stirring is the
flow pattern induced in the liquid core of the strand.
Because of high-temperature and non-transparence, it is
difficult to obtain direct experimental data on velocity
field. The main techniques used consist in reconstructing
the flow pattern from metallurgical observations taken from
the solidified ingot.

The observation of cross-sectional sulphur prints is a good example of such a technique. The presence of white bands gives a good idea of the streamline configuration and may give some indication about the value of the velocity (Fig. 5).[26]

This analysis method is useful, but only to bring some modifications or corrections in the working of a stirrer, but are unable, because of their principle, to lead to any optimization: white bands are defects to eliminate, therefore the investigation method fails as soon as the defects begin to vanish, i.e., when more accuracy is needed. A very useful experimental tool can be deduced from Takahashi's work,[27] which gives an accurate relationship between the inclination of the dendrites and the velocity. However, the same complaint is to be made of this technique as well as

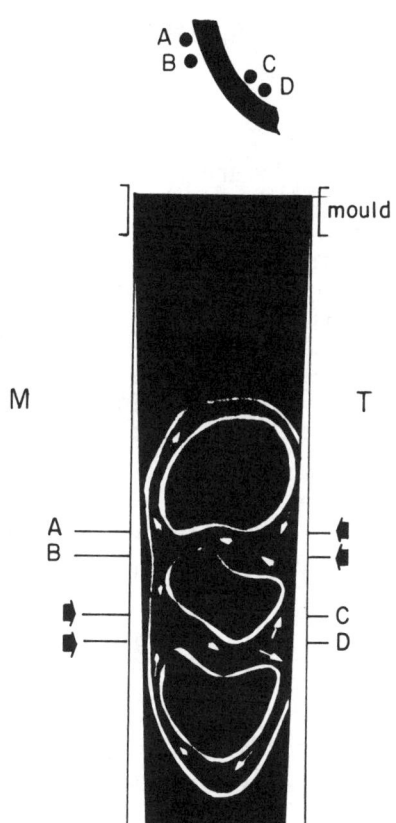

Fig. 5 Butterfly flow pattern obtained with IRSID-CEM stirring rolls.

the previous one because the diagnosis depends on the presence of defects.

Some other techniques are employed but are difficult to implement around the strand, such as radioactive tracers[28]. Simulations are possible in which electrical conducting liquid is put inside a given stirrer. But with liquid metals such as mercury, in which accurate measurements are possible, solidification is forgotten; with other molten metals, even with rather low melting points, which can be solidified, no available anenometry technique exists.

In the case of slabs submitted to horizontal traveling magnetic field, some water models can be used and give a relatively good reproduction of the flow pattern[29,30]. Propellers are used to force the liquid motion. The obtained results are good in this particular case because of the electromagnetic forces induced within a rather small region, and, at some distance from the inductor, the flow forgets the precise origin of the motion, which may then be produced by another localized cause. Fig. 6 gives the experimental flow patterns obtained in the case of an alternate stirring.

All these techniques cannot offer other information than flow patterns and, in the best cases, an order of magnitude for the velocity; however, they are quite unable to estimate the temperature field, which is, within the process, as important as the velocity field. Some rare experiments have been done and they give interesting measurements of the molten metal in the mold,[31,32] but temperature is very difficult to obtain by any other way than mathematical modeling.

The problem is much more difficult when it concerns experimental measurements of the stirring pattern in a ladle. Most of the above methods cannot be applied.

Mathematical Modeling

Today, no mathematical model exists that gives precise information about MHD phenomena when they are coupled with thermal and metallurgical effects, because of the high complexity of both mechanisms involved and geometries. However, some interesting studies offer information about the hydrodynamical and thermal behavior of the stirred molten metal and also about the solidification.

Concerning motion, the commonly used models take turbulence into account by considering a constant turbulent viscosity some orders of magnitude higher than molecular viscosity, or by using Prandtl mixing length theory, or by using a k-ε model. All these models give very approximate results and generally suffer from the lack of experimental data.

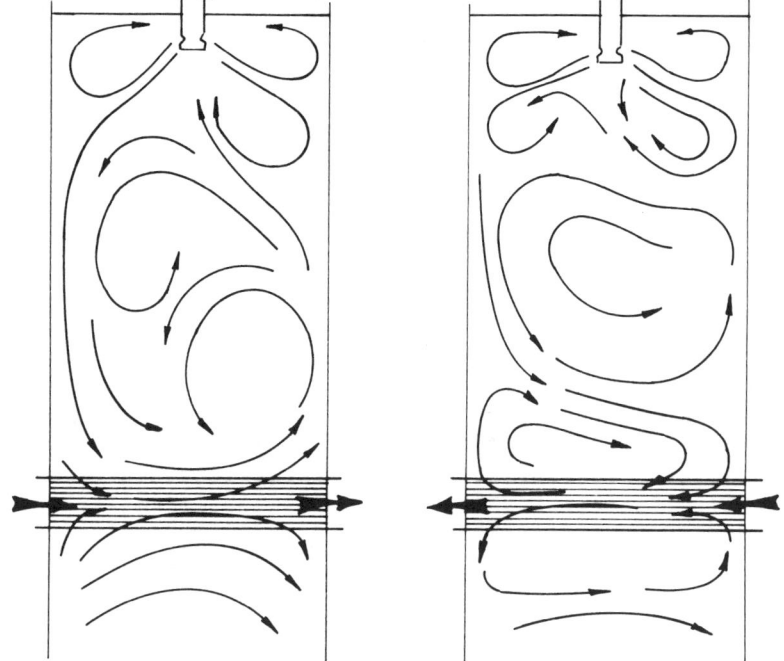

Fig. 6 Experimental flow patterns obtained in the ease of an alternating stirring.

To illustrate the efforts made in the mathematical approach of the problems, three typical and complementary works can be given as examples:

1) The work carried out by Van Den Hove[33], which gives a good modelization of the magnetic field and of the motion induced in the molten metal, in the case of the stirring below the mold of a billet. Thermal effects and solidification are not studied.
2) The work of Lesoult[34] which leads to a good modelization of the solidification in the case of slab casting. The flow pattern is not computed but schematically deduced from water models.
3) The work of Meyer and Durand[35], which is a good attempt at modeling coupled MHD and solidification phenomena. The originality of this study is to combine a numerical model and experimental simulation with Al-Cu alloys in a simple geometry, very appropriate for fundamental mechanisms analysis. The methods used and the results obtained are presented during this seminar.

In his work, Lesoult proposes a model for the growth of equiaxed crystals that are eroded away by the recircu-

lating flow induced by electromagnetic stirring. Except
for the flow pattern, the characteristic parameters of the
system are the number of crystallites eroded away per unit
of time and area surface and their radius. Heat and solute
transfer are considered as purely diffusive. The results
concern the effect of superheat, temperature field, and
solid fraction. Fig. 7 illustrates an example of computed
values for superheat and solid fraction with respect to the
factor of mass transfer between upper and lower streams.

Particular care is given by Van Den Hove in the
numerical modelization of the magnetic field generated by
the stirrer and in the definition of the stirrer itself.
The numerical model based on the finite-element method
used a mixed technique solving equations for magnetic field
scalar potential out of the metal and for the magnetic
field inside. Such a model is very appropriate for model-

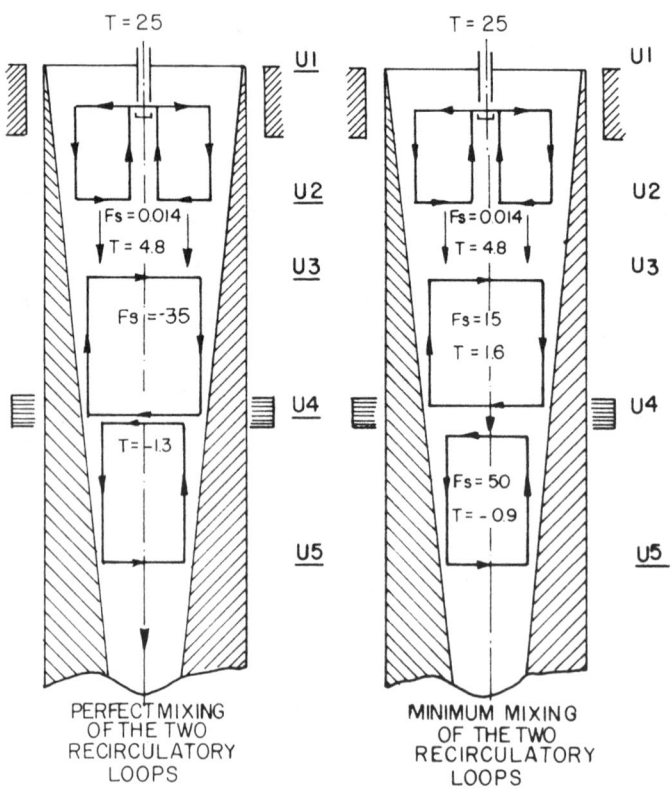

Fig. 7 An example of computed values for superheat and solid
fraction with respect to the factor of mass transfer between upper
and lower streams.

METALLURGICAL APPLICATIONS OF MHD   601

ization of the corners of the stirrer. A good representation of the fringe effect can so be deduced, with an important increase of the magnetic field near the upper and lower edges of the induction (Fig. 8).

The value of the angular velocity is computed in the case of a finite-size inductor with respect to the magnetic field intensity in the stirrer (Fig. 9).

Recirculating flows appear in a vertical plane that are very interesting concerning heat transfer and crystallites transport. These recirculating flows have two origins: first, the rotational part of the electromagnetic forces induced by fringe effects; second, the axial decreasing away from the inductor of the aximuthal component of the force: This induces high pressure within the stirrer and low pressure outside and is responsible for the axial fluid motion. The results of flow calculus are given on Fig. 10 with respect to the intensity of the magnetic field. The effect of withdrawal speed is also taken into account; the convection effect appears on Fig. 11. It should be noted that the domain of influence of the magnetic field through induced motion is not very large (four times the length of the stirrer).

Scientific Problems Connected With Metallurgical Applications of MHD

The different examples given above show how important it is for progress to be made in the analysis of MHD

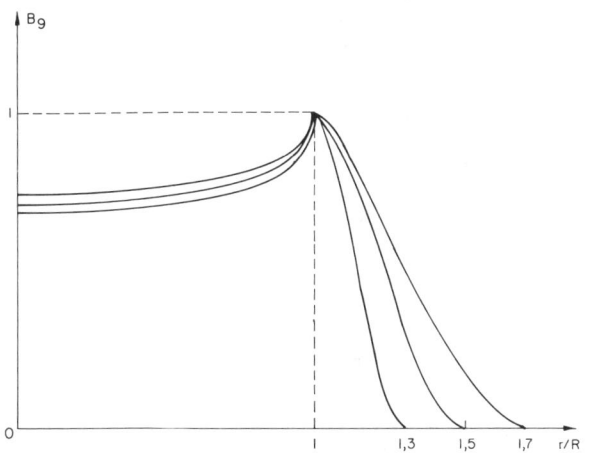

Fig. 8 Demonstration of the fringe effects by an increase of the magnetic field near the upper and lower edges of the induction.

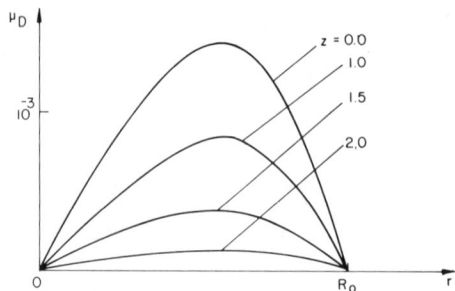

Fig. 9a Azimuthal velocity profiles for different values of distance from the stirrer center.

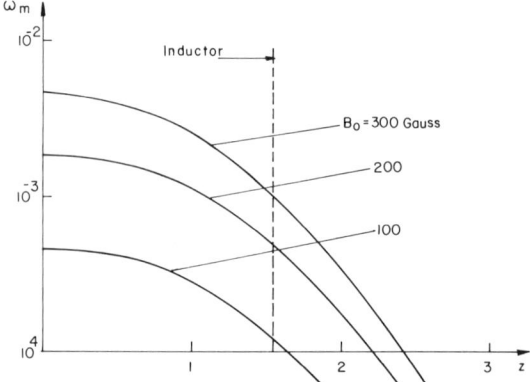

Fig. 9b Mean angular velocity profiles for different values of B_0.

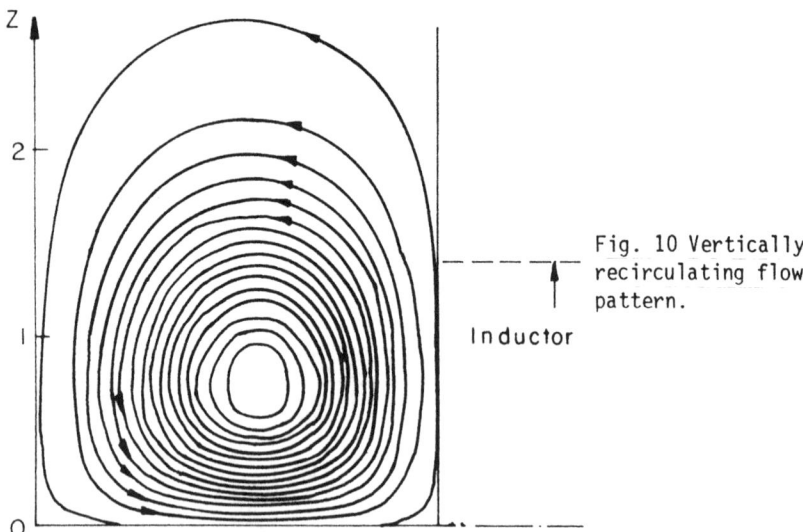

Fig. 10 Vertically recirculating flow pattern.

METALLURGICAL APPLICATIONS OF MHD 603

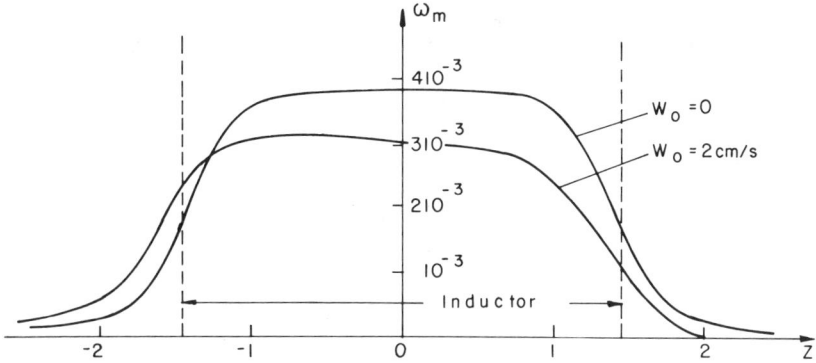

Fig. 11 Mean angular velocity for different values of withdrawal speed Wo.

phenomena and in their understanding, in order to promote a successful development of applications. Many scientific problems are to be studied for this purpose.

Electromagnetic Modelization

The problem to be solved concerns both the inductors and the generated magnetic field.
Concerning the inductors, very powerful methods exist that can give an accurate modelization when the frequency used is not very high (industrial frequencies up to 50 Hz). But if the frequency is higher, some difficulties appear: A strong interaction then exists between the different conductors of the coil and between the conductors and their magnetic environment. In this case, the current density distribution is not uniform in the cross section of the conductors, and it is wrong to consider a homogeneous skin effect, even though this is usually done. The current densities are to be considered as supplementary unknown variables, and classical methods are not valid. Recent studies demonstrate the incorrectness of the approximation of uniform current densities[36]: In the case of multiturn coils, for example, electric currents may locally flow in a direction opposed to the intuitively predicted one. A negative effect result occurs to the magnetic field and to the resistance of the coil that becomes difficult to estimate.
Computational methods are to be developed to increase the precision in the generated magnetic field calculations. Indeed, very important losses in accuracy exist in the necessary successive steps of calculus: What is interesting in MHD is not really the magnetic field itself, but the

rotational part of the Lorentz force. Usual methods need at least three derivative operations before solving the equations of motion: the first one, to derive the magnetic field from the vector potential; the second one, to deduce the current density and the Lorentz force; the third one, when taking the curl of the force, the accuracy of which depends on the validity of the resulting velocity field. Particular care is given to this way in recent works.[37,38]

Moreover, it is necessary to develop three-dimensional models for magnetic field computation, in order to take into account realistic geometries and fringe effects due to short inductors.

Velocity Field Computation

For given electromagnetic force fields, the mean velocity induced in liquid metals can be reached by using classical techniques based on finite-difference or finite-element methods. However, stirring motions induced by Lorentz forces in liquid metals are always turbulent, and all classical existing models have proved invalid in predicting MHD turbulent flows. Turbulence involved is specific for many reasons, and there is an urgent need to develop well-adapted specific models that cannot only be pure modifications or adaptions of existing ones. The presence of the magnetic field, though Joule effect, strongly modifies the properties of turbulence because of the anisotropy induced. Such anisotropy, well understood in the case of steady uniform magnetic fields,[39] acts in a more subtle and complex manner in the case of alternating magnetic fields, and simple extrapolation would be hazardous. Indeed, with alternating magnetic fields, Lorentz forces are specific and original for two reasons. First, they are not uniformly distributed in the fluid but are localized in the skin depth: An initial difficulty then appears because the electromagnetic skin depth exists in the same region as the hydrodynamical boundary layer whose dynamics are completely modified. Particular care should therefore be given in modeling these regions. Secondly, the electromagnetic forces are iron-stationary. If one neglects the pulsating part of these forces in the calculation of the mean velocity, turbulence modeling has to take this part into account, since the fluctuating force may excite in a preferred manner some scales and lead to a selective power injection into the turbulent spectrum. Moreover, in many configurations connected with metallurgical applications, the flow is recirculating and the fluid particles experience periodically the effect of the magnetic field and of the pulsating forces, when they travel within the skin depth. This periodic effect with its own typical time scale also contributes to the originality of this turbulence. It is of pressing interest to develop theoretical and experimental works about this original MHD turbulence in the presence of alternating magnetic fields, not because of its influence upon the mean velocity fields, but because of its implications on transfer mechanisms.

Thermal Effects

In many studies where the aim is only to predict the velocity field induced by electromagnetic forces, liquid metal is supposed to be isothermal. Such an approximation is valid in some particular cases such as induction furnaces, but may lead to important discrepancies between computed and real values when, for example, a phase transition occurs. If temperature gradients are generally unimportant through buoyancy effects they may induce, they have to be taken into account in the resulting variations of physical properties: magnetic permeability for example, when Curie temperature is reached somewhere in the metal, or effective viscosity for fluid motion investigation within boundary layers near a front of solidification, or in the mushy zone where crystallites are present. Some attempts are made to achieve numerical models taking into account coupled electromagnetic hydrodynamical and thermal phenomena,[40] but we are only beginning to understand them.

Transfer Phenomena

Some fundamental research work is necessary to develop and study the connections between turbulence characteristics of liquid metal flow and transfer mechanisms: momentum transfer, heat transfer, or mass transfer. In the scope of industrial applications, such work finds a justification in the answers to be given to some important questions. For example, what kind of turbulence may promote the coalescence of particles present in a bath and improve the kinetics of their removal? What kind of turbulence may, on the contrary, promote dispersion of particles or homogenization of the bath with respect to a given contaminant? Similar questions may be raised concerning transfers near the interfaces; for example, turbulence induced near the slag-metal interface highly modifies the transfers, which also have to be analyzed to define the best conditions for accelerating heterogeneous chemical reactions. Some instabilities of such interfaces have been observed not only due to Marangoni effect, but connected with turbulent intensity and a local resulting transfer rate that modifies surface tension and can lead to spontaneous emulsion of the slag near the interface. Near the walls, it is also important to control the turbulence characteristics responsible for the trapping, or, on the contrary, for the washing away of particles. The vicinity of the solidification front is also a region where the coupling between turbulence and heat and solute transfer needs to be precisely analyzed in order to make possible the control of solidification structure.

The interest of these studies, and also the origin of their complexity, is that they combine MHD with chemistry, physicochemistry, thermodynamics, and metallurgy and cannot be successful if they are not closely studied by and among the scientists of these complementary specialities.

New Possible Metallurgical Applications of MHD

Some new applications of MHD are possible that enable metallurgical operations that are difficult or impossible without electromagnetic effects.

The Electromagnetic Device for Continuous Casting of Hollow Ingots

Today, elaboration of hollow ingots or tubes is not a widespread metallurgical activity. The technique, itself, demands the solving of many problems in order to be economically valid. For the elaborated product, a very good internal and external surface quality is the basic requirement, and it is very difficult to obtain with annular geometry. Indeed, during the casting operation, non-metallic inclusions, for example, slag particles, are trapped in the first solidifying skin, and this leads to prohibitive defects on external and internal surfaces. These defects must be eliminated by a very delicate, expensive, and complicated process of scalping before using or mechanically transforming the ingot.

In the case of continuous casting of aluminum, some electromagnetic solutions have been implemented to solve the problem. Two circular inductors and shields are used inside and outside the annular volume of the molten metal to control the shape of the free surface and to maintain the stability of the equilibrium of the liquid against gravity force (Fig. 12). This process is credited to Getselev[41]. However, this technique can only be used with aluminum or with very light alloys having high thermal

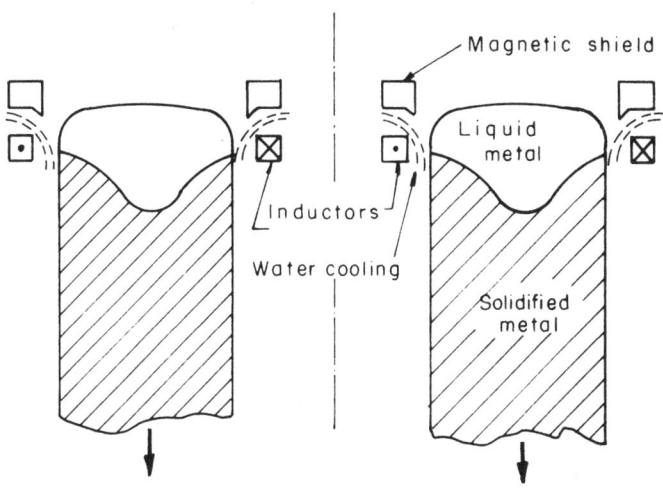

Fig. 12 Electromagnetic device for continuous casting of aluminum.

conductivity, to ensure that the depth of the molten zone does not exceed a certain number of centimeters, along with acceptable withdrawal speeds. Other metals, for example, steel, require one to cast very slowly, and this would not be profitable.

The improvement of the external surface quality can be achieved by giving the whole casting device a rotational motion. The free surface of the bath then takes on a paraboloidal shape, which prevents slag or non-metallic particles from coming into contact with the external mold and prevents them from being trapped. Yet the problem of the internal surface is not solved; on the contrary, it is intensified by this technique. We therefore propose an additional solution, which consists in using a traveling magnetic field to force the ascension of the liquid metal around the internal mold and, thus, to create a deep meniscus at the bottom of which non-metallic inclusions can accumulate and be removed. For safety reasons, an imperative condition must be respected: No electric currents or voltages can be accepted inside the internal water-cooled mold. We therefore introduce into the mold a cylindrical rotor made of iron with two strips of permanent magnets stuck on it, along helical lines. The two strips are switched and have opposite polarities. At any fixed points in space when the rotor is rotating, the magnetic field is composed of a vertically traveling magnetic field and of a rotating magnetic field. The wave-length of the helical lines and the angular velocity are chosen to favor an important synchronism velocity (~10m/s) and a rather low frequency to avoid shield effects of the copper mold.[42]

In the molten metal, the result is a recirculating flow ascending next to the internal mold and descending next to the external solidified skin. Because of the liquid momentum, the desired meniscus appears within the annular space between the two molds. In addition to the appearance of the meniscus, the recirculating flow offers two other advantages: First, non-metallic particles present in the molten metal are brought near the free surface, where they are caught by the slag; purification of the ingot is then obtained. Second, a non-negligible effect involves the improvement of the metallurgical structure of the hollow ingot because of the stirring induced near the solidification front (Fig. 13).

Such a solution is inexpensive; in the experimental device tested with mercury, the electric power needed for the rotation of the rotor was 500 W. In the industrial device, the cooling water circulation in the internal mold is used to give the desired angular velocity to a turbine fixed to the axis of the rotor.

Magnetic Shaping of Liquid Metal: A Particular Application for Metallic Glasses

Some laboratory experiments demonstrated the possibility of imposing a given shape to a liquid metal free surface

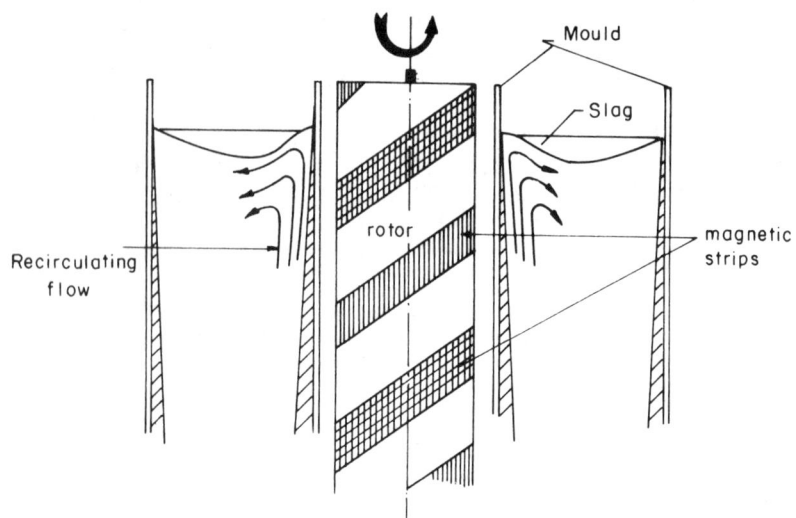

Fig. 13 Recirculating flow pattern in a hollow ingot.

by using high-frequency magnetic field: The principle used is to induce, within a very thin skin depth, a non-uniform magnetic pressure whose competition with surface tension gravity or inertial forces modify the equilibrium geometry of the initial surface. Possible metallurgical applications include the confinement or the guiding of liquid metal jet without having to resort to any wall, levitation, or continuous casting of magnetically shaped metal sections.[43]

A good example of application is the possibility of imposing a very thin ribbon-shaped cross section to an initially circular vertically falling metal jet. Such a molten metal thin ribbon can be used for amorphous metals malsing by quick cooling without temperature gradient and rapid solidifying. Today, two main techniques exist for such an elaboration "planar flow casting" and "melt spining technique."[44,45] The first one leads to amorphous 7 in.-wide ribbons but demonstrates a real technological prowess; the second one does not give good results because of the difficulties in reducing the instabilities of the liquid metal jet to be solidified. With the help of alternating magnetic fields, it is, first of all, possible to shape a circular cross section jet into a thin ribbon and secondly to stabilize the resulting ribbon until it solidifies.

The initially circular jet is submitted to the action of a uniform horizontal magnetic field. For 1-cm-diam jet, the frequency used is about 400 kHz, in order to have a very thin skin depth compared with the radius of the jet and to promote superficial electromagnetic effects. The centripetal induced electromagnetic forces are non-uniform around the jet (Fig. 14). Two stagnation points appear

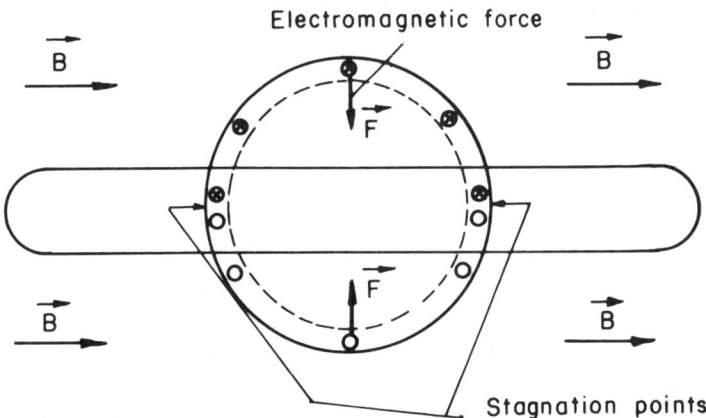

Fig. 14 A circular jet in a uniform magnetic field.

where electromagnetic fields, induced current and, therefore, electromagnetic forces are zero. On the contrary, in the perpendicular direction to the line bearing the stagnation points, electromagnetic forces are maximum. The result of the balance between this force distribution with surface tension is a stretching of liquid metal along the magnetic field lines, which leads to the desired ribbon shape. In laboratory experiments, using mercury and molten tin, ribbons less than 1 mm thick are easily obtained with moderate electric power.

An interesting effect is due to the axial velocity of the liquid jet: Because of the resulting convection, the maximum width of the ribbon does not appear between the two coils generating the magnetic field, but below this area. In view of industrial applications, this leads to the possibility of forcing two or more of such magnetically shaped ribbons to coalesce to obtain a very large ribbon whose thickness enables a rapid solidification. Yet a problem arises because of the instability resulting from the coalescence, which induced periodic vertical ripples parallel to the edges of the resulting ribbon, whose wavelength is the width of the initial ribbons. Such a perturbation cannot be reduced because it appears in a region where no magnetic fields exist and is amplified by surface tension influence. It is then necessary to use a second inductor whose effect is only to stabilize the ribbon by flattening the ripples.

This MHD application may well offer the solution to the very difficult technological problems encountered by the classical techniques. It is not an improvement of an existing technique, but an original application giving rise to a new technique. The excellent stability of the ribbon and the absence of contact between the molten metal and a material wall, which eliminates the risk of pollution and offers the possibility of working in an inert gas atmosphere, constitute the main advantages of using alternating magnetic field.

Application of Electromagnetic Levitation to Welding Operations

In metallurgical industry or in mechanical building, welding operations are often difficult because they have to be made against gravity. In this case, it is only the dexterity of the worker that makes it possible to keep the molten metal in place until it solidifies. Suitable electromagnetic forces induced in the liquid part of the metal can locally equilibrate the gravity force and, thanks to a

good design of the inductor, impose the required stable horizontal flat equilibrium shape.
Many systems are able to levitate a liquid metallic load but cannot verify the condition of the plane free surface. The system we use is composed of some horizontal conductors parallel to the axis of the molten metal volume. It is a combination of a couple of parallel conductors fed with alternating current flowing in opposed direction. This configuration presents both an advantage and an inconvenience; it gives an important restoring force that is indispensable in assuring the stability of the levitated liquid, but it leads to a stagnation point for the magnetic field, on the symmetry axis, where no magnetic force exists to prevent the liquid from linking away. To avoid the deficiency and to take advantage of the stabilizing effect of the configuration, we use two couples of conductors (Fig. 15). Yet it is necessary to obtain the superposition and the interaction of the effect induced by each couple of conductors. Two solutions are proposed to obtain this result: either by feeding the conductors with currents having the same frequency but presenting a $\pi/2$ difference of phase, or by feeding the conductors with two different frequency currents. The frequency has to verify conditions in in order to prevent the molten metal from linking through the stagnation region of the magnetic field that is traveling along the free surface. The four interaction parameters computed with the skin depth in the molten metal, the intensity of the magnetic field and the common frequency, when it is unique, the two frequencies and their difference otherwise have to be very small compared with unity. In this case, the molten metal only experiences the mean value of the magnetic pressure, which may be quasi-uniform for

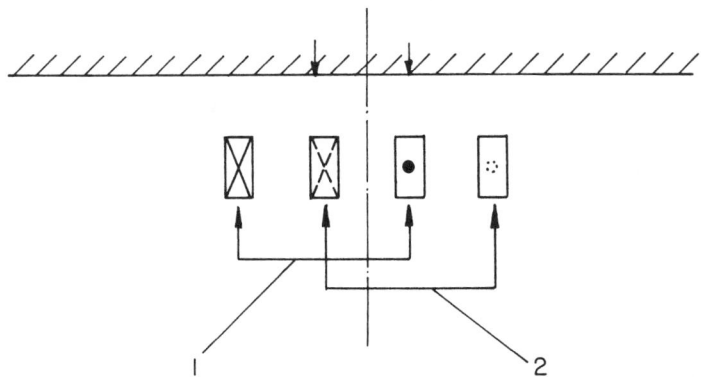

Fig. 15 Two coupled conductors used for levitation in welding operations.

suitable values of the ratio between the distance between the conductors and the distance between the conductors and the metal surface.

Two artifices are used to facilitate the implementation of the device and to increase its efficiency. A particular electric circuit composed of the active part of the inductor, capacitors, and front loops can, for very well chosen values of the capacitors and of the front loops self-inductance, lead to obtaining the two different frequencies with only one conventional induction generator with a voltage feedback (Fig. 16).

Analysis concerning the electromagnetic interaction between the different conductors of a coil[46] shows the strong unfavorable non-uniformity of the current density in the conductors cross section. To counteract this effect and to insure that the current flows as near as possible to the molten metal, we use a technological artifice: The rectangular cross section of the conductors is divided into two parts (Fig. 17): The upper face of each rectangular water-cooled rectangular conductor is made out of copper, and the three other faces are made out of a more electrical resisting metal, such as stainless steel. Measurements of the magnetic field intensity show that for a given intensi-

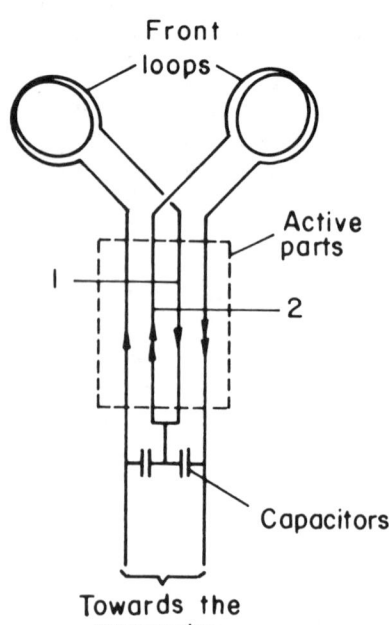

Fig. 16 The electric circuit which can yield two different frequencies with one conventional induction generator with voltage feedback.

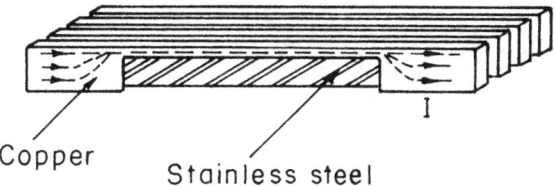

Fig. 17 The inner structure of the conductors.

ty in the conductors, the magnetic field is twice as strong when an inductor made out of two metals is used rather than when the whole cross section is made of copper.

Conclusion

Metallurgy in general, and the steel industry in particular, is faced with very important problems that are heightened in the present economic context where energy savings have become a priority. MHD can bring an important contribution to solve these problems, either by improving existing metallurgical techniques or by giving rise to innovative technologies. The application domain of alternating magnetic fields in metallurgy is very large. If MHD was, until recent years, considered as a possible solution, the time has now come to realize that, in some cases, MHD is the solution.

The success encountered by electromagnetic stirring in continuous casting is today the best example of such solutions: This success is due to the numerous metallurgical benefits that this technique offers. There is no doubt that the progress made in the comprehension of the phenomena involved has highly contributed to the widening use of continuous casting, which is now able to manufacture high quality steel.

I agree with Birat's conviction that "within a few years, electromagnetic stirring will become a key element in the generalization of the applicability of continuous casting to cover all major steel grades." However, this conviction can only become a reality on condition that knowledge continues to progress in MHD to enable the consideration of more and more refinements in modeling practical situations. This supposes very close relationships between theoreticians, scientists, and practical steelmakers or metallurgists on the one hand, and, on the other, much closer relationships between MHD specialists and chemistry, physicochemistry, and metallurgy specialists. Such a collaboration within the university setting, itself, and the collaboration between university and industry is the only solution which can effectively lead to solving the problems and bring to modern industry the innovations it needs. One of the aims of this paper is to have a catalytic effect in stimulating such a cooperative attitude.

References

[1] Johnson, A.R., "Metal Pad Velocity Measurements in Aluminum Reduction Cell," Light Metals, Vol. 1, 1978, pp. 45-48.

[2] Tarapore, E.D., "Magnetic Fields in Aluminum Reduction Cell and Their Influence on Metal Pad Circulation."

[3] Moreau, R. and Evans, J., "An Analysis of the Hydrodynamics of Aluminum Reduction Cells," Journal of the Electrochemical Society, (to be published).

[4] Szekely, J. and Nakanishik, "Stirring and Its Effects on Aluminum Deoxidation in ASEA-SKF Furnace," Transactions of the Metal Society, AIME, 1975, pp. 245-256.

[5] Sneyd, A., "Fluid Flow Induced by a Rapidly Alternating or Rotating Magnetic Field," Journal of Fluid Mechanics, 1979, pp. 209-217.

[6] Fautrelle, Y., "Analytical and Numerical Aspects of the Electromagnetic Stirring Induced by Alternating Magnetic Fields," Journal of Fluid Mechanics, 1981, pp. 405-430.

[7] Alemany, A., Barbet, J., Fautrelle, Y., and Moreau, R., "Procede de Pulverisation Electromagnetique des Metaux Liquides," French Patent No. 77.17.296, 1977.

[8] Barnier, M., and Moreau, R., "Dispositif Electromagnetique de Confinement des Metaux Liquides pour Realiser une Regulation de Debit," French Patent No. 77.21.121., 1977.

[9] Birat, J.P. and Chone, J., "Electromagnetic Stirring on Billet, Bloom, and Slab Continuous Casters: State of the Art in 1982," Ironmaking and Steelmaking, 1983, pp. 269-281.

[10] Ushijima, D., Yoshida, A., Mizutani, M., and Okajima, H., South-East Asia Iron and Steel Institute, Manilla Paper 29, presented at the Conference on Training Standards and Quality Control, 1981.

[11] Melford, D.A., Wittington, K.R., Funnel, G.D., and Armstrong, G.R., Paper presented at the Fourth International Iron and Steel Congress, London, 1982.

[12] Shiraiwa, T., Sugitani, Y., Mizutani, M., et al, Suminomo Search, 1979, pp. 97-107.

[13] Ayata, K., Mori, T., Narita, K., and Ohnishi, T., Rev. Metall. Ca. Inf. Tech., 1982, pp. 371-380.

[14] I Nouye, T., and Tanaka, H., Nippon Steel Report, 1979, pp. 1-23.

[15] Miyos, H.S., "Continuous Casting," The Metals Society, London, 1977, pp. 286-291.

[16] Lipton, J., Dacker, C.A., and Kollberg, S., Iron and Steel Engineering, 1980, pp. 66-75.

[17] AEG-Eloterm, Advertising Pamphlet, 1979.

[18] Nakatani, M., Adachi, T., Kimiya, S., and Kimura, K., "Clean Steel," The Metals Society, London, 1983, pp. 416-435.

[19] Sundberg, Y., "Principles of the Induction Stirrer," ASEA Journal, 1971, pp. 71-80.

[20] Sundberg, Y., "Metallurgical Aspects of Induction Stirring," Proceedings IUTAM Symposium on Metallurgical Applications of MHD, 1982.

[21] Ruer, J., Birat, J.P., and Alberny, R., "Brassage Electromagnetique Netinel en Lingotiere de Coulee Continue de Brames," Report EUR 707 FR, Luxembourg, 1983.

[22] Ventavoli, R., Alberny, R., and Birat, J.P., "Controle de la Proprete de Peau des Billettes de Coulee Continue par Champs Electromagnetiques Glissants Appliques au Niveau de la Lingotiere," 1978, Report EUR 6136, II FR, Luxembourg.

[23] Marr, H.S., "Electromagnetic Stirring in Continuous Casting of Steel," 1982, Proceedings of IUTAM Symposium, Cambridge, England.

[24] "Circulation Flow Rate of the Melt in the ASEA SKF Stainless Steel Refining Furnace," Mizuhima Steel Works, Kawasaki Steel Corporation, Japan, Internal Report, 1971.

[25] Marty, P., "Separation Electromagnetique Continue," These de Docteur-Ingenieur, Universite de Grenoble, France, 1982.

[26] Birat, J.P., Neu, P., Dhuyvetter, J.C., and Jeanneau, M., Proceedings of the 69th Steel Making Conference, Pittsburgh, Pa., 1982.

[27] Takahashi, T., and Hagiwara, I., Transactions of the Iron Steel Institute of Japan, 1976, pp. 283-291.

[28] Widdowson, R. and Marr, H.S., Sheffield International Congress on Solidification and Casting Metals Society Book, 1979, pp. 547-552.

[29] Sundberg, Y., ASEA Review, 1971, pp. 107-116.

[30] Szekely, J. and Yadowa, R.T., Metal Transactions, 1972, pp. 2673-2680.

[31] Offman, C., Scandinavian Journal of Metallurgy, 1981, pp. 25-28.

[32] Neu, P. and Genneson, J.C., IRSID Report No. ACI 82/321, 1982.

[33] Van Den Hove, P., "Brassage Electromagnetique a Champ Tournant dans les Puits de Coulee Continue de l'Acier," These de Docteur-Ingenieur, Universite de Grenoble, France, 1982.

[34] Lesoult, G. and Neu, P., "Modeling of Equiaxed Solidification Induced by Electromagnetic Stirring on a Steel Continuous Casting," Proceedings of the IUTAM Symposium, Cambridge, England, 1982.

[35] Meyer, J.L., "Influence de la Convection Naturelle ou du Brassage Electromagnetique sur la Solidification de Lingots d'Aluminium," These de Docteur-Ingenieur, Universite de Grenoble, France, 1983.

[36] Delage, D. and Ernst, R., "Prediction de la Repartition du Courant dans un Inducteur a Symetrie de Revolution Destine au Chauffage par Induction MF et HF," Revue Generale de l'Electricite, (to be published).

[37] Lavers, J.D., "Rotary in Mold Stirring in a Cylindrical Continuous Casting Geometry," IEEE Transactions on Industry Applications, 1983, pp. 633-639.

[38] Lavers, J.D. and Biringer, P.P., "The influence of System Geometry on Electromagnetic Stirring Forces in Induction Melting Furnaces," Proceedings of the IUTAM Symposium, Cambridge, England, 1982.

[39] Sommeria, J. and Moreau, R., "Why, How and When, MHD Turbulence Becomes Two-Dimensional," Journal of Fluid Mechanics, 1982, pp. 507-518.

[40] Masse, R., "Analyse Methodique de la Modelisation Numerique des Equations de la Physique des Milieux Continus a l'Aide de la Methode des Elements Finis -- FLUX EXPERT -- un Systeme d'Aide a la Construction de Logiciels," These de Doctorat d'Etat, Universite de Grenoble, France, 1983.

[41] Getselev, Z.N. and Marty Nov, G.I., Magnitnaya Gidrodinamika, 1975, pp. 106-111.

[42] Ernst, R., Garnier, M., Giroutru, M., Gueussier, A., and Peytavin, P., "Procede de Fabrication de Produits Longs Creux par Coulee a l'Aide d'un Champ Magnetique et Dispositif de Mise en Oeuvre du Procede," French Patent No. 82-00-763, 1982.

[43] Garnier, M., "Une Analyse des Possibilites de Controle Electromagnetique des Surfaces Libres de Metaux Fondus," These de Doctorat d'Etat, Universite de Grenoble, France, 1982.

[44] Kiebermann, H.K., "Manufacture of Amorphous Alloy Ribbons," General Electric Corporate Research and Development.

[45] Anthony, T.R. and Cline, H.E., "Dimensional Variations in Newtonian Quenched Metal Ribbons Formed by Melt Spinning and Melt Extraction," General Electric Corporate Research and Development.

[46] Biasse, J.M. and Garnier, M., "Procede de Dispositif pour Faciliter la Soudure," French Patent No. 80-052-52, 1980.

[47] Biasse, J.M., Ernst, R., and Garnier, M., "An Application of Electromagnetic Levitation to Welding," IEEE Transactions of Industry Applications, 1983, pp. 640-645.

Current Paths and MHD in Vacuum Arc Remelting

L.A. Bertram,* F.J. Zanner,† and B.M. Marder†
Sandia National Laboratories, Albuquerque, New Mexico

Abstract

In vacuum consumable arc remelting, a cylindrical ingot is formed continuously as metal is melted from a consumable electrode that is being heated by a several-kilo amp arc. For some ingot sizes and melt currents of interest in production practice, the resulting molten pool atop the ingot can have nearly equal buoyancy and Lorentz body forces. Because flow in the pool can have strong influence on the macrosegregation of the alloy components during solidification, a combined experimental and numerical simulation analysis has been undertaken. This effort seeks to determine the pool flows in the combined MHD-natural convective free-boundary problem. Since these flows will be sensitive to the current paths produced by the arcs, previous efforts have sought better characterization of the furnace arc. The present investigation is a numerical parameter study of the resulting flows and pool shapes when two extreme views of arc structure are assumed. First, a smooth current path model, suggested by the low-current arc in which the anode serves as a diffuse collector of electrons; second, an arc with coherent cathode spot motion, suggested by high-speed cinematography of the furnace arc, in which each point on the pool surface sees an intermittent charge deposition. The resulting dc currents and fields yield distinct pool sizes, but quite similar solidification conditions; components cannot significantly affect pool flows.

Paper presented at the Fourth Beer-Sheva Seminar on MHD Flows and Turbulence, Ben-Gurion University of the Negev, Beer-Sheva, Israel, Feb. 27-March 2, 1984. Copyright © 1985 by the American Institute of Aeronautics and Astronautics, Inc. All rights reserved.
*Distinguished Member Technical Staff, Applied Mathematics Division
†Applied Mathematics Division.

I. Introduction

The vacuum consumable arc remelting process is used to produce large ingots of chemically reactive or segregation sensitive alloys. A comsumable electrode having the correct overall chemical composition is melted on its lower end by a several-kilo amp electric arc (Fig. 1). The molten metal drips from the electrode into a water-cooled copper crucible where it solidifies as a larger, hopefully homogeneous and porosity-free, cylindrical ingot. The key element of the process is the control it allows of the heat extraction from the pool of molten metal atop the ingot. The ideal situation has a pool large enough to feed solidification shrinkage easily, but small enough that local solidification times are much smaller than would be achieved in any other large casting process.

This paper reports one part of an ongoing study of vacuum arc remelting: a study aimed at characterizing the solidification process[1-4]. The overall approach uses a closely coupled combination of experimental diagnostics on production furnaces and numerical simulation. Both tools are necessary because experiment cannot provide fully detailed measurements of pool liquid motions or arc conditions, and simulation cannot proceed from the first principles.

Previous results have shown that fluid motions in the pool are the result of partial cancellation of opposing Lorentz and thermal buoyancy forces for some cases of interest[3,4]. It follows that determination of solidification

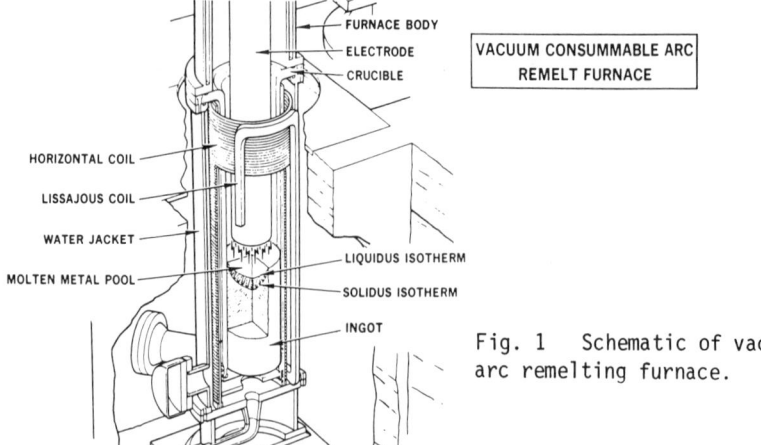

Fig. 1 Schematic of vacuum arc remelting furnace.

conditions requires accurate treatments of both the thermophysical properties of the molten alloy and of the electromagnetic environment created by the arc, if numerical simulation is to succeed. Accordingly, our specific goal here will be to examine constraints on possible structures for the vacuum arc operating in the furnace.

The approach taken involves indirect argument and simulation because of the experimental difficulties of making direct arc measurements in the furnace. In fact, where vacuum arcs are concerned, experiments even on the much more accessible breaker switch arcs have required nearly 50 years to arrive at a consistent picture of the "classical" case[5] of a single cathode spot carrying less than 150 A of current for a duration of milliseconds between cold (unmelted) Cu electrodes. Our study began by using high-speed cinematography[1,3], to get an overview of the furnace arc behavior. For analysis purposes, the arc has subsequently been treated as classical until modifications have been forced to maintain consistency with other measurable phenomena in the furnace, with the links between arc and data being provided by simulation. This development is detailed in Sec. II, where two arc models are constructed so as to be sufficiently extreme that furnace conditions must lie between them. Because the different arc models will have different Lorentz force distributions, they necessarily will have different melt pool motions as discussed in Sec. III. The implications of the differences are then considered in Sec. IV.

II. Arc Models

The classical arc operates in vacuum by creating a conduction path between two disk electrodes through a plasma of the cathode's metal vapor. The conditions required to create the plasma are achieved in a very small "cathode spot" lying close (order of a debye length) to the cathode surface. Because of its small size, this spot can sustain the very high-voltage gradients required to produce the ejected ions and electrons of the plasma, even though the total arc voltage drop is in the 20-30 V range for most electrode materials. The ions ejected by the spots are extremely energetic, with tens of electron volts of kinetic energy per ionization charge: i.e., they are not thermal, but ballistic. A cathode spot does not linger in one position, but jumps on a microsecond time scale to an adjacent position, so that it appears to do a very fast random walk over the cathode surface. Material ejected by the spot is nearly all ionized, and escapes nearly isotropically (with a slight preference for the ions to move out normal to the cathode, while the few neutrals move nearly parallel to it). A single spot can carry some material dependent maximum current in about the 50-150 A range; if the arc current exceeds this level, the spot will fission into several smaller spots, each carrying less than this ΔI_{max}. So long as the anode subtends an octant or so of solid angle at the spot, the current collection seems to be diffuse -- no anode spot will form.[5]

The furnace arc is different geometrically from this classical case, in that its anode surface is not only the pool surface, but the crucible wall surrounding the cathode (consumable electrode) as well. Further, the total current carried by the furnace, I_m, is several kiloamperes, much larger than the single-spot capacity ΔI_{max}, so there are many (say, 50) spots operating simultaneously in the furnace. Finally, the molten electrode face and molten pool surface may play a role in arc structure.[4,6]

In the high-speed movies of the furnace arc, cathode spots are visible as points of blue light moving at high speed with considerable randomness, but generally being created somewhere toward the axis of the electrode and migrating in about 10^{-3} s to its lateral surface, which they climb for some distance before being extinguished. The spots' behavior is nonclassical in that they travel in loosely bound clusters which separate ("retrograde motion"), coalesce again, etc. These ill-defined clusters make spot counting on the movie frames ineffective as a means of inferring global current distribution on the electrode face.

To different degrees for different materials, the movies indicate the presence of the plasma as a week background glow. A diffuse steady cloud, without strong spatial structure or temporal fluctuations, appears to exist. This observation suggests that the pool surface should serve, like the classical anode, as a diffuse collector of electron (and ion) current, due to the cathode spot-jet structure having been averaged out in the plasma background. This notion is supported by the tendency of the electrode face to remain flat during the whole course of melting by a "diffuse arc" of the type just described. This flatness implies macrouniform heating of the electrode face due to averaging of the intense local spot heating, again minimizing the spatial structure to be expected in the plasma.

The coarse features of these observations have been reflected in the mathematical model of pool current flow as follows.[7] Axisymmetry is assumed, since furnace construction usually emphasizes coaxial current carrying paths.

Thus, dependent variables depend on the dimensionless cylindrical coordinates (x,r) scaled to cylinder radius R. The ingot is specified to be a cylinder of aspect ratio length/R = α_r., grounded along its bottom and lateral surfaces, but with a smoothly voltage varying potential proportional to $J_0(sr)$ on its top (pool) surface, where J_0 is the Bessel function of zero order, and s is its first root: $J_0(s) = 0$ (see Ref. 8, where s = $j_{0,1}$). This results in a magnetic scaled dimensionless ($0 \geq |B| \geq |$) induction vector $\underline{B} = B(x,r) \hat{e}_\theta$, where $\hat{e}_\theta$ is the azimuthal unit vector, and

$$B(x,r) = \frac{\cosh s(\alpha_r - x)}{\cosh s\alpha_r} \frac{J_1(sr)}{J_1(s)} \qquad (1)$$

is scaled by $B_{ref} = \mu_e I_p/2\pi R$ in terms of the magnetic permeability μ_e and the total current passing through the pool I_p. The corresponding current density $\underline{j} = \nabla \times \underline{B}$ with scale $j_{ref} = I_p/2\pi R^2$, is

$$\underline{j}(x,r) = s \frac{\sinh s(\alpha_r - x)}{\cosh s\alpha_r} \frac{J_1(sr)}{J_1(s)} \hat{e}_r$$

$$+ \frac{\cosh s(\alpha_r - x)}{\cosh s\alpha_r} \frac{J_0(sr)}{J_1(s)} \hat{i} \qquad (2)$$

in terms of the first-order Bessel function J_1 and the axial and radial unit vectors $\hat{i}$ and $\hat{e}_r$, respectively.

The current streamlines for these expressions are shown in Fig. 2 for an aspect ratio α_r = 3.5. This does not differ substantially from the case α_r = 1.0, since most current has been grounded at the crucible wall within a radius of the surface in both cases.[7] Now, Eqs. (1) and (2) represent an anode model of maximal smoothness. Nevertheless, considerable support for their accuracy has come from two independent measurements while melting the U-6ω/ONb alloy at I_m = 6 kA. The first measurement, an estimate of pool volume based on macrosegregation bands in the finished ingot, was used to derive the constraint that $I_p/I_m < 0.57$. This bound was required to make the pool motions calculated as described in Sec. III yield a pool comparable to the bands in the actual ingot.[3] The second measurement, of heat flux through the crucible, indicates that the energy deposited on the inside crucible wall above the pool is nearly

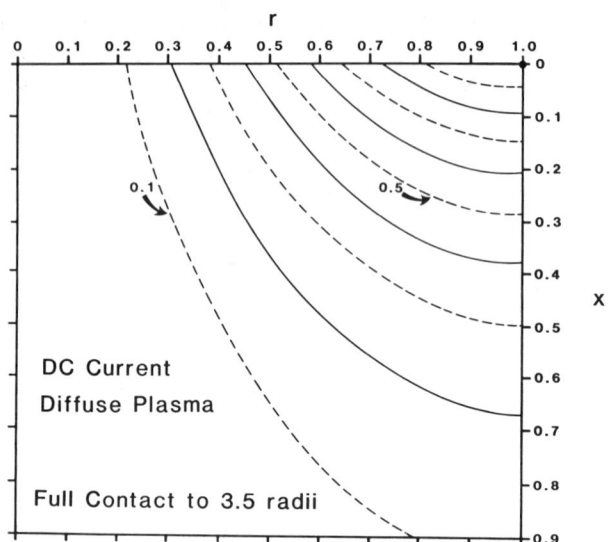

Fig. 2 Current streamlines (rB=const contours) for diffuse arc dc model given by Eqs. (1) and (2). Computational zone extends to $x = 3.5$, but only top $x = 0.9$ shown. Magnitude scaled by $(rB)_{max} = 1.122$, which occurs at dot on surface.

one-half the total coolant power.[9] Under the zero-order hypothesis that current and energy flux density are proportional in the plasma, the agreement of these two numbers supports the correctness of both.

That such zero-order arguments cannot be pushed too far is clear from another experimental result. If it is hypothesized that the energy budget of an individual spot of the furnace arc is the same as the energy budget of a classical spot, the actual furnace melt rates are seriously underestimated.[10] We have conjectured that the additional deposition on the cathode could be due to thermalized ions which are created from neutral vapor being released from the molten electrodes. This would also contribute to the evenness of the macrouniform cathode heating mentioned earlier.

More detailed, but still not fully consistent, zero-order models of the high-current arc have been discussed by Boxman et al.[11] They consider unmelted electrodes at 10-100 times the current density of a furnace arc, but, nevertheless, many features are of interest. Their multiple cathode spots also feed a quasi-steady plasma region (which they confine to the interelectrode cylindrical volume) that provides a smooth anode current deposition. In their high-density plasma, macroparticles evaporate and are ionized very rapidly, providing the thermalized ions to enhance cathode heating, which we have attributed to molten electrodes. Electron temperatures are measured at very high values 6-9 eV, as compared with classical 3 eV values.

The macroparticle-plasma interaction is particularly relevant to furnace operation because of the metal transfer mechanism. When liquid drips off the cathode, it can short the arc by forming a liquid column between electrodes.[1,2] This column is subsequently destroyed by first a pinch instability which causes a significant portion of the column to neck down by about an order of magnitude, and second by a kink instability of this necked portion, which results in a spray of droplets with high lateral velocity. At other times, cathode spots can collect on the end of a droplet and prevent its contact with the pool, but produce a spray of droplets in the process. Thus, the arc plasma quite frequently has a lot of macroparticles moving through it, most of which no doubt contribute their mass to the "skull" that forms on the crucible wall above the pool. Some, though, can contribute to the cathode heating mechanism by undergoing evaporation and ionization.

These complications indicate the difficulty of producing a complete and consistent arc model by extrapolation from the classical case. Therefore, we seek to devise a bounding ("worst-class") model to examine the most apparent of the nonclassical features not treated by Eqs. (1) and (2).

The diffuse arc model (1), (2) lacks any time-dependence. However, the movies indicate that there is a relatively strong 1-kHz component in the spot motions on the electrode. At this frequency, ac will have a skin depth of about 0.1 R in the liquid metal of the pool, a value which has been identified by e.g., Barbier et al,[12] and Koanda and Fautrelle,[13] as being near optimal for producing inductive stirring. Thus, even though weakened by the "blanketing" effect of the plasma,§ this component may be of importance to pool dynamics. Since overall pool motions are a combination of local flows, the potential importance of such inductive stirring can only be assessed by a full simulation which details all local conditions.

The diffuse arc model (1), (2) also assumes the fullest possible electrical contact with the essentially perfectly conducting OFHC*Cu crucible wall. This assumption is

---

§ Note that the very highly directed ion velocities of the classical arc support the ability of the spot's ion jet to penetrate the plasma and make this ac felt at the anode.
* Oxygen Free High Conductivity

known by direct measurement to be valid for the U-6ω/oNb alloy,[9] but is less likely to be true for stainless steels or super alloys, since they undergo considerably larger shrinkage on solidification. This shrinkage can be expected to have the effect of pulling the ingot away from the crucible wall, and thereby breaking electrical contact.

The potential importance of these two modifications of the diffuse arc model are examined together here. First, the wall contact is given an extreme form by specifying that current can flow to the crucible only at the meniscus. An equally extreme form is given to the ac component by specifying that the arc consists of a "spark ring" in which the full current I_p is deposited in a ring of radius $r_J(t)$, and that this ring continually expands until it reaches the pool edge, where it is extinguished. At this instant, a new spark appears at $r = r_0$ and repeats the motion with period T. The resulting pool currents and fields are found by solving the following magnetic diffusion problem:

$$B(x,1,t) = 0 \text{ on lateral surface } r = 1 \quad (3a)$$

$$B(\alpha_r,r,t) = 0 \text{ on bottom surface } x = \alpha_r \quad (3b)$$

$$B(0,r,t) = 0 \text{ for } r < r_J(t)$$
$$= 1/r \text{ for } r \geq r_J(t) \quad \text{on pool surface x=0 (3c)}$$

where

$$r_J(t) = r_0 + (t - [t])(1-r_0)$$

when t is scaled by T and B by B_{ref}, and t is the greatest integer function. With these boundary conditions, the magnetic diffusion equation[12]

$$\frac{\partial B}{\partial t} = \frac{2\pi}{R_\omega} \hat{e}_\theta \cdot \nabla \times \nabla \times \underline{B} \quad (3d)$$

is to be solved in the cylinder, where

$$R_\omega = 2\pi R^2/\nu_m T \quad (= 2 R^2/\delta^2)$$

in terms of the magnetic diffusivity $\nu_m = 1/(\mu_e \sigma)$ where σ is the electrical conductivity of the ingot. The skin depth is denoted by δ for frequency $\omega = 2\pi/T$.

To solve the boundary value problem (3) for its steady-state ac response, it is most convenient to separate variables and expand in series. We start with the expansion

$$B(x,r,t) = \sum_{k=1}^{\infty} B_k(x,r) \, e^{i\omega_k t} \quad (4a)$$

where $\omega_k = (k-1) 2\pi$ in dimensionless form. The Fourier coefficient B_k is then found to be

$$B_k(x,r) = \sum_{m=1}^{\infty} C_{km} J_1(s_m r) \frac{\sinh \lambda_{km}(\alpha_r - x)}{\sinh \lambda_{km}\alpha_r}$$

$$= \sum_{m=1}^{\infty} C_{km} J_1(s_m r) x(\lambda_{km} x) \quad (4b)$$

In Eq. (4b), $J_1(\cdot)$ is the Bessel function of order 1(Ref. 8), s_m is its m-th root: $J_1(s_m) = 0$, $m = 1,2,3,\ldots$, and $\lambda_{km} = (s_m^2 + i(k-1)R_\omega)^{1/2}$. The complex coefficients C_{km} are obtained from the pool surface boundary condition:

$$C_{km} = \frac{2}{J_2^2(s_m)} \int_0^1 r \, J_1(s_m r) B_k(0,r) dr \quad (4c)$$

$$B_k(0,r) = \int_0^1 e^{i\omega_k t} B(0,r,t) \, dt \quad (4d)$$

As in inductive stirring[12,13] the effective Lorentz force will be a cycle average given by

$$\int_0^1 \left[\text{Re}\{\underset{\sim}{j}\} \times \text{Re}\{\underset{\sim}{B}\} \right]_k dt$$

$$= \langle \underset{\sim}{j} \times \underset{\sim}{B} \rangle_k = -\tfrac{1}{2} \text{Re} \{\underset{\sim}{j}_k \times \overline{\underset{\sim}{B}}_k\} \quad (5)$$

where overscore denotes complex conjugate and Re $\{\cdot\}$ is "real part of."

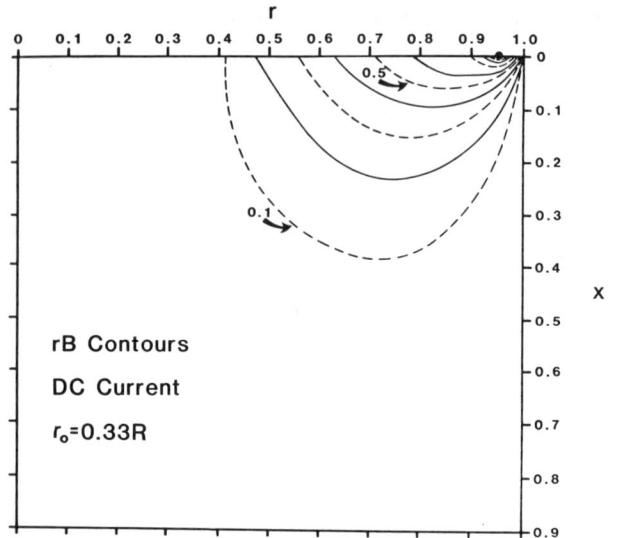

Fig. 3 Current streamlines for dc component of the spark ring arc model given by Eqs. (4) and (5). Spark starts at r_o = 0.33 and migrates to r = 1. Magnitude $(rB)_{max}$ = 1.217 occurs at dot on surface. All current is grounded at the upper right corner (pool meniscus).

The dc component (k = 1) of these fields is independent of R_ω and depends only on the dimensionless parameter r_0. From the movies, a plausible value r_0 = 0.33 is chosen, and the 20-term Bessel function expansion is used to produce the current streamlines shown in Fig. 3. By comparison with Fig. 2, it becomes clear that the effect of the restricted current flow path here is to concentrate the current (and thus Lorentz force) much nearer the meniscus.
In treating the ac terms of Eqs. (4), attention is given only to the first term, since this model was deliberately constructed to emphasize its effects. To conveniently estimate its effects, the modulus of the k = 2 term's rB values are plotted in Fig. 4 for three different values of R_ω. Since the same surface boundary condition is applied in each case, the same surface $|rB|$ values will apply; in particular, $|rB|_{max}$ = 0.29 $(rB)_{max}^{k=1}$ for all three R_ω values. Since both $\underline{j}$ and $\underline{B}$ for this component will scale with this value, the expected Lorentz force would be about $(0.29)^2 \simeq$ 0.1 times the dc Lorentz force. Therefore, except possibly in local zones where the main body forces cancel, the ac effect should not be visible except as a small correction on the basic dc-produced flows.

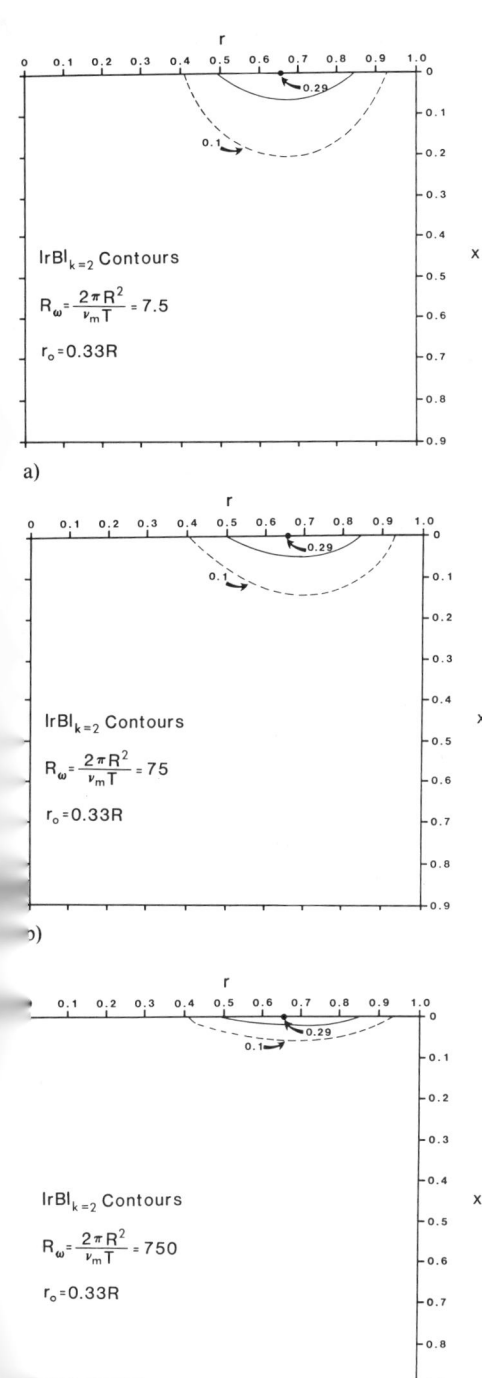

Fig. 4 Contours of rB of the first ac component given by Eqs. (4) and (5). Scale is $(rB)_{max}$ from Fig. 3. R_ω increases from top to bottom, taking on values 7.5, 75, and 750, respectively. Note skin depth increase with decreasing R_ω.

The influence of R_ω on whatever ac effect there is can be seen in Fig. 4. At R_ω = 7.5, penetration is nearly maximal (barely distinguishable from R_ω = 0 in fact). However, the phase relationships are such that the cycle averaged Lorentz force nearly vanishes, so this frequency has a very weak stirring potential. For R_ω = 75, corresponding to 1 kHz, skin depth is less, but coupling near its maximum. At 10 times higher frequency with R_ω = 750, skin depth has become so shallow that no significant stirring can be expected. These are precisely the relationships found in inductive stirring simulations.[12,13]

Given this dependence on R_ω, it is clear that a more elaborate model of the ac terms is not warranted. That is, the higher frequencies will have little or no impact on the pool flows because their associated R_ω values will be too high or low for effective coupling. Therefore, precise determination of their Fourier coefficients $B_k(x,r)$ would not improve the precision of the pool simulation.

III. Effects on Pool Dynamics

In order to connect the arc model to the metallurgically observable parameters, the free boundary phase change heat transfer problem for the ingot and pool is solved numerically. Axisymmetry is again assumed, and the equations of motion are put into the following forms; for details and boundary condition discussions, see Refs. 3, 4 and 9. Dimensionless vorticity ξ, scaled by κ_0/R^2 where κ_0 is the liquid thermal diffusivity, satisfies

$$\frac{D\xi}{Dt} = \hat{e}_\theta \cdot \nabla \times \nabla \cdot \underline{\underline{\tau}} + PrGr \frac{\partial \theta}{\partial r} - \frac{1}{A_0^2} \frac{\partial B^2}{r \partial x} \qquad (6)$$

where the Prandtl number is $Pr = \nu_0/\kappa_0$ with ν_0 = kinematic viscosity; the Grashof number is $Gr = g\alpha\Delta T_0 R^3/\nu_0^2$ based on the phase change temperature difference $\Delta T_0 = \frac{1}{2}(T_L - T_S)$, where T_L, T_S are liquidus and (nonequilibrium) solidus temperatures of the alloy, respectively. The final term on the right-hand side is the curl of the effective Lorentz force, and its coefficient is the Alfvén number based on thermal diffusion speed (κ_0/R): $A_0 = (\kappa_0/R)(B_{ref}/(\mu_e\rho_0)^{1/2})$ $D\xi/Dt$ is the Lagrangian derivative of the vorticity, and a curl of the Reynolds' stress has been neglected on the left-hand side. The viscous stress tensor is denoted by $\underline{\underline{\tau}}$: it includes a variable viscosity in this formulation.

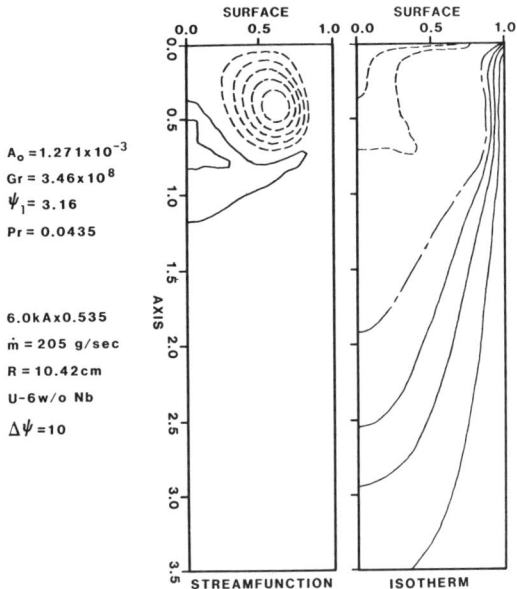

Fig. 5 Pool flows and isotherms for diffuse arc model. No-slip boundary condition applied to flow velocity on "immobilization isotherm" (chain-dashed) at dimensionless temperature of +0.33. Isotherms in solid shown as solid curves; in liquid, as broken curves. Positive streamlines (broken curves) correspond to counter-clockwise Loretnz dominated circulation; negative streamlines (solid curves), to clockwise thermal buoyancy cells.

The Boussinesq approximation has been assumed in Eq.(6), so a stream function ψ is introduced to conserve mass, giving the velocity

$$\underline{u} = (u,v,0) = \nabla \times (\psi/r)\hat{e}_\theta \quad (7a)$$

$$\xi = \hat{e}_\theta \cdot \nabla \times u = \frac{\partial^2 \psi}{\partial x^2} - \frac{\partial}{\partial r}\frac{\partial(r\psi)}{r\partial r} \quad (7b)$$

Energy conservation is modeled by solving the enthalpy (h) equation

$$\frac{Dh}{Dt} = \nabla \cdot [\kappa(T)\nabla h] + \dot{Q} \quad (8)$$

where the thermal diffusivity $\kappa(T)$ includes a specific heat factor that models the latent heat release on solidification, and where Q is the distributed heating per unit mass, due to Joule heating here.

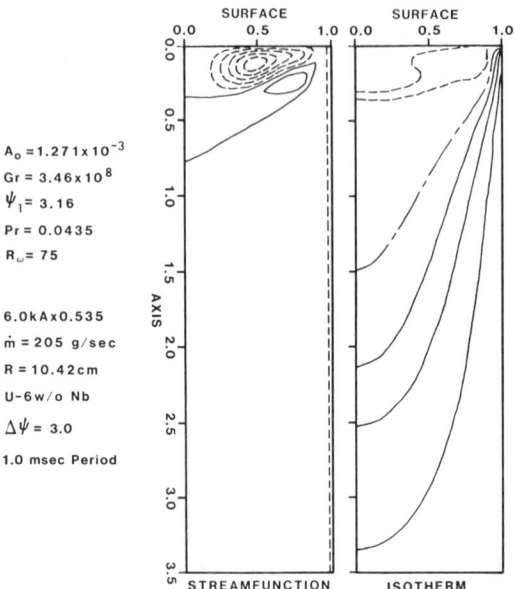

Fig. 6 Pool flows and isotherms for spark ring arc model. Note stream function is 3.0 here vs 10 in Fig. 5.

Equations (6-8) are discretized and solved in the cylinder $0 \leq r \leq 1$, $0 \leq x \leq \alpha_r$, on an analytically strained mesh with about a 4:1 expansion ratio.

To assess the differences between the two arc models, parameters are selected that correspond to the remelting of a U-6ω/oNb alloy at a melt current I_m = 6kA. For this case, the nominal parameters are Pr = 0.0435, Gr = 3.46x10^8, A_0 = 1.271x10^{-3}. In addition, the boundary conditions introduce $\Psi_I = u_I R/2_0$, a thermal Peclet number based on casting speed u_I: Ψ_1 = 3.16 here. This implies strong nonequilibrium effects in the ingot solidification. In all cases, the computational zone aspect ratio α_r = 3.5. Previous studies of pool dynamics and of energy budgets have indicated that about half the melt current passes through the pool, so I_p/I_m = 0.535 was used in the present computations. Calculations were carried out to t = 0.70-0.75; i.e. about three thermal diffusion times.

The solution for the diffuse arc model (1), (2) is shown in Fig. 5. The major features of the flow are its very large mushy zone -- the volume between the liquidus isotherm (last broken line isotherm) and the nonequilibrium solidus (first solid line isotherm) -- and the presence of counterrotating Lorentz dominated (broken streamlines) and thermal buoyancy dominated (solid streamlines) flow cells.

The spark ring model, including the dc and first ac components with $R_\omega = 75$, produces the flows in Fig. 6. Clearly, flow intensity is weaker overall, and the total pool volume is smaller. No doubt this weakening of the magnetic cell is due to the concentration of the Lorentz force near the no-slip boundary at the wall, for the spark ring model.

The pool bottoms (immobilization isotherms; chain-dashed) are of surprisingly similar shapes in the two cases and can, in fact, be overlaid on one another except for the centimeter or so near the wall. The same is true of the solidus isotherm, so the solidification process in its latter stages must be very similar indeed, making metallurgical discrimination between these two flow models very difficult. No doubt it would be necessary to do quantitative comparisons of the differences to be expected from the outer mushy zone flows in order to succeed at choosing the more metallurgically consistent flow pattern.

The effect of varying R_ω proves to be very weak, as was anticipated in the discussion in Sec. II. For $R_\omega = 7.5$, the pool appears as a very slightly shifted version of the pool in Fig. 6, due to a somewhat stronger magnetic cell. This suggests that the effect of the ac is, at worst, of a 10% order, and that it opposes the dc magnetic circulation. This opposition might be expected from the fact that the spark rings are always outward-moving, so that they would tend to entrain surface fluid in that direction and thereby create a clockwise circulation.

IV. Conclusions

From simulation results, we infer the following:

1) The effective electrical contact at the ingot lateral surface can have significant effects on the Lorentz force distribution and on pool dynamics, because of its effect on the arc dc.

2) The maximum credible amplitude of 1 kHz ac deposited on the anode due to cathode spot motions cannot have serious effects on pool dynamics.

The two extreme boundary conditions on the dc deposition on the anode result in quite different pool flows, but in such similar final solidification conditions that both refined analysis and refined metallurgical measurement would be needed to unequivocally choose between (contact boundary conditions) arc models on this basis.

Acknowledgments

The aid of J. P. Maroone of Sandia National Laboratories in data acquisition and in preparation of this paper is gratefully acknowledged. This work was performed under U.S. Department of Energy Contract No. DE-AC04-76DP00789.

References

[1] Zanner, F.J., "Observation of the Vacuum Arc and Metal Transfer During Vacuum Arc Remelting," Proceedings of the International Conference on Special Melting, edited by G.K. Bhat and R. Schlatter, American Vacuum Society, 1979, pp. 417-427.

[2] Zanner, F.J., "Metal Transfer During Vacuum Consumable Arc Remelting," Metallurgical Transactions, Vol. 10B, June 1979, pp. 133-142.

[3] Bertram, L.A. and Zanner, F.J., "Interaction Between Computational Modeling and Experiments for Vacuum Consumable Arc Remelting," Modelling of Casting and Welding Processes, edited by H.D. Brody and D. Apelian, The Metallurgical Society, Warrendale, PA, 1981, pp. 333-349.

[4] Bertram, L.A. and Zanner, F.J., "Plasma and Magnetohydrodynamic Problems in Vacuum Consumable Arc Remelting," Metallurgical Applications of MHD, edited by H.K. Moffatt and M.R.E. Proctor, The Metals Society, London, 1984, pp. 283-300.

[5] Lafferty, J.M., Ed., Vacuum Arcs, J. Wiley & Sons, New York, 1980.

[6] Zanner, F.J. and Bertram, L.A., "Behavior of Sustained High-Current Arcs on Molten Alloy Electrodes During Vacuum Consumable Arc Remelting," IEEE Transactions on Plasma Science, Vol. 11, Sept. 1983, pp. 223-232.

[7] Bertram, L.A., "A Mathematical Model for Vacuum Consumable Arc Remelt Casting," Proceedings of the First International Conference on Mathematical Modeling, Vol. 3, edited by X.J.R. Avula, University of Missouri-Rolla, 1977, pp. 1173-1182.

[8] Abramowitz, M. and Stegun, I., Handbook of Mathematical Functions, Dover Publications, New York, 1965, Chap. 9.

[9] Bertram, L.A. and Zanner, F.J., "Measurement of Ingot-Crucible Boundary Conditions During Vacuum Arc Remelting," Modeling of Casting and Welding Proceedings, Vol. 2, edited by J.A. Dantzig and J.T. Berry, The Metallurigcal Society, Warrendale, PA, 1984, pp. 33-46.

[10] Zanner, F.J., Adasczik, C., O'Brien, T.O., and Bertram, L.A., "Observations of Melt Rate as a Function of Arc Power, CO Pressure, and Electrode Gap During Vacuum Consumable Arc Remelting of Inconel 718," Metallurgical Transactions Series, Vol. 15B, March 1984, p. 1170.

[11] Boxman, R.L., Goldsmith, S., Izraeli, I., and Shalev, S., "A Model of the Multi-Cathode-Spot Arc," IEEE Transactions on Plasma Science, Vol. 11, Sept. 1983, pp. 138-145.

[12] Koanda, S. and Fautrelle, Y.R., "Modelling of Coreless Induction Furnaces: Some Theoretical and Experimental Results," Metallurgical Applications of MHD, edited by M.R.E. Proctor, The Metals Society, London, 1984, to be published.

[13] Barbier, J.N., Fautrelle, Y.R., Evans, J.W., and Cremer, P., "Simulation Numerique des Fours Chaufees par Induction," Journal de Mecanique Theorique et Appliquee, Vol. 1, 1982, pp. 533-556.

Electromagnetic Modelization of Cold Crucibles

A. Gagnoud,* D. Delage,† and M. Garnier‡

Groupement d'Interet Scientifique, Madylam, Saint Martin d'Heres, France

Abstract

In metallurgy, induction is a more and more common way of heating to produce materials by fusion at high temperature. The cold crucible is a particularly effective technique for ensuring a good degreee of purity: The watercooled copper and segmented crucible eliminates the drawbacks of the classical refractory crucible. Two applications of this technique are presented and theoretically analyzed: continuous casting and levitation melting. The casting device is studied by considering the straight cold crucible as a particular transformer in which the inductor is the primary winding, the crucible and the charge being the secondary one. For levitation melting, our theoretical approach ignores the periodicity in the azimuthal direction due to the sectors, since we consider only one very narrow split. In this model, the eddy currents are quite circumferential and may be considered as a system of a discrete number of superimposed wires of different diameters. We take advantage of an integral method based on the Biot and Savart law to calculate the vector potential of the magnetic field. Equality between internal and external currents in the crucible and Maxwell equations leads to a linear system of equations in which the intensity in each of the wires is unknown. The solution gives the eddy current distribution and the map of the magnetic field. Experimental measurements made in one split crucible show the accuracy of the modelization.

Paper presented at the Fourth Beer-Sheva Seminar on MHD Flows and Turbulence, Ben-Gurion University of the Negev, Beer-Sheva, Israel, Feb. 27-March 2, 1984. Copyright © 1985 by the American Institute of Aeronautics and Astronautics, Inc. All rights reserved.
*Doctoral Student.
†Engineer.
‡Dr.

ELECTROMAGNETIC MODELIZATION OF COLD CRUCIBLES 635

Introduction

In metallurgy, various techniques can be used to produce materials by fusion at higher temperature. Among these techniques, induction is a more and more exploited way of heating and melting, not only because of its ease and versatility, but also because interesting effects are induced in the liquid metal by electromagnetic stirring.

In many cases, melting crucibles are made from electrically conducting materials, such as graphite, or from insulating refractory oxides. Two major disadvantages appear with these kinds of crucibles: 1) chemical pollution of the melt owing to its reactivity against the crucible, and 2) mechanical erosion as the result of the stirring. One solution to this double problem consists in using a water-cooled copper crucible[1,2]. However, to prevent an electromagnetic shield effect, the crucibles must be segmented: each segment being a circular or trapezoidal copper tube cooled by an internal flow of water.

Among the different existing crucibles are the straight cylindrical crucible, used for continuous casting of metallic ingots, and the levitation crucible, which makes it possible to obtain an important overheating of the liquid metal, because there is no contact between crucible and melt[3-5].

In the laboratory or in industrial scale, some installations are working but progress must still be made in the kowledge and understanding of the phenomena involved.

Delage studied electromagnetic and thermal characteristics of cylindrical crucibles. The approach to the problem was global and led to the determination of the resistance and self-inductance equivalent to the crucible with a molten charge. The method used takes advantage of the uniformity of the magnetic field in the crucible. Although useful for tailoring a cotinuous casting system, it cannot give a precise description of other situations, such as levitation cold crucibles in which strong magnetic field gradients exist to maintain and overheat the molten charge.

The aim of this study is to obtain similar results with levitation crucibles, to be able to improve the geometry for a given effect: overheating, for example. The influence of some parameters is still unknown and needs to be made precise, for example, the number of sectors, the shape of the bottom, and the width of the splits between the sectors.

The first step considered in this paper consists of the electromagnetic modeling of the system coil-crucible-charge, taking into account the shape of the crucible.

Further, the results will be used to determine the temperature field inside the molten levitated charge, whose equilibrium shape will be computed first.

Position of the Problem

The experimental device chosen for modeling is composed of (Fig. 1) a water-cooled cylindrical copper wire inductor of four turns and a cold crucible with bottom segmented into 16 water-cooled sectors. The thickness of the space between two adjacent sectors varies from 0.5 to 1 mm. In the lower part of the bottom, a hole enables the molten metal to flow out.

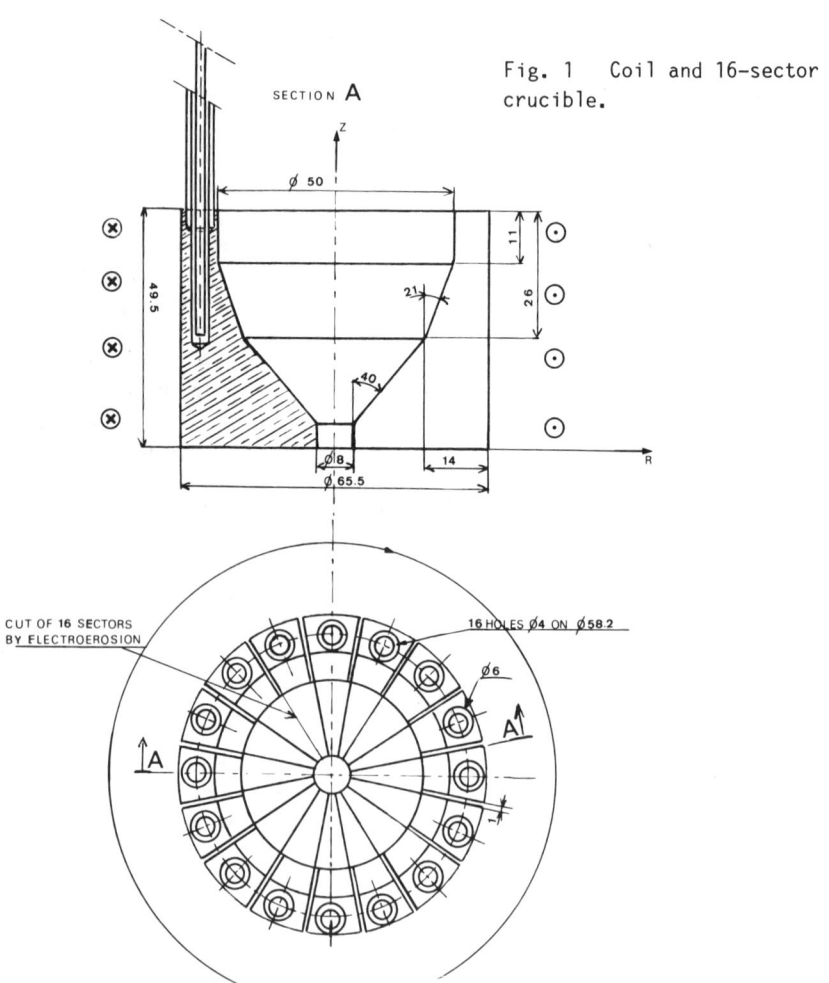

Fig. 1 Coil and 16-sector crucible.

ELECTROMAGNETIC MODELIZATION OF COLD CRUCIBLES 637

The power supply of the device is a 100 kW generator working in a frequency range from 5 to 200 kHz. Our aim is to determine the magnetic field generated by the crucible for given frequency and current intensity in the coil.

The local magnetic field, which varies periodically with time according to the frequency imposed by the generator, is

$$\vec{B}(x,y,z)\, e^{j\omega t} \qquad (1)$$

where (x,y,z) and t denote respectively spatial and temporal coordinates, and ω is the pulsation of the current in the coil. The equations to be solved are, in the air

$$\Delta \vec{B} = 0 \qquad (2)$$

and in copper,

$$\Delta \vec{B} = j\,\omega\,\mu_o\,\sigma\,\vec{B} \qquad (3)$$

with σ being the electrical conductivity and μ_o the magnetic permitivity of vacuum.

Since the crucible is segmented, the magnetic field inside is nonaxisymmetric, but presents a periodicity in the azimuthal direction Θ. The difficulty in the resolution of the equations to determine the three-dimensional magnetic field is therefore very important from an analytical point of view. So a numerical simulation of the problem has been used.

Electromagnetic Modeling

The Method

Schematization of the Problem. To simplify the modelization and to have an axisymmetric geometry, the crucible

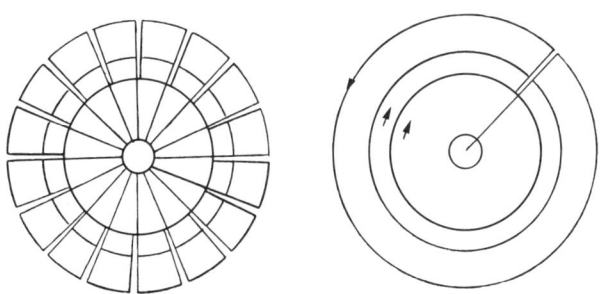

Fig. 2 In the right side a 16-sector crucible in the left side a one sector.

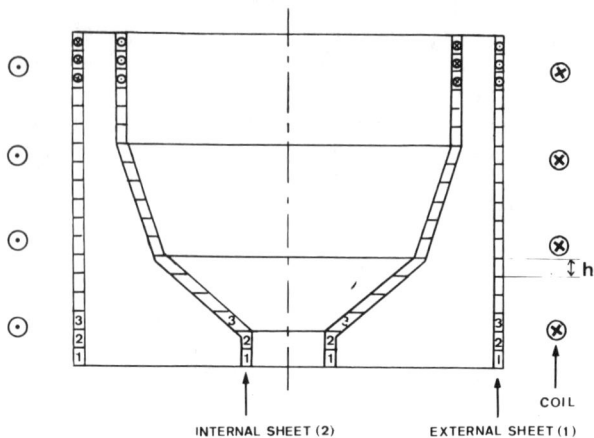

Fig. 3 Crucible considered as internal and external sheets of current is divided into elementary wires.

is assumed to be one piece with only split, whose thickness is supposed to be very small, in order to ensure the closure of the electric current paths on the internal part of the crucible (Fig. 2). Two dependent current sheets are then obtained along the internal and external faces of the crucibles; the sum of the current densities outside is opposed to the sum of the current densities inside, which results from the closure of the former (Fig. 3).

We consider that internal and external sheets of current are divided into N elementary wires with same height h (Fig. 3). The crucible is then considered as a system of 2N wires with same axis[6].

Two equations are used in our model:

$$\vec{E} = -\text{grad } V - j\omega \vec{A} \quad (4)$$

where $\vec{E}$ is the electric field, V is the electric potential, and $\vec{A}$ is the vector potential; and

$$\vec{J} = \sigma \vec{E} \quad (5)$$

where $\vec{J}$ denotes the current density.

<u>Vector Potential Relating to a Single Filiform Wire.</u>
According to the cylindrical geometry defined by Fig. 4, the vector potential A induced by the wire of radius ρ, whose center is located at $z=\xi$, has only one nonzero component, A_Θ (Ref. 7):

$$A_\Theta = (\mu_0 I/\sqrt{m\pi}) \sqrt{\rho/r} \left[(1-m/2) K(m) - E(m)\right] \quad (6)$$

ELECTROMAGNETIC MODELIZATION OF COLD CRUCIBLES

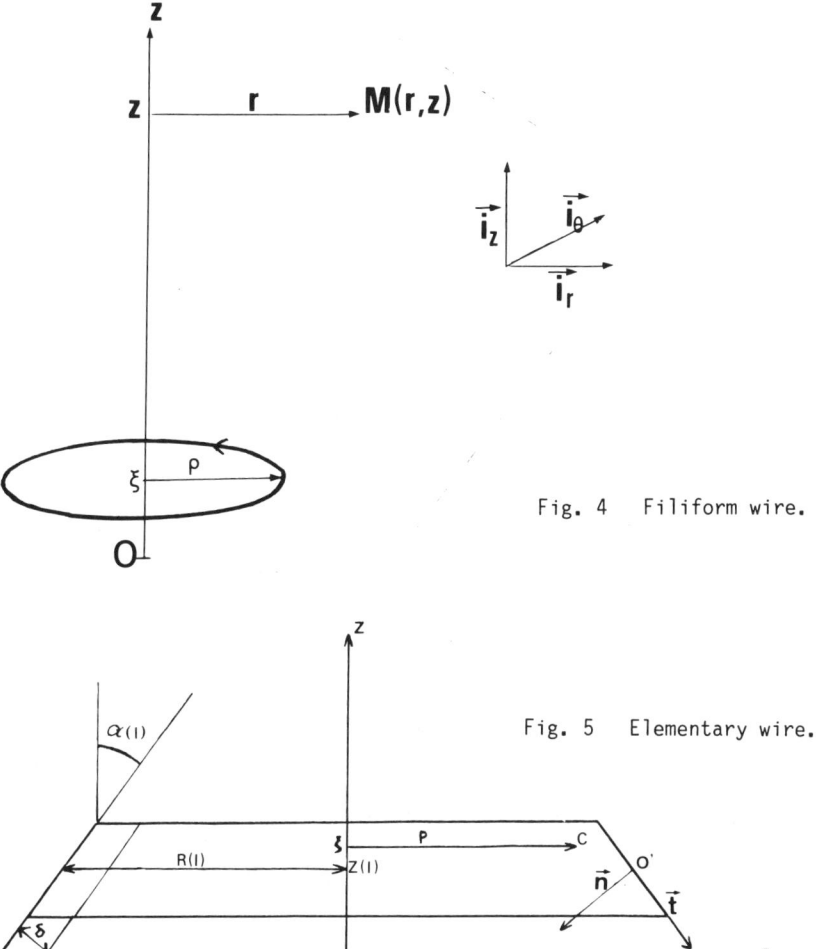

Fig. 4 Filiform wire.

Fig. 5 Elementary wire.

where

$$m = \frac{4\rho r}{(r + \rho)^2 + (z - \xi)^2}$$

K(m) and E(m) are elliptic integrals of the first and second kind.

In the following, we will take the notation:

$$f(\rho, \xi, r, z) = \sqrt{\rho/m} \; r \left[(1-m/2) \, K(m) - E(m) \right] \quad (7)$$

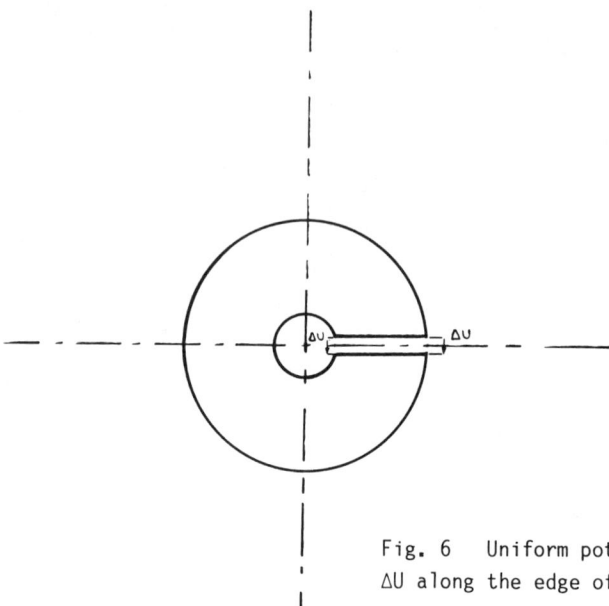

Fig. 6 Uniform potential difference ΔU along the edge of the split.

Vector Potential Relating to a Wire of Finite Size.

Let us consider a given elementary wire whose angular inclination with the vertical direction is $\alpha(i)$ (Fig. 5). We denote by n the distance of a given point C (ρ,ξ) to the wire along the normal direction; t is the distance O'C in a horizontal plane. The current in the wire (i) is written[8]:

$$I = \iint\limits_{\substack{nt \\ (\text{turn } i)}} J(t)\ e^{-\frac{(1+j)n}{\delta}}\ dt\ dn \qquad (8)$$

where δ is the skin depth

$$\delta = (2/\sigma 2\pi f\mu)^{1/2} \qquad (9)$$

J(t) is the local current density along the wire. The vector potential is then

$$A_\Theta = \frac{\mu_o}{\pi} \iint\limits_{nt} J(t)\ e^{-\frac{(1+j)n}{\delta}}\ g(n,t)\ dt\ dn \qquad (10)$$

ELECTROMAGNETIC MODELIZATION OF COLD CRUCIBLES

where

$$g(n,t) = f(\rho,\xi,r,z) \qquad (11)$$

Moreover, h is supposed to be small enough to have

$$J(t) = \text{const} = J(i) \qquad (12)$$

Because of the frequencies used, the skin depth is very small compared with the radius of the crucible (for a copper crucible $1/\sigma = 8 \times 10^{-8}$ Ωm and a frequency of 20 kHz, $\delta < 0.5$ mm). The vector potential is then given by the expression

$$A_\Theta = \frac{\mu_0 \, J(i) \, \delta \, h}{\pi \, 2(1+j) \, \cos[\alpha(i)]} \left[f\left(R(i) + \frac{h}{2} \, \text{tg} \, \alpha(i), \, Z(i) \right. \right.$$

$$\left. \left. + \frac{h}{2}, \, r, \, z\right) f\left(R(i) - \frac{h}{2} \, \text{tg}\,[\alpha(i)], \, Z(i) - \frac{h}{2}, \, r, \, z\right) \right] \qquad (13)$$

<u>Vector Potential Induced by the Crucible.</u> The resulting vector potential due to the wires of the system coil and crucible is given by the following expression:

$$A_\Theta(r,z) = \frac{\mu_0 \, \delta \, h}{2\pi(1+j)} \left\{ \sum_{i=1}^{N} J1(i) \left[f\left(R1, \, Z(i) + \frac{h}{2}, \, h, z\right) \right. \right.$$

$$+ f\left(R1, \, Z(i) - \frac{h}{2}, \, r, z\right) + \sum_{i=1}^{N} \frac{J2(i)}{\cos[\alpha(i)]}$$

$$\times \left[f\left(R2(i) + \frac{h}{2} \, \text{tg}\,[\alpha(i)], \, Z(i) + \frac{h}{2}, \, r, \, z\right) \right.$$

$$\left. + f\left(R2(i) - \frac{h}{2} \, \text{tg}\,[\alpha(i)], \, Z(i) - \frac{h}{2}, \, r, \, z\right) \right] \right\}$$

$$+ \frac{\mu_0}{\pi} I \sum_{m=1}^{4} f\left[R3, \, Z3(m), \, r, \, z\right] \qquad (14)$$

where J1(i) is the current density in the wire i of the external current sheet 1; J2(i) is the current density in the wire i of the internal current sheet 2; R1 is the radius of the external current sheet; R2(i) is the radius of the wire i of the current sheet 2; Z(i) is the vertical coordinate of the wire i of any current sheet; R3 is the radius of the coil; Z3(m) is the vertical coordinate of the wire (m) of the coil; and $\alpha(i)$ is the angle between the vertical direction and the wall of the wire i.

Electrical Potential. To obtain the values of the current density, the following equation, deduced from Eqs. (4) and (5), is to be solved:

$$\vec{J} = -\sigma \text{ grad } V - j\sigma\omega\vec{A} \qquad (15)$$

Two hypotheses can be made concerning the electrical potentital V:

In the first approach (model 1), we suppose that the induced currents are flowing along purely horizontal loops, i.e.,

$$\forall \; i \; \varepsilon \; [1,N] \quad \frac{J2(i)}{\cos[\alpha(i)]} = -J1(i) \qquad (16)$$

The resulting condition for V is then

$$\overrightarrow{\text{grad }} V = 0 \qquad (17)$$

This condition makes the vector potential (14) a linear combination of the current densities along the external sheet of the crucible and of the current in the coil. If A_i denotes the total vector potential induced in each wire of the external sheet of the crucible,

$$\forall \; i \; \varepsilon \; [1,N] \quad J1(i) = -j\sigma\omega A_i \qquad (18)$$

We then obtain a linear system of N equations with N unknown variables. The simplicity of the system is due to the restrictive hypothesis about the horizontal closure of the currents.

In the second approach (model 2), we consider that an uniform potential difference Δu exists along the edge of the split. This difference is supposed to be the same along the internal and external edges of the split (Fig. 6).

$$\forall \; i \; \varepsilon \; 1,N \quad \text{grad } V = \frac{\sigma \Delta U}{2\pi R1} \; \vec{i}_\Theta \quad \text{(external current sheet)}$$

$$\text{grad } V = \frac{\sigma \Delta U}{2\pi R2(i)} \; \vec{i}_\Theta \quad \text{(internal current sheet)} \qquad (19)$$

ELECTROMAGNETIC MODELIZATION OF COLD CRUCIBLES

The unknown current densities can be deduced from the system:

$$\forall \; i \; \varepsilon \; [1,N] \quad J1(i) = \frac{\sigma \Delta U}{2\pi R1} - j\sigma \omega A_\Theta(1,i)$$

$$J2(i) = \frac{\sigma \Delta U}{2\pi R2(i)} - j\sigma \omega A_\Theta(2,i) \quad (20)$$

$$\sum_{i=1}^{N} J1(i)S1(i) = -\sum_{i=1}^{N} J2(i)S2(i)$$

where $A_\Theta(1,i)$ is the total vector potential induced along the wire i of the external current sheet 1; $A_\Theta(2,i)$ is the total vector potential induced along the wire i of the internal current sheet 2; $S1(i)$ is the area of the wire i of the current sheet 1; and $S2(i)$ is the area of the wire i of the current sheet 2.

A linear system with 2N+1 equations and 2N+1 unknown variables (2N current densities and the potential difference) is then obtained.

<u>Possibility of Taking the Charge into Account.</u> The same method can be used to compute the electromagnetic characteristics of the crucible containing an axisymmetric charge of given position and shape. We consider that the skin depth in the charge is very small compared with its typical radius. The induced currents flow along an external sheet that is divided into Nc wires with Nc unknown current densities. In the case of the one-piece crucible with only one very thin split, because of the axial symmetry of the charge, no potential gradient exists along the external sheet of the charge. The two resulting systems to be solved in the frame of our models are model 1 with charge and model 2 with charge.

For model 1 with charge, the system supposing a horizontal closure of the induced currents is written

$$\forall \; i \; \varepsilon \; (1,N) \quad J1(i) = - j\sigma \omega A_\Theta(1,i)$$

$$\forall \; i \; \varepsilon \; (1,Nc) \quad Jc(i) = - j\sigma \omega A_\Theta(c,i) \quad (21)$$

where $Jc(i)$ is the current density in the wire i of the charge; $A_\Theta(1,i)$ is the vector potential induced by the system coil plus the crucible plus the charge along the wire i of the external current sheet of the crucible; and

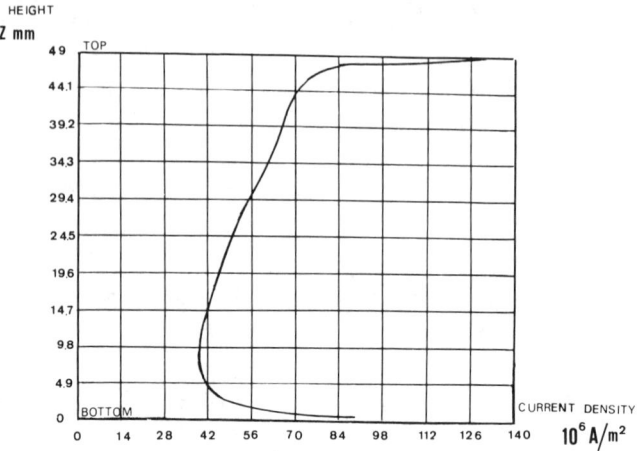

Fig. 7 Distribution of induced current density on the external sheet computed by first model.

$A_\Theta(c,i)$ is the vector potential induced by the system coil plus the charge along the wire i of the charge.
A linear system with N+Nc equations for N+Nc unknown variables is then obtained.
For model 2 with charge, a potential difference is supposed to exist along the edges of the split:

$$\forall\ i\ \varepsilon\ [1,N]\ \ J1(i) = -\frac{\sigma\Delta U}{2\pi R1} - j\ \sigma\omega\ A_\Theta(1,i)$$

$$J2(i) = \frac{\sigma\Delta U}{2\pi R2(i)} - j\ \sigma\omega\ A_\Theta(2,i) \qquad (22)$$

$$\forall\ i\ \varepsilon\ [1,\ NC]\ \ Jc(i) = -j\ \sigma\omega\ A_\Theta\ (c,i)$$

$$\sum_{i=1}^{N} J1(i) = -\sum_{i=1}^{N} \frac{J2(i)}{\cos\ [\alpha(i)]}$$

A linear system with 2N+Nc+1 equations for 2N+Nc+1 unknown variables is then obtained.

Induced Current Distribution in the Crucible.

The computed distribution is plotted on curves 1 and 2 of Figs. 7 and 8. These curves reveal strong finge effects. T

ELECTROMAGNETIC MODELIZATION OF COLD CRUCIBLES 645

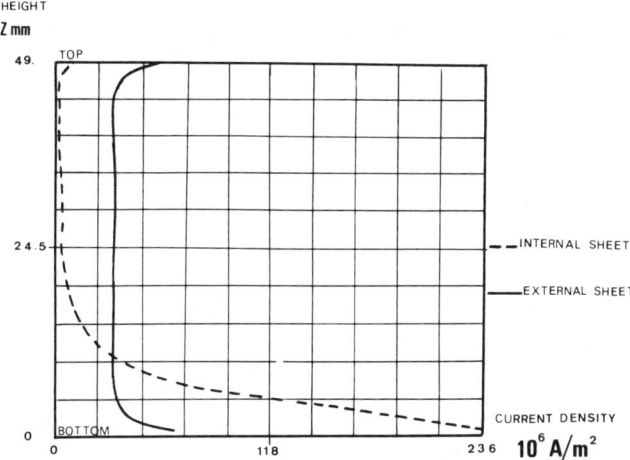

Fig. 8 Distribution of induced current density computed by second model.

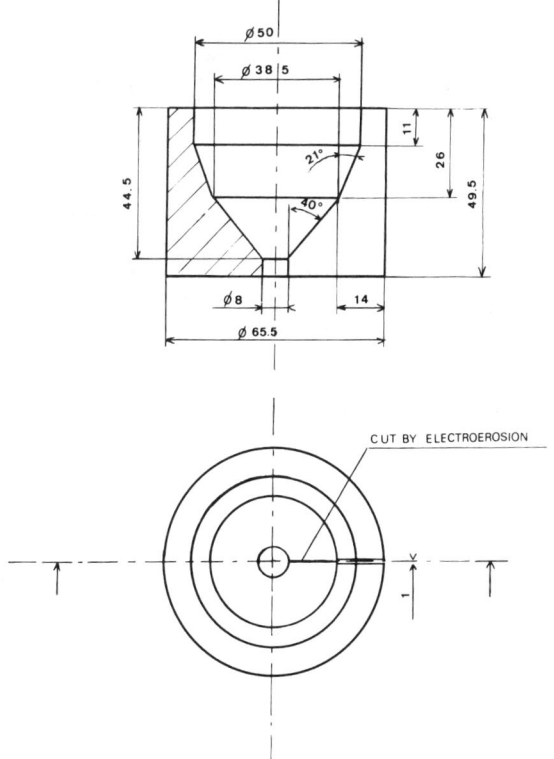

Fig. 9 A one-split crucible.

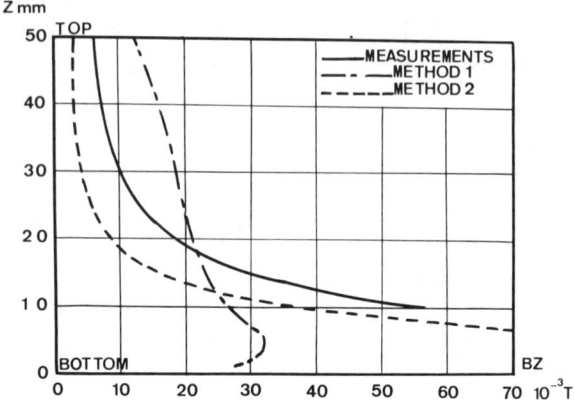

Fig. 10 Magnetic field Bz along the axis in a one-split crucible. Measurements and models.

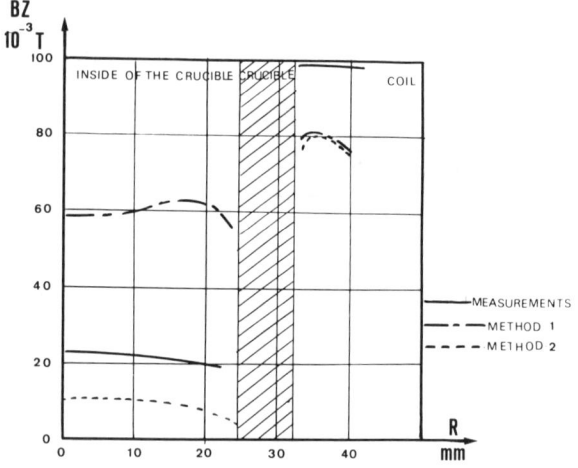

Fig. 11 Magnetic field Bz against R for z = 38 mm in a one-split crucible. Measurement and models.

first model leads to a greater current density in the upper part of the crucible. On the other hand, the second model leads to a uniform current density along the external sheet and a continuously downwards increasing current density along the internal sheet. This effect implies strong positive magnetic field gradients at the bottom of the crucible necessary to obtain good levitation.

ELECTROMAGNETIC MODELIZATION OF COLD CRUCIBLES   647

Resulting Magnetic Field in the System Coil Plus Crucible

To make a comparison between computed magnetic field and measurements[9], we use two experimental devices: a one-piece crucible with a single split (Fig. 9) and a 16-sector crucible (Fig. 1). The two crucibles have the same size and the coils are identical.

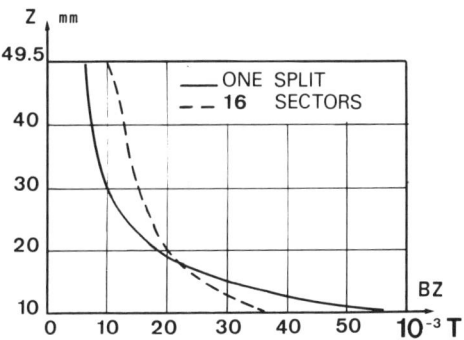

Fig. 12 Measurements of magnetic field Bz along the axis in a one-split crucible and in 16-sector crucible.

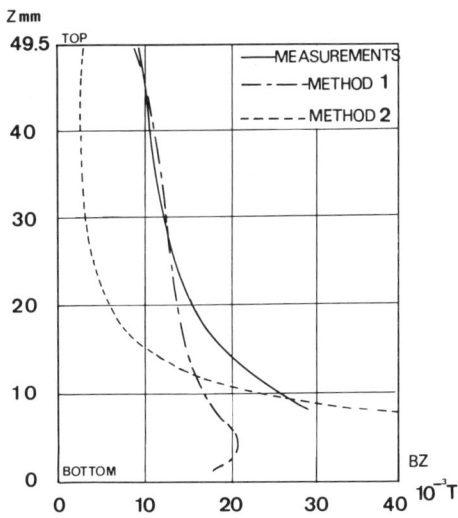

Fig. 13 Magnetic field Bz along the axis in a 16-sector crucible. Measurements and models.

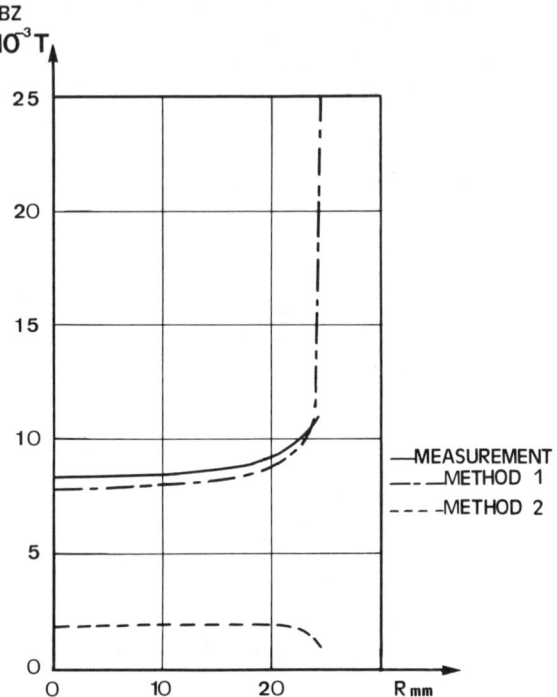

Fig. 14 Magnetic field Bz against R for z = 49.50 mm in a 16-sector crucible. Measurements and models.

A small coil made of several turns of copper wire is used to measure the magnetic field intensity. Good precision (5%) is obtained with an eight-turn coil (the diameters of coil and wire were 1 and 0.1 mm, respectively). The method consists in measuring the potential difference induced across the coil, which is proportional to the component of the magnetic field normal to the plane of the wires.

One-Split Crucible. The vertical component of the magnetic field Bz measured along the axis with respect to z is plotted on curve 3 (Fig. 10). The variation of Bz for given z with respect to the radial coordinate is plotted on curve 4 (Fig. 11).

Model 2 is in better agreement than model 1 with experimental data; curves 3 and 4 show similar behavior, with a strong increase in the magnetic field in the crucible when z decreases and a discontinuity of the magnetic field intensity when crossing the crucible wall.

Important discrepancies exist between experimental and computed values deduced from model 1.

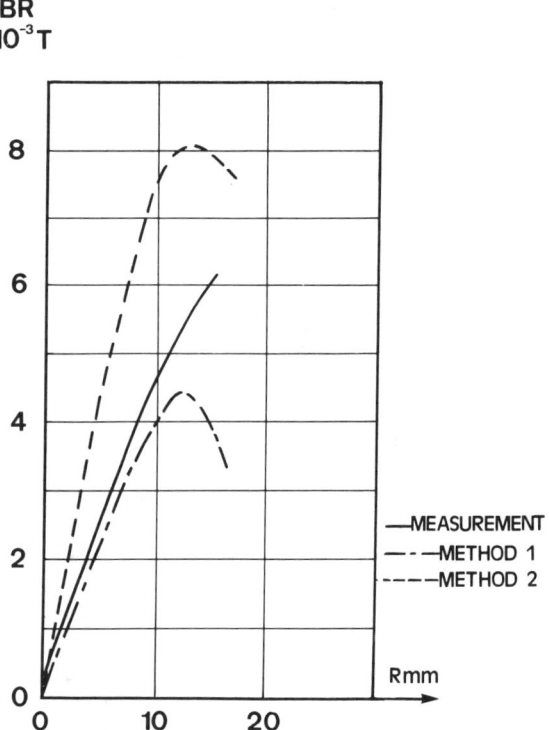

Fig. 15 Magnetic field BR against R for z = 23.5 mm in a 16-sector crucible. Measurement and models.

The 16-Sector Crucible. An important difference appears between the one-split crucible and the 16-sector crucible when considering the magnetic field intensity along the axis (curve 5) (Fig. 12). The magnetic field gradient is much more important in the case of the single split than in the case of several splits. This is due to the possibility offered to the magnetic field of escaping from the interior of the crucible through the splits. There is no doubt that the one-split crucible is better in regard to the levitation effect than any crucible with several splits[10,11].

In a comparison between experimental measurements and computed values, the value of the vertical component Bz of the magnetic field is plotted against z along the axis on curve 6 (Fig. 13). Curves 7 and 8 (Figs. 14 and 15) show respectively, for a given value of z, the variations of Bz and Br against r. Concerning curve 5, a rather good agreement is obtained between measurements and theoretical values deduced from model 1 (except near the bottom of the crucible). Model 2 is in complete disagreement with the experiment. The same conclusion is reached for curves 7 and 8. Contrary to the case of the one-split crucible, model 1 is a better way to model the magnetic field distribution. The hypothesis of horizontal closure of the induced current lines seems therefore to be physically realistic when the number of sectors increases.

Concluding Remarks

Because of the complexity of the geometry, a cold crucible is difficult to model in regard to the magnetic field distribution. The presence of the splits, which generates a periodic variation in the azimuthal direction, on the one hand, and of the bottom, which is responsible for the three-dimensionality of the magnetic field, on the other, makes it necessary to resort to simplifying hypotheses.

The two models we developed cannot describe the magnetic field configuration in any levitation cold crucible. The first model, which supposes the horizontal closure of the induced current loops, fits the 16-split crucible well. The second model, which introduces an uniform potential difference along the edges of the split, gives good results in the case of a one-piece crucible with only one split.

This leads to interesting conclusions. Indeed, we can deduce that, when the number of sectors becomes large, the induced current pathlines are horizontal, and no potential difference appears between the edges of two adjacent sectors. This fact is confirmed by experience. In a crucible with several splits it is possible to have contact between the molten charge and the sectors without modifying the electrical behavior of the crucible. On the other hand, in the case of a single one-split crucible, any such contact leads to short circuits of the crucible. Moreover, because of the singularity introduced in the axisymmetry by a single split, the induced current lines cannot be horizontal: The inclination of the current lines corresponds to a shortening of the paths along which the currents have to flow and consequently to a decrease of their resistance.

The two models are not able to describe the magnetic field in the neighborhood of the crucible wall; this is due to the influence of the splits, which are very important in this region but which very rapidly decrease toward the crucible axis.

The one-split crucible seems to be much better for levitation[12,13], since the presence of several splits tends to force the magnetic field gradients to vanish.

ELECTROMAGNETIC MODELIZATION OF COLD CRUCIBLES   651

References

[1] Delage, D.,"Aspects Electriques et Thermiques de la Fusion par Induction en Creuset Froid," (Electric and Thermic Aspects in Cold Crucible Induction Melting), Ph.D. Thesis, INPG, Grenoble, France, 1982.

[2] Delage, D., Barbier, J.N., and Fautrelle, Y.R., "Modeling of Magnetic Field, Pool Profiles, Temperature Field in Cold Crucible Induction Melting," Proceedings of the Fourth International Conference on Mathematic Modeling, Zurich, 1983.

[3] Biasse, H.M., "La levitation electromagnetique. Une Proposition d'Application a la soudure," (Electromagnetic Levitation. Proposal of Application to Soldering) Ph.D. Thesis, Grenoble, France, 1981.

[4] Okress, E.C., Wroughton, D.M., Gomenetz, G., et al, "Electromagnetic Levitation of Solid and Molten Metals," Journal of Applied Physics, Vol. 23, May 1952, pp. 545-552.

[5] Rony, P.R., "The Electromagnetic Levitation of Metals," Transactions of the Vacuum Metal Conference, edited by M.A. Cocca, American Vacuum Society, Boston, Mass., 1965, p. 55.

[6] Tarapore, E.D., "Fluid Flow and Mass Transfer in Induction Melting Furnaces," Ph.D. Thesis, University of California, Berkeley, 1976.

[7] Durand, E., Magnetostatique, edited by Masson and Cie, 1968, pp.30-38.

[8] Duperrier, S., "Pratique du Chauffage Electronique," (Application of Electronic Heating), edited Chiron, 1952, pp. 7-125.

[9] Ernst, R., "Analyse du Fonctionnement d'un Generateur a Triode Destine au Chauffage par Induction," (Working Analysis of a Triode Generator Used for Induction Heating), Revue Generale d'Electricite, Sept. 1981, pp. 667-673.

[10] Mestel, A., "Magnetic Levitation of Liquid Metals, $\underline{I}$" Journal of Fluid Mechanics, Vol. 117, 1982, pp. 27-43.

[11] Fromm, E., and Jehn, J., "Electromagnetic Forces and Power Adsorption in Levitation Melting," British Journal of Applied Physics, Vol. 16, 1965, pp. 653-663.

[12] Jones, T.B., "A Necessary Condition for Magnetic Levitation," Journal of Applied Physics, Vol. 50, July 1979, pp.5057-5058.

[13] Holmes, L.M., "Stability of Magnetic Levitation," Journal of Applied Physics, Vol. 49, June 1978, pp. 3102-3109.

Shaping of Liquid Metal Cylinders

J-P. Brancher, R. de Framond, and O. Sero-Guillaume

Groupement d'Interet Scientifique, Madylam, Saint Martin d'Heres, France

Abstract

This study deals with the shaping, by an MHD process applied to liquid metals, of cruciform section cylinders. The bidimensional problem comes from the equilibrium between the magnetic pressures created by the quadripolar inductor and the surface tension stresses. First, the equilibrium shape of the metallic liquid cyliner is calculated with a variational method, where the function of energy of the system is minimized. The results are compared with experimental measures on mercury and with a first-order approximation for low magnetic fields. Then, the jet stability inside the inductor is studied. Calculations are made with a strong surface tension.

Introduction

The shaping of liquid metal jets constitutes one of the MHD applications to metallurgy. It is obtained through the application of high-frequency magnetic field for the shaping of a metal strip, on quadripolar field for cross-shaped sections. The experiments carried out by Etay and Garnier[1] on mercury show the possibility of such a method. In an initial approximation (a weak magnetic Reynolds number and a negligible electromagnetic skin depth), the magnetostatic model is usable. It has been used by Shercliff[2], and by Brancher et al.[3] for shaping, and by Sneyd and Moffatt[4] for the "levitation melting process".
This paper is particularly interested in cruciform shaping. After considering the magnetostatic problem, the va-

---

Paper presented at the Fourth Beer-Sheva Seminar on MHD Flows and Turbulence, Ben-Gurion University of the Negev, Beer-Sheva, Israel, Feb. 27-March 2, 1984. Copyright © 1985 by author. Published by the American Institute of Aeronautics and Astronautics, Inc. with permission.

riational method was adapted to solve the free boundary problem. The calculated solutions are compared with the experimental results obtained on mercury.

Calculated configurations are stable for weak perturbations. The situation is different for perturbations of great amplitude. The problem of conditional stability is studied in case of a high superficial tension; the cast can be brought back to the center of the quadripolar inductor, even if initially it is found on the exterior of the inductor. And, if one sufficiently brings nearer two inductors to the others (rectangle case), the system is no longer stable and the shaping is no longer possible.

Shaping Calculation

Problem Definition

Given is a cylindrical jet of axis Oz, placed in a bidimensional alternative magnetic field $\vec{B_o}(x,y)\sqrt{2}\sin\omega t$. The ω pulsation is sufficiently high for the skin depth to be negligible, in order that the magnetic field does not penetrate the liquid metal.

The total field $\vec{B}(x,y)\sqrt{2}\sin\omega t$ can be separated in the following way:

$$\vec{B}(x,y) = \vec{B_o}(x,y) + \vec{b}(x,y)$$

where $\vec{B_o}$ is the applied field and $\vec{b}$ is the proper field created by the superficial induced currents.

In the metal, there is $\vec{B}(x,y)=0$ or $\vec{b}(x,y)= -\vec{B_o}(x,y)$. In the area Ω outside the metal, situated in an orthogonal plan to Oz, there is

$$\vec{\text{curl }} \vec{B} = \mu_o \vec{j_o} \quad \text{(on } Oz\text{)}$$

where $\vec{j_o}$ is the current density vector, imposed on the magnetizing circuits and μ_o is the permeability of vacuum.

In adding the equation div $\vec{B}=0$ and the boundary conditions on the frontier Γ of the area occupied by the liquid, one obtains the equations of the field by both equivalent systems:

$$\begin{array}{ll} \vec{\text{curl }} \vec{B} = \mu_o \vec{j_o} & \\ \text{div } \vec{B} = 0 & \text{in } \Omega \\ \vec{B}\cdot\vec{n} = 0 & \text{on } \Gamma \end{array} \quad \text{or} \quad \begin{array}{ll} \vec{b} = \vec{\text{grad }} u & \\ \Delta u = \dfrac{\partial^2 u}{\partial x^2} + \dfrac{\partial^2 u}{\partial y^2}=0 & \text{in } \Omega \\ \dfrac{\partial u}{\partial n} = -\vec{B_o}\cdot\vec{n} & \text{on } \Gamma \end{array} \quad (1)$$

where $\vec{n}$ is the external normal vector to Ω on Γ. $\vec{b}$ will be defined by the holomorph complex potential $F = \phi + 1\psi$ outside of the magnetizing circuits.

The equilibrium configuration in a horizontal plan is controlled by the effects of superficial tension with coefficient γ and by the action of the magnetic pressure $B^2/2\mu_0$ exerted on the interface Γ. Then, the equilibrium condition can be written as

$$\gamma/R + B^2/2\mu_0 = c^{st} \qquad (2)$$

where R is the curvature radius of Γ. (See Fig. 1).

The problem is to determine the shape of Ω, given that $\vec{B}_0$ is the imposed as well as the liquid metal flow rate. It is demonstrated in Ref. 3 that imposing the flow rate means the same thing as requiring the area A of the section Ω for the liquid metal jet, under the condition that the kinetic energy $\frac{1}{2}\rho (q_v^2/A^2)$ is great enough in relation to the magnetic energy $B^{*2}/2\mu_0$. ρ is the density of the liquid and B^* is a characteristic magnetic field. Under these conditions, one sets $A = \pi R_0^2$, where R_0 is the jet radius undeformed before it penetrates into the field zone. The problem to solve is a free boundary problem governed by the Eqs. (1) and (2), and the condition $A = \pi R_0^2$.

Variational Method

In such a problem, it is interesting to minimize a certain energetic function instead of solving Eq. (2). This point of view, adopted in Refs. 3-5, leads to a problem of minimization with constraints depending on an adimensional parameter defined from the ratio of the superficial tension

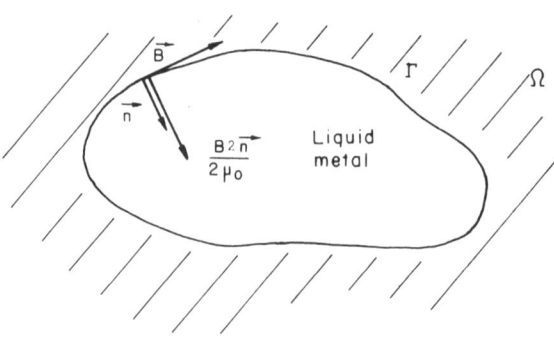

Fig. 1   A general cross-section of a liquid metal in a magnetic field

SHAPING OF LIQUID METAL CYLINDERS

and the magnetic energies. It has been shown in Refs. 4 and 5 that the energy function E to be minimized is as follows:

$$E = \gamma L - \lim_{\varepsilon \to 0} \left(\int_{\Omega\varepsilon} \frac{B^2}{2\mu_o} \, dxdy + \sum_{k=1}^{M} \frac{\mu_o I_k^2}{4\pi} \log \varepsilon \right) \quad (3)$$

in the case where the applied field is created by M filiform currents represented by M points in Ω. I_k is the value of the current in the kth circuit. $\Omega\varepsilon$ is the area Ω deprived of M disks with ε radius centered upon the M proceeding points. L is the length of the M curve.

In the case where the applied field is uniform, E is defined as

$$E = \gamma L - \frac{1}{2\mu_o} \int_{\Omega} b^2 \, dxdy + \frac{1}{\mu_o} \lim_{R \to \infty} \int_{\Gamma R} \vec{B}_o \cdot \vec{n} \, \phi \, dl \quad (4)$$

where ΓR is the circle with radius R, $\vec{n}$ is the external normal vector.

The energy function (4) is used in Ref. 3 to calculate the shaping of the liquid metal strip. Then we shall use the energy function (3), which will allow us to treat the shaping of cross-shaped sections.

The E minimization is defined in relation to the Ω area. One, then, introduces the successive derivations $\partial E/\partial \Omega$, $\partial^2 E/\partial \Omega^2$ ... of E in relation to Ω ($\partial E/\partial \Omega = 0$ gives the equilibrium Eq. (2)). The general case of this problem is studied in Ref. 5. Here, the problem is bidimensional; that is why the Ω areas group will be introduced from a series of conform transformations which forms a correspondence between Ω and the outside of the circle (o,R_o) (from the z plan to the ξ plan) defined by

$$z = C_o \xi + \sum_{n=1}^{\infty} \frac{C_n}{\xi^n}$$

where C_o, C_1, ..., C_n are complex. Then, the energy function E is to be calculated from the coefficients C_o, C_1, ..., C_n, ... When stopping the series at the Nth term, the problem of minimization of an N variables function must be solved.

Calculation of Cruciform Sections

The shaping of cross-formed sections is obtained by a quadripolar inductor where dephased currents (two by two $\phi = \pi$) with true value I. If the dimensions are reduced in re-

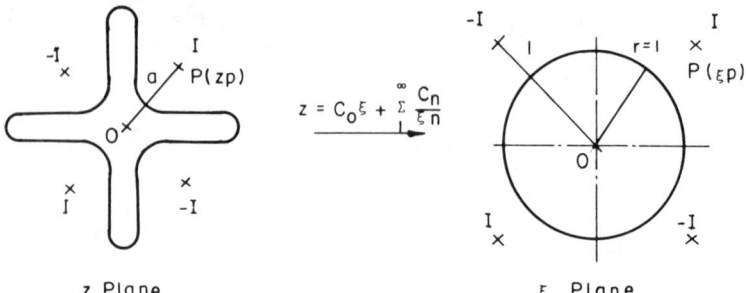

Fig. 2   The cross-section obtained by shaping with a quadripolar inductor, as seen in a z=const plane and in a ξ=const plane.

lation to R_0, one obtains, in the z plan and in the ξ plan, the configuration shown in Fig. 2.

The length calculation is obtained as in Ref. 4:

$$L = 2\pi \sum_{n=0}^{+\infty} r_n^2$$

with $r_0 = \sqrt{C_0}$, $r_1 = 0$, $r_2 = -C_1/2\sqrt{C_0}$, $r_3 = -C_3/\sqrt{C_0}$. For $n>4$, we have

$$r_n = -\frac{1}{2r_0} \left| (n-1) C_{n-1} + \sum_{k=2}^{n-2} r_k r_{n-k} \right|$$

The magnetic energy calculation uses the method of images applied to the circular configuration of the ξ plan (Fig. 2). An additional term appears due to

$$\lim_{\varepsilon \to 0} \log \varepsilon = \lim_{z \to z_p} \log |z - z_p| = \lim_{\xi \to \xi_p} \log \frac{|z-z_p|}{|\xi-\xi|} |\xi - \xi_p|$$

$$= \lim_{\varepsilon' = |\xi-\xi_p| \to 0} \left| \log \varepsilon' + \log \left|\frac{\partial z}{\partial \xi}\right|_p \right|$$

From the length and magnetic energy calculations, one obtains the nondimensional expression of E:

$$E = kL + \log \frac{1^4+1}{1(1^4-1)} - \log \left|\frac{dz}{d\xi}\right|_{\xi=e^{i\pi/4}} \tag{5}$$

SHAPING OF LIQUID METAL CYLINDERS

where $k = \pi\gamma R_0/\mu_0 I^2$, 1 satisfying for

$$ae^{i\pi/4} = C_0 \, 1 \, e^{i\pi 4} + \sum_1^\infty \frac{C_n}{1^n_1 e^{in\pi/4}}$$

The conservation of the A area imposes

$$|C_0|^2 = 1 + \sum_{n=1}^\infty n|C_n|^2$$

The desired solution must be invariant with a $\pi/2$ rotation and by symmetry of axis Ox. The C_n coefficients can therefore be chosen among real numbers, and so that only C_0 and C_{4k-1} are different from zero.

The obtained results will be compared with the experimental results in the following paragraph. However, when k is great, the measures are not significant because the jet deformations are very weak. The results are therefore compared with those obtained by a first-order approximation.

At the first order, the magnetic pressure $B^2/2\mu_0$ is approximated by that which appears for the nondeformed configuration, which is calculated by the method of images. The equilibrium condition (2) is written in a reduced form:

$$\frac{1}{R} + \varepsilon \, e(1-e)^2 \, \frac{1 - \cos 4\theta}{(e^2+1 + 2e \cos 4\theta)} = cs^{te} \qquad (6)$$

where

$$e = \frac{R_0^4}{a^4} < 1 \qquad \varepsilon = \frac{4 \mu_0 I^2}{\pi^2 R_0 \gamma} \ll 1$$

and R is the curvative radius.

The shape of the section is given in polar coordinates by

$$\rho/R_0 = 1 + \varepsilon' \, f(\theta) \quad \text{with } \varepsilon' = 0(\varepsilon) \qquad (7)$$

Developing $f(\theta)$ as a Fourier series, one obtains

$$f(\theta) = \sum_{n=2}^\infty a_n \cos n\theta$$

$$\frac{1}{R} = \sum_{n=2}^\infty (n^2-1) a_n \cos n\theta + 0(\varepsilon'^2)$$

Table 1 Comparison of numerical results and measures

k	a	Flow rate Q(cc/s)	$\Sigma\ C_n$ Measured	$\Sigma\ C_n$ Calculated
0.1	1.3	650	1.07	
0.1	1.5		1.05	1.12
0.05	1.3		1.09	1.33
0.05	1.5		1.08	1.20
0.044	1.5		1.11	
0.025	1.3		1.15	
0.025	1.5		1.14	
0.02	1.5		1.16	1.36
0.01	1.5			1.65
0.001	2.0			1.62

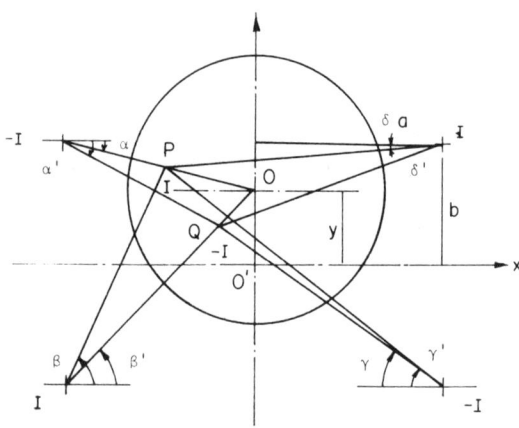

Fig. 3 The geometry used for the method of images.

The A section area is

$$\frac{1}{2} \int_0^{2\pi} \rho^2\ d\theta = \pi R_0^2 + O(\varepsilon'^2)$$

so A is conserved at the first order.
Equation (6) gives $\varepsilon' = e(1-e)^2\ \varepsilon$ and

$$\sum_{n=2}^{\infty} (n^2-1)\ a_n\ \cos n\theta + \frac{1 - \cos 4\theta}{(e^2+1 + 2e \cos 4\theta)^2} = c_s{}^{te} \quad (8)$$

SHAPING OF LIQUID METAL CYLINDERS

The a_n coefficients are calculated from the Fourier coefficients of the function

$$\frac{1 - \cos 4\theta}{(e^2+1 + 2e \cos 4\theta)^2}$$

One therefore obtains

$$a_n = 0 \text{ if } n \neq 4p$$

$$a_{4p} = \frac{-8}{\pi(16p^2-1)} \int_0^{\pi/4} \frac{1 - \cos 4\theta}{(e^2+1 +2e \cos 4\theta)^2} \cos 4p\theta \, d\theta$$

The shaping calculation thus gives

$$\frac{\rho}{R_0} = 1 + \varepsilon \sum_{p=1}^{\infty} a_{4p} \cos 4p\theta + O(\varepsilon^2)$$

Experiments and Results

The studying system consists of three different circuits, electric, cooling, and mercury, which converge to a measure center. In the latter, mercury flows by gravity from a nozzle (converging circuit) as a stable circular jet. It falls through a quadripolar inductor that, in conjunction with a capacitor battery in parallel, is crossed by an al-

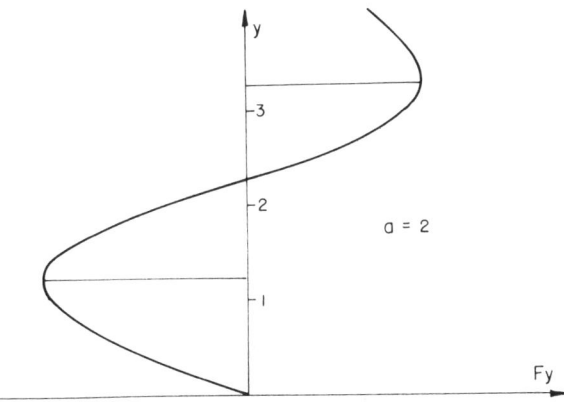

Fig. 4 The vertical strength as a function of the vertical displacement from the center of the cylinder.

ternative current with a high frequency (250 kHz) leading to an electromagnetic skin depth inside the metal of about 1 mm.

A magnetic field geometry corresponds to each geometry of the inductor and, thus, to an equilibrium shape for the metallic jet. The effect of the superficial tension, in competition with the magnetic pressure, imposes a curvature radius to the jet surface that therefore assumes the shape of a hypocycloid with round vertices.

The manipulation allows the study of the following parameters: flow rate, initial jet diameter, inductor geometry. For the measure system, we use the current between the mercury jet and a tungsten point related to a (r,θ) table. The precision of the measure is 0.1 mm. Here, we only measure the deformation on a symmetrical axis.

Results Comparison

Numerical results and the first-order approximation are given at the end of the paper. Experiments are not very easy to make, and the results are hard to use. Few experimental values can be done. Comparisons between the numerical results and measures can be seen in Table 1.

A complete discussion is not possible at this time. However, it appears that the numerical deformations are stronger than the experimental ones. This situation can be explained by the theoretical hypothesis about the filiform coils: in our experiments, the cross-sectional areas of the coils and the jet are of the same magnitude order. First of all, the currents are not concentrated on a point but are accumulated on the ring of a disk. Secondly, the induced currents act on the coils and change the current distribution on the ring so that the density is more important on the part in front of the metal.

Stability

General Problem

In the case of electromagnetic shaping on a mercury jet through a quadripolar inductor, the position of the cast at the center of the conductors is a stable equilibrium position. Let us study the evolution of this equilibrium when the jet is subjected to disturbances, that is to say, to displacements. One can see that this jet equilibrium depends on the amplitude and direction of these perturbations; so we shall call it conditional stability.

This conditional stability is studied, when superficial tension is very high, through two parallel methods. The method of images and the energetic evaluation: the latter is connected with the variational method already explained in the next section. (Fig. 3).

The $\gamma \to \infty$ Case

Let us consider the case where a cylinder with a circular section has a infinite superficial tension. We easily find the magnetic pressure around the metal (Rm<<1 and $\delta/L<<1$) and the resultant of the applied strengths through the method of images. The problem is adimensionalized in relation to the radius of the cylinder.

The problem is to study the interactions between the eight conductors. The possible electric current I_c at the center of the cylinder equals zero as the sum of the inside currents (image currents) equals zero. The next topic of interest is the resultant of the strength on axis oy when the cylinder is moved along this axis.

The Square Case. The resultant of the strengths induced by the eight conductors on axis oy is (by the fact of symmetries) as follows:

$$R = \frac{\mu I^2}{\pi} (\frac{\cos \delta}{P4} + \frac{\cos \gamma}{P_3} + \frac{\cos \beta'}{Q_2} + \frac{\cos \alpha'}{Q_1} - \frac{\cos \alpha}{P_1} - \frac{\cos \beta}{P_2}$$

$$- \frac{\cos \gamma'}{Q_3} - \frac{\cos \delta'}{Q_4})$$

So, one draws the intensity of the attractive of repulsive strength according to the vertical displacement given to the center of the cylinder (Fig. 4).

First, the two equilibriums (stable and unstable) are noted. One of them is found when the axis of the cylinder is outside the square defined by the four conductors. This means that a cylinder outside the area closed by the conductors can be brought back inside, depending on its distance from the center.

This unstable equilibrium position can be studied in relation to a displacement on x and on y, which gives the results of an equilibrium area from the geometry of the inductor. (Fig. 5)

Moreover, the intensity of the strength of attraction goes to a maximum (just as the strength of repulsion). The displacement of the cylinder corresponding to this maximum

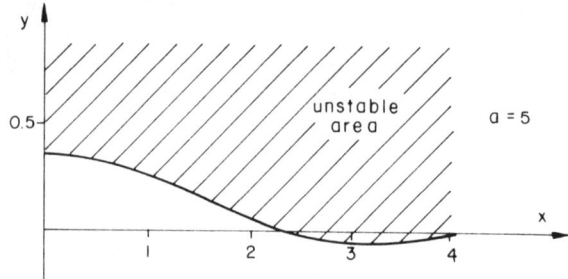

Fig. 5 The unstable area in the x-y plane where x and y are the components of the displacement of the cylinder.

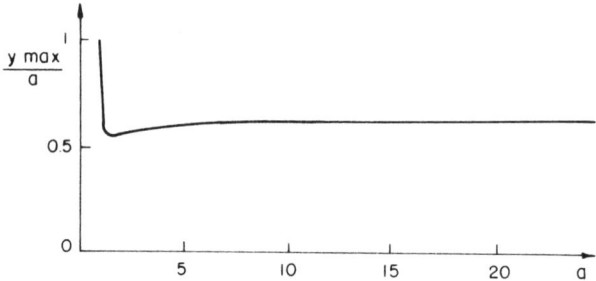

Fig. 6 The maximal displacement of the cylinder as a function of the radius of the circular cross-section.

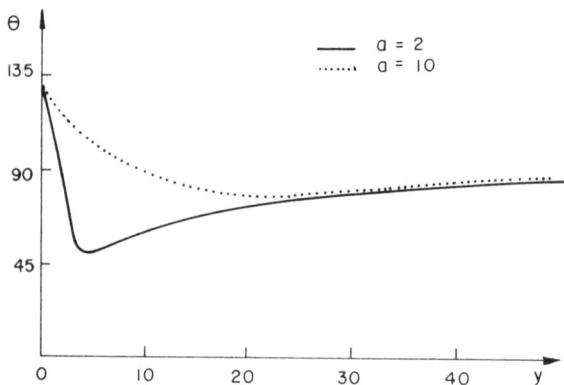

Fig. 7 The angle corresponding to the maximum magnetic pressure around the cylinder as a function of the vertical displacement, for two different values of the radius of the circular cross-section.

SHAPING OF LIQUID METAL CYLINDERS 663

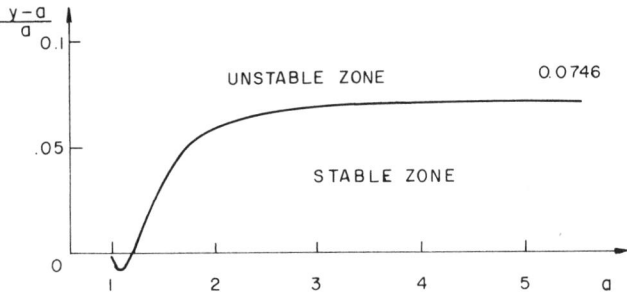

Fig. 8 Stable and unstable zones in the $(y-a)1a$, a plane where y is the vertical displacement of the cylinder and a is its circular cross-section.

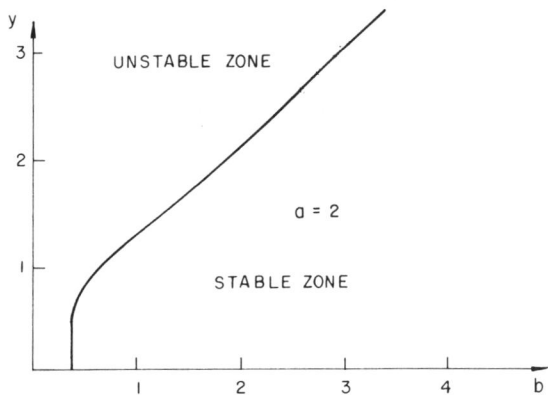

Fig. 9 Stable and unstable zones of a rectangle of length a and breadth b.

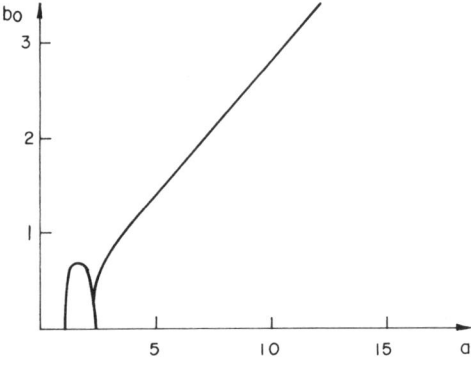

Fig. 10 The breadth, below which no stable equilibrium exists as a function of a.

depends on the space between the conductors in relation to the radius of the circular section (Fig. 6).

It is possible, in correlation, to observe the evolution of the angle corresponding t the maximum magnetic pressure around the cylinder in relation to this same geometry of the inductor (Fig. 7).

The displacement corresponding with the unstable equilibrium depends on the space between the conductors (Fig. 8). For the curves as a whole, one will notice special behaviors for inductor geometries in the same magnitude order as the cylinder diameter.

The Rectangle Case. This section deals with the unstable equilibrium points when the geometry of the conductors becomes rectangular. Given the length a, we calculate the displacements corresponding to an unstable equilibrium in relation to the breadth b of the rectangle (Fig. 9).

It is interesting to note that when b is below a certain value, the unstable equilibrium does not exist any longer. That is to say, the central point becomes an unstable equilibrium and shaping is no longer possible. Finally, one can notice that the value of b, for which stability no longer exists, varies as a in a nonlinear way (Fig. 10).

Conclusion

The calculated configurations (for square patterns) are stable for small perturbations. At every step, we must be sure that the transformation is a conformal mapping in the external part of the metal. If the solutions of $dz/d\xi = 0$ are located outside the unit disk, the method does not work. So the case without surface tension cannot be touched.

The study of the stability for disturbances of finite amplitude gives some interesting results in the strong surface tension case -- more particularly, in the situation of rectangular patterns where a cusp catastrophe is displayed. When the surface tension is weak, the method developed in Eq. (5) can be used to have informations about the sign of the second derivative of the energy and its value.

The experiments are continuing, and the apparatus will be fitted up to allow measures in all directions around the jet. Moreover, the reaction of the induced currents on the inducing ones have to be taken into account.

Appendix

Square Pattern

K = 0.100; A1 = 1.5; et N = 10

SHAPING OF LIQUID METAL CYLINDERS

K = 0.050; A1 = 1.5;   N = 10

K = 0.050; A1 = 1.3;   N = 10

K = 0.020; A1 = 1.500; et N = 10

K = 0.010; A1 = 1.5;   N = 10

K - 0.001; A1 = 2.0;   N = 10

where $K = \pi \gamma R_o / \mu_o I^2$, $A_1 = a/R_o$

Rectangular Pattern

K = 0.050; A1 = 1.50; Alfa1 = 40.00 deg; N = 10

K = 0.10;   A1 = 2.24; Alfa1 = 26.57 deg; N = 10

First-Order Approximation

$\varepsilon' = 0.32$

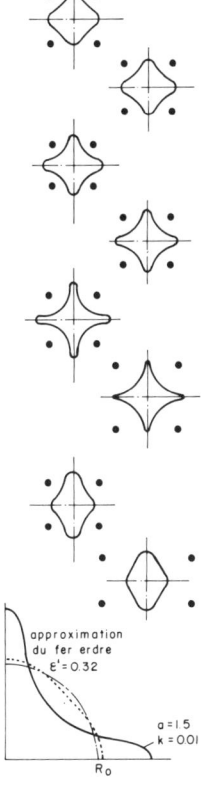

approximation du fer erdre
$\varepsilon' = 0.32$
a = 1.5
k = 0.01
R_o

References

[1] Etay, J., and Garnier, M., "Sur le Controle Electromagnetique des Surfaces," Journal de Mecanique Theorique et Appliquee, Vol. 1, No. 5, 1982.

[2] Shercliff, J.A., "Magnetic Shaping of Molten Metal Columns," Proceedings of the Royal Society of London, Series A 375, 1981.

[3] Brancher, J.P., Etay, J., and Sero-Guillaume, O., "Formage d'une Lame Metallique Liquide: Calculs et Experiences," Journal de Mecanique Theorique et Appliquee, (to be published).

[4] Sneyd, A.D., and Moffatt, H.K., "The Fluid Dynamics of the Process of Levitation Melting," Journal of Fluid Mechanics, Vol. 117, 1982.

[5] Brancher, J.P., and Sero-Guillaume, O., "Sur l'Equilibre des Liquides Magnetiques, Application a la Magnetostatique,", Journal de Mecanique Theorique et Appliquee, Vol. 2, No. 2, 1983.

Shield Effects in Continuous Electromagnetic Casting

R. Ricou* and C. Vives†
Centre Universitaire d'Avignon, Avignon, France

Abstract

The use of new local measurement techniques of velocity, magnetic field, current density, and phase difference by means of small-scale probes, which allow the experimental investigation of the flow in molten metal (up to 700°C) in the presence, or absence, of an induction magnetic field is summarily described. Next, these methods are applied to the study of electromagnetic and hydrodynamic phenomena in aluminum industrial processes, such as in electromagnetic castings, inside the sump of circular cross-section ingots. The important shield effect is also discussed as a function of the screen location by means of a mercury pool simulating the electromagnetic casting.

Introduction

The electromagnetic continuous casting process of circular and rectangular cross-section ingots of light metals and alloys has been used for a few years by principal producers of the Western world. This new process offers manifold advantages (improvement of the metal structure and of the tensile properties, very smooth surface). From the economic point of view, these properties are of great interest and, above all, allow the removal of the very costly scalping operation.

Paper presented at the Fourth Beer-Sheva Seminar on MHD Flows and Turbulence, Ben-Gurion University of the Negev, Beer-Sheva, Israel, Feb. 27-March 2, 1984. Copyright © 1985 by the American Institute of Aeronautics and Astronautics, Inc. All rights reserved.
  *Assistant, Faculté des Sciences, Laboratoire de Magnétohydrodynamique.
  †Professor, Faculté des Sciences, Laboratoire de Magnétohydrodynamique.

At first glance, the principle behind this process seems simple: The upholding effect of the conventional mold is now insured, without contact, by radial electromagnetic body forces located within the upper liquid part of the ingot. These forces are generated by a one-loop induction coil supplied with an electric current, the intensity and frequency of which are, usually, 5000 A and 2000 Hz, respectively.

A number of papers[1-3] point out the importance of the electromagnetic fringe effects due to the predominance of the characteristic horizontal dimensions of both the ingot and the inductor (the diameter, for instance) with respect to the bath and the coil heights. The Lorentz forces may be resolved into a potential component and a rotational component. The potential forces allow the upholding of the liquid metal height, ranging between 35 and 50 mm, and contribute to the shape of the meniscus. The vortical forces are the cause of an intense electromagnetic stirring.

In the attempt to damp the fluid motion, it is necessary to add an electromagnetic screen, which consists of a solid metal ring internally cooled by water. The screen is placed in such a way that it partially shields the meniscus from the action of the electromagnetic field generated by the inductor; it may be considered as a complementary coil through which the eddy currents are practically in opposite phase with respect to those of the actual inductor.

The commercial value of circular and rectangular cross-section blocks cast in electromagnetic molds is so much higher that the flatness of the ingot surface is better and the number of folds and pinholes per unit of surface is less.

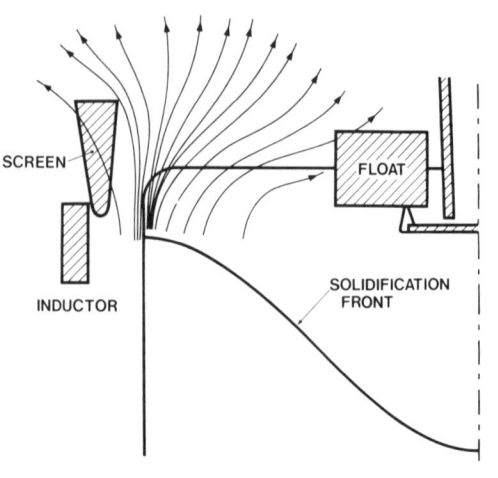

Fig. 1 Magnetic field lines.

An exact understanding of the role of the shield is necessary to eliminate these defects. Hence, we have entered upon a methodical experimental study of the effect of the screen on the electromagnetic and hydrodynamic parameters dealing with the electromagnetic casting.

The use of new local measurement techniques of velocity, magnetic field, current density, and phase difference, by means of small-scale probes, which allow the experimental investigation of the flow in molten metal (up to 700°C), is summarily described. Next, the important shield effect is discussed as a function of the screen location. Finally, the surface aspect of the solidified ingot and the reversing of the direction of the flow are both connected with the modification of the magnetic field pattern.

Measurement Techniques

Velocity Measurements

Velocity measurements have been made with an incorporated magnet probe; the working principle, the calibration technique and the behavior at high temperatures have already been related with details in two papers (Refs. 4 and 5) to which we refer the reader. Consequently, we confine ourself to summarizing its main properties. This sensor allows, by simple voltage measurements, the determination of the local and instantaneous values of the magnitude and direction of velocity up to about 700°C.

Current Density Measurements

These measurements are founded on the application of the Ohm's law for moving media:

$$\vec{J} = \sigma(\vec{E} + \vec{u} \times \vec{B})$$

which reduces to

$$\vec{J} = \vec{E} = -\sigma \frac{\partial \vec{A}}{\partial t}$$

because the displacement currents are negligible for the frequency (2000 Hz) that enters into our study. The same holds true for the product $\vec{u}\vec{B}$.

In the case of an induction electromagnetic field of axial symmetry, a potential sensor collects between its two electrode tips, AA', the emf caused by the circulation along an electric current line,

$$e_{aa'} = \int_A^{A'} \vec{E} \cdot \vec{dl} = E_\theta \ AA'$$

The rms $\varepsilon_{AA'}$ of $e_{aa'}$ being measured with a microvoltmeter, it is easy to obtain the rms of the electric current density,

$$J = \sigma \varepsilon_{AA'}/AA'$$

and of the vector potential, $A = J/\sigma\omega$.

The probe previously calibrated was made of two wires, the sheaths and cores of which, insulated by magnesia, were made of stainless steel with diameters of 1 and 0.35 mm, respectively. The electrode tips were about 1 cm apart.

Magnetic Field Measurements

The principle consists in measuring, with a microvoltmeter, the rms ε of the emf $e(t)$ that occurs at the ends of a small coil with total area S, placed perpendicularly to a time-varying magnetic field $B(t)$ (or to the components of one) of angular frequency ω. From Lenz's law,

$$||e(t)|| = \omega s \ ||B(t)||$$

or

$$\varepsilon = K' \ B$$

where K' is a calibration factor and B the rms of the magnetic field.

The coil was made up of n turns of wire; the sheaths of 0.25-mm exterior diameter were of stainless steel, and the wire core of chrome steel. The electric insulation was assured by compacted magnesia. For a frequency of 2000 Hz, the sensitivity was of the order of 2×10^{-4} T per mV, corresponding to a total area of 4 cm^2. The convolution diameter never exceeded 4 mm, and the height was about 1.5 mm. In the typical case of an electromagnetic field of axial symmetry, the coil yielded the vertical component B_z for a horizontal position and the radial component B_r for a position both vertical and perpendicular to a radius.

Phase Difference Measurements

A phasemeter (Bruël & Kjaer, type 2971) allows the measurement of the phase angle between the periodic param-

SHIELD EFFECTS IN CASTING

eters B_r, B_z, and J. Depending on the case, we operated either directly (for instance, between B_z and J) or throughout a fixed reference coil situated in the vicinity of the inductor and collecting the magnetic field leaks.

Measurements in the Sump of Aluminum Alloy Ingots

Experimental Conditions

The 345-mm-diam aluminum alloy 7049 billets were cast at a lowering rate of 50 mm/min, and the height of the liquid metal maintained by the electromagnetic force field was 35 mm. A single-phase electric current of 4970 A and 2000 Hz frequency, with a potential difference of 29 V, flowed through a single-induction coil; the electric power supplied by the alternator was 28.8 kW.

Electromagnetic Parameter Measurements

The magnetic field rms B is determined in magnitude and direction by the B_r and B_z components measured within air and liquid metal, and also by the phase angle between them. Hence, the magnetic field variation is an elliptic vibration. It follows from this that the direction of the magnetic field and the electromagnetic force are time-varying dependent. The elliptic character of the vibration has little repercussion under the meniscus, on account of the sharp predominance of B_z, with respect to B_r, in this area. On the other hand, inside the meniscus zone, the ellipticity effect, though more pronounced, is not very important. Subsequently, we agree to choose the major axis of the ellipse to represent the rms B of the magnetic field.

In Fig. 1, the modification of the inclination of the magnetic field lines may be seen comparatively to the classical pattern, which is produced by a circular winding without a screen. In general, we find, owing to the shield effect, that B_r, B_z, and J variations as a function of z are greater in the meniscus area (i.e., in the vicinity of the screen) and weaker inside the vertical zone of the electromagnetic skin situated under the screen.

In this case, where the skin depth $\delta = (2/\omega\sigma\mu)^{1/2} =$ 5.6 mm is small, with respect to the diameter of the billet, measurements show that, in the electromagnetic skin located under the meniscus, the instantaneous local values of the current density and of the axial component of the magnetic field are given by the following expressions:

$$J(r,z,t) \simeq J_{o(z)} e^{-r/\delta} \cos \left| \omega t - \frac{r}{\delta} \right| \qquad (1)$$

$$B_z(r,z,t) \simeq B_{o(z)} e^{-r/\delta} \cos |\omega t - \frac{r}{\delta} - \phi_{J,B_z}(r,z)| \quad (2)$$

Where $B_{o(z)}$, $J_{o(z)}$ are determined after extrapolation of the experimental results and with $\phi_{J,B_z}(r,z) \simeq 45$ deg.
Now, taking into account the J, B_z and $\phi(J,B_z)$ measurements conducted on the 340-mm-diam billet, we find that, after graphical integration over the volume of the electromagnetic skin, the height of aluminum alloy corresponding to the electromagnetic pressure is about 39 mm, whereas the height of liquid metal held in this case is, in fact, 35 mm.

The graphical integration of $I = \iint J \, ds$ and $W = \iiint J^2/\sigma \, dv$ (respectively, over the cross-sectional area and the volume of the electromagnetic skin) allows both the estimation of the electric current (about 750 A) and the electric power dissipated inside the sump by Joule effect (about 800 W). Finally, the power consumption inside the solid part of the metal situated just under the mushy zone may be estimated at about 1 kW.

Velocity Measurements

Figure 2 presents an illustration of local velocities taken in a temperature range from 630 to 670°C inside the sump of the 340-mm-diam billet of aluminum alloy 7049 cast in electromagnetic mold and shows the presence of two main loops. The smaller cell of about 30-mm-diameter is driven by the vortical electromagnetic forces. Near the vertical liquid wall, the fluid flows downwards and the maximum velocities are of 12 cm/s. The flow is rather stable inside these two principal vortices. On the other hand, in the areas situated above the stream jet and between the two cells, the turbulence occurs markedly.

The Shield Effect

Experimental Rig

The Study of the influence of the distance between the screen and the inductor on the liquid flows inside the sump is important, but rather difficult, especially in extreme cases. Indeed, in the absence of a screen or when the screen is deeply driven in the inductor loop, the molten metal overflows as it is too much stirred or poorly held by the electromagnetic forces. Furthermore, the hydrodynamic effect, due to the liquid metal inlet through the dispatcher, is sufficient to partially conceal the particular contribution of the electromagnetic forces, especially in the sump core.

SHIELD EFFECTS IN CASTING 673

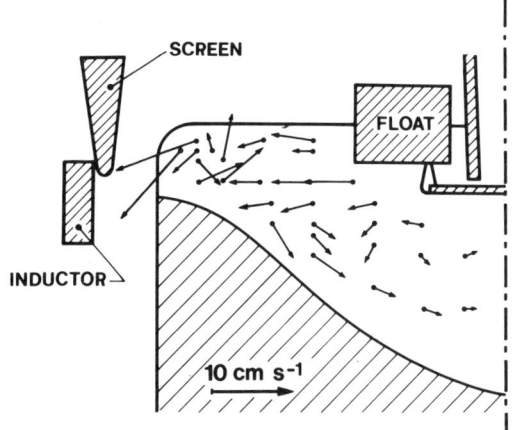

Fig. 2 Velocity measurements in the sump of an electromagnetic mold (aluminum alloy billet with a 340 mm diameter).

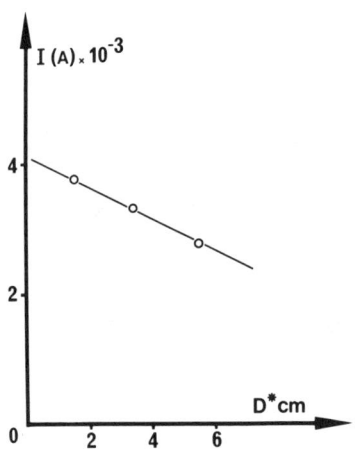

Fig. 3 Amperage inside the screen as a function of the distance D.

In order to exclusively examine the shield effect, we use a laboratory model consisting of a mercury pool. This tank is constituted by a 320-mm-diam billet made of solid stainless steel in which is made a cavity of an identical shape to the one represented in Fig. 2. The upper part of the cavity was capped by a glass fiber crown which reproduced the meniscus as well as the vertical liquid free surface contained by the electromagnetic forces. The screen was made from aluminum and an electric current of 50 Hz frequency and 2000 A-turns was flowing through the inductor.

Electric Parameter Measurements Inside the Screen

Graphical integrations of the current density J, over the cross-sectional area of the solid aluminum ring, allow the estimation of the electric current I flowing in the screen, for different values of the distance D* (D* being the difference of level between the bottom of the shield and the midheight of the actual inductor).

Current density measurements were carried out by means of a two electrode potential-difference sensor. This probe was either inserted into thin slots made inside the shield or traveled upon the surface of the ring.

The linearity between the electric current I and the distance D* is quite marked in Fig. 3, where it can be seen that I increases as D* decreases (i.e., when the screen is drawn near the inductor). Moreover, experiments show that the phase difference between the electric currents passing through the inductor and the shield lies within the range 170-190 deg, whatever D*.

Magnetic Field Patterns

Plotting of the B_r and B_z components in a radial cross section, as well as the phase angle between them, allows the display of magnetic field maps.

Figure 4a corresponds to the case where the measurements were exclusively made in the presence of the actual inductor, as it is well known the magnetic field is a linear vibration here.

Figures 4b-d show examples of magnetic field patterns solely plotted in the presence of the coil and the screen and call for the following remarks:

1) The magnetic field decreases as the proximity of the screen increases, with respect to the coil.

2) The magnetic field is generated by the electric currents which are flowing both through the actual inductor and the shield; the corresponding current densities, shifted between them in the space, are also out of phase in time. So, a rotating magnetic field may locally occur, the understanding of which is similar to the one given to those produced by the polyphase induction devices.

3) The magnetic field is oriented in the direction of the screen, in its vicinity.

A magnetic field pattern, plotted without screen and in the presence of the stainless steel block (the cavity of which being not filled with mercury), is sketched in Fig. 5. It may be seen that, once again, the magnetic field is an elliptic vibration; moreover, the field lines (represented

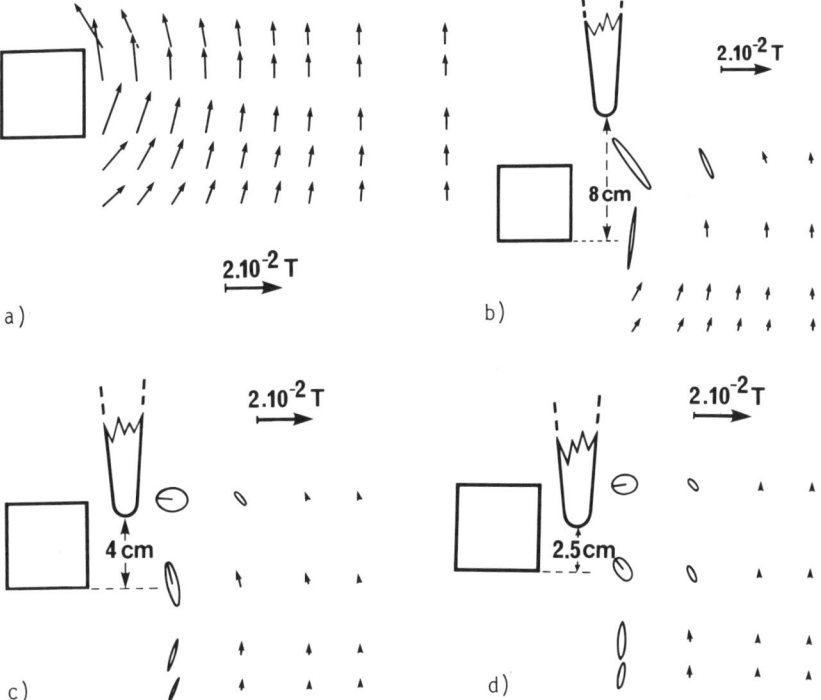

Fig. 4 Magnetic field patterns plotted in air: a) inductor without screen; b, c, d) with inductor and screen.

by the major axis of the ellipses) are now directed towards the axis of symmetry of the device. The role of the eddy currents which travel through the solid steel block is obviously analogous to those which are running inside the screen in the case of Figs. 4b-d.

The magnetic field patterns sketched in Fig. 6 have been plotted for different locations of the screen (defined by the difference of level D between the bottom of the inductor and of the shield), the cavity of the billet being always filled up with mercury. In the general case, the magnetic field must be considered as generated by all the electric currents which travel respectively through the inductor, the screen, the solid stainless steel block, and the mercury.

It may be seen in Fig. 6a that the magnetic field, plotted without screen, decreases from the wall to the center of the billet; moreover, the field lines (represented by the major axis of the ellipses) are sloped towards the axis of symmetry of the ingot. As in the induction furnace,[6] the time mean body force has a rotational part (principally ver-

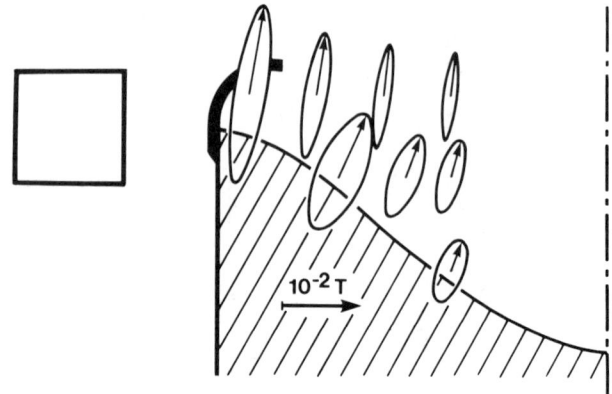

Fig. 5 Magnetic field pattern plotted in air in the absence of screen and molten metal.

tical) that is responsible for the stirring effect. The radial component of the $(\overline{JB})$ body force being centripetal (upholding effect), it follows from this that the vertical components F_z are oriented downwards; on account of the skin effect, the F_z components are strongest on the periphery of the bath and hence may contribute to develop a descending flow along the meniscus.

When the lower end of the screen is located at 8 cm of the bottom of the inductor (Fig. 6b), it may be seen that the magnetic field decreases mostly inside the sump core; on the other hand, the inclination of the field lies towards the axis of symmetry of the ingot decays.

In the case of Figs. 6c and 6d, the screen is more inserted into the actual inductor. Inside the molten metal, the magnetic field is everywhere divided by about 4, except in the peripheral zone, where the action of the screen prevails on account of the small distance which separates it from the meniscus. Above all, it is important to note the change of sign of the force field slopes, which are here directed towards the screen in the meniscus region. Therefore, the vertical components F_z are now oriented upwards and have a tendency to give rise to an ascending flow along the meniscus surface.

Velocity Patterns

Figure 7a is relative to the case where the screen is far away from the inductor and reveals the presence of a single loop, which completely fills up the half cross-section of the pool. In agreement with the comments expressed

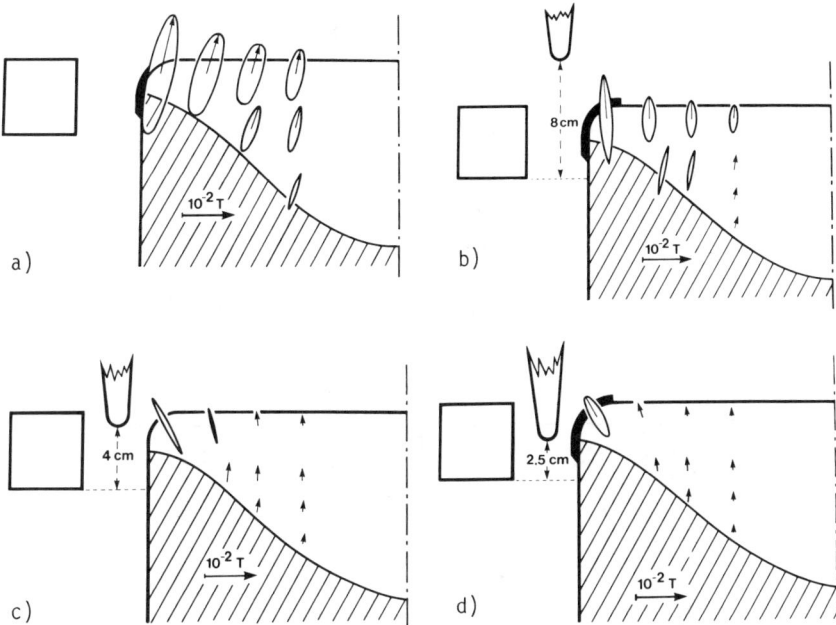

Fig. 6 Magnetic field patterns plotted in molten metal: a) with shield removed; b, c, d) for different locations of the shield.

after inspection of Fig. 6a, it is seen that the liquid flows downwards along the meniscus and upwards in the center region of the billet. Moreover, the eye of the vortex is located near the horizontal free surface and at a distance from the periphery of the order of the skin depth (71 mm).

When the bottom of the screen is on a level with the top of the inductor (Fig. 7b), it appears that the single cell, shown before in Fig. 7a, continues to exist. However, the local velocities are notably decreased, particularly in the vicinity of the meniscus and near the solid-liquid interface.

Figure 7c corresponds to the case where the screen is penetrated at midheight of the inductor. In the meniscus zone, a new loop can be seen of about 5 cm width (i.e., less than δ), where the flow is now ascending along the vertical liquid wall; this finding confirms the predictions expressed on observation of Figs. 6c and 6d. On the other hand, the direction of the stream lines remains unchanged inside the main vortex. Nevertheless, the comparison with the velocity field sketched in Fig. 7a points out that the braking effect becomes very important, the velocities being divided within a ratio ranging from 5 to 10.

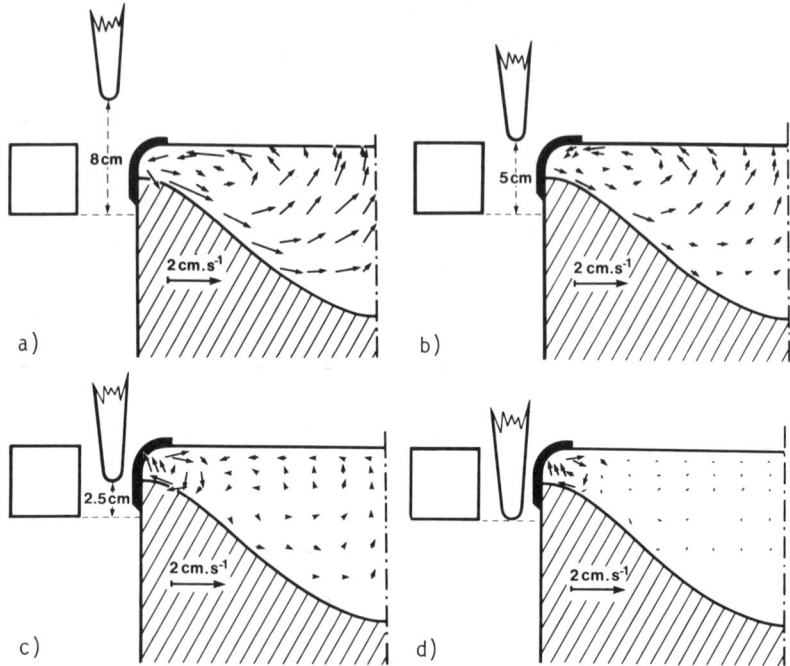

Fig. 7 Velocity maps plotted inside a mercury pool for different locations of the screen.

Finally, in the case of a total penetration of the screen (Fig. 7d), the fluid flow inside the sump is practically limited to the small peripheral loop.

The analogy between these experiments and actual electromagnetic castings is obviously far from being complete because the frequencies of the electric current, as well as both the electrical (σ) and the mechanical (ρ, ν) parameter values of the aluminum and mercury, are very different. However, it is known that the dimensions of the loop of electromagnetic origin decrease with the skin depth, i.e., when the frequency increases.[1-3] Thus, the vortex which filled the entire half cross section, in the cases of Figs. 7a and 7b, have a width of only 35 mm in the experimental conditions of the Fig. 2 (i.e., aluminum alloy, 2000 Hz frequency, 5000 A-turn and the screen only driven in at 6 mm with respect to the inductor top).

Again, in the experimental conditions relative to an aluminum ingot of 1100x300 mm^2 cross-section and with a sufficient driving of the screen into the inductor, it was possible to visually detect upward flows upon the meniscus surface establishing the presence of the small loop shown in Figs. 7c and 7d. Unfortunately, the velocity force field pattern inside this vortex was not able to be completely plotted, because its width was of the order of magnitude of the diameter (4 mm) of the probe. Nevertheless, the reversal of the direction of rotation has been discovered with certainty, by means of the incorporated magnet probe, in the vicinity of both the solidification front and vertical liquid wall.

The variations of the screen's position have a prominent effect upon the surface aspect of the ingot. In the case of a block of 1100x300 mm^2 cross section, the screen location yielding the better results corresponds with a transition rate between the upward and downward flows along the vertical liquid wall, i.e., also to a minimum of the magnitude of the velocities inside the meniscus area.

Conclusions

Obviously, the experimental results reported here do not aim to exhaust this complicated subject, but, rather, they allow one to consider a methodical experimental study of both the electromagnetic parameters and the fluid flows within the sump of ingots cast in electromagnetic molds, and also of their influence upon the crystal structure and the surface state of the as cast metal.

References

[1] Getselev, Z. N. and Martynov, G. I., "Calculation of the Main Electromagnetic Parameters of an Apparatus for Ingot Shoping During Continuous Casting," Magnetohydrodynamics, No. 4, 1973, pp. 135-138.

[2] Getselev, Z. N. and Martynov, G. I., "Calculation of the Velocity Induced in the Liquid Phase of a Casting by Electromagnetic Forces," Magnetohydrodynamics, No. 2, 1975, pp. 106-111.

[3] Getselev, Z. N., Kaeindel, D. A., Kaptilkin, A. A., and Martynov, G. I., "Experimental Study of Circulation of Liquid Metals in Electromagnetic Fields," Magnetohydrodynamics, No. 2, 1975, pp. 144-146.

[4] Ricou, R. and Vivès, Ch., "Local Velocity and Mass Transfer Measurements in Molten Metals Using an Incorporated Magnet Probe," International Journal of Heat and Mass Transfer, Vol. 25, No. 10, 1982, pp. 1579-1588.

[5] Vivès, Ch. and Ricou, R., "Velocity and Electromagnetic Parameter Measurements in the Sump of Aluminum Alloy Billets," Proceedings of the IUTAM Symposium, Cambridge. Metallurgical Applications of Magnetohydrodynamics, The Metal Society, London, 1982, pp. 24-32.

[6] Moore, D. J. and Hunt, J. C., "Electromagnetic Stirring in the Coreless Induction Furnace," Liquid-Metal Flows and Magnetohydrodynamics: Progress in Astronautics and Aeronautics, Vol. 84, AIAA, New York, 1983, pp. 359-373.

Investigation of the Turbulent Flow in an Induction Furnace Supplied with Various Frequencies

E. Taberlet* and Y.R. Fautrelle†

Institut National Polytechnique, Groupement d'Interet Scientifique, Madylam, Saint Martin d'Heres, France

Abstract

The present work deals with the numerical and experimental study of electromagnetic stirring in a mercury induction furnace supplied with various frequencies. The measurements of the mean motion show that, for a fixed coil current, stirring is maximum when the skin depth normalized by the pool radius is about 0.2 As for the turbulence, its properties are nearly homogeneous in the bath at low frequency, while the high-frequency case is characterized by an increase of turbulent fluctuations near the wall. Numerical predictions based on the k-ε model have been achieved. The results show that the confinement of the forces at high frequency leads to important inaccuracy. Such drawbacks may be suppressed by using proper wall functions.

Introduction

Electromagnetic stirring in induction furnaces has been investigated both theoretically and experimentally by many authors[1-11]. These studies have shown that there are essentially two important parameters in such a phenomenon:
1) The intensity of the inducing current that governs the magnitude of the velocity.
2) The frequency of the inductor current, or more precisely, the screen parameter R_ω, defined as

$$R_\omega = \mu \sigma \omega a^2 \quad \omega = 2\pi f$$

Paper presented at the Fourth Beer-Sheva Seminar on MHD Flows and Turbulence, Ben-Gurion University of the Negev, Beer-Sheva, Israel, Feb. 27-March 2, 1984. Copyright © 1985 by the American Institute of Aeronautics and Astronautics, Inc. All rights reserved.
*Engineer.
†Dr.

TURBULENT FLOW IN AN INDUCTION FURNACE   681

where μ, σ, f and a respectively denote the magnetic permeability, electrical conductivity of the metal, frequency of the inductor current, and radius of the pool. One may define from R_ω the so-called electromagnetic skin depth δ, namely,

$$\frac{\delta}{a} = (\frac{2}{R_\omega})^{1/2}$$

According to the value of R_ω, two flow regimes may be distinguished:
 1) The low-frequency case ($R_\omega < 1$), where the magnetic field penetrates in the whole bath, and where the velocity is an increasing function of the frequency (see Ref. 10).
 2) The high-frequency case ($R_\omega \gg 1$), where the magnetic field is confined in a thin electromagnetic wall layer and where the velocity is a decreasing function of the frequency (see, for example, Ref. 12).

The purpose of the present work is to investigate both experimentally and numerically the two asymptotic cases. The velocity measurements in a mercury induction furnace are detailed in a previous paper.[11]

We shall summarize here the main results and show how the numerical modeling (based on the k-ε turbulence model) must be adapted to take into account the skin effect.

The Experimental Apparatus

The Induction Furnace

The experimental device consists of a 200-mm diameter mercury pool located in a 15-turn coil (Fig. 1). The inductor is supplied with an electric power source (0-20kW) that can provide various frequencies in the range between 50 and 5000 Hz. In that condition, the two important parameters are $R_\omega = 2(\delta/a)^{-2}$ and $Re = u_a a/\nu$, where $u_a = B_0/\sqrt{\mu\rho}$ is the Alfven speed, B_0 is a typical induction value, and ρ is the density. The value of B_0 is related to the coil intensity I (maximum value) by

$$B_0 \sim \mu NI/H_c (1 + 0.88 \, a_c/H_c)$$

where a_c, H_c respectively denote the radius and the height of the coil. The values of R_ω and R_e are in the range

$$3 < R_\omega < 400 \quad 6 \times 10^4 < R_e < 3 \times 10^5$$

The velocity field is investigated by means of two complementary devices. Firstly, a quartz-coated hot film

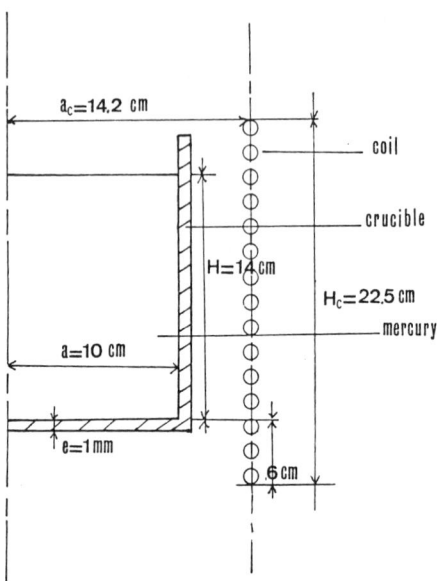

Fig. 1 Scheme of the apparatus.

probe is used with a constant temperature anemometer to get the mean velocity modulus, turbulent intensity, and time spectra. Secondly, the mean velocity streamlines and direction are obtained in a straightforward manner by locating the thermal wake close to the hot film. The thermal wake is well defined beside the great thermal diffusivity of mercury.

The Numerical Model

The predictions of the turbulent motion require the use of turbulence models. In the case considered here, turbulence is not affected either by the electromagnetic damping effects or the fluctuating part of the forces (see, for example, Ref. 11). Therefore, a classical turbulence model may be used. Among the various models, the k-ε one[13] remains the most convenient for such realistic geometries. The general procedure consists in first computing the electromagnetic forces by solving Maxwell equations that are decoupled from the motion. The Lorentz forces are then introduced in the momentum equations, which are simultaneously solved with two additional transport equations for k and ε. In such models, the influence of the boundary conditions (i.e. the wall functions) are of primary importance for the accuracy of the results. In the small skin depth case, the confinement of the electromagnetic forces requires special care near the wall. In the present work, we have studied the behavior of the k-ε model with respect to two types of boundary conditions. The main equations are given in Appendix A.

The first type, referred to hereafter as a type I model, consists in using boundary conditions based on the classical wall functions. Those wall functions are deduced from the universal velocity distribution for turbulent boundary layers without pressure gradients and body forces, i.e., the logarithmic law[13].

The secon type of boundary conditions (type II model) are obtained from the local equilibrium between the Reynolds stresses τ and the electromagnetic forces in the limit δ/a→0. The calculations are detailed in Appendix B. Assuming that the stresses are negligible far from the wall, τ may be related to the Lorentz forces as follows:

$$\tau = \rho \; \frac{\delta}{2a} \; u_a^2 \; AA' \; e^{-2y/\delta} \qquad (1)$$

where x, y, and A(x) respectively denote the curvilinear coordinate along the boundary of the pool, the transverse coordinate, and the nondimensional vector potential along the boundary. Relation (1) yields the wall stress τ_w, namely

$$\tau_w = \rho \; \frac{\delta}{2a} \; u_a^2 \; AA' \qquad (2)$$

The expression of τ_w is used to calculate the values of k and ε near the wall. It is clear that the type II model is valid only for large R_ω.

The equations have been solved by means of a finite-difference method. The mesh used in the computations was staggered, rectangular, and uniform. It is not easy to quantify the effects of the numerical diffusion for a uniform grid. Thus, various meshes have been used in order to test the convergence of the numerical scheme (22x22, 32x32, and 50x50). It is likely that the effects of numerical diffusion are mainly localized near the wall regions because of the local decay of k. The choice of a 32x32 point mesh corresponds to a compromise between the stabilization of the numerical scheme and the computation time.

Discussion

The Experimental Results

The measurements of the mean and turbulent velocities are detailed in Ref. 11. Nevertheless, the main results are

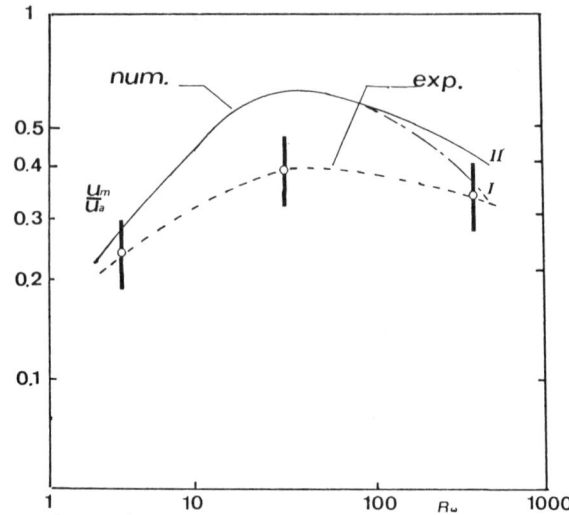

Fig. 2 Evolution of the mean velocities with respect to the frequency. (The velocities are normalized by the Alfven speed u_a). u_m/u_a: Maximum mean axial velocity component (the bars indicate the absolute error on those values).

summarized here. They concern the evolution of stirring with respect to the frequency of the supply currents.

As for the mean flow, the important feature is the experimental evidence of the maximum stirring for a fixed current intensity when the skin depth is such that $\delta/a \sim 0.2$. Figure 2 shows the evolution of the characteristic mean velocity against R_ω, i.e. the frequency.

As far as the turbulence is concerned, one may distinguish two regimes according to the frequency:

The Low-Frequency-Regime ($R_\omega \leq 1$). When the skin depth δ is of the order of the pool radius (e.g., $\delta/a = 0.7$), the turbulent properties of the flow, (i.e. the turbulent kinetic energy, the turbulent dissipation rate, and the integral scale) are nearly homogeneous in the whole bath (see Figs. 3-6, and 11). This behavior may be interpreted as the balance constraint between the driving forces and the Reynolds stresses. Since the electromagnetic forces are distributed in the whole bath, the Reynolds stresses intensity must be sufficient to equilibrate the Lorentz forces everywhere in the pool.

The High-Frequency-Regime ($R_\omega \gg 1$). For weak skin depth (e.g., $\delta/a = 0.07$), it is shown experimentally that there exist two regions, namely, the bulk and the electromagnetic layer. In the central region, turbulence properties are similar to the previous case (i.e. $R_\omega < 1$). It

TURBULENT FLOW IN AN INDUCTION FURNACE 685

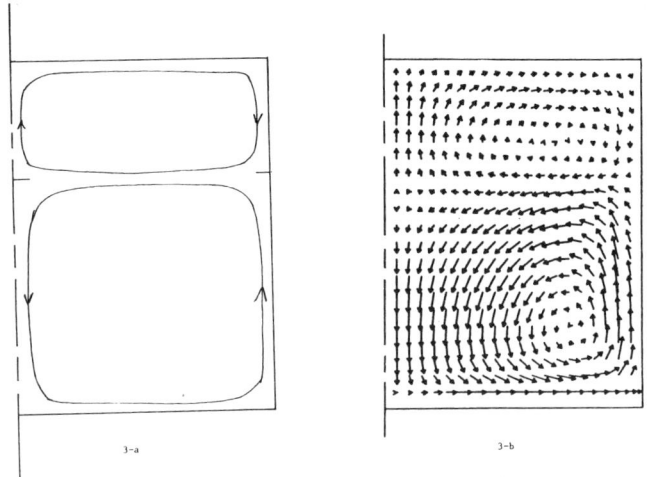

Fig. 3 Flow pattern at low frequency ($R_\omega = 3.9$). a) Experimental; b) Numerical.

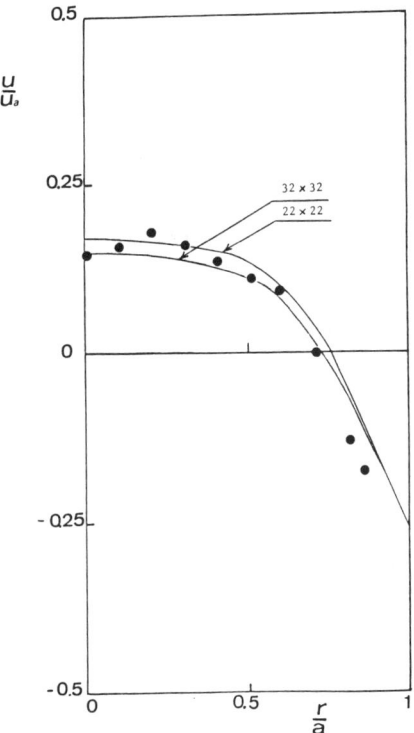

Fig. 4 Radial profile of the axial velocity component through the eye of the lower vortex at low frequency ($R_\omega = 3.9$; the velocity is normalized by the Alfven speed U_a. •:Experimental data; ---:Numerical curves.

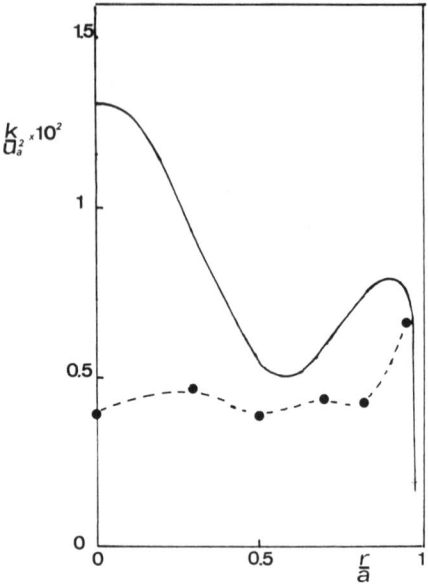

Fig.5 Radial profile of the kinetic energy of the turbulence through the eye of the vortex at low frequency ($R_\omega = 3.9$)
•: Experimental data;
—: Numerical curve.

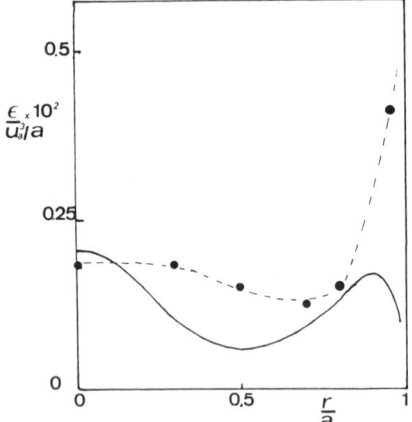

Fig.6 Radial profile of the turbulen dissipation rate through the eye of the vortex at low frequency ($R_\omega = 3.9$).
•: Experimental data;
—: Numerical curve.

is noteworthy to observe near the wall a significant growth of K and ε and a decay of the integral scale L, as shown in Figs. 7-10. The growth of k and ε is consistent with the need for sufficient Reynolds stresses to balance the Lorentz forces in the electromagnetic layer.

Numerical Results

The numerical model has been compared with the two experimental limit cases, namely, $\delta/a = 0.7$ (the low-frequency regime) and $\delta/a = 0.07$ (the high-frequency regime).

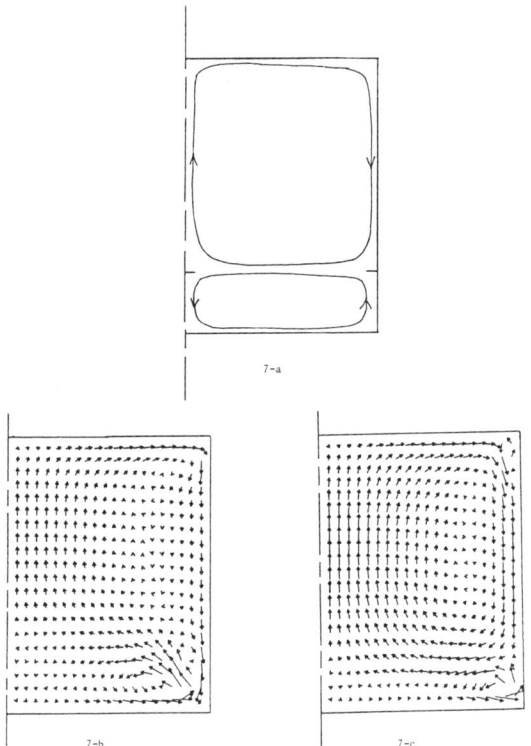

Fig. 7 Flow pattern at high frequency (R_ω = 372). a) Experimental; b) numerical with classical wall functions; c) numerical with new wall functions.

In the low-frequency regime, the results of the type I model are in fairly good agreement with the experimental ones, as shown in Figs. 3-6. It is noteworthy to observe that k and ε agree qualitatively with the experimental data. Indeed, in this case, the wall region does not intervene fundamentally in the force balance of the flow. Consequently, the classical wall functions based on the logarithmic law are quite valid. Note, however, that the numerical value of k near the axis is much greater than the experimental one. This phenomenon has not been clearly explained. Figures 4 and 8 also show the evolution of the mean velocity profiles for various mesh sizes. As expected, the high-frequency case requires a finer mesh than the low-frequency one. Note that stabilization of the profiles is almost reached for the 32x32 point grid.

In the high-frequency regime, both type I and II models have been tested. It is clear from Figs. 7-10 that modifications of the boundary conditions lead to significant im-

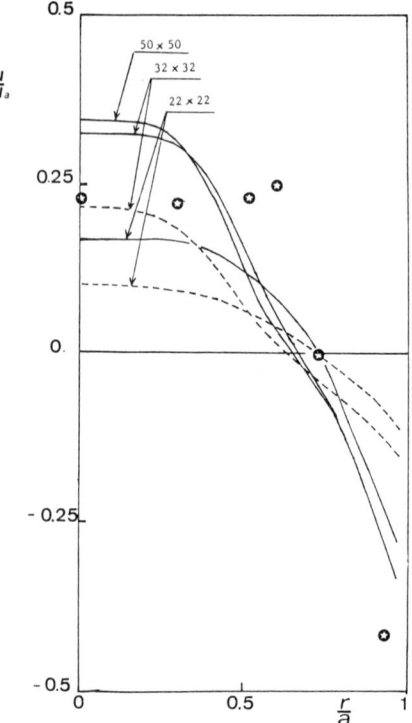

Fig.8 Radial profile of the axial velocity component through the eye of the upper vortex at high frequency ($R_\omega = 372$, the velocity is normalized by the Alfven speed u_a).
○: Experimental data; ---: numerical curves with classical functions; —: numerical curves with new wall functions.

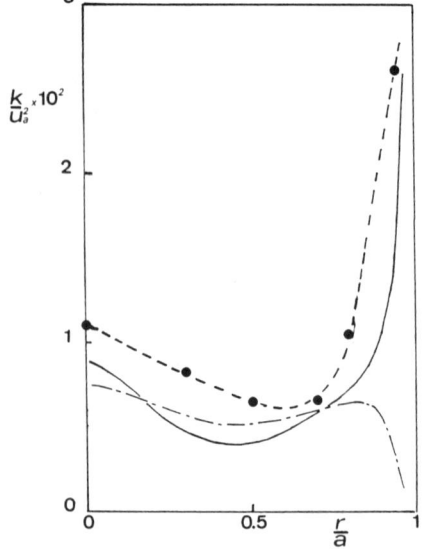

Fig.9 Radial profile of the kinetic energy of the turbulence through the eye of the vortex at high frequency $R_\omega = 372$). •:Experimental data; -.-.-.: Numerical curve with classical wall functions; ——: Numerical curve with new wall functions.

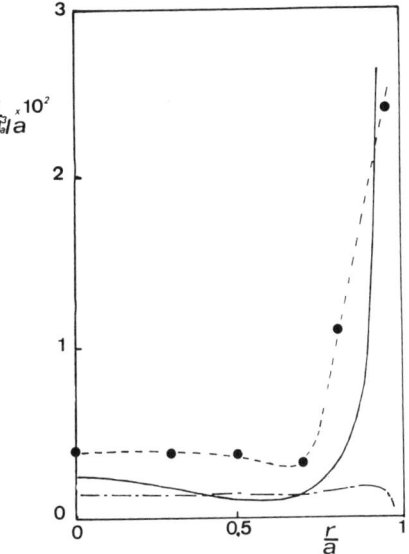

Fig.10 Radial profile of the turbulent dissipation rate through the eye of the vortex at high frequency ($R_\omega = 372$). •:Experimental data; -.-.-:Numerical curve with classical wall functions; ———: Numerical curve with new wall functions.

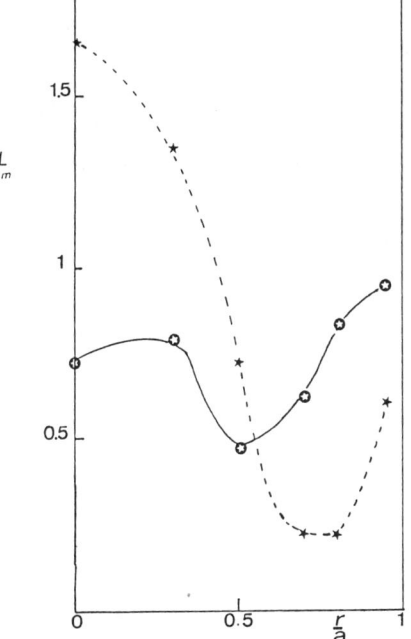

Fig.11 Radial profiles of the integral scale through the eye of the vortex. ⊙:Experimental data for low frequency $R_\omega = 3.9$; ★: Experimental data for high frequency $R_\omega = 372$.

provements of both the mean and turbulent motions, especially near the wall region. Both the mean motion patterns and the turbulence properties are noticeably improved.

The interpretation of such results is obvious. The k-ε is very sensitive to the choice of the boundary conditions and the shear stresses that determine the boundary values of k and ε. Without special mesh refinement near the wall, the type II model avoids the use of the logarithmic law and thus allows a better estimate of the wall shear stress and, hence, the turbulent parameters. The last result is very important for the knowledge of the wall turbulent transfers. Note that Eq. (1) is no longer valid in the vicinity of the corner regions. However, the typical length of that region is $O(\delta)$, and the corresponding error remains small.

In the limit case $\delta/a \to 0$, the electromagnetic body forces may be replaced by a wall stress. It is likely that type II model might be used in that case without any special mesh refinement.

Appendix A: The Governing Equations

Let $\vec{u}$, and P be the mean velocity field vector and pressure field respectively. In the Boussinesq approximation, they are governed by the mean motion equations, namely

$$\rho (\vec{u}\cdot\vec{\nabla}) \vec{u} + \vec{\nabla}p = \vec{F}_{em} + \vec{\nabla}(2\mu_t \vec{e}) \quad (3)$$

$$\vec{\nabla}\cdot\vec{u} = 0 \quad (4)$$

where $\vec{F}_{em}$, μ_t, and $\vec{e}$ respectively denote the electromagnetic body forces, turbulent viscosity, and deformation rate tensor. In the so-called k-ε model, the value of μ_t is related to the turbulent kinetic energy k and the turbulent dissipation rate ε by[13]

$$\mu_t = \rho C_\mu k^2/\varepsilon$$

where $C_\mu = 0.09$ is a constant.

The determination of k and ε requires the solution of two additional transport equations, namely,

$$\vec{u}\vec{\nabla}\phi = \vec{\nabla}\left(\frac{\mu_t}{\rho\sigma_\phi} \vec{\nabla}\phi\right) + S_\phi \quad \phi=k,\varepsilon \quad (5)$$

$$S_k = P - \varepsilon \quad \text{with} \quad P = (\mu_t/\rho)(e_{ij} e_{ij})$$

$$S_\varepsilon = (C_{1\varepsilon} P/\varepsilon - C_{2\varepsilon})\varepsilon^2/k$$

e_{ij} denoting the components of $\vec{e}$.

The values of the various constants are gathered in the following table:

$C_{1\varepsilon}$	$C_{2\varepsilon}$	σ_k	σ_ε
1.44	1.92	1.0	0.67

Appendix B: The Wall Shear Stress

Let u and v be the components of the mean velocity in a local frame (x,y), x and y respectively being the curvilinear coordinate along the boundary of the pool and the transverse coordinate. Near the wall, the Prandtl approximations hold[7] and the boundary layer equations are

$$\rho \left(u \frac{\partial u}{\partial y} + v \frac{\partial u}{\partial y} \right) + \frac{dP}{dx} = \rho \frac{u_a^2}{a} AA' e^{-2y/\delta} + \frac{\partial \tau}{\partial y} \quad (6)$$

$$\vec{\nabla} \cdot \vec{u} = 0 \quad (7)$$

where A(x) is the nondimensional vector potential along the boundary[7]. Assuming that $\tau \to 0$ as $y \to \infty$, the boundary conditions are

$$y \to \infty \quad u = U(x) \quad P = \rho (U^2/2) + a \text{ const}$$

$$y = 0 \quad u = v = 0 \quad (8)$$

Furthermore, let us assume that

$$\frac{u}{u_a} = 0 \left[(\frac{\delta}{a})^{1/2} \right] \quad (9)$$

The above assumption stems from an order of magnitude calculation given by Lillicrap and Moore[14] and may be justified a posteriori. Owing to the boundary conditions (8), it is clear that in the limit $\delta/a \to 0$ the right-hand side of Eq. (6) is of order of δ/a. The zeroth-order balance therefore is

$$\rho \frac{u_a^2}{a} AA' e^{-2y/\delta} + \frac{\partial \tau}{\partial y} = 0 \quad (\frac{\delta}{a} u_a^2) \quad (10)$$

The turbulent stresses are obtained from the integration of (10), namely,

$$\tau = \tau_w e^{-2y/\delta} \quad \text{with} \quad \tau_w = \rho u_a^2 \frac{\delta}{2a} AA' \quad (11)$$

Note that, if we assume that the mean velocities are of order of the friction velocity u_* defined from τ_w by

$$u_* = (\tau_w/\rho)^{1/2} \qquad (12)$$

the estimate (9) is consistent with (12).

References

[1] Sneyd, A., "Generation of Fluid Motion in a Circular Cylinder by an Unsteady Applied Magnetic Field," Journal of Fluid Mechanics, Vol. 49, No. 4, 1979, pp. 817-827.

[2] Tarapore, E., and Evans, J.N., "Fluid Velocities in Induction Melting Furnaces, Part I: Theory and Laboratory Experiments," Metallurgical Transactions B, Vol. 7, September 1976, pp.343-351.

[3] Tir, L.L., "Features of Mechanical Energy Transfer to a Closed Metal Circuit in Electromagnetic Systems with Azimuthal Currents," Magnitnaya Gidrodinamika, Vol. 12, No. 2, 1976, pp. 100-109.

[4] Mikelson, Y.Y., Yakovitch, A.T., and Pavlov, S.I., "Numerical Investigation of Averaged MHD-Flow in a Cylindrical Region with the Adoption of Working Hypotheses for Turbulent Stresses," Magnitnaya Gidrodinamika, Vol. 14, No. 1, 1978, pp. 51-58.

[5] Hunt, J.C.R., and Maxey, M.R., "Estimating Velocities and Shear Stresses in Turbulent Flow of Liquid Metals Driven by Low Frequency Electromagnetic Field," Proceedings of the Second Bat Sheva Seminar on MHD Flows and Turbulence, edited H. Branover, Israel University Press, Jerusâlem, 1980, pp. 249-270.

[6] Moore, D.J., and Hunt, J.C.R.,"Electromagnetic Stirring in the Coreless Induction Furnace," Proceedings of the third Beer Sheva Seminar on Liquid Metal Flows and Magnetohydrodynamics: Progress in Astronautics and Aeronautics, Vol. 84, edited by H. Branover, P.S. Lykoudis and A. Yakhot, AIAA, New York, 1983, pp. 359-373.

[7] Fautrelle, Y., "Analytical and Numerical Aspects of the Electromagnetic Stirring Induced by Alternating Magnetic Fields," Journal of Fluid Mechanics, Vol. 102, January 1981, pp. 405-430.

[8] Cremer, P., and Alemany, A., "Aspects Experimentaux du Brassage Electromagnetique en Creuset," Journal de Mecanique Theorique Appliquee, Vol. 5, No. 1, 1981, pp. 37-50.

[9] Barbier, J.N., Fautrelle, Y.R., Evans, J.W., and Cremer, P., "Simulation Numerique des Fours Chauffes par Induction," Journal de Mecanique Theorique et Appliquee, Vol. 1, No. 3, 1982, pp. 533-536.

[10] Trakas, C., Tabeling, P., and Chabrerie, J.P., "Etude Experimentale du Brassage Turbulent dans le Four a Induction," Journal de Mecanique Theorique et Appliquee, Vol. 3, No. 3, pp. 345-370.

[11]Taberlet, E., and Fautrelle, Y., "Turbulent Stirring in an Experimental Induction Furnace," Journal of Fluid Mechanics, 1985 (in press).

[12]Koanda, S., and Fautrelle, Y.,"Modelization of Coreless Induction Furnace: Some Theoretical and Experimental Results," Proceedings of the IUTAM Symposium and Metallurgical Applications of Magnetohydrodynamics, Cambridge, Great Britain, 1982, edited by H.K. Moffatt and M.R.E. Proctor, the Metals Society, London, 1984, pp. 120-128.

[13]Launder, B.E., and Spalding, D.B., "The Numerical Computation of Turbulent Flows," Computer Method in Applied Mechanics and Engineering, North Holland, Amsterdam, 1974, pp. 269-289.

[14]Lillicrap, D.C., and Moore, D.J., "Electromagnetic Stirring in Coreless Induction Furnace," Proceedings of Electroheat for Metal Conference, Cambridge, Great Britain, 1982.

The Electromagnetic Force of Narrow Stirring Inductors

F.R. Block* and E. Julius[†]
Technical University, Aachen, Federal Republic of Germany[‡]

Abstract

An analytical method is presented for calculating the electromagnetic force density distribution of finitely wide stirring cylindrical inductors. The method takes into account the actual width of the inductor as well as the stator end-windings and the finite conductivity of the liquid rotor. The current distribution of the stirring inductor is described as a surface current density and expanded in Fourier integrals. If the eddy currents and therefore the vector potential in the melt have not only axial components, as in the simple case of an infinitely wide inductor with rotatory symmetry and axially directed exciting wires, the fundamental vector field equation cannot be separated in cylindrical coordinates. Nevertheless to solve the boundary problem analytically, the solenoidal vector potential is decomposed into two solenoidal components derived from two higher vector potentials only in the fixed axial direction. The calculated three-dimensional electromagnetic force density is represented as a function of the axial and radial coordinate, and the results are compared to that of an infinitely wide stator. The validity of the model has been checked by measurements on a narrow four-pole asynchronous inductor.

Nomenclature

$\underline{a}$ = inner radius of stator core
$\underline{A}$ = vector potential
$\underline{b}$ = width of end-windings
$\underline{B}$ = magnetic induction
C_0, C_1, C_2 = integration constants

Paper presented at Fourth Beer-Sheva International Seminar on Magnetohydrodynamic Flows and Turbulence, Ben-Gurion University of the Negev, Beer-Sheva, Israel, February 27-March 2, 1984. Copyright © by the American Institute of Aeronautics and Astronautics, Inc., 1985. All rights reserved.
*Privatdozent, Dr. rer. nat.
†Dr.-Ing.
‡Im Namen und für Rechnung der VDEh-Gesellschaft zur Förderung der Eisenforschung mbH.

ELECTROMAGNETIC FORCE OF STIRRING INDUCTORS

$\vec{e}_r, \vec{e}_\phi, \vec{e}_z$ = unit vectors in radial, azimuthal, and axial direction
$\vec{E}$ = electric field strength
$\vec{f}$ = force density
f = frequency
$\vec{H}$ = magnetic field strength
i = imaginery unit
I_p, K_p = modified bessel functions of the first and the second kind of order p
$\vec{j}$ = current density
$\vec{k}$ = surface current density
L = width of stator core
p = number of pole pairs
$\vec{Q}$ = higher vector potentials
Q_0, Q_1, Q_2 = scalar potentials
r, ϕ, z = cylinder coordinates
σ = conductivity
μ = permeability
μ_0 = permeability of vacuum
φ = phase angle of magnetic induction
ψ = scalar potential
ω = angular frequency

Introduction

Today, in many continuous casting plants, liquid metals are stirred by using azimuthally traveling magnetic fields. When calculating the electromagnetic fields and forces inside the melt, the inductor is usually assumed to be infinitely wide. Normally, however, the inductors are narrow. Therefore, the electromagnetic fields depend also on the width of the stator and the form of the end-windings.

When calculating the torque of asynchronous motors, the fields of the end-windings can be neglected, because their forces and moments are self-compensating within the rigid rotor. The finite width of the stator can be taken into account by a correlation factor. In a liquid rotor, however, these forces can induce internal flows.

Because of the enormous increase in the electrical loadings of large turbogenerators, knowledge of the magnetic field in the end regions has become important. It is useful in limiting short-circuit currents and forces and in decreasing eddy current losses in the windings and adjacent metal structures.

Many efforts have been made to produce satisfactory methods of predicting the end-leakage fields, using either analytical or purely numerical methods.[1-5] In most cases, it is assumed that the rotor has high magnetic permeability and negligible electrical conductivity. Therefore, the vector field equation in cylindrical coordinates cannot be separated. The finite-elements method can be used for solving the problem.[2] The disadvantage is that it requires long calculation times. On the other hand, an analytical treatment of a simplified model identifies controlling factors rather more readily than a numerical approach, and a single basic calculation enables the field at any point to be determined.

The present paper describes an analytical approach using a vector potential derived from two scalar functions that obey elliptical differential equations.[6,7]

Basic Assumptions

The stator windings of asynchronous inductors are mounted in small slots in a usually laminated iron core. The winding is arranged in pole pairs, and the end-windings lead the current from each slot to another that is approximately one pole pitch away. The end-windings are normally shaped to lie on the surface of a truncated cone. The winding is schematically shown in Fig. 1.

The currents in the windings are approximately described by simplified surface current distributions. Although the conical shape of the end-windings could be taken into account, a simpler model has been used. The axially directed currents as well as the currents in the end-windings are represented by equivalent surface currents on a cylinder with the same radius as the inner stator. The stator core itself is assumed to be infinitely wide. In a more detailed approach, the finiteness of the stator core could also be taken into account by using the method of images.[2]

Also, the radius of the liquid rotor is the same as the radius of the stator core and the rotor is infinite in axial direction.

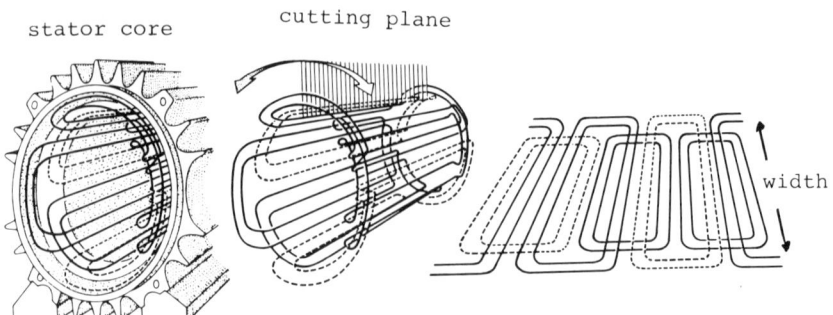

Fig. 1 Stator winding of an asynchronous motor.

ELECTROMAGNETIC FORCE OF STIRRING INDUCTORS

Because the fundamental of the rotating field is dominant, the harmonics are neglected. The angular velocity of the rotating electromagnetic field is assumed to be much higher than that of the liquid rotor.

Representation of the Inductor Currents

The surface current density has only axial components in the zone of the stator core, but has axial and azimuthal components in the end zone along the end-windings, as shown in Fig. 2. Radial components are zero everywhere. The complex axial component of the surface current density rotating with angular frequency ω, on the stator surface $r = a$, is given by:

$$K_z(\phi,z) = \hat{K}_z e^{i(\omega t - p\phi)} f(z) \quad (1a)$$

where $\hat{K}_z$ is the complex amplitude of the surface current density at $z = 0$, and p is the number of pole pairs.

Expanding the function $f(z)$ shown in Fig. 2 into a Fourier integral with respect to z gives

$$K_z(\phi,z) = 2\hat{K}_z e^{i(\omega t - p\phi)} \int_0^\infty g(m)\cos(mz)dm \quad (1b)$$

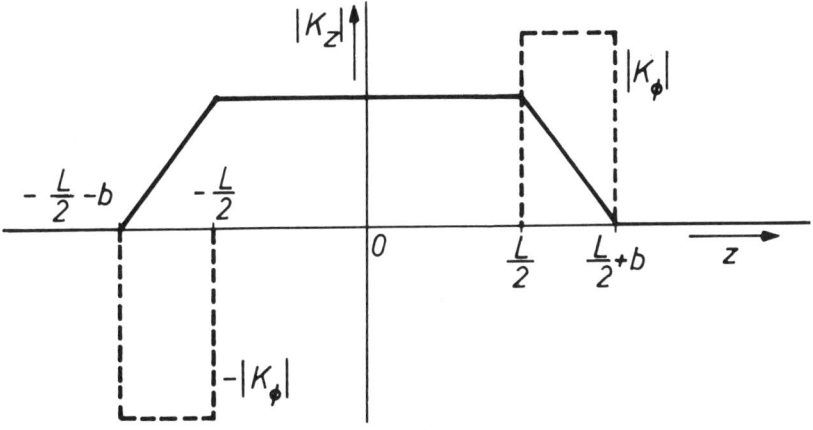

Fig. 2 Axial and azimuthal components of the surface current density along the axial coordinate z. L: width of the stator core; b: width of the end-windings.

with

$$g(m) = \frac{2}{\pi bm^2} \sin\left(m\frac{b+L}{2}\right) \sin(\frac{mb}{2}) \qquad (2)$$

As the divergence of the surface current density is zero, the expression

$$K_\phi = \frac{2ia\hat{K}_z}{p} e^{i(\omega t - p\phi)} \int_0^\infty m\, g(m) \sin(mz)\, dm \qquad (3)$$

results for the azimuthal component of the surface current density.

Calculation of Electromagnetic Field

Neglecting the movements in the liquid rotor and the displacement currents, and assuming that the field is rotating with angular frequency ω, the Maxwell equations are reduced to

$$\nabla \times \vec{B} = \mu \vec{j} \qquad \nabla \times \vec{E} = (-k^2/\mu\sigma)\vec{B} \qquad \nabla \cdot \vec{B} = 0 \qquad (4a\text{-}c)$$

$$\nabla \cdot \vec{E} = 0 \qquad k^2 = i\omega\sigma\mu \qquad \vec{j} = \sigma \vec{E} \qquad (4d\text{-}e)$$

Introducing a vector potential $\vec{A}$ in Eq. (4b), defined by the relations

$$\vec{B} = \nabla \times \vec{A} \qquad (5a)$$

and

$$\nabla \cdot \vec{A} = 0 \qquad (5b)$$

gives

$$\vec{E} + (k^2/\mu\sigma)\vec{A} = -\nabla \psi \qquad (6)$$

where ψ is a scalar potential.

Since there is no applied voltage, the scalar potential can be set at zero and, therefore, with the help of Eq. (4e), it follows that

$$\vec{j} = (-k^2/\mu)\vec{A} \qquad (7)$$

Inserting the formulas (4a) and (5a) gives

$$\nabla \times \left[\nabla \times \vec{A}(r,\phi,z)\right] = -k^2 \vec{A} \qquad (8)$$

The solenoidal field $\vec{A}$ is decomposed into two solenoidal components that are derived in a different manner from higher vector potentials, which have only components in the fixed axial direction:

$$\vec{A} = \vec{A}_I + \vec{A}_{II}$$
$$= \nabla \times \vec{e}_z Q_1 + \nabla \times (\nabla \times \vec{e}_z Q_2) \quad (9)$$

Each component of $\vec{A}$ satisfies Eq. (8) if the corresponding scalar potential Q_α obeys the equation

$$\nabla^2 Q_\alpha = k^2 Q_\alpha \quad \alpha = 1,2 \quad (10)$$

Putting

$$Q_\alpha = \hat{Q}_\alpha e^{i(\omega t - p\phi)} \quad (11)$$

where $\hat{Q}_\alpha(r,\phi,z) = \hat{Q}_\alpha(r,0,z)$, Eq. (10) is reduced in cylindrical coordinates to

$$\frac{\partial^2 \hat{Q}_\alpha}{\partial r^2} + \frac{1}{r}\frac{\partial \hat{Q}_\alpha}{\partial r} - \left(k^2 + \frac{p^2}{r^2}\right)\hat{Q}_\alpha + \frac{\partial^2 \hat{Q}_\alpha}{\partial z^2} = 0 \quad \alpha = 1,2 \quad (12)$$

Using the symmetry of the surface current density and the finiteness of the magnetic field for $r=0$, the problem admits only solutions of the form

$$\hat{Q}_1 = \int_0^\infty C_1 \sin(mz) I_p(qr) dm$$
$$\hat{Q}_2 = \int_0^\infty C_2 \cos(mz) I_p(qr) dm \quad (13)$$

where $q^2 = k^2 + m^2$.

In the region outside the liquid rotor $r > a$, the conductivity is assumed to be zero. Therefore, according to Eq. (4a), the curl of the magnetic field vanishes, and the magnetic induction can be written

$$\vec{B} = \nabla Q_0 \quad (14)$$

By virtue of Eq. (4c), Eqs. (10) and (12) hold also for Q_0 and $\hat{Q}_0$, respectively, with $k=0$.

Using the symmetry of the surface current density and the finiteness of the magnetic field for $r \to \infty$, the problem admits only solutions of the form

$$\hat{Q}_0 = \int_0^\infty C_0 \cos(mz) K_p(mr) dm \qquad (15)$$

The integration constants C_0, C_1, C_2 may be determined by applying the boundary conditions; this means the radial components of the magnetic flux must be continuous at $r=a$ and the differences in the tangential components of the magnetic field strength across the boundary are equal to the perpendicular components of the surface current density $\vec{K}$.

Once the integration constants have been determined, the final expression for the three components of the current density and the magnetic induction can be calculated from Eq. (13) using Eqs. (9), (7), and (5a).

Finally, the mean value of the electromagnetic body force $\bar{f}(r,z)$ in the liquid rotor is calculated from the relationship

$$\bar{f}(r,z) = \frac{1}{2} \text{Re}\ \{\vec{j} \times \vec{B}^*\} \qquad (16)$$

where the asterisk denotes a complex conjugate.

Theoretical and Experimental Results of the Magnetic Induction

Measurements of the magnetic field distribution on a small four-pole asynchronous inductor without rotor have been carried out to check the accuracy of the applied model. The main sizes and characteristics of the inductor are given in Table 1.

The components of the magnetic induction have been measured using small hall probes mounted on a nonconducting cylinder of radius c.

Figure 3 represents the measured and the calculated magnitudes of the magnetic induction components at $r=c=0.825a$--normalized with respect to B_0, where B_0 is the

Table 1 Main sizes and characteristics of the inductor

Inner radius of core	a = 57.5 mm
Width of core	L = 90 mm
Width of end-windings	b = 43 mm
Frequency	f = 50 Hz
Number of pole pairs	p = 2

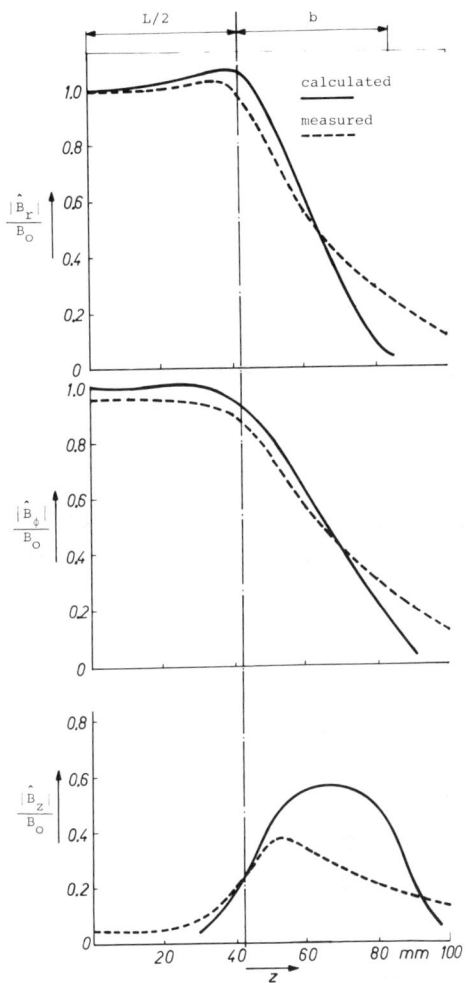

Fig. 3 Calculated and measured magnetic induction vs z at $r=c=0.825a$, $B_0 = B_r(z=0, r=c)$.

magnitude of the radial component B_r of the magnetic induction at $z=0$, $r=c$--as a function of the axial coordinate z. The curves show that the radial component has a slight maximum at the end of the stator core, while the azimuthal component of $\vec{B}$ is nearly constant along the stator core. In the zone of the end-windings the axial component B_z is approximately 50% of B_0.

In spite of the simplifications, the radial and azimuthal components show a satisfactory agreement between measured and calculated results. The measured and the calculated values of the axial component show less agreement.

Since the end-windings are normally conelike in shape with the radius of the end-windings increasing outside the stator core, the values of the actual magnetic induction are smaller than those calculated.

The highly permeable stator core concentrates the magnetic field toward the middle of the inductor.

Figure 4 shows the measured and calculated phase angles vs z. The phase of the azimuthal component at z=0 is set at zero. The agreement between calculated and measured results is good apart from the peripherical regions of the end-windings, where the field is so weak that the simplifications concerning the shape of the end-windings cause discrepancy.

In the region $z < a/2$, the magnitude of the magnetic induction is too small for measuring the phase of B_z with sufficient accuracy. But the phase distribution in the regions $z < a/2$ and $z > L/2 + b$ has no serious effect on the electromagnetic force.

Calculated Electromagnetic Force

The electromagnetic force density inside a conducting cylinder has been calculated as a function of the axial co-

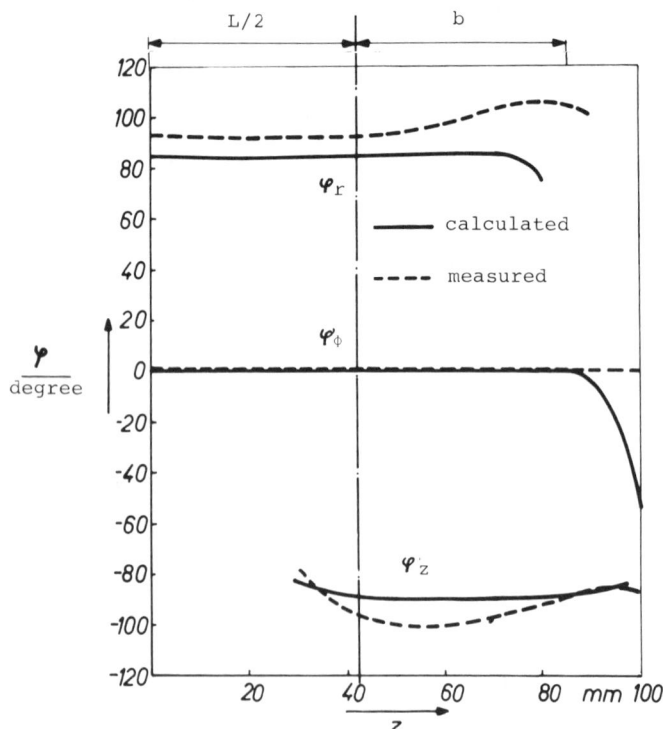

Fig. 4 Calculated and measured phase angle of the magnetic induction vs z at r=c=0.825a.

ELECTROMAGNETIC FORCE OF STIRRING INDUCTORS

Table 2 Main sizes and characteristics of the inductor

Inner radius of core	$a = 100$ mm
Width of core	$L = 200$ mm
Width of end-windings	$b = 100$ mm
Conductivity of cylinder	$\sigma = 10^6$ S/m
Frequency	$f = 50$ Hz
Number of pole pairs	$p = 1$

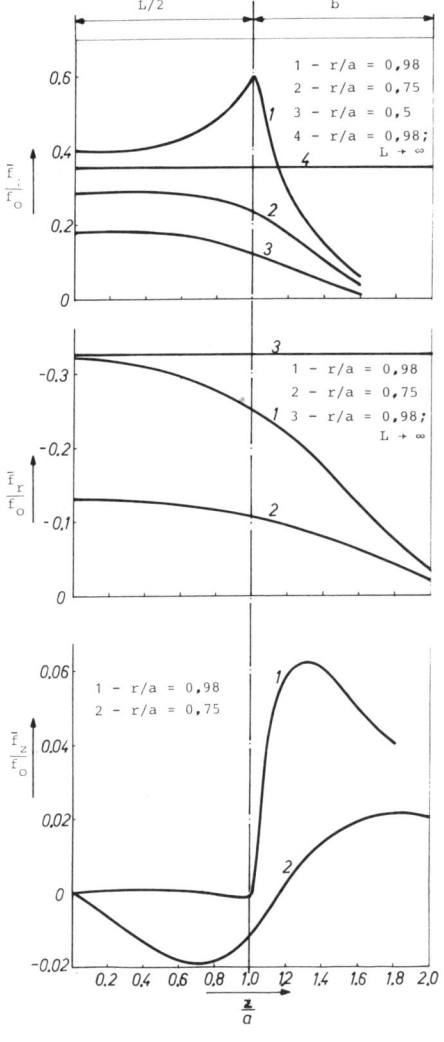

Fig. 5 Calculated force density of a finitely and an infinitely wide inductor vs z, $f_0 = \mu_0 R_z^2/a$.

ordinate z for various radii r. The main sizes and characteristics of the assumed configuration are given in Table 2.

Figure 5 shows the normalized components of the force density of a finitely and an infinitely wide inductor vs z for three different radii. Just below the surface of the cylinder, the azimuthal component of the finitely wide inductor has a sharp maximum at the end of the stator core, while near the center of the inductor, it is about 15% higher than that of an infinite one.

The radial component of the finitely wide inductor has a maximum at the middle z=0 and decreases steadily to the ends. Compared with the radial and the azimuthal components, the axial component is approximately one order of magnitude smaller and, therefore, is insignificant in most applications.

Conclusion

An analytical method of predicting the three-dimensional force density in a conducting cylinder induced by the fields of a circular asynchronous inductor is described. The fields inside the stator of a typical configuration are calculated.

Although simplification of geometrical form and magnetic behavior have been used, the calculated magnetic induction is in encouraging agreement with measured values on a small inductor.

The method presented has some advantages over finite-element methods such as flexibility, ease of computer implementation, and short calculation time. On the other hand, accuracy is limited by the assumptions regarding the boundary conditions.

In order to improve the accuracy of the method, further extensions are possible: 1) Without fundamental difficulties, the air gap between stator and rotor can be taken into account. 2) The surface current density can be described more accurately as distributed on a conical shape or arranged as a series of steps. 3) The presence of a highly permeable stator core can be taken into account by using the method of images.

But, on the other hand, calculation times thereby increase rapidly, and advantages over purely numerical methods may diminish.

Acknowledgment

This work was supported by Deutsche Forschungsgemeinschaft (DFG).

References

[1] Asworth, D. S. and Hammond, P., "The Calculation of the Magnetic Field of Rotating Machines, Part 2," Proceedings IEE, Vol. 108, Pt. A., No. 42, Dec. 1961, pp. 50-63.

[2] Lawrenson, P. J., "The Magnetic Field of the End-Windings of Turbo-Generators," Proceedings IEE, Vol. 108, Pt. A., No. 42, Dec. 1961, pp. 538-549.

[3] Reece, A. B. J. and Pramanik, A., "Calculation of the End-Region Field of A. C. Machines," Proceedings IEE, Vol. 112, No. 7, July 1965, pp. 1355-1368.

[4] Wolf, H. G., "Messung und Berechnung von Zusatzverlusten durch den Stirnstreufluβ von Drehfeldmaschinen in Konstruktionsteilen des Stators," (Measurement and Calculation of Stray Losses in the Stator Core of Rotating Machines produced by the End-Region Field), Ph.D. Thesis, TH Darmstadt, 1969.

[5] Martinelli, G. and Morini, A., "A Potential Vector Field Solution in Superconducting Magnets," IEEE Transactions on Magnetics, Vol. Mag-19, No. 4, July 1983, pp. 1537-1556.

[6] Moffat, H. K., Magnetic Field Generation in Electrically Conducting Fluids, Cambridge University Press, Cambridge, 1978, pp. 13-26.

[7] Hannakam, L., "Wirbelströme in einem massiven Zylinder bei beliebig geformter erregender Leiterschleife," (Eddy Currents in a Solid Cylinder Induced by a Coil of Arbitrary Shape), Archiv für Elektrotechnik, Vol. 55, 1973, pp. 207-215.

Study of the Electromagnetic Features in Channel Induction Furnaces

A. Moros* and J.C.R. Hunt†

University of Cambridge, Great Britain

and

D.C. Lillicrap‡

Electricity Council Research Centre, Capenhurst, Chester, Great Britain

Abstract

Recirculating flows in channel induction furnaces require investigation in order to understand and predict their performance with regard to heating and holding metals. Two idealized laboratory models for studying these flows are described. Flow indication by means of drag devices and some experimental measurements are presented. A two-dimensional finite-element model for calculating the electromagnetic forces and its curl in the channel has been developed. In this model, the equation for the vector potential $\underline{A}$ is solved at the node points of a triangular mesh. The distribution of the ($\overline{\underline{J \wedge B}}$) forces and the curl ($\overline{\underline{J \wedge B}}$) in the channel are presented and discussed.

I. Introduction

The channel induction furnace is used for heating and holding metals. A typical channel is shown in Fig. 1. The single channel furnaces are usually used for small outputs[1]. Electrically, the furnace is similar to a transformer. The molten metal in the channel forms a single turn around the primary induction coil. The large currents induced in the

---

Paper presented at the Fourth Beer-Sheva Seminar on MHD Flows and Turbulence, Ben-Gurion University of the Negev, Beer-Sheva, Israel, Feb. 27-March 2, 1984. Copyright © 1985 by the American Institute of Aeronautics and Astronautics, Inc. All rights reserved.

*Research student, Department of Engineering.

†Reader, Department of Applied Mathematics and Theoretical Physics.

‡Research officer.

FEATURES IN CHANNEL INDUCTION FURNACES

channel heat the metal, which then mixes with the larger volume of metal in the bath above. To enhance this mixing, and hence minimize the temperature rise in the loop, a unidirectional flow through the channel is advantageous. The main advantage over the coreless induction furnace is the low energy consumption.

An alternating current in the primary coil causes an electromagnetic wave front to diffuse into the metal, resulting in large induced poloidal currents $\underline{J}$ in the Z direction. This eventually melts the metal through the Joule dissipation J^2/σ.

The induced currents $\underline{J}$ interact with the magnetic flux $\underline{B}$ to give rise to an electromagnetic force ($\underline{J} \wedge \underline{B}$). For a configuration such as that of Fig. 1, at some distance from the ends of the channel the force components are primarily in the x,y plane. In any section of the channel, the force is rotational; consequently, transverse motions are set up

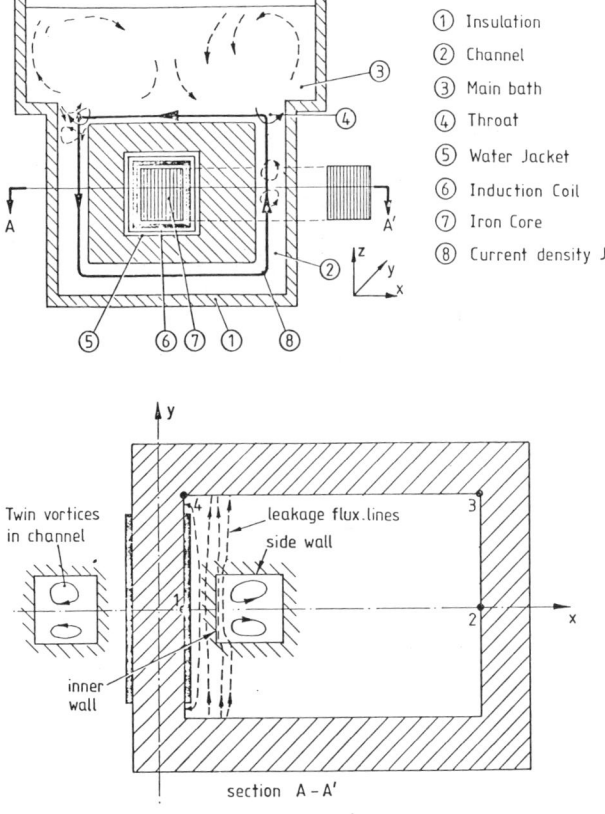

Fig. 1 Schematic view of a channel furnace

in the x,y plane, with vorticity parallel to the z axis. These are called transverse motions.

Near the channel ends where the induced currents spread out and turn the corner, the $(\underline{J} \wedge \underline{B})$ has components in the z,x plane. Since it is rotational, it can induce vorticity parallel to the Y and x axis. In Fig. 1, $\underline{\nabla} \wedge (\underline{J} \wedge \underline{B})$ is significant around the top of both legs but equal and opposite in magnitude. Consequently, there cannot be an electromagnetically driven unidirectional flow if the channel is symmetrically arranged around the coil. From the evidence of various designs, including patented designs, it is likely that unidirectional flows are induced if there is some asymmetry between the legs of the channel.

A better understanding of the nature of the stirring and how it can be related to furnace parameters that are under the control of furnace designer and operator is essential. Therefore, two idealized laboratory experiments have been planned. In the first, the primary eddy motion will be studied in a model simulating one leg of the channel, in

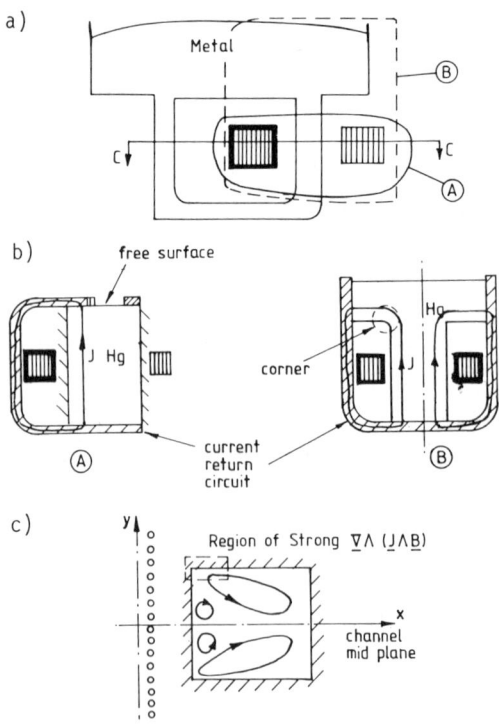

Fig. 2a Channel furnace; b) the two idealized laboratory models (vertical cross section); c) flow patterns

FEATURES IN CHANNEL INDUCTION FURNACES

which $\delta \ll x$ (region A in Fig. 2), away from the throat and close to the coils around the channel.

The second experiment will simulate the channel and the throat region (region B in Fig. 2), and the flows will be studied at the corner to obtain a better understanding of how the geometry of the throat governs unidirectional flows; different types of corner shape will be used.

II. Experimental Apparatus Simulating One Leg of the Channel

The salient features of the apparatus are shown in Fig. 3. Essentially it consists of a low-permeability ($\mu_r < 1.05$) stainless steel tank of (15x15)-cm square cross-section and 30-cm-high filled with mercury. To balance the induction heating of the mercury, the system is cooled by cold water from a gallery pouring down the outside of the tank wall. The water enters the gallery via two diametrically opposite 10-mm i.d. circular inlets and flow in the periphery of the tank. The upper and lower base of the tank and the gallery are connected with stainless steel flanges. This technique of cooling is effective and allows the apparatus to run continuously with an average temperature in the mercury of 38°C.

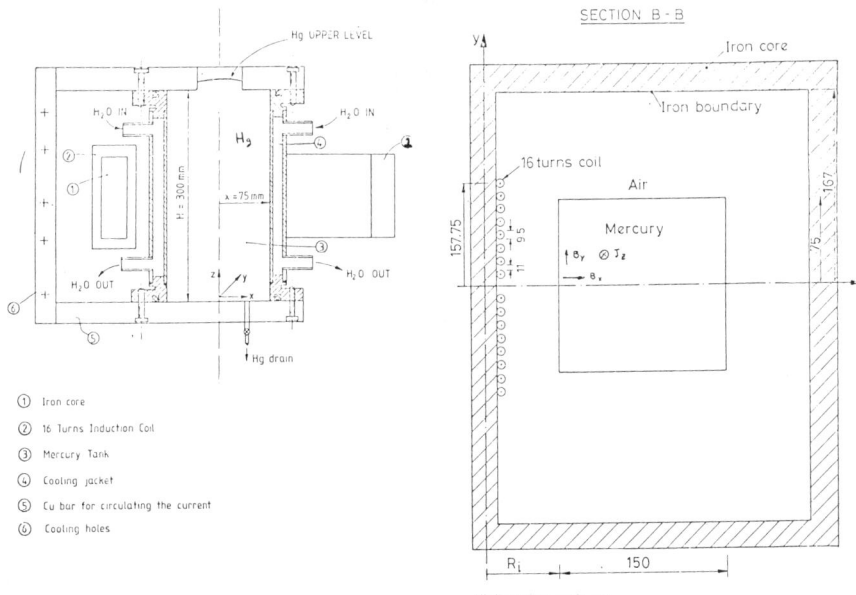

① Iron core
② 16 Turns Induction Coil
③ Mercury Tank
④ Cooling jacket
⑤ Cu bar for circulating the current
⑥ Cooling holes

Fig. 3 Laboratory model

The tank is surrounded by a laminated iron core on which a 16 turn water-cooled coil is wound around one of its legs. The coil is supplied from a single-phase 400-Hz generator. In order to recirculate the current induced in the mercury, a copper bar connects the top and bottom of the tank. The copper bar is screwed onto the top and bottom flanges and sealed with "o"rings.

The variety of the probes used to monitor the variables of the flow are supported via a stainless steel traverse mechanism located on the top of the apparatus. With this mechanism, the probes can effectively be located in the tank to within ± 1 mm.

III. Two-Dimensional Computer Model

The electromagnetic forces are computed for an infinitely long straight channel. It is assumed that the magnetic flux density and the electromagnetic forces are functions of x and y, and there is no variation in Z direction, $\partial/\partial z=0$. The coil is replaced by conductors parallel to the channel, and the iron laminations are also assumed to extend to infinity.

The governing equations are the usual Maxwell's equations:

$$\text{curl } \underline{E} = - \frac{\partial \underline{B}}{\partial t} \qquad (1)$$

$$(1/\mu) \text{ curl } \underline{B} = \underline{J} \qquad (2)$$

$$\underline{\nabla} \cdot \underline{B} = 0 \qquad (3)$$

$$\underline{J} = \sigma (\underline{E} + \underline{U} \wedge \underline{B}) \qquad (4)$$

In induction furnaces, because $Re_m \ll 1$, Eq. (4) reduces to

$$\underline{J} = \sigma \underline{E} \qquad (5)$$

With this simplification, a vector potential $\underline{A}$ can be introduced such that

$$\underline{B} = \nabla \wedge \underline{A} \qquad (6)$$

$$\underline{E} = - \frac{\partial \underline{A}}{\partial t} \qquad (7)$$

$$\underline{\nabla} \cdot \underline{A} = 0 \qquad (8)$$

FEATURES IN CHANNEL INDUCTION FURNACES

Fig. 4 Electromagnetic forces in mercury for a 150-mm square tank

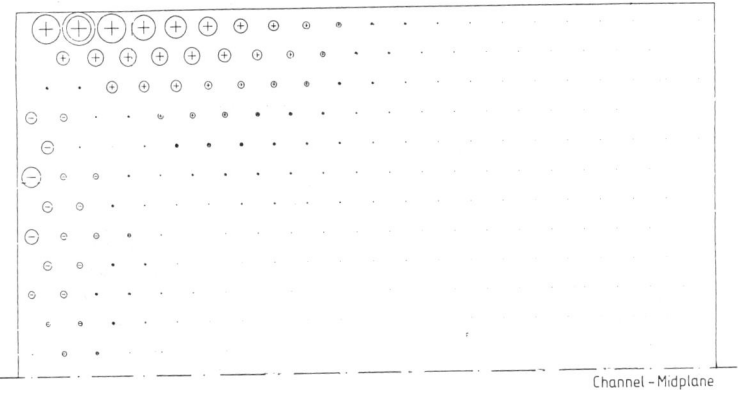

Fig. 5 Curl ($\underline{J} \wedge \underline{B}$) in the mercury for a 150-mm square tank

Then from Eqs. (1-8)

$$\underline{\nabla}^2 \underline{A} = \mu\sigma \frac{\partial \underline{A}}{\partial t} \qquad (9)$$

For a sinusoidal field $\underline{A}$ put in the form

$$A = I \ (Ao \ e^{i\omega t}) \qquad (10)$$

where I denotes the imaginary part.

Equation (9) becomes

$$\nabla^2 \underline{A}o - i\omega\sigma\mu \underline{A}o = 0 \qquad (11)$$

The solution of this equation for the appropriate boundary conditions provides all the electromagnetic quantities from which the electromagnetic force ($\underline{J}\wedge\underline{B}$) can be calculated.

To apply the boundary conditions to the problem, it is noted that the x axis in section A-A of Fig. 1 is a line of symmetry and that flux lines cross it at right angles. Furthermore, the permeability of the iron is very much greater than that of air so that flux lines enter the iron nearly at right angles to its surface. Thus, sufficient boundary conditions for a solution of Eq. (11) over the region bounded by 1,2,3,4 is that the tangential component of the magnetic field is zero along the boundary (i.e. $\partial \underline{A}/\partial Y = 0$). A solution is obtained by first creating a triangular mesh over this region and then using a finite-element method[2] to solve for the vector potential at the node points of the elements.

IV. Two-Dimensional Model Calculations

Using the described two-dimensional model, the electromagnetic force and the curl ($\underline{J}\wedge\underline{B}$) were computed for the 150-mm square channel of the described laboratory model (Sec. II), using a triangular mesh of size 5 mm. For a coil current of 1100 A (which gives $B_0 = 0.06$ T and hence $B_0/\sqrt{\mu\rho} = 0.35$ m/s), the electromagnetic forces are presented as arrows in Fig. 4 while the curl ($\overline{\underline{J}\wedge\underline{B}}$) is presented as circles in Fig. 5. In this figure, the +ve curl denotes anticlockwise rotation, while the -ve denotes clockwise rotation. The distribution of the forces shows that the maximum force occurs at the corner formed by the inner and side walls. Moving toward the midplane of the channel, the forces become nearly radial but decay rapidly with x.

For a two-dimensional model, the curl ($\overline{\underline{J}\wedge\underline{B}}$), which is the driving force for the recirculating flow, has only one component given by

$$\mathrm{curl}_Z (\overline{\underline{J}\wedge\underline{B}}) = -\frac{\partial}{\partial y}(\overline{\underline{J}\wedge\underline{B}})_X + \frac{\partial}{\partial x}(\overline{\underline{J}\wedge\underline{B}})_Y \qquad (12)$$

or

$$\mathrm{curl}_Z (\overline{\underline{J}\wedge\underline{B}}) = 2|B_x|\cdot|B_y|\omega\sigma \sin(\phi_x - \phi_y) \qquad (13)$$

The Z-axis is perpendicular to the plane of the paper.

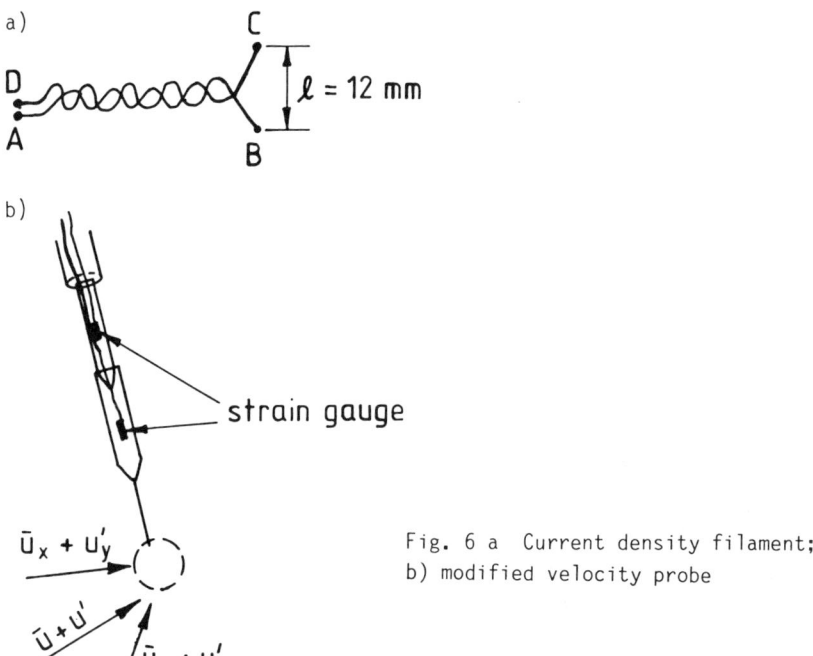

Fig. 6 a) Current density filament; b) modified velocity probe

Equation (13) is another expression that allows the curl to be calculated by avoiding differentiation of the forces.[3] Examination of the magnitude of the curl (Fig. 5) shows that the curl has its maximum value at the side wall near the corner. Moving toward the inner wall, the curl decays to zero and, near the inner wall, becomes -ve. This distribution of the curl suggests that we might have two flow regimes. The strong +ve curl along the side wall expels the metal away from the corner toward the midplane, forming a recirculating eddy as shown in Fig. 2c. The -ve curl near the inner wall suggests either another smaller recirculating eddy or a stagnant region. This suggestion is compared with measurements of the flow in the laboratory model, described in Sec. II.

V. Preliminary Experimental Results

The tank described in Sec. II has now been filled with 92 kg of mercury. With a coil current of 640 A, the Y component of the magnetic flux density $|B_Y|$ and the current density J_z were measured over a 5-mm mesh. Measurements of the current density were made with a current density probe

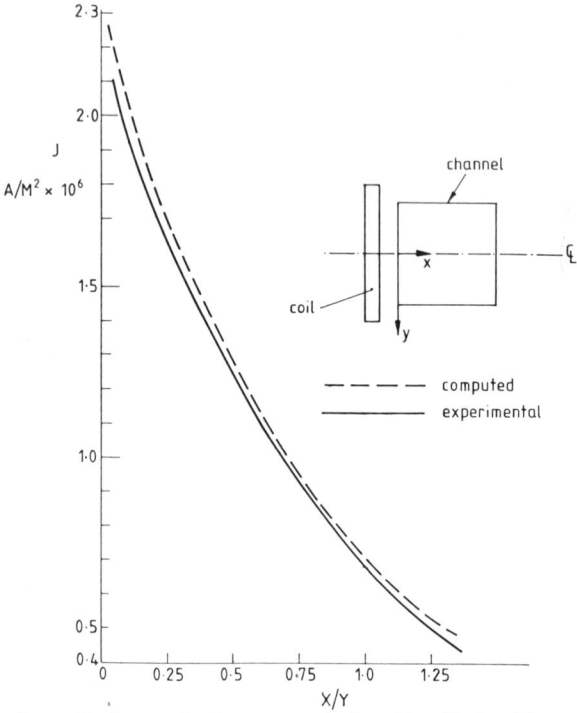

Fig. 7 Distribution of the current density JZ at $y=40$ mm from the C_L line

and measurements of the magnetic flux with a 4-mm-diameter search coil. The technique for measuring current densities in liquid MHD flows was first developed by Earlburke and Allen.[4] The probe consists of an open-circuited filament of wire that is immersed in the mercury. The entire length of the filament is electrically insulated from the mercury with the exception of the extreme two ends. The filament and its dimensions are shown in Fig. 6a. If the filament is aligned with the current direction, the voltage developed at the ends AD is proportional to the current density.

$$V_{AD} = (1/\sigma) \, 1J \qquad (14)$$

knowing the length 1 and the electrical conductivity σ, the current density J can be calculated.

Figures 7 and 8 show the distribution of the current density J and the magnetic flux $|B_Y|$, respectively. It can be seen that J as well as $|B_Y|$ decay rapidly with distance (x) from the coil.

FEATURES IN CHANNEL INDUCTION FURNACES

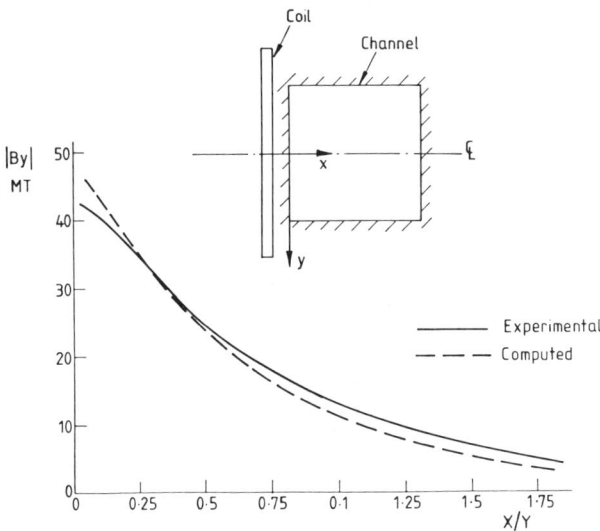

Fig. 8 Distribution of the magnetic flux density $|B_Y|$ at $y=40$ mm from the C_L line

Observations of the surface of the mercury and direction of flow inferred from a drag device show that the flow pattern is similar to the one suggested in Sec. IV and shown in Fig. 2c.

Velocity measurements will be made at different points in the x,y plane of the apparatus using the drag probe, the initial design of which has been described by Moore.[5] The modified version of the drag probe is shown in Fig. 6b. This version enables us to measure the two components of the velocity, U_x and U_y, simultaneously.

References

[1]Langman, R.D., "A Review of Engineering Development in Electric Induction Metal Melting Hold in Furnaces," British Foundryman, Sept. 1981.

[2]Winslow, A.M., "Numerical Solution of the Quasilinear Poisson Equation in a Nonuniform Triangle Mesh," Journal of Computational Physics, Vol. 2, 1967.

[3]Shercliff, J.A., Metals Technology, Vol. 11, Feb. 1984.

[4]Earlburke, P., and Alden, R.T.H., IEEE Transactions on Power Apparatus and Systems, No. 2., Feb. 1969, pp. 181-185.

[5]Moore, D.J., Ph.D. Thesis, Cambridge University, Great Britain, 1982.

Hartmann Layers in Slowly Solidifying Liquids

F.S. Hall* and G.S.S. Ludford†
Cornell University, Ithaca, New York
and
J.S. Walker†
University of Illinois, Urbana, Illinois

Abstract

Determination of the motion of a solidification front into a molten metal, a so-called Stefan problem, does not depend on details of the solidification process inside the front. Such details are needed, however, when the liquid is being stirred by a magnetic field, since they determine the structure of the coincident Hartmann layer and hence, the secondary motion in the liquid, our primary concerns here. Salient features of this novel Hartmann layer are its variable viscosity, due to variation in the concentration of solid particles, and its doubly infinite extent. In the absence of a theory in the nucleation literature, a law of growth in solid concentration is proposed for the small undercooling that must occur in such a layer. The (asymptotic) analysis is carried out for the slow solidification of a cylinder of molten metal that is being stirred by a rotating, uniform magnetic field, but the results are of general validity. The perturbation of the synchronous solid body rotation of the liquid is calculated and found to be substantially stronger than in the absence of solidification.

I. Introduction

The propagation of a solidification front into a pure molten metal at rest is a consequence of the solid sending

---

Paper presented at the Fourth Beer-Sheva Seminar on MHD Flows and Turbulence, Ben-Gurion University of the Negev, Beer-Sheva, Israel, Feb. 27-March 2, 1984. Copyright © 1985 by the American Institute of Aeronautics and Astronautics, Inc. All rights reserved.

*Graduate Research Assistant, Department of Theoretical and Applied Mechanics.
†Professor, Department of Theoretical and Applied Mechanics.

out shoots into the liquid. As a shoot grows, it sends out its own shoots, the process repeating itself to form a treelike structure known as a dendrite. The growth of a dendrite is limited only by its neighbors and the availability of liquid between its shoots. The resulting casting is coarse grained and, hence, relatively intolerant of mechanical stress.

The strength of a casting increases as the grain size is reduced, and such a reduction can be achieved by making the liquid flow along the solidification front so as to break off the dendritic shoots as soon as they form. Many nucleation sites are thereby created, the resulting grain size being related to the height of a shoot when it is broken off by the shear flow. (The strength is also increased by the more even distribution of impurities that results from smaller grain size.)

The molten metal can be made to flow along the solidification front by stirring it electromagnetically; this method is continuous and controllable. Here, we shall consider the liquid enclosed by a converging cylindrical front to be stirred by a uniform magnetic field perpendicular to and rotating about the axis of the cylinder. For best results, the rotation should be slow, so that the field (supposed to be applied by electromagnets outside the initial cylinder of molten metal) penetrates unchanged through the solid into the liquid; and the magnetic field should be strong, so as to produce a flow with high shear at the front.

Alemany and Moreau[1] have considered the problem without a solidification front, i.e., when the metal remains liquid (as it does when there is no heat loss). To leading order, the liquid is in solid body rotation, and the resulting Hartmann layer at the containing wall (assuming there is one) determines the perturbation of this flow. The object of the present paper is to show how the structure of the Hartmann layer is modified when it is also a solidification front and to determine the new perturbation of the solid body rotation in the liquid core. (The propagation of the front is determined by heat conduction considerations alone, leading to a classical Stefan problem, without reference to this structure.)

Modification of the Hartmann layer must take account of two new features: not only is the layer (doubly) infinite, instead of semiinfinite; but also it contains a mixture of solid particles (fragments of dendritic shoots) and liquid, so that the effective viscosity varies from that of the pure liquid to infinity (the value for a pure solid). A complete treatment would require theories of i) the size and shape of the fragments broken off by the shear flow, ii) the rate at which these fragments grow, and iii) the effective viscosity of the resulting suspension. Unfortunately, no such theories exist for i and ii, while there are too many for iii.

Lack of information about a theory of i would only be a handicap if there were a theory of ii. Instead, we shall propose a law of growth that does not require such information, as if it were an empirical law; indeed, we hope that it will be tested experimentally. [If the sizes of the fragments have a certain distribution, a potential justification of the law can be given (see Sec. V), but we do not consider this possibility in detail.] For iii, Mancini[2] finds an abundance of theories, each depending to a lesser or greater extent on the composition of the suspension. These mainly differ in the way that the viscosity increases as the suspension becomes saturated with solid particles. In view of our other assumptions, and because we do not expect errors in viscosity to make a qualitative difference, we have selected a simple relation between the concentration of solid particles and the viscosity of the suspension.

We propose that the rate of increase in solid concentration (1-c) is

$$k\ c\ \exp[-\alpha/(1-T)^2 T]$$

where k, α are constants, and T is the temperature measured in units of the equilibrium temperature. Such a law (with $k \gg 1$) holds in the absence of solid particles, i.e., for so-called homogeneous nucleation, when it exhibits the considerable undercooling (of the order of 10% of the equilibrium temperature) that is required to solidify a pure liquid. With particles present, the required undercooling is observed to drop to a fraction of a degree, and this is reflected by the proposed law for a much smaller α (with the same k).

Asymptotic theory of the solidification layer is based on small undercooling $\varepsilon = 1 - T_s$, where T_s is the solidification temperature. It involves a distinguished limit in which α and the small heat loss associated with slow solidification are related to ε. For the purpose of illustration, the solidification layer is supposed to have the same thickness as the Hartmann layer. (Subsequent measurements could determine whether this is reasonable or not; we merely note here that mature dendritic shoots, which could provide hundreds of fragments, are about 0.2 mm in height, whereas a typical Hartmann length is at least 0.02 mm.) The assumption of equal thicknesses for the two layers is not necessary. It amounts to relating the undercooling to the strength of the magnetic field, and these are independent parameters. Theories of i and ii would determine the undercooling, as would experiments, without reference to the magnetic field. Meanwhile, the reasonableness of the results based on this assumption suggest that the thickness of the two layers are not very different, so that, at most, minor modification of the present analysis will be needed.

The slow cooling of pure aluminum is taken as an example. The propagation of the solidification front through the cylinder of molten metal, which (as noted before) does not depend upon details of the solidification process, has been considered by Walker and Georgopoulos[3]; we use their parameter values throughout this paper. For a magnetic field of 1 t, the undercooling is found to be 0.84 K, a substantially greater amount than that found experimentally. However, only stationary melts have been examined so far; the smallness of the dendritic fragments expected in rotating melts could certainly account for the difference. (The nucleation rate depends on the curvature of any solid surface present.)

The plan of the paper is as follows. Sections II, III, and IV review the stirring problem, keeping a variable viscosity in the Hartmann layer involved. Their purpose is to develop formulas for the perturbation of the solid body rotation of the core. The investigation of the solidification layer in Secs. V, VI, and VII determines this variable viscosity. The growth law for the dendrite fragments is presented in Sec. V and used in Sec. VI to develop the structure equations. Section VII gives the solution of the structure problem and draws all elements together for the numerical calculation of the perturbation in the core.

II. MHD Stirring

The effect of a slowly rotating, uniform magnetic field on liquid metal in a cylindrical container has been considered by Alemany and Moreau[1] in the strong-field limit. An account of their work was given by Moffat[4] at the Second Beer-Sheva Seminar, but it will be convenient to have the main results derived here, especially in view of the extension to variable viscosity needed in the following section.

If the field has strength B_0 and rotates with angular velocity ω, and the cylinder has radius a, then appropriate units are length a, time ω^{-1}, velocity ωa, magnetic field B_0, and electric field $\omega a B_0$. In terms of these units, we have the electromagnetic equations

$$\nabla \times \underline{E} = -\partial \underline{B}/\partial t \quad R_m^{-1} \nabla \times \underline{B} = \underline{j} = \underline{E} + \underline{v} \times \underline{B} \quad \nabla \cdot \underline{B} = 0 \quad (1)$$

where the unit of j is the current $\sigma\omega a B_0$; here, σ is the electrical conductivity and, with μ the permeability,

$$R_m = \mu\sigma\omega a^2 \ll 1 \qquad (2)$$

is the magnetic Reynolds number, assumed small because the rotation is slow.

In the limit $R_m \to 0$, we find

$$\nabla \times B = 0 \qquad (3)$$

so that the applied magnetic field is undisturbed to leading order: In other words, the magnetic diffusion time $\mu\sigma a$ is short compared with the rotation time ω^{-1}. If the z axis is taken along the centerline of the cylinder, we may write

$$B = (\cos t, \sin t, 0) \qquad (4)$$

and then the corresponding electric field and current are

$$E = (0,0,x) \quad j = (0,0,X-V) \qquad (5)$$

where

$$X = x \cos t + y \sin t \quad V = -u \sin t + v \cos t \qquad (6)$$

are components in a frame rotating with the magnetic field (Fig. 1).

We turn now to the fluid-dynamic equations

$$\nabla \cdot v = 0 \qquad (7a)$$

$$N^{-1} Dv/Dt = -\nabla p + j \times B + M^{-2} \nabla \cdot [\eta(\nabla v + \nabla v^T)] \qquad (7b)$$

where the unit of p is the pressure $\sigma\omega a^2 B_0^2$, and allowance has been made for a variable (dimensionless) viscosity η. (The superscript T denotes the transpose of the dyadic ∇v.) Here, with ρ the density and η_r a reference viscosity,

$$N = \sigma B_0^2/\rho\omega \gg 1 \qquad (8a)$$

$$M = B_0 a (\sigma/\eta_r)^{\frac{1}{2}} \gg 1 \qquad (8b)$$

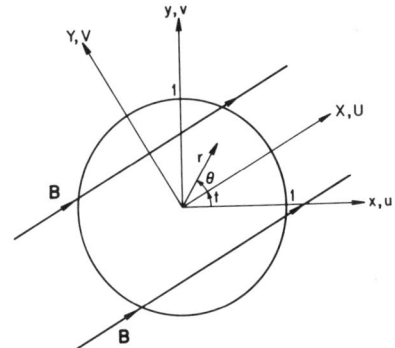

Fig. 1 Notation.

are the interaction and Hartmann numbers, assumed large because the magnetic field is strong. (The smallness of ω also contributes to N being large.)
In the limit $N \to \infty$, Eq. (7b) is replaced by its inertialess approximation

$$\nabla p = j \times B + M^{-2} \nabla \cdot [\eta(\nabla v + \nabla v^T)] \quad (9)$$

which reduces to

$$\nabla p = j \times B \quad (10)$$

wherever velocity gradients are not large enough to counterbalance the smallness of M^{-2}. Except near the wall of the container, i.e., in the so-called core, the velocity must adjust so as to make the Lorentz force a gradient field balanced by the pressure. The resulting condition is

$$\partial j/\partial X = 0 \quad (11a)$$

i.e.,

$$\partial V/\partial X = 1 \quad (11b)$$

It is now convenient to introduce the other components

$$Y = -x \sin t + y \cos t \quad U = u \cos t + v \sin t \quad (12)$$

in the rotating frame, and the streamfunction $\psi(X,Y)$:

$$U = \partial \psi/\partial Y \quad V = -\partial \psi/\partial X \quad (13)$$

The equation of continuity (7a) is automatically satisfied, while the residue (11b) of the momentum equation gives

$$\partial^2 \psi / \partial X^2 = -1 \qquad (14)$$

This must be solved under the boundary condition

$$\psi = 0 \quad \text{on} \quad X^2 + Y^2 = 1 \qquad (15)$$

showing that the velocity field in the bulk of the liquid metal is given by

$$\psi = \tfrac{1}{2}(1 - X^2 - Y^2) \qquad (16)$$

It follows that, to leading order, the core is in current-free solid body rotation synchronous with the magnetic field; the pressure there is constant.

The goal is to determine the $O(M^{-1})$ perturbation of the core flow, which involves the same balance (10), since only terms $O(M^{-2})$ were neglected in deriving it. Equation (14) still holds, but the boundary condition (15) must be replaced by a more accurate one, coming from the viscous layer between the core and the container. This modified Hartmann layer will now be examined.

III. Doubly Infinite Hartmann Layers

The viscosity will be supposed to vary across the layer from a constant value in the core, taken to be the η_r in the definition (8b) of M, to infinity at the container. Polar coordinates r, θ will be taken in the rotating frame and

$$R = M(1 - r) \qquad (17)$$

will denote the layer variable.

Since the circumferential velocity v_θ is $O(1)$, the radial velocity v_r must be $O(M^{-1})$; accordingly, we write

$$v_r = M^{-1} U \quad v_\theta = V \qquad (18)$$

To leading order, the equations of continuity (7) and momentum (9) then become

$$\frac{\partial U}{\partial R} = \frac{\partial V}{\partial \theta} \qquad (19a)$$

$$\frac{\partial p}{\partial R} = 0 \tag{19b}$$

$$\frac{\partial p}{\partial \theta} = (1 - V)\cos^2\theta + \frac{\partial}{\partial R}\left(\eta \frac{\partial V}{\partial R}\right) \tag{19c}$$

The pressure does not vary across the layer, so that, since it is constant everywhere. There is no pressure gradient to drive the circumferential flow, and V satisfies

$$\frac{\partial}{\partial R}\left(\eta \frac{\partial V}{\partial R}\right) + \cos^2\theta(1 - V) = 0 \tag{20}$$

i.e., an ordinary differential equation containing θ as a parameter. At the extremes of the range of R, we have

$$\begin{aligned}\eta &\to \infty \\ &\to 1\end{aligned} \quad \text{as} \quad R \to \pm\infty \tag{21a}$$

$$\begin{aligned}V &\to 0 \\ &\to 1\end{aligned} \quad \text{as} \quad R \to \pm\infty \tag{21b}$$

the former being a property of the given function η, and the latter coming from conditions outside the layer. Once V has been found, the continuity equation (19a) shows that

$$U = -\int_{-\infty}^{R} \frac{\partial V}{\partial \theta} \, dR \tag{22}$$

since velocity components must vanish as $R \to -\infty$.

The problem (20), (21) defines a doubly infinite Hartmann layer. When the liquid has uniform viscosity, i.e.,

$$\begin{aligned}\eta &= \infty \text{ for } R < 0 \\ &= 1 \text{ for } R > 0\end{aligned} \tag{23}$$

a classical Hartmann layer is obtained; we have

$$\begin{aligned}V &= 0 & \text{for } R < 0 \\ &= 1 - \exp(-|\cos\theta|R) & \text{for } R > 0\end{aligned} \tag{24a}$$

$$\begin{aligned}U &= 0 & \text{for } R < 0 \\ &= \tan\theta\{R\exp(-|\cos\theta|R) + |\sec\theta|[\exp(-|\cos\theta|R) - 1]\} & \text{for } R > 0\end{aligned} \tag{24b}$$

The limit

$$U \to \frac{\partial}{\partial \theta}(1 - |\sec\theta|) \quad \text{as} \quad R \to \infty \tag{25}$$

is of particular interest, since this determines the $O(M^{-1})$ velocity field in the core. The corresponding result in general is obtained by taking the limit $R \to \infty$ in formula (22):

$$U \to \frac{\partial f}{\partial \theta} \qquad (26a)$$

with

$$f(\theta) = \int_{-\infty}^{\infty} (V - V_0) dR \quad \text{as} \quad R \to \infty \qquad (26b)$$

Here the subtraction of V_0, the V profile for $\theta = 0$, ensures convergence of the integral after removal of $\partial/\partial\theta$ for reasons of accuracy (V is only known numerically).

It is not difficult to construct functions $\eta(R)$ for which the problem (20), (21) does not have a solution: The viscosity must tend to infinity fast enough as the solid side of the layer is approached. To obtain a plausible sufficient condition, we change the independent variable to

$$Z = \int_{-\infty}^{R} \frac{dR}{\eta} \qquad (27)$$

in terms of which the problem is

$$\frac{\partial^2 V}{\partial Z^2} + \eta \cos^2\theta (1 - V) = 0 \qquad \begin{array}{l} V \to 0 \text{ as } Z \to 0 \\ \to 1 \text{ as } Z \to \infty \end{array} \qquad (28)$$

As $Z \to \infty$, $\eta \to 1$, and (in general) solutions grow like $\exp(|\cos\theta|Z)$, although there is a family of solutions satisfying the right boundary condition. As $Z \to 0$, solutions satisfying the left boundary condition must have the form

$$V = \sec^2\theta \int_{-\infty}^{R} \frac{(C-R) dR}{\eta} \qquad (29)$$

as is seen by integrating the approximate equation $\partial^2 V/\partial Z^2 + \eta\cos^2\theta = 0$. Here, C is an integration constant, suitable choice of which may be expected to suppress the general exponential growth as $Z \to \infty$. Though proof is lacking, the inference is that a solution exists provided the integral

(29) converges, i.e.,

$$\int_{-\infty}^{R} \frac{R dR}{\eta} < \infty \qquad (30)$$

and that is certainly true when (as $R \to -\infty$) $\eta = O(|R|^{2+\varepsilon})$ for some $\varepsilon > 0$.

The latter theory yields a viscosity tending to infinity much more rapidly, so that we are reasonably confident of the solution that is computed numerically. Nevertheless, to emphasize the importance of condition (30), we violate it with the following example:

$$\eta = 1 + 1/Z^2 \qquad R = Z - 1/Z$$

$$1 - V = Z^{\frac{1}{2}}[AI_\nu(|\cos\theta|Z)$$

$$+ BK_\nu(|\cos\theta|Z)] \quad \text{with} \quad \nu^2 = \tfrac{1}{4} + \cos^2\theta$$

Here, A and B are integration constants, for which there is no choice satisfying the boundary conditions (21b).

IV. The Core Perturbation

To leading order, the core is at rest in a frame rotating with the applied magnetic field. To order M^{-1}, it is perturbed by the Hartmann layer that this creates near the wall of the container: Variations in the $O(1)$ flux along the layer must be compensated by an $O(M^{-1})$ flux into the layer from the core. The resulting flowfield there is calculated as follows.

We have seen that the stream function ψ still satisfies the differential equation (14), but that the boundary condition (15) must be corrected. The correction recognizes that the radial flow $\partial\psi/\partial\theta$ out of the core accounts for the velocity $M^{-1}U$ at the edge of the Hartmann layer, so that

$$\psi = M^{-1} f(\sin^{-1} Y) \quad \text{on} \quad X^2 + Y^2 = 1 \qquad (32)$$

according to result (26). It follows that

$$\psi = \tfrac{1}{2}(1 - X^2 - Y^2) + M^{-1} f(\sin^{-1} Y) \qquad (33)$$

in the core, so that the perturbation is a shear flow parallel to the magnetic field.

When the fluid has uniform viscosity, $f = 1 - |\sec\theta|$ according to result (25), so that

$$f(\sin^{-1} Y) = 1 - (1 - Y^2)^{-1/2} \qquad (34)$$

The flow is in the negative X direction when Y is positive, and in the positive X direction when Y is negative (see Fig. 2). The solution fails as $Y \to \pm 1$, i.e., in the neighborhood of the two points where the cylinder is parallel to the magnetic field. The same singularity may be expected in general, as may be seen by writing the differential equation (20) in the form

$$\frac{\partial}{\partial \overline{R}} \left(\eta \frac{\partial V}{\partial \overline{R}} \right) + (1 - V) = 0 \quad \text{with} \quad \overline{R} = |\cos\theta| R \qquad (35)$$

The effect of θ is felt in $\eta(\overline{R}/|\cos\theta|)$, which tends to ∞ for each $\overline{R} < 0$, and to 1 for each $\overline{R} > 0$ as $\theta \to \pm \pi/2$. The solution $V(\overline{R})$ may therefore be expected to tend to that for a liquid of uniform viscosity as a singularity is approached, which implies that the nature of the singularity is the same.

The general procedure for calculating the perturbation is as follows. Given $\eta(R)$ with the properties (21a), the

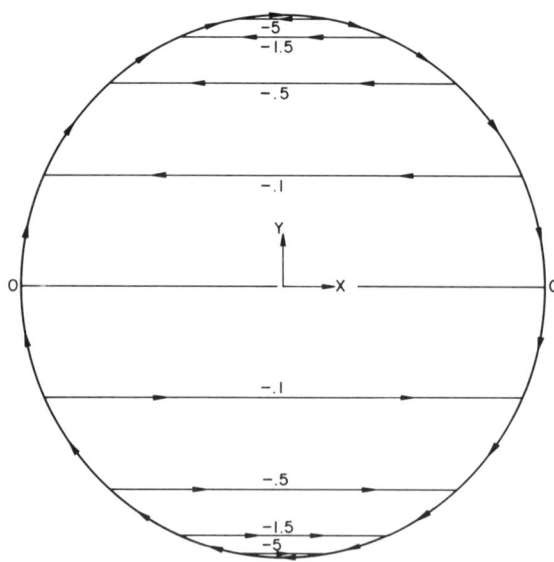

Fig. 2 Streamlines of the core perturbation, showing values of function (34).

differential equation (20) is integrated under the boundary conditions (21b) for a series of values of θ, and the corresponding f caluculated from the integral (26b). By interpolation, the values of Y for which $f(\sin^{-1}Y)$ takes on equidistant values are then found and the results displayed as in Fig. 2, which has been drawn for case (23).

We turn now to the determination of η from a structure analysis of the solidification front.

V. The Solidification Process

The model that we shall adopt is as follows. The dendritic shoots that form at the solid are supposed to be broken off by the scouring action of the shear flow while they are still small. These minute solid particles migrate through the shear layer toward the core, providing nuclei for the solidification process. The layer is therefore viewed as a gradation of neutrally buoyant particles, their concentration tending to zero on the liquid side and to 1 on the solid side.

In the absence of such particles, i.e., for so-called homogeneous nucleation, the solidification rate depends on the amount of the undercooling from T_e, the temperature at which solid and liquid are in equilibrium.[5] For each temperature less than T_e, a potential nucleus must have a critical size in order to grow, and it is a matter of determining how many of the particles, i.e., clusters of atoms that are continually forming and dissolving, are large enough. The relation between the temperature and the rate of increase in concentration of solid turns out to be

$$\frac{d}{dt}(1-c) = k\, c \exp\left(-\frac{\alpha}{(1-T)^2 T}\right) \qquad (36)$$

where c is the liquid concentration, T_e is the unit of temperature, and k is a (very large) constant. The value of α, also a constant, is 1.14 for aluminum (as may be calculated from Chalmer's formula, using the value 93 ergs/cm^2 for the interfacial energy of aluminum given by Turnbull.[6])

Near the equilibrium temperature, i.e., for 1-T small, the right side is negligibly small, corresponding to the requirement of a large critical size; as 1-T increases, the exponential slowly increases until, at some value, 1.11×10^{-1} for aluminum (corresponding to 104 K and undercooling from $T_e = 933$ K), the coefficient of c suddenly becomes appreciable, and solidification occurs in a fraction of a second. One concludes that there is a definite temperature at which nuclei of critical size are present in sufficient numbers to cause almost instantaneous solidification. Note that

$$\alpha/(1 - T)^2 T = 103 \qquad (37)$$

for aluminum, so that k must be about $\exp(103) = 10^{44.7}$.

When extraneous particles are present to act as nuclei, less undercooling is needed for solidification: The number of particles is increased, so that there are now more of critical size at any temperature. There is no theory of the nucleation rate in such circumstances, so we propose to use the same formula (36) with a smaller value for the constant α. (More precisely, we may argue that the effective change in free energies from solid to liquid is smaller because the energies of the extraneous particles are not to be counted; since the two energies, volume and surface, are reduced proportionately, the formula follows from the same analysis as for homogeneous nucleation.) The value of α is related to the number of dendritic fragments produced, which would be difficult to determine. We therefore characterize the amount by the undercooling temperature T_u at which sudden solidification now takes place. In view of the exponent (37) for homogeneous nucleation of aluminum, we then have

$$\alpha = 103 (1 - T_u)^2 T_u \qquad (38)$$

In applying formula (36) to a cylindrically converging solidification front, we take the time unit to be ω^{-1}, as for the rotating magnetic field. Then, if the speed of the solidification front is small enough for the process to be considered steady relative to the front, we may write

$$v_0 \frac{dc}{dr} = -kc \exp\left(-\frac{\alpha}{(1 - T)^2 T}\right) \qquad (39)$$

where the dimensionless speed v_0 is very small. In the aluminum problem considered by Walker and Georgopulos,[3] a cylinder of radius 5 cm solidified in 30 min, giving $v_0 = \exp(-12.3)$ for $\omega = 20$ Hz. The smallness of v_0 enhances the effect of k, so that solidification actually takes place at a temperature T_s higher than T_u.

To complete the problem, we must add the heat equation

$$v_0 \frac{dT}{dr} - Pe^{-1} \frac{1}{r} \frac{d}{dr}\left(r \frac{dT}{dr}\right) = -kv_0 \frac{dc}{dr} \qquad (40)$$

where, as before, we have replaced the time derivative by

$v_0 d/dr$; here,

$$P_e = \rho c_p a^2 \omega / \lambda \quad \kappa = h/\rho c_p T_e \quad (41)$$

with λ/c_p the thermal diffusivity, c_p the specific heat, and -h the heat of solidification per unit mass. (P_e is the Peclet number, which is supposed to have the same constant value in the solid as it has in the liquid; this assumption, equality of the two thermal diffusivities, is easily removed.)

VI. The Solidification Layer

The limiting form of the problem (39), (40) is the Stefan problem of solidifying liquids. The heat equation (40) with right-side zero has to be solved on either side of the moving interface between solid and liquid, subject to a jump $\kappa v_0 P_e$ in the temperature gradient and to a requirement that the temperature is T_s there. The remaining boundary conditions (away from the interface) provide an overdetermined problem, so that v_0 must take a particular value if there is to be a solution. The structure of the solidification front plays no role in determining its velocity. Such Stefan problems are usually tackled numerically, but an analytical treatment of the one considered here has recently been given by Walker and Georgopoulos.[3]

If, as we shall suppose, the solidification front has a (nondimensional) thickness comparable to M^{-1}, its structure must be investigated in order to determine how its effective viscosity η varies from 1 to ∞ from the liquid side to the solid side. Now, there are several theories[2] for the effective viscosity of a suspension of solid particles in a liquid, each yielding a different dependence on the concentration. Since we do not have enough information about the particles to select any one of these theories, we shall take the simple relation

$$\eta = 1/c^2 \quad (42)$$

as illustration. In view of the uncertainties in other parts of our presentation, it would be false sophistication to seek anything better at this stage.

Whenever there is a layer, there must be a small parameter ε, and the question now is to identify ε. The undercooling is clearly a candidate, but the amount that occurs in ordinary solidification processes is too small, leading to thicknesses much less than M^{-1}. However, when dendritic shoots are prevented from forming, and only small solid particles are in contact with the liquid, the undercooling may be much larger (though still small). Such is the case here, and, in the absence of experimental determinations of the amount of undercooling when the liquid is stirred, we shall proceed on the assumption that it is sufficient to make the solidification layer of thickness M^{-1}.

To determine the amount of undercooling

$$\varepsilon = 1 - T_s \qquad (43)$$

necessary to make the two layers of the same thickness, we set

$$\alpha = \tilde{\alpha}\varepsilon^\ell \qquad (44a)$$

$$T = T_s + \varepsilon^m T_1/2\tilde{\alpha} + \ldots \qquad (44b)$$

where $\tilde{\alpha}$, ℓ, m are positive constants, and T_1 represents the temperature perturbation (about T_s) in the layer. The exponent in Eq. (39) is then

$$-\frac{\alpha}{(1-T)^2 T} = -\tilde{\alpha}\,\frac{\varepsilon^{-2}}{1-\varepsilon} - \varepsilon^{\ell+m-3} T_1 + \ldots$$

The first term is to be cancelled by pre-exponential factors, i.e.,

$$\tilde{\alpha}\varepsilon^{\ell-2} = \ell n(k/v_0) \qquad (45)$$

to leading order, implying $\ell < 2$; and the second is to provide $O(1)$ variations in the layer, i.e.,

$$\ell + m - 3 = 0 \qquad (46)$$

requiring $m > 1$. The layer variable must be of the form

$$\tilde{R} = 2\tilde{\alpha}\tilde{\beta}\varepsilon^{n-m}(1-r) \qquad (47)$$

if a small temperature gradient $\tilde{\beta}\,\varepsilon^n$ ($n > 0$) outside the layer is to be matched, so that the thickness of the solidification layer is equal to that of the Hartmann layer, i.e., $\tilde{R} = R$, if

$$\varepsilon^{m-n} = 2\tilde{\alpha}\tilde{\beta}M^{-1} \qquad (48)$$

Equations (45), (46), and (48) now show that

$$\varepsilon^{1-n} = 2\tilde{\beta}M^{-1}\ln(k/v_0) \qquad (49)$$

a relation between ε and M that does not depend on α. This is an important point, since α cannot be determined directly by experiment.

For aluminum $n_r = 1.39\times 10^{-3}$ kg·m^{-1}s^{-1}, $\sigma = 4.13\times 10^6$ Ω^{-1}m^{-1}, so that a cylinder of radius 5 cm (as before) would correspond to $M^{-1} = 3.66\times 10^{-4}/B_0$, if B_0 is measured in teslas. With the values $\ln k = 10^3$ and $\ln v_0 = -12.3$ derived in the previous section, requirement (49) becomes

$$\varepsilon^{1-n} = 0.0842\tilde{\beta}/B_0 \qquad (50)$$

In order to obtain further information about ε, we write the basic Eqs. (39) and (40) to leading order in the layer variables

$$\frac{dc}{dR} = ce^{-T} \qquad (51a)$$

$$\frac{d^2T}{dR^2} = \frac{-dc}{dR} \qquad (51b)$$

where we have set

$$\frac{\tilde{\alpha}\varepsilon^{\ell-2}}{(1-\varepsilon)} - \ln(k/v_0) = (m-n)\ln\varepsilon - \ln(2\tilde{\alpha}\tilde{\beta}) \qquad (52a)$$

$$\tilde{\beta}\varepsilon^n = \kappa v_0 \text{Pe} = hv_0 a^2 \omega/T_e \lambda \qquad (52b)$$

In Eq. (51), T_1 has been replaced by T, and c now stands for its leading term in an expansion of type (44b). Equation (45) is now seen to be the first approximation of requirement (52a), which determines allowable pairs of values $\tilde{\alpha},\ell$ for the theory to be applied. Requirement (52b) provides an additional equation for ε, namely,

$$\tilde{\beta}\varepsilon^n = 0.0107 \qquad (53)$$

when the values for aluminum used before are supplemented by $h = 9.6\times 10^8$ J/m^3 and $\lambda = 134$ W/m·K (mean of solid and liquid values).

Equations (50) and (53) give

$$\varepsilon = 0.000901 \qquad (54)$$

for a magnetic field of 1 t, the strongest that is likely to be used in practice. The undercooling required for the two layers to have the same thickness is therefore about 0.84 K. Temperature measurements during the solidification of a stationary melt show considerably smaller undercooling, as would be expected for larger particles (fully formed dendritic shoots) than are envisaged here. It remains to be seen whether rotating melts do sustain larger undercoolings.

To complete the discussion in any particular case, we must check whether reasonable values of $\tilde{\alpha}$, ℓ and $\tilde{\beta}$, n can be found to satisfy the requirements (52a, 52b). Using approximation (45), we find (in the case considered here)

$$\tilde{\alpha} = 0.598 \quad \text{for} \quad \ell = 1.25 \quad (m = 1.75) \qquad (55)$$

while

$$\tilde{\beta} = 2.06 \quad \text{for} \quad n = 0.75 \qquad (56)$$

[Using these values of ℓ, $\tilde{\beta}$, n in the exact quation (52a) changes $\tilde{\alpha}$ to 0.557.] Selection of pairs of values is, however, not necessary (as the next section will show). A complete theory of the solidification process would, of course, provide $\tilde{\alpha}$, ℓ, $\tilde{\beta}$, n.

VII. Solution of the Layer Problem

If there is no temperature gradient in the melt to leading order $O(\varepsilon^n)$, the layer equations (51a, 51b) have to be solved under the boundary conditions

$$c \to 1 \quad \frac{dT}{dR} \to 0 \quad \text{as} \quad R \to +\infty \qquad (57)$$

i.e., on the liquid side. The requirement

$$\frac{dT}{dR} \to 1 \quad \text{as} \quad c \to 0 \qquad (58)$$

is then satisfied automatically, as is seen from the integral

$$S \equiv \frac{dT}{dR} = 1 - c \qquad (59)$$

of Eq. (51b). It is also automatic that the limits (58) are attained when $R \to -\infty$, as the result (64b) shows.

In view of the integral (59), the remaining layer equation (51a) becomes

$$\frac{d^2T}{dR^2} = \left(\frac{dT}{dR} - 1\right) e^{-T} \qquad (60)$$

The solution may be written in the parametric form

$$T = -\ln[-\ln(1 - S) - S] \qquad (61a)$$

$$R = \int_{1/2}^{S} \frac{dS}{(1 - S)[S + \ln(1 - S)]} \qquad (61b)$$

where we have located the origin of R at the point where S is half way between its values 0, 1 at $R = \pm\infty$. The second of these formulas is of greater interest than the first, since, according to definition (59), S is the solid concentration in the layer and, hence, directly related to the viscosity (42).

The requirement (30) prompts us to determine the asymptotic properties of T and S as $|R|$ becomes large. To determine the behavior as $R \to -\infty$, expand for small $s = 1-S$:

$$T = -\ln(-\ln s) + O(\ln s)^{-1}$$
$$R = -\ln(-\ln s) + A + O(\ln s)^{-1} \quad \text{as} \quad s \to 0 \qquad (62)$$

where

$$A = \ln \ln 2 - \int_0^{1/2} \left(\frac{1}{s-1-\ln s} + \frac{1}{\ln s}\right) \frac{ds}{s} = -2.85 \qquad (63)$$

Inversion of the second expansion and substitution in the first then give

$$T = R - A + O(e^R/R) \quad \text{as} \quad R \to -\infty \qquad (64a)$$

$$S = 1 - \exp[-\exp(A - R)][1 + O(e^R/R)] \quad \text{as} \quad R \to -\infty \qquad (64b)$$

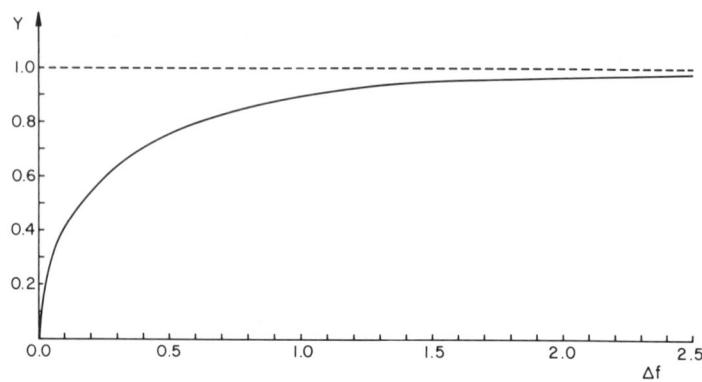

Fig. 3 Difference Δf by which shear flow in Fig. 2 is amplified.

Similarly, expansion for small S leads to

$$T = \ln(R^2/2) + O(\ln R/R) \quad \text{as} \quad R \to +\infty \quad (65a)$$

$$S = 2/R + O(\ln R/R^2) \quad \text{as} \quad R \to +\infty \quad (65b)$$

The result (64b) certainly satisfies the requirement (30) on $\eta = (1 - S)^{-2}$, and shows that the limits (58) are indeed attained as $R \to -\infty$. The result (65a) shows that the temperature perturbation becomes logarithmically infinite on the liquid side. [That it must become infinite is clear from Eq. (60) and boundary conditions (57).]

The way is now clear to determine f from the integral (26b). For each θ of interest, the differential equation (20) must be integrated (numerically) under the boundary condition (21b), with $\eta = (1 - S)^{-2}$ and S(R) given by the result (61b). Because of the singular nature of f(θ) at θ = π/2, as displayed by function (34), it is expedient to extract the singularity from the integral by writing

$$f(\theta) = \int_{-\infty}^{\infty} [(V - V_0) - (V_c - V_{c0})] dR + 1 - |\sec\theta| \quad (66)$$

where V_c is the classical function (24a). [For numerical reasons, V_c for $\eta > 0$ is actually generated as the solution of the differential Eq. (20) with $\eta \equiv 1$.]

Of most interest is the difference between f and its classical form $1 - |\sec\theta|$, i.e., the integral in expression (66). Figure 3 is a plot of this difference and shows that shear flow of Fig. 2 is substantially amplified.

Acknowledgments

This work was supported by the U.S. Army Research Office and by the National Science Foundation under Grant CPE-8108952.

References

[1] Alemany, A. and Moreau, R., "Écoulement d'un Fluide Conducteur de l'Électricité en Présence d'un Champ Magnétique Tournant," Journal de Mécanique, Vol. 16, No. 4, 1977, pp. 625-646.

[2] Mancini, F., "The Viscosity of Suspensions," M.S. Thesis, Cornell University, Ithaca, N.Y., 1984.

[3] Walker, J. S. and Georgopoulos, E ., "Slow Solidification of a Cylinder with Constant Heat Flux," International Communications on Heat and Mass Transfer, Vol. 11, 1984, pp. 45-53.

[4] Moffatt, H. K., "Rotation of a Liquid Metal Under the Action of a Rotating Magnetic Field," Proceedings of the Second Bat-Sheva Seminar on MHD Flows and Turbulence, edited by H. Branover, AIAA, New York, 1980, pp. 45-62.

[5] Chalmers, B., Principles of Solidification, John Wiley and Sons, New York, 1964; reprinted by Robert E. Krieger Publishing Company, Malabar, Fla., 1982, p. 680.

[6] Turnbull, D., "Formation of Crystal Nuclei in Liquid Metals," Journal of Applied Physics, Vol. 2 1, 1950, pp. 1022-1028.

Stirring an Aluminum Ingot Mold with a Linear Motor: Electromagnetic, Hydrodynamic, and Thermal Effects

J-L. Meyer,* R. Ernst,† and F. Durand‡

Institut National Polytechnique de Grenoble,
Groupement d'Interet Scientifique, Madylam,
Saint Martin d'Heres, France

Abstract

A three-phase linear motor allows the production of a traveling field in a rectangular vertical ingot mold used for solidification experiments on aluminum. A theoretical model of the electromagnetic and dynamic effects of a multiphased inductor is presented. It derives from a two-dimensional solution of Maxwell's equations and Ohm's law in the symmetry plane of both the inductor and the mold. The electromagnetic field is not coupled with flow phenomena. An analytical solution is developed in the case of an infinitely high inductor and mold. It describes accurately the exponential decrease of the traveling field with the distance to the inductor, depending on both the frequency through a shield parameter and the pole periodicity of the inductor. Moreover, a numerical solution based on a finite-difference scheme accurately shows the corner effects. The analytical expression of the time-averaged force is used in a theoretical model of heat and fluid flows. Results concerning temperature and velocity fields agree reasonably well with measurements performed in steady-state conditions. The model is especially suitable for treating interactions between heat and fluid flows. It provides a semiquantitative description of the overall hydrodynamic and thermal evolution during transient solidification of ingots. It can be useful for the choice of electromagnetic stirrers in relation to a required metallurgical effect.

Paper presented at the Fourth Beer-Sheva Seminar on MHD Flows and Turbulence, Ben-Gurion University of the Negev, Beer-Sheva, Israel, Feb. 27-March 2, 1984. Copyright © 1985 by the American Institute of Aeronautics and Astronautics, Inc. All rights reserved.

STIRRING AN ALUMINUM INGOT MOLD

Nomenclature

B, B_x, B_y, B_o	=	magnetic flux density: vector, its components, its value at the inductor wall
C_p	=	specific heat
$f, f_x, f_y, \bar{f}_x,$ $\bar{f}_y$	=	electromagnetic force: vector, its components, their time-averaged values
f_s	=	solid fraction of the metal
g	=	gravity vector
h, h_1, h_2, h_3	=	heat exchange coefficients of the mold walls
I	=	effective current per phase
j	=	current density
k	=	wave number
t	=	time
T	=	temperature (time-averaged value in the case of turbulent flow)
V_s	=	synchronism velocity of the traveling field
V	=	voltage drop of the linear motor
u, v	=	component of the flow velocity (time-averaged values in the case of turbulent flow)
x, y, x	=	space coordinates
$\alpha, \alpha_t, \alpha_e$	=	thermal diffusivity: molecular, turbulent, effective
β	=	coefficient of thermal expansion
α^*, β^*	=	electromagnetic forces parameters
λ	=	thermal conductivity
μ	=	magnetic permeability
ν, ν_t, ν_e	=	kinematic viscosity: molecular, turbulent, effective
ω	=	current angular frequency
ψ	=	stream function
ρ	=	density
σ	=	electrical conductivity
ξ	=	vorticity
∇	=	Nabla operator
τ	=	half-pole periodicity
ΔH	=	latent heat of solidification

Introduction

In continuous casting of steel, multiphased electromagnetic stirring has proved to be an efficient way to improve the quality of cast products.[1] Its use is considered particularly suitable in order to avoid nonuniformity of the chemical composition and of the metallurgical structure. However, different cases of application must be distinguished. For instance, stirring the liquid at the meniscus level involves principally the circulating effect of the flow, in order to eliminate the solid particles, oxide, refractory scraps, etc., which are transported by the liquid. On the other hand, when stirring in the lower solidification zone, the mixing effect is required in order to obtain a more uniform solidification structure. Consequently, it is important to design the stirring device in relation to the metallurgical effect that is expected.

At the Polytechnic Institute of Grenoble, the authors study the effects of multiphased stirring on solidification structure. But solidification by itself is a complex transient process. Therefore, a suitable theoretical modeling is helpful to understand the observed phenomena. More specifically, the use of a multiphased inductor makes it necessary to develop an adequate treatment of the magnetohydrodynamic phenomena and of their relations to thermal effects. The program described in this paper combines experiments on liquid aluminum, together with theoretical modeling of the electromagnetic field and of heat and fluid flows.

The Experimental Device

Solidification structure is sensitive to the natural convection. Therefore, the experimental device was designed so that fluid flow could be activated either by thermal convection or by electromagnetic stirring (Fig. 1).

The central part is a rectangular vertical ingot mold. Its geometric and thermal characteristics are given in Table 1. Heat is extracted through the vertical wall on the left, to simulate the wall of an ingot mold. A three-phase linear motor creates an electromagnetic field traveling along the wall, either downward, that is, in the same direction as the natural convection, or upward. Table 2 summarizes its electric characteristics, and Table 3 lists the property values used in the calculation.

The experimental program was oriented in order to get a better understanding of the hydrodynamic and thermal field in relation to the electromagnetic conditions. Temperature and velocity maps were established during a series of steady-state experiments.[2] In such cases, a furnace was added along the wall opposite to the inductor. For the velocity measurements, we used a magnetodynamic probe.[3] The flux was measured in air or in liquid aluminum by means of a flat spiral coil sensor (Fig. 2). Its internal and external dia-

STIRRING AN ALUMINUM INGOT MOLD

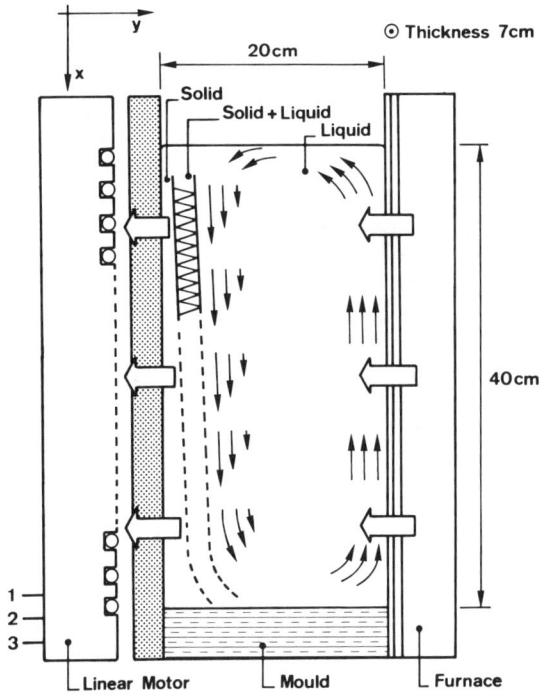

Fig. 1 Rectangular vertical ingot mold with linear motor and furnace.

Table 1 Characteristics of the ingot mold
(height: 400 mm; length: 200 mm; width: 70 mm)

	Steady-State Case $Wm^{-2}K^{-1}$	Transient Case $Wm^{-2}K^{-1}$
Chilling wall, concrete, 30 mm thick	$h_1 = 75$	120
Furnace wall, iron coated with aluminum	$h_2 = 480$	0
Free surface	$h_3 = 30$	0
Other walls, mineral wool	$h = 0$	

Table 2 Characteristics of the linear motor

Current per phase I	245–545 A
Voltage V	3.5–7.5 V
Magnetic induction B	10–23 mT
Frequency $\omega/2\pi$	50 Hz
Synchronism velocity V_s	7.2 ms^{-1}
Wave number k	43.6 m^{-1}

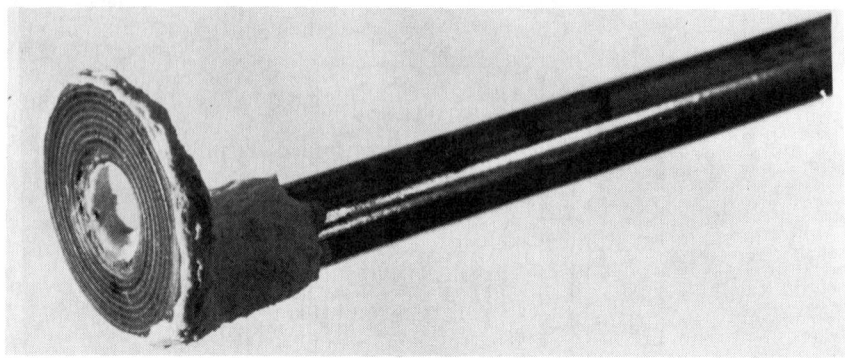

Fig. 2 Magnetic field sensor.

Table 3 Values of the properties used in the calculation

Density ρ	2389 kg m^{-3}
Viscosity ν	1.22x10^{-6} m^2 s^{-1}
Thermal expansion coefficient β	1.16x10^{-4} K^{-1}
Electrical conductivity σ	5x10^6 mho m^{-1}
Magnetic permeability μ	1.26x10^{-6} H m^{-1}
Thermal conductivity λ	103 W m^{-1} K^{-1}
Specific heat C_p	1080 J kg^{-1} K^{-1}
Thermal diffusivity α	3x10^{-5} m^2 s^{-1}

meters are 6 and 20 mm, respectively. It was protected by a stainless steel sheath and coated with alumina.

Theoretical Distribution of the Electromagnetic Forces

In the present paper, the distribution of the electromagnetic forces in the ingot mold is derived by applying the formulation of electromagnetic phenomena currently used for induction stirring.[4,5] Maxwell's equations and Ohm's law can be transformed to obtain the induction equation:

$$\mu\sigma \frac{\partial B}{\partial t} = \nabla^2 B + \mu\sigma\nabla \times (\nu \times B) \quad (1)$$

In a nondimensional form, Eq. (1) reduces to

$$\frac{\mu\sigma\tau^2}{t_c} \frac{\partial B'}{\partial t'} = \nabla'^2 B' + R_m \nabla' \times (\nu' \times B') \quad (2)$$

The prime index (') refers to nondimensional variables. τ is the half pole periodicity of the inductor. It is taken as the characteristic length of the system. $R_m = \mu\sigma\tau V_0$ is the magnetic Reynolds number. Here, V_0, the characteristic velocity, is only a few percent of the synchronism velocity $V_s = \omega/k = \omega\tau/\pi$, as it is well known in the case of asynchrone pumping. Hence, R_m is very small, typically 4×10^{-2}. Consequently, the convective transport of the magnetic field can be neglected in Eq. (2), which reduces to

$$\frac{\mu\sigma\tau^2}{t_c} \frac{\partial B'}{\partial t'} = \nabla'^2 B' \qquad (3)$$

Consequently, in our case, electromagnetic calculations can be treated beforehand, prior to the treatment of the fluid motion. The induced electromagnetic field is alternating and periodic. It is conveniently written in the form of a nondimensional complex flux density vector:

$$B' = B/B_0 = \text{Re}(\hat{B} \exp i\omega t) \qquad (4)$$

Here, $i = (-1)^{\frac{1}{2}}$. Hence, Eq. (3) reduces to Eq. (5), in which appears the shield parameter R_ω:

$$i R_\omega \hat{B}' = \nabla'^2 \hat{B}' \quad \text{with} \quad R_\omega = \mu\sigma\omega\tau^2 \qquad (5)$$

The theoretical calculation is restricted to the x-y median symmetry plane of both the inductor and the mold. This corresponds to an idealized case, where both inductor and ingot mold are indefinitely extended in the z direction.

It is convenient to express the magnetic flux density in terms of the vector potential. Here it is periodic with time, and only its z component is nonzero. It is defined by

$$\hat{B}' = \nabla' \times \hat{A}'_z \qquad (6a)$$

with

$$\nabla' \cdot \hat{A} = 0 \qquad (6b)$$

From Eq. (5), it follows:

$$i R_\omega \hat{A}'_z = \nabla'^2 \hat{A}'_z \qquad (7)$$

Boundary conditions for the problem are as follows. Far from the inductor, A'_z vanishes. At the wall of the mold close to the inductor, some approximation is needed, adapted to the complex structure of the linear motor. Assuming the magnetic yokes are perfect, the inductor is replaced by an equivalent sheet of traveling current. Hence, the tangential magnetic flux density is conveniently expressed in the form of a traveling field:

$$\text{For } y=-e \qquad B_z = B_0 \cos(\omega t - kx) \qquad (8)$$

The B_0 value can be related to I and N (the effective current intensity and the number of conducting wires per pole and phase, respectively) by applying Ampere's theorem to a circuit following the inductor surface along half a pole periodicity $\tau = n/k$ in the x direction, then returning to the origin through the magnetic yokes:

$$B_0 = \mu k \, NI \sqrt{2} \qquad (9)$$

Other boundary conditions are summarized in Fig. 3.

The magentic flux density and current density can be derived from $\hat{A}'_z$:

$$B = B_0 \, \text{Re}(\nabla' \times \hat{A}'_z \exp i\omega t) \qquad (10)$$

$$j = -\sigma\omega\tau \, B_0 \, \text{Re}(i \, \hat{A}'_z \exp i\omega t) \qquad (11)$$

Electromagnetic forces contain two terms:

$$f = j \times B = \sigma\omega\tau \, B_0^2 \, f'$$

with

$$f' = \frac{1}{2} \text{Re}\left[i \, \hat{A}'^*_z \times (\nabla' \times \hat{A}'_z) - i\hat{A}'_z \times (\nabla' \times \hat{A}'_z) \exp i2\omega t \right] \qquad (12)$$

The star (*) denotes the conjugate complex quantity. In Eq. (12), the first term is time independent; the second pulsates at a frequency double the excitating frequency. Due to the fact that inertial and viscous effects do not allow the fluid to follow such frequency motions, only the time-averaged value of the force has a dynamical effect:

$$\bar{f}' = \frac{1}{2} \text{Re}\left[i \, \hat{A}'^*_z \times (\nabla' \times \hat{A}'_z) \right] \qquad (13)$$

For numerical calculations, it is convenient to split $\hat{A}'_z$ in its modulus $|\hat{A}'_z|$ and its angular phase ϕ. Hence, the above

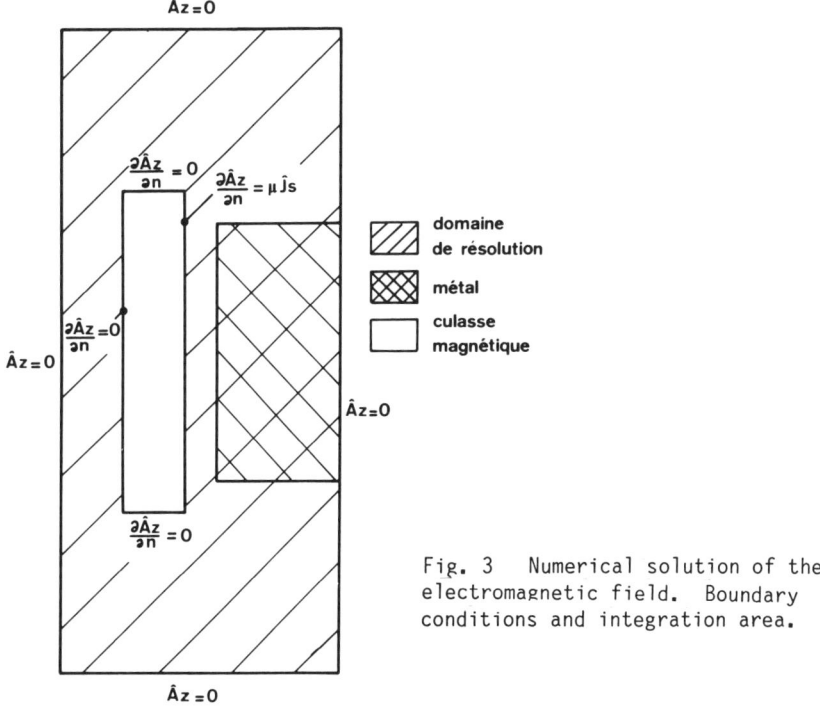

Fig. 3 Numerical solution of the electromagnetic field. Boundary conditions and integration area.

equation reduces to

$$\overline{f} = \frac{1}{2} |\hat{A}'_z|^2 \nabla' \phi \qquad (14)$$

This equation and boundary conditions are put in a finite-difference form on a nonuniform mesh, and solved using a successive over-relaxation scheme.

A completely analytical solution can be derived from Hughes and Young,[6] if it is assumed that both inductor and ingot mold are infinitely long in the x direction. Indeed, in this case, the complex value $\hat{A}'_z$ can be put in the following form:

$$\hat{A}'_z = A''_z (y') \exp(-i\pi x') \qquad (15)$$

Equations (4) and (12) now give:

$$\frac{d^2 A''_z}{d y'^2} = (\pi^2 + i R_\omega) A''_z \qquad (16)$$

The nondimensional solution is then:

$$B'_x = -B'_0 \exp(-\alpha^* y') \left(\frac{\beta^*}{\pi} \cos(\omega t - \pi x' - \beta^* y' + \phi) \right.$$
$$\left. + \frac{\alpha^*}{\pi} \sin(\omega t - \pi x' - \beta^* y' + \phi) \right) \quad (17a)$$

$$B'_y = B'_0 \exp(-\alpha^* y') \cos(\omega t - \pi x' - \beta^* y' + \phi) \quad (17b)$$

$$j'_z = -\frac{1}{\pi} B'_0 \exp(-\alpha^* y') \cos(\omega t - \pi x' - \beta^* y' + \phi) \quad (17c)$$

$$B'_0 = \pi \left| \beta^{*2} \cos h^2 \pi e' + (\alpha^* \cos h\pi e' + \pi \sin h\pi e')^2 \right|^{-\frac{1}{2}} \quad (17d)$$

$$\alpha^*, \beta^* = \left| \pm \pi^2/2 + (\pi^4 + R_\omega^2)^{\frac{1}{2}}/2 \right|^{\frac{1}{2}} \quad (17e)$$

$$e' = e/\tau \quad (17f)$$

The time-averaged components of the electromagnetic force now take the following expressions:

$$\overline{f}_x = \sigma \omega \tau B_0^2 (B'^2_0/2\pi) \exp(-2\alpha^* y') \quad (18a)$$

$$\overline{f}_y = \sigma \omega \tau B_0^2 (B'^2_0/2\pi)(\beta^*/\pi) \exp(-2\alpha^* y') \quad (18b)$$

The magnetic field decreases exponentially as a function of the distance from the wall. Equations (17) and (18) clearly show that this decrease results from both the frequency effect common to all ac inductors, and the pole periodicity, characteristic of the multiphased inductors. In the present case, the shield parameter R_ω is of the order of 10. Consequently, it is an intermediate case. It is different from the high-frequency case, in which the so-called skin effect is confined in a very narrow thickness, so the electromagnetic forces can then be treated as surface forces. It is more related to the case of asynchrone stirring, even if the frequency effect is not negligible.

Comparison Between Measurements and Calculations

The only component of the magnetic flux density that has been measured is the y component, normal to the inductor,

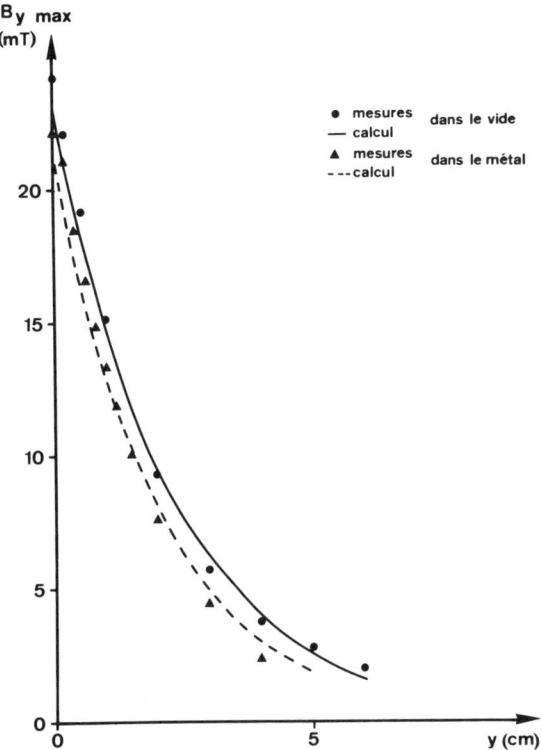

Fig. 4 Decrease of the magnetic flux density as a function of the distance to the inductor. I=545 A. Black dots, measures without metal; solid line, corresponding calculation; black triangles, measures in liquid aluminum; dotted line, corresponding calculation.

in the vertical symmetry plane. Figure 4 presents the decrease of the magnetic field as a function of the distance from the inductor wall, either when the ingot mold is empty or when it is full of liquid aluminum. Calculated values are in good agreement with the measured ones. As expressed by Eqs. (18), the magnetic field decreases exponentially as a function of distance. This effect evidences the skin effect. But the exponential decrease still holds even if the mold is empty. This is a general feature of the multiphased inductors, contrary to the single-phased inductors case. The magnetic field in the air is lower than in the metal, as expected. But in the considered case the difference is small, since the ratio is 92% at the inductor wall. Regarding the vanishing distance, which is τ/α^* from Eqs. (18), its value is 2.1 cm in the liquid aluminum, to be compared with 2.3 cm in the air. These results indicate that the

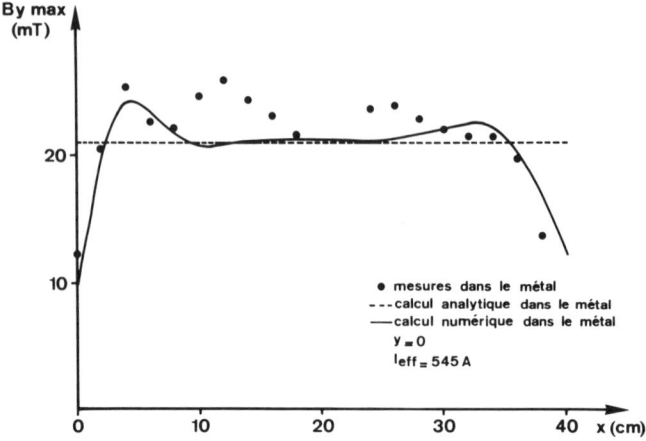

Fig. 5 Variation of the magnetic flux density along the inductor wall. I=545 A. Black dots, measures; solid line, numerical calculation; dotted line, analytical approximation.

modification of the magnetic field by the liquid metal is only slight, because the frequency used is relatively low.
 Variations of B_y along the inductor wall are shown on Fig. 5. There is a general agreement between measured values and theoretical results, both numerical or analytical ones. More precisely the numerical calculation clearly expresses the corner effect, which increases the magnetic field at the ends of the ingot. In the central part of the ingot, both numerical and analytical results are very close together. But the measured points show a variation along the x direction, which reflects the pole periodicity. This unexpected variation can be explained by a poor equilibrium between the three phases of the inductor. Indeed, measured current intensities were out of balance on the order of 10%. Numerical tests have shown that such a lack of balance can result in a relatively higher perturbation of the magnetic field. But concerning the fluid flow, it is likely that inertial effects completely damp out such variations; therefore, the analytical model gives a sufficient approximation.
 The curl of the electromagnetic forces is the parameter adequate for the analysis of the fluid flow behavior, as shown below. Figure 6 shows the variation of curl $\bar{f}$ along the inductor wall, as a result either of the numerical calculation or of the analytical expression. In the latter case, Eqs. (18) show that only the x component gives a non-zero contribution to the curl $\bar{f}$ and consequently promotes a recirculating flow. The analytical approximation assumes

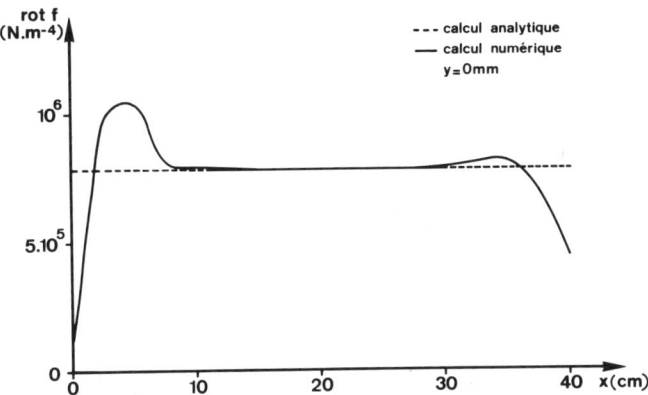

Fig. 6 Variation of curl f along the inductor wall. I=545 A. Solid line, numerical calculations; dotted line, analytical approximation.

curl $\bar{f}$ is a constant along the wall, independent of x. The value calculated by the numerical scheme is very close to the analytical one in the central part of the ingot, but becomes higher at both ends of the inductor, in relation to the increase of the electromagnetic field in these regions. But curl $\bar{f}$ never changes its sign. These results can be compared to the distribution of curl $\bar{f}$ obtained in a single-phased inductor, as obtained, for instance, by Lavers and Biringer.[7] In that case, curl $\bar{f}$ changes its sign between the upper and the lower corners and is nearly equal to zero at the center of the inductor. On the contrary, in the case of stirring by a multiphased inductor which is presently considered, curl $\bar{f}$ does not change its sign, and it is nearly constant along the ingot wall. A related remark is that the repulsion effect of the electromagnetic field is negligible as compared to the asynchrone stirring. (So fluid flow can be assumed to have only one vortex.) The difference between numerical and analytical results is low enough to allow the use of the analytical approximation in the flow modeling.

Modeling of Heat and Fluid Flows

The theoretical modeling of heat and fluid flow was described in detail in a previous paper.[8] It is written in a two-dimensional approximation, in the symmetry plane of both the mold and the inductor. Heat and fluid flows are treated simultaneously because one of the objectives is to compare electromagnetic stirring with natural thermal convection.

This model applies either to steady-state flow, the metal being entirely liquid, or to transient solidification phenomena. In the latter case, thermal effects include the latent heat release, which is assumed to be proportional to the solid fraction variation. Basic equations are conveniently expressed in terms of vorticity and stream function:

$$\xi = \frac{\partial v}{\partial x} - \frac{\partial u}{\partial y} \tag{19}$$

$$u = \frac{\partial \psi}{\partial y} \qquad v = -\frac{\partial \psi}{\partial x} \tag{20}$$

$$\frac{\partial T}{\partial t} + \frac{\partial}{\partial x}\left(T \frac{\partial \psi}{\partial x}\right) - \frac{\partial}{\partial y}\left(T \frac{\partial \psi}{\partial y}\right) - \frac{\Delta H}{C_p} \frac{\partial f_s}{\partial t}$$

$$= \frac{\partial}{\partial x}\left(\alpha_e \frac{\partial T}{\partial x}\right) + \frac{\partial}{\partial y}\left(\alpha_e \frac{\partial T}{\partial y}\right) + \frac{\bar{j}^2}{\sigma \rho C} \tag{21a}$$

$$\xi = -\left(\frac{\partial^2 \psi}{\partial x^2} + \frac{\partial^2 \psi}{\partial y^2}\right) \tag{21b}$$

$$\frac{\partial \xi}{\partial t} + \frac{\partial}{\partial x}\left(\xi \frac{\partial \psi}{\partial y}\right) - \frac{\partial}{\partial y}\left(\xi \frac{\partial \psi}{\partial y}\right)$$

$$= \frac{\partial}{\partial x}\left(\nu_e \frac{\partial \xi}{\partial x}\right) + \frac{\partial}{\partial y}\left(\nu_e \frac{\partial \xi}{\partial y}\right) + g\beta \frac{\partial T}{\partial y} + \frac{1}{\rho}\left(\frac{\partial \bar{f}_y}{\partial x} - \frac{\partial \bar{f}_x}{\partial y}\right) \tag{21c}$$

In Eq. (21c), one term expresses the influence of the buoyancy forces, and another is the curl of the electromagnetic forces, the importance of which was underlined above. Eq. (21a) contains the Joule effect, but numerical tests show it is negligible in our case, due to the relatively low frequency used here.

In stirring conditions, the flow is likely to be turbulent. This is taken into account by expressing viscosity ν_e and thermal diffusivity α_e by means of the mixing length model and Prandtl's approximation.

Concerning the hydrodynamic boundary conditions, the inductor wall needed special attention.[8] The boundary value of the vorticity was directly expressed as a function of the curl of the electromagnetic forces, which takes a high value in this area, as was mentioned above. Other boundary conditions are the usual ones: $\psi=0$ at all the walls and the free surface, and $\xi=0$ at the free surface. The heat flow boundary conditions are expressed with the help of three exchange coefficients, for the furnace wall, for the inductor wall, and for the free surface. Other walls are assumed to be insulated.

Equations (21,a,b,c) are solved simultaneously following a finite-difference scheme derived from Gosman et al.[9]

Experimental and Theoretical Results Concerning Steady-State Stirring

In order to check the theoretical modeling of heat and fluid flow, its results were compared first to a series of temperature and velocity measurements performed in steady-state conditions. The metal was entirely liquid, and a furnace placed against the left wall supplied the heat flow extracted through the inductor wall and the free surface. Figure 7 compares the theoretically calculated velocity chart with the experimental one. The theoretical model correctly represents the general distribution of the velocity. It shows the accelerated layer along the inductor wall. It gives the good position to the maximum velocity, at about 10 to 20 mm from the inductor wall. Moreover, the model

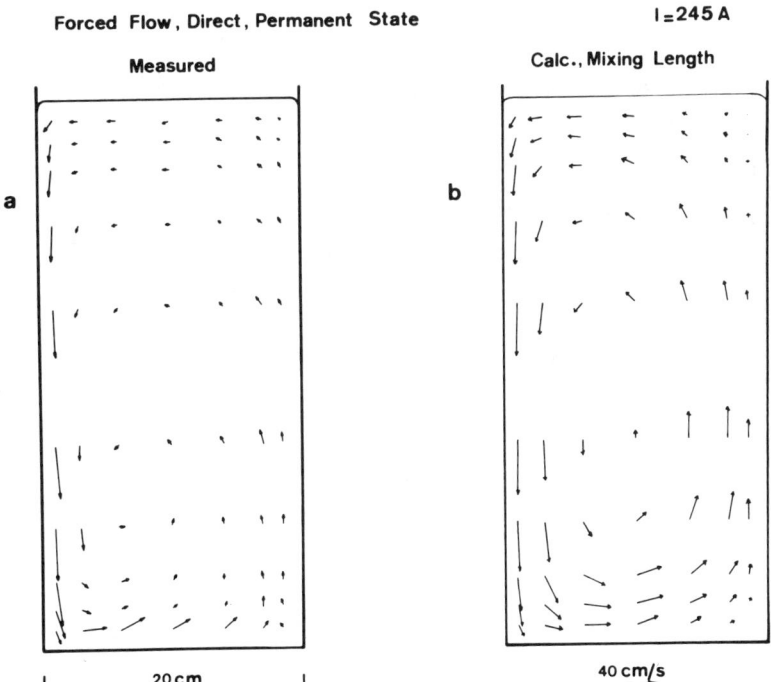

Fig. 7 Velocity field in steady-state stirring (I=545 A): a) measured, b) calculated.

gives the correct value of the maximum velocity. This means that the model gives a correct account of the glide effect common to the asynchrone stirring. Figure 8 evidences the effect of stirring on the temperature field. The latter is completely different from what could be expected in purely conductive heat transport. The bulk of the liquid is practically isothermic. The temperature gradients are confined in the boundary layers, which have a relatively limited extension.

The agreement between the modeling and the experiment is good. However, it is worth noting that the three exchange coefficients have been adjusted in order to give the same temperature at the coldest and at the hottest points (Table 1).

Similar maps were obtained in the case of natural thermal convection.[2] The comparison is very satisfactory. This is an important result in relation to solidification phenomena.

Heat and Fluid Flow During Transient Solidification

The theoretical model is now applied to the case of transient solidification. All the walls of the mold are con-

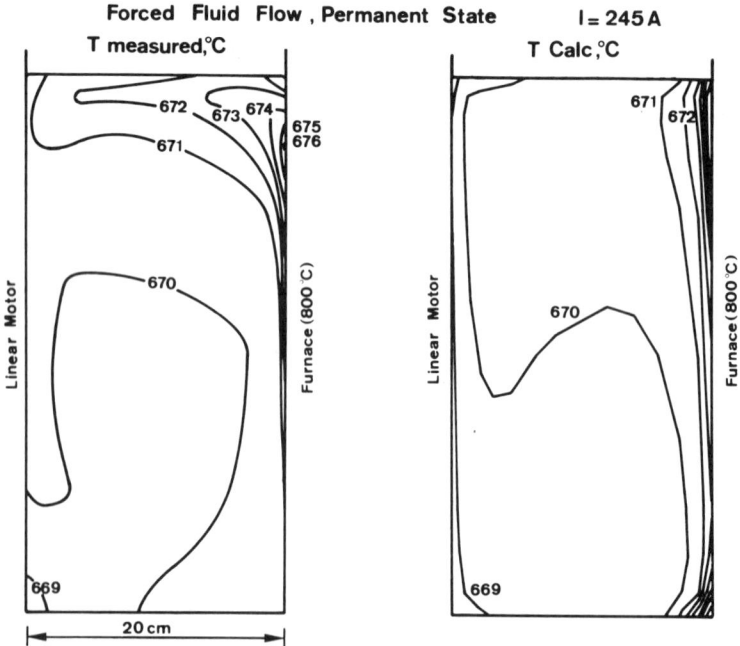

Fig. 8   Temperature field, same conditions as Fig. 7.

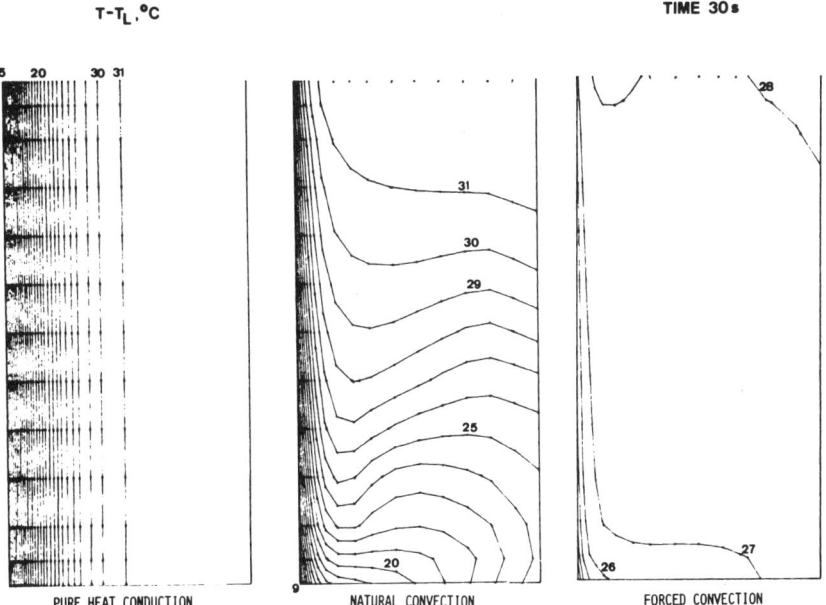

Fig. 9　Temperature field, 30 s after pouring.

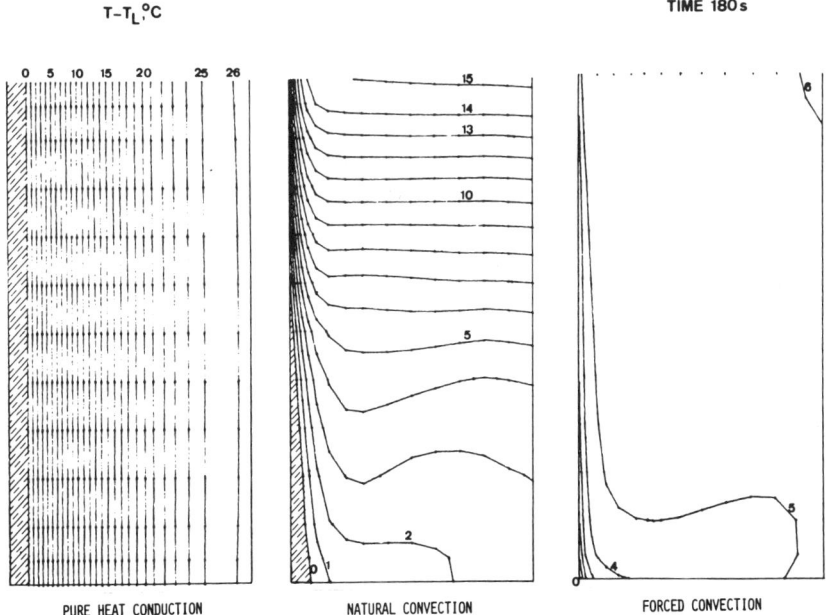

Fig. 10　Temperature field, 180 s after pouring.

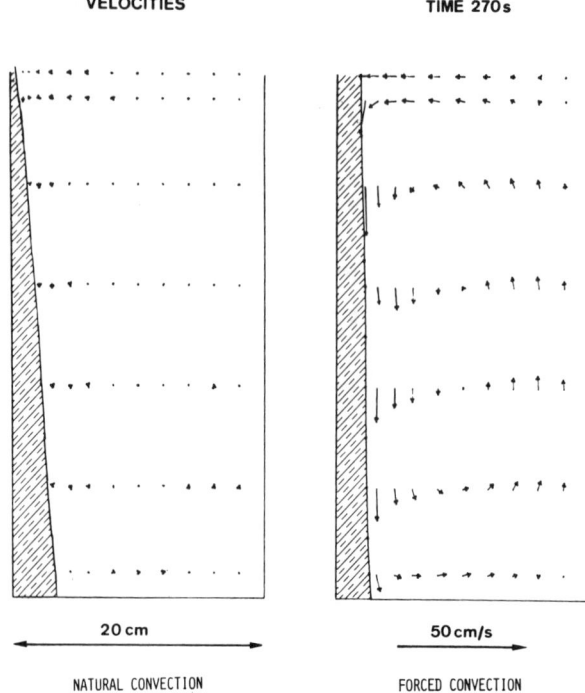

Fig. 11 Velocity field, 270 s after pouring.

sidered as insulated except for the inductor wall, for which a heat exchange coefficient is given (H=120 W m^{-2} K^{-1}). Calculations are made in three cases: melt stirred by the traveling field (I=245 A) or moved by natural thermal convection only, or motionless, the heat flow being purely conductive in the latter case, which is just a reference for comparison. Liquid is assumed to be an aluminum-2% copper solution. At the initial time it is assumed to be motionless, with a given superheat, 32°C in our case. Its liquidus temperature is 653°C. On the following figures, isotherms are referred to the residual superheat. Figure 9 depicts the temperature field 30 s after pouring. It is very different in the three cases. According to the purely conductive case, temperature gradients ought to be relatively strong in the half volume against the chill wall. In the stirred conditions, the liquid is practically isothermal, the temperature gradients are confined in a very thin boundary layer against the wall. In the natural convection case, the temperature distribution is intermediate. The convected heat has already caused a vertical temperature gradient to rise,

STIRRING AN ALUMINUM INGOT MOLD 753

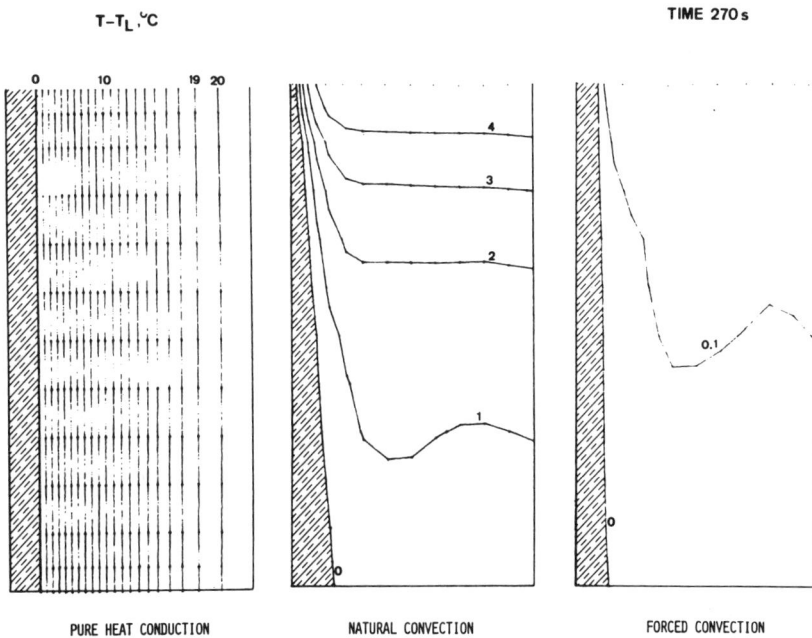

Fig. 12 Temperature field, 270 s after pouring.

which could never result from a purely conductive transport. In neither case has solidification taken place anywhere.

Solidification against the chilling wall begins 180 s after pouring in the three cases (Fig. 10). The shaded area represents the solid-liquid area. It is limited by the zero isotherm. The model accounts for an important fact: the extraction of superheat is much faster in stirring conditions. But as a counterpart, solidification begins later. This is a fact which is observed in continuous casting of steels. Superheat still exists in the bulk liquid. Consequently, it is likely that solidification proceeds in purely columnar conditions in the three cases.

Figure 11 represents the velocity distribution as calculated 270 s after pouring. The solid layer absorbs an increasing part of the electromagnetic forces, as compared to Fig. 7. But the remaining part is sufficient to move the liquid efficiently. In the natural flow case, the velocities are now very low.

Now we consider the temperature distribution at the same time, 270 s (Fig. 12). In the stirring conditions, the liquid superheat is less than a few tenths of a degree. If we take into account the undercooling existing at the columnar front, it is likely that now equiaxed crystals can form

in the whole liquid volume. In natural convection, equiaxed crystals could form only in the bottom of the liquid.
By comparison, in purely conductive conditions only columnar
solidification could be forecast.

Conclusion

Our experimental device combining a rectangular-vertical ingot mold containing liquid aluminum to a linear motor
proved to be convenient to study the effects of electromagnetic fields on solidifying melts. Electromagnetic traveling fields created by multiphased inductors must be handled
differently from the monophased induced fields. The proposed theoretical model accurately depicts the distribution
of the electromagnetic field, and of the electromagnetic
forces. The analytical approximation is sufficiently accurate to express the hydrodynamic contribution of the electromagnetic forces.

Concerning heat and fluid flows, the proposed model
provides a coherent description of the overall thermal and
hydrodynamic evolution during the solidification of ingots.
In the present state, it can be used to discuss the effects
of stirring concerning the influence of superheat, of heat
transfer coefficients, and of aspect ratio.

It can also provide a guideline for the design or the
choice of an electromagnetic stirrer in relation to a required metallurgical effect.

References

[1] Birat, J.P. and Chone, J., "Electromagnetic Stirring on Billet, Bloom and Slab Continuous Casters: State of the Art in 1982," Ironmaking and Steelmaking, 1983, pp. 269-281.

[2] Meyer, J.L., Durand, F., Ricou, R., and Vives, C., "Steady Flow of Liquid Aluminum in a Rectangular-Vertical Ingot Mold, Thermally or Electromagnetically Activated," Metallurgical Transactions Series B (in press).

[3] Ricou, R. and Vives, C., "Local Velocity and Mass Transfer Measurements in Molten Metals Using an Incorporated Magnet Probe," International Journal of Heat and Mass Transfer, Vol. 25, 1982, pp. 1579-1588.

[4] Tarapore, E.D. and Evans, J.W., "Fluid Velocities in Induction Melting Furnaces: Part I. Theory and Laboratory Experiments," Metallurgical Transactions, Series B, 1976, pp. 343-351.

[5] Fautrelle, Y.R., "Analytical and Numerical Aspects of the Electromagnetic Stirring Induced by Alternating Magnetic Fields," Journal of Fluid Mechanics, Vol. 102, 1981, pp. 405-430.

[6]Hughes, W.F. and Young, F.J., The Electromagnetodynamics of Fluids, John Wiley, New York, 1976.

[7]Lavers, J.D. and Biringer, P.P., IUTAM Symposium on Metallurgical Applications of Magnetohydrodynamics, Cambridge, U.K., 1982.

[8]Meyer, J.L. and Durand, F., "Analysis of the Transient Effects of Convection During Solidification, With and Without Electromagnetic Stirring," Proceedings of Conference "Modeling of Casting and Welding Processes," Henniker NH USA 1983, Edited by J. Berry, J. Dantzig (in press).

[9]Gosman, A.D., Pun, W.M., Runchal, A.K., et al., Heat and Mass Transfer in Recirculating Flow, Prentice Hall, Englewood Cliffs, N.J., 1962.

Author Index for Volume 100

Alemany, A........89,401,435	Jackson, W.D..............562	Papailiou, D.D............1!
Alexion, C.C.........516,533	Joussellin, F..........401,435	Petrick, M.......371,413,5€
Arts, J.G.A.................455	Julius, E......................694	Petrykowski, J.C...........)
Berry, G......................562	Kamiyama, S.I............304	Picologlou, B...............4⁹
Bertram, L.A...............617	Keeton, A.R.........516,533	Pierson, E.S.................5(
Block, F.R..................694	Khait, Y.L...................213	Reed, C.B....................4(
Brachet, M.E..............100	Laborde, R............401,435	Ricou, R......................6(
Brancher, J-P..............652	Lemnean, N................340	Roberts, J....................4⁹
Branover, H.....111,371,413	Levy, Y.......................355	Roy, P.........................1"
Caperan, P...................89	Librescu, L...................55	Sero-Guillaume, O........6!
Cristea, E-D................340	Lillicrap, D.C..............706	Sommeria, J.................
de Framond, R............652	Ludford, G.S.S............716	Sukoriansky, S.......111,3
Delage, D....................634	Lykoudis, P.S........255,280	Sulem, C.....................1
Dennis, C....................562	Malcolm, D.G.............449	Sulem, P.L..........100,1
Durand, F...................736	Marder, B.M...............617	Taberlet, E...................6
El-Boher, A................413	Merck, W.F.H.............455	Thibault, J.P.........401,4
Ernst, R.....................736	Messerle, H.K.............548	Thual, O.....................1
Fautrelle, Y.R.............680	Meyer, J-L..................736	Timnat, Y.M...............3
Flinsenberg, H.J..........475	Mihai, D.....................340	Toma, T......................3
Gagnoud, A................634	Mond, M....................329	Uhlenbusch, J..............4
Garnier, M...........589,634	Morioka, S..................317	Vives, C..................32,6
Gray, O.E., III...........533	Moros, A....................706	Walker, J.S.........3,17,7
Hall, F.S....................716	Naot, D......................202	Weil, D......................2
Herve, P.....................32	Nathenson, R.D...........533	Werkoff, F............401,4
Hunt, J......................706	Nygren, R...................496	Zanner, F.J..................€
		Zilberman, I................)

PROGRESS IN ASTRONAUTICS AND AERONAUTICS SERIES VOLUMES

VOLUME TITLE/EDITORS

*1. Solid Propellant Rocket Research (1960)
Martin Summerfield
Princeton University

*2. Liquid Rockets and Propellants (1960)
Loren E. Bollinger
The Ohio State University
Martin Goldsmith
The Rand Corporation
Alexis W. Lemmon Jr.
Battelle Memorial Institute

*3. Energy Conversion for Space Power (1961)
Nathan W. Snyder
Institute for Defense Analyses

*4. Space Power Systems (1961)
Nathan W. Snyder
Institute for Defense Analyses

*5. Electrostatic Propulsion (1961)
David B. Langmuir
Space Technology Laboratories, Inc.
Ernst Stuhlinger
NASA George C. Marshall Space Flight Center
J.M. Sellen Jr.
Space Technology Laboratories, Inc.

*6. Detonation and Two-Phase Flow (1962)
S.S. Penner
California Institute of Technology
F.A. Williams
Harvard University

---
*Out of print.

*7. Hypersonic Flow Research (1962)
Frederick R. Riddell
AVCO Corporation

*8. Guidance and Control (1962)
Robert E. Roberson
Consultant
James S. Farrior
Lockheed Missiles and Space Company

*9. Electric Propulsion Development (1963)
Ernst Stuhlinger
NASA George C. Marshall Space Flight Center

*10. Technology of Lunar Exploration (1963)
Clifford I. Cummings and Harold R. Lawrence
Jet Propulsion Laboratory

*11. Power Systems for Space Flight (1963)
Morris A. Zipkin and Russell N. Edwards
General Electric Company

*12. Ionization in High-Temperature Gases (1963)
Kurt E. Shuler, Editor
National Bureau of Standards
John B. Fenn, Associate Editor
Princeton University

*13. Guidance and Control—II (1964)
Robert C. Langford
General Precision Inc.
Charles J. Mundo
Institute of Naval Studies

*14. Celestial Mechanics and Astrodynamics (1964)
Victor G. Szebehely
Yale University Observatory

*15. Heterogeneous Combustion (1964)
Hans G. Wolfhard
Institute for Defense Analyses
Irvin Glassman
Princeton University
Leon Green Jr.
Air Force Systems Command

*16. Space Power Systems Engineering (1966)
George C. Szego
Institute for Defense Analyses
J. Edward Taylor
TRW Inc.

*17. Methods in Astrodynamics and Celestial Mechanics (1966)
Raynor L. Duncombe
U.S. Naval Observatory
Victor G. Szebehely
Yale University Observatory

*18. Thermophysics and Temperature Control of Spacecraft and Entry Vehicles (1966)
Gerhard B. Heller
NASA George C. Marshall Space Flight Center

*19. Communication Satellite Systems Technology (1966)
Richard B. Marsten
Radio Corporation of America

757

*20. Thermophysics of Spacecraft and Planetary Bodies: Radiation Properties of Solids and the Electromagnetic Radiation Environment in Space (1967)
Gerhard B. Heller
NASA George C. Marshall Space Flight Center

*21. Thermal Design Principles of Spacecraft and Entry Bodies (1969)
Jerry T. Bevans
TRW Systems

*22. Stratospheric Circulation (1969)
Willis L. Webb
Atmospheric Sciences Laboratory, White Sands, and University of Texas at El Paso

*23. Thermophysics: Applications to Thermal Design of Spacecraft (1970)
Jerry T. Bevans
TRW Systems

24. Heat Transfer and Spacecraft Thermal Control (1971)
John W. Lucas
Jet Propulsion Laboratory

25. Communication Satellites for the 70's: Technology (1971)
Nathaniel E. Feldman
The Rand Corporation
Charles M. Kelly
The Aerospace Corporation

26. Communication Satellites for the 70's: Systems (1971)
Nathaniel E. Feldman
The Rand Corporation
Charles M. Kelly
The Aerospace Corporation

27. Thermospheric Circulation (1972)
Willis L. Webb
Atmospheric Sciences Laboratory, White Sands, and University of Texas at El Paso

28. Thermal Characteristics of the Moon (1972)
John W. Lucas
Jet Propulsion Laboratory

29. Fundamentals of Spacecraft Thermal Design (1972)
John W. Lucas
Jet Propulsion Laboratory

30. Solar Activity Observations and Predictions (1972)
Patrick S. McIntosh and Murray Dryer
Environmental Research Laboratories, National Oceanic and Atmospheric Administration

31. Thermal Control and Radiation (1973)
Chang-Lin Tien
University of California at Berkeley

32. Communications Satellite Systems (1974)
P.L. Bargellini
COMSAT Laboratories

33. Communications Satellite Technology (1974)
P.L. Bargellini
COMSAT Laboratories

34. Instrumentation for Airbreathing Propulsion (1974)
Allen E. Fuhs
Naval Postgraduate School
Marshall Kingery
Arnold Engineering Development Center

35. Thermophysics and Spacecraft Thermal Control (1974)
Robert G. Hering
University of Iowa

36. Thermal Pollution Analysis (1975)
Joseph A. Schetz
Virginia Polytechnic Institute

37. Aeroacoustics: Jet and Combustion Noise; Duct Acoustics (1975)
Henry T. Nagamatsu, Editor
General Electric Research and Development Center
Jack V. O'Keefe, Associate Editor
The Boeing Company
Ira R. Schwartz, Associate Editor
NASA Ames Research Center

38. Aeroacoustics: Fan, STOL, and Boundary Layer Noise; Sonic Boom; Aeroacoustic Instrumentation (1975)
Henry T. Nagamatsu, Editor
General Electric Research and Development Center
Jack V. O'Keefe, Associate Editor
The Boeing Company
Ira R. Schwartz, Associate Editor
NASA Ames Research Center

39. Heat Transfer with Thermal Control Applications (1975)
M. Michael Yovanovich
University of Waterloo

40. Aerodynamics of Base Combustion (1976)
S.N.B. Murthy, Editor
Purdue University
J.R. Osborn, Associate Editor
Purdue University
A.W. Barrows and J.R. Ward, Associate Editors
Ballistics Research Laboratories

41. **Communications Satellite Developments: Systems** (1976)
Gilbert E. LaVean
Defense Communications Agency
William G. Schmidt
CML Satellite Corporation

42. **Communications Satellite Developments: Technology** (1976)
William G. Schmidt
CML Satellite Corporation
Gilbert E. LaVean
Defense Communications Agency

43. **Aeroacoustics: Jet Noise, Combustion and Core Engine Noise** (1976)
Ira R. Schwartz, Editor
NASA Ames Research Center
Henry T. Nagamatsu, Associate Editor
General Electric Research and Development Center
Warren C. Strahle, Associate Editor
Georgia Institute of Technology

44. **Aeroacoustics: Fan Noise and Control; Duct Acoustics; Rotor Noise** (1976)
Ira R. Schwartz, Editor
NASA Ames Research Center
Henry T. Nagamatsu, Associate Editor
General Electric Research and Development Center
Warren C. Strahle, Associate Editor
Georgia Institute of Technology

45. **Aeroacoustics: STOL Noise; Airframe and Airfoil Noise** (1976)
Ira R. Schwartz, Editor
NASA Ames Research Center
Henry T. Nagamatsu, Associate Editor
General Electric Research and Development Center
Warren C. Strahle, Associate Editor
Georgia Institute of Technology

46. **Aeroacoustics: Acoustic Wave Propagation; Aircraft Noise Prediction; Aeroacoustic Instrumentation** (1976)
Ira R. Schwartz, Editor
NASA Ames Research Center
Henry T. Nagamatsu, Associate Editor
General Electric Research and Development Center
Warren C. Strahle, Associate Editor
Georgia Institute of Technology

47. **Spacecraft Charging by Magnetospheric Plasmas** (1976)
Alan Rosen
TRW Inc.

48. **Scientific Investigations on the Skylab Satellite** (1976)
Marion I. Kent and Ernst Stuhlinger
NASA George C. Marshall Space Flight Center
Shi-Tsan Wu
The University of Alabama

49. **Radiative Transfer and Thermal Control** (1976)
Allie M. Smith
ARO Inc.

50. **Exploration of the Outer Solar System** (1976)
Eugene W. Greenstadt
TRW Inc.
Murray Dryer
National Oceanic and Atmospheric Administration
Devrie S. Intriligator
University of Southern California

51. **Rarefied Gas Dynamics, Parts I and II (two volumes)** (1977)
J. Leith Potter
ARO Inc.

52. **Materials Sciences in Space with Application to Space Processing** (1977)
Leo Steg
General Electric Company

53. **Experimental Diagnostics in Gas Phase Combustion Systems** (1977)
Ben T. Zinn, Editor
Georgia Institute of Technology
Craig T. Bowman, Associate Editor
Stanford University
Daniel L. Hartley, Associate Editor
Sandia Laboratories
Edward W. Price, Associate Editor
Georgia Institute of Technology
James G. Skifstad, Associate Editor
Purdue University

54. **Satellite Communications: Future Systems** (1977)
David Jarett
TRW Inc.

55. **Satellite Communications: Advanced Technologies** (1977)
David Jarett
TRW Inc.

56. **Thermophysics of Spacecraft and Outer Planet Entry Probes** (1977)
Allie M. Smith
ARO Inc.

57. **Space-Based Manufacturing from Nonterrestrial Materials** (1977)
Gerard K. O'Neill, Editor
Princeton University
Brian O'Leary, Assistant Editor
Princeton University

58. **Turbulent Combustion** (1978)
Lawrence A. Kennedy
State University of New York at Buffalo

59. **Aerodynamic Heating and Thermal Protection Systems** (1978)
Leroy S. Fletcher
University of Virginia

60. **Heat Transfer and Thermal Control Systems** (1978)
Leroy S. Fletcher
University of Virginia

61. **Radiation Energy Conversion in Space** (1978)
Kenneth W. Billman
NASA Ames Research Center

62. **Alternative Hydrocarbon Fuels: Combustion and Chemical Kinetics** (1978)
Craig T. Bowman
Stanford University
Jorgen Birkeland
Department of Energy

63. **Experimental Diagnostics in Combustion of Solids** (1978)
Thomas L. Boggs
Naval Weapons Center
Ben T. Zinn
Georgia Institute of Technology

64. **Outer Planet Entry Heating and Thermal Protection** (1979)
Raymond Viskanta
Purdue University

65. **Thermophysics and Thermal Control** (1979)
Raymond Viskanta
Purdue University

66. **Interior Ballistics of Guns** (1979)
Herman Krier
University of Illinois at Urbana-Champaign
Martin Summerfield
New York University

67. **Remote Sensing of Earth from Space: Role of "Smart Sensors"** (1979)
Roger A. Breckenridge
NASA Langley Research Center

68. **Injection and Mixing in Turbulent Flow** (1980)
Joseph A. Schetz
Virginia Polytechnic Institute and State University

69. **Entry Heating and Thermal Protection** (1980)
Walter B. Olstad
NASA Headquarters

70. **Heat Transfer, Thermal Control, and Heat Pipes** (1980)
Walter B. Olstad
NASA Headquarters

71. **Space Systems and Their Interactions with Earth's Space Environment** (1980)
Henry B. Garrett and Charles P. Pike
Hanscom Air Force Base

72. **Viscous Flow Drag Reduction** (1980)
Gary R. Hough
Vought Advanced Technology Center

73. **Combustion Experiments in a Zero-Gravity Laboratory** (1981)
Thomas H. Cochran
NASA Lewis Research Center

74. **Rarefied Gas Dynamics, Parts I and II (two volumes)** (1981)
Sam S. Fisher
University of Virginia at Charlottesville

75. **Gasdynamics of Detonations and Explosions** (1981)
J.R. Bowen
University of Wisconsin at Madison
N. Manson
Université de Poitiers
A.K. Oppenheim
University of California at Berkeley
R.I. Soloukhin
Institute of Heat and Mass Transfer, BSSR Academy of Sciences

76. **Combustion in Reactive Systems** (1981)
J.R. Bowen
University of Wisconsin at Madison
N. Manson
Université de Poitiers
A.K. Oppenheim
University of California at Berkeley
R.I. Soloukhin
Institute of Heat and Mass Transfer, BSSR Academy of Sciences

77. **Aerothermodynamics and Planetary Entry** (1981)
A.L. Crosbie
University of Missouri-Rolla

78. **Heat Transfer and Thermal Control** (1981)
A.L. Crosbie
University of Missouri-Rolla

79. **Electric Propulsion and Its Applications to Space Missions** (1981)
Robert C. Finke
NASA Lewis Research Center

80. **Aero-Optical Phenomena** (1982)
Keith G. Gilbert and Leonard J. Otten
Air Force Weapons Laboratory

81. **Transonic Aerodynamics** (1982)
David Nixon
Nielsen Engineering & Research, Inc.

82. **Thermophysics of Atmospheric Entry** (1982)
T.E. Horton
The University of Mississippi

83. **Spacecraft Radiative Transfer and Temperature Control** (1982)
T.E. Horton
The University of Mississippi

84. **Liquid-Metal Flows and Magnetohydrodynamics** (1983)
H. Branover
Ben-Gurion University of the Negev
P.S. Lykoudis
Purdue University
A. Yakhot
Ben-Gurion University of the Negev

85. **Entry Vehicle Heating and Thermal Protection Systems: Space Shuttle, Solar Starprobe, Jupiter Galileo Probe** (1983)
Paul E. Bauer
McDonnell Douglas Astronautics Company
Howard E. Collicott
The Boeing Company

86. **Spacecraft Thermal Control, Design, and Operation** (1983)
Howard E. Collicott
The Boeing Company
Paul E. Bauer
McDonnell Douglas Astronautics Company

87. **Shock Waves, Explosions, and Detonations** (1983)
J.R. Bowen
University of Washington
N. Manson
Université de Poitiers
A.K. Oppenheim
University of California at Berkeley
R.I. Soloukhin
Institute of Heat and Mass Transfer, BSSR Academy of Sciences

88. **Flames, Lasers, and Reactive Systems** (1983)
J.R. Bowen
University of Washington
N. Manson
Université de Poitiers
A.K. Oppenheim
University of California at Berkeley
R.I. Soloukhin
Institute of Heat and Mass Transfer, BSSR Academy of Sciences

89. **Orbit-Raising and Maneuvering Propulsion: Research Status and Needs** (1984)
Leonard H. Caveny
Air Force Office of Scientific Research

90. **Fundamentals of Solid-Propellant Combustion** (1984)
Kenneth K. Kuo
The Pennsylvania State University
Martin Summerfield
Princeton Combustion Research Laboratories, Inc.

91. **Spacecraft Contamination: Sources and Prevention** (1984)
J.A. Roux
The University of Mississippi
T.D. McCay
NASA Marshall Space Flight Center

92. **Combustion Diagnostics by Nonintrusive Methods** (1984)
T.D. McCay
NASA Marshall Space Flight Center
J.A. Roux
The University of Mississippi

93. **The INTELSAT Global Satellite System** (1984)
Joel Alper
COMSAT Corporation
Joseph Pelton
INTELSAT

94. **Dynamics of Shock Waves, Explosions, and Detonations** (1984)
J.R. Bowen
University of Washington
N. Manson
Universite de Poitiers
A.K. Oppenheim
University of California
R.I. Soloukhin
Institute of Heat and Mass Transfer, BSSR Academy of Sciences

95. **Dynamics of Flames and Reactive Systems** (1984)
J.R. Bowen
University of Washington
N. Manson
Universite de Poitiers
A.K. Oppenheim
University of California
R.I. Soloukhin
Institute of Heat and Mass Transfer, BSSR Academy of Sciences

96. **Thermal Design of Aeroassisted Orbital Transfer Vehicles** (1985)
H.F. Nelson
University of Missouri-Rolla

97. **Monitoring Earth's Ocean, Land, and Atmosphere from Space — Sensors, Systems, and Applications** (1985)
Abraham Schnapf
Aerospace Systems Engineering

98. **Thrust and Drag: Its Prediction and Verification** (1985)
Eugene E. Covert
Massachusetts Institute of Technology
C.R. James
Vought Corporation
William F. Kimzey
Sverdrup Technology AEDC Group
George K. Richey
U.S. Air Force
Eugene C. Rooney
U.S. Navy Department of Defense

99. **Space Stations and Space Platforms — Concepts, Design, Infrastructure, and Uses** (1985)
Ivan Bekey
Daniel Herman
NASA Headquarters

100. **Single- and Multi-Phase Flows in an Electromagnetic Field** *Energy, Metallurgical, and Solar Applications* (1985)
Herman Branover
Ben-Gurion University of the Negev
Paul S. Lykoudis
Purdue University
Michael Mond
Ben-Gurion University of the Negev

(Other Volumes are planned.)

RAYMOND H. FOGLER LIBRARY
DATE DUE